Handbook of Humidity Measurement

Methods, Materials and Technologies

Volume 2: Electronic and Electrical Humidity Sensors

**Handbook of Humidity Measurement: Methods, Materials
and Technologies, 3-Volume Set**
(ISBN: 978-1-138-29787-6)

**Handbook of Humidity Measurement,
Volume 1: Spectroscopic Methods of Humidity Measurement**
(ISBN: 978-1-138-30021-7)

**Handbook of Humidity Measurement,
Volume 2: Electronic and Electrical Humidity Sensors**
(ISBN: 978-1-138-30022-4)

**Handbook of Humidity Measurement,
Volume 3: Sensing Materials and Technologies**
(ISBN: 978-1-138-48287-6)

Handbook of Humidity Measurement

Methods, Materials and Technologies

Volume 2: Electronic and Electrical Humidity Sensors

Ghenadii Korotcenkov

CRC Press
Taylor & Francis Group
Boca Raton London New York

CRC Press is an imprint of the
Taylor & Francis Group, an **informa** business

CRC Press
Taylor & Francis Group
6000 Broken Sound Parkway NW, Suite 300
Boca Raton, FL 33487-2742

© 2019 by Taylor & Francis Group, LLC
CRC Press is an imprint of Taylor & Francis Group, an Informa business

No claim to original U.S. Government works

Printed on acid-free paper

International Standard Book Number-13: 978-1-138-30022-4 (Hardback)

This book contains information obtained from authentic and highly regarded sources. Reasonable efforts have been made to publish reliable data and information, but the author and publisher cannot assume responsibility for the validity of all materials or the consequences of their use. The authors and publishers have attempted to trace the copyright holders of all material reproduced in this publication and apologize to copyright holders if permission to publish in this form has not been obtained. If any copyright material has not been acknowledged please write and let us know so we may rectify in any future reprint.

Except as permitted under U.S. Copyright Law, no part of this book may be reprinted, reproduced, transmitted, or utilized in any form by any electronic, mechanical, or other means, now known or hereafter invented, including photocopying, microfilming, and recording, or in any information storage or retrieval system, without written permission from the publishers.

For permission to photocopy or use material electronically from this work, please access www.copyright.com (http://www.copyright.com/) or contact the Copyright Clearance Center, Inc. (CCC), 222 Rosewood Drive, Danvers, MA 01923, 978-750-8400. CCC is a not-for-profit organization that provides licenses and registration for a variety of users. For organizations that have been granted a photocopy license by the CCC, a separate system of payment has been arranged.

Trademark Notice: Product or corporate names may be trademarks or registered trademarks, and are used only for identification and explanation without intent to infringe.

Visit the Taylor & Francis Web site at
http://www.taylorandfrancis.com

and the CRC Press Web site at
http://www.crcpress.com

Contents

Preface xiii
Acknowledgments xv
Author xvii

SECTION I Introduction in Humidity Sensors

Chapter 1 Introduction to Electronic and Electrical Humidity Sensors 3

    1.1   Introduction 3
    1.2   Classification of Humidity Sensors 5
    1.3   Materials Acceptable for Application in Humidity Sensors 7
    1.4   Requirements for Humidity Sensors 10
    1.5   Purpose and Content of Book 11
    References 12

SECTION II Conventional Methods and Brief History of Humidity Measurements

Chapter 2 Gravimetric Method of Humidity Measurement 17

    2.1   History of Gravimetric-Based Hygrometers 17
    2.2   Gravimetric Measurement of Air Humidity 18
    References 21

Chapter 3 Mechanical (Hair) Hygrometer 23

    3.1   History of Hair Hygrometers 23
    3.2   Features of Humidity Control Using Hair Hygrometers 25
    3.3   Other Types of Mechanical Hygrometers 26
    3.4   Hygrographs and Hygrothermographs 27
    3.5   Electronic Expansion Hygrometer 28
    References 28

Chapter 4 Psychrometer 31

    4.1   History 31
    4.2   Principles of Operation 31
    4.3   Realization 33
    4.4   Accuracy of Measurements 36
    4.5   Advantages and Limitations 38
    References 40

Chapter 5 Chilled-Mirror Hygrometer or Mirror-Based Dew-Point Sensors 41

    5.1   Condensation Methods of Humidity Measurement 41
    5.2   Chilled-Mirror Hygrometers 42

	5.3 Surface Conductive Dew-Point Hygrometer	46
	5.4 Dew-Point Hygrometers and Their Performances	47
	References	48

Chapter 6   Heated Salt-Solution Method for Humidity Measurement 51

     6.1   The Dunmore Cell 51
     6.2   Limitations of LiCl-Based Humidity Sensors 52
     6.3   Planar LiCl-Based Humidity Sensors 53
     References 55

Chapter 7   Electrolytic or Coulometric Hygrometers 57

     7.1 P_2O_5-Based Coulometric Hygrometers 57
           7.1.1   Principles of Operation 57
           7.1.2   Sensor Configuration 57
           7.1.3   Advantages and Limitations 58
           7.1.4   Modern P_2O_5-Based Humidity Sensors 61
     7.2   Pope Cell 62
           7.2.1   Advantages and Limitations of Pope Cell 63
     7.3   Other Possible Coulometric Methods of Humidity Measurement 63
     References 63

Chapter 8   Humidity Measurement Based on Karl Fischer Titration 65

     8.1   KF Titration 65
     8.2   Principles of KF Titration Method 65
     8.3   Advantages and Limitations 67
     References 69

Chapter 9   Other Conventional Methods of Humidity Measurement 71

     9.1   Chemical Method 71
           9.1.1   Calcium Carbide Method 71
     9.2   Constant-Pressure Hygrometer 71
     9.3   Constant-Volume Hygrometer 71
     9.4   Pneumatic Bridge Method of Humidity Measurement 72
     9.5   Diffusion Hygrometer 72
     9.6   Cloud or Fog Chamber Hygrometer 73
     9.7   Mass Spectrometric Measurement of Air Humidity 73
     References 74

SECTION III   Electronic and Electrical Humidity Sensors and Basic Principles of Their Operation

Chapter 10   Capacitance-Based Humidity Sensors 79

     10.1   Basic Principles of Operation 79
           10.1.1   Humidity Sensors of Permittivity-Type 79
                   10.1.1.1   Parallel Plate Structure 81
                   10.1.1.2   Planar Capacitive Sensors with IDE 83
           10.1.2   Requirements for Materials Suitable for Use in Capacitive Humidity Sensors 85

Contents

- 10.2 Polymer-Based Capacitance Humidity Sensors ... 86
  - 10.2.1 Mechanisms of Water Interaction with Polymers ... 86
  - 10.2.2 Implementation of Polymer Humidity Sensors ... 87
- 10.3 Metal Oxide–Based Capacitive Humidity Sensors ... 93
  - 10.3.1 Metal Oxides in Humidity Sensors ... 93
  - 10.3.2 Mechanisms of Water Interaction with Metal Oxides ... 94
  - 10.3.3 Al_2O_3-Based Humidity Sensors ... 96
  - 10.3.4 Other Metal Oxide-Based Humidity Sensors ... 100
- 10.4 Summary ... 103
- References ... 105

Chapter 11 Resistive Humidity Sensors ... 113

- 11.1 General Consideration ... 113
- 11.2 Polymer-Based Resistive Humidity Sensors ... 114
  - 11.2.1 Introduction ... 114
  - 11.2.2 Semiconductor or Conductive Polymers ... 115
  - 11.2.3 Polymers Containing Inorganic Salts ... 117
  - 11.2.4 Polyelectrolyte-Based Humidity-Sensitive Materials ... 119
  - 11.2.5 Nanocomposite-Based Humidity Sensors ... 126
  - 11.2.6 Summary ... 131
- 11.3 Metal Oxide–Based Resistive Humidity Sensors ... 133
  - 11.3.1 Mechanisms of Conductivity in Metal Oxides in Humid Air ... 133
    - 11.3.1.1 Grotthuss-Type Transport of Protons ... 133
    - 11.3.1.2 Conductivity in Heated Semiconducting Metal Oxides ... 135
    - 11.3.1.3 Conductivity in Metal Oxides Doped with Alkali Ions ... 135
  - 11.3.2 Metal Oxides in Resistive Humidity Sensors ... 136
  - 11.3.3 Examples of Realization ... 137
    - 11.3.3.1 General Consideration ... 137
    - 11.3.3.2 RT Humidity Sensors ... 138
  - 11.3.4 Mechanism of Sensor Response and the Role of Structural Factor in the Sensor Performance ... 142
  - 11.3.5 The Role of Alkaline Ions in Sensing Effect Observed in Metal Oxide-Based Humidity Sensors ... 145
  - 11.3.6 Heated Sensors with Semiconducting and Ionic Metal Oxides ... 148
  - 11.3.7 Summary ... 150
- 11.4 Other Materials Acceptable for Application in Resistive Humidity Sensors ... 150
- References ... 151

Chapter 12 Gravimetric Humidity Sensors ... 161

- 12.1 Introduction ... 161
- 12.2 QCM-Based Sensors ... 162
  - 12.2.1 General Consideration ... 162
  - 12.2.2 FBAR Sensors ... 164
  - 12.2.3 SMR Structures ... 166
- 12.3 Surface Acoustic Wave Sensors ... 167
  - 12.3.1 Propagation Modes of the Acoustic Wave ... 168
  - 12.3.2 SAW Oscillator as Sensing Element ... 170
- 12.4 Advantages and Shortcomings of Acoustic Mass-Sensitive Sensors ... 171

12.5 Sensing Layers in Gravimetric Humidity Sensors 173
    12.5.1 General Requirements 173
    12.5.2 Humidity-Sensitive Materials Used in Gravimetric Humidity Sensors 175
        12.5.2.1 Polymers 175
        12.5.2.2 Metal Oxides 176
        12.5.2.3 Other Materials 177
12.6 Acoustic Humidity Sensors' Performance 177
    12.6.1 QCM Humidity Sensors 177
        12.6.1.1 Conventional QCM Humidity Sensors 177
        12.6.1.2 FBAR Humidity Sensors 181
    12.6.2 SAW-Based Humidity Sensors 183
        12.6.2.1 General Consideration 183
        12.6.2.2 Polymer SAW-Based Humidity Sensors 187
        12.6.2.3 SAW Humidity Sensors Based on Inorganic Humidity-Sensitive Materials 190
        12.6.2.4 Stability of SAW Humidity Sensors 193
        12.6.2.5 New Approaches to Development of SAW-Based Humidity Sensors 194
    12.6.3 Dew-Point Hygrometers 194
12.7 Summary 197
References 199

Chapter 13 Cantilever- and Membrane-Based Humidity Sensors 205

13.1 General Consideration 205
    13.1.1 Membrane-Based Sensors 205
    13.1.2 Microcantilevers 205
13.2 Sorption-Induced Effects and Their Influence on Cantilever Operation 207
    13.2.1 Sorption Models 207
    13.2.2 Microcantilevers and Their Modes of Operation 209
        13.2.2.1 Static Mode 209
        13.2.2.2 Dynamic Mode 209
13.3 Microcantilever Deflection Detection Methods 211
    13.3.1 Optical Methods 211
    13.3.2 Capacitive Method 212
    13.3.3 Piezoelectric Method 212
    13.3.4 Interferometry Method 212
    13.3.5 Optical Diffraction Grating Method 213
    13.3.6 CCD Detection Method 213
    13.3.7 Hard-Contact/Tunneling 213
    13.3.8 Piezoresistive Method 213
13.4 Resonant Operating Mode 215
    13.4.1 Mechanical Properties of Microcantilevers 215
    13.4.2 Mass Resolution Limitations 216
13.5 Humidity-Sensitive Materials 216
    13.5.1 Polyimide 216
    13.5.2 Other Humidity-Sensitive Polymers 218
13.6 Humidity Sensor Implementation 218
    13.6.1 Functionalization Methods 218
    13.6.2 Capacitive Humidity Sensors 221

		13.6.3	Piezoresistive Humidity Sensors	224
			13.6.3.1 Piezoresistors	224
			13.6.3.2 Sensor Performance	227
	13.7	Microresonator-Based Humidity Sensors		233
	13.8	Summary		235
	References			236

Chapter 14 Thermal Conductivity-Based Hygrometers 243

- 14.1 Principle of Operation 243
- 14.2 Approaches to the Design of Thermal Conductivity-Based Hygrometers 244
- 14.3 Micromachined Humidity Sensors 247
- 14.4 Summary 250
- References 251

Chapter 15 Field Ionization Humidity Sensors 253

- 15.1 Principles of Operation: Corona Discharge 253
- 15.2 Gas Ionization Sensors Based on 1D Structures 253
- 15.3 Summary 256
- References 256

Chapter 16 Humidity Sensors Based on Thin-Film and Field-Effect Transistors 257

- 16.1 Thin-Film and Field-Effect Transistors..257
- 16.2 Humidity-Sensing Characteristics of Organic-Based Transistors 259
  - 16.2.1 TFT-Based Sensors 259
  - 16.2.2 MISFET-Based Sensors 267
- 16.3 Other Materials in TFT and FET-Based Humidity Sensors 267
  - 16.3.1 MISFET-Based Sensors 268
  - 16.3.2 GasFET or FET with Air Gap 269
  - 16.3.3 TFT-Based Sensors 275
- 16.4 Summary 277
- References 278

Chapter 17 Hetero-Junction-Based Humidity Sensors 283

- 17.1 Schottky Barrier-Based Humidity Sensors 283
  - 17.1.1 Principles of Operation 283
  - 17.1.2 Sensor Performance 283
- 17.2 p–n Hetero-Contact-Type Humidity Sensors 286
- 17.3 Summary 289
- References 289

Chapter 18 Kelvin Probe as a Humidity Sensor 291

- 18.1 Work Function 291
- 18.2 Kelvin Probe 291
- 18.3 Sensor Performance 293
- 18.4 Summary 296
- References 297

Chapter 19 Solid-State Electrochemical Humidity Sensors 299

    19.1 Introduction 299
    19.2 Principles of Operation 299
    19.3 Sensors Performance 300
        19.3.1 Potentiometric Humidity Sensors 300
        19.3.2 Amperometric Humidity Sensors 304
        19.3.3 Impedance Sensors 308
    19.4 Summary 312
    References 312

SECTION IV New Trends and Outlook

Chapter 20 Microwave-Based Humidity Sensors 317

    20.1 Introduction 317
    20.2 Microwave Sensors 318
        20.2.1 Transmission Sensors 319
        20.2.2 Resonator Sensors 320
        20.2.3 Impedance Meters 320
    20.3 Examples of Humidity Sensor Realization: Humidity Sensor Performance 321
        20.3.1 Transmission Sensors 321
        20.3.2 Resonant and Impedance Sensors 323
    20.4 Summary 335
    References 336

Chapter 21 Integrated Humidity Sensors 339

    21.1 Humidity Sensors Integrated with Heater 339
    21.2 Monolithic Integration of the Humidity Sensors with the Readout Circuitry 341
    21.3 Summary 344
    References 345

Chapter 22 Humidity Sensors on Flexible Substrate 347

    22.1 Flexible Electronics 347
    22.2 Flexible Platforms 347
    22.3 Humidity Sensors on Flexible Substrates 350
        22.3.1 Capacitive Humidity Sensors on Flexible Substrates 351
        22.3.2 Resistive Humidity Sensors on Flexible Substrates 353
        22.3.3 Multiparameter Sensing Platform 356
        22.3.4 Features of Fabrication Technology 357
    22.4 Paper-Based Humidity Sensors 359
    22.5 Summary 362
    References 362

Chapter 23 Nontraditional Approaches to Humidity Measurement 365

    23.1 Humidity Detection Using Triboelectric Effect 365
    23.2 Humidity Influence on the Breakdown Voltage 366
    23.3 Humidity Measurement Using Heat Pipe 366

	23.4	Self-Powered Active Humidity Sensor	368
	23.5	Humidity Sensor Using a SMT	372
	References		373

Chapter 24 Summary and Outlook 375

    24.1 Summary 375
    24.2 Smart Sensors 376
        24.2.1 Architecture of Smart Sensors 377
        24.2.2 Advantages and Limitations 379
    References 380

Index 381

Preface

On account of unique water properties, humidity greatly affects living organisms, including humans and materials. The amount of water vapor in the air can affect human comfort, and the efficiency and safety of many manufacturing processes, including drying of products, such as paint, paper, matches, fur, and leather; packaging and storage of different products, such as tea, cereal, milk, and bakery items; and manufacturing of food products, such as plywood, gum, abrasives, pharmaceutical powder, ceramics, printing materials, and tablets. Moreover, the industries discussed above are only a small part of the industries for which humidity should be controlled. In agriculture, the measurement of humidity is important for crop protection (dew prevention), soil moisture monitoring, and so on. In the medical field, humidity control should be used in respiratory equipment, sterilizers, incubators, pharmaceutical processing, and biological products. Humidity measurements at the Earth's surface are also required for meteorological analysis and forecasting, climate studies, and special applications in hydrology, aeronautical services, and environmental studies, because water vapor is the key agent in both weather and climate.

This means that the determination of humidity is of great importance. Therefore, humidity control becomes imperative in all fields of human activity, from production management to creating a comfortable environment for our living, and for understanding the nature of the changes happening to the climate. Humidity can change in a wide range, making devices and sensors capable of carrying out its measurement across the entire range of possible changes are necessary. It is clear that these sensors and measurement systems must be able to work in a variety of climatic conditions, ensuring the functioning of their control and surveillance systems for a long time.

In the past decade, much development took place of new methods for measuring humidity, as well as improvements in and optimization of the manufacturing technology of already-developed humidity sensors, and for the development of different measuring systems with an increased efficiency. As a result, the field of humidity sensors has broadened and expanded greatly. At present, humidity sensors are used in medicine, agriculture, industry, environmental control, and other fields. Humidity sensors are widely used for the continuous monitoring of humidity in diverse applications, such as the baking and drying of food, cigar storage, civil engineering to detect water ingress in soils or in the concrete in civil structures, medical applications, and many other fields. However, the process of developing new humidity sensors and improving older types of devices used for humidity measurement is ongoing. New technologies and the toughening of ecological standards require more sensitive instruments with faster response times, better selectivity, and improved stability. It is therefore time to resume the developments carried out during this time and identify methods for further development in this area. This is important, as too many approaches and types of devices that can be used for measuring humidity are proposed. They use different measuring principles, different humidity-sensitive materials, and various configurations of devices, making it difficult to conduct an objective analysis of these devices' capabilities. We hope that the detailed data presented in this book on various types of humidity sensors that are developed by different teams, accompanied by an analysis of their strengths and weaknesses, will allow cross-comparison and the selection of suitable sensing methods for specific applications. As a result, conditions will be created for the development of devices to ensure accurate, reliable, economically viable, and efficient humidity measurements.

This series is organized as follows. Considering current trends in development of instruments for measuring humidity, this publication is divided into three parts: The first volume focuses on the review of devices based on optical principles of measurement, such as optical UV and fluorescence hygrometers, and optical and fiberoptic sensors of various types. As indicated, atmospheric water plays a key role in the climate. Therefore, various methods for monitoring the atmosphere have been developed in recent years, on the basis of measuring electromagnetic field absorption in different spectral ranges. All these methods, covering the optical (FTIR and Lidar techniques), microwave, and THz ranges, are discussed in this volume, and analysis of their strengths and weaknesses is given. The role of humidity-sensitive materials in optical and fiberoptic sensors is also detailed. This volume also describes reasons that cause us to control humidity, features of water and water vapors, and units used for humidity measurement. This information will certainly be cognitive and interesting for readers.

The second volume is entirely devoted to the consideration of different types of solid-state devices, the operating principles of which are based on other physical principles. Detailed information is provided, including advantages and disadvantages about the capacitive, resistive, gravimetric, hygrometric, field ionization, microwave, solid-state electrochemical, and thermal conductivity-based humidity sensors, followed by a relevant analysis of the properties of humidity-sensitive materials used to develop such devices. Humidity sensors based on thin-film and field-effect transistors, heterojunctions, flexible substrates, and integrated humidity sensors are also discussed in this volume. This is an age of automation and control. Therefore, in addition to interest in sensor properties, such as accuracy and long-term drift, there is also interest in durability in different environments, component size, digitization, simple and quick system integration, and last, but not least, price. This means that modern humidity sensors should be able to integrate all these demands into one sensor. The experiment showed that these capabilities can be fully realized in electric and electronic sensors manufactured using semiconductor solid-state technology. Such humidity sensors can be fabricated with low cost and are more convenient for moisture control. Great attention is also paid to consideration of conventional devices that were used to measure humidity for several centuries. It is important to note that many of these methods are widely used today.

The third volume focuses on considering properties of various materials suitable for the development of humidity sensors. Polymers, metal oxides, porous semiconductors (Si, SiC), carbon-based materials (black carbon, carbon nanotubes, and graphene), zeolites, silica, and some others are included. Features of humidity sensor fabrication and related materials are also considered. Market forces naturally lead to ever-more-specialized and innovative products; sensors should be smaller in size, cheaper, more robust, and accurate in measurement; they should have better sensitivity and stability. This challenge requires new technological solutions, some of which (e.g., integration and miniaturization) are considered in this volume. Specificity of the humidity sensor calibration and humidity sensor market analysis also are covered.

These books show that materials play a key role in humidity sensor functioning, and that the range of materials that can be used in their development is broad. Each material has its advantages and disadvantages; therefore, selection of optimal sensing materials for humidity sensors is complicated and multivariate. However, the number of published books or reviews providing an analysis of all possible humidity sensor materials through field application is limited. Therefore, it is difficult to conduct a comparative analysis of various materials and to choose a humidity-sensing material optimal for particular applications. This book contributes to the solution of this problem. Readers of these three volumes, including scientists, can find a comparative analysis of all materials acceptable for humidity sensor design and can estimate their real advantages and shortcomings. Moreover, throughout these books, strategies are described for the fabrication and characterization of humidity-sensitive materials and sensing structures employed in sensor applications. One can consider the present books as a selective guide to materials for humidity sensor manufacture.

Thus, these books provide an up-to-date account of the present status of humidity sensors, from understanding the concepts behind them to the practical knowledge necessary for their development and manufacture. In addition, these books contain a large number of tables with information necessary for humidity sensor design. The tables alone make these books helpful and convenient to use. Therefore, this issue can be utilized as a reference book for researchers and engineers, as well as graduate students who are either entering the field of humidity measurement for the first time, or who are already conducting research in these areas but are willing to extend their knowledge in the field. In this case, these books will act as an introduction to the world of humidity sensors that may encourage further study, and estimate the role that humidity sensors may play in the future. I hope they will also be useful to university students, postdoctoral candidates, and professors. Their structure offers the basis for courses in the field of material sciences, chemical sensors, sensor technologies, chemical engineering, semiconductor devices, electronics, medicine, and environmental monitoring. We believe that practicing engineers, measurement experts, laboratory technicians, and project managers in industries and national laboratories who are interested in looking for solutions to humidity measurement tasks in industrial and environmental applications, but do not know how to design or select optimal devices for these applications, will also find useful information to help them to make the right choices concerning technical solutions and investments.

Ghenadii Korotcenkov

Acknowledgments

My sincere gratitude goes to CRC Press for the opportunity to write this book. I also acknowledge the Gwangju Institute of Science and Technology, Gwangju, South Korea, and Moldova State University, Chisinau, Republic of Moldova, for inviting me and supporting my research in various programs and projects. I thank my wife and the love of my life, Irina Korotcenkova, for always being there for me, inspiring me and supporting all my endeavors. She gives me purpose, motivates me to continue my work, and makes my life so much more exciting. Also, I am grateful to my daughters, Anya and Anastasia, for being a part of my life and encouraging my work. Special thanks go to my friends, colleagues, and coauthors for their support and collaboration over the years. Great thanks to all of you; this would not be possible without you by my side.

Author

Ghenadii Korotcenkov earned his PhD in material sciences from the Technical University of Moldova, Chisinau, Moldova, in 1976 and his doctor of science degree (doctor habilitate) in physics from the Academy of Science of Moldova in 1990 (Highest Qualification Committee of the USSR, Moscow, Russia). He has more than 45 years of experience as a scientific researcher. For a long time, he was the leader of the gas sensor group and manager of various national and international scientific and engineering projects carried out in the Laboratory of Micro- and Optoelectronics, Technical University of Moldova. His research had financial support from international foundations and programs such as the CRDF, the MRDA, the ICTP, the INTAS, the INCO-COPERNICUS, the COST, and NATO. From 2007 to 2008, he was an invited scientist in the Korea Institute of Energy Research, Daejeon, South Korea. Then, until the end of 2017, he was a research professor at the School of Materials Science and Engineering at the Gwangju Institute of Science and Technology, Gwangju, South Korea. Currently, he is the chief scientific researcher at the Department of Physics and Engineering at the Moldova State University, Chisinau, Republic of Moldova.

Specialists from the former Soviet Union know G. Korotcenkov's research results in the field of study of Schottky barriers, metal oxide semiconductor structures, native oxides, and photoreceivers on the basis of III–Vs compounds, such as InP, GaP, AlGaAs, and InGaAs. His present scientific interests, dating from 1995, include material sciences, focusing on metal oxide film deposition and characterization; surface science; and the design of thin-film gas sensors and thermoelectric convertors. These studies were carried out in cooperation with scientific teams from Ioffe Institute (St. Petersburg, Russia), University of Michigan (Ann Arbor, MI, USA), Kiev State University (Kiev, Ukraine), Charles University (Prague, Czech Republic), St. Petersburg State University (St. Petersburg, Russia), Illinois Institute of Technology (Chicago, IL, USA), University of Barcelona (Barcelona, Spain), Moscow State University (Moscow, Russia), University of Brescia (Brescia, Italy), Belarus State University (Minsk, Belarus), South-Ukrainian University (Odessa, Ukraine).

G. Korotcenkov is the author or editor of 38 books, including the 11-volume *Chemical Sensors* series published by the Momentum Press (USA), 15-volume *Chemical Sensors* series published by Harbin Institute of Technology Press (China), three-volume *Porous Silicon: From Formation to Application* published by CRC Press (USA), two-volume *Handbook of Gas Sensor Materials* published by Springer (USA), and three-volume *Handbook of Humidity Measurement*, which is published by CRC Press (USA). In addition, at present, G. Korotcenkov is a series editor of the *Metal Oxides* series, which is published by Elsevier.

G. Korotcenkov is author and coauthor of more than 600 scientific publications, including 25 review papers, 38 book chapters, and more than 250 articles published in peer-reviewed scientific journals (h-factor = 41 [Scopus] and h-factor = 47 [Google Scholar citation]). He is a holder of 17 patents. He has presented more than 200 reports at national and international conferences, including 15 invited talks. G. Korotcenkov was co-organizer of several international conferences. His name and activities have been listed by many biographical publications, including *Who's Who*. His research activities have been honored by an Award of the Supreme Council of Science and Advanced Technology of the Republic of Moldova (2004); Prize of the Presidents of the Ukrainian, Belarus, and Moldovan Academies of Sciences (2003); and National Youth Prize of the Republic of Moldova in the field of science and technology (1980), among others. G. Korotcenkov also received a fellowship from the International Research Exchange Board (IREX, Washington, DC, 1998), Brain Korea 21 Program (2008–2012), and Brainpool Program (Korea, 2007–2008 and 2015–2017).

Section I

Introduction in Humidity Sensors

1 Introduction to Electronic and Electrical Humidity Sensors

1.1 INTRODUCTION

In Volume 1, the importance of controlling the humidity level of the surrounding atmosphere was shown, and spectroscopic methods were considered to allow realizing such control. This volume will show that electrical and electronic methods for measuring air humidity are the same effective methods as optical methods. Moreover, in many cases, their use is preferable, since it makes this process simpler, faster, cheaper, and more comprehensive.

Today is an age of automation and control. Therefore, in addition to interest in the properties of sensors, such as sensitivity, accuracy, and stability, there is also interest in durability of operation in different environments, component size, digitization, simple and quick system integration, and, last but not least, price (Romig and Smith 1998; Christian 2002). This means that modern humidity sensors should be able to integrate all these demands into one device. The experiment described in Chapters 10 through 24 showed that these capabilities can be fully realized in electric and electronic sensors manufactured using semiconductor, solid-state technology. Such humidity sensors can be fabricated with low cost and used for moisture control more conveniently. As a result, this class of solid-state devices, based on a variety of principles and materials, has become the solution of choice for mainstream applications, because technology advances are turning it into an accurate, compact, stable, and low-power alternative (see Chapters 10 through 22). Manufacturers of consumer products, such as home appliances, and residential heating and air conditioning systems, are the main consumers for such solid-state humidity sensors (Fenner and Zdankiewicz 2001). By integrating such sensors into various devices and continuously monitoring the level of humidity, it is possible to significantly increase an appliance's energy efficiency while optimizing its ability to preserve or prepare food and increase the efficiency of the work of heating, ventilation, and air conditioning (HVAC) systems designed for materials drying, creating comfortable living conditions, and air handling. The automotive industry is also a field of application of such sensors. Automotive applications currently include HVAC control inside the automobile (climate control), with new applications emerging for controlling the fuel efficiency of internal combustion engines by monitoring combustion air quality. Other applications of humidity sensors are described in Volume 1 of this series. The main applications of humidity sensors are shown in Figure 1.1. It should be noted that, in spite of the fact that humidity sensors on the sensor market have been present for several decades, interest in this problem is not reduced; this is illustrated by a constant increase in the number of articles published on this topic (see Figure 1.2).

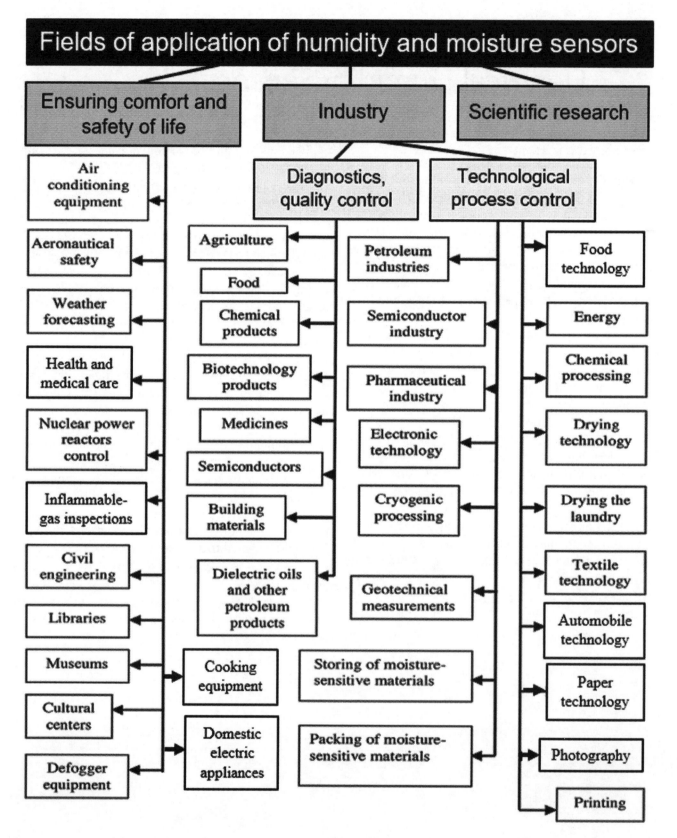

FIGURE 1.1 Fields of humidity sensors application. (Reprinted with permission from Blank, T.A. et al., Recent trends of ceramic humidity sensors development: A review, *Sens. Actuators B*, 228, 416–442, 2016, Copyright 2016, Elsevier.)

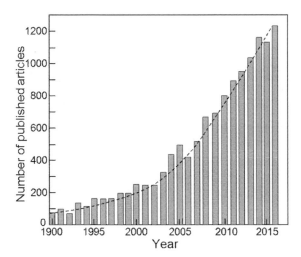

FIGURE 1.2 Diagram illustrates the growth in the number of articles published on topics related to the development and testing of humidity sensors. The diagram is based on Scopus data.

1.2 CLASSIFICATION OF HUMIDITY SENSORS

At present, there is a variety of devices based on different materials and operating on diverse principles, which can be applied for humidity measurement (Sager et al. 1996; Rittersma 2002; Xu 2004; Wang et al. 2005; Okcan and Akin 2007; Yeo et al. 2008; Khanna 2012; Alwis et al. 2013; Huang et al. 2016). For their classification, various approaches can be used (Traversa 1995; Fenner and Zdankiewicz 2001; Chen and Lu 2005; Farahani et al. 2014). For example, considering transduction mechanisms, we can distinguish five general categories of humidity sensors, which are the following: (1) optical, (2) electrochemical, (3) electrical (4) mass sensitive; and (5) thermometric (see Table 1.1). *Electrochemical sensors* (Chapter 19) are based on the detection of electroactive species involved in chemical recognition processes and use a charge transfer from a solid or liquid sample to an electrode, or vice versa. *Electrical sensors*, operating due to a surface interaction with water vapor, combine a large group of humidity sensors, such as polymer- and metal oxide–based capacitance (Chapter 10), semiconductor-resistive (Chapter 11), work-function type (Chapter 18), Schottky barrier (Chapter 17), and Thin-Film and Field-Effect Transistor (FET)-based sensors (Chapter 16) Rittersma 2002; Wang et al. 2005; Khanna 2012). *Mass-sensitive gas sensors* (Chapters 12 and 13), such as microcantilever (Chapter 13), quartz crystal microbalance (QCM)–and surface acoustic wave (SAW)–based sensors (Chapter 12), rely on disturbances and changes to the mass of the sensor surface during interaction with chemicals. *Optical gas sensors* transform changes in optical phenomena that result from an interaction of the water vapor with the receptor part (Xu 2004; Yeo et al. 2008; Alwis et al. 2013). Such sensors were considered in detail in Volume 1. *Thermometric sensors* (Chapter 14) convert the temperature changes into electrical signals, such as the change of the resistance, current, and voltage (Sager et al. 1994; Okcan and Akin 2004). Of course, humidity sensors may exhibit characteristics that fall into more than one of these five broad categories. For example, some mass sensors may rely on electrical excitation or optical settings. Given that the optical sensors have been analyzed in detail in Volume 1 of our text, naturally they will not be considered in this volume.

We need to point out that our classification represents but one of the possible alternatives. Detection principles can also be used for humidity sensor classification. According to detection principles, commonly used humidity sensors can be classified into the following three groups: (1) sensors based on reactivity of water vapor, (2) sensors based on physical properties of water vapor, and (3) sensors based on water sorption (see Table 1.2).

No doubt, humidity sensors can be graded according to the approach used for design of humidity monitoring devices. According to this approach, humidity monitoring devices come in two main types: portable devices and fixed instruments for humidity measurement. Fixed-type detectors are generally mounted near the process area or in the room that should be controlled. Generally, they are installed for continuous monitoring; therefore, they are connected with a system dealing with the collection and processing of information received.

According to the technology used for humidity sensor fabrication, they can be classified as ceramic, polymer, thin-film, and thick-film sensors. Technologies of

TABLE 1.1
Classification of Humidity Sensors

Class of Humidity Sensors	Operating Principle
Electrochemical	Changes in current, voltage, capacitance/impedance
Electrical	Changes in conductivity, work function, and electrical permittivity (capacitance)
Mass-sensitive	Changes in weight
Optical devices	Changes in light intensity, color, or emission spectra
Thermometric	Heat effects of a specific chemical reaction. Changes in temperature, heat flow, heat content

TABLE 1.2
Classification of Humidity Sensors According to Detection Principle

Detection Principle	Examples
Sensors based on reactivity of gas	Electrochemical sensors
	Semiconductor sensor
	Colorimetric paper tape
	Chemoluminescence
	Schottky barrier/hetero contact sensor
	FET-based sensors
Sensor based on physical properties of gas	Nondispersive infrared absorption
	UV absorption
	Thermal conductive sensor
	Gas ionization
Sensors based on gas sorption	Polymer sensors (swelling)
	Fiberoptic sensors
	Capacitive sensors
	Mass-sensitive (quartz crystal microbalance [QCM], surface [SAW], and bulk acoustic wave [BAW], microcantilevers)

humidity sensors fabrication will be considered in detail in Volume 3 of our series. Micromachined humidity sensors (Section 13.6), designed during last decades, can also be referred to this principle of classification. One should note that bulk and surface micromachining processes are the next step towards microminiaturization of sensors aimed at further improving sensors' parameters (Fenner and Zdankiewicz 2001; Qui et al. 2001; Nizhnik et al. 2011; Zaghloul and Voiculescu 2011; Huang et al. 2016). A new generation of sensors fabricated using micromachining technology should have small size, rapid response time, low operating voltages that are compatible with low-power complementary metal–oxide–semiconductor (CMOS) devices, and greatly reduced power consumption. In this case, a humidity microsensor becomes just another part on the circuit board. As for the current state, most of the humidity sensors on the market are still manufactured using traditional thin-film and thick-film technologies and materials borrowed from hybrid microelectronics.

Humidity sensors can also be clustered according to their working temperature. This approach to classification gives two large groups: "hot sensors" and "cold sensors." The group of so-called "hot sensors" mainly clusters the different types of metal oxide–heated humidity sensors. This also includes thermal conductive sensors (Chapter 11). "Cold sensors" operate at ambient temperature. Most humidity sensors belong to this group. This group of sensors clusters the capacitive sensors, piezoelectric crystal sensors (bulk acoustic wave sensors, surface acoustic wave sensors), optical sensors, and conducting polymer sensors. However, it is important to note that devices related to this group can also work at elevated temperatures.

The atmosphere in which humidity sensors work can also be used to classify sensors. Currently, one can find sensors that were developed and tested in the atmospheres of N_2 (Korsah et al. 1998; Saha et al. 2008), Ar (Basu et al. 2001; Li et al. 2001), H_2 (Neumeier et al. 2008), O_2 (Basu et al. 2001), O_2/N_2 (Maskell and Page 1999), and even such gas mixtures as $Ar/He/N_2/O_2$ (Saha et al. 2008).

A classification could also consider materials used as humidity-sensitive materials (Schubert and Nevin 1985; Traversa 1995; Pokhrel et al. 2003; Chen et al. 2005; Kuang et al. 2007; Kassas 2013; Wang et al. 2013). In accordance with this classification, there are polymer (Sections 10.2, 11.2), ceramic (Section 10.3), ionic salt (Section 11.2), metal oxide (Section 10.3, 11.3), carbon-based (Section 12.5), and semiconductor humidity sensors (Chapter 16). Humidity sensors using nanowires, nanofibers, and *p-n* hetero-junctions are subclasses of the ceramic (inorganic) type. Regarding the intrinsic properties of sensing elements, ceramic and polymeric types can be designed by utilizing their conducting or dielectric properties. Materials acceptable for application in humidity sensors will be considered in detail in Volume 3 of our text.

Humidity sensors can also be classified according to the approach to collecting information. Some devices carry out the measurement discretely, others permanently. Discrete methods are those that require a finite sample of the test gas, which yields a single measurement after execution of one or more operational procedures. Continuous methods are those that continually sample the test gas and yield a continuous indication or recording of the humidity.

It should be recognized that the classification of humidity sensors according to the measurement unit is the most common approach. According to the measurement unit, humidity sensors can be divided into two types: relative humidity (RH) sensors and absolute humidity (trace moisture) sensors. At the most, humidity sensors are RH sensors—that is, they measure the ratio of moisture in the air to the highest amount of moisture at a particular air temperature, which is RH (see Chapter 1 in Volume 1). RH sensors are the most commonly used because they are generally simpler and thus cheaper and therefore extensively applied in applications involving indoor air quality and human comfort issues (Kulwicki 1991). Accordingly, in research laboratories and public applications, RH sensors are ubiquitously applied to simplify the design process and further use as secondary sensors. Absolute humidity sensors are mostly used for traceable purposes (trace moisture measurement as dew/frost point [D/F PT], parts per million by weight [PPMw] or volume [PPMv]). These sensors can be used as primary sensors, because they describe the absolute amount of water vapor in gaseous environments. It is important to note that every type of humidity sensor mentioned above has different humidity-sensing behavior. More detailed description of humidity sensors, including construction and principles of operation, will be presented in the following chapters.

In closing, some definitions of sensor characteristics are summarized in Table 1.3. Most of these characteristics are given in manufacturers' data sheets. However, information on the reliability and robustness of a sensor is rarely given in a quantitative manner.

1.3 MATERIALS ACCEPTABLE FOR APPLICATION IN HUMIDITY SENSORS

The results of numerous studies have shown that, in theory, any material can be used in the design of a gas sensor, regardless of its physical, chemical, structural, or electrical properties (Korotcenkov 2010, 2011). Prototypes of humidity sensors based on polymers, metal oxides, semiconductors, solid electrolytes, ionic membranes, carbon nanotubes, graphene, organic semiconductors, and ionic salts have been already tested. It was shown that these materials may be used successfully in the design of humidity sensors of various types. It was also established that some of these materials show excellent adaptability for application in humidity sensors. This means that in the present book, we must consider a variety of humidity-sensing layers to form a real conception of materials suitable for application in humidity sensors.

TABLE 1.3
Summary of Main Sensor Characteristics

Parameter	Definition
Sensitivity	The slope of the output characteristic curve ($\Delta y/\Delta x$)
Sensitivity error	A departure from the ideal slope of the characteristic curve
Range	Maximum minus minimum value of the measured stimulus
Dynamic range	Total range of the sensor from minimum to maximum
Full-scale output	Algebraic difference between the electrical output with maximum and minimum input stimuli
Resolution	Smallest measurable increment in measured stimuli
Detection limit	The lowest concentration value that can be detected by the sensor in question, under definite conditions. Whether the analyte can be quantified at the detection limit is not determined
Sensing frequency	Maximum frequency of the stimulus that can be detected
Response time	Time required for a sensor output to change from its previous state to a final, settled value within an error tolerance band of the correct new value
Recovery time	Time required for recovery initial state of sensors after interaction with analyte
Accuracy	Error of measurement, in percent full-scale deflection
Hysteresis	Capability to follow the changes in the input parameter, regardless of the direction
Precision	Degree of reproducibility of a measurement (repeatability, reproducibility)
Drift	Long-term stability (deviation of measurement over a time period)
Zero drift	Percentage change in the zero point or baseline of a gas sensor or gas detection system over a specific period of time
Linearity	Extent to which the actual measured curve of a sensor departs from the ideal curve
Dynamic linearity	Ability to follow rapid changes in the input parameter, amplitude and phase distortion characteristics, response time
Monotonicity	The dependent variable always either increases or decreases as the independent variable increases
Saturation	No desirable output with further increase in stimuli (span-end nonlinearity)
Offset	Output exists when it should be zero
Dead band	Insensitivity of a sensor in a specific range of input signals
Size	Leading dimension of sensors
Weight	Weight of sensors

Source: Reprinted with permission from Atashbar, M.Z. et al., Basic principles of chemical sensors operation, In: G. Korotcenkov (ed.), *Chemical Sensors: Fundamentals of Sensing Materials, Vol. 1, General Approaches*, pp. 1–62. Copyright 2010, Momentum Press, New York.

In Volume 1 of this text, the main phenomena responsible for humidity-sensing effects in polymers and metal oxides have already been considered. It has been shown that the principle of operation of the humidity devices is based on the effect of water vapor on the properties of humidity-sensitive materials through five mechanisms: diffusion, adsorption, coordination chemistry interactions, chemisorption, and capillary condensation. At low temperatures, adsorption involves a weak interaction between water molecules and the surface of the sensing film through van der Waals forces or acid-base interactions, depending upon the type of film material used. Van der Waals forces are a low energy balance between molecular attractive and repulsive forces involving reaction energies on the order of 0–10 kJ/mole (0–0.1 eV). Acid-base interactions involve proton or electron pair interaction between a target gas and molecules of the sensing film, with reaction energies usually less than 40 kJ/mole (0.4 eV) (Nieuwenhuizen and Barendsz 1987). Coordination chemistry interactions involve films containing metal-ligand complexes, such as metallophthalocyanines. Because of the low energies involved, adsorption and coordinate chemistry reactions are fully reversible (Fenner and Zdankiewicz 2001). Reversible reactions are essential if the sensor is to continue working after its first exposure. As for chemisorption processes, they can also be involved in humidity-sensitive effects. However, the dominant role in sensing effects they play in ceramics-based humidity sensors at temperatures above 200°C, because chemisorption involves strong vapor/film molecular interactions where chemical bonds are broken and formed with reaction energies around 300 kJ/mole (Traversa 1995). Detailed descriptions of humidity sensor operation and approaches to humidity sensor simulation can be found in many reviews, including those of Yamazoe and Shimizu (1986), Traversa (1995), and Khanna (2012), and on the sites of various companies (e.g., ww6.hygrometrix.com).

Regarding capillary condensation, this phenomenon in sensing effects is involved when considering the fine structure of the porous matrix. According to the basic theory of adsorption on porous matrix (Adamson and Gast 1997), when the vapor molecules are first physicosorbed onto the porous material, capillary condensation will occur if the micropores are narrow enough. The critical size of pores for a capillary condensation effect is characterized by Kelvin radius. In the case of water, the condensation of vapor into the pores can be expressed with a simplified Kelvin equation (Ponec et al. 1974):

$$r_K = \frac{2\gamma V_M cos\theta}{\rho RT ln(\%RH/100)} \quad (1.1)$$

where γ is the surface tension of vapor in the liquid phase; V_M is molecular volume; θ is contact angle; ρ is the density of vapor in liquid phase. In this equation, the thickness of the adsorbed layer has been ignored. While V_M and surface tension γ are constants at room temperature, the possibility to control the condensation by simply changing the contact angle θ becomes attractive. It was shown that the Kevin radius increases with the relative humidity, and the rate of change (the slope) also increases with relative humidity. This means that the pores with smaller diameter are filled first, while bigger pores are filled later.

The result of the interaction of humidity-sensitive material with water vapors, as stated in Volume 1, may be the change in the parameters, such as effective dielectric permittivity, mass, resistance, and volume (swelling effect). The latter effect is realized in polymers. When using nanocomposites, such as a polymer-conductive additive, the swelling effect can be accompanied by a strong change in the conductivity, which can also be used to measure humidity. Swelling affects the conductive pathways between conductive particles such as metals, carbon nanotubes (CNTs), graphene, and conductive metal oxides, resulting in an increase in the film's bulk resistance (see Section 11.2).

Experience shows that materials aimed for using in humidity sensors developed for sensors market should meet tough requirements. For example, based on the analysis carried out by Wolfbeis (2005), before incorporating a humidity-sensing material in a real device, its quality should be estimated by the number of yes or no answers to the following questions:

- Does the tested material give a high signal-to-noise ratio?
- Does the response of the sensing material strongly change with the analyte?
- Is the sensor material stable over time in storage?
- Do the components used for making the sensor material not leach or deteriorate on exposure to the (often complex or aggressive) analyte?
- Is the humidity indicator stable during longtime operation?
- Does the sensor material adequately adhere to the support?
- Can the analytical signal be referenced to another signal?
- Can the optical response of the material be described (or at least modeled) by a fairly uncomplicated mathematical equation?
- Are the materials used affordable, and do they come (or can they be made) in constant quality?

- Can the sensor material, be made at adequate cost?
- Has it been demonstrated that the materials used are nontoxic?

According to Wolfbeis (2005), some of the above criteria are so-called "killer criteria."

Without a doubt, parameters of sensing materials can be optimized with regard to their use in humidity sensors. The problem of optimization exists while designing or manufacturing any electronic device. However, in the case of humidity sensors of many types, this problem has some specific peculiarities due to the absence of strict quantitative theory that would describe their operation. Since the number of physical and chemical parameters that characterize sensor properties is large, and some of these parameters are difficult to control, the problem of optimization is largely empirical and remains a kind of art. At present, therefore, the field of humidity sensors is characterized by a search for optimal sensing materials and design of technology that allows achieving the optimal parameters of those humidity-sensing materials. At present, the optimal parameters of humidity-sensing materials are understood as follows (Yamazoe and Shimizu 1986; Randin and Zullig 1987; Geisslinger 1997; Fenner and Zdankiewicz 2001; Chen and Lu 2005): (1) high response to the water vapor, (2) low cross-sensitivity with gases present in the atmosphere, (3) long operating lives, (4) fast and reversible interaction with analytes, (5) absence of long-term drift, (6) effective low-cost technology, (7) high reproducibility, and (8) uniform and strong binding to the surface of the substrate.

The specified set of parameters indicates that the materials for humidity sensors have to possess a specific combination of physical and chemical properties, and not every material can fulfill these requirements. A systematic consideration of the required properties of materials for humidity sensor applications indicates that the key properties that determine a specific choice include: (1) adsorption ability; (2) electronic, electrophysical, and chemical properties; (3) catalytic activity; (4) permittivity; (5) thermodynamic stability; (6) crystallographic structure; (7) interface quality; (8) compatibility with current or expected materials to be used in processing; and (9) reliability. Many different materials appear to be favorable in terms of some of these properties, but very few materials are promising with respect to all of these requirements. Despite the naming of many humidity sensitive materials in this text, only a few can be considered as materials for use in real humidity sensors aimed for market. For example, at present, a majority of humidity sensors available on the sensors market are built with sensing materials such as metal oxides, polymers, and polyelectrolytes (PEs) only (see Chapters 10 through 12 and 19), which can be referred to as traditional or conventional sensing materials. PEs are a special class of modified polymers in which one type of an ionic chemical radical group is fixed to the repeat units of the polymer backbone to form a single-ion conducting material. The introduction of water vapor to a PE film under a voltage bias will hydrolyze the ionic groups, resulting in a flow of ions. Film conductivity can be measured as ionic impedance (i.e., AC resistance) and will vary in proportion to the water vapor concentration present. Table 1.4 provides examples of materials that are most widely used in the development of humidity sensors.

TABLE 1.4

The Most Suitable Materials for the Development of Humidity Sensors

Type of Sensing Material	Candidate Materials	Measurement Scheme
Porous ceramic	$MgCr_2O_4$-TiO_2; TiO_2-V_2O_5; $ZnCr_2O_4$-$LiZnVO_4$	Impedance (ionic)
	$Sr_{1-x}La_xSnO_3$; ZrO_2-MgO	Impedance (electronic)
	Al_2O_3; Ta_2O_4-MnO_2	Capacitive
Polymer	Cellulose acetate; cellulose acetate buthyrate	Capacitive; mass-sensitive
	Polyimide	Capacitive; mass-sensitive; piezoresistive
	Phthalocyanine	Mass-sensitive
	Crosslinked hydrophobic acrylic polymer-carbon nanoparticles	Impedance (electronic)
Polyelectrolyte	Polyvinyl acetate-LiCl; sulfonated polysulfone; 2-hydroxy-3-methacryl-oxypropyl trimethyl ammonium chloride + methacrylic ester; polystyrene sulfonate + vinyl polymer + N, N'-methylene-bis-acrylamide	Impedance (ionic)

Source: Data extracted from Fenner, R. and Zdankiewicz, E., *IEEE Sensor J.*, 1, 309–317, 2001.

Note that, in addition to specific electronic, electrophysical and chemical properties, humidity-sensitive materials should also have specific structural properties. The experiment has shown that the performance of a humidity sensor is largely controlled by its nano- and microscopic structure, including the pore size, thickness of the porous layer, size distribution of the surface structural unit, regularity of the surface morphology, and electrode distance (Kim et al. 2000; Di Francia et al. 2002). It has been found that porous films exhibit higher humidity sensitivity than the nonporous counterparts. This fact indicates that the regularity and controllability of porous structures are of great importance in sensor design (Shah et al. 2007). For example, due to a water condensation in nanosize pores, it is difficult to prepare rapid humidity sensors with absence of hysteresis on the basis of nanoporous material (Bjorkqvist et al. 2004). Water transport in nanoporous material is described by Knudsen diffusion, which slows with decreasing pore size. At the same time, reducing pore size in many cases contributes to an increase in sensitivity of the humidity sensor. Experiment has shown that the characteristics of humidity sensors are also strongly affected by the hydrophilic/hydrophobic properties of the walls (Yarkin 2003). Based on the Kelvin equation, the Kelvin radius becomes smaller when the surface becomes more hydrophobic. It is also important to note that dopants, which can be added to the humidity-sensitive materials as catalysts to promote the dissociation of adsorbed water into hydrogen and hydroxyl ions, can also have a significant impact on humidity-sensing properties of these materials (Chapters 10 and 11).

The processes mentioned in this chapter as applied to various humidity-sensitive materials will be considered in subsequent chapters in more detail.

1.4 REQUIREMENTS FOR HUMIDITY SENSORS

In the conclusion of this chapter, we list the most important requirements for humidity sensors. According to the findings of numerous developers and humidity sensor users, sensors designed for real applications should have the following:

- *Good reproducibility*: Reproducibility is a very important characteristic in regards to reliability. Reproducibility is the precision of a set of measurements taken over a long time interval, or performed by different operators, or in different laboratories. For a sensor to be useful, its output should be consistent with time and not depend on the operator.
- *Good repeatability*: The precision of a set of measurements taken over a short time interval. A high measurement accuracy cannot be achieved with poor repeatability.
- *High sensitivity*: An ideal sensor will have a large and constant sensitivity. The sensors with high sensitivity may relieve the design complications of support electronic circuits. If the sensitivity is too low, large signal amplification and a large number of parts are required to meet a satisfactory sensor output. In relation to output signal quality, the signal-to-noise ratio may also degrade.
- *Small hysteresis*: Hysteresis is the difference of sensor output between increasing and decreasing humidity. Sensors with small hysteresis are useful for precise reading.
- *Good linearity*: Linearity is the closeness of the calibration curve to a specified straight line. Linearity may not be a major issue, because compensation is possible with a microprocessor. However, good linearity reduces the demand for extra electronic circuits, parts, and program effort.
- *Wide measuring range*: A wide measuring range allows limiting the number of devices needed to solve various tasks during the process of atmosphere monitoring.
- *Fast response speed*: Response time may become important in the application, such as industrial processes and medical facilities.
- *Short time to operational status*: After storage in inappropriate conditions, some types of devices may lose the ability to measure, and special long-term treatments are often required to restore their working functions. Short time to operational status provides a quick transition of the sensor to the working state.
- *High durability*: Durability can be classified into mechanical and chemical aspects. Mechanical durability is determined by the type of substrate, sensor structure, and packaging method. Chemical durability is determined by the type of sensing materials.
- *Negligible temperature dependence*: Negligible temperature dependence is useful to simplify the sensor design and support electronics and programming for compensation.
- *Resistance against contamination*: Contaminations, such as smoke, oil, organic solvent, and acidic/

alkaline chemicals, degrade the performance of the sensor, resulting in misleading sensor outputs. Selection of humidity-sensing material should reflect intended application area of humidity sensors. Recovery methods from contaminations may be employed in certain applications for more reliable sensors.
- *Low cost*: Although performance of the sensor may be excellent, if the production cost is very high, this may become a disincentive to entering markets. To avoid this problem, design and production should be reviewed carefully.
- *Easy connection to control units*: This sensor feature reduces the demand for extra electronic circuits, parts, and program effort.

1.5 PURPOSE AND CONTENT OF BOOK

As shown in previous sections, many types of devices can be used for humidity control (Wexler 1965; Nitta 1981; Fleming 1981; Spomer and Tibbitts 1997; Wiederhold 1997; Rittersma 2002; Srivastava 2012). Currently, there are a large number of published books on chemical sensors in which one can find information regarding the development and use of humidity sensors (Nenov and Yordanov 1966; Fraden 2004; Comini et al. 2009; Korotcenkov 2011–2012, 2012–2013, 2013). However, as a rule, in such books, humidity sensors can be described in a particular chapter, considering some specific types of sensors or the properties of specific humidity-sensitive materials. On the basis of these publications, it is difficult to form a general idea about developments in the field of humidity sensors. The collection of books devoted to consideration of the direct features of humidity measurement is very limited (Wexler 1965; Teweles and Giraytys 1970; Wiederhold 1997; Shallcross 1997; Bentley 1998; Harriman III 2002; Bas 2003; Purushothama 2010; Castillo 2011; Okada 2011; Liu et al. 2013; Kämpfer 2013; Thomson 2013; Wernecke and Wernecke 2014). Moreover, most of these books were published long ago and may not reflect the current state of research and development. Also, typically, these books do not encompass the variety of approaches that can be used to develop humidity sensors. In some cases, the consideration is too superficial. Therefore, to obtain comprehensive information about humidity sensors, we must study several books and reviews written by different authors with various views on the approaches to the analyzed object. In this publication, we have tried to close this gap and present a more general view. We offer a book that considers, in detail, different transduction methods and all possible approaches to the development of humidity sensors, including the possible configurations, materials sensitive to humidity, and technologies that can be used in the manufacture of these devices. Consideration of these topics includes the achievements of recent years. At the same time, we do not forget that conventional devices of humidity measurements, so-called hygrometers, are widely used. Their operation is based, in many cases, on somewhat different principles in comparison with electronic and electric humidity sensors (see Table 1.5). A general review of the state of the art in the field of conventional hygrometry is given by Wexler (1965), Fleming (1981), Yamazoe and Shimizu (1986), Sonntag (1994), Wiederhold (1997), Spomer and Tibbitts (1997) and Visscher (1999). Of course, not all of these devices have a sufficiently high sensitivity and operation speed. Some of them are simple, while others are characterized by increased complexity. Some devices are easy to use, while others require a considerable amount of skills. However, it seems that, along with information on modern approaches, it is of interest for the reader to obtain information about devices, some of which were developed several centuries ago. In addition, some of these devices are mandatory in the calibration of electronic and electrical humidity sensor (see, for example, Chapters 2 and 9).

TABLE 1.5
Principles of Humidity Measurement in Conventional Hygrometers

Principle of Operation	Examples of Instruments
Reduction of temperature by evaporation	Psychrometer
Dimensional changes due to absorption of moisture, based on hygroscopic properties of materials	Hygrometers with sensors of hair, wood, natural and synthetic fibers
Chemical or electrical changes due to absorption or adsorption	Electrical hygrometers such as Dunmore cell, lithium chloride, carbon and aluminum oxide strips
Formation of dew or frost by artificial cooling	Cooled mirror surfaces
Diffusion of moisture through porous membrane	Diffusion hygrometers
Absorption spectra of water vapor	Infrared and UV absorption; Lyman-alpha radiation hygrometers

Source: Data extracted from Middleton, W.E.K. and Spilhaus, A.F., *Meteorological Instruments*, University of Toronto Press, Toronto, Canada, 1953.

This book is thus organized as follows. It begins with brief review of conventional methods of humidity measurements (Section I). In Section II, modern humidity sensors are discussed. Here, we analyze principles and construction of various electric and electronic humidity sensors. The main humidity-sensitive materials, namely polymers and metal oxides, that can be used in various electric and electronic humidity sensors are reviewed in this part of this book. Other humidity-sensitive materials are analyzed in Volume 3 of our series. There will also be a description of the technologies that are used or can be applied for humidity sensor fabrication. A general comparison of the methods acceptable for humidity control, description of calibration process, and analysis of the market of electronic humidity sensors can be found in Volume 3 as well. We do not analyze the methods intended for measuring soil moisture, which is a special topic for consideration. A description of these methods can be found in the reviews of Robinson et al. (2008), Robens et al. (2011), Alwis et al. (2013), and Lekshmi et al. (2014).

Analyzing the contents of these books, we conclude that our books can justifiably be regarded as an encyclopedia of techniques, materials, and technologies used in the measurement of humidity, as well as a handbook that one can use as a guide when developing or refining humidity sensors. Despite the fact that these books are devoted to humidity sensors, they can be used as a basis for studying the principles of functioning and development of any gas sensors.

REFERENCES

Adamson A.W., Gast A.P. (1997) *Physical Chemistry of Surface*. Wiley, New York.

Alwis L., Sun T., Grattan K.T.V. (2013) Optical fibre-based sensor technology for humidity and moisture measurement: Review of recent progress. *Measurement* 46, 4052–4074.

Atashbar M.Z., Krishnamurthy S. and Korotcenkov G. (2010) Basic principles of chemical sensors operation. In: G. Korotcenkov (ed.) *Chemical Sensors: Fundamentals of Sensing Materials*, Vol. 1: General Approaches, pp. 1–62. Copyright 2010, Momentum Press.

Bas E. (ed.) (2003) *Indoor Air Quality: A Guide for Facility Managers*. Fairmont Press, Lilburn, Georgia, 383 p.

Basu S., Chatterjee S., Saha M., Bandyopadhay S., Mistry K.K., Sengupta K. (2001) Study of electrical characteristics of porous alumina sensors for detection of low moisture in gases. *Sens. Actuators B* 79, 182–186.

Bentley R.E. (ed.) (1998) *Handbook of Temperature Measurement, Vol. 1: Temperature and Humidity Measurement*. Springer Verlag, Singapore, p. 240.

Bjorkqvist M., Salonen J., Paski J., Laine E. (2004) Characterization of thermally carbonized porous silicon humidity sensor. *Sens. Actuators A* 112, 244–247.

Blank T.A., Eksperiandova L.P., Belikov K.N. (2016) Recent trends of ceramic humidity sensors development: A review. *Sens. Actuators B* 228, 416–442.

Castillo J.M. (ed.) (2011) *Relative Humidity: Sensors, Management, and Environmental Effects*. Nova Science Publishers, New York, p. 242.

Chen Z., Lu C. (2005) Humidity sensors: A review of materials and mechanisms. *Sens. Lett.* 3, 274–295.

Chen Y.S., Li Y., Yang M.J. (2005) Humidity sensitive properties of NaPSS/MWNTs nanocomposites. *J. Mater. Sci.* 40, 5037–5039.

Christian S. (2002) New generation of humidity sensors, *Sensor Rev.* 22 (4), 300–302.

Comini E., Faglia G., Sberveglieri G. (eds.) (2009) *Solid State Gas Sensing*. Springer, New York, p. 337.

Di Francia G., Noce M.D., Ferrara V.L., Lancellotti L., Morvillo P., Quercia L. (2002) Nanostructured porous silicon for gas sensor application. *Mater. Sci. Technol.* 18, 767–771.

Farahani H., Wagiran R., Hamidon M.N. (2014) Humidity sensors principle, mechanism, and fabrication technologies: A comprehensive review. *Sensors* 14, 7881–7939.

Fenner R., Zdankiewicz E. (2001) Micromachined water vapor sensors: A review of sensing technologies. *IEEE Sensor J.* 1 (4), 309–317.

Fleming W.J. (1981) A physical understanding of solid state humidity sensors. *SAE Technical* 1656–1667. Paper 810432.

Fraden J. (2004) *Handbook of Modern Sensors*. Springer, New York.

Geisslinger C. (1997) Choosing the right humidity sensor. *Sensors* 1997, 38–41.

Harriman III L.G. (ed.) (2002) *The Dehumidification Handbook*, Munters Corporation, Amesbury, MA.

Huang J.-Q., Li F., Zhao M., Wang K. (2016) A surface micromachined CMOS MEMS humidity sensor. *Micromachines* 6, 1569–1576.

Kämpfer N. (ed.) (2013) *Monitoring Atmospheric Water Vapour. Ground-Based Remote Sensing and In-situ Methods*. Springer, New York, 242 p.

Kassas A. (2013) Humidity sensitive characteristics of porous Li-Mg-Ti-O-F ceramic materials. *Am. J. Anal. Chem.* 04, 83–89.

Kim S.J., Park J.Y., Lee S.H., Yi S.H. (2000) Humidity sensors using porous silicon layer with mesa structure. *J. Phys. Appl. Phys.* 33, 1781–1784.

Khanna V.K. (2012) Detection mechanisms and physicochemical models of solid-state humidity sensors, In: Korotcenkov G. (ed.) *Chemical Sensors: Simulation and Modelling, Vol. 3: Solid State Devices*. Momentum Press, New York, pp. 137–190.

Korotcenkov G. (ed.) (2010–2011) *Chemical Sensors*, Vols. 1–6, Momentum Press, New York.

Korotcenkov G. (ed.) (2012–2013) *Chemical Sensors: Simulation and Modeling*. Vols. 1–5, Momentum Press, New York.

Korotcenkov G. (2013) *Handbook of Gas Sensor Materials*, Vols. 1 and 2, Springer, New York.

Korotcenkov G. (ed.) (2011) *Chemical Sensors*. Vol. 4. Momentum Press, New York.

Korsah K., Ma C.L., Dress B. (1998) Harmonic frequency analysis of SAW resonator chemical sensors: Application to the detection of carbon dioxide and humidity. *Sens. Actuators B* 50, 110–116.

Kuang Q., Lao C., Wang Z.L., Xie Z., Zheng L. (2007) High-sensitivity humidity sensor based on a single SnO_2 nanowire. *J. Am. Chem. Soc.* 129, 6070–6071.

Kulwicki B. (1991) Humidity sensors. *J. Am. Ceram. Soc.* 74, 697–708.

Lekshmi S.U.S., Singh D.N., Baghini M.S. (2014) A critical review of soil moisture measurement. *Measurement* 54, 92–105.

Li Y., Yang M.J., Camaioni N., Casalbore-Miceli G. (2001) Humidity sensors based on polymer solid electrolytes: Investigation on the capacitive and resistive devices construction. *Sens. Actuators B* 77, 625–631.

Liu X., Jiang Y., Zhang T. (2013) *Temperature and Humidity Independent Control (THIC) of Air-conditioning System*, Springer-Verlag, Berlin, Germany, p. 356.

Maskell W.C., Page J.A. (1999) Detection of water vapour or carbon dioxide using azirconia pump-gauge sensor. *Sens. Actuators* B 57, 99–107.

Middleton W.E.K., Spilhaus A.F. (1953) *Meteorological Instruments*, University of Toronto Press, Toronto, Canada.

Nenov T., Yordanov S.P. (1966) *Ceramic Sensors: Technology and Applications*. CRC Press, Boca Raton, FL.

Neumeier S., Echterhof T., Bölling R., Pfeifer H., Simon U. (2008) Zeolite based trace humidity sensor for high temperature applications in hydrogen atmosphere. *Sens. Actuators B* 134, 171–174.

Nieuwenhuizen M.S., Barendsz A.W. (1987) Processes involved at the chemical interface of a SAW chemosensor. *Sens. Actuators* 11, 45–62.

Nitta T. (1981) Ceramic humidity sensor. *Ind. Eng. Chem. Prod. Res. Dev.* 20, 669–674.

Nizhnik O., Higuchi K., Maenaka K. (2011) A standard CMOS humidity sensor without post-processing. *Sensors* 11, 6197–6202.

Okada C.T. (ed.) (2011) *Humidity Sensors: Types, Nanomaterials and Environmental Monitoring*. Nova Science, New York, p. 187.

Okcan B., Akin T. (2004) A thermal conductivity based humidity sensor in a standard CMOS process, In: *Proceedings of the 17th IEEE International Conference on Micro Electro Mechanical Systems (MEMS)*, 25–29 2004, Maastricht, the Netherlands. doi:10.1109/MEMS.2004.1290644.

Okcan B., Akin T. (2007) A low-power robust humidity sensor in a Standard CMOS Process. *IEEE Trans. Electron. Dev.* 54 (11), 3071–3078.

Pokhrel S., Jeyaraj B., Nagaraja K.S. (2003) Humidity-sensing properties of $ZnCr_2O_4$-ZnO composites. *Mater. Lett.* 22–23, 3543–3548.

Ponec V., Knor Z., Cerný S. (1974) *Adsorption on Solids*. Butterworth, London, UK, p. 405.

Purushothama B. (2010) *Humidification and Ventilation Management in Textile Industry*. CRC Press, Boca Raton, FL, p. 410.

Qui Y.Y., Azeredo-Leme C., Alcacer L.R., Franca J.E. (2001) A CMOS humidity sensor with on-chip calibration. *Sens. Actuators A* 92 (1–3), 80–87.

Randin J.-P., Zullig F. (1987) Relative humidity measurements using a coated piezoelectric quartz crystal sensor. *Sens. Actuators* 11, 319–328.

Rittersma Z.M. (2002) Recent achievements in miniaturised humidity sensors—A review of transduction techniques. *Sens. Actuators A* 96, 196–210.

Robens E., Rübner K., Klobes P., Balköse D. (2011) Water vapour sorption and humidity—A survey on measuring methods and standards, In: Okada C.T. (ed.) *Humidity Sensors: Types, Nanomaterials and Environmental Monitoring*. Nova Science Publishers, Hauppauge, NY, pp. 1–54.

Robinson D.A., Campbell C.S., Hopmans J.W., Hornbuckle B.K., Jones S.B., Knight R. (2008) Soil moisture measurement for ecological and hydrological moisture shed-scale observatories: A review. *Vadose Zone J.* 7 (1), 358–389.

Romig Jr. A.D., Smith J. (1998) The coming revolution in ICs: Intelligent, integrated microsystems. *Micromach. Dev.* 3 (2), 4–6.

Sager K., Gerlach G., Schroth A. (1994) A humidity sensor of a new type. *Sens. Actuators B* 18, 85–88.

Sager K., Schroth A., Nakladal A., Gerlach G. (1996) Humidity-dependent mechanical properties of polyimide films and their use for IC-compatible humidity sensors. *Sens. Actuators A* 53, 330–334.

Saha D., Das S., Sengupta K. (2008) Development of commercial nanoporous trace moisture sensor following sol–gel thin film technique. *Sens. Actuators B* 128, 383–387.

Schubert P., Nevin J. (1985) A polyimide-based capacitive humidity sensor. *IEEE Trans. Electron Dev.* ED-32, 1220–1223.

Shah J., Kotnala R.K., Singh B., Kishan H. (2007) Microstructure-dependent humidity sensitivity of porous $MgFe_2O_4$-CeO_2 ceramic. *Sens. Actuators B* 128, 306–311.

Shallcross D. (1997) *Handbook of Psychrometric Charts. Humidity Diagrams for Engineers*. Springer, Dordrecht, the Netherlands.

Sonntag D. (1994) Advancements in the field of hygrometry. *Zeitschrift Meteorologie* 3 (2), 51–66.

Spomer L.A., Tibbitts T.W. (1997) Humidity. In: Langhans R.W., Tibbitts T.W. (eds.), *Plant Growth Chamber Handbook*. Committee on Controlled Environment Technology and Use, NCERA-101, Iowa Agricultural and Home Economics Experiment Station, Ames, IA. pp. 43–64.

Srivastava R. (2012) Humidity sensor: An overview. *Intern. J. Green Nanotechnol.* 4 (3), 302–309.

Teweles S., Giraytys J. (eds.) (1970) *Meteorological Monographs* Vol. 11, No. 33. American Meteorological Society, Lancaster Press, Lancaster, PA., 455 p.

Thomson G. (2013) *The Museum Environment*, 2nd edn. Butterworth-Heinemann, Oxford, UK, 312 p.

Traversa E. (1995) Ceramic sensors for humidity detection: The state-of-the-art and future developments. *Sens. Actuators B* 23, 135–156.

Visscher G.J.W. (1999) Humidity and moisture measurement. In: Webster J.G. (ed.) *The Measurement, Instrumentation and Sensors Handbook*. Boca Raton, FL, CRC Press, Ch. 72.

Wang J., Wang X.-H., Wang X.-D. (2005) Study on dielectric properties of humidity sensing nanometer materials. *Sens. Actuators B* 108, 445–449.

Wang K., Qian X., Zhang L., Li Y., Liu H. (2013) Inorganic-organic p-n heterojunction nanotree arrays for a high-sensitivity diode humidity sensor. *ACS Appl. Mater. Interfaces* 5, 5825–5831.

Wernecke R., Wernecke J. (2014) *Industrial Moisture and Humidity Measurement: A Practical Guide*. Wiley, Weinheim, Germany, 520 p.

Wexler A. (ed.) (1965) *Humidity and Moisture. Measurement and Control in Science and Industry*. Vols. 1–3. Reinhold Pub. Corp., New York.

Wiederhold P.R. (1997) *Water Vapor Measurement: Methods and Instrumentation*. CRC Press, Boca Raton, FL, 384 p.

Wolfbeis O.S. (2005) Materials for fluorescence-based optical chemical sensors. *J. Mater. Chem.* 15, 2657–2669.

Xu L. (2004) Optical fiber humidity sensor based on evanescent wave scattering, PhD Thesis, Mississippi State University, Starkville, MS.

Yamazoe N., Shimizu Y. (1986) Humidity sensors: Principles and applications. *Sens. Actuators* 10, 379–398.

Yarkin D.G. (2003) Impedance of humidity sensitive metal/porous silicon/n-Si structures. *Sens. Actuators A* 107, 1–6.

Yeo T.L., Sun T., Grattan K.T.V. (2008) Fibre-optic sensor technologies for humidity and moisture measurement. *Sens. Actuators A* 144, 280–295.

Zaghloul M.E., Voiculescu I. (2011) Integrated chemical sensors, In: Korotcenkov G. (ed.) *Chemical Sensors: Comprehensive Sensor Technologies, Vol. 4: Solid State Devices*. Momentum Press, New York, pp. 485–514.

Section II

Conventional Methods and Brief History of Humidity Measurements

2 Gravimetric Method of Humidity Measurement

2.1 HISTORY OF GRAVIMETRIC-BASED HYGROMETERS

It should be noted that gravimetric-based sensors were first used to measure air humidity. It is considered that the first prototype hygrometers were devised and developed during the Western Han dynasty (206 BCE–220 CE) in Ancient China to study weather (Hamblyn 1965). The Chinese used a bar of charcoal and a lump of earth: Its dry weight was taken and then compared with its damp weight after being exposed to the air. The differences in weight were used to tally the humidity level. However, the earliest record of humidity measurement, indicating the use of the instrument called *hygrometer*, was done in the mid-fifteenth century by the Germans (Selin 2008; Robens et al. 1994; 2011). In 1450, a German cardinal, philosopher, and administrator, Nicolaus de Cusanus (Nicolas Cryfts) (1401–1464) (Gerland and Traumuller 1899), described in his book *Idiota de Staticis Experimentis* the first hygrometer with the following: "If anyone hangs on one side of a big balance with dry wool and loads of stone on the other side until equilibrium is established, at a place and in air of moderate temperature he could observe that with increasing humidity the weight of the wool increases and with increasing dryness of the air it decreases. By these differences it is possible to weigh the air and it is likely that one might perform weather forecasting" (Cusanus 1450). About 20 years later, the Italian architect and painter Leon Battista Alberti (1404–1472) recommended a balance with a sponge: "We know that a sponge becomes wet from the humidity of the air and using this fact we make a balance with which we weigh the weight of the air and the dryness of the winds" (Alberti 1485). Later, in 1481, Italian artist, scientist, and inventor Leonardo Da Vinci (1452–1519) drew Nicolas Cryft's hygrometer in his *Codex Atlanticus*, using a sponge or cotton instead of wool (see Figure 2.1) (da Vinci 1955, 1986). The purpose of the hygrometer, according to Leonardo Da Vinci, was to know the qualities and thickness of the air, and when it was going to rain. Robens et al. (1994), after conducting experiments, showed that the hygrometers described in fifteenth-century literature could have

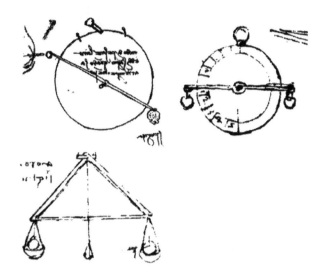

FIGURE 2.1 Sketches of hygrometers by Leonardo da Vinci. (Reprinted from www.museogalileo.it.)

been used, although hair would have been a much better absorbent material, in comparison with cotton.

According to da Vinci, his hygrometer consisted of scales containing a hygroscopic substance (sponge, cotton) in one pan and wax in the other (wax does not absorb water) (Aretin 1872). Models of these instruments can be seen in the London Science Museum. On dry weather conditions, the scales mark zero. However, as the moisture in the air increases, the weight of the hygroscopic substance rises accordingly, and the scales will tip toward the pan where the hygroscopic substance is (Figure 2.2a). Of course, the accuracy of such measurements was very low. It is believed that the most accurate instrument of this group was the paper disc hygrometer (Figure 2.2b). This hygrometer was invented by John Coventry and made by Adams. The paper disc hygrometer includes a stack of paper discs, serving as a hygroscopic substance, a balance and a scale inscribed on it. The variations in atmospheric humidity cause the weight of the paper discs to change. This alters the position of the mobile arm, which can be returned to equilibrium by shifting the cursor.

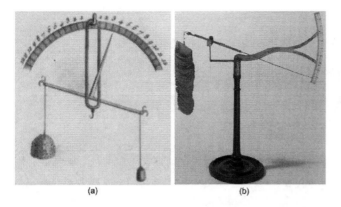

FIGURE 2.2 Illustration of the gravimetric hygrometer operation (a) and Adams paper disc hygrometer (b) from the eighteenth century. Museo Galilei, Florence, Italy. (Reprinted from http://catalogue.museogalileo.it.)

Naturally, these materials (sponge, cotton, paper) are currently not used for the development of humidity sensors. However, in some cases, it is being applied to the gravimetric method, based on the change in the weight of the adsorbent, such as anhydrous phosphorous pentoxide (P_2O_5), calcium chloride ($CaCl_2$), or magnesium perchlorate ($Mg[ClO_4]_2$) after interaction with the air (Till 1959; Wexler and Hyland 1965; Wiederhold 1997). For the first time, such an approach based on measurement of the variation in the weight of a hygroscopic salt was suggested by Santorio (1561–1636). According to Till (1959), phosphorous pentoxide, although very efficient, is messier and more difficult to handle than the other two desiccants. Calcium chloride has poor efficiency in comparison with the other two. The maximum rate of flow that will ensure a complete removal of the water from the air is less for magnesium perchlorate than for phosphorous pentoxide. When the gas flow does not exceed this maximum, however, magnesium perchlorate has proved to be as efficient a drying agent as phosphorous pentoxide. Also, the weight of water absorbed per unit weight of desiccant is several times greater with anhydrous magnesium perchlorate than with phosphorous pentoxide. Nearly 60% of its own weight of moisture can be absorbed before any trace of water vapor is detected in the effluent gas. In addition, it does not form channels, and it contracts in the volume after absorbing moisture.

2.2 GRAVIMETRIC MEASUREMENT OF AIR HUMIDITY

In principle, the gravimetric method is the most fundamental way of accessing the amount of water vapor in the moist gas, because this method uses the absorption of water vapor by a desiccant from a known volume of air (Regnault 1845; Picard et al. 2008). The mass of the water vapor is determined by weighing the drying agent before and after absorbing the vapor. The mass of the dry sample is determined either by weighing (after liquefaction, to render the volume of the sample manageable) or by measuring its volume (and having knowledge of its density). As a rule, gravimetric hygrometer measures the mass ratio r, defined by

$$r \equiv m_w/m_g \quad (2.1)$$

where m_w is the mass of the water in the gas and m_g is the mass of the dry gas.

Currently, the gravimetric method is the most common method for determining the material moisture content (see Figure 2.3). In a gravimetric hygrometer, the water vapor can be also frozen out by a cold trap and weighed, while the volume or mass of dry gas is

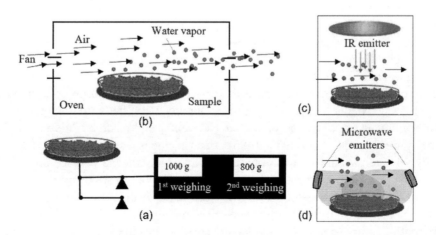

FIGURE 2.3 (a) Diagram illustrating gravimetric method. (b) Drying samples in drying cabinet. (c) With infrared radiation. (d) With microwave radiation. Humidity is removed by circulating air.

measured directly. In principle, the separation of gas and water vapor may be also conducted by using a semipermeable membrane, through which only one of the moist gas constituents passes. For example, such a solution is possible for the hydrogen–water vapor system using a palladium membrane. At certain temperatures, a palladium membrane is permeable to hydrogen.

Also, since the method involves a direct measurement of a fundamental property of moist air, it has often been accepted as a standard for a humidity measurement against which other instruments may be calibrated. Unfortunately, although the method is simple and attractive in principle, it is awkward in application; investigators have stressed, or their results show, the great difficulty involved in obtaining a reliable record of high accuracy (Till 1959). For example, certain problems arise in the actual weighing of the absorbent tubes. Also, if the tubes are at a different temperature than their surroundings on the balance, convection currents may cause trouble. Therefore, the application of this method is limited by the laboratory environment.

The details of the gravimetric apparatus and the techniques used depend somewhat upon the desired accuracy of the results, but in each case, the principle is the same. The moist air sample to be analyzed is passed through a series of tubes containing an absorbent. The air sample must be clean initially or filtered so that the absorbent picks up nothing but water vapor. The filter and the tubing used to conduct the sample from the test space to the absorbent tubes must not alter the water vapor content of the sample. The sorption of water on the walls of the tubing, for example, must not be significant. It was found that rubber is unsuitable; even the small amounts of rubber used to connect glass or metal tubes introduced a significant error at low temperatures, at which the concentrations of water vapor are very small. The dried air sample is then metered in some way to permit the calculation of the quantity of air involved. The measured quantities of major interest, therefore, are the increase in the weight of the absorbent tubes and the metered quantity of air. Before these quantities can be used to give the humidity ratio of the sample, however, a large number of extraneous factors must be considered and their effects upon the records accurately evaluated.

No doubt, such instruments are not useful for day-to-day measurements: The operation of such a standard requires high skill and sophisticated hardware. This method is also unsuitable for in situ application. This is another disadvantage of the gravimetric method. Since the result of a measurement gives the average value over an extended time, the instrument is used in combination with a humidity generator, capable of producing a gas of constant humidity. In addition, a substantial volume sample of air is required for accurate measurements to be taken, and a practical apparatus requires a steady flow of the humid gas for a number of hours, depending upon the humidity, in order to remove a sufficient mass of water vapor for an accurate weighing measurement. As a consequence, the method is restricted to providing an absolute calibration reference standard. Such an apparatus is found mostly in the national calibration standards laboratories. In these labs (the National Institute of Standards and Technology, NIST [United States], the National Physical Laboratory, NPL [United Kingdom], and The National Research Laboratory of Metrology, NRLM [Japan]), among others, this method is being used for primary standard (Wexler 1948; Wexler and Hyland 1965; Scholz 1984; Visscher 1999; Picard et al. 2008). Achievable accuracies are approximately 0.1%–0.2% in mixing ratio, or 0.04°C in the range of −35°C to +50°C dew point, increasing to 0.08°C at +80°C and 0.15°C at −75°C.

The device shown in Figure 2.4a has better accuracy (Meyer et al. 2010). This is a second-generation gravimetric hygrometer designed by the NIST Thermodynamic Metrology Group (United States). A block diagram of the NIST gravimetric hygrometer is shown in Figure 2.5. A gravimetric hygrometer measures humidity by separating the water from the gas and subsequently determining the masses of the water and the gas separately. For these purposes, the gravimetric hygrometer employs water-collection tubes and an automated, continuous-flow gas-collection system. This makes it far easier to operate than its predecessor, which was rarely used because of its onerous nature. The mass of the water is determined from the increase in mass of the water-collection tubes and trap (if used). The mass of the dry gas is determined by measuring the volume and density of the gas that has filled a gas-collection tube (Figure 2.4). Determined masses of the water and the gas are then used to calculate the water mole fraction in the gas. Other humidity quantities, such as the dew point and relative humidity, can be easily determined from the mole fraction using additional measurements of the pressure and temperature of the gas made before it enters the hygrometer.

The design of the NIST gravimetric hygrometer allows gas collection to be limited only by the patience of the operator. As the gas flows through the gravimetric hygrometer, desiccant-filled water-collection tubes trap the water in the gas before the gas enters the gas collection system. For gas with mass ratios greater than 15 mg/g, the gas first flows through an additional water trap that condenses out most of the water, and the rest is trapped by the desiccant. The water-collection system consists of three desiccant-filled tubes connected in series. A diagram of one of these tubes is shown in Figure 2.6a. The first and second tubes that the gas passes through contain anhydrous $Mg(ClO_4)_2$,

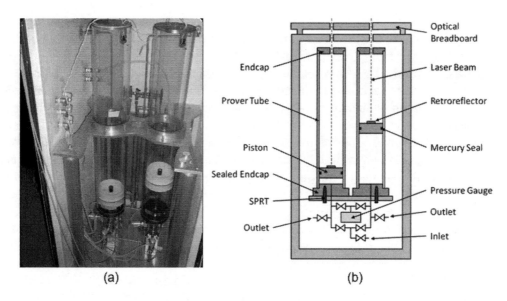

FIGURE 2.4 The image (a) and schematic diagram (b) of the gas collection system of the NIST Gravimetric Hygrometer, comprising two glass cylindrical tubes with moveable prover Teflon pistons, whose positions indicate the collected amount of gas. Mercury O-rings seal the pistons against the tubes but allow vertical motion of the pistons. Hygrometer contains two standard platinum resistance thermometers (SPRTs). (Reprinted from www.nist.gov.)

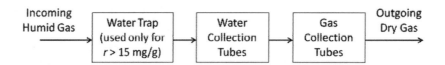

FIGURE 2.5 Block diagram of the NIST gravimetric hygrometer. (Reprinted with permission from Meyer, C.W. et al., The second-generation NIST standard hygrometer, *Metrologia*, 47, 192–207, 2010. Copyright 2010, Institute of Physics.)

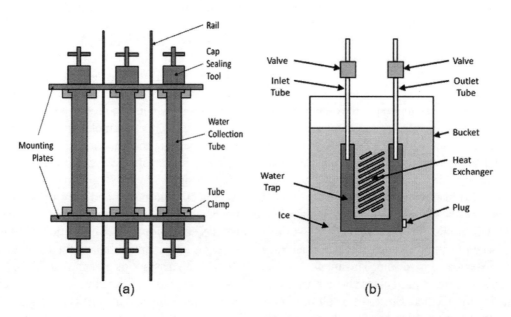

FIGURE 2.6 (a) Mounting of water-collection tubes in the manifold of the gravimetric hygrometer and (b) Water trap for the gravimetric hygrometer, which is used when the dew-point temperature of the gas is above the ambient temperature. (Reprinted with permission from Meyer, C.W. et al., The second-generation NIST standard hygrometer, *Metrologia*, 47, 192–207, 2010. Copyright 2010, Institute of Physics.)

and the third tube contains anhydrous P_2O_5. The latter desiccant is more powerful, but the former is recyclable and more convenient to use. The first tube collects the vast majority of the water, and the second tube collects the remaining amount. The third tube exists to verify that all water in the gas has been removed. An external cold trap (Figure 2.6b) allows the hygrometer to measure a mole fractions factor of 10 higher than before. The hygrometer uses a laser interferometry method, in conjunction with a prover piston technique to increase the accuracy of the gas mass measurement (see Figure 2.4b). With the new design, uncertainty of the new gravimetric hygrometer is 35% lower than before. The hygrometer can measure the mole fractions from 2.2×10^{-4} (a frost point of $-60°C$) to 0.31 (a dew point of $70°C$). Under optimal conditions, its relative uncertainty is within 0.09% over most of its range.

However, we must understand that achieving this accuracy requires an implementation of special conditions. As is known, the accuracy of the gravimetric method depends on the accuracy of the weight measurement. This means that the difficulty of obtaining accurate determinations increases rapidly with decreasing moisture content. In the gravimetric method, this problem is being solved by pumping a large amount of gas (i.e., the larger samples of test gas are used as the humidity decreases). For example, in order to obtain an accuracy of 0.1% using a scale with the typical accuracy of 0.2 mg, the sample must contain 200 mg of water, which is, for example, 20,000 mg for a sample with 1% water and 200,000 mg for a sample with 0.1% water. This corresponds to 15 and 155 liters of gas, respectively. If the accuracy of measurement is lower, then the required amount of gas is even more. When using the volumetric and pressure methods, this approach is not applicable, since the sampling volume is fixed. Therefore, in this case, the relative error of measurement increases with decreasing humidity (Wexler 1970). One of the major sources of error is the adsorption of water vapor on the interior surfaces of the apparatus. Another source of error arises from the change in the temperature during a measurement. Corrections can be applied, or the entire apparatus can be immersed in a thermostat camera. The time required to make a measurement depends on the apparatus used and the skill of the operator.

REFERENCES

Alberti L.B. (1912) L'architettura, Padua 1483/Firenze 1485, Zehn Bucher uber die Baukunst. Translated by M. Theurer. Wissenschaftliche Buchgesellschaft, Darmstadt, Germany, 1975. (Reprint of the 1st ed., Heller, Wien 1912, p. 357.)

Aretin (ed.) (1872) *L. da Vinci, Codex Atlanticus—Saggio del Codice Atlantico*. Vol. fol. 249, Milano, Italy.

Cusanus N. (1942) Idiota de Staticis Experimentis, Dialogus, Strasburg, 1450. German transl.: H. Menzel-Rogner, Der Laie uber Versuche mit der Waage, Philosophische Bibliothek, Vol. 220, Meitner, Leipzig, Germany.

Gerland E., Traumuller F. (1965) *Geschichte der Physikalischen Experimentierkunst*. Leipzig, Germany, 1899. Reprint: Olms, Hildesheim, Germany, pp. 83–84.

Hamblyn R. (1965) *The Invention of Clouds: How an Amateur Meteorologist Forged the Language of the Skies*. Picador, New York.

Leonardo da Vinci (1955) *Das Lebensbild Eines Genies*. Vollmer, Wiesbaden, Germany, p. 211.

Leonard da Vinci (1986) *Catalogue "Les Mots dans le Dessin."* Cabinet des Dessin, Louvre, Paris, pp. 68–69.

Meyer C.W., Hodges J.T., Hyland R.W., Scace G.E., Valencia-Rodriguez J., Whetstone J.R. (2010) The second-generation NIST standard hygrometer. *Metrologia* 47, 192–207.

Picard A., Davis R.S., Gläser M., Fujii K. (2008) Revised formula for the density of moist air (CIPM-2007). *Metrologia* 45 (2), 149–155.

Regnault H.V. (1845). Etude sur l'hygrométrie. *Ann. Chim. Phys. Ser.* 3, 15, 129–236.

Robens E., Massen C.H., Hardon J.J. (1994) Studies on historical gravimetric hygrometers. *Thermochim. Acta*, 235, 125–133.

Robens E., Rübner K., Klobes P., Balköse D. (2011) Water vapour sorption and humidity—A survey on measuring methods and standards, In: Okada C.T. (ed.) *Humidity Sensors: Types, Nanomaterials and Environmental Monitoring*. Nova Science Publishers, Hauppauge, NY, pp. 1–54.

Scholz G. (1984) A standard calibrator for air hygrometers. *Bull. Oiml.* 97, 18–27.

Selin H. (ed.) (2008) *Encyclopaedia of the History of Science, Technology, and Medicine in Non-Western Cultures*, 2nd edn. Springer, Heidelberg, Germany.

Till C.E. (1959) Use of a gravimetric technique for humidity measurement, *Internal Report, Division of Building Research, National Research Council Canada*, 1959-02-01. doi:10.4224/20338270.

Visscher G.J.W. (1999) Humidity and moisture measurement, In: Webster J.G. (ed.) *The Measurement, Instrumentation, and Sensors: Handbook*. CRC Press, Boca Raton, FL, Ch. 72.

Wexler A. (1948) Divided flow, low-temperature humidity test apparatus. *J. Res. NBS* 40, 479–486.

Wexler A. (1970) Measurement of humidity in the free atmosphere near the surface of the earth, In: Teweles S., and Giraytys J. (eds.) *Meteorological Monographs*, Vol. 11, No. 33, American Meteorological Society. Lancaster Press, Lancaster, PA, pp. 262–281.

Wexler A., Hyland, R. W. (1965) The NBS standard hygrometer, In: *Humidity and Moisture*, Wexler A. (ed.), Vol. III. Reinhold Publication, New York.

Wiederhold P.R. (1997) *Water Vapor Measurement, Methods and Instrumentation*. Marcel Dekker, New York.

3 Mechanical (Hair) Hygrometer

3.1 HISTORY OF HAIR HYGROMETERS

According to historical sources (Pfaundler 1907; Brodgesell and Liptak 1993; Robens et al. 1994; Camuffo et al. 2014), the first mechanical hygrometers appeared in 1783: Swiss Horace Bénédict de Saussure invented the first hygrometer using a human hair to measure humidity (de Saussure 1783). A diagram illustrating the principle of the hair hygrometer operation is shown in Figure 3.1. Other variants of hair hygrometers are shown in Figure 3.2.

However, attempts to develop such hygrometers had been undertaken before. For example, in 1626, for medical purposes, Sanctorius (1626) invented hygrometers based on the change in length of a ballasted cord (Figure 3.3). A cord was stretched horizontally on a wall, and from its center, a ballast ball was suspended. When the relative humidity increased, the cord was tightened and the ball was lifted. This model is very similar to the string hygrometer developed by Folli and Viviani (Figure 3.4). This model was developed around 1664 (D'Alancé 1707).

The hair hygrometer designed by de Saussure is based on a sorption method. It uses the characteristic of the hair so that its length expands or shrinks the response to the relative humidity (Sonntag 1994). The hair shrinks when humidity drops and swells when humidity increases; similarly, cordage and catgut are shortened and untwisted by moisture. For example, the length of a human hair from which liquid is removed increases by 2.0%–2.5% when relative humidity changes from 0% to 100%. A humidity change takes an effect on the moisture content in such materials. The hair is made from keratin,

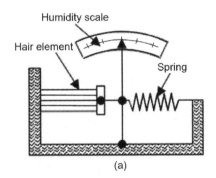 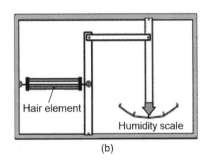

FIGURE 3.1 (a, b) Diagrams illustrating the principle of hair hygrometers operation. (Reprinted from https://learn.weatherstem.com.)

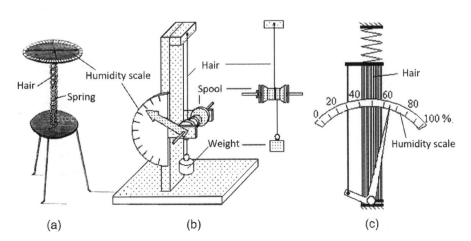

FIGURE 3.2 (a, b, c) Variants of old retro hair hygrometers. (Reprinted from http://americanhistory.si.edu and www.pinterest.com/sapunkovan/masons-hygrometer.)

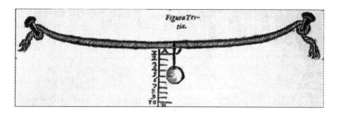

FIGURE 3.3 Sanctorius hygrometer based on the change in length of a cord ballasted in the middle. (Reprinted from www.imss.fi.it.)

FIGURE 3.4 String hygrometer by Folli and Viviani with rotating pointer. This model is presented in Museo Galileo-Institute and Museum of History of Science, Florence. (Reprinted from http://imss.fi.it.)

a protein that is wound into a coil. The turns of the coil are held together by hydrogen bonds. These bonds break in the presence of water, allowing the coil to stretch and the hair to lengthen. The bonds reform when the hair dries, which allows people to style their hair simply by wetting it, shaping it, then drying it. Different types of human hair show different changes in length. However, there is still a relationship between the length of hair and relative humidity. Therefore, manufacturers generally use a bundle of hairs. This averages the individual responses of each strand, since different hairs respond at slightly different rates of expansion and contraction. One should note that the dimensions of various organic materials also vary with their moisture content.

Variants of more advanced hair hygrometers are shown in Figure 3.5. The hair is hygroscopic (tending toward retaining moisture); therefore, in the late 1700s, such devices were called *hygroscopes* by some scientists. That word is no longer in current use, but *hygroscopic* and *hygroscopy*, which derive from it, still are. The traditional folk art device known as a *weather house* works on this principle. The length change in hair hygrometers may be magnified by a mechanism (see, for example, Figure 3.6).

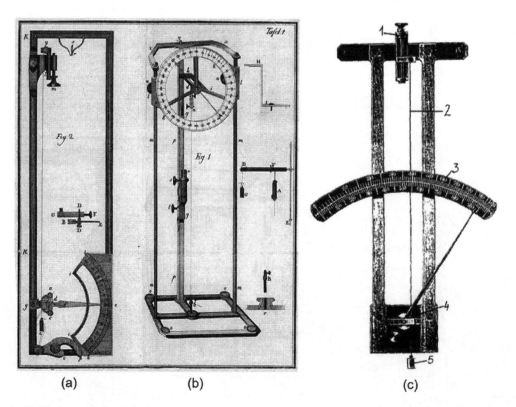

FIGURE 3.5 (a, b) Variants of the hair hygrometer designed by Horace Bénédict de Saussure (1783): (a) Hand hygrometer; (b) Large hygrometer; (c) Museum exhibit, hand hygrometer: 1, zero adjustment; 2, hair; 3, nonlinear graduation in relative humidity; 4, roller with pointer; 5, prestressing weight. (a, b - reprinted from Saussure 1783/1900, www.galltec-mela.de/histoire; c - reprinted from http://americanhistory.si.edu.)

Mechanical (Hair) Hygrometer

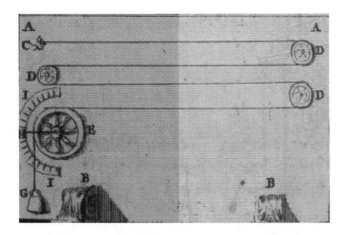

FIGURE 3.6 The drawings present in the museums describing the work of a hair hygrometer. A long cord is fixed at one extreme (C) and ballasted (G) at the other one. A number of pulleys (D) fixed to the wall (A) reduce the total length from linear to bidimensional. The ballasted rope drives a rotating pointer (H, E). (Reprinted from www.imss.fi.it.)

3.2 FEATURES OF HUMIDITY CONTROL USING HAIR HYGROMETERS

The hair hygrograph or hygrometer is considered to be a satisfactory instrument for use in situations or during periods in which extreme and very low humidity are seldom or never found. The hair hygrometer is not expensive, and its usage is simple. Some types record on a chart driven by clockwork or batteries. Very basic types are not powered at all. Electronic sensor-based hygrometers are usually preferred now, but many mechanical hygrometers remain in use for room monitoring (see Figure 3.7a, b). However, the response of the hair to humidity is nonlinear. In addition, the response tends to drift over time, and this response is characterized by hysteresis: The hair length changes more when the humidity increases than when it decreases. The change of the hair length observed when the humidity increases is up to 5%–6% larger than that observed when the humidity decreases. The response time of the hair hygrometer also depends on the air temperature. The time constant of the hair hygrometer is approximately 10 seconds at 20°C and approximately 30 seconds at −30°C. The hair is also highly sensitive to contamination, such as dust, ammonia, oil, and exhaust gas.

It was established that, by rolling the hairs to produce an elliptical cross-section and by dissolving out the fatty substances with alcohol, the ratio of the surface area to the enclosed volume increases and yields a decreased lag coefficient, which is particularly relevant for use at low air temperatures. This procedure also results in a more linear response function, although the tensile strength is reduced. For accurate measurements, a single hair element is preferred, but a bundle of hairs is commonly used to provide a degree of ruggedness. Chemical treatment with barium (BaS) or sodium (Na_2S) sulfide yields further linearity of response.

Testing has shown that 20%–90% relative humidity (RH) is the best range for hair hygrometer usage. The accuracy attainable by hygroscopic techniques in this range of humidity is typically within 5%–10% relative humidity, but like many other instruments, hair hygrometers can be more accurate in the middle of the relative humidity scale than at the very high or low ranges. Accuracy may be closer to ± 2–3% RH between 40% and 60% RH at room temperatures. Outside of that range, accuracy will decline. Accuracy depends on many factors. Sources of measurement error include the response of different hairs and the response of the mechanical linkage that connects the hair to the indicating scale. Accuracy depends also on the conditions of exploitation. For example, after the hair hygrometer is exposed in low temperature and low humidity for a long time, reading error increases due to the increasing of delay. If a hair hygrometer is left in a low-humidity condition for a long time, its reading changes cause large errors as well. To improve accuracy, the device should be calibrated in the room where it will be used, and calibrated at a condition in the humidity range expected for the room. Thus, users should maintain a healthy skepticism concerning a hair hygrometer's readings, including those devices that are marketed as "certified." Often, a small tap with a finger is enough to change a humidity reading by 3%, as the mechanical linkage can seize up over time. They are best used as a general indication of humidity rather than for important readings.

Hair hygrometers, similar to the wet and dry bulb hygrometers discussed in Chapter 4, require good circulation of the measured gas. In air ducts, there is generally an adequate gas velocity to ensure a

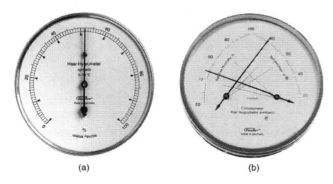

FIGURE 3.7 The hair tension dial hygrometers fabricated by Fischer Scientific: (a) hygrometer and (b) hygrometer-thermometer. (Reprinted from https://www.fishersci.com.)

dependable measurement; however, if the instrument is to be mounted in a room, the location must be carefully chosen. The instrument should not be mounted near the doors or other openings where it will be exposed to spurious drafts; flush mounting on a panel should be avoided, because the atmosphere in the back of the panel is stagnant. The hair element can be mounted on the top or back of the instrument case, depending on the installation. The element can also be mounted on an extension in the back of the instrument so that the sensing portion is in the room or compartment where relative humidity is to be measured, while the readout device is surface mounted on the wall outside (Brodgesell and Liptak 1993).

3.3 OTHER TYPES OF MECHANICAL HYGROMETERS

Besides hair in the earlier versions of mechanical hygrometers, some other materials have been used. Paper, for example, also changes length at air humidity change (see Figure 3.8). However, unlike hair, paper does not have the necessary reproducibility and stability parameters. In addition, its properties are strongly dependent on the manufacturer and composition.

There have also been attempts to develop wood-based hygrometers using volumetric changes of the material. As is known, wood shrinks at low RH and swells at increasing levels. Following this principle, a hygrometer was developed to monitor such dimensional changes and transform them into a rotation of a pointer on a circular scale (Figure 3.9).

In museums, one can also find hygrometers, using humidity-sensitive materials such as leather, horn, and ivory (Camuffo et al. 2014). One was developed in 1687 by Guillaume Amontons. A hygrometer was composed of a vertical glass tube, 3 feet long, and at the bottom, a

FIGURE 3.8 Hygrometer of the type invented or perfected by Vincenzo Viviani (second half seventeenth century). The hygroscopic substance is a paper ribbon. The brass bar has a small, turned column at each end. The columns carry two small rolls, on which the ends of the paper ribbon are wrapped. The ribbon is weighted at the center by a wooden staff carrying a small pointer (the ribbon and staff are modern restorations). The changes in atmospheric humidity cause variations in the length of the paper ribbon, which are registered on a scale. (Reprinted from www.imss.fi.it.)

FIGURE 3.9 Volumetric hygrometer based on wood shrinkage and swelling. (Reprinted from www.imss.fi.it.)

leather bag filled of mercury was applied. When air was moist, the leather bag expanded and mercury descended in the tube (Amontons 1695; Berryat 1754). The bag with the mercury moved the interface level of two immiscible liquids up and down in the tube.

It should be noted that the mechanical hygrometers can be designed based on principles different from those shown in Figures 3.1 through 3.9. For example, a laminate hygrometer or the metal–paper coil hygrometer shown in Figure 3.10 is made by attaching thin strips of wood or paper to thin metal strips, forming a laminate. The laminate is formed into a spiral or a helix. As the humidity changes, the spiral flexes, due to the change in the length of the wood or paper. One end of the spiral is anchored, and the other is attached to a pointer (similar

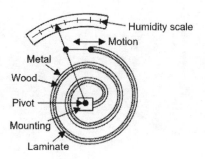

FIGURE 3.10 Hygrometer using metal/wood or metal/paper laminate. (Reprinted from https://learn.weatherstem.com.)

to a bimetallic strip used in temperature measurements), and the scale is graduated in percentage of humidity. Such a configuration is useful for giving a dial indication of humidity changes. Hygrometers, which use indicated approach, appear most often in very inexpensive devices, and their accuracy is limited, with variations of 10% or more.

A contemporary adaptation of the hair hygrometer is the plastic expansion hygrometer. In this popular and economical instrument, the hair is replaced by a hygroscopic polymer, such as nylon, polyimide plastic, or cellulose. Like human hair, synthetic polymer materials change length as the water vapor concentration varies. This elongation has been used to build dial and digital indicators and also relatively inexpensive industrial thermostats, humidistats, and residential furnace humidifiers. The operation of polymer-based element instruments is similar to the previously described human hair hygrometers. While the hygroscopic polymer is more uniform than human hair, the same advisory cautions apply—do not expect accuracy greater than ±7% RH, calibrate them regularly in the environment where they are used, and do not expect accuracy if wide swings in RH are common. In addition, because of the long lag time for synthetic fibers, such sensors should never be used below 10°C.

3.4 HYGROGRAPHS AND HYGROTHERMOGRAPHS

In addition to devices that provide information on air humidity, there is a large group of instruments, hygrographs, recording these changes in time (see Figure 3.11). This is the most commonly used hair hygrometer. This

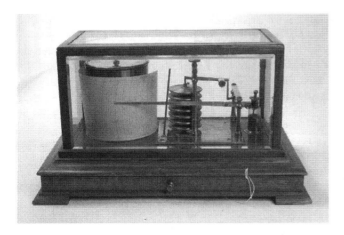

FIGURE 3.11 Old retro hygrograph with the function of recording the change in humidity over time. (Reprinted from www.shutterstock.com.)

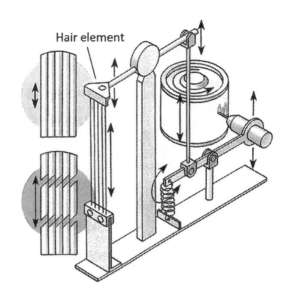

FIGURE 3.12 Hair hygrograph. Diagram illustrating the mechanism of recording the change in humidity over time. (Reprinted from www.daviddarling.info.)

uses a bundle of hairs held under slight tension by a small spring and connected to a pen arm in such a way as to magnify a change in the length of the bundle. A pen at the end of the pen arm is in contact with a paper chart that has been fitted around a metal cylinder and registers the angular displacement of the arm. The cylinder rotates about its axis at a constant rate determined by a mechanical clock movement (see Figure 3.12). The rate of rotation is usually one revolution per week or per day. The chart has a scaled time axis that extends around the circumference of the cylinder and a scaled humidity axis parallel to the axis of the cylinder. The cylinder normally stands vertically. The mechanism connecting the pen arm to the hair bundle may incorporate specially designed cams that translate the nonlinear extension of the hair in response to humidity changes into a linear angular displacement of the arm.

Modern humidity recorders based on a mechanical hygrometer are generally available as two-pen instruments, with the second pen recording temperature (see Figure 3.13). Such hygrothermographs have been the basic monitoring tool in museums for some time. They provide a continuous record of the temperature and humidity variations over a period of 1, 7, 31, or 62 days. Hygrothermographs are accurate within ±3% to 5% when properly calibrated. They are the most accurate within the range of 30%–60% RH. Transmitters are available in both digital and analog designs, and the analog ones can be electronic or pneumatic (Brodgesell and Liptak 1993).

FIGURE 3.13 Old retro hygrometer with humidity and temperature recording function. (Reprinted from www.shutterstock.com.)

3.5 ELECTRONIC EXPANSION HYGROMETER

In the mechanical hygrometer, the change in expansion of material is measured and indicated by a mechanical linkage of gears, levers, and dials. A modification of this concept replaces the linkage with electronics. The hair, plastic, and, in one case, a desert plant seed case are connected to an electronic strain gauge that measures the pressure exerted as the sensing element contracts. This is often an improvement over mechanical hygrometers, since electronics tend to be more repeatable than linkages, particularly over long periods of time.

Another approach was realized by Ha et al. (2000). They proposed so-called mechanical-optoelectronic humidity sensors (see Figure 3.14). This device consists of a light emitting diode (LED), a very sensitive photodiode, and a mechanical system. The sensor has a bunch of human hair at one end, and the other end has a thin metal sheet with a fittable window with respect to the LED and photodiode. A spiral spring is connected with the metal sheet. When the humidity concentration changes, the contraction and expansion occur in the hair; this hair pulls the metal sheet up or down. Thus, the window opening area changes and varies the light intensity to the photo detector from the LED, resulting in a photocurrent change with respect to humidity on the output side. Developers argue that this sensor has good linearity, long life, small hysteresis, stable operation, and less temperature dependency.

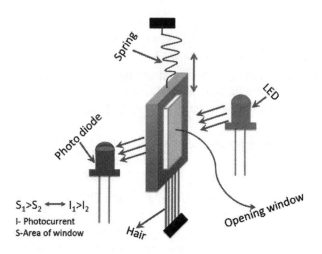

FIGURE 3.14 Schematic representation of a humidity sensor that operates by the mechanical-optoelectronic principle. LED, light-emitting diode. (Reprinted from Tripathy, A. et al., Role of morphological structure, doping, and coating of different materials in the sensing characteristics of humidity sensors, *Sensors*, 14, 16343–16422, 2014. Published by Multidisciplinary Digital Publishing Institute (MDPI) as open access.)

REFERENCES

Amontons G. (1695) Remarques & Expériences Physiques sor la construction d'une nouvelle Clepsidre, sur les Barometres, Termometres & Hygrometres. Jombert, Paris.

Berryat (1754) Anneé MDCLXXXVII. Physique Générale. In: Histoire de l'Académie des Sciences de Paris, Tome II, page 22. In Recueil de mémoires ou Collection de pieces Académiques, Tome I, p. 131. Fournier, Auxerre et Desventes, Dijon.

Brodgesell A., Liptak B.G. (1993) Moisture in air: Humidity and Dew point, In: Liptak B.G. (ed.) *Analytycal Instrumentation*, Chilton Book Company, Radnor, PA, pp. 215–226.

Camuffo D., Bertolin C., Amore C., Bergonzini A., Cocheo C. (2014) Early hygrometric observations in Padua, Italy, from 1794 to 1826: The Chiminello goose quill hygrometer. Versus the de Saussure hair hygrometer. *Clim. Change*, 122, 217–227.

D'Alancé J. (1707) *Traittéz des Barométres, Thermométres et Notiométres ou Hygrométres*. Maurret, Amsterdam, the Netherlands.

Ha N.T.T., An D.K., Phong P.V., Hoa P.T.M., Mai L.H. (2000) Study and performance of humidity sensor based on the mechanical–optoelectronic principle for the measurement and control of humidity in storehouses. *Sens. Actuators B* 66, 200–202.

Pfaundler L. (ed.) (1907) *Müller-Pouillet, Lehrbuch der Physik*, Vol. III. Vieweg, Braunschweig, Germany, pp. 832–833.

Robens E., Massen C.H., Hardon J.J. (1994) Studies on historical gravimetric hygrometers. *Thermochim. Acta*, 235, 125–133.

de Saussure H.B. (1783/1900), Essais sur l'hygrometrie, Neuchatel, 1783. Ostwald's Klassiker der exakten Wissenschaften, No. 115/1 19, Engelmann, Leipzig, Germany, 1900.

Sonntag D. (1994) Advancements in the field of hygrometry. *Zeitschrift Meteorologie* 3 (2), 51–66.

Tripathy A., Pramanik S., Cho J., Santhosh J., Osman N.A.A. (2014) Role of morphological structure, doping, and coating of different materials in the sensing characteristics of humidity sensors. *Sensors* 14, 16343–16422.

4 Psychrometer

4.1 HISTORY

The best-known instrument for humidity measurement is the psychrometer, although it is more commonly known by its use as the "wet-bulb/dry-bulb" method. This wet and dry bulb hygrometer is also called a *psychrometer—psychros* means "cold" in Greek. Psychrometry has been a popular method for monitoring humidity for a long time, primarily due to its simplicity and inherent low cost. Psychrometry is one of the oldest and the most accepted methods of measuring humidity (see Figure 4.1). In 1825, August constructed the first psychrometer (Rübner et al. 2008). According to other data, Leslie (France) was the first to measure humidity using a differential thermometer with a dry bulb and a wet bulb. He did it in 1799 (Berzelius 1841). Leslie observed that, when a piece of muslin was moistened with water and exposed to the atmosphere, evaporation accompanied by cooling generally occurred. If the muslin was wrapped around a thermometer, an indication of the cooled temperature could be observed. Leslie theorized that the amount of cooling was dependent on the hydrometric state of the atmosphere, and he assembled the first psychrometer. According to other data, the psychrometer was first developed by James Hutton (1726–1797) during the late 1700s (Middleton 1969). It measured the cooling effect of evaporating moisture on a thermometer bulb. There is also information that the formula for calculating humidity from readings of the dry- and wet-bulb thermometers was derived by (France) in 1815. Further improvements were proposed in 1887 by Richard Assmann (Germany), who invented the aspirated psychrometer. Assmann added a ventilator, and this aspiration psychrometer is regarded as a standard instrument for air humidity measurement.

4.2 PRINCIPLES OF OPERATION

A psychrometer generally consists of two thermometers of the same specifications. At that one thermometer with an ordinary dry bulb and the other one, called the wet

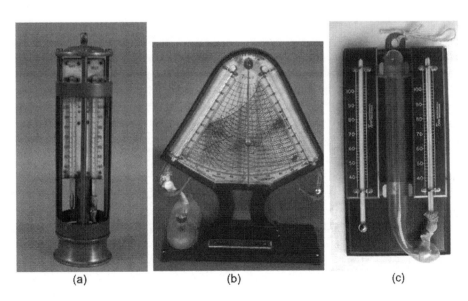

FIGURE 4.1 (a, b, c) Examples of the first wet-bulb hygrometers or psychrometers. The hygrometer at the middle (*b*) is a hygrodeik. Like the hygrometer, it uses wet- and dry-bulb thermometers. However, it has a nomograph, which enables the user to set the wet- and dry-bulb temperatures and then read off the relative humidity. The wick around the wet bulb was supplied with water from the bottom of a glass tube. (Reprinted from https://www.pinterest.com and www.explainthatstuff.com.)

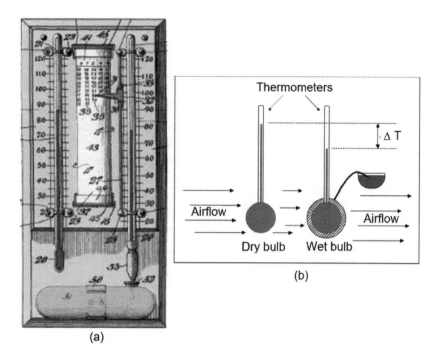

FIGURE 4.2 (a) Schematic diagram of psychrometer; (b) Diagram showing the principle of dry- and wet-bulb measurement. (Reprinted from https://www.pinterest.com.)

bulb, is enclosed in a porous medium (wick) that is maintained wet by capillary action from a reservoir of water (Bindon 1965; Wiederhold 1997). The wet-bulb thermometer (T_{WB}) will show a lower temperature than the dry-bulb thermometer (T_{DB}) if the air was not saturated with water vapor (see Figure 4.2). When water changes phase from liquid to vapor, an amount of heat equal to the latent heat of vaporization must be supplied from the environment for each unit mass of water evaporated. This means that evaporation from the moist cloth lowers the wet-bulb thermometer's temperature. The vaporization, in addition to its dependence on the amount of heat available, is also a function of the degree of saturation or relative humidity (RH) of the atmosphere surrounding the water: At low humidity, water evaporates more actively, so that the wet-bulb temperature decreases more. Thus, the aspirated psychrometer measures humidity by measuring the difference between the dry-bulb temperature and the wet-bulb temperature. The difference between the dry-bulb temperature and the wet-bulb temperature is the "wet-bulb depression."

The humidity can be estimated using tables (Table 4.1) or by calculation (sometimes internally calculated and displayed directly by modern psychrometers). For example, the humidity can be calculated by using the psychrometer equation (Sprung's formula [Sprung 1888]).

$$p_w = p_{ws}(T_{DB}) - B \cdot P \cdot (T_{DB} - T_{WB}), \quad (4.1)$$

where P is the total barometric pressure in kPa and B is the psychrometer coefficient that varies between $B = 6.4 \cdot 10^{-4}$ °C^{-1} and $B = 6.8 \cdot 10^{-4}$ °C^{-1} depending on the ventilation of the psychrometer. This coefficient has to be determined for each specific design of psychrometer and, in particular, for each design of the wet bulb. This means that individual values are assigned by calibration of the psychrometer. The RH can be calculated based on the Equation 4.2:

$$T_{WB} = T - \frac{\alpha}{B \cdot P} \cdot exp\left(\frac{\beta \cdot T}{\lambda + T}\right) \cdot \left(1 - \frac{U_W}{100\%}\right), \quad (4.2)$$

where α, β, and λ are Magnus parameters given in Table 4.2 (P, total pressure; U_w, RH;%, T – °C).

In the version adopted by the World Meteorological Organization (WMO) (1989a, 1989b), this basic formula for RH calculation is as follows:

$$\text{RH} = 100 \cdot \frac{p_{ws}(T_{WB}) - A \cdot P \cdot (T_{DB} - T_{WB})}{p_{ws}(T_{DB})}, \quad (4.3)$$

where $p_{ws}(T_{WB})$ and $p_{ws}(T_{DB})$ are the saturation vapor pressures corresponding to the temperature of the web bulb and the dry bulb, correspondingly; P is the pressure

TABLE 4.1
Psychrometric Chart of Relative Humidity (in percent)

Depression of the Wet Bulb (Dry-Bulb Temperature Minus Wet-Bulb Temperature in °C)

Dry-Bulb Temperature (°C)	0	0.5	1.0	1.5	2.0	2.5	3.0	3.5	4.0	4.5	5.0	7.5	10.0	12.5	15.0	17.5	20
−20	100	70	41	11													
−15	100	79	58	38	18												
−10	100	85	69	54	39	24	10										
−7.5	100	87	73	60	48	36	22	10									
−5	100	88	77	66	54	43	32	21	11	1							
−2.5	100	90	80	70	60	50	42	37	22	12	3						
0	100	91	82	73	65	56	47	39	31	23	15						
2.5	100	92	84	76	68	61	53	46	38	31	24						
5	100	93	86	78	71	65	58	51	46	38	32	1					
7.5	100	93	87	80	74	68	62	56	50	44	38	11					
10	100	94	88	82	76	71	65	60	54	49	44	19					
12.5	100	94	89	84	78	73	68	63	58	53	48	25	4				
15	100	95	90	85	80	75	70	66	61	57	52	31	12				
17.5	100	95	90	86	81	77	72	68	64	60	55	36	18	2			
20	100	95	91	87	82	78	74	70	66	62	58	40	24	8			
22.5	100	96	92	87	83	80	76	72	68	64	61	44	28	14	1		
25	100	96	92	88	84	81	77	73	70	66	63	47	32	19	7		
27.5	100	96	92	89	85	82	78	75	71	68	65	50	36	23	12	1	
30	100	96	93	89	86	82	79	76	73	70	67	52	39	27	16	6	
32.5	100	97	93	90	86	83	80	77	74	71	68	54	42	30	20	11	1
35	100	97	93	90	87	84	81	78	75	72	69	56	44	33	23	14	6
37.5	100	97	94	91	87	85	82	79	76	73	70	58	46	36	26	18	10
40	100	97	94	91	88	85	82	79	77	74	72	59	48	38	29	21	13
42.5	100	97	94	91	88	86	83	80	78	75	72	61	50	40	31	23	16
45	100	97	94	91	89	86	83	81	78	76	73	62	51	42	33	26	18
47.5	100	97	94	92	89	86	84	81	79	76	74	63	53	44	35	28	21
50	100	97	95	92	89	87	84	82	79	77	75	64	54	45	37	30	23

TABLE 4.2
Magnus Parameters for Equation 4.2

Condition	T Range (°C)	α (hPa)	B	λ (°C)
Above water	−45 to 60	6.112	17.62	243.12
Above ice	−80 to 0.01	6.112	22.46	272.62

of the air; T_{DB} and T_{WB} are the temperatures of the dry and wet bulbs; and A is the psychrometer coefficient. Psychrometer coefficient depends on the air flow speed (>2.2 m/s), thermodynamic properties of water, vapor pressure, and geometry of the wet-bulb thermometer, so it is difficult to determine precisely. Where the moisture content of gases other than air is to be determined, the psychrometric constant A must be modified, to account for the physical properties of the particular gas.

4.3 REALIZATION

At present, there are three types of psychrometers: simple psychrometers for home use (Figure 4.3), sling psychrometers (Figure 4.4), and aspirating psychrometers (see Figure 4.5). The whirling psychrometer, today commonly known as the *sling psychrometer*, was invented in 1853 by August Bravais. For a dependable measurement, the sample velocity should be well in excess of 3 m per second. The psychrometer should, therefore, be mounted where there is an adequate circulation. To take readings with a sling psychrometer, whirl it around for 1 minute to pass air over the wet and dry bulbs (Figure 4.4).

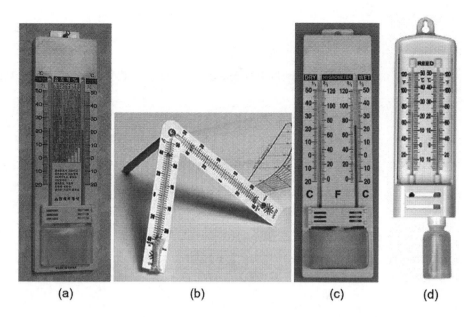

FIGURE 4.3 Psychrometers for home use fabricated by different companies. (a - Reprinted from http://kotzur.com; b – from https://megadepot.com, c – from www.psscientific.com, and d – from http://zhiling.en.hisupplier.com.)

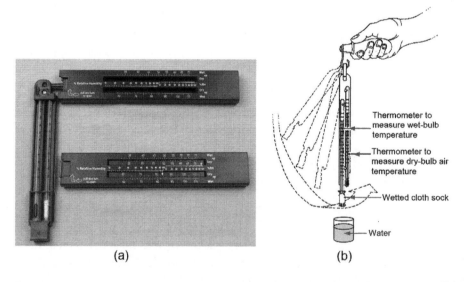

FIGURE 4.4 (a) Sling psychrometer fabricated by Avagadro's Lab Supply, Inc. (http://www.avogadro-lab-supply.com). (b) Principle of use. Idea from Gust J. An error analysis of psychrometers, In: *Proceedings of NCSL Workshop & Symposium*, Session 7C, July 16–20, Fairmont Hotel, Dallas, TX, pp. 583–606, 1995.)

The aspirating psychrometer uses a battery-powered fan to steadily blow air over the bulb at a set speed.

The same approaches can be used for realization of psychrometers with thermocouples or thermistors (see Figure 4.6). Replacement of thermometers to the thermocouples and thermistors has also allowed implementation of this principle of humidity measurement with modern, thin-film technologies. In particular, Berlicki et al. (1998) designed a thin-film psychrometer based on thermopile. Configuration of this sensor is shown in Figure 4.7.

Berlicki et al. (1998) believe that this sensor could be used in two different measurement procedures. In the first case, the thermopile sensor can give a voltage signal directly proportional to the temperature difference between the wet and dry junctions of the thermocouples, whereas the ambient temperature could be measured with a thermoresistor. Due to the small mass of the thin-film sensor, a relatively fast (below 1 minute) response can be expected. In the second measurement procedure, known as the *compensation method*, the electrical power supplied to the wet end of the thermopile, in order to

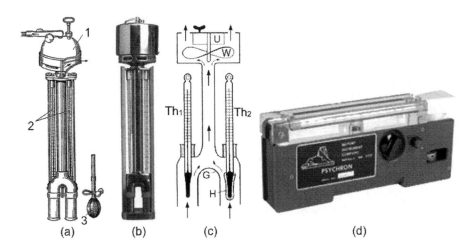

FIGURE 4.5 (a, b) Aspiration psychrometer by Assmann: *1*, fan; *2*, psychrometric thermometers; *3*, eyedropper to wet the wet thermometer (http://www.th-friedrichs.de). (c) Schematic diagram illustrating operation of Assmann's aspiration psychrometer: *H*, cotton envelope; *G*, air channel; *Th*, thermometer; *W*, ventilator; *U*, motor. (d) Portable battery-powered and motor-aspirated psychrometer, Psychron, designed by Belfort Instrument (Reprinted from http://belfortinstrument.com/).

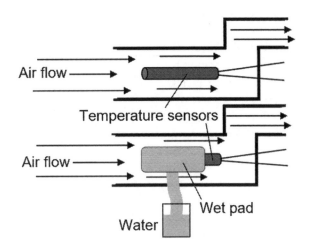

FIGURE 4.6 Wet- and dry-bulb psychrometer with temperature sensors.

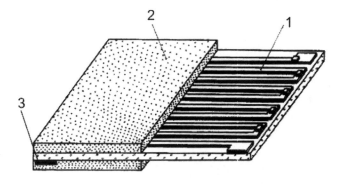

FIGURE 4.7 Sensor wafer: *1*, thermopile; *2*, heating thermoresistor; and *3*, filter paper wetted with water. The sensor was designed in the form of a wafer on which a NiCr-Ni thermopile and thermoresistor were fabricated by thin-film technology. The substrate was in the form of a glass foil of 100 μm thickness (Corning 7059), and the films were deposited by the magnetron sputtering method. Part of the wafer was covered with a filter paper wetted with water. A thermoresistor could be applied both for the ambient temperature measurements and to heat the active (wet) part of the thermopile. (Reprinted with permission from Berlicki, T.M. et al., Thermoelement humidity sensor, *Sens. Actuators A*, 64, 213–217, Copyright 1998, Elsevier.)

bring the temperature difference between the thermopile junctions back to zero, could be measured. According to Berlicki et al. (1998), contrary to sorption sensors, the measurement is free of undesirable drift (compensation method) and similar to the classical psychrometric method; this sensor is not sensitive to disadvantageous environmental conditions (for example, oil vapor). If the sensor was used for the direct readout of temperature difference between the wet and dry areas, the drift caused by aging processes depends only on such processes inside the thermopile, which usually can be neglected if thermopile materials with good long-term stability are used. The rate of time response depends on the sensor mass (in this model, it is approximately 60 s). The preferred RH measurement range is 10%–90%. There exists a possibility to increase the resolution of readout in the range of approximately 80%–100% RH. A disadvantage of the sensor is the necessity of wetting it before every measurement.

Other approaches to the development of psychrometers have also been proposed. For example, Montanini (2007) suggested an optical psychrometer in which temperature measurements are performed by means of two fiber Bragg grating sensors, used as dry-bulb and wet-bulb thermometers. The adopted design exploits both the high accuracy of psychrometric-based RH measurements with acknowledged advantages of wavelength-encoded fiberoptic sensing. The measurement accuracy was estimated to be within 2% RH. Undoubtedly, the sensors developed by Berlicki et al. (1998) and Montanini (2007) are only prototypes, and they may never be commercialized humidity sensors. The need to have a water tank to be replenished substantially narrows the application area of such hygrometers. However, their development has shown that the emergence of microminiature psychrometers is possible.

Gregory (1947) has shown that, instead of direct temperature measurements, the temperature of the thermometers may be equalized. In this method, the temperature of the wet bulb is raised to that of the dry bulb and the heat measured, which is required to maintain equilibrium. This is accomplished by winding a manganin wire heater around the wet bulb and under the wick and measuring the current when the wet- and dry-bulb thermometers read alike. A convenient arrangement utilizes a differential thermocouple psychrometer whose output is read on a galvanometer. The current to the wet-bulb heater is adjusted until the galvanometer no longer deflects. However, this approach was not used in the development of psychrometers intended for the market.

4.4 ACCURACY OF MEASUREMENTS

It is worth noting that this is a fairly accurate method of measuring humidity. For both sling and aspirating psychrometers, uncertainties of around 2%–7% RH are obtainable (Wiederhold 1997). Its accuracy depends on a number of factors, such as the length of the thermometer and how accurately it can be read; thoroughness of wetting of the web of the bulb sensor; exposure to the gas; purity of distilled water; degree of thermal conduction along the thermometer; shielding from the radiant heat exchange; degree of ventilation (volumetric flow rate); diameter of the wire; shape, size (thickness and length), and material of the wick covering the bulb; relative positions of the wet and dry bulb; and accuracy of the psychrometric formula used (Bindon 1965; Gust 1995; Lemay et al. 2001; Toida et al. 2006; Ustymczuk and Giner 2011; d'Ambrosio Alfano et al. 2012).

The physical location of the test and amount of time spent whirling or blowing air over the wet bulb are additional factors that directly affect the accuracy of the test result. Therefore, while using the sling psychrometer, the thermometers should be whirled rapidly for 15 or 20 seconds, stopped, and quickly read—the wet bulb first, because it will begin to change when the air movement stops. The test should be repeated until two or more wet-bulb readings equal the lowest reading obtained (Gust 1995). An aspirating psychrometer is more accurate. The aspirating psychrometer is more reliable because it minimizes possible operator errors and ensures a constant air flow past the wick. The accuracy of aspirating and sling psychrometers can also be affected by altitude, especially at lower RH. At lower atmospheric pressure, water evaporates faster, lowering the temperature of the dry bulb more. It should also be borne in mind that, when the temperature is near or below the freezing point, the psychrometer is not a very reliable instrument for measuring humidity.

As indicated before, a psychrometer does not directly measure RH and dewpoint temperature. These values are calculated using a formula such as Equation 4.3, into which the dry- and wet-bulb temperatures are inserted. However, graphs and psychrometric slide-rule calculators are available for this. Indeed, reference tables can be employed to find these partial pressures, but most tables only have one-degree resolution at best, leaving the metrologist to interpolate between integer degrees. Other common instruments or charts include calibration curves or sliding scales, which can introduce errors if one incorrectly reads the scale or curve. Interpolation error can also occur. In addition, the calculation is valid only for air at ambient conditions and should not be applied to other gases or if the temperature/pressure conditions differ significantly from ambient. Charts such as the U.S. Weather Bureau Psychrometric Tables (Figure 4.8) make this determination a little easier. However, to use these tables, one must know the local atmospheric pressure for that day. However, with the aid of a computer or advanced calculator, these equations and calculations can be expediently computed, yielding much more accurate and reliable measurements than the charts. Advances in computing technology provide an

15

Table I.—*Temperature of dew-point in degrees Fahrenheit.*

Pressure = 30.0 inches.

Air temp. t	Vapor press. e	Depression of wet-bulb thermometer $(t-t')$. .2	.4	.6	.8	1.0	1.2	1.4	1.6	1.8	2.0	2.2	2.4	2.6	2.8	3.0
−40	.0019	−52														
−39	41	−50														
−38	44	−49														
−37	46	−48														
−36	48	−46														
−35	.0051	−45														
−34	54	−43	−59													
−33	57	−42	−56													
−32	61	−40	−53													
−31	65	−38	−50													
−30	.0069	−36	−47													
−29	74	−35	−45													
−28	78	−33	−42	−56												
−27	83	−32	−40	−54												
−26	89	−30	−38	−50												
−25	.0094	−29	−36	−46												
−24	.0100	−28	−34	−43	−60											
−23	106	−27	−32	−40	−54											
−22	112	−26	−31	−38	−49											
−21	119	−25	−29	−35	−45											
−20	.0126	−23	−28	−33	−42	−57										
−19	133	−22	−26	−31	−39	−51										
−18	141	−21	−25	−29	−36	−46										
−17	150	−20	−23	−28	−33	−42	−59									
−16	159	−19	−22	−26	−31	−38	−51									
−15	.0168	−18	−21	−24	−29	−35	−45									
−14	178	−16	−19	−23	−27	−32	−40	−56								
−13	188	−15	−18	−21	−25	−30	−36	−48								
−12	199	−14	−17	−20	−23	−27	−33	−43	−59							
−11	210	−13	−16	−18	−22	−25	−30	−38	−50							
−10	.0222	−12	−14	−17	−20	−24	−28	−33	−44							
− 9	234	−11	−13	−16	−18	−22	−26	−30	−38	−51						
− 8	247	−10	−12	−14	−17	−20	−24	−28	−34	−44						
− 7	260	− 9	−11	−13	−16	−18	−22	−26	−31	−38	−51					
− 6	275	− 8	−10	−12	−14	−17	−20	−23	−28	−33	−44					
− 5	.0291	− 7	− 8	−10	−13	−15	−18	−21	−25	−30	−37	−50				
− 4	307	− 6	− 7	− 9	−11	−14	−16	−19	−22	−27	−32	−42	−59			
− 3	325	− 4	− 6	− 8	−10	−12	−14	−17	−20	−24	−29	−35	−47			
− 2	344	− 3	− 5	− 7	− 8	−10	−13	−15	−18	−21	−25	−30	−38	−53		
− 1	363	− 2	− 4	− 5	− 7	− 9	−11	−13	−16	−19	−22	−27	−32	−42	−60	
0	.0383	− 1	− 3	− 4	− 6	− 7	− 9	−12	−14	−17	−20	−23	−28	−35	−46	
+ 1	403	± 0	− 2	− 3	− 4	− 6	− 8	−10	−12	−15	−17	−20	−25	−30	−37	−60
2	423	+ 1	− 1	− 2	− 3	− 5	− 6	− 8	−10	−13	−15	−18	−21	−26	−31	−40
3	444	2	+ 1	− 1	− 2	− 4	− 5	− 7	− 9	−11	−13	−16	−19	−22	−27	−32
4	467	3	2	± 0	− 1	− 2	− 4	− 5	− 7	− 9	−11	−14	−16	−19	−23	−28
5	.0491	4	3	+ 1	± 0	− 1	− 3	− 4	− 6	− 7	− 9	−12	−14	−17	−20	−24
6	515	5	4	3	+ 1	± 0	− 1	− 3	− 4	− 6	− 8	−10	−12	−15	−17	−21
7	542	6	5	4	2	+ 1	± 0	− 1	− 3	− 4	− 6	− 8	−10	−12	−15	−18
8	570	7	6	5	4	3	+ 1	± 0	− 2	− 3	− 5	− 6	− 8	−10	−13	−15
9	600	8	7	6	5	4	3	+ 1	± 0	− 2	− 3	− 5	− 6	− 8	−10	−13
10	.0631	9	8	7	6	5	4	3	+ 1	± 0	− 2	− 3	− 5	− 6	− 8	−10
11	665	10	9	8	7	6	5	4	3	+ 1	± 0	− 1	− 3	− 4	− 6	− 8
12	699	11	10	9	8	7	6	5	4	3	+ 2	± 0	− 1	− 3	− 4	− 6
13	735	12	11	11	10	9	8	7	6	4	3	+ 2	± 0	− 1	− 2	− 4
14	772	13	12	12	11	10	9	8	7	6	5	3	+ 2	+ 1	− 1	− 2
15	.0810	14	13	13	12	11	10	9	8	7	6	5	4	2	+ 1	± 0
16	850	15	14	14	13	12	11	10	9	8	7	6	5	4	3	+ 1
17	891	16	15	15	14	13	12	12	11	10	9	8	7	6	4	3
18	933	17	17	16	15	14	13	13	12	11	10	9	8	7	6	5
19	.0979	18	18	17	16	15	15	14	13	12	11	10	9	8	7	6
20	.1026	19	19	18	17	16	16	15	14	13	12	12	11	10	9	8

t	e	t	e	$(t-t')$ .1	.2	.3	.4	.5
−60	.0010	−50	.0021	−60				
−59	11	−49	22	−58				
−58	12	−48	24	−56				
−57	13	−47	26	−55				
−56	13	−46	27	−53				
−55	.0015	−45	.0029	−51				
−54	16	−44	31	−50	−60			
−53	17	−43	33	−49	−58			
−52	18	−42	35	−48	−56			
−51	.0019	−41	37	−46	−54			
		−40	.0039	−45	−52			
		−39	41	−44	−50			
		−38	44	−43	−49	−59		
		−37	46	−42	−48	−56		
		−36	48	−41	−46	−54		
		−35	.0051	−40	−45	−51		
		−34	54	−38	−43	−49	−59	
		−33	57	−37	−42	−47	−56	
		−32	61	−35	−40	−46	−53	
		−31	65	−34	−38	−44	−50	−60
		−30	.0069	−33	−36	−42	−47	−55

FIGURE 4.8 U.S. Weather Bureau psychrometric tables use to calculate relative humidity. (Reprinted from www.nist.gov.)

opportunity to properly utilize humidity-determining calculations, free of human errors inherent in massive, complex mathematics and interpolation from charts.

4.5 ADVANTAGES AND LIMITATIONS

According to WMO (1989a, 1989b), usage of the psychrometer for humidity measurement has both advantages and disadvantages, which are listed in Table 4.3.

According to Gust (1995), in addition to limitations listed in Table 4.3, the wet- and dry-bulb technique has other limitations that may be important in the case of environmental chambers:

- No measurement below the freezing point.
- The sample can be humidified, because the wet sock adds water to the environment (a problem with chambers operating at low humidity). This means that this hygrometer is not suitable for operation in small, enclosed areas (where the moisture from the wet bulb significantly changes the water vapor content in the environment).
- Sluggish response, thus poor control of characteristics: The wet-bulb temperature reacts slowly to changes in humidity because of the mass of the wet-bulb thermometer and the wick. Slow reaction to changes in temperature is due to the time required by the water supply to adapt.
- Requires a water supply and, therefore, can support the growth of micro-organisms.
- Cannot be easily calibrated and trouble-shot.

As to the merits of the psychrometer, the operating experience showed that a prominent advantage of these devices is its adaptability to measurement environments. For example, Costello et al. (1991) have shown that the aspirated psychrometer could work reliably in a dusty environment with minor maintenance required only every 5–10 days. The same results were reported by Duuren (1991). He used four types of equipment for measuring the moisture content of air for 7 months in heavy clay dryers that included the wet- and dry-bulb psychrometers. Thus, compared to a modern electronic RH sensor, such as capacitive RH sensors or resistive RH sensors, the psychrometer is one of the RH measurement devices that can be used in a dirty environment, withstand state changes of water, and provide a large range of measurement values. As a result, the psychrometer is widely used in different measurement environments in many current applications (Yin et al. 1996; Maskan et al. 2002; Swami et al. 2007; Tanaka et al. 2010).

As indicated before, the psychrometer is difficult to operate at temperatures below freezing. A wick cannot be used to convey water from a reservoir to the wet-bulb sleeve by capillary action when the wet-bulb temperature is below 0°C. However, in reality, a psychrometer in many cases is used in climates where such temperatures occur. Under these conditions, care should be taken to form only a thin layer of ice on the sleeve. It is an absolute necessity that the thermometers be artificially ventilated; if they are not, the management of the wet bulb will be extremely difficult. The water should, as far as possible, have a temperature close to a freezing point. If a button of ice forms at the lowest part of the bulb, it should be immersed in water long enough to melt the ice. The time required for the wet bulb to reach a steady reading after the sleeve is wetted depends on the ventilation rate and the actual wet-bulb temperature. An unventilated thermometer usually requires from 15 to

TABLE 4.3
Advantages and Disadvantages of the Psychrometer for Humidity Measurement

Advantages	Disadvantages
Simple, cheap, reliable, and robust	Slow response. The time constant of the psychrometer is about 40 seconds
Can have good stability	Large physical size
Wide range of humidity can be measured	Some skill is required to use and maintain the instrument: keeping one thermometer bulb wet and ensuring good airflow around it
Tolerate high temperatures and condensation	A large air sample is required for measurement
The simplicity of the device affords easy repair at minimum cost	Measurement is complicated below 10°C: The cooling of the wet bulb to its full depression becomes difficult. As a result, the accuracy below 20% RH seriously deteriorates
	Wick can become contaminated
	Results have to be calculated from tables or software
	Whirling types are prone to serious errors

RH, relative humidity.
Source: WMO World Meteorological Organization: Assmann Aspiration Psychrometer Intercomparison (Sonntag D.), (Instruments and Observing Methods Report No. 34), (WMO/TD-No. 289), WMO, Geneva, Switzerland, 1989a; WMO World Meteorological Organization: International Hygrometer Intercomparison (Skaar J., Hegg K., Moe T., Smedstud K.). (Instruments and Observing Methods Report No. 38), (WMO/TDNo.316), WMO, Geneva, Switzerland, 1989b.

45 minutes, while an aspirated thermometer will require a much shorter period. It is essential that the formation of a new ice film on the bulb be made at an appropriate time. If hourly observations are being made with a simple psychrometer, it will usually be preferable to form a new coating of ice just after each observation. If the observations are made at longer intervals, the observer should visit the screen sufficiently in advance of each observation to form a new ice film on the bulb. The wet bulb of the aspirated and sling psychrometers should be moistened immediately before use. The evaporation of an ice film may be prevented or slowed by enclosing the wet bulb in a small glass tube, or by stopping the ventilation inlet of the wet bulb between intervals. (Note that the latter course should not be taken if the circumstances are such that the ventilating fan would be overheated.)

While the wet- and dry-bulb measurement technique has a sound theoretical basis, the problem is that it is simple only in appearance. As shown previously, many factors influence the accuracy of the measurements. Therefore, to achieve accurate results, one should follow these guidelines (Gust 1995):

- The wick used must be of the type supplied with the instrument, fit properly, and be absorbent.
- Do not touch the wick. Wicks should be cleaned before use by boiling in water with little detergent for approximately 10 minutes.
- The wick should be closely fitted to the thermometer bulb so as to minimize errors due to heat conduction along the stem of the thermometer.
- Water used must be highly pure—either deionized or distilled.
- The wick must be changed daily or at least weekly depending on the contamination level in the environment (e.g., dust, pollen, salt spray).
- Be sure that the aspirating psychrometer has a good battery.
- Use the correct psychrometric tables; a psychrometer is said to be aspirated if the air velocity past the bulbs is greater than 3 m per second.
- Psychrometer coefficient, used for establishing the psychrometric chart that converts the wet- and dry-bulb temperature readings into RH, should be determined for each specific design of psychrometer and in particular for each design of the wet bulb.
- Failure to ensure the conditions above will usually lead to the psychrometer's overestimating humidity.
- The wet- and dry-bulb thermometers should not only be accurate, but they should also be matched so as to minimize the error on the temperature depression readings (or temperature difference).
- Make sure the thermometers are mounted at a location where conditions are fairly representative of the average conditions inside the chamber.
- The errors of measurement can result due to poor choice in the mounting location of the wet- and dry-bulb thermometers. This is the case when the thermometers are installed too closely to a source of moisture (e.g., water supply for the wet bulb, steam injector).

However, the use of a psychrometer also showed that, even if all requirements were met, the accuracy of the measurement in some cases may not be sufficient. As a rule, conventional psychrometers have low measurement accuracy compared to the chilled-mirror hygrometer (Chapter 5) and electronic sensors (read Chapters 10–22 in Sections III and IV). Therefore, substantial efforts have been made to improve the accuracy of the psychrometer readings. The most dramatic changes in the structure of psychrometers occurred when replacing the thermometer on the thermocouples and thermistors (Powell 1936; Kawata and Omori 1953; Nantou 1979; Nichols 1992), which permitted reduction in the size of the psychrometer, to provide continuous monitoring and simplify the recording of readings and the processing of results. The accuracy of the digital ventilated psychrometer was less than 2% RH (Nantou 1979; Nichols 1992; Bhuyan and Bhuyan 1995). Thermocouples and thermistors are of special interest where low lag is important, where there is little or no ventilation, or where very low RHs are to be measured. Even higher measurement accuracy was reported by Zhang et al. (2016). By optimizing the distance between the wet and dry bulb, the wet cloth size, the distance between the wet bulb and water, the distance from blow fan to dry bulb and wet bulb probe, the air flow speed, the accuracy of resistance temperature detectors, and the signal conversion, it was managed to achieve measurement accuracy of ± 0.6%. In addition, psychrometers with electronic temperature sensors can provide the highest accuracy, near 100% RH. Since the dry bulb and wet bulb sensors can be connected differentially (Slatyer and Bierhuizen 1964), this allows the wet-bulb depression (which approaches zero as the RH approaches 100%) to be measured with minimal error.

REFERENCES

Berlicki T.M., Murawski E., Muszynski M., Osadnik S.J., Prociow E.L. (1998) Thermoelement humidity sensor. *Sens. Actuators A* 64(3), 213–217.

Berzelius J.J. (1841) *Chemische Operationen und Geräthschaften*. 3 edn. Lehrbuch der Chemie, Vol. 10. Arnoldsche Buchhandlung, Dresden, Leipzig, Germany.

Bhuyan M., Bhuyan R. (1995) An on-line method for monitoring of relative humidity using thermal sensors, In: *Proceedings of the IEEE International Conference on industrial Automation and Control*, Hyderabad, India, pp. 7–11.

Bindon H.H. (1965) A critical review of tables and charts used in psychrometry, In: Wexler A. (ed.) *Humidity and Moisture*, Vol. 1. Reinhold, New York, pp. 3–15.

Costello T.A., Berry I.L., Benz R.C. (1991) A fan-actuated mechanism for controlled exposure of a psychrometer wet bulb sensor to a dusty environment. *Appl. Eng. Agric.* 7, 473–477.

d'Ambrosio Alfano F.R., Palella B.I., Riccio G. (2012) On the problems related to natural wet bulb temperature indirect evaluation for the assessment of hot thermal environments. *Ann. Occup. Hyg.* 56, 1063–1079.

Duuren V. (1991). Process control in the dryers of the heavy clay industry. Practical investigation working of ten commercially obtainable moisture sensors. *Rev. Sci. Instrum.* 14(4), 77–81.

Gregory H.S. (1947) The measurement of humidity. *Instrum. Pract.* 1, 307, 447.

Gust J. (1995) An error analysis of psychrometers, In: *Proceedings of NCSL Workshop & Symposium*, Session 7C, July 16–20, Fairmont Hotel, Dallas, TX, pp. 583–606.

Kawata S., Omori Y. (1953) An investigation of thermocouple psychrometer. *I. J. Phys. Soc. Jpn.* 8, 768–776.

Lemay S.P., Guo H., Barber E.M., Zyla L. (2001) A procedure to evaluate humidity sensor performance under livestock housing conditions. *Biosyst. Eng.* 43, 14–21.

Maskan A., Kaya S., Maskan M. (2002) Hot air and sun drying of grape leather (pestil). *J. Food Eng.* 54, 81–88.

Middleton W.E.K. (1969) *Catalog of Meterological Instruments in the Museum of History and Technology*. Smithsonian Institution Press, Washington, DC.

Montanini R. (2007) Wavelength-encoded optical psychrometer for relative humidity measurement. *Rev. Sci. Instrum.* 78(2), 025103.

Nantou Y. (1979) Digital ventilated psychrometer. *IEEE Trans. Instrum. Meas.* 28, 42–45.

Nichols E.L. (1992) An automatic psychrometer sensor for poultry farming, In: *Proceedings of the IEEE International Conference on Southeast*, Birmingham, AL, pp. 750–752.

Powell R.W. (1936) The use of thermocouples for psychrometric purposes. *Proc. Phys. Soc. London* 48, 406–414.

Rübner K., Balköse D., Robens E. (2008) Methods of humidity determination. Part I: Hygrometry. *J. Thermal Anal. Calorimetry* 94(3), 669–673.

Slatyer R.O., Bierhuizen J.F. (1964) A differential psychrometer for continuous measurements of transpiration. *Plant Physiol.* 39(6), 1051–1056.

Sprung A. (1888) Über die Bestimmung der Luftfeuchtigkeit mit Hilfe des Assmannschen Aspirationspsychrometers. *Z. Angew. Meteorol., Das Wetter* 5, 105–108.

Swami S.B., Das S.K., Maiti B. (2007) Texture profile analysis of cooked sun dried nuggets (bori) prepared with different levels of moisture content and percent air incorporation in its batter. *Int. J. Food Eng.* 3, 1–15.

Tanaka F., Ide Y., Kinjo M., Genkawa T., Hamanaka D., Uchino T. (2010) Development of thick layer re-wetting model for brown rice packaged with LDPE and PBT films. *J. Food Eng.* 101, 223–227.

Toida H., Ohyama K., Kozai T., Hayashi M. (2006) A method for measuring dry-bulb temperatures during the operation of a fog system for greenhouse cooling. *Biosyst. Eng.* 93, 347–351.

Ustymczuk A., Giner S.A. (2011) Relative humidity errors when measuring dry and wet bulb temperatures. *Biosyst. Eng.* 110, 106–111.

Wiederhold P.R. (1997) *Water Vapor Measurement, Methods and Instrumentation*. Marcel Dekker, New York.

WMO (1989a) World Meteorological Organization: Assmann Aspiration Psychrometer Intercomparison (Sonntag D.). (Instruments and Observing Methods Report No. 34), (WMO/TD-No. 289), WMO, Geneva, Switzerland.

WMO (1989b) World Meteorological Organization: International Hygrometer Intercomparison (Skaar J., Hegg K., Moe T., Smedstud K.). (Instruments and Observing Methods Report No. 38), (WMO/TDNo.316), WMO, Geneva, Switzerland.

Yin X.Z., Zhou Y.J., Li Y.X. (1996) The test of parameterized psychrometric coefficient with wind speed. *J. Arid Meteorology* 14, 48–53.

Zhang W., Ma H., Yang S.X. (2016) An inexpensive, stable, and accurate relative humidity measurement method for challenging environments. *Sensors* 16, 398.

5 Chilled-Mirror Hygrometer or Mirror-Based Dew-Point Sensors

5.1 CONDENSATION METHODS OF HUMIDITY MEASUREMENT

Mirror-based dew-point sensors are based on the condensation method. It is known that every object in a moist atmosphere has water molecules on its surface. The concentration on these molecules is related to the temperature of the object and to the dew point of the atmosphere. If the temperature of the surface is above the dew point of the atmosphere, the thin layer of the water molecules will be invisible. However, as the surface is cooled to the dew point, the density of water molecules at the surface becomes so great that the water condenses on the surface, and the dew can be seen. Mirror-based dew-point sensors have better accuracy than the methods described in previous chapters, but these devices are costlier in fabrication and exploitation, because they require maintenance by qualified personnel (Wiederhold 1997; Chen and Lu 2005).

It is worth noting that the invention of the first condensation hygrometer is ascribed to Ferdinand II, the Grand Duke of Tuscany (1650–1670) (see Figure 5.1) (Magalotti 1667; Targioni Tozzetti 1780). The truncated cone in the lower part was lined with glass, to make it impermeable, and filled with "snow or ice very finely ground." The humidity in the air, in contact with the iced glass, condensed, causing the formation of water drops that descended toward the apex of the cone, gathering in the graduated glass below. The higher the atmospheric humidity, the greater is the condensation. The quantity of water gathered in the glass over a certain period of time was therefore a relative measure of the humidity.

Regarding a dew-point hygrometer, it was invented in 1854 by H.V. Regnault (France). In Regnault's hygrometer, a silver tube containing ether was cooled by blowing air through it with a rubber bulb. When condensation appeared on the outside, the air was saturated with water at that temperature (dew point) (see Figure 5.2a and b). From a graph of water content of saturated air against temperature, the water content at dew point and that at room temperature may be found. According to other sources (Rübner et al. 2008), John Frederic Daniell invented a dew-point hygrometer (see Figure 5.3) in 1820.

There are two possible sources of serious error with instruments designed by Daniell and Regnault (Middleton 1969). One is the gradient of temperature between the surface of the bulb and the thermometer;

FIGURE 5.1 The condensation hygrometer, which was probably invented by Grand Duke Ferdinand II de' Medici. The date was about 1655 (Middleton 1969). The truncated cone in the lower part was lined with glass, to make it impermeable, and filled with "snow or ice very finely ground." The humidity in the air, in contact with the iced glass, condensed, causing the formation of water drops that descended toward the apex of the cone, gathering in the graduated glass below. The higher the atmospheric humidity, the greater is the condensation. The quantity of water gathered in the glass over a certain period of time was therefore a relative measure of the humidity. However, this way of measuring humidity was time-consuming and probably very sensitive to the speed of the air past the cone. (Reprinted from https://siarchives.si.edu/history/national-museum-american-history.)

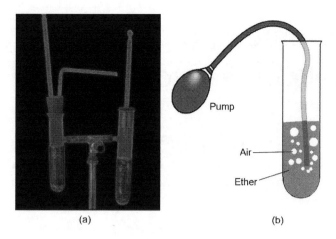

FIGURE 5.2 (a) Regnault's hygrometer. and (b) Diagram illustrating its operation. (Reprinted from https://siarchives.si.edu/history/national-museum-american-history.)

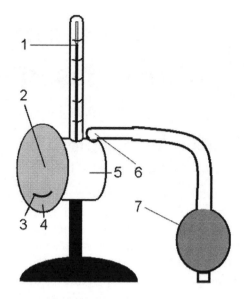

FIGURE 5.4 Lambrecht's dew-point hygrometer: 1, thermometer; 2, metal mirror; 3, gap; 4, reference mirror; 5, container filled with ether; 6, air supply; 7, air pump. (Reprinted from https://siarchives.si.edu/history/national-museum-american-history.)

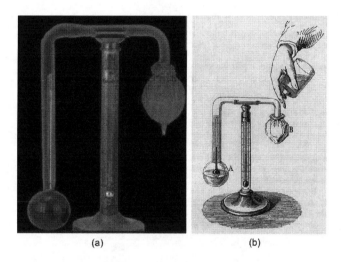

FIGURE 5.3 (a) The condensation instruments developed by John Frederic Daniell in 1820. The most usual means of cooling the surface in the nineteenth century was by the evaporation of ether. Daniell built a closed-glass system terminating in two bulbs, one covered with muslin, the other of black glass, and containing ether and a thermometer. To use it, the instrument is tilted so that most of the ether is in the uncovered bulb; the bulb that is covered with muslin is then moistened with ether from a bottle (b). The evaporation of this causes the ether in the instrument to distill over, reducing the temperature of the bulb until dew begins to form on its surface. At this moment, the internal thermometer is read (Middleton 1969). Thus, the inside thermometer gives the dew point. From the dew point and the air temperature, one calculates the relative humidity. This device was the ancestor of the modern chilled-mirror dew-point meter (i.e., the most accurate laboratory instrument to measure humidity and to perform calibrations of other instruments). (Reprinted from https://siarchives.si.edu/history/national-museum-american-history.)

however, this error can be greatly reduced by noting the temperature at which the dew disappears, as well as that at which it forms. The other is less tractable and results from the frequent presence of small amounts of water in the ether used to cool the bulb. To get away from this, it is better to evaporate the ether in the chamber that is to be cooled by drawing air through it and discharging the resulting vapor at a distance. The best design for a hygrometer operating in this way is probably a dew-point hygrometer, which was designed by Lambrecht in 1881. This hygrometer is shown in Figure 5.4. The Lambrecht dew-point hygrometer is a polished metal mirror that is cooled by evaporation of ether until moisture just begins to condense onto it. This point was controlled optically. A distinctive feature of these hygrometers is that a comparison surface, not cooled, is next to the surface on which the dew is formed, which significantly increases the accuracy of determining the dew-point temperature.

5.2 CHILLED-MIRROR HYGROMETERS

Compared to the previously described devices, a chilled-mirror hygrometer is an improved device. The chilled-mirror hygrometer is an absolute humidity sensor. In this device, the sample air is drawn to the metallic mirror surface through piping to determine directly the dew-point temperature. As the mirror cools, condensation forms when its surface temperature falls below the dew-point temperature, but it evaporates and disappears at higher temperatures. The temperature of the metallic mirror, when condensation forms, is measured using a platinum resistance thermometer properly embedded in the mirror, and the result is taken as the dew-point temperature.

Historically, the cooling of the mirror surface has been accomplished with acetone and dry ice, liquid carbon dioxide (CO_2), mechanical refrigeration, and, more recently, thermoelectric heat pumps. It is important to note that the first commercial applications of the hygrometer started in the early 1960s, after it became practical to use it in thermoelectric coolers. Prior to 1960, the use of chilled-mirror hygrometers was primarily confined to laboratories because of the difficulties in cooling the mirror with cryogenic liquids, controlling its temperature, and detecting the onset of dew. The detection of condensation was observed visually, and the equilibrium cooling was manually controlled.

The manually cooled, visually observed hygrometer is commonly known as the *dew cup*. A dew cup is the simplest instrument for the measurement of dew points, which duplicates the experiments of Charles LeRoy in 1751. The gas sample is drawn from the furnace or generator across the outside of a polished cup made of chromium-plated copper. The cup is enclosed in a glass container so the moisture can be seen condensing on the cup surface when the dew point is reached. Dew is observed by the operator, and measurement differences can occur due to the operator's interpretation. The cup surface is cooled progressively by dropping small pieces of dry ice in acetone (or methanol) until the dew point is reached, as indicated by condensation on the cup surface at a temperature indicated by a thermometer in the acetone. It is a relatively inexpensive technique, and, when operated by an experienced and skilled technician, it is quite accurate. However, it does suffer from some limitations (Brodgesell and Liptak 1993, 1995), which are listed below:

1. It is not a continuous measurement.
2. It is operator dependent, so the readings may vary from operator to operator.
3. Versions using expendable coolants require a replacement of supplies.

All of these difficulties can be overcome by the thermoelectrically cooled, optically observed dew-point hygrometer. An instrument of this type is shown in Figure 5.5. These versions utilize optical phototransistor detection designs to control automatically the surface at the dew point or the frost point. The temperature instrumentation has included the entire spectrum, from the glass bulb thermometers to all types of electrical

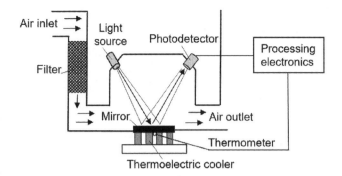

FIGURE 5.5 Schematic of the dew-point sensor.

temperature elements. The mirror surface is chilled to the dew point by a thermoelectric cooler, while a continuous sample of the atmospheric gas is passed over the mirror. In the chilled-mirror technique, a mirror is constructed from a material with good thermal conductivity, such as silver or copper, and properly plated with an inert metal such as iridium, rubidium, nickel, or gold to prevent tarnishing and oxidation. The mirror is illuminated by a light source—for example, a light-emitting diode (LED), and observed by a photodetector bridge network. As condensate forms on the mirror, the change in reflectance is detected by a reduction in the direct reflected light level received by the photodetector because of the light-scattering effect of the individual dew molecules. This light reduction forces the optical bridge toward a balance point, reduces the input error signal to the amplifier, and proportionally controls the drive from the power supply to the thermoelectric cooler. This continuously maintains the mirror at a temperature at which a constant thickness of the dew layer is retained. In other words, the temperature of the mirror is carefully controlled to maintain an equilibrium between condensation and evaporation of water as the humidity changes. Independently, embedded within the mirror, a temperature-measuring element measures the dew-point temperature directly.

The 1990s were a time of research aimed at the improvement of the mirror-based dew/frost hygrometers (Jachowicz 1992; Jachowicz and Makulski 1993; Matsumoto and Toyooka 1992, 1995a, 1995b; Pascal-Delannoy et al. 1998; Sorli et al. 2002; Chen and Lu 2005). The improvement was focused on the sensing element (i.e., how to accurately detect the temperature at which water begins to condense on the mirror surface). For example, in order to stabilize the mirror temperature, additional heat was injected into the mirror (Jachowicz 1992; Jachowicz and Makulski 1993). The experimental results showed that temperature was many times faster, and overcondensation was minimized. The temperature fluctuations around the dew point usually were not over 0.03 K. During operation of the hygrometer, the heat pump cools the mirror and simultaneously gives out a huge amount of heat into the heat housing. This temperature change influences the sensitivity of the optical dew detector, which decreases the hygrometer's accuracy. In order to solve this problem, optical fibers were used to separate the optical dew detector from the mirror area, as shown in Figure 5.6. The accuracy of the hygrometer was significantly improved at elevated temperatures. At 5°C, the error of the fiber optical dew detector is approximately −0.10°C to −0.11°C, while the regular optical dew detector is approximately −0.63°C to −0.65°C (Jachowicz and Zalewski 1994). Kimura (1996) proposed to measure the absolute humidity independently of the ambient temperature, using a thermal humidity sensor with only a single micro-air-bridge heater, which is heated up to two level temperatures by pulse currents.

It is necessary to add that the optical dew-point mirror hygrometers are still available in large quantities. It is important that chilled-mirror hygrometers can be made very compact using a solid-state optics and thermoelectric cooling and with extensive capabilities using microprocessors. A view of modern devices

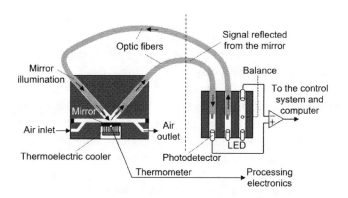

FIGURE 5.6 Schematic of the hygrometer measurement head with fiberoptic dew-point detector. (Reprinted with permission from Jachowicz, R.S. and Zalewski, D., Hygrometer with fibre optic dew point detector, *Sens. Actuators A*, 42, 503–507, Copyright 1994, Elsevier.)

using the method of humidity measurement is shown in Figure 5.7. The main advantages of this design include its wide range; its accuracy, which is limited only by the quality of the thermometer used to detect the temperature of the cooled surface; and its self-calibrating nature. Self-calibration is achieved by allowing the surface to heat up and then checking whether, upon

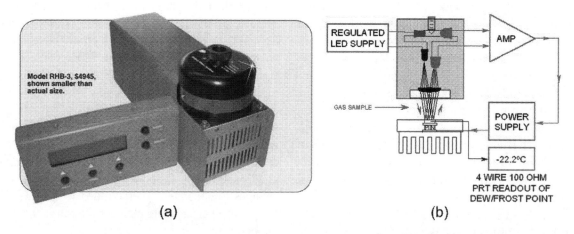

FIGURE 5.7 (a) Chilled-mirror dew-point optical hygrometer designed by Omega Engineering (www.omega.com) CMH Block Diagram (b), Figure illustrates how the dew-point hygrometers detect and measure the dew point. The condensate mirror is illuminated with a high-intensity, solid-state, light-emitting diode (LED). A photodetector monitors the LED light reflected from the mirror. The photodetector is fully illuminated when the mirror is clear of dew, and it receives less light as dew forms. A separate LED and photodetector pair are used as a known reference to compensate for any thermally induced changes in the optical components. The photodetectors are arranged in an electrical bridge circuit, the output current of which is proportional to the light reflected from the mirror. The bridge output controls the electrical current to the thermoelectric cooler. A large bridge current develops when the mirror is dry, causing the mirror to cool toward the dew point. As dew begins to form on the mirror, less light is reflected, and the bridge output decreases. This, in turn, causes a decrease in cooling current. A rate feedback loop within the amplifier ensures critical response, causing the mirror to stabilize quickly at a temperature that maintains a thin dew or frost layer on the mirror surface. (Reprinted from www.omega.com.)

recooling it, the dew point detected is the same value as before. At present, the dew-point mirror hygrometer is recognized as the most precise method of determining the water vapor content of a gas above 5% relative humidity (RH). Unlike polymer RH sensors, lithium chloride dew cells, and other chemically based sensors, dew-point mirror hygrometers do not lose their calibration. Cooling or heatup rates are reasonably fast—approximately 3°F (1.5°C) per second. The instrument can be used with all common atmospheric gases, such as nitrogen, oxygen, and carbon dioxide. The ability of a chilled-mirror instrument to measure the values of low humidity depends both on the power available for cooling the mirror and on the evacuation of heat away from the mirror.

Despite their wide use, the optical dew-point hygrometers have several drawbacks, listed in Table 5.1. In addition, the instrument is unsuitable for the samples containing components that react with water, such as chlorine or sulfur dioxide, and components with a dew point higher than that of water. It is also necessary to monitor the presence of air supplements such as natural gas, such as a methane. One should keep in mind that the chilled mirror will tend to control the first dew point it encounters as it cools. In such instances, the methane dew/frost point temperature can be above the water dew point, causing an erroneous reading.

In recent years, studies carried out have allowed to optimize significantly the performance of dew-point hygrometers and to reduce the influence of the factors distorting their readings. These improvements included so-called *automatic balance* and *continuous balance*, used in the measurement process. In all of these, the mirror is continuously at the dew-point temperature and has at all times, except during the short balancing cycle, a dew or frost layer on it. Other approaches to optimize dew-point hygrometers that have substantially improved their performance have also been used. For example, in the hygrometer, unique integral cylindrical filters were incorporated that were not consumed by the sample air, as in conventional systems, but that caused most of the sample air to bypass the mirror and filter, and thereby further reduced mirror contaminants. Mirror cleaning cycles go as long as 1 year in some cases. They were also developed Dual-Mirror Twin-Beam Sensors, which virtually eliminated the need for regular recalibration of optics, were also developed. All this increased the attractiveness of these hygrometers for various applications. For example, despite the bulkiness of equipment needed to conduct research using the chilled-mirror principle, this method was adapted to measure the distribution of humidity in height (Hall et al. 2016). The need for such studies was discussed in Volume 1 of this text. The National Oceanic and Atmospheric Administration (NOAA) frost-point hygrometer (FPH), the schematic diagram of which is shown in Figure 5.8, is the latest model of a balloon-borne instrument to measure the water vapor profiles up to 28 km. The instrument has an uncertainty in the stratosphere that is lower than 4%–6% and up to 10%–12% in the troposphere. A digital microcontroller version of the instrument improved upon the older versions in 2008 with sunlight filtering, better frost control, and resistance to radiofrequency interference (RFI). A new thermistor calibration technique was implemented in 2014, decreasing the uncertainty in the thermistor calibration fit to less than 0.01°C, covering the full range of frost- or dew-point temperatures ($-93°C$ to $20°C$) measured during a profile. The description of previous models of balloon-borne frost-point hygrometers, used to collect vertical profiles of the water vapor abundance and distribution in the atmosphere since the late 1950s, can be found in Vomel et al. (2007) and Hall et al. (2016).

TABLE 5.1
Advantages and Disadvantage of Dew-Point Mirror Hygrometers

Advantages	Disadvantages
• The principle is simple; place a temperature-controlled mirror in the air stream and cool it until dew forms on the mirror.	• Expensive.
	• Large physical size. Not a compact system.
	• Sufficiently large power consumption.
• This is a direct, fundamental measurement of dew point.	• Some skill is required to use and maintain the instrument.
• Can provide precise measurement. Uncertainty around 0.2°C.	• A large air sample is required for measurement.
• Good long-term performance. Inert wetted components. Five to 20 years of service without noticeable drift.	• Frequent mirror contamination. Other gases may condense prior to water, such as heavy hydrocarbons.
• Wide measurement range.	• Instability under continuous use.
• Some models can measure at process pressure.	• Dew points below 0°C require careful interpretation.
	• Can be slow in response.

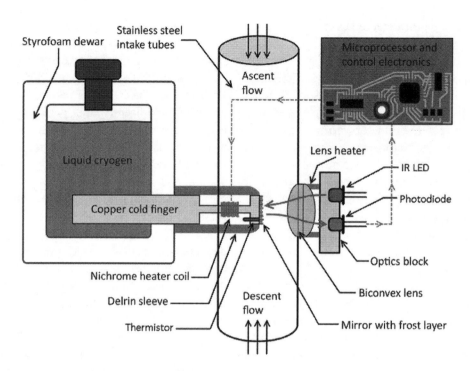

FIGURE 5.8 Schematic of the NOAA FPH instrument. A copper cold finger immersed in a Dewar containing liquid cryogen provides cooling power throughout the profile. A polished mirror disk resides at the opposite end of the cold finger, with ambient air passing over it at 3–6 m/s. A nichrome heater wrapped around the narrow shaft of the continuous cold finger and the mirror piece provides heat to the mirror. An optical source and detector, composed of an infrared light emitting diode (LED) and a photodiode, is used to monitor the mirror's reflectivity as condensate accumulates in the form of dew or frost. A biconvex lens focuses the light returned off the mirror into the photodiode. A calibrated thermistor embedded in the mirror accurately measures the frost-point temperature. The frost-point temperature is achieved when a stable frost layer is present, showing equilibrium between the water vapor in the air sample and the condensate on the mirror. Due to the material and geometry, the surface of the mirror is considered to be uniform in temperature across the 7.1 mm diameter. The electronics board and optics block are enclosed by insulating foam, along with the battery pack, while the lens, intake tubes, and mirror head are all exposed to ambient air. The microcontroller uses the photodiode signal to regulate the mirror temperature such that the reflectivity of the frost on the mirror is constant over the flight. (Reprinted with permission from Hall E.G., et al., Advancements, measurement uncertainties, and recent comparisons of the NOAA frost point hygrometer, *Atmos. Meas. Tech.*, 9, 4295–4310, 2016. Copyright 2016, European Geosciences Union as open access.)

5.3 SURFACE CONDUCTIVE DEW-POINT HYGROMETER

It should be noted that the dew-point sensors, based on condensation method can be designed on the basis of devices controlling the surface conductivity change occurring during the condensation of water vapor (Brodgesell and Liptak 1995). It is known that the appearance of water adsorbed on the surface of many solid bodies is accompanied by a sharp increase in their surface conductivity. This means that the surface condensation can be detected electrically, although the water vapor is not visible to the eye. These condensed water molecules permit a current to flow on the surface of even an excellent insulator. Moreover, this current flow is a function of the surface material and moisture density at the surface. As a rule, for a given material and fixed applied potential, the current increases logarithmically as the surface temperature decreases to the dew point of the atmosphere. Below the dew point, the logarithmic relationship does not hold, and thus the surface conductivity can be related to a dew point. As it was shown before, a dew formation can be observed optically; however, it is necessary to recognize that a visual observation is more difficult, because dew will form at or below the dew-point temperature, whereas the conductivity is a continuous function of temperature. Besides, the optical dew-point mirror hygrometers have significantly larger sizes and more complex configuration.

The measuring element in the surface conductive dew-point hygrometer consists of a highly polished inert surface inlaid with an intermeshed gold grid and a

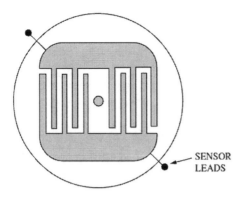

FIGURE 5.9 Sensing surface and housing of the surface conductivity type dew-point detector. (Reprinted with permission from Brodgesell A., Liptak B.G., Moisture in air: Humidity and dew point, In: Liptak B.G. (ed.), *Instrument Engineers' Handbook. Vol. 1: Process Measurement and Analysis*, 1420–1433, CRC Press, Boca Raton, FL, 1995. Copyright 1995, Taylor & Francis Group.)

thermocouple imbedded in the surface (see Figure 5.9). A fixed potential is maintained across the gold grids, and the current flow is compared to the reference current flow at dew point. This signal is amplified and used to modulate a cooler so that the surface is maintained at the dew point of the sample. Usually, the cooler is a Peltier element that pumps heat away from the sensor when electric power is supplied to it. From the thermocouple imbedded in the surface, the output is normally presented as dew-point temperature; however, the readout in terms of relative or specific humidity is also possible.

It is worth noting that the sampling system (see Figure 5.10) is needed for both types of dew-point hygrometers. At that, it is important to have an adequate flow through the sensor. Too little flow can slow the response (particularly at very low frost points). Too much flow can cause instability of the control system at high dew points and can reduce the depression capability of the cooling pump at very low dew points. Too much flow also accelerates the rate of system contamination. A flow rate of a little over 1 liter per minute is ideal for most applications. In many cases, flow rates between 0.1 and 2.5 liter per minute may be used.

RH is determined from the dew/frost point measurements by the following formula:

$$\mathrm{RH} = 100 \cdot \frac{p_{\mathrm{ws}}(T_{\mathrm{DF}})}{p_{\mathrm{ws}}(T_a)}, \tag{5.1}$$

where $p_{\mathrm{ws}}(T_{\mathrm{DF}})$ is a saturation vapor pressure at the dew/frost point temperature T_{DF}, and $p_{\mathrm{ws}}(T_a)$ is a saturation vapor pressure at the air temperature T_a.

5.4 DEW-POINT HYGROMETERS AND THEIR PERFORMANCES

In principle, the dew-point hygrometer allows determining the moisture content in a very wide range, including very low humidity levels. For example, the limitation on the optical hygrometer is the absolute amount of moisture available in a gas sample at −100°F (−73°C), or 1.5 ppm water vapor. However, this requires a more careful approach to the choice of materials for the sampling system components and to create the measuring system without leakage. For example, of equal importance is the effect that material absorption/desorption characteristics have on overall system response. Stainless steel and nickel alloy tubing are the best nonhygroscopic materials and should be used for dew points from 0 to −100°F (−17°C to −73°C). Copper, aluminum alloys, Teflon®, and polypropylene are suitable above −20°F (−29°C) dew point. Most other plastic and rubber tubing is unacceptable in all ranges. The temperature stability of the sampling system components is also quite important. For any given equilibrium sampling condition, a specific amount of moisture will be absorbed onto the sampling system's wetted surface, so control of the sample line temperature may be necessary to ensure an equilibrium condition. In the same time, the problem of leakage is relative; that is, if the dew point being measured is close to the ambient dew point, leakage into the system may not bias the reading substantially. If the system is pressurized above atmospheric pressure so as to create a leakage out of rather than into the system, the error introduced will be less (Brodgesell and Liptak 1993).

The *depression capability* of the instrument, which refers to the disadvantages of a dew-point hygrometer, is a function of the size and efficiency of the thermoelectric cooler. Recent advances in the technology of the thermoelectric coolers have made possible a practical

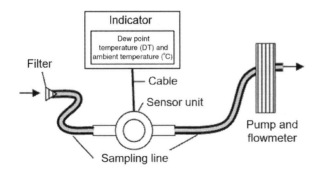

FIGURE 5.10 Structure of sampling system. (Reprinted from http://www.jma.go.jp.)

instrument utilizing three serial stages of thermoelectric cooling, capable of depression in excess of 160°F (71°C). Thus, new instruments are capable to monitor virtually all dew-point ranges encountered in current industrial processes.

With regard to the measurement accuracy, it is determined by the accuracy of the temperature measurement, the lag effect of the mirror to heating and cooling, and the airflow rate. Nonsoluble and soluble contaminants such as salt, dust, and oil misted on the mirror surface can also be sources of uncertainty. Therefore, World Meteorological Organization (WMO) recommends to filter the air supply to the device to avoid contamination of the mirror with dust, droplets, or mist. The mirror should be regularly cleaned daily or at least weekly with deionized, distilled water and alcohol to remove the oil-based contaminants.

In addition, care should be taken, because sometimes dew is not dew. When measuring the dew-point or frost-point temperatures that are below the freezing point, it is desirable to know the phase (liquid or solid) of the condensate. When the temperature of the mirror surface is slightly below 0°C, the condensate may be a super-cooled liquid water, or it may be in the form of solid ice. Generally, if the dew point has been dropping, the condensate will initially be in the form of super-cooled water. In time, it will usually change to the ice phase, because that has a lower energy state. At very low temperatures, the change to ice is more rapid, and at −40°C, it generally exists only in the ice form. Therefore, it is desirable to know the state of the condensate because, for a given mirror temperature, the vapor pressure in equilibrium with the condensate is higher for water and lower for ice. For a given vapor pressure over the mirror, if the condensate is ice, then the mirror temperature at equilibrium is slightly higher than it would be if the condensate was a liquid. Thus, the mirror should be monitored at dew points lower than 0°C to define either a frost or dew point so the correct formula is used in the RH calculation. In some industrial and metrology applications, the hygrometer has a microscope providing direct observation in such a way that the user can accurately discern the phase state on the mirror surface.

A response time of dew-point hygrometers depends on many factors. According to General Eastern Instruments operator's manual for GE General Eastern OptiSonde Chilled-Mirror Hygrometer (www.general eastern.com), at dew points above 0°C, the system stabilizes within a few seconds at a consistent dew or frost layer. Once the system is stable, valid readings may be taken. When the system is operating at very low frost points (below −40°C), extra care may be required when interpreting readings because of the slower response of the system. Time response depends on a number of factors, including the dew/frost point, slew rate, upstream filtering, and flow rate. As the dew/frost point becomes lower, the water molecules in the air sample become scarcer, and it takes longer to condense a frost layer on the mirror sufficiently thick to establish an equilibrium condition. The temperature slew rate is dependent on the dew point and depression (the temperature difference between the mirror and the sensor body); at higher dew points and moderate depressions, it is typically 1.5°C per second. At lower dew points and/or larger depressions, the slew rate becomes progressively slower. The flow rate affects the response by determining the rate at which water vapor is supplied or carried off.

REFERENCES

Brodgesell A., Liptak B.G. (1993) Moisture in air: Humidity and dew point, In: Liptak B.G. (ed.) *Analytical Instrumentation*. Chilton Book Company, Radnor, PA, pp. 215–226.

Brodgesell A., Liptak B.G. (1995) Moisture in air: Humidity and dew point, In: Liptak B.G. (ed.) *Instrument Engineers' Handbook. Vol. 1: Process Measurement and Analysis*, CRC Press, Boca Raton, FL, pp. 1420–1433.

Chen Z., Lu C. (2005) Humidity sensors: A review of materials and mechanisms. *Sensor Lett.* 3 (4), 274–295.

Hall E.G., Jordan A.F., Hurst D.F., Oltmans S.J., Vömel H., Kühnreich B., Ebert V. (2016) Advancements, measurement uncertainties, and recent comparisons of the NOAA frost point hygrometer. *Atmos. Meas. Tech.* 9, 4295–4310.

Jachowicz R.S. (1992) Dew point hygrometer with heat injection—principle of construction and operation. *Sens. Actuators B* 7, 455–459.

Jachowicz R.S., Makulski W.J. (1993) Optimal measurement procedures for a dew point hygrometer system. *IEEE Trans. Instrum. Measur.* 42, 828–833.

Jachowicz R.S., Zalewski D. (1994) Hygrometer with fibre optic dew point detector. *Sens. Actuators A* 42, 503–507.

Kimura M. (1996) A new method to measure the absolute—Humidity independently of the ambient temperature. *Sens. Actuators B* 33, 156–160.

Magalotti L. (1667) *Saggi di naturali esperienze fatte nell'Accademia del Cimento*. Giuseppe Cocchini, Florence, Italy.

Matsumoto S., Toyooka S. (1995a) Laser dew-point hygrometer. *Jpn. J. Appl. Phys. Part 1*, 34, 316.

Matsumoto S., Toyooka S. (1995b) Estimation of initial and response times of laser dew-point hygrometer by measurement simulation. *Jpn. J. Appl. Phys. Part 1*, 34, 5847.

Matsumoto S., Toyooka S. (1992) Determination of the dew point using laser light and a rough surface. *Opt. Commun.* 91 (1–2), 5–8.

Middleton W.E.K. (1969) *Catalog of Meterological Instruments in the Museum of History and Technology*, Smithsonian Institution Press, Washington, DC.

Pascal-Delannoy F., Sackda A., Giani A., Foucaran A., Boyer A. (1998) Fast humidity sensor using optoelectronic detection on pulsed Peltier device. *Sens. Actuators A* 65, 165–170.

Rübner K., Balköse D., Robens E. (2008) Methods of humidity determination. Part I: Hygrometry, *J. Therm. Anal. Calorim.* 94 (3), 669–673.

Sorli B., Pascal-Delannoy F., Giani A., Foucaran A., Boyer A. (2002) Fast humidity sensor for high range 80%–95% RH. *Sens. Actuators A* 100, 24–31.

Targioni Tozzetti G. (1780) *Notizie degli aggrandimenti delle Scienze Fisiche accaduti in Toscana nel corso di anni LX del secolo XVII*, Tomo I. Bouchard, Florence, Italy.

Vomel H., David D.E., Smith K. (2007) Accuracy of tropospheric and stratospheric water vapor measurements by the cryogenic frost point hygrometer: Instrumental details and observations. *J. Geophys. Res.* 112, D08305.

Wiederhold P.R. (1997) *Water Vapor Measurement, Methods and Instrumentation*. Marcel Dekker, New York.

6 Heated Salt-Solution Method for Humidity Measurement

The method of heated salt-solution is also based on the condensation method (Wiederhold 1997). The saturated salt lithium chloride sensor has been one of the most widely used dew-point sensors. Its popularity stems from its simplicity, low cost, and durability, and the fact that it provides a fundamental measurement. The equilibrium vapor pressure at the surface of a saturated salt solution is less than that for a similar surface of pure water at the same temperature. This effect is exhibited by all salt solutions, but particularly by lithium chloride, which has an exceptionally low equilibrium vapor pressure. As a consequence, a solution of lithium chloride is extremely hygroscopic under typical conditions of surface atmospheric humidity; if the ambient vapor pressure exceeds the equilibrium vapor pressure of the solution, the water vapor will condense over it and change its conductivity. This means that the conductivity of the solution is an indication of the relative humidity (RH) in the air.

6.1 THE DUNMORE CELL

The Dunmore or LiCl dew-point sensor was invented in 1938 and named after its inventor (Dunmore 1938). It was used for over 40 years, being the only electrical moisture-sensing device available at the time. In particular, in 1944, the electric resistance of cotton impregnated with LiCl or $CaCl_2$ solution was used to measure air humidity, and in 1946 this method was applied to the measurement of humidity on human skin and clothing (Ogden and Rees 1946).

The first sensing elements consisted of a thin-walled, hollow metal socket, wrapped with the tape impregnated with salt crystals (lithium chloride). Two wires were wrapped over the tape and connected to a regulated AC voltage source (50 or 60 Hz) (Figure 6.1A). The salt crystals completed the electric circuit between the wires. For example, the humidity-resistance characteristics A–E in Figure 6.2 are obtained when the amounts of impregnated lithium chloride are 0% (A), 0.25% (B), 0.5% (C), 1% (D), and 2.2% (E), respectively. As a rule, polyvinyl acetate was used for impregnation. It is seen that a wide RH range from 10% to 100% can be measured with a set of units with different sensing characteristics. A small-sized, lightweight humidity sensor has been made using plant pith as the porous binder (Figure 6.1B)

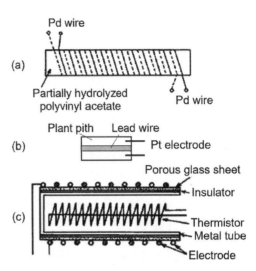

FIGURE 6.1 Various configurations of lithium chloride humidity sensors. (a) Dunmore-type humidity sensor; (b) A small-sized, lightweight humidity sensor developed using plant pith as the porous binder; (c) A dew indicator developed using a lithium chloride humidity sensor and a thermistor. Pd–palladium, Pt–platinum. (Reprinted with permission from Yamazoe, N. and Shimizu, Y., Humidity sensors: Principles and applications, *Sens. Actuators*, 10, 379–398, 1986. Copyright 1986, Elsevier.)

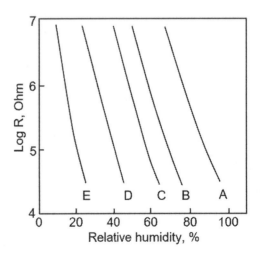

FIGURE 6.2 Resistance-humidity characteristics for several units of a Dunmore-type humidity sensor with different amounts of lithium chloride. (Reprinted with permission from Yamazoe, N. and Shimizu, Y., Humidity sensors: Principles and applications, *Sens. Actuators*, 10, 379–398, 1986. Copyright 1986, Elsevier.)

(Shiba 1956; Yamada 1979). These sensors show rather slow responses to humidity but fairly good stability, and have been widely used in radiosonde circuits, as well as in instruments for medical services and other applications (Holvo 1981).

More recent sensors, designed as a dew indicator (Hickes 1947), consist of a tube, or bobbin, with a resistance thermometer fitted axially inside (Figure 6.1C). The external surface of the tube is covered with a glass fiber material (usually a tape wound around and along the tube) that is soaked with an aqueous solution of lithium chloride, sometimes combined with potassium chloride. Bifilar silver or gold wire is wound over the covering of the bobbin, with equal spacing between the turns. Current flows between adjacent bifilar windings, which act as electrodes, and through the solution.

As mentioned above, lithium chloride is hygroscopic and therefore can take up moisture from the air. When the sensing element is exposed to the sample atmosphere, water, condensing on the crystals, forms an ionic solution, which permits an electric current to flow between the wires. This current, in turn, heats the solution and raises its vapor pressure. As more water condenses, more current flows, resulting in a further increase of the solution vapor pressure until equilibrium is reached. At decreasing moisture content of the measured gas, water evaporates from the element, decreasing the current flow and resulting in a new equilibrium at a lower-solution vapor pressure. A temperature sensor inside the hollow socket is used to detect the sensor temperature and to provide a signal for readout (Brodgesell and Liptak 1993). Thus, an operational equilibrium temperature of the instrument depends on the ambient water-vapor pressure. Above the equilibrium temperature, evaporation will increase the concentration of the solution, and the electrical current and the heating will decrease and allow heat losses to cause the temperature of the solution to fall. Below the equilibrium temperature, condensation will decrease the concentration of the solution, and the electrical current and heating will increase and cause the temperature of the solution to rise. At the equilibrium temperature, neither evaporation nor condensation occurs, because the equilibrium vapor pressure and the ambient vapor pressure are equal. This means that the humidity measurement is achieved by the measurement of the dew-point temperature. In practice, the equilibrium temperature measured is influenced by individual characteristics of sensor construction and has a tendency to be higher than that predicted from equilibrium vapor-pressure data for a saturated solution of lithium chloride. However, reproducibility is sufficiently good to allow the use of a standard transfer function for all sensors constructed to a given specification (WMO 1992, 2011). Nelson and Amdur (1965) found that measuring dew points using salt-solution, phase-transition hygrometry can be as precise as $\pm 0.15°C$. This is exceptional for any type of hygrometer. Additionally, no water supply is needed, no wet-bulb freezing problem would appear, and the elevated temperature of the element not only promotes a natural air flow past it, but also reduces errors due to spurious thermal loads.

Several efforts have been made to improve this device (e.g., by incorporation of an additional heater), which is connected to a control circuit to keep the vapor pressure of the saturated lithium chloride solution equal to the vapor pressure of the atmosphere, or by measuring the conductance with a pair of interdigital electrodes of noble metals (Dunmore 1939; Mathews 1963). Experiment has shown that improved accuracy may be obtained when a solution of lithium chloride is heated indirectly. The conductance of the solution is measured between two platinum electrodes and provides control of a heating coil. The latest development in resistive humidity sensors uses a ceramic coating to overcome limitations in environments where condensation occurs. The sensors consist of a ceramic substrate with noble metal electrodes deposited by a photoresist process. The substrate surface is coated with a conductive polymer/ceramic binder mixture, and the sensor is installed in a protective plastic housing with a dust filter.

6.2 LIMITATIONS OF LiCl-BASED HUMIDITY SENSORS

Varying the concentration of lithium chloride, a wide RH range up to 100% can be measured using this device. However, this sensor cannot be used to measure RH below 11%, because the saturation vapor pressure of lithium chloride occurs at approximately 11% RH. Also, the sensors cannot be used in the presence of gases in the atmosphere, interacting with the salt crystals or water. Specifically, gases such as ammonia, sulfur dioxide, sulfur trioxide, and chlorine react with water, and these sensors therefore cannot be applied for operation in an atmosphere containing such gases. The sensors are also not suitable for the samples containing diethylene glycol or triethanolamine (Brodgesell and Liptak 1993). Relatively slow response is also one of the disadvantages of heated salt solution–based sensors. Strong ventilation affects the heat transfer characteristics of the sensor, and fluctuations in ventilation also lead to unstable operation.

It must be borne in mind that the use of lithium chloride sensors also has a temperature limit (WMO 1992, 2011). The equilibrium vapor pressure for saturated lithium chloride depends upon the hydrate's being in equilibrium with the aqueous solution. In the range of solution temperatures corresponding to the dew-points of −12°C to 41°C, monohydrate normally occurs. Below −12°C, dihydrate forms, and above 41°C, anhydrous lithium chloride forms. Close to the transition points indicated in Eq. 6.1, the operation of the hygrometer is unstable and the readings are ambiguous, because the chemically bound water content of the undissolved salt changes abruptly, which is associated with an enthalpy change. However, the −12°C lower dew-point limit may be extended to −30°C by the addition of a small amount of potassium chloride (KCl).

$$\begin{aligned} LiCl \cdot 3H_2O &\rightarrow LiCl \cdot 2H_2O \; (-20°C) \\ LiCl \cdot 2H_2O &\rightarrow LiCl \cdot H_2O \; (20°C) \\ LiCl \cdot H_2O &\rightarrow LiCl \; (100°C) \end{aligned} \quad (6.1)$$

One should note that lithium chloride–based sensors can have very high resolution. However, such resolution can be achieved for sensors operated in humidity spans of 20%–40% RH. For very low RH, a control function in the 1%–2% RH range and accuracies of 0.1% can be achieved. These sensors show rather slow responses to humidity but a fairly good stability, and have been widely used in the radiosonde circuits, as well as in the instruments for medical services. However, their use in very humid environments should be avoided so as not to lower the accuracy and lifetime. The major advantage of this sensor is that it is suitable for batch fabrication. Saturated salt sensors are an attractive proposition when a low-cost, rugged, slow-responding, and moderately accurate sensor is required. However, it requires regular maintenance of the lithium chloride solution to guarantee the stable operation of the device, which is an important disadvantage (Yamazoe and Shimizu 1986).

It should also be considered that the instrument does not require a high-velocity sample to perform satisfactorily; in fact, the sample velocity in excess of 0.3 m per second will result in a poor measurement and shortened element life. At high flow rates, it is difficult to establish equilibrium between the sensing element and the sample. As a result, the feedback is lost, which causes the work to be at higher power and a reduction of element life. The sensing element therefore must be located in a relatively quiescent zone or must be protected from direct impingement of the sample. In ducts, a sheet metal hood installed over the element and open on the downstream side is adequate. In pipelines, the element can be installed through the side outlet of a tee or can be mounted separately in a sampling chamber piped to the process line. The latter installation is preferred for samples under pressure, because the element can be services without shutting down the pipeline (Brodgesell and Liptak 1993).

6.3 PLANAR LiCl-BASED HUMIDITY SENSORS

With regard to the miniaturization of such sensors in planar version, the attempts are also available. In particular, Sakai et al. (1999) proposed the planar sensor, combining at one ceramic platform the lithium chloride dew-point sensor, the temperature sensor, and the heater (see Figure 6.3). One should note that such platform is typically used for fabrication various gas sensors.

It should be noted that the lithium chloride–based humidity sensors in the planar version are still manufactured. Moreover, they occupy a large proportion of humidity sensors of resistive or impedance type. Research carried out in this area is mainly related to the search for the optimal materials used for impregnation by lithium chloride (Geng et al. 2007; Li et al. 2008; Song et al. 2009; Jiang et al. 2014). For example, Jiang et al. (2016) has suggested hierarchically porous polymeric microspheres (HPPMs) to use for this purpose. Porous polymers with abundant pores are capable of physically immobilizing functional molecules, meanwhile small molecules are able to transport into the interior of the polymers freely.

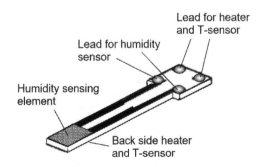

FIGURE 6.3 A prototype of the planar dew-point hygrometer. (Idea from Sakai, Y. et al., A new type LiCl dew point hygrometer probe fabricated with a composite of porous polymer and the salt, In: *Proceedings of the 10th International Conference on Solid-State Sensors and Actuators, Transducers*, 99, June 7–10, Sendai, Japan, pp. 1664–1667, 1999.)

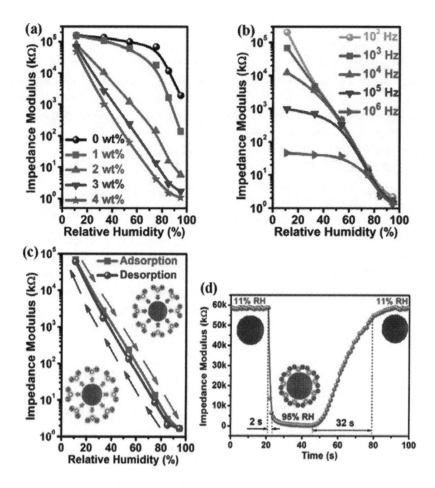

FIGURE 6.4 (a) The impedance modulus of 0, 1, 2, 3, 4 wt % LiCl/HPPMs sensors in different humidity atmospheres. (b) The relationship of impedance modulus to relative humidity (RH) for 3 wt % LiCl/HPPMs-1 sensor with working frequencies of 10^2, 10^3, 10^4, 10^5, and 10^6 Hz. (c) The impedance modulus of 3 wt % LiCl/HPPMs-1 sensor in the continuous adsorption and desorption processes. (d) The impedance modulus of 3 wt % LiCl/HPPMs-1 sensor in the RH atmosphere change between 11% and 95% RH. The HPPMs were synthesized by the polymerization of phloroglucin/dimethoxymethane in a mixture solution of methanol and 1,2-dichloroethane without any catalyst under hydrothermal conditions (150°C for 72 h). (Reprinted with permission from Jiang, K. et al., Excellent humidity sensor based on LiCl loaded hierarchically porous polymeric microspheres, *ACS Appl. Mater. Interfaces*, 8, 25529–25534, 2016. Copyright 2016, American Chemical Society.)

Furthermore, the cross-linked structures of porous polymers can enhance the durability of humidity sensors. Hierarchically porous polymers (HPPs), as a special kind of porous polymers, contain pores across multiple scales (Seo et al. 2014). For HPPs, micropores can offer the function of adsorption, while mesopores and macropores can enormously increase accessibility of the microporous surface by enhanced diffusion effect. Therefore, when HPPs are used as humidity-sensitive materials, the mesopores or macropores can accelerate a diffusion rate of water molecules. The parameters of the sensors that implement this approach are shown in Figure 6.4. It is seen that the 3 wt % LiCl/HPPMs-1 sensor shows high sensitivity, small hysteresis, enhanced durability, and rapid response/recovery. The sensors also show a high stability of the characteristics (see Figure 6.5).

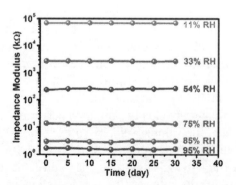

FIGURE 6.5 Durability of 3 wt% LiCl/HPPMs sensor in 95% relative humidity atmosphere with a period of 30 days. (Reprinted with permission from Jiang, K. et al., Excellent humidity sensor based on LiCl loaded hierarchically porous polymeric microspheres, *ACS Appl. Mater. Interfaces*, 8, 25529–25534, 2016. Copyright 2016, American Chemical Society.)

REFERENCES

Brodgesell A., Liptak B.G. (1993) Moisture in air: Humidity and dew point, In: Liptak B.G. (ed.) *Analytycal Instrumentation*. Chilton Book Company, Radnor, PA, pp. 215–226.

Dunmore F.W. (1938) An electrometer and its application to radio meteorography. *J. Res. Nat. Bur. Std.* 20, 723–744.

Dunmore F.W. (1939) An improved electric hygrometer. *J. Res. Nat. Bur. Std.* 23, 701–714.

Geng W.C., Wang R., Li X.T., Zou Y.C., Zhang T., Tu J.C., He Y., Li N. (2007) Humidity sensitive property of Li-doped mesoporous silica SBA-15. *Sens. Actuators B* 127, 323–329.

Hickes W.F. (1947) Humidity measurement by a new system. *Refrig. Eng.* 54, 351–354.

Holvo H.R. (1981) A dew-point hygrometer for field use. *Agr. Meteorol.* 24, 117–130.

Jiang K., Fei T., Zhang T. (2014) Humidity sensor using a Li-loaded microporous organic polymer assembled by 1,3,5-Trihydroxybenzene and Terephthalic Aldehyde. *RSC Adv.* 4, 28451–28455.

Jiang K., Zhao H., Dai J., Kuang D., Fei T., Zhang T. (2016) Excellent humidity sensor based on LiCl loaded hierarchically porous polymeric microspheres. *ACS Appl. Mater. Interfaces* 8, 25529–25534.

Li Z., Zhang H., Zheng W., Wang W., Huang H., Wang C., MacDiarmid A.G., Wei Y. (2008) Highly sensitive and stable humidity nanosensors based on LiCl doped TiO_2 electrospun nanofibers. *J. Am. Chem. Soc.* 130, 5036–5037.

Mathews D.A. (1963) Review of the Lithium Chloride radiosonde hygrometer, In: *Proceedings of the Conference on Humidity and Moisture*, Vol. VI, Washington, DC, pp. 219–227.

Nelson D.E., Amdur E.J. (1965) The mode of operation of saturation temperature hygrometers based on electrical detection of a salt-solution phase transition. In: Wexler A. (ed.) *Humidity and Moisture*, Vol. I. Reinhold, New York, p. 617.

Ogden L.W., Rees W.H. (1946) A method of measuring temperature on the skin and clothing of a human subject. *Shirley Inst. Mem.* XX, 163.

Sakai Y., Matsuguchi M., Makihata H. (1999) A new type LiCl dew point hygrometer probe fabricated with a composite of porous polymer and the salt, In: *Proceedings of the 10th International Conference on Solid-State Sensors and Actuators, Transducers*, Vol. 99, June 7–10, Sendai, Japan, pp. 1664–1667.

Seo M., Kim S., Oh J., Kim S.J., Hillmyer M.A. (2014) Hierarchically porous polymer from a hypercrosslinked block polymer precursor. *J. Am. Chem. Soc.* 137, 600–603.

Shiba K. (1956) Measurement of detailed distribution of humidity. *Asahi-kagaku* 16, 79–83.

Song X.F., Qi Q., Zhang T., Wang C. (2009) A humidity sensor based on KCl-doped SnO_2 nanofibers. *Sens. Actuators B* 138, 368–373.

Wiederhold P.R. (1997) *Water Vapor Measurement, Methods and Instrumentation*. Marcel Dekker, New York.

WMO (1992) *World Meteorological Organization: Measurement of Temperature and Humidity*. (Wylie R.G. and Lalas T.) (Technical Note No. 194), (WMO-No. 75), WMO, Geneva, Switzerland.

WMO (2011) World Meteorological Organization: *Technical Regulations*, Vol. I – General Meteorological Standards and recommended Practices, (WMO-No. 49), WMO, Geneva, Switzerland.

Yamada Y. (1979) Lithium chloride humidity sensor. *Denshigyutsu* 21 (9), 26–30.

Yamazoe N., Shimizu Y. (1986) Humidity sensors: Principles and applications. *Sens. Actuators* 10, 379–398.

7 Electrolytic or Coulometric Hygrometers

7.1 P$_2$O$_5$-BASED COULOMETRIC HYGROMETERS

7.1.1 Principles of Operation

It should be noted that anhydrous phosphorous pentoxide (P$_2$O$_5$), which can be used in the gravimetric hygrometers, could also be a base for *electrolytic* or *coulometric* hygrometers (Brodgesell and Liptak 1993, 2003; Jefferies 1993; Ma et al. 1995; Brodgesell et al. 2003). The coulometric method was invented by Keidel (1959). Later, it was established that the coulometric sensor is a very robust and cost-effective tool for determination of trace humidity. Several models of such sensors were designed by MEECO Inc. (Warrington, PA, http://meeco.com). The electrolytic sensor utilizes a cell coated with a thin film of phosphorous pentoxide (P$_2$O$_5$), which absorbs water from the gas under measurement (see Figure 7.1). When an electrical potential (voltage) is applied to the electrodes, incorporated in the measurement cell, the water vapor absorbed by the P$_2$O$_5$ is dissociated into hydrogen and oxygen molecules (according to reactions 7.1 and 7.2), generating a finite current (Smith and Mitchell 1984). This reaction takes place at a voltage of at least 2 V (DC). The resulting electrolysis current is a measure of the sample's moisture content based on Faraday's law. The amount of current required to dissociate the water is proportional to the number of water molecules presented in the sample gas.

$$P_2O_5 + H_2O \rightarrow 2HPO_3 \quad (7.1)$$

$$2HPO_3 \rightarrow H_2 + 1/2 O_2 + P_2O_5 \quad (7.2)$$

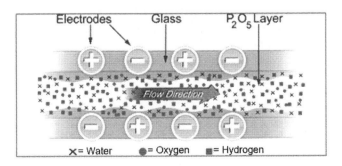

FIGURE 7.1 Longitudinal section of sensing element illustrates the operation of electrolytic hygrometer. (Reprinted from http://meeco.com.)

Under steady-state conditions, the following relationship is established between the electrolysis current and the moisture content:

$$I = C_{H_2O} \frac{Q \cdot n \cdot F}{M_{H_2O}} \quad (7.3)$$

where I is the current for electrolysis of the moisture, A; C_{H_2O} is the concentration of the moisture by weight at the CEC outlet, g/cm^3; Q is the gas flow rate through the CEC, cm^3/s; n is the number of elementary charges needed for electrolysis of a single water molecule; F is Faraday's number, coulombs per mole; M_{H_2O} is the molecular weight of water, g/mol. Thus, the electrolytic current is a measure of the rate at which water is being electrolyzed, which in equilibrium conditions equals the rate at which it is being absorbed at the phosphoric acid surface. Since two electrons are required for electrolysis of each water molecule, the electrolysis current is a measure of the water present in the sample. If the volumetric flow rate of the sample gas into the electrolysis cell is controlled at a fixed value, then the electrolysis current is a function of water concentration in the sample. Thus, a knowledge of the gas flow rate, together with this current, gives an absolute and continuous measure of the humidity mixing ratio of the sampled gas. The moisture volume fraction in this case can be determined from the formula (Pirog et al. 2011)

$$C_{H_2O} = I \cdot \frac{M_{H_2O}}{Q \cdot n \cdot F} = 7.4794 \cdot 10^6 \left(\frac{I}{Q} \right) \quad (7.4)$$

where C_{H_2O} is the moisture volume fraction, ppm; $7.4794 \cdot 10^6$ is a factor due to the choice of units for the physical quantities, (ppm·cm^3)/(A·min). Thus, while keeping the voltages at the electrodes constant, the sensor can be seen as a variable resistor.

7.1.2 Sensor Configuration

The typically used equipment for a simple electronic data processing unit is shown in Figure 7.2. It consists of a dc-regulated power supply, a high precision amperemeter, and a microcontroller unit with special functions that are important for service times.

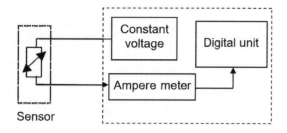

FIGURE 7.2 Equipment for the coulometric sensor.

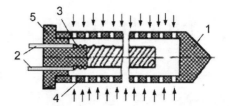

FIGURE 7.3 The device of sensing elements of the coulometric sensor. 1, housing; 2, electrodes; 3, P_2O_5 film; 4, diffusion barrier; 5, insulating base.

The main element of the hygrometer (the coulometric sensor) has various design versions. Usually the coulometric method utilizes a cell that consists of a channel, small in a cross-sectional area but long in the length, usually a capillary tube with electrodes embedded on its interior surface. The oldest and most common modification consists of a cylindrical bushing made of plastic (usually Teflon®-fluoroplastic) (Nakamura et al. 1973). The fluoroplastic tube is fixed in a housing (plastic or metal) with contacts for connecting the sensor to the measuring device. The geometric dimensions of the sensing element and the gas flow are selected so as to ensure complete removal of moisture from the gas for a given upper limit of measurement. In "glass"-sensitive elements, the moisture-absorbing substance is applied to a glass substrate (Semchevskii et al. 2009). Silica cylindrical rods can also be used (Goldsmith and Cox 1967). The use of a glass base reduces the possibility of short circuits between the electrodes and the penetration of water vapor through a plastic sheath having a certain porosity. In addition, the use of a glass tube makes it possible to visually monitor the state of the moisture-absorbing film and electrodes.

Electrodes, between which the P_2O_5 film is applied, are fused into the inner channel of the glass tube (see Figure 7.1); the electrode leads are also fused into the glass. One should note that the reliability of glass-based sensors is significantly higher than the reliability of sensors on a plastic base. Sensors on a plastic basis are characterized by a reduction in the mean time between failures after each regeneration. Sensors on a glass base, in which multiple regeneration is permissible, do not have this disadvantage.

There are also variants of sensors with a diffusion barrier (Figure 7.3), in order to reduce the influence of fluctuations in the gas flow during meteorological measurements of air humidity. In these sensors, some of the moisture from investigated gas stream diffuses through the porous hydrophobic barrier and then undergoes electrolysis. The diffusion rate does not depend on the air velocity; therefore, in the diffusion hygrometer, there is no need for a gas flow regulator, since its function is performed by the diffusion barrier. In the diffusion-sensitive element, the electrodes and the P_2O_5 film are located on the outer surface of a glass or fluoroplastic rod, and the diffusion barrier is a coaxial perforated tube made of fluoroplastic with a rod.

Typically, electrodes in the coulometric electrolytic cell are made of platinum (Goldsmith and Cox 1967; Nakamura et al. 1973; Semchevskii et al. 2009). The electrodes, made of rhodium, Rh + Pt and Rh + Ir alloys instead of the conventional platinum electrodes, were also used (Pirog et al. 2011). It was assumed that the rhodium electrodes will increase the life of the sensors by preventing the occurrence of platinum black in the interelectrode space and eliminate the errors from the recombination of hydrogen with oxygen into the water. In the process of recombination, which takes place especially at high hydrogen concentrations in the test gas, platinum can play the role of an active catalyst; rhodium has less catalytic activity. However, studies have shown that this error, when measuring the humidity of hydrogen within a wide range by a sensor with platinum electrodes, can be neglected.

After coating with phosphoric acid, the sensing elements usually are mounted in a Pyrex glass mantle. However, it is believed that for the low humidity measurement, stainless steel is better material for the housing. The geometric dimensions of the coulometric electrolytic cell (CEC) are selected so that moisture is entirely extracted from a gas being analyzed during its passage through the inner channel of the cell. A voltage that is fed to the electrodes depends on the geometry of the cell but usually is of the order of several tens of volts (Semchevskii et al. 2009).

7.1.3 Advantages and Limitations

Experiments showed that the electrolytic hygrometer is ideally suited for usage in very dry applications, such as ultra-pure gas, cryogenic engineering, special catalysis, fine chemicals, and vacuum heat treating, as well as in scientific research. This hydrometer provides a reliable performance for a long period only in the low ppm range, up to 1000 ppm by volume—that is, from a frost point of −78°C to −20°C at standard atmospheric pressure (Hibbs and Nation 1964; Jefferies 1993; Ma et al. 1995). The limitation at higher humidity is caused by the overheating of the absorbent due to the higher currents involved. This limit can be improved by applying the diffusion method. Another way is to dilute the test gas with a dry one (for example, nitrogen) in a constant ratio. With recirculation dilution, the analyzable gas at the sensor inlet is continuously diluted with dry gas taken from the sensor output.

Undoubtedly, the listed limitations when using the electrolytic hygrometer are a shortcoming of these sensors. However, at the same time, it is their advantage, since this method permits hygrometry with established electrolytic cell technology to achieve very high sensitivity to air humidity. Modifications in the Keidel cell allowed achieving a reliable moisture determination in the 1- to 10-ppb range (Nakamura et al. 1973; Mettes et al. 1993; Ma et al. 1995). The results of electrolytic hygrometer testing in this range are listed in Table 7.1. Moreover, the controlled addition of moisture to the sample gas, combined with the sample flow variations and the signal integration, permits the moisture determination below the normal background signal of the sensor (Ma et al. 1995). The configuration of such a sensor and the gas system used are shown in Figure 7.4. In addition, the elimination of polymeric materials from the sample flow path minimizes undesirable outgassing effects. It is recognized, though, that the work at ppb level is not easy, because even a very small amount of spurious water from outgassing has a significant effect on the instrument readings. This means that the hygrometer can operate over a wide range of absolute humidity and these sensors can control humidity in the range, where operation of other sensors and hydrometers is problematic. The majority of the trace-moisture measurements requires calibration, which at low-ppb levels is difficult to achieve. In contrast, the coulometric method is an absolute technique (Hulanicki 1995) that relies solely on physical principles and does not require calibration against moisture standards. This means that

TABLE 7.1
Hygrometer Responses to a Series of Gas Samples with Known Generated-Moisture Values

Humidity, ppb	
Generated Value	**Hygrometer Reading**
0	0.05
0.02	0.08[b]
2.0	1.7
2.5	2.5
5.0	5.0
	7.0[b]
10.0	13.0

Source: Data extracted from Ma, C. et al., *Micro*, 13, 43–49, 1995.

b – duplicate run

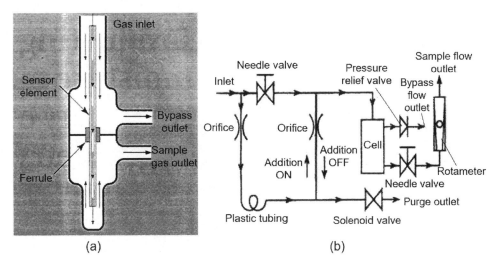

FIGURE 7.4 (a) Measurement cell of electrolytic hygrometer. *Arrows* show the direction of the sample stream. Mass-flow controllers to regulate the sample and bypass flows are located downstream of the sensor to eliminate their effects on moisture and (b) the flow schematics of the instrument. In this device, as an inexhaustible moisture source, atmospheric moisture generates the additions. More detailed description of such hygrometer operation can be found in Ma, C. et al., *Micro*, 13, 43–49, 1995. (Reprinted from http://meeco.com.)

no additional calibration is required when replacing the unit. This is particularly important for online applications, where recalibration interrupts service and proves costly.

It is worth noting that electrolytic hygrometers, as well as all moisture detectors, are affected by contaminants in natural gas. In particular, the electrolytic hygrometer cannot be used for humidity measurements in the gases that are corrosive (chlorine, etc.), react with a phosphoric acid solution, or readily combine with P_2O_5 to form water. Some other gases for which this instrument is unsuitable are unsaturated monomers, alcohols, amines, ammonia, hydrogen fluoride, and $CHClF_2$ (Freon) refrigerant (Brodgesell et al. 2003). Alcohols are seen by the cell as water. Amines and ammonia usually react with the desiccant. Hydrogen fluoride can corrode the internals of the cell. The data collected on the $CHClF_2$ refrigerant indicates an anomaly, although the reason for this is not fully understood. Unsaturated hydrocarbons, such as butadiene, or monomers with a strong tendency to polymerize cannot be monitored, as the cell will be quickly coated with polymer. In addition, the instrument should not be used with samples whose components may deposit in the cell (condensable vapors). This means that applications must be selected carefully, and for a long sensor life, the purity of the sample gas should be kept very low. However, if misapplied, the electrolytic cell is easily ruined. Contaminants such as glycols and particulates also shorten a cell life. When a cell becomes contaminated, it will show a memory for polarity; that is, the outputs under forward and reverse flow through the cell will not be equal. Therefore, filters are required to reduce contamination. A filter is normally located upstream to prevent contaminants from reaching the cell. It should also be borne in mind that a cell will lose sensitivity when exposed to moisture levels of a few ppm over a period of weeks. This sensitivity loss is due to the elution of desiccant with the sample. However, this process occurs over a long period of time, and the cell can be recoated fairly easily in the field during periodic maintenance (Brodgesell et al. 2003).

If the efficiency of absorption and electrolysis of water vapor is 100%, then the accuracy of the method is limited largely by the constancy of the mass flow rate and the current measuring circuitry. This means that the electrolytic method, based on Faraday's law, requires a strong control of the sample flow through the cell. Therefore, a typical sampling system used in coulometric hygrometers to maintain a constant flow maintains a constant pressure inside the cell. Sample gas enters the inlet, passes through a stainless steel filter, and enters a stainless steel manifold block. It is most important that all components prior to the sensor be made of an inert material, such as stainless steel, to minimize the water absorption. After passing through the sensor, the sample gas pressure is controlled by a differential pressure regulator that compares the pressure of the gas leaving the sensor to the gas venting to the atmosphere through a preset valve and flow meter. Thus, a constant flow is maintained in spite of nominal pressure fluctuations at the inlet port. One should bear in mind that after electrolysis, some of the resultant gaseous oxygen and hydrogen can recombine into water, which then can be reabsorbed and electrolyzed again by the cell. When this occurs, there is an increase in electrolysis current and a consequent error in indicated humidity.

In both academic studies and commercially, electrolytic hygrometers were traditionally used with a sample flow rate through the cell of 100 sccm. This appeared optimum, since a lower flow reduced sensitivity (less water enters the cell per minute). By contrast, a higher flow may not allow complete absorption of water from the sample. According to Goldsmith and Cox (1967), the maximum flow rate at which the hygrometer will operate may be determined by two factors:

i. The physical dimensions that determine the rate at which the water molecules diffuse to the absorbed surface
ii. The accommodation coefficient of the water molecules on the phosphoric acid surface

In summary, while the electrolytic hygrometer can provide a primary, reliable measurement at low moisture, the accuracy of the device is dependent on maintaining a controlled and monitored sample flow. Recent studies at MEECO showed that much lower cell flow rates, just 10 sccm rather than 100 sccm, can serve natural gas applications without loss of sensitivity or response time. This reflects two important design features: (1) cell materials that do not absorb moisture and (2) a high ratio of bypass to sample flow rates. The use of lower flow rates greatly increases the contamination resistance and cell life by reducing the amount of contaminants entering the cell by 90%. Because contaminants enter the cell at only one-tenth the rate, the cell life can be extended 10-fold by reduction of flow to 10 sccm. Thus, the need to change cells for cleaning and resensitizing is greatly reduced.

Estimates show that, by using special techniques and special cells, an accuracy within 1%–2% of the reading can be attained. This is more than adequate for most industrial applications. Although this instrument will operate satisfactorily with a variety of samples, a phenomenon called the *recombination effect* introduces large errors at low moisture levels in hydrogen-rich or oxygen-rich samples (Brodgesell et al. 2003). Recombination is the reversion to water of the electrolysis products; it introduces an error into the measurement when the recombined oxygen and hydrogen are re-electrolyzed. Apparently, all electrodes catalyze this reaction, although some electrode materials do so more than others. The use of rhodium as the electrode material has been found to minimize recombination. However, when monitoring very low moisture levels in oxygen-rich or hydrogen-rich atmospheres, the recombination produces a relatively large error, even with the best choice of electrode materials. For such an application, two sensors are used; one measures at the sample flow rate X and the other at the sample flow rate 2X. Because the error due to recombination is a constant, subtraction of the two sensor outputs yields a signal that is independent of recombination.

As for operation speed of the hygrometer, it must be recognized that the response time is rather slow. The response time varies with water vapor concentration, the cell voltage, and the thickness of absorption layer. Usually, a response time of 15 minutes was required to reach 80% of final value for low-level intrusions. Electrolytic hygrometers fabricated by MEECO have slightly better parameters. The transducer reached 50% of final response in less than 2 minutes and 90% of final response in less than 5 minutes. Drying time was also impervious to the sample flow. Instruments were dried in less than 5 minutes in both cases. According to Goldsmith and Cox (1967), the response time of the hygrometer depends on three factors:

i. The condition of the pipework
ii. The thickness and continuity of the absorbent
iii. The presence of contaminants in the absorbent film

If the pipework leading to the hygrometer is slow in equilibrating with the water content of the gas being passed to the hygrometer, then this will determine the apparent response time of the instrument. At low humidity, glass, plastics (other than polytetrafluoroethylene), hard-soldered joints, or dirty surfaces in the pipes will always control the response in this way. With stainless-steel pipes carefully degreased and acid cleaned, this difficulty is overcome once the pipework is dry.

7.1.4 Modern P_2O_5-Based Humidity Sensors

Modern coulometric humidity sensors in principle do not differ from the sensors developed several decades ago. They are only generally performed in a planar version, using thick-film or thin-film technology for manufacturing interdigitized electrodes on a ceramic substrate (Lorek et al. 2010; Tiebe et al. 2012). As a result, it was possible to significantly reduce the dimensions of the sensors and significantly expand the field of application. The extreme dimension differences between a typical industrial sensor with a cylindrical body and the planar sensor are shown in Figure 7.5. The electrodes are arranged on a plane (flat) ceramic substrate (see Figure 7.5b). There is also the capability for the implementation of other sensors, such as for pressure or temperature. Humidity sensors of different types also can be fabricated on the same substrate. Experiments with different electrode spacing have shown that a smaller electrode spacing between anode and cathode leads to faster response times.

A certain drift of the characteristics associated with the material accumulation on the anode was also discovered. This material is probably P_2O_5. It should be noted that this effect is also peculiar to conventional coulometric humidity sensors. This problem is solved by using appropriate additional coatings and special

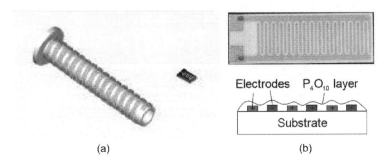

FIGURE 7.5 (a) Comparison of conventional and planar coulometric sensors. (b) Plan view (the view from above) and scheme of the planar sensor.

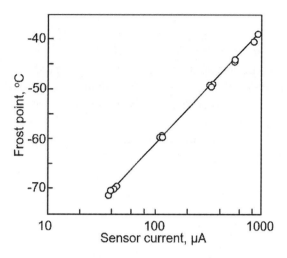

FIGURE 7.6 Sensor signal of a planar coulometric humidity sensor at 20 Nl/h. (Data extracted from Lorek A. et al. Development of a gas flow independent coulometric trace humidity sensor for aerospace and industry. In: *Proceedings of the First European Conference on Moisture Measurement*, Weimar, Germany, October 5–7, pp. 289–296, 2010.)

measurement modes. Typical characteristics of planar sensors are shown in Figure 7.6. The presence of a porous membrane also contributes to reducing the influence of a fluctuating gas flow rate.

However, it is necessary to recognize that coulometric humidity sensors in the traditional design are still present on the market. In Figure 7.7, the sensor offered by AMETEK (www.ametekpi.com) and the cell used for such sensors are shown.

7.2 POPE CELL

The polystyrene surface resistivity-based sensor, the so-called Pope cell, named after its inventor, Dr. D.H. Pope, can also be attributed to electrolytic or *coulometric* hygrometers. The Pope cell is structurally similar to the Dunmore element but, instead of lithium chloride solution, it uses sulfuric acid. The Pope cell is comprised of an insulating ceramic substrate on which a grid of interdigitated electrodes is deposited. These electrodes are overlayed with a humidity-sensitive salt imbedded in a polymer resin. Usually, a polystyrene that has been treated with sulfuric acid was used in such sensors. This acid treatment causes ulphonation of the polystyrene molecules. The sulfate radical SO_4 becomes very mobile in the presence of hydrogen ions (from the water vapor) and readily detaches to take on the H^+ ions. The humidity-sensitive layer is then covered by a protective coating that is permeable to the water vapor. Water vapor is able to proceed throughout this surface layer and enter the polymer region. This infiltration of water vapor allows the imbedded salt to ionize and become very mobile within the polymer resin. This alters the impedance of the sensor as a function of humidity. The ac excitation is provided to the electrodes, and the ac-excited Wheatstone bridge circuit can be used to measure the impedance (Brodgesell and Liptak 1993). This type of humidity sensor is shown in Figure 7.8.

The sulphonated polystyrene sensors have a large signal change with humidity. A typical sensor can change its resistance four orders of magnitude with relative humidity, from 1 kΩ at 90% relative humidity (RH) to 10 MΩ at 10% RH. This large signal change enhances

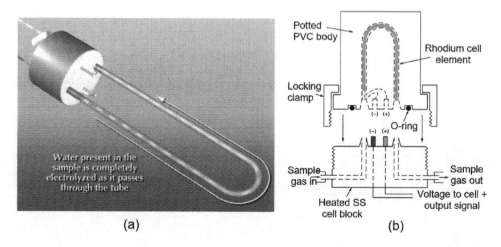

FIGURE 7.7 (a) Basic construction of an electrolytic sample cell developed by AMETEK. (Reprinted from https://www.ametek.com.) (b) Electrolytic hygrometer cell. PVC, polyvinyl chloride. (Reprinted with permission from Brodgesell, A. et al., Moisture in gases and liquids, In: Lipták B. (ed.), *Instrument Engineers' Handbook*, Vol. 1, 1434–1449, 2003. Copyright 2003, CRC Press.)

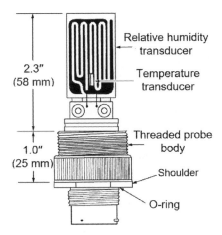

FIGURE 7.8 Polystyrene surface conductivity humidity element. (Courtesy of General Eastern, formerly Phys-Chem Scientific Corp.) (Reprinted with permission from Brodgesell, A. et al., Moisture in gases and liquids, In: Lipták B. (ed.), *Instrument Engineers' Handbook*, Vol. 1, 1434–1449, 2003. Copyright 2003, CRC Press.)

the RH resolution. Because the operating portion of the sensor is only its surface, its speed of response is fast—only a few seconds. These sensors are generally suited to the higher RH ranges due to the extremely high impedance below 10% RH and are most often used at RH levels above 15%. Typically, such sensors cover a range of 15%–99% RH.

7.2.1 Advantages and Limitations of Pope Cell

The wide range of RH measurements and a large change in the impedance of the humidity-sensitive layer under the influence of the air humidity are important advantages of these sensors. However, due to the extremely active surface of the sensor, the sensing surface can be easily washed off or contaminated to cause errors in the readings (Soloman 2010), which is a disadvantage of these sensors. In addition, the resistance of these sensors has a nonlinear dependence on the humidity that is difficult to linearize electronically with good accuracy. The Pope cell is also sensitive to contaminant errors, and the older sensors exhibit a considerable amount of hysteresis. The more advanced resistive and capacitance polymer sensors offer better performance with virtually no hysteresis and far better resistivity to contaminants. Pope cells are therefore finding fewer and fewer applications.

7.3 OTHER POSSIBLE COULOMETRIC METHODS OF HUMIDITY MEASUREMENT

Weaver and Riley (1948) developed a type of electric hygrometer that had application in routine checking of water vapor content of gases, particularly those in high-pressure cylinders. The sensitive element was a film of electrolyte, usually phosphoric and sulfuric acid, with suitable electrodes for use in measuring the electric resistance of the film, that mounted in a small case incorporated in high-pressure cylinder similarly to an aviation engine spark plug after appropriate change. Since the electric resistance of the film was unstable, a comparison procedure was resorted to, in which the resistance of the film in equilibrium with the atmosphere to be measured is immediately matched by exposing to an atmosphere, the moisture content of which can be controlled in a known way.

In the primary use of the instrument, measuring the dryness of an aviator's oxygen, pressure control was found to be most convenient and is described here only in its most elemental form. Gas, usually nitrogen, was humidified 100% while at high pressure. A sample of this nitrogen, at atmospheric pressure, was passed through the cell and the resistance noted, usually as the reading of a galvanometer in an unbalanced ac Wheatstone bridge. Then a sample from the compressed gas under test was passed through the cell, at a pressure that was reduced until the same reading was obtained. From the measured pressures and the known water content of the saturated sample, either the weight of water vapor per unit volume, vapor pressure, or relative humidity of the test sample can be computed. The instrument was characterized by speed, the fuse of a small sample, and greater sensitivity than was possessed by other instruments or methods comparable in these respects. This device was available commercially.

The variation in electric resistance of cotton wool and human hair has also been investigated by Burbidge and Alexander (1927). They established that the logarithm of the resistance of these materials was proportional to relative humidity.

REFERENCES

Brodgesell A., Liptak B.G. (1993) Moisture in air: Humidity and dew point, In: Liptak B.G. (ed.) *Analytical Instrumentation.* Chilton Book Company, Radnor, PA, pp. 215–226.

Brodgesell A., Liptak B.G. (2003) Moisture in air: Humidity and dew point, In: Liptak B.G. (ed.) *Instrument Engineers' Handbook, Vol. 1, Process Measurement and Analysis.* CRC, Boca Raton, FL, pp. 1420–1433.

Brodgesell A., Liptak B.G., Tatera J.F. (2003) Moisture in gases and liquids, In: Lipták B. (ed.) *Instrument Engineers' Handbook, Vol. 1: Process Measurement and Analysis*, 4th edn. CRC Press, Boca Raton, FL, pp. 1434–1449.

Burbidge P.W., Alexander N.S. (1927) Electrical methods of hygrometry. *Trans. Phys. Soc. London* 40, 149.

Goldsmith P., Cox L.C. (1967) An improved electrolytic hygrometer. *J. Sci. Instrum.* 44, 29–36.

Hibbs J.M., Nation G.H. (1964) The application of an electrolytic hygrometer to the determination of oxide in lead. *Analyst* 89, 49–54.

Hulanicki A. (1995) Absolute methods in analytical chemistry. *Pure Appl. Chem.* 67 (11), 1905–1911.

Jefferies J. (1993) Product quality improvement with correct moisture measurement in thermal processes using electrolytic hygrometers, Industrial Heating (http://meeco.com).

Keidel F.A. (1959) Determination of water by direct amperometric measurement. *Analyt. Chem.* 31, 2043–2048.

Lorek A., Koncz A., Wernecke R. (2010) Development of a gas flow independent coulometric trace humidity sensor for aerospace and industry, In: *Proceedings of the First European Conference on Moisture Measurement*, October 5–7, Weimar, Germany, pp. 289–296.

Ma C., Shadman F., Mettes J., Silverman L. (1995) Evaluating the trace-moisture measurement capability of coulometric hygrometry. *Micro* 13 (4), 43–49. http://meeco.com/meeco/technology.

Mettes J., Haggerty C., Zoladz E. (1993) Moisture monitoring in high-purity gas distribution systems, In: *Proceedings of SEMI Ultraclean Manufacturing Conference 93*, Austin, TX.

Nakamura K.-I., Ono K., Kawada K., Mitsui T. (1973) Electrochemical characteristics of a $Pt-P_2O_5$ electrolytic hygrometer. *Electroanal. Chem. Interfacial Electrochem.* 47, 175–179.

Pirog V.P., Gaba A.M., Rudykh I.A., Semchevskii A.K. (2011) Application of coulometric electrolytic cells in absolute humidity hygrometers. *Meas. Techn.* 53 (12), 1411–1416.

Semchevskii A.K., Gaba A.M., Nosenko L.F., Pirog V.P., Klopotov K.I. (2009) Experience of joint development of a coulometric hygrometer. *Meas. Techn.* 52 (4), 424–426.

Smith D.M., Mitchell J. Jr. (1984) Coulometric hygrometry, In: Smith D.M. and Mitchell J. (eds.) *Aquametry* (*part 2*), 2nd edn. John Wiley & Sons, New York, Chapter 3.

Soloman S. (2010) *Sensor Handbook*, 2nd edn. McGraw Hill, New York.

Tiebe C., Hübert T., Lorek A., Wernecke R. (2012) New planar trace humidity sensor, In: *Proceedings of the 14th International Meeting on Chemical Sensors, IMCS 2012*, May 20–23, Nuremberg, Germany, pp. 294–297.

Weaver E.R., Riley R. (1948) Measurement of water in gases by electrical conduction in a film of hygroscopic material and the use of pressure changes in calibration. *J. Research NBS* 40, 169–214, RP1865.

8 Humidity Measurement Based on Karl Fischer Titration

8.1 KF TITRATION

It is worth noting that, in addition to the instrumental methods of measuring humidity, there is a chemical method based on titration; the so-called Karl Fischer (KF) titration method (Scholz 1984; Schöffski 2000; Bruttel and Schlink 2003; Rübner et al. 2008). KF titration is a classic titration method in analytical chemistry that uses coulometric or volumetric titration to determine the trace amounts of water in a sample. The measurement cell for KF titration is shown in Figure 8.1. It was invented in 1935 by the German chemist Karl Fischer. Water is titrated using the KF reagent, which consists of iodine (I_2), sulfur dioxide, a basic buffer, and a solvent. The basis of this method is the reaction described by R.W. Bunsen (1853):

$$SO_2 + I_2 + 2H_2O \rightarrow H_2SO_4 + 2HI \quad (8.1)$$

Karl Fischer discovered that this reaction can be used for the water determinations in a nonaqueous system if sulfur dioxide was present in excess and the acids produced were neutralized by a base such as a pyridine. One should note that the selection of pyridine as the base for the KF reaction was completely at random. This led to the establishment of the classical KF reagent, a solution of iodine and sulfur dioxide in a mixture of pyridine and methanol. Further studies on the subject of the KF reaction have revealed that pyridine was not directly involved in the reaction (i.e., it only acted as a buffering agent and could therefore be replaced by other bases).

8.2 PRINCIPLES OF KF TITRATION METHOD

The KF titration method is mainly used for determination of moisture in the solids and liquids, where it is important to know the percentage of water (Bruttel and Schlink 2003; Isengard 1991, 2003; GFS 2004). However, the determination of the moisture content in the air and other gases is also possible (Archer and Hilton 1974; Miyake and Suto 1974; Davies 1975;

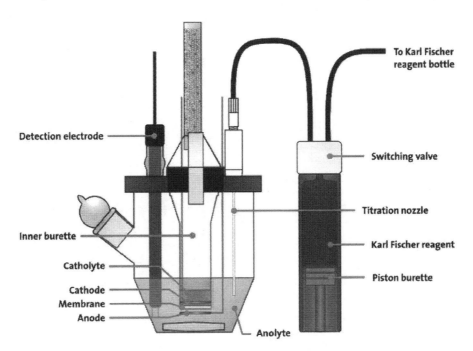

FIGURE 8.1 Schematic diagram illustrating the measurement cell for Karl Fischer titration. (Reprinted from www.metrohm.com.)

Bruttel and Schlink 2003; Rübner et al. 2008; Schaller 2007). The original composition was modified and adapted to these purposes. As a rule, alcohol (mostly methanol) was used as the solvent. However, other alcohols, such as either diethylene glycol monoethyl or imidazole, can also be used. To obtain a quantitative reaction, the ester is preferably neutralized by imidazole to yield alkyl sulfite. In a second step, alkyl sulfite is oxidized by iodine to give alkyl sulfate in a reaction that requires water.

As it follows from equation 8.1, the KF titration requires water to be in direct contact with the reagent. Therefore, for determination of the water vapor in the gas, a measured volume of gas runs through the titration cell or adsorber that physically or chemically retains water. The absorber can be a dry solvent, KF reagent, desiccant, or cold trap. A liquid of low vapor pressure is preferred to keep evaporation loss to a minimum, particularly if a large volume of gas must be passed to accumulate sufficient water for accurate measurement. Ethylene glycol, therefore, is sometimes also employed in mixture with other solvents or KF reagent absorber. The amount of water retained is then determined by titration with KF reagent or with a standard solution of water in methanol if KF reagent was employed as absorber. It is best to introduce the gas directly into the titration cell (see Figures 8.1 and 8.2b) through a capillary that is immersed as deeply as possible in the KF solution. A flow meter is installed in the gas line and the flow rate adjusted with a control valve. It was established that the gas flow should be between 50 mL/min and a maximum of 250 mL/min (3–15 liter/h).

The consumption of iodine is measured either coulometrically or volumetrically (Keidel 1959; Möhlmann 2004). Coulometric titration is particularly effective for this purpose because of its high precision and the low level of moisture in most gases. In the coulometric procedure, the iodine required for the determination of water by the KF reaction is produced by electrolysis of the reagent containing the iodide, and then the water content in a sample is determined by measuring the quantity of electricity that is required for the electrolysis (i.e., for the production of iodine), based on the quantitative reaction of the generated iodine with water. In the volumetric titration method, iodine required for reaction with water is previously dissolved in the water used for titration, and the water content is determined by measuring the amount of iodine solution consumed as a result of reaction with water in a sample (see Figure 8.2a).

The end point is detected most commonly by a bipotentiometric method in both cases. For this purpose, a second pair of platinum electrodes are immersed in the anode solution. Platinum electrodes placed into the working medium are polarized either by constant current or by constant voltage. Prior to the equivalence point, the solution contains I^- but little I_2. At the equivalence point, the excess of I_2 appears and the abrupt voltage drop marks the end point. The amount of charge needed to generate I_2 and reach the end point can then be used to calculate the amount of water in the original sample.

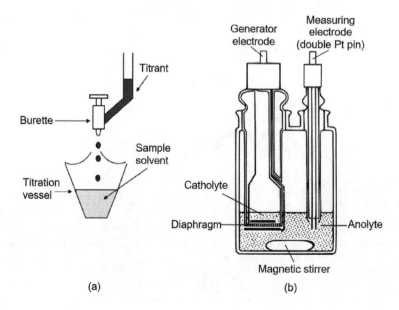

FIGURE 8.2 (a) Volumetric Karl Fischer titration: Iodine is added by a burette during titration. (b) Coulometric Karl Fischer analysis: Iodine is generated electrochemically during titration. The terms *generator electrode* and *measuring electrode* mean an arrangement of two electrodes (anode and cathode) to form an electrolytic cell.

The quantity of electricity is directly proportional to the mass of iodine generated and hence to the mass of water determined. As gas samples normally have a low water content, coulometric titration is to be preferred. If the volumetric method is used, then a solvent mixture made up of equal amounts by volume of methanol and ethylene glycol should be used as the absorption solution.

Because many gases include the amounts of water of only few g/dm^3, the admission of the gas sample should be made carefully in order to avoid contamination or adsorption of water by the tubing system (Schaller 2007). Subsequently, the determination of humidity is made according to the International Organization for Standardization (ISO) (1993a, 1993b, 1993c, 1997). It is often advisable to set up the titrator directly at the sampling point or to introduce the sample directly into the detached but conditioned titration vessel. In order to achieve a stable equilibrium, the gas line must be thoroughly purged with the gas sample (during 10–30 min). This is done by installing a three-way stopcock in the line and not passing the gas sample through the titration cell during the purging phase. The time during which the sample is passed into the titration cell must be measured in order to subsequently carry out the necessary calculation of the water content. The amount of sample depends primarily on the water content, the method used (coulometric or volumetric titration), and the required accuracy. At lower moisture concentration in the gas, large volumes of gas are required for accurate measurement. With larger amounts of the sample, one should consider that the methanol that evaporates must be topped up from time to time.

The determination of humidity usually is made according to the ISO (ISO 1993a, 1993b, 1993c). The water content is normally calculated in µg per liter or mg per liter:

$$\text{Gas volume} = \text{flow-through time} \times \text{gas flow}$$

$$\text{Water content, coulometry} = A/V$$

$$\text{Water content, volumetry} = (A \cdot t)/V$$

where A is a titrator reading (µg H_2O or mL KF reagent [KFR]), t is a titer of KF solution (mg H_2O/mL), and V is a gas volume. Simple calculations show that 1 mg of water corresponds to a consumption of 10.712°C electrical current. If the water content is to be calculated in percentages, then the mass of the gas introduced into the titration vessel must be known. This can be obtained by differential weighing (sample container before and after removing the gas) or by calculation.

8.3 ADVANTAGES AND LIMITATIONS

It is important that this method can be applied to natural gas (Davies 1975) and other gases that do not react with KF reagents. According to ISO, this method is applicable to water concentrations between 5 mg/cm^3 (5000 ppm) and 5000 mg/cm^3 (5·10^6 ppm). Undoubtedly, the indicated sensitivity of this method is greatly inferior to the sensitivity of other instrumental methods, especially considering modern electronic moisture sensors. However, in many cases, the specified measurement range is sufficient, and this method is therefore still in use.

Besides that, Metrohm (www.metrohm.com), Mettler Toledo (http://www.mt.com/), and Sigma-Aldrich (www.sigmaaldrich.com) claim that this method actually permits determination of water content in a much wider range, from a few ppm up to 100%. The volumetric KF titration is suitable for the samples in which water is present as a major component from 100 ppm to 100%, while coulometric KF analysis is suitable for the samples in which water is present in a trace amounts, from 1 ppm to 5%. According to their estimates, the measurement accuracy can be within 1% of available water. However, this accuracy depends on the volume of tested gas pumped through the solution and the amount of water contained in this gas. Therefore, the volume of tested gas must be of such size that the mass of water vapor extracted is large enough to permit a determination with suitable accuracy. An especially large amount of sample for analysis is required for the samples with less than 0.05% of water. The result is, therefore, an average for the duration of the test. For example, at a sampling rate of 50 mL per minute, which permits reasonably complete extraction of the water vapor from the test gas, the time required to collect 0.001 g of water vapor from the air with a frost point of −40°C is approximately 20 minutes.

Generally, KF is conducted using a separate KF titrator or, for volumetric titration, a KF titration cell installed into a general-purpose titrator. However, KF is also suitable for automation (Kelley et al. 1959; Miyake and Suto 1974). For example, automatic KF titrators fabricated by Mettler Toledo are shown in Figure 8.3. The automatized KF titrator designed by Metrohm (www.metrohm.com) for gas analysis is shown in Figure 8.4. Generally, such an apparatus consists of an automatic burette, a back titration flask, a stirrer, and equipment for amperometric titration at constant voltage or potentiometric titration at constant current. The analysis is typically complete within a minute.

The main advantages of this method usually include the accuracy, speed, and selectivity. KF is selective for water, because the titration reaction itself consumes water.

FIGURE 8.3 Automatic Karl Fischer titrators fabricated by Mettler Toledo. (Reprinted from http://www.mt.com/.)

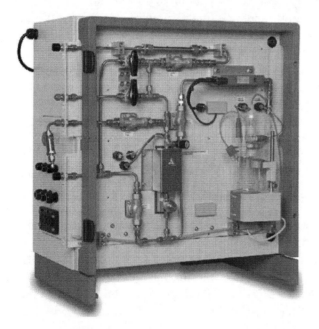

FIGURE 8.4 Karl Fischer gas analyzer designed by Metrohm. The 875 KF gas analyzer is the robust solution for water determination in gases by automated coulometric KF titration. The system consists of a control unit and an analysis module. The analysis module is equipped with a mounting board for the gas handling system and the water analysis cell, as well as an integral coulometer that performs all the necessary analysis steps fully automatically. (Reprinted from www.metrohm.com.)

If no side reactions occur, only water will be determined. The KF response is linear. Therefore, a single-point calibration using a calibrated 1% water standard is sufficient, and no calibration curves are necessary. The method can be validated and therefore fully documented. However, the strong redox chemistry (SO_2/I_2) means that the redox-active sample constituents may react with the reagents.

For this reason, KF is unsuitable for solutions containing, for example, dimethyl sulfoxide.

KF suffers also from an error called *drift*, which is an apparent water input that can confuse the measurement. Atmospheric humidity represents the most relevant source of error in the KF titration. Moisture can enter the *sample*, the *titrant*, and the *titration stand*. This problem is particularly relevant in tropical climates or in coastal regions, where the relative humidity can achieve values of more than 80%. The glass walls of the vessel adsorb water, and if any water leaks into the cell, the slow release of water into the titration solution can continue for a long time. Therefore, the titration stand must be sealed as tightly as possible against atmospheric moisture. In addition, before measurement, it is necessary to carefully dry the vessel and run a 10- to 30-minute "dry run" in order to calculate the rate of drift. The drift is then subtracted from the result.

It is also necessary to bear in mind that certain reactive substances can interfere in the KF determination of water (GFS 2004). Samples known to contain such substances can be given special treatment, or special conditions can be employed to minimize or eliminate interferences. In general, substances that interfere are of three types: reductants that are oxidized by iodine, oxidants that are reduced by iodide, and substances that form water in reaction with components of the KF reagent or its reaction products.

Reductants that interfere include the following:

- Certain phenols, such as hydroquinone and aminophenols (some of them can still be titrated by volumetric KF, adding salicylic acid to the titration solvent)
- Thiols (mercaptans [RSH]) oxidize iodine as follows:

$$2\ RSH + I_2 \rightarrow RSSR + 2\ HI \quad (8.2)$$

- Thioacetate and thiosulfate

Because these reductants consume iodine, they cause high results in the KF titration.

Oxidants that interfere are fewer in number. This is because KF reagents commonly contain a large excess of sulfur dioxide, which is capable of reducing many oxidants. Therefore, the only oxidants that interfere, assuming sufficient sulfur dioxide is present, are those that fail to be reduced by sulfur dioxide but are strong enough to oxidize iodide to iodine. An example is sodium dichromate. Its presence causes low results in KF titrations, because any iodide formed in the titration will react with dichromate to form iodine to thus regenerate the titrant in situ. Potassium permanganate is a strong oxidant, but it is relatively insoluble in methanol and does not interfere. Substances that interfere by water-forming reactions include the following: esters, active carbonyl compounds, basic oxides and hydroxides, and inorganic carbonates.

Aldehydes and ketones react with methanol to form acetals and ketals. Aldehydes are much more reactive than ketals. The longer the chain length of an aliphatic ketone, the lower its reactivity, and aromatic ketones are less reactive than aliphatic ones. The above reaction can be suppressed by using methanol-free KF solvents and titrants. For the other interfering compounds, proper choice of conditions and/or prior treatment (addition of a buffer) suffices in most applications to overcome interferences.

REFERENCES

Archer E.E., Hilton J. (1974) The determination of small amounts of water in gases using Karl Fiescher reagent. *Analyst* 99, 547–550.

Bruttel P., Schlink R. (2003) *Water Determination by Karl Fischer Titration*. Metrohm Ltd., Herisau, Switzerland.

Bunsen R.W. (1853) Ueber eine volumetrische Methode von sehr allgemeiner Anwendbarkeit. *Liebigs Ann. Chem.* 86, 265–291.

Davies R.J. (1975) The determination of water in natural gas using a modified Karl Fischer titration apparatus. *Analyst* 100, 163–167.

Fischer K. (1935) Neues Verfahren zur maßanalytischen Bestimmung des Wassergehaltes von Flüssigkeiten und festen Körpern. *Angew. Chem.* 48 (26), 394–396.

GFS (2004) *Moisture Measurement by Karl Fischer Titrimetry*. GFS Chemicals, Inc., Powell, OH.

Isengard H.-D. (1991) Bestimmung von Wasser in Lebensmitteln nach Karl Fischer. ZFL-Internat. Zeitschr. f. Lebensmittel-Technik, Marketing, Verpackung und Analytik. *Eur. Food Sci.* 42, 1–6.

Isengard H.-D. (2003) How to determine water in foodstuffs? *Analytix* 3, 11–15.

ISO (1993a) ISO 10110-2: Natural gas: Determination of water by the Karl Fischer method. Part 2: Titrimetric method. ISO, Genève.

ISO (1993b) ISO 10110-3: Natural gas: Determination of water by the Karl Fischer method. Part 3: Coulometric method. ISO, Genève.

ISO (1993c) ISO 10110-9: Natural gas. Determination of water. ISO, Genève.

ISO (1997) ISO 11541: Natural gas. Determination of water content at high pressure. ISO, Genève.

Keidel F.A. (1959) Determination of water by direct amperometric measurement. *Anal. Chem.* 31 (12), 2043–2048.

Kelley M.T., Stelzner R.W., Laing W.R., Fisher D.J. (1959) Automatic coulometric titrator for the Karl Fischer determination of water. *Anal. Chem.* 31 (2), 220–221.

Miyake S., Suto T. (1974) An automatic colulometric titrator for the determination of water by the Karl Fischer method. *Bunseki Kagaku* 23, 482–490.

Möhlmann D., Wernecke R., Schwanke V. (2004) Measurement principle and equipment for measuring humidity contents in the upper martian surface and subsurface, In: *Proceedings of the 37th ESLAB Symposium on Tools and Technologies for Future Planetary Exploration, ESA: Noordwijk*, pp. 163–168.

Rübner K., Balköse D., Robens E. (2008) Methods of humidity determination. Part I: Hygrometry. *J. Thermal Anal. Calorimetry* 94 (3), 669–673.

Schaller E. (2007) Wassergehalt in gasen. *LABO* 38(2), 24–26.

Schöffski K. (2000) Die wasserbestimmung mit Karl-Fischer-titration. *Chem. Unserer Zeit* 34, 170–175.

Scholz E. (1984) *Karl Fischer Titration: Determination of Water*. Springer-Verlag, Berlin, Germany.

9 Other Conventional Methods of Humidity Measurement

9.1 CHEMICAL METHOD

Undoubtedly, there are many other humidity control methods (Wexler 1970). For example, there is *chemical method* when a sample of the test gas is passed continuously through a chemical reactant, such as calcium hybride, sodamine, or calcium carbide, which converts the water vapor into a more easily detectable gas, such as hydrogen, ammonia, and acetylene, respectively. The appearance of hydrogen and acetylene and their concentration can be determined using, for example, a simple gas sensor, such as pellistor, or a pressure sensor. The method is capable of continuously recording the water vapor content in the air over a frostpoint range of −30°C to 0°C with an accuracy within 0.25°C–2.50°C, with a response time of a few minutes. The apparatus is simple to construct and relatively rugged. As an example, consider in more detail the calcium carbide method.

9.1.1 Calcium Carbide Method

The calcium carbide method is a fast but destructive method for measuring the moisture quotient of materials. When calcium carbide gets in contact with water, acetylene gas is released (Eq. 9.1). Cameron Hugh patented in 1930 a "process and apparatus for detecting and determining the quantity of percentage of moisture in a substance" based on this reaction (Cameron 1930). The original patent was mostly intended for measuring moisture in flour; however, the reaction can also be used with respect to air humidity. The sample that should be studied is weighed and then placed in a gas pressure vessel with a calcium carbide ampoule and some steel balls. When the vessel is shaken, the steel balls break the ampoule. As a consequence, the calcium carbide reacts with the water in the sample. A gauge at the top of the vessel can be used to measure the resulting gas pressure. The quantity of the generated gas is directly proportional to the moisture content of the sample. The air humidity corresponding to the measured pressure can be determined from conversion tables. The calcium carbide method is a relatively fast technique for measuring moisture content. However, this method is not characterized by high accuracy: The problem with the method is the indirect measurement of moisture quotient through pressure. In some countries, this method is used for determining moisture in concrete in special applications.

$$CaC_2 + 2\ H_2O \rightarrow C_2H_2\uparrow + Ca(OH)_2 \qquad (9.1)$$

9.2 CONSTANT-PRESSURE HYGROMETER

The volume of water vapor in a gas sample can also be measured if its change in volume is measured at constant pressure before and after the water vapor is absorbed (Wexler and Brombacher 1972). Conversely, if the volume is held constant, the change in pressure gives the pressure of the water vapor. These methods are useful only in laboratory investigations. The difficulty of obtaining accurate determinations increases rapidly with decrease in temperature of the gas sample. In one form of constant pressure apparatus, a manometer and also a graduated tube containing an absorbing liquid, such as sulfuric acid, are connected to a gas container (Blackie 1936). In operation, sulfuric acid is slowly admitted to the container to absorb the water vapor in the sample and at the same time in sufficient volume to maintain constant the absolute pressure of the gas, as indicated by the manometer and a barometer. The volume of acid admitted is the volume of water vapor in the sample, subject to corrections for lack of constancy of temperature or of the reference pressure. If the apparatus just described is modified so that the sulfuric acid forms part of the original volume, a constant volume apparatus results. In this case, the change in pressure as the water vapor is absorbed is the water-vapor pressure.

9.3 CONSTANT-VOLUME HYGROMETER

The tilting form of absorption hygrometer described by Mayo and Tyndall (1921) is essentially a constant volume instrument (Wexler and Brombacher 1972). Here, the absorbing material (P_2O_5) is installed in a piston that moves from one end of the cylinder containing the gas sample to the other as the cylinder is oscillated, thus forcing the gas through the absorbing chemical. This piston action reduces the time required for complete absorption of the water vapor. The fall in gas pressure that is measured is the water-vapor pressure. The device is proposed as a working standard in calibration of hygrometers. It is not available commercially (Wexler and Brombacher 1972).

One version of constant-volume hygrometer (Okada and Tamura 1940) completely dispenses with an absorbing chemical and, instead, uses low temperatures to condense part of the water vapor content of a gas sample and, from a measurement of the temperature and reduction in pressure, permits the determination of the initial vapor pressure. In this instrument, two identical vessels, one containing dry gas and the other the gas sample of unknown water vapor content, are sealed and connected through a differential manometer. The vessels are then immersed in a liquid bath and gradually cooled until the differential manometer registers a pressure difference, indicating that condensation of water vapor has taken place in the gas sample. At this point, the temperature and pressure difference are read. From the saturation pressure at the observed temperature and from the pressure difference, the initial water-vapor content is then computed.

By employing a liquid-air trap, the moisture in a large volume of gas can be condensed and then suitably measured. The known volume of gas is passed through the trap and, while the low temperature is maintained, the trap is evacuated. The apparatus is then allowed to warm up, preferably in a thermostatted bath, and the vapor pressure measured. When using this method with gases containing carbon dioxide, the temperature of the air trap is raised above −78°C before evacuation. This permits any condensed carbon dioxide to be vaporized and removed from the trap.

For additional information on absorption hygrometers and descriptions of a number of these old instruments, read (Shaw 1916; Dowling 1930; Kleinschmidt 1935; Thornthwaite 1941).

9.4 PNEUMATIC BRIDGE METHOD OF HUMIDITY MEASUREMENT

There is also the so-called *pneumatic bridge method* (Greenspan 1965). The pneumatic bridge is roughly analogous to a Wheatstone bridge, in which four small, critical flow nozzles are the analogue of the resistors and a differential pressure gauge is the analogue of the galvanometer (see Figure 9.1). The bridge is balanced when the ratio of the volumetric flow in the downstream nozzle to that in the upstream nozzle is identical for both arms of the bridge, and the pressure difference across the bridge will then be zero. When an absorber, which removes the

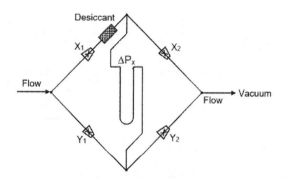

FIGURE 9.1 Simplified schematic of pneumatic bridge hygrometer: X_1, X_2, Y_1, and Y_2 are nozzles.

water from the gas flowing through the bridge, is incorporated into one branch of a balanced bridge between the upstream and downstream nozzles, then there appears an unbalance of pressures. The decrease in the air humidity reduces the mass flow and thus affects the balance of the bridge, which can be measured by the differential pressure gauge. This pressure unbalance is approximately proportional to the partial pressure of the water vapor constituent. The general principle of this instrument was first described by Wildhack (1950). Although the theory predicts the output of the instrument in term of the humidity of the test gas, a calibration is required to obtain the highest accuracy.

9.5 DIFFUSION HYGROMETER

At the time, a *diffusion hygrometer* was also proposed (Greinacher 1944, 1954). This hygrometer was based upon the diffusion of water vapor through a porous membrane. The diffusion hygrometer used the property of certain porous materials, such as clay, marble, gypsum, alabaster, cellophane, and gelatin, which behaved as semipermeable membranes with respect to the water vapor and the air. In this case, one can create conditions when the pressure difference between the vessel with desiccant and the atmosphere can serve as a basis for determining the humidity of the air. In practice, such a hygrometer consists of a closed chamber having porous walls and containing a hygroscopic compound, whose absorption of water vapor causes a pressure drop within the chamber that is measured by a manometer (see Figure 9.2). A differential manometer, communicating with the enclosed vessel and the ambient atmosphere (whose humidity is being measured), registers a pressure drop Δp_1 that is directly proportional to the partial pressure e of the water vapor in the ambient atmosphere. In order to avoid consideration of the constant of proportionality of the apparatus, a similar vessel

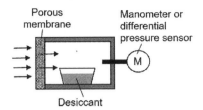

FIGURE 9.2 Diagram illustrating the principle of operation of diffusion hygrometer.

with an identical porous clay plate and manometer, but containing water instead of the desiccant, is employed. The pressure drop Δp_2 indicated by the latter arrangement is directly proportional to the difference between the saturation vapor pressure e_s and the ambient partial pressure at the ambient temperature. The relative humidity is given by the relation in Equation 9.2. Further theoretical consideration of this hygrometer was presented by Spencer and Rourke (1947).

$$\text{RH} = 100\frac{e}{e_s} = 100\frac{\Delta p_1}{\Delta p_1 + \Delta p_2} \qquad (9.2)$$

Thus, diffusion hygrometer is a laboratory method, which determines the relative humidity, basing on the difference in pressure in a closed chamber and the atmosphere. A diffusion hygrometer can be a small and low-cost indicator. However, the instrument is very sensitive to rapid temperature change. In addition, it is relatively sluggish in its response to the humidity change. This method is also not distinguished by high accuracy.

9.6 CLOUD OR FOG CHAMBER HYGROMETER

The *cloud or fog chamber hygrometer* uses another effect. With this type of hygrometer, the formation of fog is used as a criterion for saturation. It was used in the industry because of its portability, simplicity, and ability to cover a broad range of dew points without the need for cooling. Samples of the moist gas are compressed to successively higher pressures. The atmosphere sample is held in the observation chamber for several seconds to stabilize the temperature, after which the quick opening valve is made. This is accompanied by releasing the pressure and creating adiabatic cooling, which causes a visible condensation or formation of the fog in the chamber. The procedure is repeated to find the end point, the point at which the fog disappears. The temperature is indicated by a thermometer that extends into the observation chamber. The dew point is then determined by referring to a chart based on the initial temperature reading of the thermometer and the pressure ratio gage reading at the point where the fog disappeared (Griffiths 1933). Though simple in construction, the method has the same drawback as the dew cup in that it is subject to mistake during interpretation, and this method requires the skill of the operator. Measurement errors of 5°C are not uncommon. It is also a one-time measurement.

However, at present, when a large number of advanced electronic and electrical sensors of humidity appeared, hygrometers mentioned above are almost never used. Thus, we are not going to consider them in more detail.

9.7 MASS SPECTROMETRIC MEASUREMENT OF AIR HUMIDITY

Various mass spectrometers can also be used to determine air humidity. For example, studies performed by Kaufmann et al. (2016) have shown that mass spectrometers can be used even for accurate quantification of low water vapor concentrations в upper troposphere and lower stratosphere (UTLS). The airborne mass spectrometer (AIMS)-H_2O, developed by Kaufmann et al. (2016), consists of a linear quadrupole mass spectrometer (Huey et al. 1995), which was designed and built by THS instruments at the Georgia Institute of Technology (Greg Huey, Atlanta, USA). This spectrometer allows measurement of water vapor by direct ionization of ambient air. A schematic diagram of the flight configuration of AIMS is shown in Figure 9.3.

The developers claim that the new configuration of AIMS was developed in response to the large discrepancies between different airborne water vapor measurements that have been found in the past device (Kley et al. 2000). According to Kaufmann et al. (2016), with the mass spectrometer AIMS-H_2O, which includes in-flight calibration, it was achieved a significant progress in the field of airborne water vapor measurements especially in the case of the low H_2O mixing ratios of the UTLS. A comprehensive and in-depth error analysis allowed the developers to achieve accuracy between 7% and 15% in the measurement range between 1 and 500 ppmv, depending on the specific humidity and time resolution of the measurement. However, it should be recognized that this device for measuring humidity is expensive and cumbersome, and requires special conditions (high vacuum) and highly qualified specialists, which significantly limits the area of possible application (see Figure 9.4).

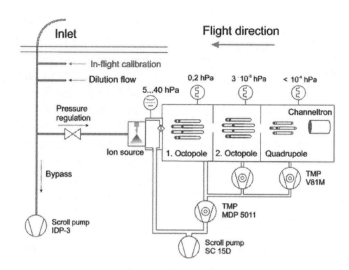

FIGURE 9.3 Schematic of the flight configuration of AIMS. Ambient air enters via a backward-faced inlet and passes through a pressure regulation valve before entering the ion source. The ion beam is then focused by two adjacent octopoles and finally separated by mass-to-charge ratio in the quadrupole. Additionally, connections for an optional dilution of ambient air and background measurements and for addition of trace gases for inflight calibration are mounted right beneath the inlet. (Reprinted with permission from Kaufmann S. et al., The airborne mass spectrometer AIMS – Part 1: AIMS-H_2O for UTLS water vapor measurements, *Atmos. Meas. Technol.*, 9, 939–953, 2016. Copyright 2016, European Geoscience Union as open access.)

FIGURE 9.4 The front view of instrument rack in AIMS-H_2O configuration integrated in the HALO (High Altitude and Long Range Research Aircraft) standard rack. The inlet line is connected to a trace gas inlet (TGI) mounted at the top fuselage of the aircraft. (Reprinted with permission from Kaufmann S. et al., The airborne mass spectrometer AIMS – Part 1: AIMS-H_2O for UTLS water vapor measurements, *Atmos. Meas. Technol.*, 9, 939–953, 2016. Copyright 2016, European Geoscience Union as open access.)

REFERENCES

Blackie A. (1936) A sulphuric acid hygrometer. *J. Sci. Instr.* 13, 6.

Cameron H.F. (1930) Process and apparatus for detecting and determining the quantity of percentage of moisture in a substance. GB Patent 335,308.

Dowling J.J. (1930) A vapor pressure hygrometer. *J. Sci. Inst.* 13, 214.

Greenspan L. (1965) A pneumatic bridge hygrometer for use as a working humidity standard, In: Wexler A (Ed.) *Humidity and Moisture*, Vol. 3. Reinhold, New York, pp. 433–444.

Greinacher H.S. (1944) Ein neuer Feuchtigkeitsmesser: Das diffusion hygrometer. *Helv. Phys. Acta* 17, 437–454.

Greinacher H.S. (1954) Diffusion hygrometer mit direkter Ablesung. *Schweis. Arch. Angew. Wiss. Technol.* 20, 198–200.

Griffiths E. (1933) *The Measurement of Humidity in Closed Spaces*. Spec. Rept. No.8 Rev. Ed., Food Investigations Board, D.S.I.R., London, UK.

Huey L.G., Hanson D.R., and Howard C.J. (1995) Reactions of SF_6 and I with atmospheric trace gases. *J. Phys. Chem.* 99, 5001–5008.

Kaufmann S., Voigt C., Jurkat T., Thornberry T., Fahey D.W., Gao R.-S., Schlage R., Schäuble D., Zöger M. (2016) The airborne mass spectrometer AIMS – Part 1: AIMS-H_2O for UTLS water vapor measurements. *Atmos. Meas. Technol.* 9, 939–953.

Kleinschmidt E. (1935) *Handbuch der Meteorologischen Instrumente*. Julius Springer, Berlin, Germany.

Kley D., Russell III J.M., Phillips C. (Eds.) (2000) SPARC report no. 2: Upper tropospheric and stratospheric water vapour. http://www.sparc-climate.org.

Mayo H.G., Tyndall A.M. (1921) The tilting hygrometer; a new form of absorption hygrometer. *Proc. Phys. Soc. Lond.* 34 (1), xvii.

Okada T., Tamura M. (1940) Studies with the condensation hygrometer. *Proc. Imp. Acad. (Tokyo)* 16, 141 and 208.

Shaw A.N. (1916) Improved methods of hygrometry. *Trans. Roy. Soc. Can. Ser.* 3 (10), 85.

Spencer G., Rourke E. (1947) Theoretical basis for the diffusion hygrometer. *Phil. Mag.* 38, 573–580.

Thornthwaite C.W. (1941) Chemical absorption hygrometer as meteorological instrument. *Trans. Am. Geophys.* Union, Pt. II, p. 417.

Wexler A. (1970) Measurement of humidity in the free atmosphere near the surface of the earth, In: Teweles S., Giraytys J. (Eds.) *Meteorological Observations and Instrumentation*, Meteorological Monographs Vol. 11, No. 33. American Meteorological Society, Lancaster Press, Lancaster, PA, pp. 262–281.

Wexler A., Brombacher W.G. (1972) Methods of measuring humidity and testing hygrometers. A review and bibliography, In: Bloss R.L., Orloski M.J. (Eds.) *Precision Measurement and Calibration: Mechanics*. National Bureau of Standard, Washington, DC, pp. 261–280.

Wildhack W.A. (1950) A versatile pneumatic instrument based on critical flow. *Rev. Sci. Instr.* 21 (1), 25–30.

Section III

Electronic and Electrical Humidity Sensors and Basic Principles of Their Operation

10 Capacitance-Based Humidity Sensors

10.1 BASIC PRINCIPLES OF OPERATION

Capacitance-type humidity sensors are a huge part of existing sensor types in both research and industry, as they offer significant advantages in terms of simplicity of fabrication, sensitivity, and low-power operation (Ishihara and Matsubara 1998; Kummer et al. 2004; Lee and Lee 2005; Chatzandroulis et al. 2011). They dominate in atmospheric and process measurements and are the only types of full-range relative humidity (RH) measuring devices capable of operating accurately down to 0% RH. According to estimations (Rittersma 2002), capacitive RH sensors represent more than 75% of the available humidity sensors on the market. Capacitive humidity sensors are commercially available from, for example, Sensirion (www.sensirion.com), Vaisala (www.vaisala.com), and Humirel (www.humirel.com).

In the simplest case, a capacitance-type sensor is made of two parallel plates. In such structure, the capacitance between the two electrodes is given by

$$C = \varepsilon_r \varepsilon_0 \frac{A}{d} \quad (10.1)$$

where ε_r and ε_0 are the relative and vacuum permittivity constants, respectively, A is the plate surface area, and d is the plate distance. From this equation, it is evident that only three ways exist to effect a change in the capacitance of that device: (1) alter the distance d between the two plates, (2) alter the overlapping area A between the two plates, and (3) change the dielectric permittivity between the plates (Ishihara and Matsubara 1998). This means that capacitive sensors can detect only those gases and vapors that affect these parameters. Water vapor can exert such influence and therefore, by measuring the change in the capacitance, the presence of the water vapors in the air can be detected. The principles of operation and more detailed description of the constructions of capacitance gas and chemical sensors can be found in Ishihara and Matsubara (1998) and Chatzandroulis et al. (2011).

10.1.1 Humidity Sensors of Permittivity-Type

It is worth noting that the most common humidity sensors of the capacitive type are sensors of the permittivity type (i.e., sensors in which the variable parameter is the permittivity of the space between the electrodes).

The simplest version of such sensors is air-gap sensors (Ford 1948; Fraden 2004; Zarnik and Belavic 2012; Choi and Kim 2013). As is known, the dielectric constant of air increases with increasing air humidity (see Table 10.1). According to Lea N. (1945), moisture in the atmosphere changes the electrical permittivity in air according to the equation:

$$\kappa = 1 + \frac{211}{T}\left(P + \frac{48 P_s}{T} \cdot H\right) 10^{-6} \quad (10.2)$$

where T is the absolute temperature (in K), P is the pressure of moist air (in mmHg), P_s is the pressure of saturated water vapor (in mmHg) at temperature T, and H is the RH (in %).

An analysis of the possibilities of designing a humidity sensor based on this effect was carried out in Choi and Kim (2013). Simulation has shown that humidity sensors can be developed that they will be high-speed and reliable. However, the same analysis indicates that the sensors of this type do not have a high sensitivity, since the change in permittivity under the influence of humidity is insignificant (Table 10.1). With the growth of temperature, this influence increases, but to a very small degree. It must be taken into account that the temperature change is also accompanied by a change in the permittivity of the air (see Table 10.2), commensurate with the change in ε under the influence of humidity. All this means that the development of humidity measurement devices based on such sensors sharply increases the requirements of stability and sensitivity of devices, as well as minimization of parasitic capacitances. Another problem of such sensors is the low dielectric permeability of air. Thus, to achieve a capacitance value acceptable for measurement, the

TABLE 10.1
Dependence of the Dielectric Constant of Air on the Relative Humidity at Normal Temperature and Pressure

Relative Air Humidity, %	Dielectric Constant, ε
0	1.00058
50	1.00060
100	1.00064

TABLE 10.2
Temperature Influence on the Dielectric Constant of Air

Temperature		
°C	K	Dielectric Constant, ε
−60	213	1.00081
+20	293	1.00058
+60	333	1.00052

TABLE 10.3
Dielectric Constants of Selected Materials

Material	Dielectric Constant, ε
Vacuum	1.0000 00
Air (1 atm)	1.0000 54
Air (100 atm)	1.0548
H_2O (20°C)	80
Polymers	
Polytetrafluoroethylene (Teflon)	2–2.1
Polyethylene	2.2–2.4
Polyamide	2.8
Polystyrene	2.6–3
Polyvinyl chloride	3.2
Nylon	4–5
Conductive metal oxides	
SnO_2	9.9
TiO_2	86–173
$SrTiO_3$ (Strontium titanate)	310
Other materials	
Paper	2–4
Wood, dry	2–6
Si	11–12
Glass	3.7–10
SiO_2	3.9–4.5
Alumina	9.1–11.5
Ethanol (25°C)	24.3

Source: Blythe, T. and Bloor, D., *Electrical Properties of Polymers*, Cambridge University Press, Cambridge, NY, 2005.

gap between the electrodes must be very small and the area very large, which creates significant difficulties in the implementation of such sensors. Another option to increase the sensitivity is to increase the pressure (Choi and Kim 2013).

In the case of using porous dielectrics between electrodes capable of accumulating water, the situation is much better. As is known, due to the polar structure of the H_2O molecule, water exhibits a very high permittivity $\varepsilon_w = 80$ at room temperature (Grange and Delapierre 1991). Dielectric materials usually have considerably smaller permittivity (see Table 10.3). This means that a change in the water content in the dielectric can give a much larger capacity change than a change in air humidity, especially if the dielectric used has a low dielectric constant and is porous (Khanna and Nahar 1984; Kim et al. 2000). The smaller the dielectric constant of the dielectric used is, and the greater is the proportion of space between the electrodes that the water occupies, the greater will be the effect, appearing in the increase in capacity. Based on this requirement, it becomes clear that high adsorption capacity and high porosity are important parameters of materials suitable for the development of capacitive humidity sensors. As can be seen from Table 10.3, polymers and some inorganic dielectrics meet these requirements.

At equilibrium conditions, the amount of moisture present in a hygroscopic material depends on both the ambient temperature and the ambient water vapor pressure. So, there is a relationship between RH, the amount of moisture present in the sensor, and the sensor capacitance. Capacitive sensors, as well as other absorption-based humidity sensors, typically show a nonlinear behavior as a function of RH (Rittersma 2002). This behavior can be described by the phenomenological equation:

$$\frac{C_S}{C_0} = \left(\frac{\varepsilon_w}{\varepsilon_d}\right)^n \quad (10.3)$$

where ε_d and ε_w are the permittivity of the dielectric at a dry and wet state and n is a factor related to the morphology of the dielectric. This relationship is the basis of the operation of a capacitive humidity instrument. In real polymer-based humidity sensors, the change in a dielectric constant can achieve 30% when humidity changes within 0%–100% RH. In the absence of moisture, the dielectric constant of the hygroscopic dielectric material and the sensor geometry determine the value of capacitance C_0. It is important to note that the measurement is made from a large base capacitance; thus the 0% capacitance readings are made at a finite and measurable RH capacitance level. For example, the typical capacitance variability of the humidity-sensitive films is 0.2–0.5 pF for a 1% RH change, while this value is between 100 and 500 pF at 50% RH for the bulk capacitance at room temperature (Fontes 2005).

It is important to note that, in order to achieve a noticeable change in capacity, water in dielectric material must be in a free state, because only the free-water molecules have dielectric properties similar to those of liquid water, while the bound water exhibits icelike dielectric properties. According to Evans (1965) and Matzler and Wegmuller (1987), freshwater ice has a permittivity of 3.17–3.19, which is significantly less than that of water. With increasing temperature, the permittivity of ice increases in accordance with expression (Matzler and Wegmuller 1987):

$$\varepsilon' = 3.1884 + 0.00091 \cdot T \ (T \text{ in } °C) \quad (10.4)$$

Free water is usually understood as water held in capillaries. Taking into account the above, we come to the conclusion that the processes of capillary condensation should play an important role in the effects responsible for the sensitivity of capacitive humidity sensors. More detail description of the mechanism of water vapor interaction with polymers and ceramics one can find in Chapter 10 (Volume 1) and Chapter 2 (Volume 3) of our series.

10.1.1.1 Parallel Plate Structure

In classical capacitive sensors, a humidity-sensitive material is placed between the top and bottom electrodes (see Figure 10.1). This is the so-called parallel plate structure, which can be implemented on typical substrates of ceramic, glass, or silicon. The sensor material is made very thin to achieve a large signal change

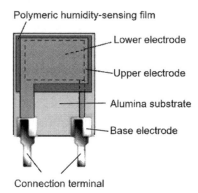

FIGURE 10.1 Capacitive-type humidity sensors with parallel plate structure. (Reprinted from www.vaisala.com)

with humidity. This permits the water to enter and leave easily and also promotes fast drying and easy calibration of the sensor. As a rule, inert materials, stable in the presence of water vapor, are used to form these electrodes. In order to ensure the access of water vapor to the humidity-sensitive material, the upper electrode must be water vapor permeable (i.e., porous). Implementation of this requirement is being achieved in various ways. The simplest solution is the deposition of a thin film of gold or platinum, which at a thickness of 10–20 nm has sufficient gas permeability. This approach is used in particular in Vaisala Corporation (Helsinki, Finland, www.vaisala.com).

Perforation of a metal film can also be used. This approach was used in the works of Jachowicz and Sentura (1981) and Korvink et al. (1993). Possible versions of electrode geometry for capacitive humidity sensors have been considered by Korvink et al. (1993). Some of these variants are presented in Figure 10.2. However, the most common version is the version with an "interdigital" electrode (Mamishev et al. 2004). The term *interdigital* refers to a digit-like or finger-like periodic pattern of parallel in-plane electrodes, used to build up the capacitance associated with the electrical fields that penetrate into a material sample. These designs allow for a uniform electrical distribution in the dielectric and give possibility for the vapor to diffuse freely into the dielectric. An intermediate solution is to provide the top electrode with small meshes.

Of course, in addition to interdigitated ones, there may be other options for forming the upper electrodes of capacitive sensors (see Figure 10.2). For example, Kim et al. (2009) have compared the characteristics of Al_2O_3-based humidity sensors with interdigitated and rectangular spiral-shaped electrode types (see Figure 10.3) and found that the rectangular spiral-shaped electrode type had a little more capacitance than the interdigitated type because the interdigitated type had an ineffective zone at the branching corners. Hysteresis of the rectangular spiral-shaped type tended to be larger than that of the interdigitated type, but the sensitivity of the rectangular

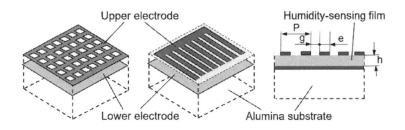

FIGURE 10.2 Possible configurations of top electrodes in parallel plate capacitors. (Idea from Korvink, J.G. et al., *Sens. Mater.*, 4, 323–335, 1993.)

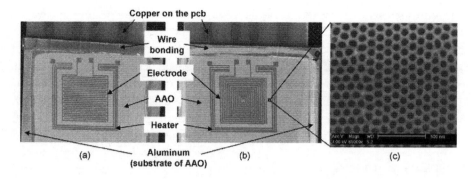

FIGURE 10.3 Humidity sensor devices with (a) interdigitated type and (b) rectangular spiral-shaped type, and (c) SEM micrograph of anodic aluminum oxide (AAO). (Reprinted with permission from Kim, Y. et al., Capacitive humidity sensor design based on anodic aluminum oxide, *Sens. Actuators B*, 141, 441–446, 2009. Copyright 2009, Elsevier.)

spiral-shaped type was higher than that of the interdigitated type. Therefore, Kim et al. (2009) believed that the sensors with rectangular spiral-shaped type electrodes could be efficient if hysteresis and nonlinearity were minimized by controlling the design factors. However, the indicated approach to contact forming complicates the technology. In addition, these devices have longer response times, because the humidity-sensitive material is partially covered by the upper capacitor electrode.

To solve this problem, Kang and Wise (1999, 2000) proposed to abandon the formation of a continuous layer of humidity-sensitive material. The capacitor sensor configuration option proposed by Kang and Wise (2000) is shown in Figure 10.4. In this configuration, the water vapor can penetrate into columnar-shape polyimides circumferentially, due to the higher surface area of the film; therefore, a high-speed adsorption occurs (Figure 10.5). Kang and Wise (2000) reported that a response time of 1.0 second with polyimide column diameters of 5 μm and a sensitivity of 30.0 fF/%RH have been obtained. According to the developers, this sensor also had linear humidity response. In this design, the recovery time was reduced due to presence of the heater. In addition, it has been verified that the integrated heater gives possibility to estimate the high RH values with an error of 2% RH.

The same approach was used by Laville and Pellet (2002b), Kim et al. (2010) and Choi et al. (2015). To improve the access of water vapor to polyimide (PI), they used the etching of the PI layer between electrodes on different depths, thus exposing a greater surface area of PI to moisture. As a result, Choi et al. (2015) managed to increase the sensitivity from 640 fF/%RH to 1500 fF/%RH, to decrease hysteresis from 1.60% to 0.37% and to improve the rate of response. Choi et al. (2015) have also shown that the most effective is etching not on the entire thickness of the PI film. The same effect was observed by Kim et al. (2010). After etching the PI layer, the sensitivity increased from 351 fF/%RH to 506 fF/%RH, and the response time decreased from 33 to 6 seconds. The sensors demonstrated also good linearity, weak temperature dependence (Figure 10.6a), and high stability of the parameters (see Figure 10.6b). Typically, plasma etching was used to etch the PI.

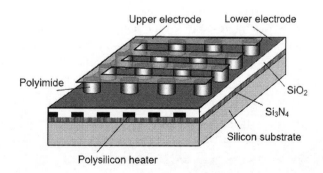

FIGURE 10.4 The capacitive humidity sensor with cylindrical polyimide columns with diameters of a few microns integrated with polysilicon heater. The polysilicon heater has been fabricated on the top of the Si_3N_4 layer underneath the sensor. (Idea from Kang, U. and Wise, K.D., *IEEE Trans. El. Dev.*, 47, 702–710, 2000.)

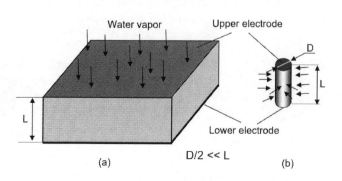

FIGURE 10.5 (a) Conventional structure and (b) high-speed structure.

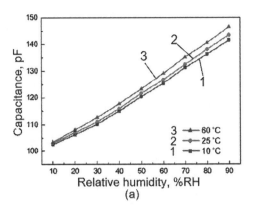
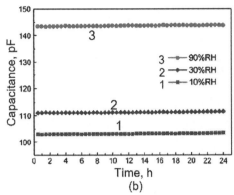

FIGURE 10.6 (a) Temperature dependence of the humidity sensor measured at 10, 25, and 60°C; (b) Stability of the humidity sensor at 10% relative humidity (RH), 30% RH, and 90% RH of the environment. (Reprinted with permission Kim, J.-H. et al., High-performance capacitive humidity sensor with novel electrode and polyimide layer based on MEMS technology, *Microsyst. Technol.*, 16, 2017–2021, 2010. Copyright 2010, Springer Science + Business Media.)

An interesting option of the parallel plate, capacitive humidity sensor was suggested by Lazarus et al. (2010, 2011, 2012). It is known that parallel plate, capacitive sensors have not been integrated with electronics, primarily because the fabrication of electrodes above the polymer layer, while making contact down to the metals in complementary metal–oxide–semiconductors (CMOS), significantly complicates the fabrication process. Lazarus et al. (2010, 2011, 2012) developed technology that allows resolving this problem. They proposed to remove the underlying substrate and arrange both interdigitated electrodes perpendicular to the main surface. To implement such a configuration, a standard CMOS-MEMS (micro-electro-mechanical systems) process was used. Sensitive polymer, polyimide, was added between two parallel electrodes by using a custom drop-on-demand inkjet system to deposit a polymer in the solution to an attached well, with polymer drawn into the released structure using capillary forces. Wicking allows well-controlled, repeatable material deposition without damaging the fragile released structure. The image of the sensor developed by Lazarus and Fedder (2012) is shown in Figure 10.7. However, the technological process of manufacturing such sensors is being significantly complicated, and there is no noticeable effect associated with the improvement of the sensor parameters.

10.1.1.2 Planar Capacitive Sensors with IDE

Experiment and modeling showed that capacitive sensors can be also designed based on an interdigitated electrode (IDE) structure in which interdigitated electrodes are under humidity-sensitive materials and form a planar structure (Lin et al. 1991; Park et al. 2001;

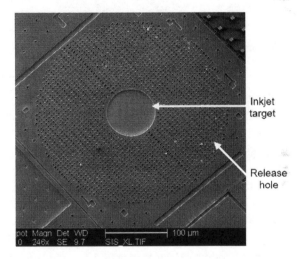

FIGURE 10.7 Sensor before inkjet deposition. (Reprinted with permission from Lazarus, N. and Fedder, G.K., Integrated vertical parallel-plate capacitive humidity sensor, *J. Micromechan. Microeng.*, 21, 065028, 2012. Copyright 2012, Institute of Physics.)

Chatzandroulis et al. 2011). In this case, humidity-sensitive material can be applied directly to the top of this IDE structure (Figure 10.8).

The operating principle of the planar interdigital sensor basically follows the rule of the two–parallel plate capacitor, in which electrodes open up to provide a one-sided access to the sensing material. The electrical filed lines of a parallel plate capacitor and an interdigital sensor are shown in Figure 10.9. Of course, in such a system, the sensor behavior becomes a function of system properties. In order to get a strong signal, the electrode pattern of the interdigital sensor can be repeated many times. The conventional interdigital

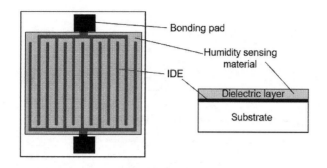

FIGURE 10.8 Capacitive humidity sensors with interdigitated electrodes. Typically, capacitive humidity sensors are composed of two interdigitated electrodes (IDEs) covered by a dielectric layer, which is sensitive to the humidity change.

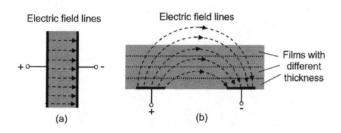

FIGURE 10.9 Electric field lines of (a) parallel plate capacitors, and (b) planar interdigital sensor.

sensor is shown in Figure 10.8. Models for IDE capacitors have been presented by Jachowicz and Sentura (1981). It was shown that the penetration depth of the electric filed lines varies for different pitch lengths. The pitch length of the interdigital sensor is the distance between two consecutive electrodes. The penetration depth can be increased by increasing the pitch length, but the electrical field strength generated at the neighboring electrodes will be weak.

The usage of IDE structure greatly simplifies the technology of manufacturing sensors and contributes to an increase in sensitivity, since the absence of an upper electrode facilitates diffusion and interaction of water vapor with humidity-sensitive material. In addition, the manufacturing technology of interdigitated capacitance-type sensors is easier to integrate in the standard CMOS processing. One good example of such sensors is found in the work of Dai (2007), who presented the fabrication of a humidity sensor with a microheater using 0.35 μm CMOS process and a postprocess.

However, for the effective operation of such sensors, a certain ratio between the thickness of the sensitive element and the geometry of the IDE is necessary (Hagleitner et al. 2001; Kummer et al. 2004; Igreja and Dias 2006) (see Figure 10.10), since the response of a capacitance-type IDE chemical sensor is influenced

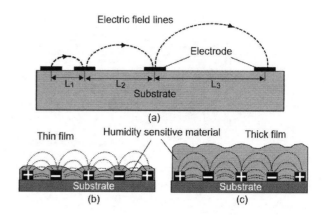

FIGURE 10.10 (a) Electrical field formed between positive and negative electrodes for different pith lengths; (b, c) the configuration of electric field lines in capacitance sensors with different thickness of sensitive (polymer) films. (Reprinted with permission from Kummer, A.M. et al., *Anal. Chem.*, 76, 2470–2477, 2004. Copyright 2004, American Chemical Society.)

by the periodicity of the electrodes and the thickness of the sensitive coating layer (Hagleitner et al. 2001; Kummer et al. 2004; Igreja and Dias 2006; Kummer and Hierlemann 2006). For example, it was established that, for a layer thickness of less than half the periodicity, swelling of the polymer on analyte absorption always results in a capacitance increase, regardless of the dielectric constant of the absorbed analyte. This is due to the increased polymer/analyte volume within the field-line region, exhibiting a larger dielectric constant than that of the air it displaces. On the other hand, the capacitance change for a polymer layer thickness larger than half the periodicity of the electrodes is determined by the ratio of dielectric constants of the analyte and polymer. If the dielectric constant of the polymer is lower than that of the analyte, the capacitance will be increased, and if the polymer dielectric constant is larger, the capacitance will be decreased. This means that, in the most optimal form, the thickness of the sensitive layer must be greater than the distance between the electrodes. It was also concluded that, for maximum effect, it is desirable that the electrodes have the maximum possible thickness with the smallest width.

Rivadeneyra et al. (2014, 2016) have shown that, in addition to interdigitated contacts, other electrode cofigurations can be used to fabricate capacitive sensors in the planar version. They are shown in Figure 10.11. As a result of numerical simulations and experimental studies, Rivadeneyra et al. (2016) have concluded that, in terms of sensitivity to RH, all sensors showed similar trends in frequency, but the highest value was achieved by serpentine electrodes (SREs), followed by spiral electrodes, IDE, and

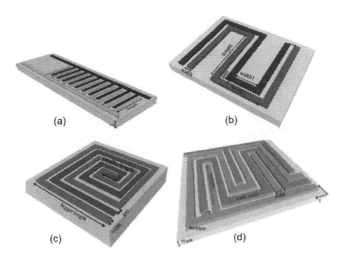

FIGURE 10.11 Layout of the designed electrodes: (a) IDE sensor, (b) meandered capacitor, (c) spiral electrodes, and (d) serpentine electrodes (SRE) capacitor. (Reprinted with permission from Rivadeneyra, A. et al., Printed electrodes structures as capacitive humidity sensors: A comparison, *Sens. Actuators A*, 244, 56–65, 2016. Copyright 2016, with permission from Elsevier.)

meandered electrodes, in this order. Looking at the thermal drift, meandered lines, IDE and SRE present values around ±0.5 fF/(mm$^2 \cdot$°C) whereas spiral electrodes show higher temperature sensitivities in frequency. Rivadeneyra et al. (2016) believe that this fact can be associated with the anisotropic behavior of this polyimide with temperature, which manifests a higher thermal dependence in this structure because its asymmetric electrodes compared to the other three layouts, whose electrodes are symmetric. Therefore, Rivadeneyra et al. (2016) have offered, depending on the specific application, a various configuration. For example, SRE and spiral electrodes should be selected when sensitivity is a critical factor. However, these two designs are more complex than the other two; therefore, their yield rates are lower than the other two configurations. Regarding IDE electrodes, experience has shown that these electrodes are the best compromise between sensitivity and a manufacturing yield. At the same time, meandered electrodes can be easily adapted to more complex designs. Sensor frequency is also important to choose the suitable electrode design. For example, Rivadeneyra et al. (2016) have established that spiral electrodes showed higher thermal dependence above 1 MHz.

It is of note that, despite the importance of the geometry of the electrodes and the configuration of the humidity sensor, its main parameters are controlled by the properties of the humidity-sensitive material itself. The same conclusion was made by Wang and co-workers (2005) while studying the capacitive-type humidity sensors based on nanometer barium titanate.

In addition to elements mentioned above, capacitive humidity sensors can have an additional layer applied to the upper electrode. Such an upper layer performs a protective function. Usually it is made from a porous polymer film and acts as a mechanical filter to prevent contamination by dust, impurities, and oils. This layer should be a water vapor–permeable.

Cantilever, metal-insulator-semiconductor (MIS), Schottky diode and *p-n* junction–based humidity sensors, as well as microwave-based humidity sensors and sensors on flexible substrate can also be of capacitance type. Their features will be described later in Chapters 13, 17, 20 and 22 of present volume.

10.1.2 Requirements for Materials Suitable for Use in Capacitive Humidity Sensors

One should note that the selection of the most suitable sensing material for a capacitance-type humidity sensor application is rather difficult, because this material must have a certain set of properties (Kummer et al. 2004; James et al. 2005; Chatzandroulis et al. 2011). As it follows from the consideration, the optimal humidity-sensitive materials designed for application in capacitive sensors should be a dielectric capable of absorbing the water vapor (i.e., to be hygroscopic and hygrophilic, to be stable even with water condensation, to have low dielectric permeability and high permeability to water vapor, and a porous structure with an appropriate set of pore sizes). It is also desirable that this material has a selectivity in the interaction with water vapor and does not react to the appearance of other gases and vapors.

Some of these requirements can be met by a large group of materials with dielectric properties. As a result, at present, such materials as polymers (Delapierre et al. 1983; Sakai et al. 1996; Harrey et al. 2002; Kraus et al. 2003; Dabhade et al. 2004; Matsuguchi et al. 2004), metal oxides (Seiyama et al. 1983; Khanna and Nahar 1984; Erson et al. 1990; Traversa 1995; Nahar 2000), porous silicon (Furjes et al. 2003), porous silicon carbide (Connolly et al. 2004), zeolites (Urbiztondo et al. 2011), and carbon-based materials (Llobet 2013) have been utilized as humidity-sensitive materials.

However, in practice, the main focus is on the development of polymer- and metal oxide–based capacitive humidity sensors. Therefore, in this chapter, only these materials are considered, while other humidity-sensitive materials, such as silica, zeolites, carbon, graphene, carbon nanotubes (CNTs), phosphates, black phosphorus, metal organic frameworks (MOFs), and porous semiconductors will be considered in special chapters (Chapters 1–9 in Volume 3 of our series) devoted specifically to these materials.

In general, polymers, in comparison with ceramics, cannot withstand high humidity levels and elevated temperatures for a long time (Yamazoe and Shimizu 1986), but the response characteristics are more linear that those of porous ceramics. In addition, from a technological point of view, polymers have the great advantage that they can be added at the end of the fabrication sequence of a CMOS process used during fabrication of modern humidity sensors.

10.2 POLYMER-BASED CAPACITANCE HUMIDITY SENSORS

The most widely used materials in capacitance-type humidity sensors are polymers (Chatzandroulis et al. 2011). They are particularly suitable for sensing water vapors because they exhibit a rapid reversible water vapor sorption and are easy to apply as thin or thick films by a variety of techniques. With reference to the polymer coating, the mechanisms of water interaction with polymer, which can be accompanied by the change in the capacitance, are shown in Figure 10.12. These changes are the result of changing the volume (swelling) and dielectric constant, taking place as a result of the incorporation of the water molecules into the polymer matrix during water absorption. This means that, in polymeric humidity sensors, all possible approaches to the development of humidity sensors can be realized.

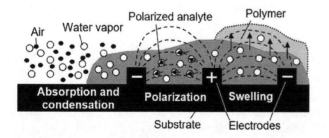

FIGURE 10.12 Schematic of sensing principle showing analyte absorption and the two relevant effects changing the sensor capacitance: the change of the dielectric constant and swelling. The interdigitated electrodes (+, −) on the substrate (*black*) are coated with a polymer layer (*gray*). *Big and small globes* represent analyte and air molecules, respectively. Analyte molecules are polarized in the electric field (*solid lines*). The analyte-induced polymer swelling is indicated with the *dashed lines* (right side). (Reprinted with permissions from Kummer, A.M. et al., *Anal. Chem.*, 76, 2470–2477, 2004. Copyrights 2004, American Chemical Society.)

10.2.1 MECHANISMS OF WATER INTERACTION WITH POLYMERS

The relation between the technology of making polymers and their hygroscopic properties has been presented in a number of publications (Mercer and Goodman 1991; Lin et al. 1993; Sakai 1993). Moisture absorbed in a polymer material modifies many physical material properties, including the dielectric constant, conductivity, modulus, impact strength, ductility, and toughness. Of course, the change in the dielectric constant is the most important for capacitive humidity sensors. As indicated before, the swelling effect can also be present in the interaction of polymers with water vapor. The mechanism of such interaction was considered earlier in Volume 1 (Chapter 10) of present series. However, when designing sensors for the market, permittivity-type sensors are preferred, since the thermoset polymers used for their development have significantly better parameter stability than thermoplastic polymers, in which the swelling effect can take place. The swelling and/or geometrical deformation of the polymer film causes the large temperature coefficient and lack of long-term stability and reproducibility of the sensing behavior. It seems that the cross-linked hydrophobic polymer films are adequate for a practical capacitive-type humidity sensors (Sadaoka 2009). Thermoset polymer-based capacitive sensors, as opposed to thermoplastic-based capacitive sensors, provide better resistivity against chemical liquids and vapors such as isopropyl, benzene, toluene, formaldehydes, oils, common cleaning agents, and ammonia vapor. This type of capacitive sensors also has a wider temperature range (up to 200°C), lower hysteresis and temporal drift, and no problems with high humidity measurement or saturation. However, cantilever- and membrane-based humidity sensors for effective operation nevertheless require thermoplastic polymers, since their work is based on exactly the swelling effect.

Moisture is exchanged between the material and the environment until a steady state is reached. In steady state, the net gain and loss of moisture becomes zero. The equilibrium moisture content is a function of humidity, temperature, the type of material, and the moisture history of the material. As is known, the absorbed water in a material exists in several different forms: chemisorbed, physisorbed, and condensed states. In the first state, water molecules are chemically bound to the constituents of the material; in the second state, they are held by the surface forces; and in the third state, water is condensed inside small pores present inside the material.

In polymers, the water molecules in the polymer are bonded to the hygroscopic groups of the polymer molecules by weak van der Waals's interactions

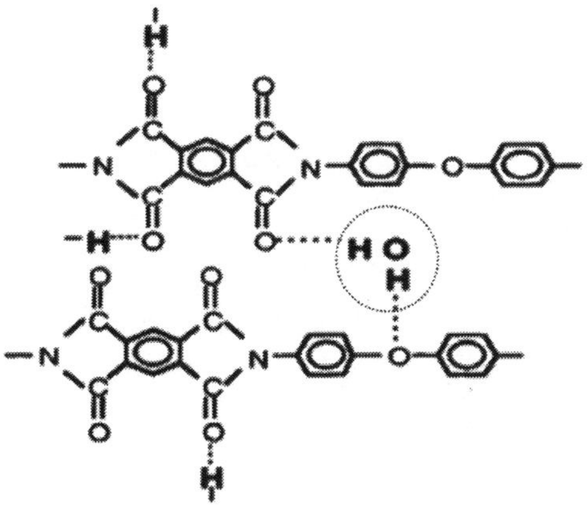

FIGURE 10.13 Possible bonding sites in polyimides for water molecules. (Idea from Melcher, J. et al., *IEEE Trans. Electron. Insulation*, 24, 31–38, 1989.)

(Kämpfner 2013). That uptake of water molecules by the hydroactive sponge-like polymer structure results in a change of the relative dielectric permittivity ε_r of the polymer layer. Water molecules can also condense in microvoids, depending on the humidity level. Figure 10.13 shows possible bonding sites for water molecules in polyimides; they are bounded either to the carbonyl group or to the oxygen of the ether linkage (Melcher et al. 1989).

The response of the polymer-based, capacitive-type sensor to the change in the amount of water vapor in its ambient gas is mainly linearly proportional to the change in relative humidity of the gas, because the driving force of the process of water vapor solubility in the polymer is the free energy for adsorption (Majewski 2016), G:

$$G = RT \cdot \ln P/P_s \qquad (10.5)$$

where:
> R is the universal gas constant, T is the absolute temperature, P is the partial pressure of water vapor in gas, and P_s is the saturation pressure of water vapor in the gas.

As the RH is defined as $\varphi = P/P_s$, it is directly related to the amount of water molecules adsorbed in the polymer.

It has been shown by measuring the equilibrium moisture content in a material as a function of the water vapor pressure (the plot of this measurement result is called the *equilibrium moisture content isotherm*) that moisture starts to condense as the RH level becomes higher (Yang et al. 1985, 1986). Experimental results show that the isotherm curve is concave with respect to the vapor pressure axis at the lower vapor pressures, while it is convex to the axis at higher pressures. The transition from the concave to the convex form is explained to be due to clustering of water molecules inside the material; the transition point at which moisture clustering starts to occur depends on the material and the temperature.

It is important that polymer-based capacitive humidity sensors are characterized by the response that is close to linear. This linear response is due to a very simple principle, described as follows: For insulating polymers, the absorbed water, the weight of which is proportional to RH, occupies the free space between the polymeric molecules. Therefore, the change of the dielectric constant of the hygroscopic polymer is linearly proportional to the amount of water absorbed. The description of such models can be found in Grange et al. (1987), Matsuguchi et al. (1993), Liu et al. (1995), Ralston et al. (1996), Dokmeci and Najafi (2001), Harrey et al. (2002), and Ingram et al. (2003).

Moisture transport inside a material is due to diffusion as well as reaction. At higher temperature, however, reaction occurs at a very slow rate so that the transport due to this mechanism can be neglected. In the temperature range at which humidity sensors are normally operated (> −50°C), moisture transport is due to diffusion through the microvoids.

10.2.2 Implementation of Polymer Humidity Sensors

At present, when developing capacitive humidity sensors, a large number of different polymers have been tested. Some of the polymeric materials used for capacitance sensor design are listed in Table 10.4. The thickness of polymeric films can vary from 100 nm to hundreds of micrometers.

As indicated, capacitive sensors require polymers, which should have dielectric properties and be porous and stable in the presence of water. The porosity of the humidity-sensitive polymer is important for obtaining quick responses. Simplicity of processing is also one of the criteria for selecting polymers for use in sensors. The important requirements for fabricating capacitive-type humidity sensors are also a low hygroscopic and the rigid structure of the sensing polymer thin film. It is also believed that polymers for capacitive humidity sensors should preferably be hydrophobic but partly hygroscopic materials in order to absorb moisture (Sakai et al. 1996). In other words, the polymers for capacitive sensors should be both nonionic and highly polar macromolecules.

The experiment showed that polyimides best correspond to these requirements. Figure 10.14 illustrates the chemical formula of a Kapton, a typical polyimide.

TABLE 10.4
Capacitance-Type, Humidity Polymer-Based Sensors

Sensor Type	Sensitive Material	References
Permittivity-based sensors		
IDE	Polyimide	Chandran et al. (1991); Boltshauser and Baltes (1991); Boltshauser et al. (1991); Shibata et al. (1996); Park et al. (2001); Laville and Pellet (2002a), (2002b); Dai (2007); Lee et al. (2008)
	Benzocyclobutene (BCB)	Laville and Pellet (2002a, 2002b)
	Poly CA; poly CAB; MMA; PVP, and mixtures thereof	Oprea et al. (2008)
	BEHP-co-MEH:PPV+PAASS	Sajid et al. (2017)
	BBCB	Pecora et al. (2008)
	Polar PEUT; PDMS	Kummer et al. (2006)
	PHEMA; PMMA; PHEMA; EPR	Kitsara et al. (2006)
Parallel plate	Polyimide	Kang and Wise (2000); Laville and Pellet (2002a, 2002b); Kim et al. (2010); Choi et al. (2015)
	Heteropolysiloxane	Endres and Drost (1991)
Bimorph-based sensors		
Cantilevers	Polyimide	Chatzandroulis et al. (2002)
Membranes	PVAc; PHEMA; PDMS; PMMA; EPN	Chatzandroulis et al. (2004)
	Alkane thiol coatings with COOH and NH_3 end groups	Lim et al. (2007)
	PAAM	Park et al. (2007)
	PIB	Park et al. (2008)
MEMs and CMOS-devices	Polyimide	Dokmeci and Najafi (2001); Qui et al. (2001); Gu et al. (2004).

Note: APTMOS, 3-amino-propyl-trimethoxysilane; AuHFA, fluoroalcohol-coated gold nanospheres; BEHP-co-MEH:PPV, Poly{[2-[2_,5_-bis(2__-ethylhexyloxy)phenyl]-1,4-phenylenevinylene]-co-[2-methoxy-5-(2_-ethylhexyloxy)-1,4-phenylenevinylene]}; BBCB, bis-benzocyclobytene; CA, cellulose acetate; CAB, cellulose acetate-butyrate; CEE, chloroethyl ether; DIMP, diisopropyl methylphosphanate; DMMP, dimethyl methylphosphanate; EPR, epoxidized novolac; IPA, isopropanol; OV-225, cyanopropyl methyl phenylmethyl silicone; OV-275, dicyanoallyl silicone; PAAM, polyallylamine hydrochloride; PAASS, Poly(acrylic acid) partial sodium salt; PAPPS, propylaminopropyl polysiloxane; PDMS, poly(dimethyl siloxane); PECH, polyepichlorohydrin; PEG, polyethylene glycol; PEI, polyethyleneimine; PEO, polyethylene oxide; PEVA, polyethylene-co-vinyl acetate; PEUT, poly(etherurethane); PHEMA, poly(2-hydroxyethyl methacrylate); PIB, polyisobutylene; PMMA, poly(methylmethacrylate); PTMOS, propyl-trimethoxysilane; PVA, polyvinyl alcohol; PVP, polyvinylpyrrolidone; SXFA, siloxanefluoroalcohol.

Therefore, the first capacitance-type polymer humidity sensors were designed in the early 1980s on the base of polyimide films (Sheppard et al. 1982; Delapierre et al. 1983; Denton et al. 1985). Subsequently, most commercial sensors were also manufactured on the basis of this material. As is known, these polymers are insoluble in water. The polyimides are chemically inert in cured form and have a good resistance to chemical corrosion (Delapierre et al. 1983; Ralston et al. 1996). In addition,

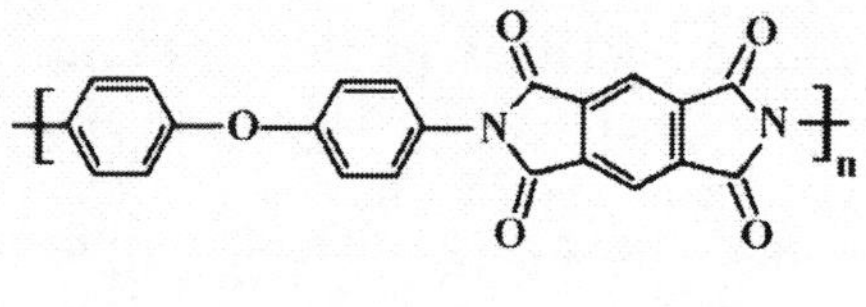

FIGURE 10.14 Kapton, a typical polyimide.

polyimides possess required toughness and thermal stability. Polyimides are thermally stable up to temperatures around 450°C. Moreover, it is a perfect planarizer used to planarize irregular surfaces. It also has a low relative permittivity and high breakdown voltage. Experiments show that the dielectric constant of polyimide films changes from approximately 3 to 4 as the RH changes from 0%RH to 100%RH (Melcher et al. 1989; Noe et al. 1991). In addition, the dielectric constant change with respect to the humidity change is almost linear, especially within the 20%RH–70%RH range. Polyimides also absorb water reversibly, with little or almost no hysteresis. The diffusion constant is often very large, leading to the fast response time. According to Okamoto et al. (1992) and Xu et al. (2007), the diffusion coefficient of water in the polyimide films is ranged from 3×10^{-8} to

5×10^{-9} cm²/sec, depending on the type of polyimide. Being fully compatible with silicon processing technology, polyimide deposition can be performed at the end of a CMOS fabrication process; therefore, humidity sensors, which are monolithically integrated with the readout circuit, can be obtained (Qui et al. 2001).

It thus becomes clear why various polyimides in the form of films and resins have been used extensively in RH sensors of capacitance type (Yamazoe and Shimizu 1986; Mercer and Goodman 1991; Kuroiwa et al. 1993; Ralston et al. 1996; Matsuguchi et al. 1998b; Jachowicz and Weremczuk 2000; Kang and Wise 2000; Laville and Pellet 2002a, 2002b). For example, Ralston et al. (1996) characterized the performance of 13 polyimide films in RH-sensing applications. In Figure 10.15, the hysteresis loop for the Kapton® HN (from DuPont™)-based sensor is shown as a typical example. The humidity-sensing characteristics of polyimide-based devices are excellent, except for instability at high humidities and high temperatures (Ralston et al. 1991; Sakai et al. 1996). Sakai et al. (1996) believed that these disadvantages are due to the following reasons: (1) Unreacted sites are likely to be present, and (2) microvolds may arise from volatile byproducts such as water.

Hysteresis is another serious problem in polymer-based capacitive sensors. As a rule, serious hysteresis occurs at high values of RH. One should note that hysteresis is a common problem for virtually all kinds of capacitive humidity sensors, due to the slower diffusion time of moisture-sensitive films while dehumidifying.

According to Sakai et al. (1996) and Matsuguchi et al. (1998a), hysteresis comes from the clusters of absorbed water inside the bulk polymer. The formation of clusters indicates that hygroscopicity of some polymers is too high, and relatively large voids exist in the polymeric structures. The water clusters may also deform the polymers due to formation of ink-bottled-type pores in polymer films and shorten the lifetimes of the sensors. New designs based on polyimide oligomers improved the parameters of the sensors (Matsuguchi et al. 1993). It was established that these oligomers polymerize, forming a cross-linking bond without evolving byproducts. The cross-linking leads to a rigid hydrophobic film, which does not contain microvoids and is chemically and thermally stable. As a result, improved polyimides have been successfully applied on Si chips as humidity-sensitive dielectrics with high mechanical strength, temperature capability, and resistance to chemical attack. The best sensors have almost linearity in the signal with RH over the range 10%–90% RH, hysteresis less than 1%, and response time less than 10–30 seconds. The change of 25%–28% normalized capacitance has been reported as the RH value was increased from 0%–100%. Long-term stability (see Figure 10.16) and a good durability against high RH (90% RH–95% RH) and dew drops have also been claimed. In addition, the sensor proved to be robust to chemicals (Matsuguch et al. 1998b).

Some modified polyimide films, capable of enlarging the active surface and improving the measuring sensitivity, have also been reported (Yang et al. 2002; Laconte et al. 2003; Kim et al. 2004; Lei et al. 2015). For example, Kim et al. (2004) developed a humidity sensor using a

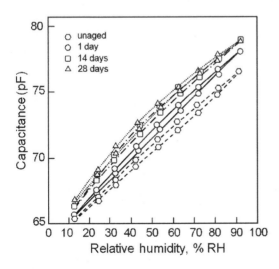

FIGURE 10.15 Capacitance versus relative humidity (RH) of Kapton HN-based humidity sensor in the unaged state, and after aging for 1, 14, and 28 days aging at 85°C/85% RH. (Reprinted with permission from Ralston, A.R.K. et al., A model for the relative environmental stability of a series of polyimide capacitance humidity sensors, *Sens. Actuators B*, 34, 343–348, 1996. Copyright 1996, Elsevier.)

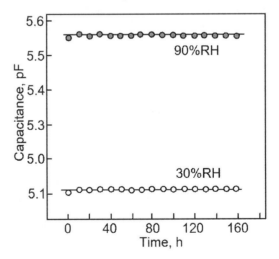

FIGURE 10.16 Measured capacitance as a function of time in 30% and 90% RH at 27°C in a week. (Reprinted with permission from Gu, L. et al., A novel capacitive-type humidity sensor using CMOS fabrication technology, *Sens. Actuators B*, 99, 491–498, 2005. Copyright 2005, Elsevier.)

polyimide film without hydrophobic elements. Polyimide is a polymer material that can be obtained by thermal or chemical polymerization of polyamic acid (PAA) composed of two monomers, diamine and dianhydride, and aprotic solvent. Polyimide has been widely used for interlayer dielectrics in electronic industry where a low dielectric constant and insensitivity to moisture are imperative. Therefore, such elements as fluorine or chlorine have been included in diamine or dianhydride. To exclude these elements from the PI, Kim et al. (2004) suggested a different route. The polyimide film was obtained by synthesizing and thermally polymerizing polyamic acid composed of m-pyromellitic dianhydride, phenelenediamine, and dimethylacetamide. The synthesis procedure and chemical structure of the polyimide film is shown in Figure 10.17. Kim et al. (2004) reported that synthesized PI had excellent properties, such as thermal stability (glass transition temperature > 450°C) and chemical resistance, which make it fully compatible with general microfabrication processes. Regarding the humidity-sensitive properties, the measurement has shown that the sensitivity of these sensors was 0.75 pF/%RH over a range from 10% to 90%RH, hysteresis of 0.77% over the same %RH range, and maximum drift of 0.25% at 50% RH (see Figure 10.18). Lei et al. (2015) suggested forming a humidity-sensitive PI layer as nanofiber forests. Lei et al. (2015) formed nanofibers with super hydrophilicity by using a plasma-stripping technique. But the sensitivity of these sensors did not exceed 0.14 pF/%RH and was nonlinear.

Polyimide layers can be patterned photolithographically by using a photoresist. More important, however, is that there are polyimide precursors or soluble polyimides that possess photoresistant properties (Boltshauser et al. 1991). Moreover, these photosensitive polyimides

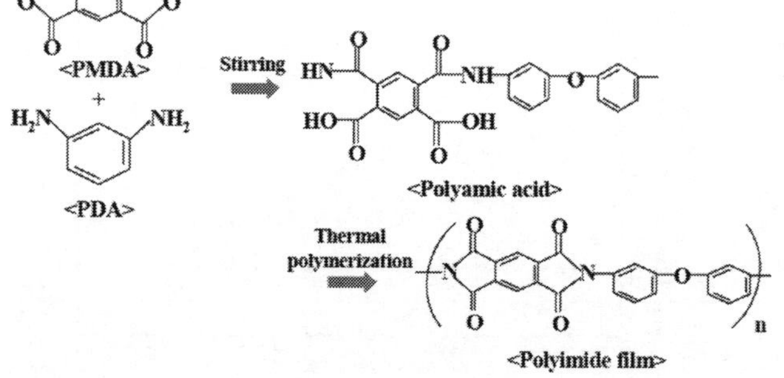

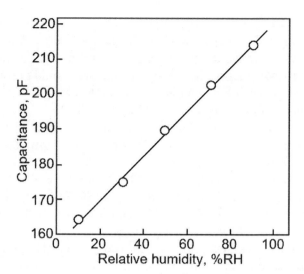

FIGURE 10.18 Capacitance versus relative humidity (RH) of humidity sensor with modified polyimide (PI). (Reprinted with permission from Kim, Y.-H. et al., A highly sensitive capacitive-type humidity sensor using customized polyimide film without hydrophobic elements, *Sens. Mater.*, 16, 109–117, 2004. Copyright 2004, Myu Scientific Publishing as open access.)

FIGURE 10.17 The synthesis sequence and chemical structure of the modified polyimide film. (Reprinted with permission from Kim, Y.-H. et al., A highly sensitive capacitive-type humidity sensor using customized polyimide film without hydrophobic elements, *Sens. Mater.*, 16, 109–117, 2004. Copyright 2004, Myu Scientific Publishing as open access.)

provide higher yields at lower cost because of fewer and safer processing steps. They can easily be employed in common photo techniques. Just as in the case of conventional photoresists of the negative type, light exposure through a mask gives rise to large solubility differences by crosslinking in the exposed area directly in the layer to be patterned (i.e., direct patterning) (Horie and Yamaskito 1995). This is important for high resolution. After development with the use of a solvent and subsequent curing, the appropriate polyimide patterns are obtained for the described application. Thus, the usage of nonphotosensitive polyimide, which should be patterned by dry or wet etching using a photoresistive layer as a mask, requires the additional process steps of spin-coating and patterning of photoresist. The use of photosensitive polyimide requires only two steps of the process: ultraviolet exposure and development. Since the photosensitive type needs considerably fewer processing steps than for the nonphotosensitive type, this type of polyimide seems to be the most promising for use as the sensitive layer in a humidity sensor fabricated using semiconductor and micromachining technologies.

With regard to adhesion properties of polyimide films, excellent adhesion of the polyimide to metals (and vice versa) and to the cured polyimide film beneath is an important advantage of such coatings. Difficulties arise only when good mechanical properties and good adhesion are required, since polyimides with rigid and linear molecular structure show low stress but have weaker adhesive forces than flexible ones, which, in turn, have poorer mechanical properties (Numata et al. 1991). The exposure of polyimide layers to water at elevated temperatures also lowers the adhesion of polyimides, because water can hydrolyze chemical bonds between the polyimide and the substrate surface. However, adhesive forces can be increased by using special adhesion prompters for various kinds of polyimides and substrates (e.g., amino-organosilane), or by plasma treatment of the substrate surface (Toray 1992). Some new types of commercially available polyimides already contain an integrated adhesion promoter: Pimel G-X Grade (Asahi 1994), Pyralin PI 2700 (DuPont 1994), and probimide 7000, 7500 (OCG 1994). New generations of polyimides have passed the tape test, even when wafers coated with the polyimide on a silicon nitride surface were boiled at 121°C and 2 atm for 400 hours (DuPont 1994) and 500 hours (Asahi 1994; OCG 1994). Of course, these data depend on the interface polyimide/substrate.

Experimentation has shown that cellulose acetate (CA) or cellulose acetate butyrate (CAB) (Thoma et al. 1979; Matsuguchi et al. 1998; Sadaoka et al. 1988), polymethyl methacrylate (PMMA) and polyethylene terephthalate (PETT) (Roman et al. 1995; Meanna Pérez and Freyre 1997), polyphenylacetylene (PPA) (Furlani et al. 1992), polydimethylphosphazene (PDMP) (Anchisini et al. 1996), polyethersulphone (PES) (Harrey et al. 2002), polysulfone (PSF) (Kuroiwa et al. 1995), hexamethyldisilazane (HMDSN) (Kraus et al. 2003), poly{[2-[2',5'-bis(2"-ethylhexyloxy)phenyl]-1,4-phenylenevinylene]-co-[2-methoxy-5-(2'-ethylhexyloxy)-1,4-phenylene vinylene]}+Poly(acrylic acid) partial sodium salt (BEHP-co-MEH:PPV+PAASS) (Sajid et al. 2017), and [bis(benzo cyclobutene)] (BCB) (Tetelin et al. 2003; Zampetti et al. 2009) also can be successfully used for these purposes (Sakai et al. 1996; Yang et al. 2006). The relationship between the permittivity of cellulose derivatives and humidity is shown in Figure 10.19. It is seen that the change in permittivity is strong. For example, Sajid et al. (2017), using BEHP-co-MEH:PPV+PAASS (see Table 10.4) as sensitive material, designed a humidity sensor with sensitivity 34 pF/%RH in the range of 0% and 80% RH. The sensor response was also very fast. The response and recovery time did not exceed 9 seconds. However, the capacitive response was highly nonlinear. The main changes in capacity occurred at a humidity above 50% RH.

However, many problems such as hysteresis, nonlinearity of characteristics, stability at high humidities and at high temperature, and durability on exposure to some kinds of organic vapors have yet to be solved. It is very difficult to find a compromise between these parameters, since very often, optimization of technology

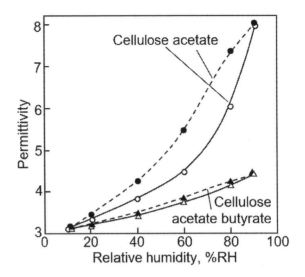

FIGURE 10.19 Humidity dependence of permittivity: (○, ●) cellulose acetate, (△, ▲) cellulose acetate butyrate. The humidity dependence of the permittivity observed at 1 kHz. *Open symbols*, humidification. *Closed symbols*, desiccation. (Reprinted with permission from Sadaoka, Y., Capacitive-type relative humidity sensor with hydrophobic polymer films, In: Comini, E. et al. (eds.), *Solid State Gas Sensing*, 109–151, 2009. Copyright 2009, Springer Science+Business Media.)

aimed at improving one parameter is accompanied by a deterioration of the other. For example, as shown in Figure 10.20, increasing the sensitivity of sensors based on BEHP-co-MEH:PPV+PAAPSS is accompanied by increased hysteresis and nonlinearity of characteristics.

Sadaoka et al. (1988) have also established that the hysteresis for the various cellulose derivatives–based sensors was caused by the formation of clusters of sorbed water molecules. As with polyimides, in some cases, cross-linking of the polymer chains was found to depress the clustering of water. For example, Matsuguchi et al. (1995), using this approach, strongly reduced hysteresis in polyvinyl chloride acetate (PVCA)-based humidity sensors. In addition, the cross-linked polymers were durable in the presence of organic vapors. Roman et al. (1995) established that the copolymerization of PMMA with cross-linking agents enhanced the sensitivity and response time of the films; on the other hand, higher durability and lower hysteresis were achieved against acetone vapor. Matsuguchi et al. (2004) developed a capacitive humidity sensor based on a PMMA cross-linked with divinylbenzene (DVB). Thanks to the rigid cross-linked structures in the polymer, the irreversible increase in the volume of the sensing layers caused by swelling was prevented, which improved the long-term stability of the capacitive humidity sensors. However, it should be recognized that, as a rule, in terms of their parameters, the above sensors are inferior to polyimide-based sensors.

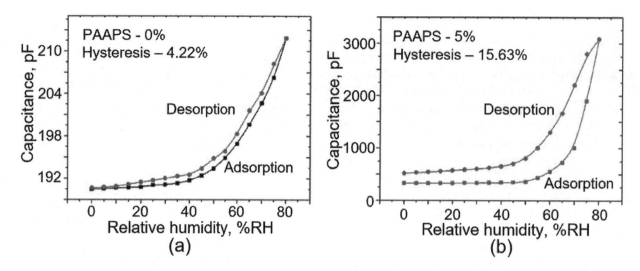

FIGURE 10.20 Capacitive response of sensors at 1 kHz showing hysteresis effect for adsorption and desorption cycles in dependence on the concentration of poly(acrylic acid) partial sodium salt (PAAPS) in the polymer: (a) 0%; (b) 5%. (Reprinted with permission from Sajid, M. et al., Highly sensitive BEHP-co-MEH:PPV + Poly(acrylic acid) partial sodium salt based relative humidity sensor, *Sens. Actuators B*, 246, 809–818, 2017. Copyright 2017, Elsevier.)

Earlier, porosity and good permeability for water vapor were noted to be important conditions for the development of high-speed sensors (Comyn 1985). Various methods are used to solve this problem. For example, Delapierre et al. (1983) have proposed the tensile-stressed fracture technique. A porous chromium electrode was evaporated under conditions such that the sensitive film was tensile stressed, and these stresses caused the creation of a high volume of cracks in the film, resulting in higher water vapor permeability rates by several orders of magnitude compared to nonporous polymers.

As for the selectivity of capacitive humidity sensors, it must be recognized that all sensors, regardless of their type, are not selective. This means that they can react to the appearance of other gases and vapors. The sensors react especially strongly to solvent vapors, which, like water, are inclined to capillary condensation. For example, Figure 10.21 shows the effect of water vapor and various solvents on the readings of poly(ether urethane) (PEUT)-based CMOS capacitive sensors (Kummer et al. 2004). Of course, due to the abnormally large dielectric constant of the water, water

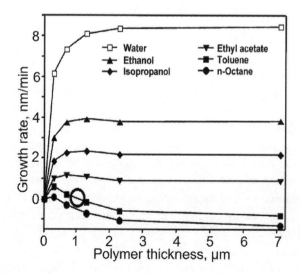

FIGURE 10.21 Measured dependence of the physical sensitivity, $\Delta C/\varphi_A$, on the layer thickness, h, for analytes with various dielectric constants. Low ε_A: n-octane (1.93) and toluene (2.36). High to very high ε_A: ethyl acetate (5.88), 2-propanol (18.5), ethanol (24.3), and water (76.6). (Reproduced with permission from Kummer, A.M. et al., *Anal. Chem.*, 76, 2470–2477, 2004. Copyright 2004, American Chemical Society.)

has the strongest influence on the sensor readings. But the effect of other gases that may appear in the surrounding atmosphere should also be considered.

Discussions on more general problems associated with the use of polymer films in humidity sensors can be found in Harsanyi (1995) and Osada and De Rossi (2000).

10.3 METAL OXIDE–BASED CAPACITIVE HUMIDITY SENSORS

10.3.1 Metal Oxides in Humidity Sensors

At present, numerous metal oxides were tested as humidity-sensitive materials for capacitive humidity sensors (see Table 10.5). It was shown that usually, metal oxides designed for application in capacitive sensors are characterized by high mechanical strength, resistance to chemical attack, and thermal stability. In addition, because of the immense surface-to-volume ratio and the abundant void fraction, very high sensitivities can be obtained with porous metal oxide ceramics. Generally, ceramic humidity sensors are more chemically and thermally stable than the polymeric types. Reviews of ceramics for humidity

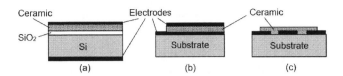

FIGURE 10.22 Possible configurations of metal oxide–based capacitive humidity sensors: (a) metal-insulator-semiconductor (MIS) structure; (b) conventional parallel plate structure; (c) structure with interdigitated electrodes.

sensors have been presented by Seiyama et al. (1983), Fagan and Amarakoon (1993), Traversa (1995), Chen and Lu (2005), and Blank et al. (2016).

As a rule, capacitive sensors are made in the form of a parallel plate structure, which can have two configurations (see Figure 10.22), either top electrode-porous ceramic-SiO_2-silicon-ohmic contact electrode (Li et al. 1995, 1997, 1999, 2000), or the top electrode-porous ceramic-bottom electrode. At present, sensors with interdigitated electrodes are also being developed and tested. The upper electrode, as in the case of polymeric materials, must be porous (i.e., permeable to water vapor).

TABLE 10.5
Metal Oxides Used in Capacitive Humidity Sensors

Type of Oxides	Oxides	References
Binary oxides	ZnO	Qi et al. (2008); Li et al. (2013)
	Al_2O_3	Nahar and Khanna (1998); Mai et al. (2000); Nahar (2000); Basu et al. (2001); Mistry et al. (2005); Saha et al. (2005a, 2008); Islam et al. (2012); Kim et al. (2012); Mahboob et al. (2016)
	WO_3	Wenmin and Wlodarski (2000)
	SnO_2	Wang et al. (2016)
	CeO_2	Xie et al. (2015)
	SiO_2	Wang et al. (2005a)
Mixed oxides and composites	ZnO/TiO_2	Gu et al. (2011)
	TiO_2–WO_3	Faia et al. (2009); Rocha and Zanetti (2011)
	$ZnWO_4$	You et al. (2012)
Perovskite	$BaTiO_3$	Li et al. (1995); Yuk and Troczynski (2003); Wang et al. (2005b); He et al. (2010)
	$LaFeO_3$	Zhao et al. (2013)
Layered perovskite	Bi_2MO_6	Zheng et al. (2010)
Composites based on perovskites	$BaMO_3$	Viviani et al. (2001)
	$BaTiO_3$:M	Viviani et al. (2001); Agarwal and Sharma (2002)
	$LaCo_{0.3}Fe_{0.7}O_3$	Wang et al. (2011)
	$Sr_{1-x}La_xTiO_3$	Li et al. (1997)
	$SrNb_xTi_{1-x}O_3$	Li et al. (1999)
Spinel-type	$MgAl_2O_4$	Ahn et al. (2005)
Composites based on spinel-type oxides	$ZnCr_2O_4$–K_2CrO_4	Bayhan and Kavasoglu (2006)
	$MgCr_2O_4$–TiO_2	Saha et al. (2005b)
Other metal oxides	$CaCu_3Ti_4O_{12}$	Li (2016)
	Ca, Mg, Fe, Ti-oxides	Mahboob et al. (2016)

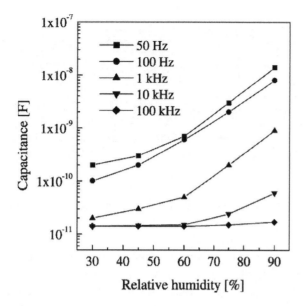

FIGURE 10.23 Capacitance of BaTiO$_3$ thin film vs relative humidity (RH) for various measurement frequencies. (Reprinted with permission from Yuk, J. and Troczynski, T., Sol–gel BaTiO$_3$ thin film for humidity sensors, *Sens. Actuators B*, 94, 290–293, 2003. Copyright 2003, Elsevier.)

As for the measurement modes, as indicated earlier, measurements are made at low frequencies, typically 0.1–1.0 kHz (Figure 10.23). The capacitance decrease with increasing frequency at the same RH is attributed to the increased difficulty for dipoles of water molecules to re-orient with increasing frequency (Callister 1994).

10.3.2 Mechanisms of Water Interaction with Metal Oxides

It is worth noting that, based on metal oxides, it is possible to develop humidity sensors of permittivity type only, since the interaction with water is not accompanied by a change in the size of the sensing element. Such humidity sensors belong to the permittivity-type sensors.

Moreover, in such sensors, unlike polymeric sensors, water can be found only on the surface of crystallites or in intercrystalline space. Taking this into account, it becomes clear that the main processes controlling the capacitive properties of metal oxides are the adsorption of water vapor followed by their condensation in the pores (Fleming 1981). It was experimentally confirmed that the dielectric constant of the ceramic is proportional to the water content (Li et al. 2000).

The principle of adsorption mechanism of H$_2$O on activated metal oxide sites has received abundant attention since as early as the 1960s (Yates 1961; Fripiat and Uytterhoeven 1962). When studying the surface behavior and the existence of hydroxyl functional groups, the infrared spectroscopy technique was frequently applied by several researchers to disks of iron oxide and silica (Blyholder and Richardson 1962; Fripiat et al. 1965). The mechanism of water molecule interaction on metal oxide surfaces was discovered and explained in the studies of silica gel and hematite (α-Fe$_2$O$_3$) from approximately 1968 to 1971 (Anderson and Parks 1968; Hertl and Hair 1968; Hair and Hertl 1969; Morimoto et al. 1969; McCafferty and Zettlemoyer 1971).

Proposed are the following mechanism models of water adsorption (Traversa 1995). As dry oxides are kept in contact with humid air, in the first stage of the interaction, a few water vapor molecules are chemically adsorbed (chemisorption) at the neck of the crystalline grains on activated sites of the surface, which is accompanied with a dissociative mechanism of vapor molecules to form hydroxyl groups (two hydroxyl ions per water molecule). As an interaction between the surface ions of the grain necks and the adsorbed water, the hydroxyl group of each water molecule is adsorbed on metal cations that are present in the grains' surfaces and possess high charge carrier density and strong electrostatic fields, thus providing mobile protons. The protons migrate from site to site on the surface and react with the neighbor surface O$_2^-$ groups (oxygen) to form a second

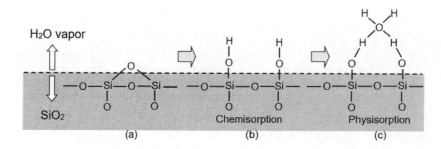

FIGURE 10.24 Adsorption sites on the silica surfaces and schematic presentation of water-vapor adsorption mechanisms on the SiO$_2$ surface. (Idea from Fripiat, J.J. and Uytterhoeven J., *J. Phys. Chem.*, 66, 800–805, 1962; Voorthuyzen, J.A. et al. *Surf. Sci.*, 187, 201–211, 1987.)

hydroxyl (OH−) group (McCafferty and Zettlemoyer 1971; Nitta and Hayakawa 1980). As a result, the surface of most metal oxides is covered with hydroxyl groups when exposed to humid atmospheres (Figure 10.24). Thus, the chemisorbed layer is the first formed layer, so once it has formed on the surface, it will not change further by exposure to humid air.

As a second stage, after chemical completion of the first layer, subsequent water vapor layers are physically adsorbed (physisorption) on the first formed hydroxyl layer via hydrogen bonding. The physisorbed water easily dissociates to form H_3O^+ because of the high electrostatic fields in the chemisorbed layer. As water vapor continues to increase in the surface, another water molecule adsorbs and an extra layer forms on the first physisorbed layer via double hydrogen bonding to two neighboring hydroxyl groups (Hair and Hertl 1969; Thiel and Madey 1987). This means that the physisorption changes from monolayer to multilayer. Finally, by forming more layers, a large amount of water molecules is physisorbed on the necks and flat surfaces, hence singly bonded water vapor molecules become mobile and able to form continuous dipoles and electrolyte layers between the electrodes, resulting in an increased dielectric constant and bulk conductivity (Kurosaki 1954). Water molecules in the succeeding physisorbed layers are only singly bonded and form a liquid-like network (McCafferty and Zettlemoyer 1971). Therefore, multilayer formation due to the physisorption of water vapor can be certified by observing the increase in the permittivity (Nitta and Hayakawa 1980). Figure 10.25 shows a multilayer structure of adsorbed water vapor molecules on the surface of iron oxide (McCafferty and Zettlemoyer 1971). A similar mechanism of hydroxylation, and hence multilayer formation on the surface of silica, was reported by Hair and Hertl in 1969.

It is important to consider that physisorption of water vapor molecules can only be accomplished at temperatures below 100°C, while at higher temperatures up to approximately 400°C, only chemisorption is responsible for the surface interaction of hydroxyls in ceramics (Egashira et al. 1978, 1981). Surface hydroxyls start to desorb at approximately 400°C.

As mentioned earlier, the condensation of water vapor occurs as a result of a capillary effect. The humidity easily adsorbs throughout the open porosities and leads to water condensation within the capillary pores, which are distributed between the grains. The behavior of this condensation is a function of ceramic pore size and its distribution. According to the basic theory of adsorption on a porous matrix (Adamson and Gast 1997), when the vapor molecules are first physicosorbed onto the porous material, capillary condensation will occur if the micropores are narrow enough. The critical size of pores for a capillary condensation effect is characterized by the Kelvin radius (Equation 10.6). In the case of water, the condensation of vapor into the pores can be expressed by a simplified Kelvin equation (Ponec et al. 1974):

$$r_K = \frac{2\gamma V_M \cos\theta}{\rho RT ln(\%RH/100)} \quad (10.6)$$

where γ is the surface tension of vapor in the liquid phase, V_M is molecular volume, θ is contact angle, and ρ is the density of vapor in the liquid phase. Equation 10.6 was obtained for case of cylindrical pores (Foster 1932). In this equation, the thickness of the adsorbed layer has been ignored. While V_M and surface tension γ are constants at room temperature, the possibility of controlling the condensation by simply changing the contact angle θ becomes attractive.

Thus, the condensation occurs in all pores with radii up to r_k, the Kelvin radius, and under a constant water vapor pressure or RH. The smaller the r_k or the lower temperature, the more easily condensation occurs (i.e., with a lower air humidity, free water forms in the pores of the dielectric).

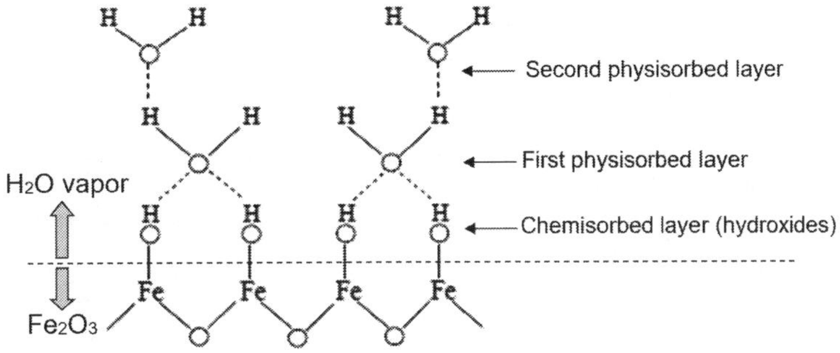

FIGURE 10.25 Multilayer structure of adsorbed water vapor molecules on the surface of iron oxide. (Reprinted with permission from McCafferty, E. and Zettlemoyer, A.C., Adsorption of water vapour on α-Fe_2O_3. Discuss. *Faraday Soc.*, 52, 239–254, 1971. Copyright 1971, The Royal Society of Chemistry.)

Taking into account that condensed water has the maximum effect on capacity, it becomes evident that it is the pore size and the porosity of the metal oxide that controls the sensitivity of the capacitive sensor. The smaller the pore size is, the lower is the sensitivity threshold. The greater the porosity and open pores' volume, the greater is the amount of condensed water and the greater is the range of capacity change when interacting with water vapor. It is important that the pores are open. The presence of open porosity permits the condensation of water in the capillary pores. It is belived that the volume fraction of pores in metal oxides designed for capacitive sensors should be as high as 45%. The pore size also determines the response rate. Sensors with pore size above 100 nm in diameter exhibit a fast response, but in the case when the pore size is smaller than 10 nm in diameter, the response and especially recovery may be long.

The pore size distribution is also an important parameter, as it seems to determine the nature of the dependence of the change in capacitance on the RH. All this indicates that the parameters of the sensors largely depend on the structure of the material, and not on the nature of the metal oxide itself. The nature of the metal oxide appears only in the stability of the parameters of the sensors and the features of interaction with other vapors that can be present in the atmosphere. Considering the general nature of the effect of the structure of metal oxides on the characteristics of capacitive and resistive humidity sensors (Gusmano et al. 1990, 1993), a special chapter will be devoted to this issue in Volume 3 (Chapter 2).

It should be noted that, in pure form, capacitive sensors based on metal oxides are realized only for metal oxides with dielectric properties such as Al_2O_3. For semiconductor metal oxides, such as ZnO, SnO_2, TiO_2, which are also used to develop humidity sensors of capacitive type, the situation is more complicated, as along with the change in capacitance, the conductivity of structure changes as well.

Therefore, such sensors would be more appropriate to call *impedance-based sensors*. In more detail, the mechanisms of interaction of water vapor with metal oxides, which are accompanied by a change in conductivity, will be examined in Chapter 11, analyzing resistive humidity sensors.

10.3.3 Al_2O_3-Based Humidity Sensors

One of the first ceramics to be used in humidity capacitance-type sensors was porous Al_2O_3, obtained by electrochemical etching of aluminum under anodic bias. The first humidity-sensitive Al_2O_3 layer, formed through anodization on the Al metal surface, was reported in 1953. Al_2O_3 exhibits a pore size distribution that depends on the etching parameters such as current density (Masuda et al. 1993). Today, the material is used in many commercial humidity sensors (www.systechinst.com; www.amsystems.co.uk), on the one hand because the etching technology is well established, and on the other hand, because Al_2O_3 has proven to be stable at elevated temperature and high humidity level (Nahar and Khanna 1982; Khanna and Nahar 1984; Sberveglieri et al. 1994, 1995; Nahar and Khanna 1998; Mai et al. 2000; Nahar 2000; Varghese and Grimes 2003; Timar-Horvath et al. 2008; Cheng et al. 2011; Li et al. 2012). The Al_2O_3-based humidity sensor is a volume-effect device based on physical adsorption discussed above. At low humidity, the walls of the pores are lined with one molecular-thick liquid layer. As the humidity increases, after saturating the walls, due to capillary condensation effect, the water starts condensing in the pores and changing the dielectric constant (Boucher 1976; Neimark and Ravikovitch 2001). The number of water molecules absorbed determines the electrical impedance of the capacitor, which is proportional to the water vapor pressure.

A diagram illustrating the structure of Al_2O_3-based capacitive humidity sensors offered on the sensor market is shown in Figure 10.26. The basic construction of such

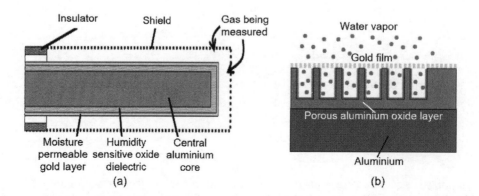

FIGURE 10.26 (a, b) Cross-sections of aluminum oxide sensor. (Adapted from http://www.systechinst.com/.)

sensors consists of a high-purity aluminum wire, the surface of which was anodically etched and then chemically oxidized to produce a pore-filled insulating layer of partially hydrated aluminum oxide. On a microscopic level, the aluminum oxide appears as a matrix with many parallel pores. When exposed to even small amounts of water vapor, the superstructure enables water molecules to permeate into the matrix, where microcondensation occurs. Since the dielectric constant of dry gases is significantly lower than in gases containing moisture, each pore acts as a microcapacitor. As the microcapacitors are in a parallel arrangement, the total capacitance is additive. Thus, the capacitance, measured between the sensor's aluminum core and the gold film, fluctuates as the water vapor content in the air changes. The sensor is excited with a low-voltage alternating current at a fixed frequency.

Based on an equivalent circuit model (Kovac et al. 1978; Nahar and Khanna 1982; Khanna and Nahar 1984; Nahar et al. 1984), the porous Al_2O_3 film capacitive sensor usually is represented by parallel resistance and capacitance (see Figure 10.27). In the diagram, C_s and R_s represent the capacitance and resistance of the solid walls; C_a and R_a are the capacitance and resistance of the pore area filled with air; C_b and R_b are the capacitance and resistance of the barrier layer below the air portion. C_{eff} denotes the capacitance of the multiple-dielectric capacitor comprising the sublayers: barrier layer, chemisorbed layer, and water condensed in the pore, as given by the Maxwell-Wagner effect. R_{eff} is the parallel resistance of the capacitor.

Based on the equivalent circuit, the capacitance and resistance can be obtained as (Nahar and Khanna 1982):

$$C = C_s + \frac{C_a \cdot C_b}{C_a + C_b} + C_{eff} \quad (10.7)$$

and

$$\frac{1}{R} = \frac{1}{R_s} + \frac{1}{R_a + R_b} + \frac{1}{R_{eff}} \quad (10.8)$$

Because of its low cost and easy process, anodic etching and oxidation of Al_2O_3 has great priority over other technologies such as evaporation, sputtering, and spray pyrolysis (Chen and Lu 2005). The oxide layer that forms the dielectric separating layer of the capacitor is in the form of a mass of tubular pores running up from the aluminum core to the exposed surface. The small pore radius makes Al_2O_3 sensitive to very low water vapor pressure. The aluminum oxide is then coated with a thin, permeable layer of gold. The gold layer and the aluminum probe form the anode and cathode (see Figure 10.26). The pore size of the aluminum oxide layer is specific to water vapor and smaller molecules, but due to the dielectric constant of water compared to that of other vapors that may enter the pores, such as hydrogen, the sensor response is resistant to many contaminants and specific to changes in water vapor, regardless of the matrix gas.

Aluminum oxide sensors offer many advantages (Nahar 2000; Chen and Lu 2005): wide dynamic range of humidity measurement, good linearity of $C = f(RH)$ dependences, high sensitivity and rapid response, relatively stable with low hysteresis and temperature coefficients, low or modest maintenance requirements, available in small sizes, and capable of measuring very low dew-point levels without the need for cooling, as chilled-mirror hygrometers need. However, sensor parameters of Al_2O_3-based devices (sensitivity, humidity range, response time, stability, etc.) strongly depend on the thickness and the density of the porous layer, and the size of the pores as well (Banerjee and Sengupta 2002; Varghese et al. 2002). For example, the response time of ceramic humidity sensors is generally limited by diffusion in the pores (Cunningham and Williams 1980; Seiyama et al. 1983). This means that the response and recovery times are controlled by the pore size (Sun et al. 1989). It is also clear from the empirical point of view that, for design devices with highly humidity-sensitive properties in wide range of humidity, it is necessary to use ceramics with a large pore volume and a wide

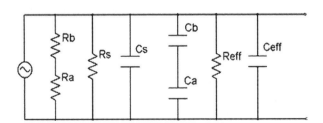

FIGURE 10.27 Schematic of the electrical equivalent circuit model of an Al_2O_3 moisture sensor. (Reprinted with permission from Nahar, R.K. and Khanna, V.K., A study of capacitance and resistance characteristics of an Al_2O_3 humidity sensor, *Int. J. Electron.*, 52, 557–567, 1982. Copyright 1982, Taylor & Francis Group.)

dispersion of pore sizes, from nano- to macropores (Traversa 1995). The diameters and depths of the pores can be controlled by the technological parameters such as concentration and temperature of the electrolyte and the current density in the anodizing cell. Therefore, the detection limit could be set very low by shrinking the pore size (the minimum detectable humidity decreases as the pore radius decreases). In particular, Chen et al. (1990), optimizing the pore size, achieved sensitivity as low as 1 ppmv. Thus, control of microstructure of the film is again important for sensing characteristics, and the film preparation conditions should be carefully chosen (Almasi Kashi et al. 2012). To obtain quick responses, the upper electrode should be as thin and porous as possible. At the same time, in order to obtain a linear dependence of the capacitance on humidity, the Al_2O_3 layer must have a definite distribution in pore size. If the pores had the same size, then, as a rule, the change in capacity is close to stepped (see Figure 10.28), when at a certain humidity, capillary condensation of water and a sharp increase in capacity occur (Kim et al. 2009).

The primary problem of anodized Al_2O_3-based humidity sensors is that, when exposed for a long duration in high humidity, a significant degradation in sensitivity and drift in the capacitance characteristics would be expected (Chen and Lu 2005). In this case, the accuracy of the aluminum oxide moisture sensor is low, each sensor requires a separate nonlinear calibration curve, and the unit must be periodically recalibrated to compensate for aging and contamination. This is the major drawback for the aluminum oxide sensor (Emmer et al. 1985). Many researchers try to solve this problem by aging the alumina films in boiling water or macerating the film in some ion solutions; however, the drift cannot be completely eliminated. This means that these sensors are designed for low-dewpoint measurements and should be disabled if exposed to high humidity or wetted. In other words, aluminum oxide sensors are good for very dry and clean conditions where quick measurement is not required.

Nahar (2000) believed that strong drift of sensor characteristics was attributed to the widening of the pores due to diffusion of the adsorbed water. However, other research found that the basic reason for long-term drift of sensors using porous, anodic aluminum oxide films is related to the structure transformation of Al_2O_3 during exploitation (Chen and Lu 2005). It was established that anodic aluminum oxide used in humidity sensors contains α-Al_2O_3, γ-Al_2O_3, and amorphous phases. At that, α-phase of Al_2O_3 is stable, while γ-phase and amorphous Al_2O_3 suffer from degradation. When the structure is exposed to a humid atmosphere, the γ-phase or amorphous Al_2O_3 changes to γ-$Al_2O_3 \cdot H_2O$ (boehmite) (Young 1961). This irreversible phase change causes volume expansion of aluminum oxide, resulting in the gradual decrease of surface area and porosity (Emmer et al. 1985). This further gradually decreases the adsorption ability of the film and causes the long-term calibration drift. Many researchers tried to improve the anodic aluminum oxide sensors by optimization of anodization process (Varghese et al. 2002; Varghese and Grimes 2003), aging the aluminum oxide in boiling water or macerating the films in some ion solutions (Emmer et al. 1985). Mai et al. (2000) proposed to use thermal annealing at about 400°C. However, the drift cannot be

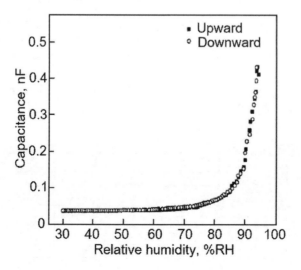

FIGURE 10.28 Capacitance characteristics of interdigitated type Al_2O_3-based sensor with 30 μm porous layer thickness and 60 nm pore diameter. (Reprinted with permission from Kim, Y. et al., Capacitive humidity sensor design based on anodic aluminum oxide, *Sens. Actuators B*, 141, 441–446, 2009. Copyright 2009, Elsevier.)

completely eliminated. Even the commercial aluminum oxide humidity sensors have to be calibrated twice a year to assure their accuracy (Chen and Lu 2005). This problem seriously hinders the widespread use of aluminum oxide humidity sensors. Chen and Jin (1992a) suggested a different approach to improving stability. As mentioned before, α-Al_2O_3 is a very stable phase. However, it is very difficult to form α-Al_2O_3, because the temperature for phase change from γ to α is at least 900°C (Chen and Jin 1992b). Fortunately, there is an electrochemical approach to deposit α-Al_2O_3 without a high-temperature process, which is based on the using the anodic spark deposition process. This is a unique process for forming certain ceramic coatings (McNeill 1958; Gruss and McNeill 1963; Koshkarian and Kriven 1988), including α-Al_2O_3 (Brown et al. 1971). With respect to Al_2O_3-based humidity sensors, this process was implemented by Chen et al. (1991, 1992a), who successfully fabricated humidity sensors with the capability of detecting moisture level as low as 1 ppmv using α-Al_2O_3 films formed by anodic spark deposition in the melt of $NaHSO_4$-$KHSO_4$ mixture. The anodic spark–deposited α-Al_2O_3 films exhibited a continuous open pore structure. The average pore size in formed α-Al_2O_3 films was in the range from 1 to 2 μm, and the porosity was approximately 30%. In order to overcome a short circuit at high AC voltage (> 0.5 V), the α-Al_2O_3 film was re-anodized in diluted sulfuric acid or borax solution for a short time to form a thin barrier layer. The fabricated sensors showed high sensitivity for moisture level from 1000 to 1 ppmv or −20°C to −76°C dew/frost point (D/F PT) (see Figure 10.29). The sensors also demonstrated excellent long-term stability. For more than 6 months, there was no drift for data. The response was also very fast. From the low to the high moisture level, the response time was approximately 5 seconds, and from the high to the low moisture level, approximately 20 seconds. Therefore, the long-term calibration-drift occurring in γ-Al_2O_3 film sensors was eliminated in the α-Al_2O_3 film sensors.

One should say that the technology of anodic etching for the formation of porous Al_2O_3 is not very consistent with modern methods of manufacturing devices used for mass production. AAO films are mostly fabricated by anodization of aluminum sheets of a few hundred micrometers in thickness (Jessensky et al. 1998; Lee et al. 2006; Bai et al. 2008). Because of that, in recent years, the Al_2O_3-based humidity-sensitive generation methods have been widely used in the production of Al_2O_3-based humidity sensors without the use of anodic etching and oxidation. In this case, the layer of porous aluminum oxide on a conductive substrate is formed using the methods of thick-film technology and then is coated with a thin film of gold (Chatterjee et al. 2001). For example, Chatterjee et al. (2001) have used the tape casting technique, which was adopted to improve the porous structure of Al_2O_3 layer. The sensors fabricated were sensitive to gas moisture in the range of

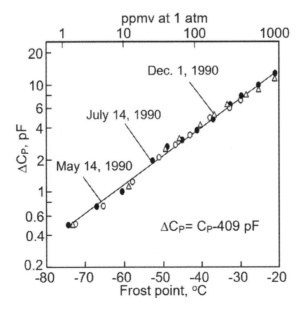

FIGURE 10.29 Operating characteristics of α-Al_2O_3-based capacitance humidity sensors during long-term stability test. (Data extracted from Chen, Z. and M.-C. Jin, An alpha-alumina moisture sensor for relative and absolute humidity measurement, In: *Proceedings of the 27th annual conference of IEEE Industry Application Society*, Houston, TX, 2, 1668–1675, 1992a.)

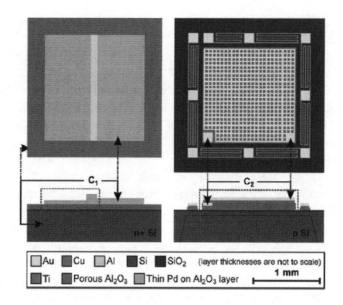

FIGURE 10.30 Top view and cross-section of the first (C1) and the second (C2) capacitive relative humidity (RH) sensor structures. (Reprinted with permission from Juhasz, L. and Mizsei, J., Humidity sensor structures with thin film porous alumina for on-chip integration, *Thin Solid Films*, 517, 6198–6201, 2009. Copyright 2009, Elsevier.)

100–1000 ppmv. Experiments have shown that humidity sensors can also be manufactured using an anodic etching of an aluminum film deposited on a silicon surface or SiO_2 (Westcott and Rogers 1985; Sberveglieri et al. 1994; Juhász and Mizsei 2009). In this case, the manufacturing process can also be integrated into standard microelectronic technology. Versions of humidity sensors made using the anodization of sprayed Al films are shown in Figure 10.30.

10.3.4 Other Metal Oxide-Based Humidity Sensors

Porous ceramic films of other metal oxides can also be used (Yagi and Nakata 1992; Fenner et al. 2001; Wang et al. 2005b). However, most of them are laboratory samples. As a rule, porous ceramic films of other metal oxides are formed on the substrates using a thick-film technology. Thick films are usually printed onto an alumina substrate as a paste or conductive ink with film thickness greater than 10 μm. Thin-film technology can also be used.

Most of the metal oxides used to develop capacitive humidity sensors have electrical conductivity. This means that, in reality, all these sensors are impedance sensors, because simultaneously with the change in capacitance, there is a change in conductivity. Therefore, the common structure of metal oxide-based capacitive humidity sensors is the MIS, in which the insulator layer is SiO_2. As an example of capacitive sensors, manufactured basing on the mentioned approach, one can consider sensors developed on the basis of perovskites, such as strontium lanthanum titanate ($Sr_{1-x}La_xTiO_3$) on a silicon substrate (Li et al. 1997), $SrNb_xTi_{1-x}O_3$ (Li et al. 1999), and $BaMO_3$ (Viviani et al. 2001; Agarwal and Sharma 2002; Wu and Gu 2009). Experiments conducted by Viviani et al. (2001) have shown that, in the $BaMO_3$ system, highly porous $BaTiO_3$ with the addition of 0.3 at% of La had maximum sensitivity. However, according to Agarwal and Sharma (2002), an optimum additive to $BaTiO_3$ is Sr. These experiments have shown that the major factors affecting the sensitivity and response time were porosity and pore size distribution. For example, Viviani et al. (2001) have established that a decrease in the grain size gives a significant increase in sensor sensitivity, but this growth is accompanied by an increase in the response time (see Table 10.6). It should be noted that the use of composites for the development of capacitive sensors is a general trend in the development of humidity sensors, since this approach provides additional possibilities for controlling the structure of the materials being formed and creating the necessary porosity of the structure (Korotcenkov and Cho 2017).

Undoubtedly, capacitive sensors can be manufactured without an additional insulating layer. In particular, Steele et al. (2007, 2008) have manufactured capacitive-based humidity sensors using coplanar gold IDEs coated with nanostructured TiO_2, SiO_2, and Al_2O_3 thin films produced by the glancing angle deposition. It was established that the capacitance response of the TiO_2 film was the greatest. The devices were sensitive over a wide range of RH levels (< 1%–92%) and exhibited extremely fast, subsecond response times. The capacitance in this RH range changed from approximately 1 nF to 1000 nF (Steele et al. 2007). Typical adsorption and desorption

TABLE 10.6
Parameters of BaMO$_3$-Based Capacitive Humidity Sensors

Metal oxide	Grain size, μm	Volume of pores with diameter <20 nm, %	$\Delta C/C_m$, %	Time constant of response τ_1, s	τ_2, s
BaTiO$_3$	2	32	154	12	650
BaTiO$_3$	10	35	11	15	250
BaTiO$_3$:La	0.2	55	3560	220	1000
BaZrO$_3$	0.6	16	65	8	45
BaHfO$_3$	0.6	14	84	25	190
BaSnO$_3$	9	0	5	7	170

Source: Data extracted from Viviani, M. et al., *J. Eur. Ceram. Soc.*, 21, 1981–1984, 2001.
Note: Samples were obtained by cold isostatic pressing and sintering of fine powders prepared by wet chemical synthesis.

response times were measured to be 220 and 400 milliseconds, respectively. It was shown that sensitivity (nF/%RH) can be increased by decreasing the electrode periodicity or by increasing the planar area of the electrodes, or both. It was concluded that these parameters are a consequence of the specific structure of the deposited films. All films used in this study were composed of individual columns (100–200 nm in diameter) that are oriented along the substrate normal and separated by voids, with an overall film porosity of 60%–70%.

An even higher sensitivity was demonstrated by ZnO/TiO$_2$ core/shell, nanorod-based, capacitive thin-film humidity sensors with interdigitated electrodes fabricated on the glass substrates by Gu et al. (2011).

Vertically aligned ZnO nanorod arrays were grown hydrothermally and subsequently coated with a TiO$_2$ layer in a sol–gel process. Morphological characterization showed that the primarily formed zinc oxide nanorods were coated with anatase titanium oxide shells as a second layer. Compared with those sensors based on individual ZnO and TiO$_2$, the ZnO/TiO$_2$ nanocomposite (ZTNA) sensors exhibited a considerably enhanced sensitivity at 95% RH (31 and 1,380 times greater than the ZnO nanorod arrays and TiO$_2$ thin films, respectively) (Gu et al. 2011). Further, the core/shell arrays' capacitance varied from 10^1 to 10^6 pF over the whole humidity range of 11%–95% RH at room temperature (see Figure 10.31). Gu et al. (2011) believed

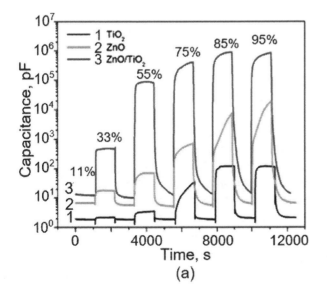
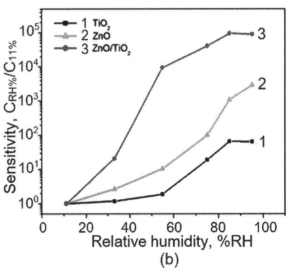

FIGURE 10.31 (a) Time-dependent capacitance of sensors based on the TiO$_2$ thin film (*black*) and pristine ZnO nanorod arrays (*green*) compared to ZnO/TiO$_2$ nanocomposite (ZTNA) sensors (*blue*) at various humidities; (b) comparison of the concentration-dependent sensitivity of the different sensor types. (For interpretation of the references to color in the figure caption, the reader is referred to the web version of the article.) (Reprinted with permission from Gu, L. et al., Humidity sensors based on ZnO/TiO$_2$ core/shell nanorod arrays with enhanced sensitivity, *Sens. Actuators B*, 159, 1–7, 2011. Copyright 2011, Elsevier.)

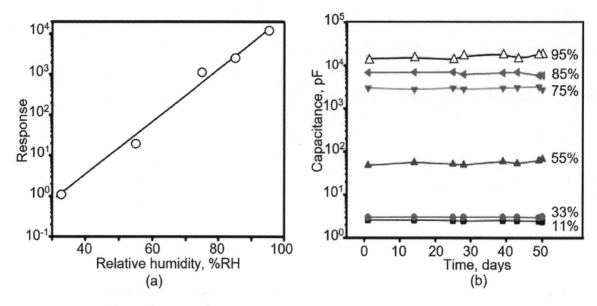

FIGURE 10.32 (a) Logarithmic response behavior vs relative humidity (RH) for cubic bismuth phosphate following the expression $y = -2.2 + 0.06695x$ (correlation coefficient $R = 0.99544$). (b) Long-term stability of cubic bismuth phosphate at various RH values over 50 days. (Reprinted with permission from Sheng, M. et al., Humidity sensing properties of Bismuth Phosphates, *Sens. Actuators B*, 166–167, 642–649, 2012. Copyright 2012, Elsevier.)

that achieved parameters are the result of a specific surface morphology and the high degree of hydrophilicity, which lead to the enhanced adsorption facilities and thus better humidity-sensing properties of the ZnO/TiO$_2$ nanorod arrays.

Extremely high sensitivity was also achieved for capacitive humidity sensors based on bismuth phosphates, such as cubic sillenite and monoclinic types synthesized through a hydrothermal method (see Figure 10.32). These sensors were investigated by Sheng et al. (2012). The sillenite family encompasses a variety of compounds having the general formula $Bi_{12}(Bi_{4/5-nx}M_{5x}{}^{n+})O_{19.2+nx}$ (M = M^{2+}, M^{3+}, M^{4+} and M^{5+} [only V, As, and P]) in common. Cubic bismuth phosphate has mostly been accessed with solid-state reactions in the Bi_2O_3/$BiPO_4$ system. It was established that sensors with the cubic bismuth phosphate structure revealed linear capacitance variations of four orders, namely from 1.1 to 12,908 pF over the RH range from 30% to 95%, while the monoclinic bismuth phosphate sample showed only three orders of magnitude change from 1.2 to 1,097 pF in a nonlinear fashion. Sheng et al. (2012) assumed that structurally flexible, sillenite-related cubic bismuth phosphate is a promising humidity sensor material due to a polarizable oxide framework of Bi^{3+} cations with neighboring oxygen atoms that can easily form hydrophilic OH– bonds and adsorb water molecules efficiently. Moreover, the high amount of polarizable Bi^{3+} in the sillenite structure enhances the number of active sites for the surface adsorption of water in comparison with monoclinic $BiPO_4$.

Clearly, not all metal oxide sensors have such a high sensitivity to humidity. For a large group of sensors developed on the basis of SnO_2 (Wang et al. 2016), ZnO (Narimani et al. 2016), Ca, Mg, Fe, Ti-oxides (Mahboob et al. 2016), $CaCu_3Ti_4O_{12}$ (Li 2016), the change in capacitace during the measurement did not exceed 1.5–10 times, which was apparently the manifestation of the structure of the films used and the specificity of the interaction of these metal oxides with water vapor. For example, the structure of a ZnO film used by Narimani et al. (2016) does not contribute to capillary condensation of water vapor; the pore size is too large (Figure 10.33). However, even this change significantly exceeds the change in the capacitance observed in polymer humidity sensors.

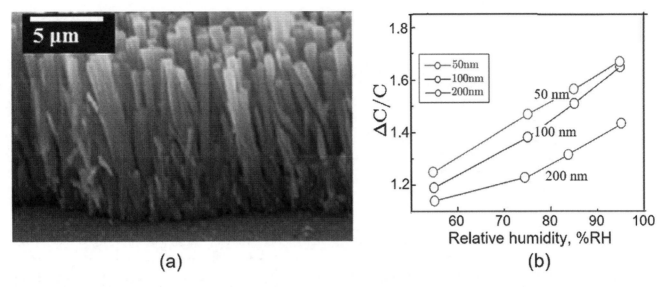

FIGURE 10.33 (a) Cross-sectional view of the grown ZnO nanorods. (b) The capacitance measured for each capacitor at different relative humidities (RHs). (Reprinted with permission from Narimani, K. et al., Fabrication, modeling and simulation of high sensitivity capacitive humidity sensors based on ZnO nanorods, *Sens. Actuators B*, 224, 338–343, 2016. Copyright 2016, Elsevier.)

10.4 SUMMARY

Examination carried out in this chapter shows that humidity sensors of capacitive type have a certain set of advantages and disadvantages. Typically, the following advantages are attributed to capacitive humidity sensors (Wilson 2005):

- Near-linear voltage output
- Wide RH range
- Interchangeable
- Stable over long-term use with minimal long-term drift and hysteresis
- Condensation tolerance
- Highly resistant to chemicals
- Because of their low temperature effect, they are often used over wide temperature ranges up to 200°C without active temperature compensation. For metal oxide–based sensors, the temperature dependence can be observed
- Very low-power consumption and high output signals
- Fast response at optimal design
- Low cost

It is these advantages that have provided capacitive RH sensors for wide application in industrial, commercial, and weather telemetry applications. Only this type of devices guarantees almost full-range RH measuring with acceptable accuracy.

As for the general shortcoming of capacitive humidity sensors, they can be attributed to the following:

1. The change in capacitance is small compared with the capacitance of even a few meters of cable. This means that it is necessary to reduce stray capacitance, which requires the use of a special configuration of sensors, and the electronic processing has to be completed close to the sensor. If one data logger is connected to several RH sensors, each will need its own power supply (extra wires in the cable) and relatively bulky electronics. This is an irritating extra expense, because some data loggers have all the necessary processing power built into them. In addition, in order to reduce the effect of parasitic capacitances, the sensors must be sufficiently large, which limits the possibility of their microminiaturization.

2. The operation of the sensors requires electronics to convert capacitance to RH.
3. The possibility of significant drift.
4. The loss of relative accuracy at low end (< 5%). The last remark is mainly related to polymer sensors. For example, on the contrary to polymers, for Al_2O_3-based sensors, very dry environments (as low as −75°C to −100°C dew-point temperatures) is the area recommended for use. As an option to increase the sensitivity of a polymer capacitive sensor of parallel type in the region of a low RH, an increase in area and a decrease in the thickness of the sensor could be used. These changes would increase the change in sensor's capacitance at low humidity, but can cause extension of the response time. The low humidity level can also be made measurable when increased by compression of a sample of humid gas in a bypass pipe containing measurement cell with sensor, or by cooling the gas in a capacitive dew-point hygrometer (Majewski 2016). However, this complicates the measuring circuit.

One should note that, despite the weak dependence of the parameters of capacitive sensors on temperature, temperature is one of the main sources of errors in measuring the humidity (Rotronic 2014). Sensor hygroscopic properties vary with temperature. An RH instrument relies on the assumption that the relationship between the amount of moisture present in the sensor hygroscopic material and RH is constant. However, in most hygroscopic materials, this relationship varies with temperature. In addition, the dielectric properties of the water molecule are affected by temperature. At 20°C, the dielectric constant of water has a value of approximately 80. This constant increases by more than 8% at 0°C and decreases by 30% at 100°C. Sensor dielectric properties also vary with temperature. The dielectric constant of most dielectric materials decreases as temperature increases. Fortunately, the effect of temperature on the dielectric properties of most plastics is usually more limited than in the case of water. However, they exist. All this can lead to errors in the measurement. Therefore, humidity values reported by the electronics must compensate for the impact of temperature on the sensor. Failure to do so can result in large measurement errors, sometimes up to 8% RH or more. It is also necessary to consider that any difference between the ambient temperature and the sensor temperature also causes an error. For example, at 20°C and 50% RH, a difference of 1°C between the ambient temperature and the sensor temperature results in an error of approximately 3%.

If we compare polymer and metal oxide humidity sensors, it should be noted that, despite the general nature of the phenomena responsible for the humidity-sensitive effect in these devices, which include physisorption, chemisorption of water vapor molecules, and their capillary condensation (Steele et al. 2006), polymer-based humidity sensors have the advantage of providing a faster desorption mechanism of water molecules than ceramic sensors, which need a heater to accelerate the desorption process. Polymer sensors also have a weaker temperature dependence of parameters and better stability in the condensation of water vapor. However, below RH of 5%–10%, the polymer-based sensors often exhibit an increasing drop in accuracy together with nonlinearity of $C = f(RH)$ characteristics.

In this respect, metal oxide sensors, especially on the basis of conductive oxides, significantly exceed polymer sensors and sensors based on dielectric materials. This is due to the peculiarity of the equivalent circuit of such sensors and a large surface area, which significantly exceeds the geometric dimensions of the sensor. When interacting with water vapor, in addition to the change in the dielectric constant, the width of the space charge region on the grain surface changes significantly, which, due to the large surface area, makes a significant contribution to the change in the capacitance of the sensor than the change in the effective permittivity. Moreover, the modification of the microstructure and the chemical composition of ceramic materials permit both a performance optimization in sensors, exploiting their electrical properties and tailoring to specific requirements. But these advantages do not compensate for the poor stability and reproducibility of metal oxide sensors, which is most important for practical applications. It was found that, due to porous structure, metal oxide ceramics are highly sensitive to contaminants such as dust, dirt, oil, smoke, alcohol, solvents, and smoke (Clayton et al. 1985; Nakajima et al. 1999). The adhesion or adsorption of these compounds on the ceramic surface causes irreversible changes in the sensor's response. This means that they require maintenance (e.g., by heating them from time to time and evaporating condensed vapor). Temporary heating could also be a solution for drift due to the formation of chemisorbed OH-groups. Prolonged exposure to humid environments leads to the gradual formation of stable chemisorbed OH^- groups on the surface, causing a progressive drift in the resistance of the ceramic humidity sensor (Visscher and Kornet (1994). Experimentation has shown that the hydroxyl ions are removed by heating to temperatures higher than 400°C (Morimoto et al. 1969; Egashira et al. 1978, 1981). This means that the surface-related phenomena of humidity sensing by ceramics make these materials less resistant than polymers to the surface. It should also be noted that the nonlinearity of the characteristics, long stabilization time, slow response time, and frequent calibration are also characteristic for many of the tested metal oxide–based capacitive humidity sensors (Li 2016; Tripathy et al. 2016).

REFERENCES

Adamson A.W., Gast A.P. (1997) *Physical Chemistry of Surface*. John Wiley & Sons, New York.

Agarwal S., Sharma G.L. (2002) Humidity sensing properties of (Ba, Sr)TiO$_3$ thin films grown by hydrothermal-electrochemical method. *Sens. Actuators B* 85, 205–211.

Ahn K., Wessels B.W., Sampath S. (2005) Spinel humidity sensorsprepared by thermal spray direct writing. *Sens. Actuators B* 107, 342–346.

Almasi Kashi M., Ramazani A., Abbasian H., Khayyatian A. (2012) Capacitive humidity sensors based on large diameter porous alumina prepared by high current anodization. *Sens. Actuators A* 174, 69–74.

Anchisini R., Faglia G., Gallazzi M., Sberveglieri G., Zerbi G. (1996) Polyphosphazene membrane as a very sensitive resistive and capacitive humidity sensor. *Sens. Actuators B* 35, 99–102.

Anderson J.H., Parks G.A. (1968) Electrical conductivity of silica gel in the presence of adsorbed water. *J. Phys. Chem.* 72, 3662–3668.

Asahi Breweries. (1994) Technical data sheet for Asahi Chemical "Pimel TL-500, G-7000 and IX Grade. Asahi Chemical Co., Ltd.," Functional Products Division, Tokyo, Japan.

Bai A., Hu C.-C., Yang Y.-F., Lin C.-C. (2008) Pore diameter control of anodic aluminum oxide with ordered array of nanopores. *Electrochim. Acta* 53, 2258–2264.

Banerjee G., Sengupta K. (2002) Pore size optimisation of humidity sensor—A probabilistic approach. *Sens. Actuators B* 86, 34–41.

Basu S., Chatterjee S., Saha M., Bandyopadhay S., Mistry K.K., Sengupta K. (2001) Study of electrical characteristics of porous alumina sensors for detection of low moisture in gases. *Sens. Actuators B* 79, 182–186.

Bayhan M., Kavasoglu N. (2006) A study on the humidity sensing properties of ZnCr$_2$O$_4$–K$_2$CrO$_4$ ionic conductive ceramic sensor. *Sens. Actuators B* 117, 261–265.

Blank T.A., Eksperiandova L.P., Belikov K.N. (2016) Recent trends of ceramic humidity sensors development: A review. *Sens. Actuators B* 228, 416–442.

Blyholder G., Richardson E.A. (1962) Infrared and volumetric data on the adsorption of ammonia, water, and other gases on activated iron(III) oxide 1. *J. Phys. Chem.* 66, 2597–2602.

Blythe T., Bloor D. (2005) *Electrical Properties of Polymers*. Cambridge University Press, Cambridge, NY.

Boltshauser T., Baltes H. (1991) Capacitive humidity sensors in SACMOS technology with moisture absorbing photosensitive polyimide. *Sens. Actuators A* 26, 509–512.

Boltshauser T., Chandran L., Balks H., Bose F., Steiner D. (1991) Humidity sensing properties and electrical permittivity of new photosensitive polyimides. *Sens. Actuators B* 5, 161–164.

Boucher E.A. (1976) Review porous materials: Structure, properties and capillary phenomena. *J. Mater. Sci.* 11, 1734–1750.

Brown S.D., Kuna K.J., Van T.B. (1971) Anodic spark deposition from aqueous solution of NaAlO$_2$ and Na$_2$SiO$_3$. *J. Am. Ceram. Soc.* 54, 384–390.

Callister W.D. (1994) *Material Science and Engineering*. John Wiley & Sons, New York, pp. 623–632.

Chandran L., Baltes H., Korvink J. (1991) Three-dimensional modeling of capacitive humidity sensors. *Sens. Actuators A* 25, 243–247.

Chatterjee S., Basu S., Chattopadhyay D., Mistry K.Kr., Sengupta K. (2001) Humidity sensor using porous tape cast alumina substrate. *Rev. Sci. Instrum.* 72 (6), 2792–2795.

Chatzandroulis S., Tserepi A., Goustouridis D., Normand P., Tsoukalas D. (2002) Fabrication of single crystal Si cantilevers using a dry release process and application in a capacitive-type humidity sensor. *Microelectron. Eng.* 61–62, 955–961.

Chatzandroulis S., Tegou E., Goustouridis D., Polymenakos S., Tsoukalas D. (2004) Capacitive-type chemical sensors using thin silicon-polymer bimorph membranes. *Sens. Actuators B* 103, 392–396.

Chatzandroulis S., Tsouti V., Raptis I., Goustouridis D. (2011) Capacitance-type chemical sensors, In: Korotcenkov G. (ed.) *Chemical Sensors: Comprehensive Sensor Technologies, Solid State Devices*, Vol. 4. Momentum Press, New York, pp. 229–260.

Chen Z., Jin M.-C. (1992a) An alpha-alumina moisture sensor for relative and absolute humidity measurement. In: *Proceedings of the 27th Annual Conference of IEEE Industry Application Society*, Houston, TX, Vol. 2, pp. 1668–1675.

Chen Z., Jin M.-C. (1992b) Effect of high substrate temperatures on crystalline growth of Al$_2$O$_3$ films deposited by reactive evaporation. *J. Mater. Sci. Lett.* 11, 1023–1025.

Chen Z., Jin M.-C., Zhen C. (1990) Humidity sensors with reactively evaporated Al$_2$O$_3$ films as porous dielectrics. *Sens. Actuators B* 2, 167–171.

Chen Z., Jin M.-C., Zhen C., G. Chen G. (1991) Properties of modified anodic-spark-deposited alumina porous ceramic films as humidity sensors. *J. Am. Ceram. Soc.* 74, 1325–1330.

Chen Z., Lu C. (2005) Humidity sensors: A review of materials and mechanisms. *Sens. Lett.* 3 (4), 274–295.

Cheng B., Tian B., Xie C., Xiao Y., Lei S. (2011) Highly sensitive humidity sensor based on amorphous Al$_2$O$_3$ nanotubes. *J. Mater. Chem.* 21, 1907–1912.

Choi J.M., Kim T.W. (2013) Humidity sensor using an air capacitor. *Trans. Electrical Electron. Mater.* 14 (4), 182–186.

Choi K.S., Kim D.S., Yang H.J., Ryu M.S., Chae J.S., Chang S.P. (2015) Comparison and analysis of capacitive humidity sensors with water vapor inlet holes of different depths. *J. Micro/Nanolith. MEMS MOEMS* 14 (2), 025001.

Clayton W.A., Preud P.J., Baxter R.D. (1985) Contamination resistant capacitive humidity sensor. In: *Proceedings of the Conference on Humidity and Moisture*, Washington, DC, pp. 535–544.

Comyn J. (Ed.) (1985) *Permeation of Gases and Vapors in Polymers*. Elsevier, London, UK.

Connolly E.J., Pham H.T.M., Groeneweg J., Sarro P.M., French P.J. (2004) Relative humidity sensors using porous SiC membranes and Al electrodes. *Sens. Actuators B* 100, 216–220.

Cunningham R.E., Williams R.J.J. (1980) *Diffusion in Gases and Porous Media*. Plenum, New York.

Dabhade R.V., Bodas D.S., Gangal S.A. (2004) Plasma-treated polymer as humidity sensing material—A feasibility study. *Sens. Actuators B* 98, 37–40.

Dai C.-L. (2007) A capacitive humidity sensor integrated with micro heater and ring oscillator circuit fabricated by CMOS–MEMS technique. *Sens. Actuators B* 122, 375–380.

Delapierre G., Grange H., Chambaz B., Destannes L. (1983) Polymer based capacitive humidity sensor—Characteristics and experimental results. *Sens. Actuators A* 4, 97–104.

Denton D.D., Day D.R., Priore D.F., Senturia S.D. (1985) Moisture diffusion in polyimide films in integrated circuits. *J. Electron. Mater.* 14, 119–136.

Dokmeci M. Najafi K.A (2001) High-sensitivity polyimide capacitive relative humidity sensor for monitoring anodically bonded hermetic micropackages. *J. Microelectromechan. Syst.* 10, 197–204.

DuPont (1994) *Technical Data Sheet for DuPont "Pyralin PI 2700" Series*. Dupont Co., Wilmington, DE.

Endres H.E., Drost S. (1991) Optimization of the geometry of gas sensitive interdigital capacitors. *Sens. Actuators B* 4, 95–98.

Egashira M., Kawasumi S., Kagawa S., Seiyama T. (1978) Temperature programmed desorption study of water absorbed on metal oxides: I. *Bull. Chem. Soc. Jpn.* 51, 3144–3149.

Egashira M., Nakashima M., Kawasumi S., Seiyama T. (1981) Temperature programmed desorption study of water absorbed on metal oxides. 2. Tin oxide surfaces. *Z Phys. Chem.* 85, 4125–4130.

Emmer I., Hajek Z., Repa P. (1985) Surface adsorption of water vapour on hydrated layers of Al_2O_3. *Surf. Sci.* 162, 303–309.

Erson R.C., Muller R.S., Tobias C.W. (1990) Investigations of porous silicon for vapor sensing. *Sens. Actuators A* 23, 835–839.

Evans S. (1965) Dielectric properties of ice and snow—A review. *J. Glaciology.* 5 (42), 773–779.

Fagan J.G., Amarakoon R.W. (1993) Reliability and reproducibility of ceramic sensors. III: Humidity sensors. *Am. Ceram. Soc. Bull.* 72, 119–130.

Faia P.M., Ferreira A.J., Furtado C.S. (2009) Establishing and interpreting an electrical circuit representing a TiO_2–WO_3 series of humidity thick film sensors. *Sens. Actuators B* 140, 128–133.

Fenner R. Zdankiewicz E. (2001) Micromachined water vapor sensors: A review of sensing technologies. *IEEE Sens. J.* 1, 309–317.

Fleming W.J. (1981) A physical understanding of solid-state humidity sensors, In: *Proceedings of Int. Automotive Meet.*, SAE, Detroit, MI, Paper no. 810432, pp. 51–62.

Fontes J. (2005) *Sensor Technology Handbook*. Elsevier, New York, pp. 271–284.

Ford L.H. (1948) The effect of humidity on the calibration of precision air capacitors. *J. IEE, Part II: Power Eng.* 95 (48), 709–712.

Foster A.G. (1932) The sorption of condensible vapours by porous solids. Part I. The applicability of the capillary theory. *Trans. Faraday Soc.* 28, 645.

Fraden J. (2004) *Handbook of Modern Sensors: Physics, Designs, and Applications*. Springer Verlag, New York.

Fripiat J.J., Jelli A., Poncelet G., André J. (1965) Thermodynamic properties of adsorbed water molecules and electrical conduction in montmorillonites and silicas. *J. Phys. Chem.* 69, 2185–2197.

Fripiat J.J., Uytterhoeven J. (1962) Hydroxyl content in silica gel "Aerosil". *J. Phys. Chem.* 66, 800–805.

Furjes P., Kovacs A., Ducso Cs., Adam M., Muller B., Mescheder U. (2003) Porous silicon-based humidity sensor with interdigital electrodes and internal heaters. *Sens. Actuators B* 95, 140–144.

Furlani A., Iucci G., Russo M.V., Bearzotti A., D'Amico A. (1992) Thin films of Iodine—Polyphenylacetylene as starting materials for humidity sensors. *Sens. Actuators B*7, 447–450.

Grange H., Beith C., Boucher H., Delapierre G. (1987) A capacitive humidity sensor with every fast response time and very low hysteresis. *Sens. Actuators* 12, 291–296.

Grange H., Delapierre G. (1991) Polymer-based capacitive hygrometers, In: Yamazoe N. (Ed.) *Chemical Sensor Technology*, vol. 3. Elsevier, Amsterdam, the Netherlands, pp. 147–162.

Gruss L.L., McNeill W. (1963) Anodic spark reaction products in Aluminate, Tungstate, and Silicate solutions. *Electrochem. Technol.* 1, 283–287.

Gu L., Huang Q.A., Qin M. (2004) A novel capacitive-type humidity sensor using CMOS fabrication technology. *Sens. Actuators B* 99, 491–498.

Gu L., Zheng K., Zhou Y., Li J., Mo X., R. Patzke G., Chen G. (2011) Humidity sensors based on ZnO/TiO_2 core/shell nanorod arrays with enhanced sensitivity. *Sens. Actuators B* 159, 1–7.

Gusmano G., Nunziante P., Traversa E., Montanari R. (1990) Microstructural characterization of $MgFe_2O_4$ powders. *Mater. Chem. Phys.* 26, 513–526.

Gusmano G., Montesperelli G., Traversa E., Mattogno G. (1993) Microstructure and electrical properties of MgAl$_2$O$_4$ thin films for humidity sensing. *J. Am. Ceram. Soc.* 76, 743–750.

Hagleitner C., Hierlemann A., Lange D., Kummer A., Kerness N., Brand O., Baltes H. (2001) Smart single-chip gas sensor microsystem. *Nature* 414, 293–296.

Hair M.L., Hertl W. (1969) Adsorption on hydroxylated silica surfaces. *J. Phys. Chem.* 73, 4269–4276.

Harrey P.M., Ramsey B.J., Evans P.S.A., Harrison D.J. (2002) Capacitive-type humidity sensors fabricated using the offset lithographic printing process. *Sens. Actuators B* 87, 226–323.

Harsanyi G. (1995) *Polymer Films in Sensor Applications*. Technomic Publishing Company, Inc., Basel, Switzerland.

He Y., Zhang T., Zheng W., Wang R., Liu X., Xia Y., Zhao J. (2010) Humidity sensing properties of BaTiO$_3$ nanofiber prepared via electrospinning. *Sens. Actuators B* 146, 98–102.

Hertl W., Hair M.L. (1968) Hydrogen bonding between adsorbed gases and surface hydroxyl groups on silica. *J. Phys. Chem.* 72, 4676–4682.

Horie K., Yamashita T. (1995) *Photosensitive Polyimides*. Technomic Publishing Company, Lancaster, PA.

Igreja R., Dias C.J. (2006) Dielectric response of interdigital chemocapacitors: The role of the sensitive layer thickness. *Sens. Actuators B* 115, 69–78.

Ingram J.M., Greb M., Nicholson J.A., Fountain A.W. (2003) Polymeric humidity sensor based on laser carbonized polyimide substrate. *Sens. Actuators B* 96, 283–289.

Ishihara T., Matsubara S. (1998) Capacitive type gas sensors. *J. Electrocer.* 2 (4), 215–228.

Islam T., Kumar L., Khan S.A. (2012) A novel sol-gel thin film porous alumina based capacitive sensor for measuring trace moisture in the range of 2.5–25 ppm. *Sens. Actuators B* 173, 377–384.

Jachowicz R.S., Sentura S.D. (1981) A thin film capacitance humidity sensor. *Sens. Actuators* 2, 171–186.

Jachowicz R.S., Weremczuk J. (2000) Sub-cooled water detection in silicon dew point hygrometer. *Sens. Actuators A* 85, 75–83.

James D., Scott S.M., Ali Z., O'Hare W.T. (2005) Chemical sensors for electronic nose systems. *Microchim. Acta* 149, 1–17.

Jessensky O., Müller F., Gösele U. (1998) Self-organized formation of hexagonal pore arrays in anodic alumina. *Appl. Phys. Lett.* 72, 1173.

Juhász L., Mizsei J. (2009) Humidity sensor structures with thin film porous alumina for on-chip integration. *Thin Solid Films* 517, 6198–6201.

Kämpfner N. (Ed.) (2013) *Monitoring Atmospheric Water Vapour*. Springer, New York.

Kang U., Wise K.D. (1999) A robust high-speed capacitive humidity sensor integrated on a polysilicon heater. In: *Proceedings of Transducers'99*, Sendai, Japan, pp. 1674–1677.

Kang U., Wise K.D. (2000) A high speed capacitive humidity sensor with on-chip thermal reset. *IEEE Trans. El. Dev.* 47, 702–710.

Khanna V.K., Nahar R.K. (1984) Effect of moisture on the dielectric properties of porous alumina films. *Sens. Actuators* 5, 187–198.

Kim J.-H., Hong S.-M., Moon B.-M., Kim K. (2010) High-performance capacitive humidity sensor with novel electrode and polyimide layer based on MEMS technology. *Microsyst. Technol.* 16, 2017–2021.

Kim Y.-H., Kim Y.-J., Lee J.-Y., Kim J.-H., Shin K.H. (2004) A highly sensitive capacitive-type humidity sensor using customized polyimide film without hydrophobic elements. *Sens. Mater.* 16 (3), 109–117.

Kim Y., Jung B., Lee H., Kim H., Lee K., Park H. (2009) Capacitive humidity sensor design based on anodic aluminum oxide. *Sens. Actuators B* 141, 441–446.

Kitsara M., Goustouridis D., Chatzandroulis S., Beltsios K., Raptis I. (2006) A lithographic polymer process sequence for chemical sensing arrays. *Microelectron. Eng.* 83, 1192–1196.

Korotcenkov G., Cho B.K. (2017) Metal oxide based composites in conductometric gas sensors: Achievements and challenges. *Sens. Actuators B* 244, 182–210.

Korvink J.G., Chandran L., Boltshauser T., Baltes H. (1993) Accurate 3D capacitance evaluation in integrated capacitive humidity sensors. *Sens. Mater.* 4 (6) 323–335.

Koshkarian K.A., Kriven W.M. (1988) Investigation of a ceramic-metal interface prepared by anodic spark deposition. *J. Phys. Colloq.* 49, C5–213.

Kovac M.C., Chleck D., Goodman P. (1978) A new moisture sensor for in situ monitoring of sealed packages. *Solid State Technol.* 21, 35–39.

Kraus F., Cruz S., Muller J. (2003) Plasma-polymerized silicon organic thin films from HMDSN for capacitive humidity sensors. *Sens. Actuators B* 88, 300–311.

Kummer A.M., Hierlemann A., Baltes H. (2004) Tuning sensitivity and selectivity of complementary metal oxide semiconductor-based capacitive chemical microsensors. *Anal. Chem.* 76, 2470–2477.

Kummer A.M., Hierlemann A. (2006) Configurable electrodes for capacitive-type sensors and chemical sensors. *IEEE Sensors J.* 6 (1), 3–10.

Kummer A.M., Burg T.P., Hierlemann A. (2006) Transient signal analysis using complementary metal oxide semiconductor capacitive chemical microsensors. *Anal. Chem.* 78, 279–290.

Kurosaki S. (1954) The dielectric behavior of sorbed water on silica gel. *J. Phys. Chem.* 58, 320–324.

Kuroiwa T., Hayashi T., Ito A., Matsuguchi M., Sadaoka Y., Sakai Y. (1993) A thin film polyimide based capacitive type relative humidity sensor. *Sens. Actuators B* 13(14), 89–91.

Kuroiwa T., Miyagishi T., Ito A., Matsuguchi M., Sadaoka Y., Sakai Y. (1995) A thin-film polysulfone-based capacitive-type relative-humidity sensor. *Sens. Actuators B* 25, 692–695.

Laconte J., Wilmart V., Flandre D., Raskin J.P. (2003) High-sensitivity capacitive humidity sensors using 3-layer patterned polyimide sensing film. *Proc. IEEE Sens.* 1, 372–377.

Laville C., Pellet C. (2002a) Interdigitated humidity sensors for a portable clinical microsystem. *IEEE Trans. Biomed. Eng.* 49, 1162–1167.

Laville C., Pellet C. (2002b) Comparison of three humidity sensors for a pulmonary function diagnosis microsystem. *IEEE Sensors J.* 2 (2), 96–101.

Lazarus N., Bedair S.S., Lo C.-C., Fedder G.K. (2010) CMOS-MEMS capacitive humidity sensor. *J. Microelectromech. Syst.* 19 (1), 183–191.

Lazarus N., Fedder G.K. (2011) Integrated vertical parallel-plate capacitive humidity sensor. *J. Micromech. Microeng.* 21, 065028.

Lazarus N., Fedder G.K. (2012) Integrated vertical parallel-plate capacitive humidity sensor. *J. Micromechan. Microeng.* 21 (6), 065028.

Lazarus N., Fedder G.K (2012) Designing a robust high-speed CMOS-MEMS capacitive humidity sensor. *J. Micromech. Microeng.* 22, 085021.

Lea N. (1945) Notes on the stability of LC oscillators. *J. IEE, Part III* 92, 261–274.

Lee C-Y., Lee G.-B. (2005) Humidity sensors: A review. *Sensor Lett.* 3, 1–14.

Lee C.-Y., Wu G.-W., Hsieh W.-J. (2008) Fabrication of micro sensors on a flexible substrate. *Sens. Actuators A* 147, 173–176.

Lee W., Ji R., Gösele U., Nielsch K. (2006) Fast fabrication of long-range ordered porous alumina membranes by hard anodization. *Nat. Mater.* 5, 741–747.

Lei C., Tang L.C., Mao H.Y., Wang Y., Xiong J.J., Ou W. et al. (2015) Nanofiber forests as a humidity-sensitive material. In: *Proceedings of IEEE International Conference on MEMS 2015*, Estoril, Portugal, January 18–22, pp. 857–860.

Li G.Q., Lai P.T., Zeng S.H., Huang M.Q., Cheng Y.C. (1997) Photo-, thermal and humidity sensitivity characteristics of $Sr_{1-x}La_xTiO_3$ on SiO_2/Si substrate. *Sens. Actuators A* 63, 223–226.

Li G.Q., Lai P.T., Zeng S.H., Huang M.Q., Li B. (1999) A new thin-film humidity and thermal micro-sensor with Al/ $SrNb_xTi_{1-x}O_3/SiO_2$/Si structure. *Sens. Actuators A* 75, 70–74.

Li G.Q., Lai P.T., Zeng S.H., Huang M.Q., Liu B.Y. (1995) Effects of chemical composition on humidity sensitivity of Al/ $BaTiO_3$/Si structure. *Appl. Phys. Lett.* 66, 2436–2438.

Li G.Q., Lai P.T., Zeng S.H., Huang M.Q., Zeng S.H., Li B., Cheng Y.C. (2000) A humidity-sensing model for metal-insulator-semiconductor capacitor with porous ceramic films. *J. Appl. Phys.* 87 (12), 8716–8720.

Li J., Lin X., Li J., Liu Y., Tang M. (2012) Capacitive humidity sensor with a coplanar electrode structure based on anodised porous alumina film. *Micro. Nano Lett.* 7, 1097–1100.

Li M. (2016) Study of the humidity-sensing mechanism of $CaCu_3Ti_4O_{12}$. *Sens. Actuators B* 228, 443–447.

Lim S.H., Jaworski J., Satyanarayana S., Wang F., Raorane D., Lee S.-W., Majumdar A. (2007) Nanomechanical chemical sensor platform. In: *Proceedings of the 2nd IEEE International Conference on Nano/Micro Engineered and Molecular Systems*, Bangkok, Thailand, January 16–19, pp. 886–889.

Lin J., Heurich M., Obermeier E. (1993) Manufacture and examination of various spin-on glass films with respect to their humidity-sensitive properties. *Sens. Actuators B* 13(14), 104–106.

Lin J., Miiller S., Obermeier E. (1991) Two-dimensional and three-dimensional as basic elements for chemical sensors interdigital capacitors. *Sens. Actuators B* 5, 223–226.

Liu X., Eriksent G.F., Leistikot O. (1995) A new water permeability sensor for I 1 testing thin films. *J. Micromech. Microeng.* 5, 147–149.

Llobet E. (2013) Gas sensors using carbon nanomaterials: A review. *Sens. Actuators B* 179, 32–45.

Mahboob Md. R., Zargar Z.H., Islam T. (2016) A sensitive and highly linear capacitive thin film sensor for trace moisture measurement in gases. *Sens. Actuators B* 228, 658–664.

Mai L.H., Hoa P.T.M., Binh N.T., Ha N.T.T., An D.K. (2000) Some investigation results of the instability of humidity sensors based on alumina and porous silicon materials. *Sens. Actuators B* 66, 63–65.

Majewski J. (2016) Polymer-based sensors for measurement of low humidity in air and industrial gases. *Prz. Electrotechniczn.* 92 (8), 74–77.

Mamishev A.V., Sundara-Rajan K., Zahn M. (2004) Interdigital sensors and transducers. *Proc. IEEE* 92, 808–845.

Masuda H., Nishio K., Baba N. (1993) Fabrication of a one-dimensional microhole array by anodic oxidation of aluminium. *Appl. Phys. Lett.* 63, 3155–3157.

Matsuguchi M., Sadaoka Y., Nosaka K., Ishibashi M., Sakai Y., Kuroiwa T., Ito A. (1993) Effect of sorbed water on the dielectric properties of acetylene-terminated polyimide resins and their application to a humidity sensor. *J. Electrochem. Soc.* 140, 825–829.

Matsuguchi M., Shinmoto M., Sadaoka Y., Kuroiwa T., Sakai Y. (1995) Effect of cross-linking degree of PVCA film on the characteristics of capacitive-type humidity sensor. In: *Proceedings of the International Solid-State Sensors and Actuators Conference (TRANSDUCERS '95)*, Stockholm, Sweden, 25–29 June, Vol. 2, pp. 825–828.

Matsuguchi M., Umeda S., Sadaoka Y., Sakai Y. (1998a) Characterization of polymers for a capacitive-type humidity sensor based on water sorption behavior. *Sens. Actuators B* 49, 179–185.

Matsuguchi M., Kuroiwa T., Miyagishi T., Suzuki S., Ogura T., Sakai Y. (1998b) Stability and reliability of polyimide humidity sensors using cross-linked polyimide films. *Sens. Actuators B* 52, 53–57.

Matsuguchi M., Yoshida M., Kuroiwa T., Ogura T. (2004) Depression of a capacitive-type humidity sensor's drift by introducing a cross-linked structure in the sensing polymer. *Sens. Actuators B* 102, 97–101.

Matzler C., Wegmuller U. (1987) Dielectric properties of freshwater ice at microwave frequencies. *J. Phys. D: Appl. Phys.* 20, 1623–1630.

McCafferty E., Zettlemoyer A.C. (1971) Adsorption of water vapour on α-Fe_2O_3. *Discuss. Faraday Soc.* 52, 239–254.

McNeill W. (1958) The preparation of Cadmium Niobate by an anodic Spark reaction. *J. Electrochem. Soc.* 105, 544–547.

Meanna Pérez J.M., Freyre C. (1997) A poly(ethyl eneterephthalate)-based humidity sensor. *Sens. Actuators B* 42, 27–30.

Melcher J., Daben Y., Arlt G. (1989) Dielectric effects of moisture in polyimide. *IEEE Trans. Electron. Insulation* 24 (1), 31–38.

Mercer F.W., Goodman T.D. (1991) Effect of structural features and humidity on the dielectric constant of polyimides. *High Perform. Polym.* 3(4), 297–310.

Mistry K.K., Saha D., Sengupta K. (2005) Sol–gel processed Al_2O_3 thick film template as sensitive capacitive trace moisturesensor. *Sens. Actuators B* 106, 258–262.

Morimoto T., Nagao M., Tokuda F. (1969) Relation between the amounts of chemisorbed and physisorbed water on metal oxides. *J. Phys. Chem.* 73, 243–248.

Nahar R.K. (2000) Study of the performance degradation of thin film aluminum oxide sensor at high humidity. *Sens. Actuators B* 63, 49–54.

Nahar R.K., Khanna V.K. (1982) A study of capacitance and resistance characteristics of an Al_2O_3 humidity sensor. *Int. J. Electron.* 52, 557–567.

Nahar R.K., Khanna V.K. (1998) Ionic doping and inversion of the characteristic of thin film porous Al_2O_3 humidity sensor. *Sens. Actuators B* 46, 35–41.

Nahar R.K., Khanna V.K., Khokle W.S. (1984) On the origin of the humidity-sensitive electrical properties of porous aluminium oxide. *J. Phys. D: Appl. Phys.* 17, 2087–2095.

Nakajima A., Miazusawa Y., Yokoyama T., Chakraborty S., Hara K. (1999) Highly durable humidity sensor fabricated on a sapphire substrate. In: *Proceedings of Transducers'99*, Sendai, Japan, pp. 1672–1673.

Narimani K., Nayeri F.D., Kolahdouz M., Ebrahimi P. (2016) Fabrication, modeling and simulation of high sensitivity capacitive humidity sensors based on ZnO nanorods. *Sens. Actuators B* 224. 338–343.

Neimark A.V., Ravikovitch P.I. (2001) Capillary condensation in MMS and pore structure characterization. *Micropor. Mesopor. Mat.* 44–45, 697–707.

Nitta T., Hayakawa S. (1980) Ceramic humidity sensors. *IEEE Trans. Components Hybrids Manuf. Technol.* 3, 237–243.

Noe S.C., Pan J.Y., Senturia S.D. (1991) Optical waveguiding as a method for characterizing the effect of extended cure and moisture on polyimide films, In: *Proceedings of Annual Technical Conference-Society of Plastic Eng.*, Vol. 49, pp. 1598–1601.

Numata S., Tawata R., Ikeda T., Fujisaki K., Shimanoki H., Miwa T. (1991) Preparation of aromatic acid anhydride complexes as cross linking agents, *Japan Patent JP 03090076*.

OCG (1994) *Technical Data Sheet for Ciba Geigy "Probimide 7000, 7500 and 400"*. OCG Microelectronics Materials AG, Basel, Switzerland.

Okamoto K.I.,Tanihara N., Watanabe H.,Tanaka K., Kita H.,Nakamura A., et al. (1992) Sorption and diffusion of water vapor in polyimide films. *Polymer Phys.* 30(11), 1223–1231.

Oprea A., Bârsan N., Weimar U., Bauersfeld M.L., Ebling D., Wöllenstein J. (2008) Capacitive humidity sensors on flexible RFID labels. *Sens. Actuators B* 132, 404–410.

Osada Y., De Rossi D.E. (Eds.) (2000) *Polymer Sensors and Actuators*. Springer, Berlin, Germany.

Park K.K., Lee H.J., Kupnic M., Oralkan Ö., Khuri-Yakub B.T. (2008) Capacitive micromachined ultrasonic transducer as a chemical sensor. In: *Proceedings of the 7th IEEE Conference on Sensors, IEEE Sensors*, Lecce, Italy, October 26–29, 2008, pp. 5–8.

Park K.K., Lee H.J., Yaralioglu G.G., Ergun A.S., Oralkan Ö., Kupnic M. et. al. (2007) Capacitive micromachined ultrasonic transducers for chemical detection in nitrogen. *Appl. Phys. Lett.* 91, 094102.

Park S., Kang J., Park J., and Mun S. (2001) One-bodied humidity and temperature sensor having advanced linearity at low and high relative humidity range. *Sens. Actuators B* 76, 322–326.

Pecora A., Maiolo L., Cuscunà M., Simeone D., Minotti A., Mariucci L., Fortunato G. (2008) Low-temperature polysilicon thin film transistors on polyimide substrates for electronics on plastic. *Solid State Electron.* 52, 348–352.

Ponec V., Knor Z., Cerný S. (1974) *Adsorption on Solids*. Butterworth, London, UK, p. 405.

Qi Q., Zhang T., Yu Q., Wang R., Zeng Y., Li L., Yang H. (2008) Properties of humidity sensing ZnO nanorods-base sensor fabricated by screen-printing. *Sens. Actuators B* 133, 638–643.

Qui Y.Y., Azeredo-Leme C., Alcacer L.R., Franca J.E. (2001) A CMOS humidity sensor with on-chip calibration. *Sens. Actuators A* 92, 80–87.

Ralston A.R.K., Buncick M.C., Denton D.D. (1991) Effects of aging on polyimide: A model for dielectric behavior, In: *Proceedings of the 6th International IEEE Conference on Solid-State Sensors and Actuators (TRANSDUCERS'91)*, pp. 759–763.

Ralston A.R.K., Klein C.F., Thoma P.E., Denton D.D. (1996) A model for the relative environmental stability of a series of polyimide capacitance humidity sensors. *Sens. Actuators B* 34, 343–348.

Rittersma Z.M. (2002) Recent achievements in miniaturized humidity sensors – a review of transduction techniques. *Sens. Actuators A* 96, 196–210.

Rivadeneyra A., Fernández-Salmerón J., Agudo-Acemel M., López-Villanueva J.A., Capitan-Vallvey L.F., Palma A.J. (2016) Printed electrodes structures as capacitive humidity sensors: A comparison. *Sens. Actuators A* 244, 56–65.

Rivadeneyra A., Fernández-Salmerón J., Banqueri J., López-Villanueva J.A., Capitan-Vallvey L.F., Palma A.J. (2014) A novel electrode structure compared with interdigitated electrodesas capacitive sensor. *Sens. Actuators B* 204, 552–560.

Rocha K.O., Zanetti S.M. (2011) Structural and properties of nanocrystalline WO$_3$/TiO$_2$-based humidity sensors elements prepared by high energy activation. *Sens. Actuators B* 157, 654–661.

Roman C., Bodea O., Prodan N., Levi A., Cordos E., Manoviciu I. (1995) A capacitive-type humidity sensor using cross-linked poly(methyl methacrylate-Co-(2 hydroxypropyl)-methacrylate). *Sens. Actuators B* 25, 710–713.

Rotronic (2014) Measurement solutions. Part One: Theory. Humidity Academy. https://www.instrumart.com/assets/Rotronic-humiditytheory.pdf.

Sadaoka Y. (2009) Capacitive-type relative humidity sensor with hydrophobic polymer films, In: Comini E., Gaglia G., Sberveglieri G. (Eds.) *Solid State Gas Sensing*. Springer, New York, pp. 109–151.

Sadaoka Y., Matsuguchi M., Sakai Y., Takahashi K. (1988) Effects of sorbed water on the dielectric constant of some cellulose thin films. *J. Mater. Sci. Lett.* 7, 121–124.

Saha D., Das S., Sengupta K. (2008) Development of commercial nanoporous trace moisture sensor following sol–gel thin film technique. *Sens. Actuators B* 128, 383–387.

Saha D., Giri R., Mistry K.K., Sengupta K. (2005b) Magnesium chromate–TiO$_2$ spinel tape cast thick film as humidity sensor. *Sens. Actuators B* 107, 323–331.

Saha D., Mistry K.K., Giri R., Guha A., Sensgupta K. (2005a) Dependence of moisture absorption property on sol–gel process of transparent nano-structured γ-Al$_2$O$_3$ ceramics. *Sens. Actuators B* 109, 363–366.

Sajid M., Kim H.B., Yang Y.J., Jo J., Choi K.H. (2017) Highly sensitive BEHP-co-MEH:PPV + Poly(acrylic acid) partial sodium salt based relative humidity sensor. *Sens. Actuators B* 246, 809–818.

Sakai Y. (1993) Humidity sensors using chemically modified polymeric materials. *Sens. Actuators B* 13(14), 82–85.

Sakai Y., Sadaoka Y., Matsuguchi M. (1996) Humidity sensors based on polymer thin films. *Sens. Actuators B* 35, 85–90.

Sberveglieri G., Murri R., Pinto N. (1995) Characterisation of porous Al$_2$O$_3$-SiO$_2$/Si sensor for low and medium humidity ranges. *Sens. Actuators B* 23, 177–180.

Sberveglieri G., Rinchetti G., Groppelli S., Faglia G. (1994) Capacitive humidity sensors with controlled performances, based on porous Al$_2$O$_3$ thin films grown on SiO$_2$-Si substrate. *Sens. Actuators B* 18/19, 551–553.

Seiyama T., Yamazoe N., Arai H. (1983) Ceramic humidity sensors. *Sens. Actuators* 4, 85–96.

Sheng M., Gu L., Kontic R., Zhou Y., Zheng K., Chen G., Mo X., Patzke G.R. (2012) Humidity sensing properties of Bismuth Phosphates. *Sens. Actuators B* 166–167, 642–649.

Sheppard N.F., Day D.R., Lee H.L., Senturia S.D. (1982) Microdielectrometry. *Sens. Actuators A* 2, 263–274.

Shibata H., Ito M., Asakursa M., Watanabe K. (1996) A digital hygrometer using a polyimide film relative humidity sensor. *IEEE Trans. Instrum. Meas.* 45 (3), 564–569.

Steele J.J., Fitzpatrick G.A., Brett M.J. (2007) Capacitive humidity sensors with high sensitivity and subsecond response times. *IEEE Sens. J.* 7, 955–956.

Steele J.J., Gospodyn J.P., Sit J.C., Brett M.J. (2006) Impact of morphology on high-speed humidity sensor performance. *IEEE Sens. J.* 6, 24–27.

Steele J.J., Taschuk M.T., Brett M.J. (2008) Nanostructured metal oxide thin films for humidity sensors. *IEEE Sens. J.* 8, 1422–1429.

Sun H.T., Ming-Tang W., Ping L., Xi Y. (1989) Porosity control of humidity-sensitive ceramics and theoretical model of humidity-sensitive characteristics. *Sens. Actuators* 19, 61–70.

Tetelin A., Pellet C., Laville C., N'Kaoua G. (2003) Fast response humidity sensors for a medical microsystem. *Sens. Actuators B* 91, 211–218.

Thiel P.A., Madey T.E. (1987) The Interaction of water with solid surfaces: Fundamental aspects. *Surf. Sci. Rep.* 7, 211–385.

Thoma P., Colla J., Stewart R. (1979) A capacitance humidity-sensing transducer. *IEEE Trans. Components, Hybrids, Manuf. Technol.* 2, 321–323.

Timar-Horvath V., Juhasz L., Vass-Varnai A., Perlaky G. (2008) Usage of porous Al$_2$O$_3$ layers for RH sensing. *Microsyst. Technol.* 14, 1081–1086.

Toray (1992) *Technical Data Sheet for Toray "Photoneece UR-5100"*. Toray Ind., Tokyo, Japan.

Traversa E. (1995) Ceramic sensors for humidity detection: The state of the art and future developments. *Sens. Actuators B* 23, 135–156.

Tripathy A., Pramanik S., Manna A., Bhuyan S., Shah N.F.A., Radzi Z., Osman N.A.A. (2016) Design and development for capacitive humidity sensor applications of lead-free Ca, Mg, Fe, Ti-oxides-based electro-ceramics with improved sensing properties via physisorption. *Sensors* 16, 1135.

Varghese O.K., Gong D., Paulose M., Ong K.G., Grimes C.A., Dickey E.C. (2002) Highly ordered nanoporous alumina films: Effect of pore size and uniformity on sensing performance. *J. Mater. Res.* 17 (5), 1162–1171.

Varghese O.K., Grimes C.A. (2003) Metal oxide nanoarchitectures for environmental sensing. *J Nanosci. Nanotechnol.* 3 (4), 277–293.

Visscher G.J.W., Kornet J.G. (1994) Long-term tests of capacitive humidity sensors. *Meas. Sci. Technol.* 5, 1294–1302.

Viviani M., Buscaglia M.T., Buscaglia V., Leoni M., Nanni P. (2001) Barium perovskites as humidity sensing materials. *J. Eur. Ceram. Soc.* 21, 1981–1984.

Voorthuyzen J.A., Keskin K., Bergveld P. (1987) Investigations of the surface conductivity of silicon dioxide and methods to reduce it. *Surf. Sci.* 187, 201–211.

Urbiztondo M., Pellejero I., Rodriguez A., Pina M.P., Santamaria J. (2011) Zeolite-coated interdigital capacitors for humidity sensing. *Sens. Actuators B* 157, 450–459.

Wang C.-T., Wu C.-L., Chen I.-C., Huang Y.-H. (2005a) Humidity sensors based on silica nanoparticle aerogel thin films. *Sens. Actuators B* 107, 402–410.

Wang J., Wang X.-H., Wang X.-D. (2005b) Study on dielectric properties of humidity sensing nanometer materials. *Sens. Actuators B* 108, 445–449.

Wang L.L., Kang L.P., Wang H.Y., Chen Z.P., Li X.J. (2016) Capacitive humidity sensitivity of SnO_2:Sn thin film grown on siliconnanoporous pillar array. *Sens. Actuators B* 229, 513–519.

Wang L.L., Wang H.Y., Wang W.C., Li K., Wang X.C., Li X.J. (2013) Capacitive humidity sensing properties of ZnO cauliflowers, grown on silicon nanoporous pillar array. *Sens. Actuators B* 177, 740–744.

Wang Z., Shi L., Wu F., Yuan S., Zhao Y., Zhang M. (2011) Structure and humidity sensing properties of $La_{1-x}K_xCo_{0.3}Fe_{0.7}O_{3-\delta}$ perovskite. *Sens. Actuators B* 158, 89–96.

Wenmin Q., Wlodarski W. (2000) A thin-film sensing element for ozone, humidity and temperature. *Sens. Actuators B* 64, 42–48.

Westcott L., Rogers G. (1985) Humidity sensitive MIS structure. *J. Phys. E: Sci. Instrum.* 18, 577–580.

Wilson J.S. (Ed.) (2005) *Sensor Technology Handbook*. Elsevier, Burlington NJ.

Wu Y., Gu Z. (2009) Metal-insulator-semiconductor $BaTiO_3$ humidity sensor, In: *Proceedings of 2009 Symposium on Photonics and Optoelectronics*, Wuhan, China, 14–16 August, pp. 1–4.

Xie W.Y., Liu B., Xiao S.H., Li H., Wang Y.R., Cai D.P. et al. (2015) High performance humidity sensors based on CeO_2 nanoparticles. *Sens. Actuators B* 215, 125–132.

Xu Y., Chen C., Li J. (2007) Sorption and diffusion characteristics of water vapor in dense polyimide membranes. *J. Chem. Eng. Data* 52(6), 2146–2152.

Yagi H., Nakata M. (1992) Humidity sensor using Al_2O_3, TiO_2 and SnO_2 prepared by sol-gel method. *J. Ceram. Soc. Jpn.* 100, 152–156.

Yamazoe N., Shimizu Y. (1986) Humidity sensors: principles and applications. *Sens. Actuators* 10, 379–398.

Yang B., Aksak B., Lin Q., Sitti M. (2006) Compliant and low-cost humidity nanosensors using nanoporous polymer membranes. *Sens. Actuators B* 114, 254–262.

Yang D.K., Koros W.J., Hopfenberg H.B., Stannett V.T. (1985) Sorption and transport of water in kapton polyimide. *J. Appl. Polymer Sci.* 30, 1035–1047.

Yang D.K., Koros W.J., Hopfenberg H.B., Stannett V.T. (1986) The effects of morphology and hygrothermal aging on water sorption and transport in Kapton polyimide. *J. Appl. Polymer Sci.* 31, 1619–1629.

Yang Y.L., Lo L.H., Huang I.Y., Chen H.J.H., Huang W.S., Huang S.R.S. (2002) Improvement of polyimide capacitive humidity sensor by reactive ion etching and novel electrode design, In: *Proceedings of the IEEE Sensors Conference*, June 12–14, Orlando, FL, Vol. 1 (1), pp. 511–514.

Yates D.J.C. (1961) Infrared studies of the surface hydroxyl groups on titanium dioxide, and of the chemisorption of carbon monoxide and carbon dioxide. *J. Phys. Chem.* 65, 746–753.

You L., Cao Y., Sun Y.F., Sun P., Zhang T., Du Y., Lu G.Y. (2012) Humidity sensing properties of nanocrystalline $ZnWO_4$ with porous structures. *Sens. Actuators B* 161, 799–804.

Young L. (1961) *Anodic Oxide Films*. Academic Press, New York, NY.

Yuk J., Troczynski T. (2003) Sol–gel $BaTiO_3$ thin film for humidity sensors. *Sens. Actuators B* 94, 290–293.

Zampetti E., Pantalei S., Pecora A., Valletta A., Maiolo L., Minotti A., Macagnano A., Fortunato G., Bearzotti A. (2009) Design and optimization of an ultrathin flexible capacitive humidity sensor. *Sens. Actuators B* 143, 302–307.

Zarnik M.S., Belavic D. (2012) An experimental and numerical study of the humidity effect on the stability of a capacitive ceramic pressure sensor. *Radioengineering* 21 (1), 201–206.

Zhao J., Liu Y., Li X., Lu G., You L., Liang X. et al. (2013) Highly sensitive humidity sensor based on high surface area mesoporous $LaFeO_3$ prepared by a nanocasting route. *Sens. Actuators B* 181, 802–809.

Zheng K., Zhou Y., Gu L., Mo X., Patzke R.G., Chena G. (2010) Humidity sensors based on Aurivillius type Bi_2MO_6(M = W, Mo) oxide films. *Sens. Actuators B* 148, 240–246.

11 Resistive Humidity Sensors

11.1 GENERAL CONSIDERATION

In resistive sensors, the humidity measure is a change in the resistance/conductivity of humidity-sensitive film caused by interaction with water vapor. Modulation of electrical conductivity (or electrical conductance) by the presence of water vapor takes place via reactions such as adsorption, chemical reactions, diffusion, and swelling, taking place on the surface or in the bulk of the sensing layer. This modulation can be measured as a change in the current, which is correlated to the concentration of the water vapor in the surrounding atmosphere.

Figure 11.1 shows the typical structure of conductometric sensors, which is very simple. These sensors are suitable for mass production and can be characterized by small size and low cost. Conductometric gas sensors consist of two elements, a sensitive conducting layer, usually deposited on insulated substrate, and contact electrodes. The design configuration of most resistive sensors is based on interdigitated (interdigital) electrodes (Mamishev et al. 2004). These electrodes generally are fabricated using noble metal deposited on a glass, ceramic, or silicon substrate by the thick-film printing techniques (Traversa et al. 2000) or thin-film deposition (Kunte et al. 2009). A photoresist process can also be used. The humidity-sensitive films are deposited between them such that they touch both electrodes. One should note that in the hybrid structures, frequently the thick-film printed layer is the bottom layer, and the electrodes are deposited on the top of humidity-sensitive layer. To make the measurement, a dc or ac voltage is applied to the device, and the current, flowing through the electrodes, is monitored as the response. Depending on the configuration of the sensor (distance between electrodes) and humidity-sensitive material used, the impedance range of typical resistive elements can vary from 1 kΩ to 100 MΩ.

It is important to note that there are two opinions regarding the effect of the configuration of the electrodes on the parameters of the sensors. According to the first opinion, a change in the configuration of the interdigitated electrodes (IDEs) within reasonable limits has little effect on the sensor parameters. Cha and Gong (2013) fabricated polymer-based sensors using platforms with a different number of electrode fingers in IDE (3, 4, and 5) and gap sizes (310, 360, 410, and 460 mm), and they did not observe a noticeable change in the sensitivity of the sensors. IDEs were prepared using screen-printing technology. At the same time, Yang et al. (2002) believe that the distance between the electrodes is of utmost importance: One would think that a distance decrease could be convenient anyway, because of the lowering of the film impedance. However, as the distance between the electrodes decreases, the influence of the processes that occur on the electrode increases sharply. According to Yang et al. (2002), for each system, a target distance between the electrodes should be found at which the impedance is not too high at low relative humidity (RH)% and not too low at the higher values of RH%.

Ceramics, polymers and polymeric and solid electrolytes are commonly used humidity-sensitive materials for resistive-type sensors (Moneyron et al. 1991; Chakraborty 1995; Chakraborty et al. 1998; Wang et al. 2009a, 2009b; Anbia et al. 2012; Kim et al. 2012; Farahani et al. 2014). Carbon nanotubes and graphene can be also used for fabricating such sensors (Llobet 2013). The change in the resistance occurs due to the change in the number of free charge carriers or movable ions in humidity-sensitive film caused by water absorption and dissociation of ionic functional groups in polymers. The resistance or impedance

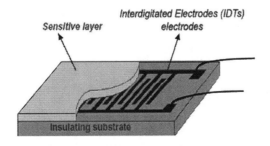

FIGURE 11.1 Structure and appearance of a typical planar thick/thin film–based humidity sensor (Model HR202, Open Impulse Inc. www.openimpulse.com) based on the interdigitated structure with the porous sensing element.

of the resistive-type sensor usually decreases as the RH increases. In some cases, the film-based sensors are formed by applying both printing techniques (e.g., screen or inkjet printing), and coating techniques such as chemical vapor deposition (CVD), spin coating and dip coating, or vacuum physical vapor deposition (PVD) techniques, such as thermal evaporation and cold sputtering (Tai et al. 2005). Among the mentioned deposition methods, electrochemical deposition is mostly operative when coating of a minuscule area with prepared polymers is required. However, there are rare works in which different deposition methods, such as spray techniques (Racheva et al. 1994) or combination of spray pyrolysis with the other techniques (Niranjan et al. 2001), were applied.

The main considerations in the fabrication of a stable resistive humidity sensor are careful selection of the sensitive layer, film-coating technique, bias potential, and measurement scheme. The sensitive layer should have affinity to the water vapor, and the film-coating technique needs to be compatible with the device fabrication process. In addition, its chemical composition, thickness, and morphology should be reproducible.

The main advantages of resistive humidity sensors are easy fabrication, simple operation, and low production cost, which means that well-engineered resistive sensors can be mass produced at reasonable cost. Moreover, these sensors are compact and durable. As a result, they are amenable to being placed in situ in monitoring wells.

11.2 POLYMER-BASED RESISTIVE HUMIDITY SENSORS

11.2.1 Introduction

Currently, one can find a huge number of works devoted to the development of humidity sensors of resistive type based on various polymers (Otsuka et al. 1980; Hijikagawa et al. 1983; Yamazoe and Shimizu 1986; Sakai et al. 1996; Osada and De Rossi 2000; Lee and Lee 2005; Fratoddi et al. 2016). Studies have shown that polymers from various groups can be used as humidity-sensitive materials (Huang 1985; Harsanyi 1995; Sakai 2000). The first group of such materials is the semiconductor or conductive polymers (Antoniadis et al. 1994; Yang et al. 1999; Chen and Lu 2005; Mogera et al. 2014; Squillaci et al. 2015; Wu and Hong 2016; Zafar and Sulaiman 2016). Polyaniline (PANI), poly(p-diethynylbenzene) (PDEB), and polyethylenedioxythiophene (PEDOT) represent this group of polymers.

The second group is formed by polyelectrolytes, in which the polymer itself contains an anionic or cationic group, usually on a side chain. Nafion, a per fluorinated sulfonated ionomer, is an example of this type of polymers. Humidity-sensitive copolymers based on conducting polymers and polyelectrolytes also can be prepared (Ogura et al. 1997; Liu et al. 2008). Conducting polymers are partly hydrophobic and thus exhibit higher durability due to a lower water uptake (Rauen et al. 1993). It was also established that, for sensors based on conducting/semiconducting polymers, dispersing some ions inside the materials leads to reduction in resistivity at low RH and thus generates greater absolute signals (Ogura et al. 1996; Fei et al. 2014a). One should note that polyelectrolyte-based humidity sensors form the largest group of resistive sensors. It is believed that polymers based on microporous hydrophilic polymers with ionic conductivity (polymer electrolytes or polyelectrolytes) are the best material for application in the resistance-type humidity sensors (Yamazoe and Shimizu 1986; Sakai 2000; Casalbore-Micelia et al. 2005).

The third type of polymer applied in resistive sensors is the polymer that does not itself possess charged moieties along its chain. Rather, the polymer acts as the solvent for electrolyte ions, which are able to move through the polymer matrix much as in a liquid electrolyte. Thus, the polymer serves as a solid ionic conductor. An example of this type of polymer is polyethylene oxide, in which lithium and other small cations have high mobility (Londono et al. 1997).

Hydrogels are actually a fourth type of polymer acceptable for application (Tierney 1996). However, in hydrogels, the solvent for the ions is water; the water-soluble polymer serves only to increase the viscosity of the solution. Hydrogel electrolytes are formed when the dissolved polymer is cross-linked into a rigid, three-dimensional matrix that can contain up to 98 vol.% water. To this extent, the hydrogel can be considered as a "solid" and, thus, a more easily processed form of liquid water. An example of a polymer that can form hydrogels is poly(vinylalcohol) (PVA). PVA can be cross-linked by a variety of agents and can form gels containing as little as 2–3 wt% PVA. However, considering the physical properties of hydrogels, their use in solid-state resistive sensors is very limited. However, hydrogels have found application as replacements for liquid electrolytes in electrochemical sensors. Because of their viscosity, they resist leakage from sensor housings, and many sensing components can be incorporated into them, such as ionic salts and electrochemical mediators.

A common thing about all types of polymer-based, resistive humidity sensors is the strong frequency dependence of conductivity (Figure 11.2). Therefore, all measurements are usually carried out at low frequencies of 20–100 Hz.

Resistive Humidity Sensors

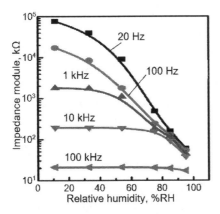

FIGURE 11.2 The impedance modules of cross-linked, amphiphilic, polymer-based sensor at different relative humidities (RHs) under different frequencies (1 V ac). (Reprinted with permission from Fei, T. et al., Stable cross–linked amphiphilic polymers from a one–pot reaction for application in humidity sensors, *Sens. Actuators B*, 227, 649–654, 2016. Copyright 2016, Elsevier.)

The phenomenon of ion transport in the abovementioned polymer electrolyte materials could not be understood clearly as yet, due to the lack of knowledge of the exact structural property correlations (Agrawal and Pandey 2008; Hallinan and Balsara 2013). However, the macroscopic studies on the basic ionic parameters and their temperature variations provide a wealth of information regarding ion dynamics. On the basis of temperature-dependent conductivity studies, it has been observed that these systems usually exhibited two dominant conduction mechanisms (Ratner et al. 2000) that divided these materials further into two separate groups. One group of polymer electrolytes obeys the Vogal–Tamman–Fulcher (VTF) type relationship, expressed by the following equation:

$$\sigma = A \cdot T^{-1/2} \cdot exp\left[-\frac{E_a}{T-T_0}\right] \quad (11.1)$$

where A is the pre-exponential factor, E_a is the activation energy, and T_0 is the equilibrium glass transition temperature close to T_g of the polymer electrolyte material. E_a can be computed from the nonlinear-least-square fit of the data from log σ versus $1/T$ plots. The other group follows the usual Arrhenius-type equation:

$$\sigma = \sigma_0 \cdot exp\left(-\frac{E_a}{kT}\right) \quad (11.2)$$

where the activation energy E_a can be computed from the linear-least-square fit of the data from log σ versus $1/T$ plots. The VTF conductivity versus reciprocal temperature plot is typically nonlinear, which is indicative of a conductivity mechanism involving ionic hopping motion coupled with the relaxation/breathing and/or segmental motion of polymeric chains. The materials exhibiting linear Arrhenius variations indicate ion transport via a simple hopping mechanism decoupled from the polymer-chain breathing.

11.2.2 Semiconductor or Conductive Polymers

The conducting mechanism of conjugated polymers could be interpreted by the principles of electron–hole pair traveling under electric field (Munn et al. 1997). However, the traveling is one-dimensional rather than three-dimensional due to the structure of conjugated polymers. Similar to inorganic semiconductors, the conductivity of polymers can be considerably enhanced by doping. The doping of polymers is actually to oxidize (*p*-type doping) or reduce (*n*-type doping) the backbone by chemical agents (Schopf and Kobmehl 1997; Nalwa 1997). The oxidation/reduction also generates by-products, like positive or negative ions. These ions become part of the polymer to keep the net charge at zero. They are usually called *counter ions*. It is expected that a *p*-type polymer semiconductor may contain negative counter ions, and an *n*-type one may contain positive counter ions.

In case that sufficient amount of water is absorbed on the doped polymers, one may expect that polymers may show some ionic conduction with the counter ions as the carriers. Water is well known for its protonation, and the released proton interacts with universally conjugated C=C double bonds. This effect was discovered and used for humidity sensing. In particular, study of the humidity-sensing property of PANI to water vapor has shown that this effect can be regarded as electron hopping assisted by proton exchange. Its conduction is both electronic and ionic. The ionic conduction is favourable, as long as mobile counter ions (for example, Cl⁻) exist in the polymer (Angelopoulos et al. 1987). Although it is verified that PANI and its derivative are sensitive to humidity, the response is very low due to weak hygroscopicity, at most one order of magnitude change in conductivity (Nechtschein et al. 1987; Traversa and Nechtschein 1987). An additional drawback of PANI is its poor processability. However, it is reported that converting poly(anthranilic acid) (PANA) into PANI by heat treatment is a convenient method for fabricating PANI with good humidity-sensing property (Ogura et al. 1999). Wu and Hong (2016) have also studied PANI-based humidity sensors. They have shown that the oxidation state of the interfacial PANI films (which closely relies on the conductivity of PANI films) could be optimized by controlling the concentration of aniline in the interfacial. At optimal conditions, the conductivity response of the thin-film humidity sensor attained 550% when the RH changed from 0% to 100% (Figure 11.3). At that, the sensor exhibited response (5–7 s) and recovery times (4–7 s) under dynamic tests that correspond to excellent performance data among the PANI-based humidity sensor systems reported previously. Furthermore, the sensitivity of the PANI film could be increased to 1900% and 14,000% when the PANI film was doped with acetic acid and hydrogen chloride, respectively (see Figure 11.3). But this improvement in sensitivity was accompanied by the increase in two times in the response and recovery times. Wu and Hong (2016) believe that the negative effect of the acid doping agents on the sensor's response and recovery times could be ascribed to the enhanced hydrophilicity of the interfacial PANI film after doping. In this case, the sensor requires the prolonged time period to absorb higher amount of water until establishing a new equilibrium state.

PDEB is also a conducting polymer due to its long-chain, conjugated structure. As reported by Yang et al. (1999), PDEB synthesized by some organic nickel catalyst is also sensitive to RH. A humidity sensor based on PDEB exhibited impedance change from 10^3 to 10^7 Ohm in the range of 15%–92% RH, small hysteresis (< 3% RH), and small temperature dependence. Sensors were prepared using Langmuir–Blodgett (LB) technology. The authors believed that the good humidity-sensing characteristics of the PDEB-based sensor are concerned with the interaction between hydrogen protons and super π-conjugate orbits in PDEB. They also supposed that the low hysteresis is conditioned by the fact that the highly hydrophobic structure of PDEB results in only weak interaction with moisture, and therefore reversible absorption/desorption can be easily achieved. This research has also shown that all stages of the formation of humidity-sensing material, including solvent and additive used for making sensors, have a strong effect on the sensing characteristics (see Figure 11.4). For example, when manufacturing the most sensitive sensors, a solution of PDEB incorporating a solution of epoxy chloropropane-co-epoxyethane was used. More recently, the same research group reported other conducting polymers, such as poly(propargyl benzoate) (PPBT), that also could be used as humidity-sensitive materials.

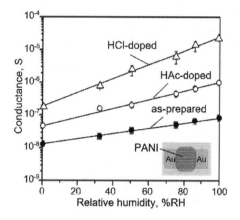

FIGURE 11.3 The response in conductance of the polyaniline (PANI) thin-film sensor to humidity (*inset*: schematic representation of the PANI thin-film sensor for electrical testing). All tests were conducted under 1 V bias voltage at room temperature (RT). (Reprinted with permission from Wu, T.-F. and Hong, J.-D., Humidity sensing properties of transferable polyaniline thin films formed at the air–water interface. *RSC Adv.*, 6, 96935–96941, 2016. Copyright 2016, Royal Society of Chemistry.)

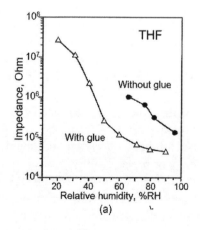 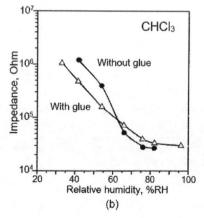

FIGURE 11.4 Effect of solvents and additive on humidity response of poly(*p*-diethynylbenzene) (PDEB)-based sensor. Solvent: (a) Tetrahydrofuran (THF) and sol-gel CHCl$_3$; additive: (Δ) with glue, ($\bullet$) without glue. (Reprinted with permission from Yang, M. et al., A novel resistive–type humidity sensor based on poly(p–diethynylbenzene), *J. Appl. Polym. Sci.*, 74, 2010–2015, 1999. Copyright 1999, Wiley-VCH Verlag GmbH & Co. KGaA.)

In terms of developing effective humidity sensors based on conductive polymers, special attention should be paid to the results obtained by Mogera et al. (2014). Using supramolecular nanofibers, Mogera et al. (2014) managed to produce highly sensitive humidity sensors with ultrafast response. The charge transfer (CT) nanofibers were prepared by the self-assembly of coronene-viologen-based donor and acceptor (D and A) pairs in water. Briefly, the potassium salt of coronene tetracarboxylate (CS) and the dodecyl substituted unsymmetrical viologen derivative (DMV) were used as D and A pairs that interact via ground-state CT interactions to form a hierarchical self-assembly. In water, these two components stack themselves to form cylindrical micelles (diameter, < 6 nm) following a surfactant-like assembly. Testing the sensors showed that, while a film of nanofibers exhibited high sensitivity ($\approx 10^4$), devices with few nanofibers showed extremely fast response (≈ 10 ms). Using UV-vis, XRD, and atomic force microscopy (AFM) measurements, it was found that the π-π interaction between the donor and acceptor molecules depends sensitively on the surrounding humidity influencing electrical conduction across the nanofiber. It was also established that the fabricated devices were found to be stable over 8 months during the study.

11.2.3 Polymers Containing Inorganic Salts

Since the first humidity sensor based on LiCl proposed by F.W. Dunmore in 1938 (Dunmore 1938, 1939), the type of sensor using LiCl dispersed in polyvinyl acetate has been used for humidity measurement for a long period of time. Sadaoka and Sakai (1984, 1986) have found that polyethylene oxide (PEO) doped with LiClO$_4$ had also a good impedance variation in the range 10^7–10^3 Ω without any hysteresis over the whole humidity range. Sadaoka and Sakai (1986) have also studied the humidity dependence on impedance of PEO doped with various alkali salts, such as LiF, LiCl, LiBr, LiClO$_4$, NaF, NaCl, KCl, and KI (see, for example, Figure 11.5). They assumed that the salts, composed of a small alkali ion and a large anion, form a complex with PEO, and the formation of these complexes is responsible for low resistivity and low activation energy for electrical conduction. Sadaoka and Sakai (1986) believe that the sorption of water results in an increase in the dielectric constant of the hybrid film, an expansion of the free volume by the plasticizer effect, and the formation of a shallow trapping site. The increase in the dielectric constant is responsible for the decrease in the dissociation energy of the ion pair. The expansion of free volume and the formation of shallow trapping sites leads to an increase in ion mobility (i.e., a decrease in the potential barrier height for ion migration). An et al. (1987) have also analyzed the influence of salts on the conductivity of polymers. They prepared a humidity sensor using mixtures of poly(styrene-co-quaternized-vinylpyridine) and perchlorate, such as HClO$_4$, LiClO$_4$, and KClO$_4$. It was found that the resistivity was the smallest in the HClO$_4$ complex and successively became large in the order of the KClO$_4$ complex and the LiClO$_4$ complex. From the electrochemical aspect, it was cleared that the degree of dissociation of the counterion exerted influence upon the resistivity of these complexes. It was also reported that

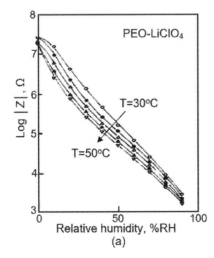
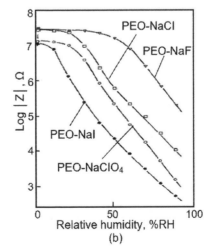

FIGURE 11.5 (a) Humidity dependence of impedance for polyethylene oxide (PEO)-LiClO$_4$ (Li/O = 0.2) at 10^3 Hz: (○) 30°C, (●) 35°C, (△) 40°C, (▲) 45°C, (□) 50°C, and (b) Humidity dependence of impedance at 10^3 Hz and 30°C: (●) PEO NaI, (○) PEO-NaClO$_4$, (□) PEO-NaCl, (△) PEO-NaF. (Reprinted with permission from Sadaoka, Y. and Sakai, Y., A humidity sensor using alkali saltpoly(ethylene oxide) hybrid films, *J. Mater. Sci.*, 21, 235–240, 1986. Copyright 1986, Springer Science+Business Media.)

the amount of sorbed water and the conductivity was in the order of $HClO_4$>$KClO_4$>$LiClO_4$.

Later, Sakai et al. (1987, 2001), studying a copolymer containing poly(2-acrylamido-2-methylpropane sulfonic acid) (AMPS) and hydrophobic poly(ethylene glycol dimethacrylate) and its alkali salts, have established that the majority of the carriers in these materials are the alkali cations. The alkali salts of AMPS were synthesized by neutralization with LiOH, NaOH, KOH, RbOH, or CsOH in an aqueous solution cooled in an ice bath. Results related to RH influence on the impedance of the films are plotted in Figure 11.6. The impedance

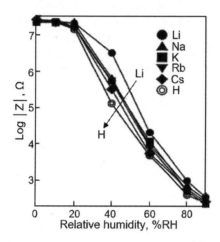

FIGURE 11.6 Plot of impedance as a function of relative humidity for alkali salts and acid: (•) Li, (Δ) Na, (□) K, (▲) Rb, (◊) Cs salt and (◎) acid form of poly(AMPS), measured at 1 kHz and 30°C. (Reprinted with permission from Sakai, Y. et al., Humidity sensor based on alkali salts of poly(2–acrylamido–2–methylpropane sulfonic acid), *Electrochim. Acta*, 46, 1509–1514, 2001. Copyright 2001, Elsevier.)

values are in the order of H<Cs<Rb<K<Na<Li salt. This order is the same as that of the equivalent conductance of the alkali ions at an infinite concentration. However, as can be seen in Figure 11.6, the nature of the alkali ions has almost no effect on the sensitivity of the sensors to RH. In addition, it was concluded that the Cs salt of poly(AMPS) is the most moisture-resistive material among the alkali salts and the acid form of poly(AMPS). The sensor, based on such material, did not deteriorate at all, even after dipping in water for 120 minutes.

The same approach, applied to nanoporous polymer based on Poly divinyl benzene (PDVB), was used by Fei et al. (2014a). The hydrophobic porous polymer was used as the host to prepare composited humidity-sensitive materials. Different levels of LiCl were loaded into the nanoporous polymer, and their humidity-sensing properties were studied. It was shown that alth0ough the polymer was not sensitive to humidity, the LiCl/PDVB composites showed excellent humidity-sensing properties (Figure 11.7a). At that, the sensor based on 5 wt% LiCl/PDVB showed the best sensing properties to RH. The impedance of the optimum sensor changed three orders of magnitude over the whole humidity range (10%–95% RH) with a good linearity, small hysteresis, and rapid response. The response time was 6 seconds when RH increased from 11% to 95% RH, and the recovery time was 30 seconds when RH decreased from 95% to 11% RH. The humidity-sensing properties of the composites comes from the interaction between the loaded LiCl and the water molecules, which decreased the impedances of the sensors as the RH increased. Fei et al. (2014a) assumed that

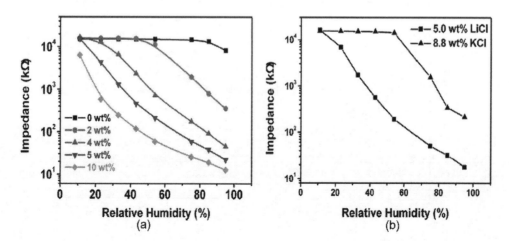

FIGURE 11.7 (a) The impedances of 0, 2, 4, 5, 10 wt% LiCl/PDVB sensors at different relative humidities (RHs) (1 V ac, 100 Hz) and (b) The impedances of 5.0 wt% LiCl/PDVB and 8.8 wt% KCl/PDVB sensors at different RH (1 V AC, 100 Hz). (Reprinted with permission from Fei, T. et al., Humidity sensors based on Li–loaded nanoporous polymers, *Sens. Actuators B*, 190, 523–528, 2014. Copyright 2014, Elsevier.)

the protons are the domination carriers in low humidity and ions (Li+ and Cl−) in a high humidity range. According to Fei et al. (2014a), the hydrophobic property of the polymer is beneficial for the stability of the sensors. Since the interaction of pure polymer and water molecules is weak, the stability of the polymer at high RH is much better than amphiphilic polymers used for humidity sensors. It is important that the addition of KCl instead of LiCl does not give such a significant effect (Figure 11.7b). Fei et al. (2014a) believe that sensing properties of LiCl/PDVB sensor at a wide humidity range come from the stronger hydrophilic properties of LiCl than KCl.

Another approach was used by Huang et al. (1991). They have proposed a humidity sensor using a composite film of perfluorosulfonic ionomer (PFSI)-H_3PO_4. It was reported that humidity from very low values (≈ 2 ppm) up to 100%RH can be measured with this sensor (see Figure 11.8). In addition, the sensors exhibited fast response times to moisture surge ($\tau \leq 10$ s), have no significant hysteresis (< 3% RH) or flow-rate dependence, and were immune to prolonged exposure to very high humidity levels. The sensor was exposed continuously for 10 hours to an atmosphere of 95% RH, and after that there was no irreversible effect from prolonged exposure to a highly humid atmosphere, and the response remained unaltered. It is possible that the intrinsic water affinity is increased by the presence of $H_3PO_4(P_2O_5)$, resulting in conductivity increase. In addition, the current must also be related to the proton activity of the film. It was assumed that the increase in the activity of the proton from H_3PO_4 is associated with the influence of perfluorosulfonic acid.

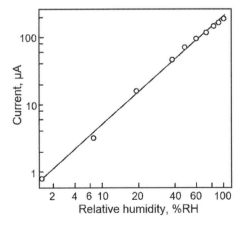

FIGURE 11.8 Logarithmic plot of sensor current as a function of relative humidity for a 5:1 PFSI:H_3PO_4 film sensor at 10 V applied voltage. (Reprinted with permission from Huang, H. et al., *Anal. Chem.*, 63, 1570–1573, 1991. Copyright 1991, American Chemical Society.)

11.2.4 Polyelectrolyte-Based Humidity-Sensitive Materials

Polyelectrolytes are polymers, whose repeating units bear an electrolyte group. Polycations and polyanions are polyelectrolytes. The most common acid groups are –COOH, –SO_3H, and –PO_3H_2, and the most common basic group is –NH_2. This means that the counterions for these groups are typically small, inorganic ions that are mobile within the polymer matrix. Based on functional groups, humidity-sensitive polyelectrolytes can be fundamentally divided into three major categories: quaternary ammonium salts (Rauen et al. 1993; Sakai et al. 1995, 2000; Gong et al. 2001; Lee et al. 2001; Park et al. 2002; Lee et al. 2003a, 2003b), sulfonate salts (Tsuchitani et al. 1988; Sakai et al. 1991), and phosphonium salts (Lee et al. 1999, 2003; Gong et al. 2002). Ammonium and sulfonate salts are traditional polyelectrolytes used in moisture sensing. Phosphonium salts were developed during the last decades. Since phosphorous is just below nitrogen in the periodic table, the chemical properties of phosphonium are nearly identical to those of ammonium. Sometimes phosphonium salts are favored in humidity sensing due to the easy formation of organic quaternary phosphonium with vinyl monomers (Son and Gong 2002).

It is important that polyelectrolyte properties are similar to both electrolytes (salts) and polymers, and are sometimes called *polysalts*. Like salts, their solutions are electrically conductive. Like polymers, their solutions are often viscous. Charged molecular chains, commonly present in soft-matter systems, play a fundamental role in determining structure, stability, and the interactions of various molecular assemblies. In particular, the type of dissociating groups in the polyelectrolyte determines its solubility in water and in other polar and hydrogen-bonding liquids (alcohols, etc.). For example, a sulfonated linear polystyrene readily dissolves in water, whereas polystyrene itself is one of the most water-resistant polymers. Both natural and synthetic polyelectrolytes are manufactured on a large scale. Common natural polyelectrolytes are pectin (polygalacturonic acid), alginate (alginic acid), carboxymethyl cellulose, and polypeptides. Examples of common synthetic polyelectrolytes are polyacrylic acid, polystyrene sulfonate, and their salts. PVA is also one such polymer that has an OH group at every other carbon in the backbone chain.

Polymers, having noted above strong acidic or basic groups, easily sorb water molecules. Moreover, these functional groups dissociate in aqueous solutions (water), making the polymers charged. As a result, their ionic conductivity increases with an increase in water adsorption, due to increases in the ionic mobility and/or charge carrier concentrations. The more the humidity in the atmosphere increases, the greater the ionization becomes. Conversely, when humidity decreases, ionization is reduced, and the concentration of mobile ions decreases. In this way, electrical resistance of the humidity-sensitive film responds to changes in humidity absorption and desorption. This means that the resistance of the polymer film can be measured as a value proportional to the RH (%RH). Such behavior of polyelectrolytes in humid air indicates that polyelectrolyte is an excellent material for electrical resistive-type humidity sensors: Its electrical conductivity varies with water sorption.

Contrary to the Pope cells, which are surface-resistance elements, the sensors analyzed are based on effects of a volumetric nature. Therefore, such sensors often are called *bulk polymer-resistive sensors*. One of the first polyelectrolytes used in developing humidity sensors was organic polymers, having constituent ionic monomers such as sodium styrenesulphonate (Tsuchitani et al. 1988; Rauen et al. 1993). Sodium styrenesulfonate (NaSS) is a highly reactive vinyl monomer containing a hydrophilic sulfonate group. Significant advantages of NaSS as compared with other monomers are its low toxicity and high thermal stability.

An important problem encountered in adopting these materials to humidity sensors is the water resistivity. Polymer electrolytes are generally hydrophilic and soluble in water, so that they have a poor durability against water or dew condensation (Sakai et al. 1996; Sakai 2000). Additionally, due to their high solubility, the conductivity reaches a very high value even if humidity is little. Their sensitivity to high RH is quite weak. A large hysteresis is also among the other significant disadvantages of these polymers. Therefore, improving their resistance to water is one of the main projects for this type of sensor. For resolving this problem, the hydrophilic sites or hydrophilic polymers were introduced by graft polymerization or impregnation in hydrophobic polymers, such as polyethylene. Random or block copolymerization also can be used. It was shown that this method is proved to be effective for lowering the conductivity of polyelectrolytes at low RH and enhancing sensitivity at high RH. As a result, humidity sensors, based on polyethylene film or a sintered alumina plate, both impregnated with a hydrophilic polymer, having a quaternary ammonium group or a sulfonate group, such as poly-N,N-dimethyl-3,5-dimethylene piperidium chloride or polystyrene sulfonate, were designed (Sakai et al. 1996). Humidity sensors based on hydrophilic polymers such as poly-(2-acrylamido-2-methylpropane sulfonate) or poly-(2-hydroxy-3-methacryloxypropyl trimethylammonium chloride) grafted on the surface of the pores of porous polyethylene and polytetrafluoroethylene (PTFE) films also were tested and shown parameters acceptable for real application (Sakai et al. 1985, 1996). The humidity dependence of the impedance measured at 1 kHz for the porous polyethylene films sulfonated at various reaction times is shown in Figure 11.9. Sakai et al. (1985) have shown that the grafted copolymer films are stable for long periods of time. These sensors, based on a three-dimensional thermosetting resin, demonstrated excellent water resistivity and high sensitivity to humidity. Their characteristics as humidity sensors did not change, even after the sensors have been immersed in water. One disadvantage of this type of sensor is the rather complicated chemical procedures for preparing the material.

Takaoka et al. (1983) and Kinjo et al. (1983) proposed another approach to fabrication of humidity sensors. A pair of interdigitated gold electrodes was printed and heated on the ceramic plate. The plate was coated with a copolymer of hydrophilic and hydrophobic monomers. The procedure for fabrication of this sensor is simpler than that described for the previous one. Many reports describing this type of device using either polyelectrolyte (Tsuchitani et al. 1988; Noguchi et al. 1989; Huang et al. 1991; Rausen et al. 1993; Xin and Wang 1994) or doped acetylene-type polymers have been published (Furlani et al. 1992a,

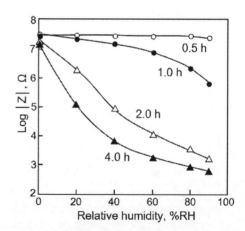

FIGURE 11.9 Impedance of the sulfonated porous polyethylene films as a function of humidity. The sulfonation reaction time is indicated in the figure. (Reprinted with permission from Sakai, Y. et al., Humidity sensors based on polymer thin films, *Sens. Actuators B*, 35, 85–90, 1996. Copyright 1996, Elsevier.)

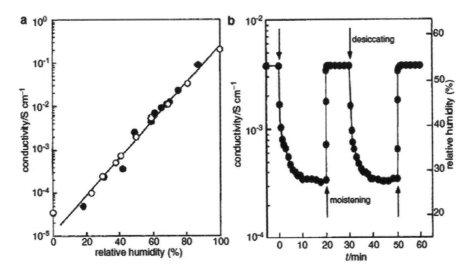

FIGURE 11.10 (a) Relationship between the conductivity of polyaniline (PANI)- poly(vinylalcohol) (PVA) composite and relative humidity in the (*open circle*) desiccating and (*filled circle*) moistening processes and (b) Conductivity and relative humidity as a function of time. Desiccating and moistening processes were begun at the point indicated by the *arrow*. The composite film was prepared with 0.28 vol% PANI and 99.72 vol% PVA. (From Ogura, K. et al., The humidity dependence of the electrical conductivity of a soluble polyaniline–poly(vinyl alcohol) composite film, *J. Mater. Chem.*, 7, 2363–2366, 1997. Copyright 1997, The Royal Society of Chemistry.)

1992b; Goldenberg and Krinichnyi 1993; Hwang et al. 1993; Taka 1993). However, synthesized copolymers were not sufficiently resistive to water. It was established that *p*-diethynylbenzene-co-propargyl alcohol (Li et al. 2002) and ethynylbenzene-copropargyl alcohol (Li and Yang 2002) copolymers are also candidates for humidity sensing. However, these two copolymers only respond to RH over 30%, and this response is nonlinear (Li and Yang 2002; Li et al. 2002).

Using a similar methodology, some researchers combine PANI (Ogura et al. 1997) and o-phenylenediamine (PoPD), a close structure to PANI with hygroscopic polymers like PVA to enhance the response (Ogura et al. 1996, 2001; Tonosaki et al. 2002). The conductivity response of PANI/PVA copolymers is shown in Figure 11.10. As reported, the copolymer PoPD/PVA was able to detect RH below 10% (Ogura et al. 1996, 2001). Some hysteresis was observed in this type of sensor after long-term operation or short exposure to high humidity. The doped copolymer film of poly(o-anisidine)(PoAN)/PVA was reported to be humidity sensitive as well (Ogura et al. 2000). In addition, Ogura et al. (2000) have shown that time of response strongly depends on the nature of PoAN doping (see Table 11.1)—in particular, the response time increases with an increase in the size of the protonic acid. It was suggested that this effect takes place because the rate of the acid–base transition of the conducting polymer

TABLE 11.1
Response Time of Acid-Doped Poly(o-anisidine)/Poly(vinylalcohol) Composites to Humidity Variation between 22% and 60% Relative Humidity

Acid-Doped Poly(o-anisidine) (wt%)[a]	Dopants[b]	Response Time (s)[c] Moistening[d]	Desiccating[e]
20	SA	6	12
	TSA	30	18
	CSA	36	27
	DBSA	60	24
1	SA	30	233
	TSA	60	315
	CSA	120	696
	DBSA	180	1140

Note: Measurements were performed at 25 ± 1°C.
SA, camphorsulfonic acid; DBSA, p-dodecylbenzenesulfonic acid; SA, sulfuric acid; TSA, p-toluenesulfonic acid.
[a] The weight percentage of acid-doped PoAN in the PVA composite was 20 or 1.
[b] Dopants used for the protonation of PoAN.
[c] The time required to reach a steady conductivity after exposure to humidity of 22% or 60%.
[d] Humidity was changed from 22% to 60%.
[e] Humidity was changed from 60% to 22%.

is limited by the diffusion of protonic acid. Hence, the conducting polymer should be prepared with a dopant anion as small as possible in size to make a highly sensitive humidity sensor with these composites.

Yang et al. (2006) have shown that nanoporous polymer membranes of polycarbonate (PC), cellulose acetate (CA), and polyester (nylon) also can be used for humidity sensor design (see Figure 11.11). There are also reports regarding humidity sensors based on poly(ethyleneterephtalate) (Meanna Perez and Freyre 1997), polyphospazene (Anchisini et al. 1996), polyether block amide (PEBA) (Yatsuzuka et al. 1993), naphthalene-based poly (arylene ether ketone) containing sulfobutyl pendant groups (Qi et al. 2016), poly(amide-sulfone)s (Jeon and Gong 2009), sulfonated polycarbonate (Rubinger et al. 2013), sulfonated tetrafluoroethylene-based fluoropolymer-copolymer (Nafion) (Wang et al. 1998), and many other polymers (Lee et al. 2003b, 2005a; Zhang et al. 2009; Takamuku et al. 2015).

In particular, Wang et al. (1998) studied sulfonated tetrafluoroethylene-based fluoropolymer-copolymer (Nafion) in different ionic forms (H^+, Li^+, and Na^+) and examined their impedance changes under the influence on RH. It was found that the content of the cation in the Nafion has a significant influence on the sensing properties of the film. The comparison of operating properties of H-, Li-, and Na-Nafion film–based humidity sensors has shown that the Li-Nafion film exhibited the largest sensitivity and stability (see Figure 11.12). Moreover, it was established that Nafion-based films have good stability against the various

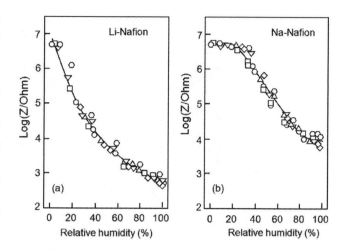

FIGURE 11.12 The relative humidity (RH) responses of (a) Li-Nafion and (b) Na-Nafion at different steps of the stability tests: ○ original untreated sample, immediately after preparation; □ – after being treated with 100% RH for 3 days at room temperature; Δ – after being put in water for 5 h at room temperature; – after being heated at 120°C for 2 h; ◊ after being heated at 100°C with 100% RH for 3 h; and ○ – after being treated with C_2H_5OH/CH_3OH water solution for 2 wk. (Reprinted with permission from Wang, H. et al., Comparison of conductometric humidity-sensing polymers, *Sens. Actuators B*, 40, 211–216, 1998. Copyright 1998, Elsevier.)

environments, such as high temperature (up to 473 K), high RH, and alcohol vapor (Tailoka et al. 2003).

The introduction of cross-linkers in polymers is another approach to stabilize polymer polyelectrolytes during their interaction with water (Sakai et al. 1988, 1995, 1996, 2000; Lee et al. 2004; Suna et al. 2010; Fei et al. 2014b, 2016). It is known that the polyelectrolytes at a certain degree of cross-linking form three-dimensional structures that swell in water rather than dissolving in it. Mechanism of polymer polyelectrolyte's swelling was discussed in Volume 1 (Chapter 10). The cross-linked polyelectrolytes can retain (extremely) large amounts of liquid relative to their own mass through hydrogen bonding with water molecules. They are called *hydrogels* and *superabsorbent polymers* (SAPs) when (slightly) cross-linked. The SAP's ability to absorb water is a factor of the ionic concentration of the aqueous solution. In deionized and distilled water, a SAP may absorb water up to 500 times its weight and from 30 to 60 times its own volume—that is, the hydrogel can consist of more than 99% liquid. The total absorbency and swelling capacity of SAPs is controlled by the type and degree of cross-linking. However, when developing humidity sensors, as a rule, it is necessary to avoid the presence of a swelling effect.

As a result of numerous studies aimed at searching for the optimal compositions of humidity-sensitive

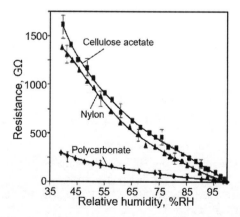

FIGURE 11.11 The dc resistance measurements at 25°C for various relative humidity values for devices fabricated on (1) polycarbonate, (2) cellulose acetate, and (3) nylon membranes, respectively. (Reprinted with permission from Yang, B. et al., Compliant and low-cost humidity nanosensors using nanoporous polymer membranes, *Sens. Actuators B*, 114, 254–262, 2006. Copyright 2006, Elsevier.)

materials and the optimum degree of cross-linking (in many cases, simple cross-linking cannot assure satisfactory insolubility, lifetime, and intensity), polymer polyelectrolytes have been developed that have high stability in wet atmosphere, and the sensors based on them are highly sensitive (Su et al. 2001; Gong et al. 2002; Lee et al. 2005b; Li et al. 2005; Gong et al. 2010; Fei et al. 2014b, 2016; Qi et al. 2016; Park and Gong 2017). In such polymer polyelectrolytes, the cross-linker forms a "bridge" that connects two polymer chains together so that a dense, insoluble, and intensive polymer network is resulted. In particular, using this approach, sensors based on the using either polyelectrolyte or doped acetylene-type polymers have been improved. As is known, the above-mentioned polymers are not sufficiently resistive to water. Forming the cross-linked polymer networks on the surface of this type of substrate allowed for making the polymer coating insoluble in water but sensitive to humidity. The indicated approach was applied for poly(2-hydroxy-3-methacryloxypropyl-trimethyl ammonium chloride), organoalkoxysilanes [RSi(OC$_2$H$_5$)$_3$] poly(vinylpyridine), and poly(chloromethyl styrene) (Sakai et al. 2000). In the case of poly(vinylpyridine) and poly(chloromethyl styrene), the degree of reaction should be as high as possible in order to reduce the hysteresis. It was found that the cross-linking of poly(methyl methacrylate) (PMMA) also promoted the improvement of PMMA-based humidity sensors (Matsuguchi et al. 1991). Matsuguchi et al. (1991) observed the decrease of the hysteresis and the temperature coefficient. The durability in the presence of organic vapors was also much improved. The cross-linking of PMMA was carried out using divinylbenzene or ethylene glycol dimethacrylate. For humidity-sensing polyelectrolytic films, the same method (cross-linking) is also used to enhance the mechanical properties of the sensors, as well as strengthening the adherence to the substrate (Chen and Lu 2005).

Gong et al. (2002) proposed humidity sensors using cross-linked polyelectrolyte prepared from mutually reactive copolymers containing phosphonium salt (Figure 11.13). These sensors, which changed resistance from 725 kΩ to 3 kΩ at RH change from 30% to 90% RH, were also found to be comparatively resistant to water. Lee et al. (2001, 2003a, 2003b) proposed two types of resistive humidity sensors using epoxy resin containing quaternary ammonium salts and polyelectrolytes based on new-type mutually cross-linkable copolymers. The impedance of the latter type varies from 755 kΩ to 2.52 kΩ in the humidity range of 30% to 90% RH. Cross-linking techniques provide an efficient means of improving the water durability and of providing a long-term

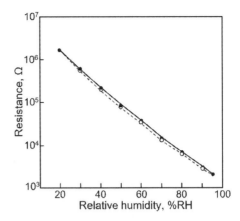

FIGURE 11.13 Dependence of resistance on the relative humidity (RH) sensor using the cross-linked copolymers VP/VTBPC and CEVE/IVE: 4/1; (○) humidification process, and (●-black) desiccation process at 25°C, 1 kHz, and 1 V. VP, 4-vinylpyridine; VTBPC, vinylbenzyl tributyl phosphonium chloride; CEVE, 2-chloroethylvinyl ether; and IVE, isobutyl vinyl ether. (Reprinted with permission from Gong, M.S. et al., Humidity sensor using cross–linked polyelectrolyte prepared from mutually reactive copolymers containing phosphonium salt, Sens. Actuators B, 86, 160–167, 2002. Copyright 2002, Elsevier.)

stability (300 days). In addition, these devices have a rapid response. Response time typically was less than 75 seconds when the humidity was changed abruptly from 33% to 94% RH. Su et al. (2001) presented a resistive humidity sensor fabricated by thick-film techniques using poly(2-acrylamido-2-methylpropane sulfonate) (poly-AMPS) modified with tetraethyl orthosilicate (TEOS) as the sensing material. This sensor was simple to fabricate and exhibited a reduced degree of hysteresis (< 2%), good linearity at humidity levels in the range 30%–90% RH, good-enough long-term stability (Figure 11.14), and satisfactory resistance to high-humidity atmospheres (e.g., 95% RH). Fratoddi et al. (2004) investigated the resistive-type humidity sensors based on two poly(monosubstituted) acetylenes, namely poly(N,N-dimethylpropargylamine) (Pd-PDMPA) and poly(propargylalcohol) (PPOH). These sensors exhibited detectable responses to RH as low as 2% and a variation of five orders of magnitude in the RH range 0–90%. Li et al. (2005) have used the copolymer of 4-binylpyridine and butyl methacrylate, which was simultaneously cross-linked and quaternized by reaction with 1,4-dibromobutane. It was found that this polymer is good humidity-sensitive material. The humidity sensor based on the cross-linked and quaternized copolymer showed high sensitivity (S > 10^2 in the range of 35%–95% RH), good response linearity in semilogarithmic scale, quick response, and, especially, a very small hysteresis

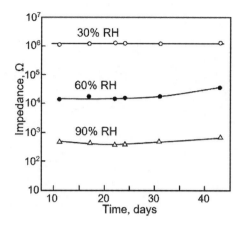

FIGURE 11.14 Effect of long-term stability on the sensor response with humidity-sensing layer of 20 μm thickness prepared using the addition of TEA (0.4 mL) and TEOS (16.25%, w/w) into poly(2-acrylamido-2-methylpropane sulfonate) (poly-AMPS): (○) 30% RH, (●) 60% RH, (△) 90% RH. RH, relative humidity; TEOS, hydrolytic tetraethyl orthosilicate; TEA, triethylamine. (Reprinted with permission from Su, P.G. et al., Use of poly(2–acrylamido–2–methylpropane sulfonate) modified with tetraethyl orthosilicate as sensing material for measurement of humidity, *Anal. Chim. Acta*, 449, 103–109, 2001. Copyright 2001, Elsevier.)

(< 1 % RH). In addition, it also exhibited a resistance to humid environment and ethanol vapor.

Park and Gong (2017) have developed a water-durable, polymeric-resistive humidity sensor. It was designed using rigid sulfonated polybenzimidazole (SPBI) with tetramethylammonium (Me_4N^+) as the counter cation. Sensors exhibited a linear response with a resistivity, varying by four orders of magnitude, between 20% and 95% RH. In addition, they have shown that cross-linking and anchoring the film to the substrates with a coupling reagent significantly improved the water durability and stability at a high humidity and high temperature. When the device was kept at 120°C for 480 hours, the resistance changed slightly.

It was also found that, under certain conditions and compliance with the manufacturing technology, polymer-resistive sensors can show very high reproducibility in electrical conductivity under various measurement regimes. Therefore, such sensors can be highly interchangeable. For commercial sensors, the variation in humidity readings when measured under the same conditions does not exceed ± 2%–5% RH. As a rule, these sensors can operate in the temperature range from –40° to 100°C. At higher temperatures, as a rule, the degradation of characteristics occurs.

Studies have shown that sensitivity, stability, reproducibility, and reliability of polymer-based resistive humidity sensors depend on the chemical structure of the polymers. However, there is a common concept for designing polymeric films for humidity sensors, which is independent of the polymer used. These investigations have provided useful information concerning sensitivity, response time, and sensing mechanism. The response time is determined by the time it takes for water vapor to diffuse into or out from a polymer film to reach an equilibrium, so that the film should be as thin as possible for quick responses. Besides the film thickness, an increase in hydrophobicity of the constituent ionic monomer appears to shorten the response time, while the mole fraction of the constituent ionic monomer to nonionic monomer strongly affects the sensing characteristics (Tsuchitani et al. 1985). In addition, Sakai et al. (1995), studying cross-linked proton conductive membranes (PCMS) using N,N,N,N-Tetramethyl-1,6-hexanediamine (TMHDA) as cross-linking reagent, established that as the degree of quaternization (that is to say, the density of the hydrophilic group $-N(CH_3)_2^-$) increases, the response time decreases (see Figure 11.15).

It was also found that the hysteresis becomes smaller as the density of hydrophilic sites in the film increases (Sakai et al. 1989) (see Figure 11.16). It was confirmed that the decrease in hysteresis and the response time takes place due to increase in the diffusion coefficient of water molecule in the polymer as the quaternization proceeds (Sakai et al. 2000). At a fixed RH and temperature, the electric resistance decreases

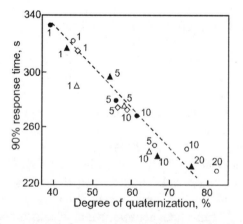

FIGURE 11.15 Plot of 90% response time in the dehumidifying run, from 80% to 50% relative humidity (RH), against the degree of quaternization. The reaction time in hours is indicated near the plots. The PCMS/TMHDA mole ratios are: ● – 8; ◊ – 4; △ – 2.7; ▲ – 2; ○ – 1.25. PCMS, poly(chloromethyl styrene); TMHDA (crosslinking reagent), N,N,N',N'-tetramethyl-1,6 hexanediamine. (Reprinted with permission from Sakai, Y. et al., Humidity sensor durable at high humidity using simultaneously crosslinked and quaternized poly(chloromethyl styrene), *Sens. Actuators B*, 24, 689–691, 1995. Copyright 1995, Elsevier.)

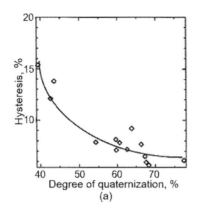 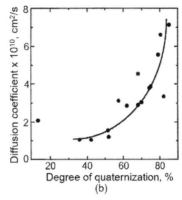

FIGURE 11.16 (a) Hysteresis of the response of cross-linked poly(chloromethyl styrene) (PCMS) film to relative humidity (RH) at 40% RH as a function of degree of conversion and (b) Diffusion coefficient of water molecules in the film of cross-linked poly(chloromethyl styrene) (PCMS) film as a function of degree of conversion. (Reprinted with permission from Sakai, Y. et al., Humidity sensor using cross–linked poly(chloromethyl styrene), *Sens. Actuators B*, 66, 135–138, 2000. Copyright 2000, Elsevier.)

when the fraction of the constituent ionic monomer increases. It seems that the ion pour of the polymer electrolytes is dissociated with water adsorption to liberate hydrated ions as dominant charge carriers. Wang et al. (1998) prepared films of differently modified PVA and phthalocyaninosilicon (TA), and examined their impedance changes under influence on RH. They established that the content of the hydrophilic group, as well as its hydrophilicity, really plays a crucial role in determining the response of the polymer film to RH. At that, a significant difference in sensing behavior was observed between molecular and ionic forms of TA film—namely, the molecular TA film showed almost no response, while the ionic form TA film showed a large conductometric response to RH. Lee and Chiu (2000), when studying poly(butyl acrylate [BA]-acrylic acid [AA]) synthesized by soapless emulsion copolymerization, have found that the sensitivity to humidity of the polymer films increased, with an increase of hydrophilic functional groups. They concluded that, with more hydrophilic functional groups, the dissociation of ions (Na^+) is much more possible, and the mobility of ions (e.g., Na^+) increases due to adsorption of many more water molecules at higher RH.

Regarding the role of polymer type in sensor response, one can state that the results depend on many factors and vary in a wide range. For example, according to Chatzandroulis et al. (2004), among poly(hydroxyethymethacrylate) (PHEMA), poly(methylmethacrylate) (PMMA), poly(vinyl acetate) (PVAc), epoxidized novolac (EPN), and polydimethylsiloxane (PDMS) layers, PHEMA presented the highest response to water vapor. This can be explained by the formation of hydrogen bonds between the OH group of PHEMA and water molecules. On the other hand, both PMMA and PVAc, lacking a high hydrogen-bonding capability, showed considerably lower sensitivity, whereas EPN and PDMS, being more hydrophobic, presented the lowest response to humidity. A detailed analysis of various ion-conducting polymers promising for application in sensors can be found in Sakai (2000).

It is important to note that, in comparison to capacitance measurement, the measurement of resistance is very simple and straightforward. This change can be measured by an electronic circuit. However, the most resistive sensors use symmetrical ac excitation voltage with no dc bias to prevent polarization of the sensor. Nominal excitation frequency is from 30 Hz to 10 kHz. This means that a resistive sensor also needs an alternating excitation voltage, not for the measurement but to avoid destroying it, caused by one-way electrolytic ion movement in humidity-sensitive materials. Then, the resulting current flow is converted and rectified to a dc voltage signal for additional scaling, amplification, linearization, or A/DR conversion. The response time for most resistive sensors ranges from 10 to 60 seconds for a 63% step change (Sakai et al. 1996).

Typical characteristics of resistive polymer humidity sensors are shown in Figures 11.9 through 11.13. It is seen that resistive sensors exhibit a strong nonlinear response to changes in humidity: As a rule, the resistance has an exponential dependence on humidity resistance. However, this response may be linearized by analog or digital methods.

11.2.5 Nanocomposite-Based Humidity Sensors

Other popular polymer-based materials for resistive humidity sensors are polymer-based nanocomposites (Hatchett and Josowicz 2008). Developers, through the use of various additives such as black carbon (Ishida et al. 1978; Barkauskas 1997; Shim and Park 2000), carbon nanotubes (CNTs) (Yoo et al. 2010; Çiğil et al. 2017), graphene (Lin et al. 2013), metal oxides (Sun et al. 2009; Zhang et al. 2017), and metals (Çiğil et al. 2017), are trying to improve the parameters of the sensors being developed. Currently, all composites used in the development of humidity sensors can be divided into three groups:

- composites based on polymers with insulating properties
- composites based on polymers with semiconductor properties
- composites based on polyelectrolytes

In composites belonging to the first group, the polymer mainly acts as the insulating matrix, while dispersed conducting particles (black carbon, CNTs, graphene, conductive metal oxides, and metals) provide the conducting path for sensing (Cho et al. 2004). Well-known nonconducting polymers, such as PVA, polystyrene (PS), PMMA, and PVAc, can be used in such composites. Due to adsorption of water vapors in such nanocomposites, there are volumetric changes of the matrix polymer (swelling), which can lead to a distinct change in percolation-type conductivity. With that, maximum effect takes place around a certain critical composition of the material, which is called the *percolation threshold*. When the concentration of conducting particles is very high, the particles pack closely in the composite and form conductive pathways, which impart a low-resistance response to the sensor. As the concentration of conducting particles decreases (e.g., in a swollen polymer), the distances between particles increase and the resistance of the composite gradually increases. When the concentration of conducting particles decreases to a point at which the conductive pathways are disrupted, the resistance of the composite increases sharply (by many orders of magnitude). This point is the percolation threshold.

Surely, the percolation threshold is dependent on the shape of the conducting particle. As a rule, a composite consisting of particles with higher aspect ratio shows a lower threshold and higher sensitivity (Abraham et al. 2004). Conducted experiments showed that CNTs, with almost one-dimensional threadlike structure and good conductivity, are ideal as the dispersed conducting particles in the insulating matrix for gas-sensing systems. It was established that addition of CNTs to a polymer matrix leads to a very low electrical percolation threshold. For example, Hu et al. (2006) reported that the percolation threshold of 0.9 wt% for electrical conductivity in CNT/poly(vinylidene fluoride) (PET) composite has been found (see Figure 11.17). In this research, pristine multiwall nanotubes (MWNTs) with diameter 10–20 nm and length 5–15 μm were used.

From the presented results, it becomes clear that if the concentration of conducting particles is only slightly

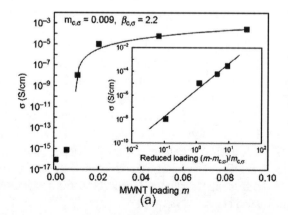
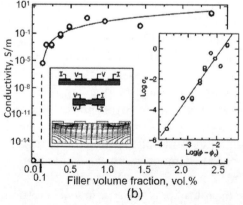

FIGURE 11.17 Electrical conductivity(-ies) of the (a) poly(vinylidene fluoride) (PET)/MWNT and (b) Reduced graphene oxide (RGO)/epoxy nanocomposites as a function of MWNT and graphene loading. *Inset*: a log–log plot of electrical conductivity vs reduced MWNT and graphene loading. The *solid lines* are fits to a power law dependence of electrical conductivity on the reduced fillers loading. ([a] Reprinted with permission from Hu, G. et al., Low percolation thresholds of electrical conductivity and rheology in poly(ethylene terephthalate) through the networks of multi–walled carbon nanotubes, *Polymer*, 47, 480–488, 2006. Copyright 2006, Elsevier; [b] Reprinted with permission from Potts, J.R. et al., Graphene–based polymer nanocomposites, *Polymer*, 52, 5–25, 2011. Copyright 2011, Elsevier.)

higher than the percolation threshold, a small amount of swelling may cause a dramatic change in the sensor resistance. In particular, increases in the resistance by factors of more than 10^6 can be achieved (Tsubokawa et al. 2001). Thus, in such sensors, conductive nanoparticles provide electrical conductivity and the polymer (any polymer) provides the sensor function. Therefore, these sensors mainly have electronic mechanism of conductivity. Considering that the amount of swelling corresponds to the concentration of the water vapor in contact with polymer, and therefore this effect, the change in conductivity measured with a dc voltage can be used to determine the concentration of water vapor in the air. Because weak Van der Waals forces between the polymer and water vapor molecules are responsible for the swelling of the sensing layer, the change is purely physical and reversible, which makes the sensor reusable.

The first humidity sensor using this principle of operation was developed in 1978 (Ishida et al. 1978). The humidity-sensitive film in this case was a cross-linked hydrophilic acrylic polymer in which carbon particles are dispersed. In the following, the same approach was used by Barkauskas (1997). In particular, Barkauskas (1997) developed a resistive humidity sensor using a sensing film prepared from PVA and graphitized carbon black disperse phase. The experimental data revealed a sensitivity of 8 Ohm/%RH at room temperature and 5 Ohm/%RH at 100°C. The response time of this sensor was 45 seconds. The same situation is typical for other humidity sensors being investigated. Experiments related to humidity sensor testing have shown that, while the absolute response was large, the response times tended to be slow, typically on the order of minutes. Thin films of composite could respond more quickly.

The same approach was used by Yoo et al. (2010). To enhance the sensitivity of polyimide-based resistive sensors, Yoo et al. (2010) prepared a nanocomposite film in which plasma-treated multiwall CNTs were blended with polyimide. It was established that when the threshold concentration of CNTs (around 0.1 wt%) is reached, a sharp drop in resistance occurs and the change of the character of the resistance dependence versus humidity takes place (Figure 11.18). Yoo et al. (2010) observed that, for small concentrations of CNTs, the resistance decreases with increasing humidity, while for large concentrations the resistance growth is observed. At that, when the concentration was higher than a critical value, the sensor exhibited a very good linearity over a wide humidity range. A similar change in resistance with increasing air humidity was observed by Shim and Park (2000) in the study of poly(1,5-diaminonaphthalene) (DAN)/carbon composite.

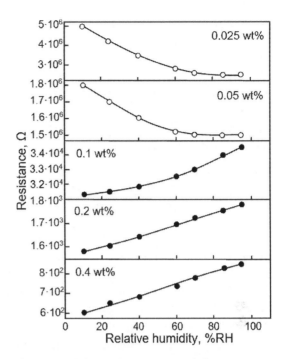

FIGURE 11.18 The resistance of polyimide (PI)-carbon nanotubes (CNTs) composite as a function of the relative humidity (RH) for different values of the carbon nanotube concentration. Data taken at room temperature at the frequency of 20 kHz. (Reprinted with permission from Yoo, K.P. et al., Novel resistive type humidity sensor based on multiwall carbon nanotube–polyimide composite film, *Sens. Actuators B*, 145, 120–125, Copyright 2010, Elsevier.)

However, one should consider that high sensitivity is realized only in the concentration range of conductive additives corresponding to the percolation threshold. In addition, the sensor response in this concentration range has a threshold character with a strong nonlinearity. If the concentration of conductive additives significantly exceeds this concentration, then the sensitivity would be much lower, but the change in conductivity depending on air humidity in this case can be described by a linear expression. Therefore, practical resistive sensors have generally been fabricated using polymers loaded with conducting particles at levels substantially above the percolation threshold. Exactly such an approach was used by Barkauskas (1997) and Yoo et al. (2010).

Certainly, resistors with different polymer matrices will respond differently to various vapor. Belmaraes et al. (2004) established that the swelling of the polymer matrix is the greatest when there is a match between the solubility parameter of the polymer and that of the vapor. This means that the selectivity and the sensitivity of the resistive sensors designed on the basis of the swelling effect will be governed by conducting particles and the polymer selected for sensor design. For example, poly(N-vinyl-pyrrolidone) is hydrophilic, so it swells substantially (and increases the sensor resistance) in water vapor, but not in toluene vapor, whereas polyisobutylene is hydrophobic, so it swells (and increases the sensor resistance) in toluene vapor, but not in water vapor (Belmaraes et al. 2004).

As a disadvantage of such sensors, the following ones are being emanated (Korotcenkov 2013):

- Low reproducibility of the sensor parameters, especially near the percolation threshold and their strong instability due to strong dependence of sensor characteristics on the concentration and parameters of the conducting nanoparticles, the nonuniform distribution of particles all over the polymer volume, and the agglomeration of conducting particles in polymers during operation. All this creates significant difficulties in the development of sensors suitable for the sensor market. All the factors listed above require more careful control in a technological process. As a result, the cost of sensors could grow.
- It is also noted that some hysteresis can occur when the sensor is exposed to high concentrations of water vapor.
- It should also be considered that the polymers possessing the swelling effect do not have a high stability of characteristics, since they are subjected to degradation under UV radiation and during interaction with ozone.

Composites belonging to the second group (i.e., composites based on polymers with semiconductor properties) can also work in accordance with the mechanisms described earlier. However, the presence of a conductive polymer in the matrix sharply reduces the conduction jump in the region of the percolation threshold. In addition, the conductivity of the polymer itself can be higher than the conductivity through the conductivity network formed by conductive additives (Lin et al. 2013). Therefore, in such composites, the sensor response is controlled by the polymer, and the main task of solid additives is to improve the structure of the humidity-sensitive material—namely, increasing the film porosity and improving the stability of their parameters. For example, Parvatikar et al. (2006) studied WO_3/PANI nanocomposites and found that the additive of WO_3 (10–50 wt%) had little effect on the characteristics of the sensors. Chaluvaraju et al. (2016) synthesized Ta_2O_5/polypyrrole (PPY) composite and also did not observed a considerable improvement in the sensor performance (Figure 11.19). Tandon et al. (2006) synthesized iron oxide-polypyrrole (Fe_3O_4-PPY) nanocomposites and found that the composite structure, sensitivity behavior, and conductivity value were greatly influenced by the amounts of PPY, and the resulting polymer composites showed lower resistivity compared to pure PPY. TiO_2/PPY and TiO_2/PPY/poly-[3-(methacrylamino)propyl] trimethyl ammonium chloride) (PMAPTAC) composites have also been synthesized and tested as humidity sensors (Su and Wang 2008). By characterization of the sensors at 30% to 90% RH, it was found that the sensors made of the TiO_2/PPy/PMAPTAC composite thin films showed the highest sensitivity, smallest hysteresis, and greatest linearity. Shukla et al. (2012) also claimed that ZnO/PPY composite-based humidity sensors had a better sensitivity and linearity, and shorter response time compared to PPY-based sensors. But experimental confirmation was not given.

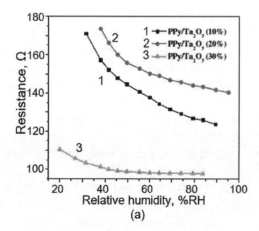
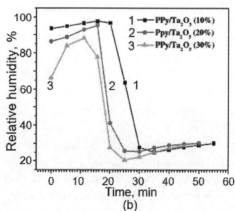

FIGURE 11.19 (a) Resistance as a function of humidity response for the PPy/Ta_2O_5 composites and (b) Humidity response as a function of time for the PPy/Ta_2O_5 composites. (Reprinted with permission from Chaluvaraju, B.V. et al., Thermo–electric power and humidity sensing studies of the polypyrrole/tantalum pentoxide composites, *Mater. Sci. Mater. Electron.*, 27, 1044–1055, 2016. Copyright 2016, Springer Science+Business Media.)

As for the third group of polymer-based composites (i.e., composites based on polyelectrolytes), this group, considering their properties considered in Section 11.2.4, seems to be the most promising for use in humidity sensors of resistive type. It is important to note that, in composites based on polyelectrolytes, a carbon is practically not used. The addition of carbon does not give an improvement in sensitivity (Li et al. 2001). Li et al. (2001) believe that carbon, which competes with polymer phase in water adsorption, can lower the formation inside the polymer of the "solution medium" created by the dissolution of the ions in the adsorbed water. In most studies, metal oxide nanoparticles are used as additives in such nanocomposites.

For example, Feng et al. (1997) prepared SiO$_2$/Nafion composite thin films using casting and dip coating methods. The response of SiO$_2$/Nafion composite to RH is shown in Figure 11.20. It is seen that sensors developed demonstrated high sensitivity, small hysteresis, and high stability. Su and Tsai (2004) used a composite material of nanosized SiO$_2$ and poly(AMPS) to fabricate a resistive humidity sensor. The sensor showed a negligible hysteresis in the range of 30%–90% RH and fast response upon humidification and dehumidification. Chou et al. (1999) investigated a humidity sensor with porous ceramics fabricated from ceramic fiber, kaolin, and sodium salt of carboxylmethyl cellulose (CMC). The conductivity was shown to change by 4–5 orders of magnitude when the RH varied from 10% to 90%. Furthermore, the response time was reported to be 5–8 minutes for RH values of 10% to 90%. Wang et al. (2002, 2003b) reported a humidity-sensing composite material composed of nanocrystalline BaTiO$_3$ and polymer polystyrene sulfonic sodium (PSS). BaTiO$_3$/PSS humidity sensors possess good sensitivity and linearity characteristics at RH ranges of 33%–98% (Figure 11.21) and exhibit a maximum humidity hysteresis of 8% RH. Additionally, the response and recovery times of these sensors are 50 and 120 seconds, respectively. Zhang et al. (2017) have used the same polymer. They have

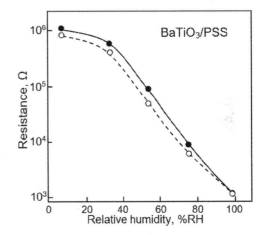

FIGURE 11.21 Resistance–RH curves of the BaTiO$_3$/PSS humidity sensor. The polystyrene sulfonic sodium (PSS) was soaked in 10% NaCl solution for 48 h, then ground into powder. (Reprinted with permission from Wang, J., Preparation and electrical properties of humidity sensing films of BaTiO3/polystrene sulfonic sodium, *Mater. Chem. Phys.*, 78, 746–750, 2003. Copyright 2003, Elsevier.)

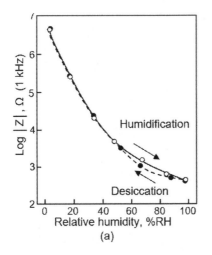
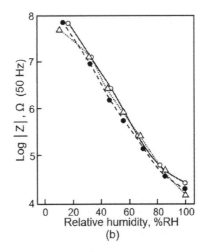

FIGURE 11.20 (a) The response calibration curve of the Nafion film and (b) The response calibration curves of the SiO$_2$/Nafion film in origin (O), and after treated at 100% RH and room temperature for 24 h (●), and in the vapor of a 10% CaHsOH and 10% CH$_3$OH water solution for 3 days (Δ). (Reprinted with permission from Feng, C.D. et al., Humidity sensing properties of nation and sol–gel derived SiO$_2$/Nafion composite thin films, *Sens. Actuators B*, 40, 217–222, 1997. Copyright 1997, Elsevier.)

shown that Co_3O_4/PSS composite-based humidity sensors prepared using a layer-by-layer method had not only high sensitivity at room temperature ($S(R_D/R_W) > 10^4$), but also fast response/recovery time ($\tau \approx 25$–30 s) and outstanding repeatability, as well as stability. Lee et al. (2013) also state that the addition of oxide nanoparticles such as SiO_2 and aluminum zinc oxide (AZO) in PEDOT:PSS improves the sensitivity and stability of the humidity sensor parameters. In addition, PEDOT:PSS/SiO_2-based sensors had higher chemical resistance, while PEDOT:PSS/AZO sensors had better stability in comparison with PEDOT:PSS-based humidity sensors.

Using in-situ synthesized inorganic/organic nanocomposites of sodium polystyrenesulfonate (NaPSS) and ZnO, Li et al. (2004) developed thin-film humidity sensors with four orders of magnitude sensitivity over the 11%–97% RH range, which had better linearity (see Figure 11.22), small hysteresis ($\approx$ 2% RH), and quicker response (2 s for absorption and desorption, respectively) than sensors prepared with NaPSS. Sun et al. (2009) have also used NaPSS to prepare the metal oxide/polymer composite. TiO_2/NaPSS-based sensors have been fabricated by the dip-coating method onto an alumina substrate. By testing the sensors over different RH ranges (11%–95% RH), Sun et al. (2009) have found that the TiO_2/NaPSS hybrid films had higher sensitivity, better linearity, and faster response and recovery in all RH ranges in comparison with polymer films.

However, the mechanism of the optimizing effect observed in the parameters of such polymer-based humidity sensors after addition of inorganic particles has not been studied in most cases. The explanations are usually a set of assumptions and speculations. For example, according to Zhang et al. (2017), at low RH, the chemisorbed water molecules on the surface of Co_3O_4/PSS film will form a layer of hydroxyl groups on the oxide surface. With RH increasing, the physical adsorption of water vapor will occur on the chemisorbed layer, and H_3O^+ will produce charge carriers due to the ionization of water molecules. As the adsorption of water molecules continues, the multilayer physisorption of water molecules exhibits a liquid-like behavior, and the protons hopping-transport via ionic conductivity takes place. At high RH conditions, the free water can interpenetrate into the interlayer of Co_3O_4/PSS film, which is contributed to a large enhancement in the dielectric constant and the sensor response. According to Lee et al. (2013), the improvement of PEDOT:PSS/AZO-based sensor performance takes place because the AZO particles in the sensor film would generate Zn^+ and Al^+ ions as the chemisorption of initial water molecules (Tai and Oh 2002). These ions locally induced electric fields to promote dissociation of water molecules, which offer additional protons. This phenomenon provides additional protons to maintain the carrier transportation of blended material in longtime scale.

Other developers of the metal oxide/polymer composite-based humidity sensors believe that such composite materials simultaneously exhibit both ionic and electronic conduction mechanisms. So, the conduction by both ions and electrons (with ions being the dominant carriers) should be responsible under high-moisture atmospheric conditions, while in a low-moisture environment, the dominant charge carriers should be electrons (Raj et al. 2002; Farahani et al. 2014). Therefore, for explanation of humidity-sensitive characteristics, they offered the following mechanism:

1. At low-RH levels, electronic conduction based on the electron donation from water molecules is the dominant responsible mechanism.
2. At medium-RH levels, based on the number of physisorbed layers in different intervals, electrostatic fields are due to the first chemisorbed layer; both electronic and ionic conduction are responsible with the dominant transition being from electronic to ionic conduction mechanism.
3. At high-RH levels, ionic conduction based on proton hopping between water molecules is the dominant responsible mechanism.

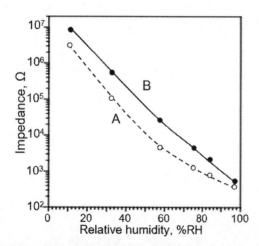

FIGURE 11.22 Electrical response of NaPSS and NaPSS/ZnO film to humidity. Composition of precursor solution: (a) NaPSS/CMC/H_2O = 10 mg/10 mg/1 cm³ and (b) NaPSS/CMC/$Zn(OH)_2$/H_2O = 10 mg/5 mg/0.0167 cm³/1 cm³. NAPSS, sodium polystyrenesulfonate; CMC, sodium carboxymethylcellulose. (Reprinted with permission from Li, Y. et al., Humidity sensors using in situ synthesized sodium polystyrenesulfonate/ZnO nanocomposites, *Talanta*, 62, 707–712, 2005. Copyright 2005, Elsevier.)

However, based on a detailed examination of the characteristics of various metal oxide/polymer composites, it can only be noted that the use of such composites does not give a noticeable improvement in sensitivity. See, for example, Figure 11.22. Usually, an improvement in linearity is observed, either a decrease in hysteresis or a decrease in the response and recovery time. In the case of PSDA/MPTMS/ZnO composite (Zor and Cankurtaran 2016), only an improvement in linearity was observed (see Figure 11.23). Çiğil et al. (2017), when studying polyethylene glycol acrylate (PEGA)/CNTs/Au composites, observed only an improvement in thermal stability and a decrease in resistance during the incorporation of CNT and Au into the polymeric network. At the same time, the humidity sensitivity decreased.

It is difficult to expect that simple mixing of a polymer and a chemically inert material can give a noticeable optimizing effect, especially at room temperature. All these effects can be explained by simply improving the structure of the humidity-sensitive material, in particular by improving its water vapor permeability and optimizing the pore size distribution. Exactly this conclusion was drawn by Najjar and Nematdoust (2016) in the process of studying the humidity sensors based on ZnO/PPy composite. The increase in the sensor response by increasing the ZnO content in the hybrid polymers was attributed to the increase in the surface area available for the water molecules to absorb and interact with PPy polymer. As revealed by the SEM images (Figure 11.24), the size of ZnO–PPy hybrid polymer nanoparticles was decreased by increasing the ZnO content in the composite. By decreasing the size of hybrid polymer nanoparticles, their surface-to-volume ratio increased dramatically. Consequently, more water molecules could be absorbed on the surface of polymer nanoparticles, and as a result, a bigger decrease in the sensor resistivity, and hence a higher increase in its response were observed.

11.2.6 Summary

As for the advantages and disadvantages of resistive polymer humidity sensors, of course they are not the same for all types of polymer humidity sensors. Some sensors are more sensitive, others are more stable, et cetera. Therefore, the advantages and disadvantages mentioned in this section are, to some extent, averaged over all types of sensors. They can be formulated as follows (Wilson 2005; Soloman 2010):

Advantages:
- No calibration standards. Therefore, resistive polymer humidity sensors are highly interchangeable and field replaceable.
- Long-term stability. By virtue of their structure, bulk polymer-resistive sensors are relatively immune to surface contamination. Life expectancy is near 5 years.
- Usable from remote locations.
- Small, compact size.
- Low cost.
- Low power input.
- Fast response.
- Excellent reproducibility.
- High accuracy.
- Broad humidity range.
- Sensors are suitable for many industrial, commercial, and residential applications, especially control and display products.

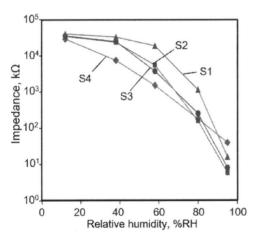

FIGURE 11.23 Impedance values versus relative humidity (RH) for poly(diphenylamine sulfonic acid) (PSDA)/3-mercaptopropyltrimethoxysilane (MPTMS)/ZnO composite measured at 0.2 V, 1 kHz. Sensors were fabricated using drop-coating method and interdigitated electrodes. PDSA:MPTMS = 80:2; S1, 0% ZnO; S2, 10 wt% ZnO ($d < 100$ nm); S3, 30 wt% ZnO; S4, 50 wt% ZnO. (Reprinted with permission from Zor, F.D. and Cankurtaran, H., Impedimetric humidity sensor based on nanohybrid composite of conducting poly(diphenylamine sulfonic acid), *J. Sensors*, 2016, 5479092, 2016. Copyright 2016, Published by Hindawi Publishing Corporation as open access.)

Disadvantages:
- Although surface buildup does not affect the accuracy of the sensor, the surface contamination does have an adverse effect on the response time.
- Exposure to chemical vapors and contaminants may cause premature failure.
- Like all RH sensors, when exposed to extreme environmental conditions, accuracy degradation can result. For example, in some resistive sensors, there is a tendency to shift values when exposed to

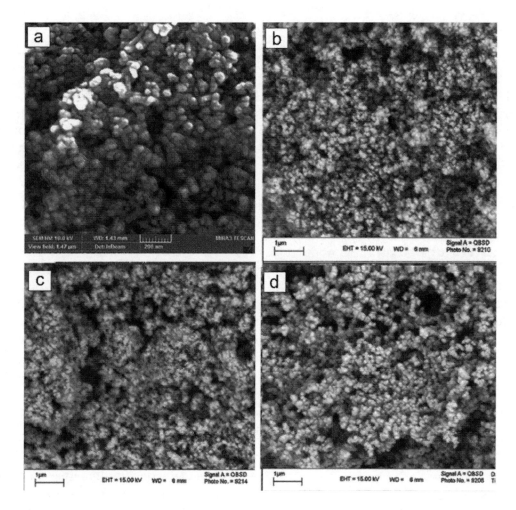

FIGURE 11.24 SEM images of (a) ZnO nanoparticles surface modified with toluene 2,4-diisocyanate (TDI) (ZnO–TDI); (b) ZnO–TDI–Py nanoparticles; (c) ZnO–TDI–PPy containing 46% of ZnO; and (d) ZnO–TDI–PPy containing 14% ZnO. (From Najjar, R. and Nematdoust, S., *RSC Adv.* 6, 112129–112139, 2016. Reproduced by permission of The Royal Society of Chemistry.)

condensation if a water-soluble coating was used. However, polymer-type sensors have proven to be more resistant to such errors or drift than the older Pope cells and Dunmore cells. As a rule, the shift in calibration takes place after long exposure to either high (> 95%) RH or continuous condensation.

- Due to the extremely high resistance at RH values of less than 5%–10%, this sensor is generally better suited to the higher RH ranges. However, too high a level (> 95%) is not desirable. As a rule, sensors do not like water condensation.
- Unlike capacitive sensors, resistive sensors are characterized by a much stronger temperature dependence of the parameters (see Figure 11.25a). Resistive humidity sensors have significant temperature dependencies when installed in an environment with large (> 3°C) temperature fluctuations. This means that temperature control, temperature stabilization, and temperature compensation are required for correct operation of resistive humidity sensors. Although it should be recognized that there are developments (Sakai et al. 2001), according to which the temperature dependence of the parameters of resistive sensors is practically absent (see Figure 11.25b). True, it is difficult to find an explanation for such a result, since the adsorption, diffusion, and mobility of ions are temperature-dependent phenomena. However, it is difficult to use the advantages of such sensors, since polymers used for the fabrication of these sensors were not stable in the highly humid atmosphere (Sakai et al. 2001).
- And finally, the change in impedance is too high and hence it is difficult to control the dynamics

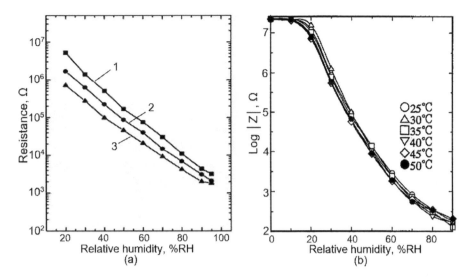

FIGURE 11.25 (a) The resistance dependence on relative humidity (RH) of humidity sensor using the cross-linked copolymers VP/VTVPC and CEVE/IVE: 4/1 at (1, □) 15°C, (2, ○) 25°C, and (3, Δ) 35°C at 1 kHz, and 1 V. Gong et al. (2002). (b) Plot of impedance of acid-type poly(AMPS) as a function of RH at various temperatures. AMPS, poly(2-acrylamido-2-methylpropane sulfonic acid). The AMPS homopolymer was prepared by the conventional polymerization of the AMPS monomer. (Reprinted with permission from Sakai, Y. et al., Humidity sensor based on alkali salts of poly(2-acrylamido-2-methylpropane sulfonic acid), *Electrochim. Acta*, 46, 1509–1514, 2001. Copyright 2001, Elsevier.)

of humidity change in wide range, especially if these changes are accompanied by the change in temperature.

11.3 METAL OXIDE–BASED RESISTIVE HUMIDITY SENSORS

11.3.1 Mechanisms of Conductivity in Metal Oxides in Humid Air

As was shown in previous chapter, the unique structures of ceramic materials, comprising grains, grain boundaries, surface areas, and controlled porous microstructures, makes them suitable candidates for application in various sensors, including humidity sensors of resistive type (Nitta and Hayakawa 1980; Shimizu et al. 1985a, 1985b, 1985c Yamazoe and Shimizu 1986; Traversa 1995; Rittersma 2002; Lee and Lee 2005; Farahani et al. 2014). It was established that these sensors in dependence on the nature and structure of metal oxides can have different mechanisms of humidity sensing. But all these sensing mechanisms rely on the three main processes: chemical adsorption (chemisorption), physical adsorption (physisorption), and capillary condensation of water vapor in a porous metal oxide matrix (Foster 1932; Hertl and Hair 1968; Seiyama et al. 1983; Thiel and Madey 1987; Traversa 1995). All these processes were described in previous Volume 1 (Chapter 10) of our series. Therefore, in this chapter, only the mechanisms of humidity sensing, namely how humidity can influence the conductivity of metal oxides, are considered.

11.3.1.1 Grotthuss-Type Transport of Protons

As known, water molecules on the surface of oxides are in chemisorption and physisorption states (Thiel and Madey 1987; Henderson 2002; Traversa 1995). The chemisorption, which is the stronger process, causes dissociation of water molecules to form surface hydroxyls.

This process is accompanied by a change in the concentration of chemisorbed oxygen and hydroxyl groups on the surface of metal oxides, hence the height of the potential barrier at the grain boundary responsible for the resistance of the polycrystalline materials (see the following section). It was found that after hydroxyl formation, the next water molecule will be physisorbed through hydrogen double bonds on the two neighboring hydroxyl groups, and a proton may be transferred from a hydroxyl group to the water molecule to form a H_3O^+ ion (see Figure 11.26). The appearance of H_3O^+ takes place due to dissociation of physisorbed water in the high electrical field in the chemisorbed layer.

As for the transport of protons in such structures, it is usually described in the framework of the so-called Grotthuss mechanism (Grotthuss 1806). One should note that the conductivity described by the Grotthuss mechanism refers to the so-called surface conductivity. Initially, the Grotthuss mechanism was acknowledged

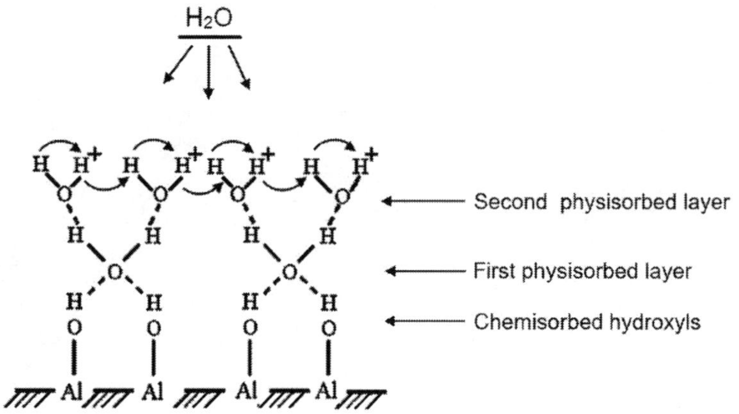

FIGURE 11.26 Schematic illustration of buildup of adsorbed water layers upon an alumina surface. (Idea from Fleming, W. J., *Soc. Automot. Eng. Trans. Section 2*, 90, 1656–1667, 1981.)

during the aquatic electrolysis of water for the positive charge migrations. In this conduction mechanism, which exists in all liquid water, protons are tunneled (proton dancing) from one water vapor molecule to the subsequent one through hydrogen bonding (Wraight 2006). It was established that this charge transport on the surface of metal oxides occurs when H_3O^+ releases a proton to a nearby water molecule, ionizing it and forming another H_3O^+ group, resulting in the proton hopping from one water molecule to another. This process is known as the *Grotthus chain reaction* and is illustrated in Figure 11.26 (Agmon 1995). Hence the dominant charge carrier in a high-moisture atmosphere is the H^+ (proton). It is important that a higher carrier concentration was found when more than one layer of physisorbed water molecule was presented on the surface. According to Anderson and Parks (1968), in this case, water molecules form a liquid-like network, which greatly increases the dielectric constant and, therefore, the proton concentration. Thus, the Grotthuss-type or "free-proton" mechanism requires close proximity of water molecules, which are firmly held but able to rotate. Studies have shown that the water molecules in the first physisorbed layer cannot move or rotate freely, while the water molecules in the second and succeeding layers are freer to rotate. In other words, from the second physisorbed layer, when water molecules become mobile and finally almost identical to the bulk liquid water, the Grotthuss mechanism can become dominant. This means that this mechanism of conductivity requires the presence of superficial liquid-like layers (aqueous layers) that appear with high humidity and condensation (capillary condensation) of water on the metal oxide surface. Thus, this mechanism indicates that sensors based purely on water-phase protonic conduction would not be quite sensitive to low humidity, at which the water vapor could rarely form continuous mobile layers on the sensor surface. In addition, such mechanism of conductivity testifies that the RH sensitivity of a ceramic oxide is related to the number of water-molecule adsorption sites, which may be increased by the presence of defect lattice sites and exogenous oxygen atoms at the oxide surface (Thiel and Madey 1987). Note that the sensor mentioned above based on mechanism of conductivity can operate at low temperature (< 100°C) and is widely accepted for many ionic ceramics, also of the nonoxide type (Traversa 1995). In addition to the protonic conduction in the absorbed layer, electrolytic conduction occurs in the liquid layer of water condensed within capillary pores, thereby resulting in an enhancement of conductivity (Kulwicki 1984).

Thus, to summarize what was said earlier, the humidity sensitivity at low RH is controlled by chemisorption, which causes dissociation of water vapor molecules and establishment of hydroxyl ions at the surface of metal oxides. When only hydroxyl ions are present on the oxide surface, the charge carriers are protons. The protons hop between adjacent hydroxyl groups (Traversa 1995). The electron tunneling between donor water sites is also possible in humidity-sensitive materials (Yeh et al. 1989). The tunneling effect, along with the energy induced by the surface anions, facilitates electrons to hop along the surface that is covered by the immobile layers and therefore contributes to the conductivity. This mechanism can be used for explaining sensitivity to low humidity levels, at which there is not effective protonic conduction (Traversa 1995). Increased humidity level causes water physisorption and formation of the first layer of hydronium ions, which promotes the transfer of protons between neighboring sites and increases the conductivity (ionic or protonic conduction). At higher RH, the water vapor condensation occurs in the capillary quasi apertures to form liquid-like layers. The electrolytic conduction leads to the occurrence of further conductivity (Nitta and Hayakawa 1980).

Transport by any other species is termed *vehicle mechanism*. Vehicle mechanism is most frequently encountered in aqueous solution and other liquid/melts (Conway et al. 1956; Agmon 1995). In solids, vehicle mechanism is usually restricted to materials with open structures (channels, layers) to allow passage of the large ions and molecules. Compounds with less amount of water would be expected to conduct by vehicle mechanism, in which a nucleophilic group such as H_2O or NH_3 acts as a proton carrier.

11.3.1.2 Conductivity in Heated Semiconducting Metal Oxides

In semiconducting metal oxides, which operate at increased temperatures, the mechanism of sensitivity to humidity is different. It is known that water molecules act as an electron donating gas on semiconductive oxides, causing the decrease of resistance of *n*-type oxides and the increase in resistance of *p*-type oxides. In fact, semiconductor gas sensors, which are used widely to detect reducing gases, are known to suffer from interfering effects of water vapor under certain conditions (Barsan and Ionescu 1983; Korotcenkov et al. 1999). Because the conductivity is caused by the surface concentration of electrons, this sensing mechanism is usually called *electronic type*. However, at low temperatures, nothing prevents the formation of water layer at the surface of metal oxides due to physical adsorption of water vapor and capillary condensation (Morrison 1982). This means that the situation described earlier is being realized on the surface of conductive metal oxides. Therefore, at room temperatures, the conductivity of ceramic semiconducting materials is actually due to present of both electrons and protons (ionic). Only at temperatures exceeding 100°C, when the moisture cannot effectively condense on the surface, transport of protons can be neglected. Therefore, in this temperature range, chemisorption of water molecules is only responsible for the changes in the electrical conductivity of metal oxides. The electrical behavior of semiconducting metal oxides in the presence of water vapor is thus similar to that observed when they are in contact with reducing gases. Some attributed this effect to the donation of electrons from the chemically adsorbed water molecules to the ceramic surface (Boyle and Jones 1977; Avani and Nanis 1981). It has also been reported that water chemisorption on the semiconductive metal oxides caused changes in the state and reactivity of chemisorbed oxygen (O^-, O^{2-}, etc.), which was accompanied by the donation of electrons in the conduction zone and changes in oxide conductivity (Yamazoe et al. 1979; Korotcenkov et al. 1999). As a result of above processes, the changes in the parameters such as concentration of oxygen chemisorbed at the surface, the height of potential barrier at intergrain contacts and film conductivity/resistivity take place. Oxide resistivity varies, decreasing or increasing according to the type of semiconducting oxides (*n*- or *p*-type). However, during the chemisorption reaction, when two hydroxyl groups are formed, with H^+ bonded to an oxide ion and OH to a metal ion (McCafferty and Zettlemoyer 1971), there is no electron transfer to the oxide (Shimizu et al. 1989a, 1989b).

A water molecule interaction with ionized oxygen vacancies (V_O) can also be responsible for conductivity change (Seiyama et al. 1983). Such a mechanism is usually realized in perovskite-type oxides (ABO_3) at high temperatures (see Section 11.3.6). Electrons trapped by ionized oxygen are liberated by the adsorption of water molecules (Traversa 1995).

As indicated before, electronic mechanism is responsible for the need to operate semiconductor humidity sensors at elevated temperatures, where physical adsorption of water is minimized. In the same way, the humidity sensitivity of semiconducting oxides is affected by the presence of oxidizing and reducing gases, as well as vapors of organic solvents, which also cause resistivity changes. This is one of the major problems related to the use of electronic conductivity-based humidity sensors.

11.3.1.3 Conductivity in Metal Oxides Doped with Alkali Ions

Seiyama et al. (1983) established that the introduction of small amounts of alkali ions has a favorable effect on ceramic humidity sensors. The alkali ions, such as K^+ (Kim et al. 1992), Na^+ (Kim and Gong 2005), and Li^+ (Yokomizo et al. 1983; Kazemzadeh et al. 2008), added act as dominant charge carriers instead of protons and thus often endow the sensors with higher humidity sensitivity as well as the base resistance stability (see Figure 11.27). This approach to the formation of humidity-sensitive metal oxides is almost identical to the approach used in the manufacture of polymeric humidity sensors described in the previous section. Recently, this approach was used by Song et al. (2009) for SnO_2, by Qi et al. (2009a) for TiO_2, by Wang et al. (2000) for $BaTiO_3$, by Kazemzadeh et al. (2008) for Mn-Co-Ni-O oxide, and by Anbia and Fard (2011) for $Ti_{0.9}Sn_{0.1}O_2$-based humidity sensors. It was shown that this approach can give the decrease in resistance in the range of low RH, and sensing characteristics can be really improved. For example, Kazemzadeh et al. (2008) reported that the addition of Li^+ to the nanostructured Mn-Co-Ni-O metal oxide has increased the sensitivity to humidity 10 times in comparison with the undoped one.

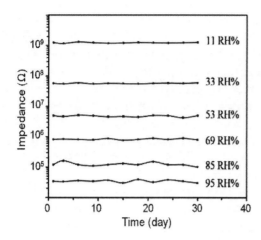

FIGURE 11.27 Stability test of $Ti_{0.9}Sn_{0.1}O_2$-based humidity sensors doped by KCl. (Reprinted with permission from Anbia, M. and Fard, S. E. M., Improving humidity sensing properties of nanoporous TiO_2–10mol% SnO_2 thin film by co-doping with La+ and K+, *Sens. Actuators B*, 160, 215–221, 2011. Copyright 2011, Elsevier.)

11.3.2 Metal Oxides in Resistive Humidity Sensors

For application in resistive humidity sensors, we do not have so pronounced a leader among metal oxides (Yamazoe and Shimizu 1986; Traversa 1995; Rittersma 2002; Lee and Lee 2005; Farahani et al. 2014). In the literature, one can found numerous studies focused on finding appropriate conducting materials (see Table 11.2), which can be ether ionic or electron conductors. For example, Ying et al. (2000) investigated TiO_2-K_2O-$LiZnVO_4$ in the range of 10%–99% RH. Qu et al. (2000) studied $Ba_{1-x}Sr_xTiO_3$. Feng et al. (1998) tested Nafion™ and NafionTM/SiO_2 composite films. Shimizu et al. (1985c) reported about humidity sensors based on $MgAl_2O_4$ and $MgFe_2O_4$. Qu and Meyer (1997) and Qu et al. (2000) presented a thick-film, porous $MnWO_4$ ceramic layer sandwiched by interdigitated metal films. Sundaram and Nagaraja (2004a, 2004b) tested MoO_3 and WO_3 and composites based on these materials, such as of $MMoO_4$ (M = Ni, Cu, Pb) and $PbWO_4$. Sol-gel $BaTiO_3$ (Yuk and Troczynski 2003), sol-gel silica films synthesized by block copolymers (Bearzotti et al. 2004), sol-gel strontium-doped lead-zirconium titanate ($PbSr_xZr_{0.3}Ti_{0.7}O_3$, x = 0.1–0.5) (Ansari et al. 2004), and sol-gel chromium titanate oxide (CTO) ($Cr_{2-x}Ti_xO_3$, x =0.05–0.4) (Neri et al. 2004) were also used as humidity-sensing layers. Arshak et al. (2002) developed a thick-film resistive humidity sensor using a combination of MnO, ZnO, and Fe_2O_3. Kotnala et al. (2008) proposed to use $Mg_{1-x}Li_xFe_2O_4$ ($0.0 \leq x \leq 0.6$) as humidity material. Yadav et al. (2011) analyzed morphological and humidity-sensing characteristics of SnO_2-CuO, SnO_2-Fe_2O_3, and SnO_2-SbO_2 nanocomposites. Li^+-Modified $Ca_xPb_{1-x}TiO_3$ films for humidity sensor were tested by Jingbo et al. (2001). $BaTiO_3$ as a humidity-sensitive material was studied in Caballero et al. (1999), Kim (2002), Yuk and Troczynski (2003), Wang et al. (2003a, 2003b), and Ertug et al. (2010). Humidity-sensitive characteristics

TABLE 11.2
Metal Oxides Tested as Humidity-Sensitive Material in Resistive Humidity Sensors

Type of Oxides	Oxides	Ref.
Binary oxides	CeO_2, SnO_2, CuO, ZnO, WO_3, TiO_2, VO_2, V_2O_5, MnO_2, Mn_3O_4, Mn_2O_3	Xu and Miyazaki (1993); Xu et al. (2009); Yadav et al. (2008, 2009); Kannan et al. (2010); Yin et al. (2011); Peng et al. (2012); Pawar et al. (2015); Xie et al. (2015); Feng et al. (2017); Jadkar et al. (2017); Ramkumar and Rajarajan (2017)
Mixed oxides, complex oxides, and composites	$Sr_{0.95}La_{0.05}SnO_3$; $MnWO_4$; $Ba0_{.99}La_{0.01}SnO_3$; MgO-Al_2O_3; MnO-ZnO-Fe_2O_3; $Zn_{1-x}Co_xMoO_4$; SnO_2:MgO; $PbWO_4$; La_2O_3-TiO_2-V_2O_5; Fe_2O_3/SiO_2; $ZnMoO_4$-ZnO; La_2CuO_4	Shimizu et al. (1987, 1989a, 1989b); Qu and Meyer (1997) and Qu et al. (2000); Arshak et al. (2002); Raj et al. (2002); Sundaram and Nagaraja (2004a, 2004b); Upadhyay (2008); Yuan et al. (2011); Hayat et al. (2016); Klym et al. (2016); Jeseentharani et al. (2017); Sabarilakshmi and Janaki (2017)
Perovskites	$BaTiO_3$; $CoTiO_3$; $KNbO_3$; $SrTiO_3$	Caballero et al. (1999); Kim (2002); Yuk and Troczynski (2003); Wang et al. (2003); Ertug et al. (2010); Wang et al. (2011a); Li et al. (2015); Ganeshkumar et al. (2017); Lu et al. (2017)
Perovskite-based oxides	$BaPbTiO_3$; $Ba_{0.5}Sr_{0.5}TiO_3$; $PbSr_xZr_{0.3}Ti_{0.7}O_3$; $CaCu_3Ti_4O_{12}$	Yeh et al. (1989); Tian et al. (2005); Ahmadipour et al. (2016)
Metal oxides doped by alkali ions	$BaTiO_3$:Na,K; (Mn-Co-Ni-O):Li; $Ca_xPb_{1-x}TiO_3$:Li; Fe_2O_3:Li	Wang et al. (2000); Bonavita et al. (2001); Jingbo et al. (2001); Kazemzadeh et al. (2008)

of $Ba_{0.99}La_{0.01}SnO_3$ were presented by Upadhyay (2008). Bauskar et al. (2012) reported humidity-sensing properties of $ZnSnO_3$ cubic crystallites synthesized by a hydrothermal method. Sadaoka et al. (1987) fabricated resistive humidity sensors using KH_2PO_4-doped, porous (Pb, La)(Zr, Ti)O_3. One should note that there are many other materials that have been tested as moisture-sensitive materials (Farahani et al. 2014). However, the results presented in the literature do not give possibility to select the most acceptable metal oxide, because the unified basis for comparison is absent. Moreover, every developer believes that his or her material is better.

11.3.3 Examples of Realization

11.3.3.1 General Consideration

Fabrication of humidity sensors based on metal oxides does not differ from the technologies developed for manufacturing gas sensors and described in numerous articles and books (Korotcenkov 2010, 2011). This means that humidity sensors can be realized both in a ceramic version (Nitta and Hayakawa 1980; Nitta 1981; Fagan and Amarakoon 1993), and in thin-film (Ahmadipour et al. 2016) and thick-film versions (Arshak and Twomey 2002; Sabarilakshmi and Janaki 2017). It is important that the different structural properties due to the fabrication and film thickness lead to different sensor performance. In thick-film sensors, the film thickness is typically in the range of 2–300 μm. Thick-film technology, based mainly on a screen printing, is the most common fabrication technique for such sensors. Other methods, such as spin and dip coating, also can be used. Thick-film technology is desirable for the mass production of resistive humidity sensors on account of its cost efficiency, robustness, and flexibility in device design (Golonka et al. 1997). In addition, the thick-film technology, in comparison with thin-film technology, provides a higher porosity of the formed film (Korotcenkov and Cho 2009). Films prepared by thick-film technology have microstructural properties similar to those of sintered bodies but can reduce the dimensions of the sensing devices, which can be used in hybrid circuits. Due to high porosity, the inner surface of metal oxide matrix also becomes a working surface. Therefore, the gases diffuse into it, leading to good sensitivity. The microstructure of the sensing layer is a function of temperature parameters of grains sintering. The conductance usually is controlled by the contact resistance between the grains. In thin-film sensors, the film thickness is typically 5–500 nm, in special cases even 1 μm. The most common preparation process is thin-film technology based on sputtering, evaporation, chemical vapor deposition, spray pyrolysis, chemical deposition, or laser ablation methods of forming the sensing layer. Sensing layers in thin-film sensors have denser structure in comparison with thick-film sensors. This feature is especially marked for films deposited using magnetron sputtering methods (Ahmadipour et al. 2016). As a result, sensors based on such films usually have a low sensitivity, although there may be exceptions (Kannan et al. 2010). In addition, at certain thicknesses in these sensors, diffusion limitation begins to affect the humidity-sensitive effects and slow down the response and recovery processes during interaction with water vapor.

When developing resistive humidity sensors, first of all, the requirements coming from the operating conditions and the field of application are taken into account. As a rule, these requirements are aimed at reducing power consumption, decreasing the size, increasing stability, et cetera. As a result, the developers of humidity sensors focus on sensors operating at room temperature. These sensors have the widest range of applications. Heated humidity sensors are also being developed (see Section 11.3.6), but the field of their application is substantially smaller.

Regarding the measurement modes used for the operation of resistive humidity sensors, as well as for capacitive sensors, measurements are made at low frequencies of the order of 0.1–1.0 kHz. At higher frequencies, the dependence of the resistance on air humidity decreases significantly (Figure 11.28). It was established that the measured voltage level does not influence the sensing properties of the sensor, such as sensitivity, and response and recovery times (Wang et al. 2003a).

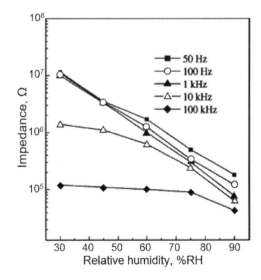

FIGURE 11.28 Impedance of $BaTiO_3$ thin film vs relative humidity (RH) for various frequencies. (Reprinted with permission from Yuk, J. and Troczynski, T., Sol-gel $BaTiO_3$ thin film for humidity sensors, *Sens. Actuators B*, 94, 290–293, 2003. Copyright 2003, Elsevier.)

11.3.3.2 RT Humidity Sensors

When developing metal oxide–based RT humidity sensors, a large number of different metal oxides belonging to different groups and having various conductivity were tested, from dielectrics to semiconducting and ionic conductive ones (see Table 11.2). These studies have shown that a large group of tested metal oxides makes it possible to manufacture resistive sensors that, under suitable conditions, are stable at room temperatures and allow for a wide range of humidity measurements with the required reproducibility (see Figure 11.29).

For example, the first commercial sample of the humidity sensor was developed in 1978 on the basis of $MgCr_2O_4$-TiO_2 composites (Nitta and Hayakawa 1980; Nitta 1981; Fagan and Amarakoon 1993). This sensor was developed for practical use in microwave ovens. The sensing element was a small, porous (35% porosity, with an average pore size of 300 nm) rectangular wafer made of a $MgCr_2O_4$-TiO_2(35 mol%) with porous RuO_2 electrodes and a coil heater for self-cleaning. The sensing element was heat cleaned at 500°C before each operation in order to eliminate the surface hydroxyl groups, which impede Grotthuss-type conduction, resulting in a drift of the resistance of the element. Figure 11.30a shows the typical resistance-humidity characteristics of this sensor (Nitta 1988). The response time of the sensor was less than 10 seconds.

More stable sensors based on $ZnCr_2O_4$-$LiZnVO_4$ have been developed by Yokomizo et al. (1983) and Uno et al. (1983) for use in air conditioners. Its ceramic body structure consisted of $ZnCr_2O_4$ spinel grains 2–3 μm in size covered by the $LiZnVO_4$ glassy phase. Its porosity was nearly 12%, with an average pore size of 300 nm. The sensing element is a circular wafer 8 mm in diameter and 0.2 mm thick, with Au electrodes applied to the opposite faces of the disc. Figure 11.30b shows the resistance-humidity characteristics of this sensor (Uno et al. 1983). The response time was approximately 150 seconds. The interesting feature of this sensor is that it can be operated without heat cleaning, because it does not show drifts in the resistance as a result of exposure to wet environments. The resistivity of the sensor is mainly determined by water adsorbed on the amorphous phase, which has stable OH radicals on the surface on which water is physisorbed (Ichinose 1985).

Many studies were devoted to TiO_2-based humidity-sensitive materials (see Table 11.3). In general, the results obtained for pure TiO_2 were not completely satisfactory because of the high resistivity of TiO_2 and its poor long-term stability (Katayama et al. 1990a). However, Yeh et al. (1989) and Katayama et al. (1983) reported that TiO_2 doped with V_2O_5 had much better stability. For example, sintered TiO_2 with 35% open porosity can be reversibly operated without repeated heat cleaning, showing a conductance versus RH sensitivity at 25°C of more than four orders of magnitude in the RH range of 15%–90% at 400 Hz (Yeh et al. 1989). Doping TiO_2 with V_2O_5 (2 mol%), with the aim of lowering the resistivity, was found to be

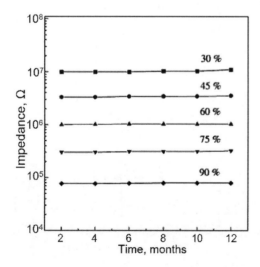

FIGURE 11.29 Impedance variations with aging time of $BaTiO_3$ thin films at various relative humidity (RH) levels. (Reprinted with permission from Yuk, J. and Troczynski, T., Sol–gel $BaTiO_3$ thin film for humidity sensors, *Sens. Actuators B*, 94, 290–293, 2003. Copyright 2003, Elsevier.)

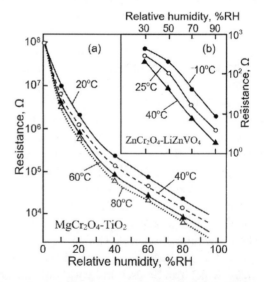

FIGURE 11.30 The relative humidity (RH) dependence of the resistance of (a) $MgCr_2O_4$-TiO_2- and (b) $ZnCr_2O_4$-$LiZnVO_4$-based sensors at different temperatures. (Data extracted from Nitta, T. Development and application of ceramic humidity sensors. In: Seiyama, T., Ed. *Chemical Sensor Technology, Vol. 1*. Kodansha, Tokyo/Amsterdam, the Netherlands: Elsevier, 1988;57–78 and Uno S., et al. $ZnCr_2O_4$–$LiZnVO_4$ ceramic humidity sensor. In: Seiyama, T., et al., Eds. *Analytical Chemistry Symposia Series: Chemical Sensors, Vol. 17*. Kodansha, Tokyo/Amsterdam, the Netherlands: Elsevier, 1983; 375–380.)

TABLE 11.3
Comparison of TiO$_2$-Based Humidity Sensors

Type of Materials	Humidity Range	Sensor Response, S	Response/Recovery Time (s)	Humidity Hysteresis	Ref.
TiO$_2$:V$_2$O$_5$ ceramics	15%–95%	> 10^4			Yeh et al. (1989)
TiO$_2$:Nb$_2$O$_5$ ceramics	10%–90%	> 10^2	60–180/60–180	NA	Katayama et al. (1990)
TiO$_2$:LiCl nanofibers	11%–95%	10^4	3/7	< 2.5% RH	Li et al. (2008)
TiO$_2$ nanofibers	11%–95%	10^4	3/3	NA	Qi et al. (2009a)
TiO$_2$ ceramics	09%–95%	10^5	5/8	< 1% RH	Wang et al. (2011a)
TiO$_2$/graphene composites	12%–90%	10^2	128/68	< 0.39% RH	Lin et al. (2015)
TiO$_2$ ceramics	11%–95%	10^4	32/131	≈ 4% RH	Wang et al. (2016)
TiO$_2$:Co mesoporous	2%–90%	≥ 10^5	24/400	< 1% RH	Li et al. (2017)

NA, not applicable; RH, relative humidity; S, sensor response calculated as R_D/R_W or I_W/I_D, where R_D and I_D are resistance and current measured in dry atmosphere, and R_W and I_W are resistance and current measured in wet atmosphere.

effective in increasing the RH sensitivity of the material. V$_2$O$_5$ influenced not only the intrinsic resistance of titania, but also its total porosity and pore-size distribution. A commercial sensor has been developed with this material, using a small rectangular wafer, which needs a heater for self-cleaning (Katayama et al. 1983). The addition of Nb$_2$O$_5$ (0.5%–1%) to titania was later investigated by the same authors (Katayama et al. 1990a). The humidity sensitivity of Nb$_2$O$_5$-doped TiO$_2$ was greatly affected by the microstructure, which was varied by sintering at different temperatures (Lee et al. 1994). Figure 11.31 shows the pore size distribution curves for the samples sintered at various temperatures, and the corresponding impedance-RH characteristics, respectively (Katayama et al. 1990a). The response time, which was rather fast for all the specimens, also depended on the microstructure of the specimens, being faster for the samples with large pores. This material showed an impedance drift during exposure to moisture, but an appropriate aging treatment at high temperature and high humidity was effective in reducing or eliminating the drift. The similar results were obtained for TiO$_2$-SnO$_2$-based sensors fabricated by Yamamoto and Murakami (1989) using the sedimentation method. These sensors also exhibited a degradation due to the formation of hydroxyl groups on the sensitive surface, but their properties were recoverable with heat treatment at approximately 500°C.

A humidity sensor using Al$_2$O$_3$, TiO$_2$, and SnO$_2$ powders prepared by the sol–gel method was developed by Yagi and Nakata (1992) for ordinary air conditioning and conditioning for special purposes at high humidity levels. An ultrafine powder of the mixed oxides was obtained from a solution in ethanol of aluminium, titanium, and tin alcoholates in the proportions 60, 20, and 20 mol%. The

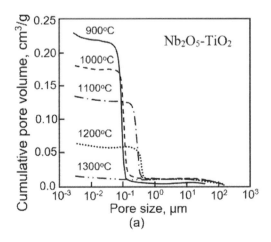
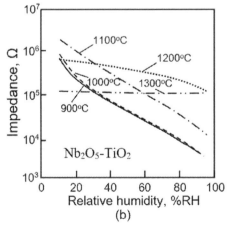

FIGURE 11.31 (a) Pore-size distribution curves of Nb$_2$O$_5$-doped TiO$_2$ samples, sintered at 900°C, 1000°C, 1100°C, 1200°C, and 1300°C and (b) Humidity sensitivity measured at 25°C of the Nb$_2$O$_5$-doped TiO$_2$ samples in (a). (Reprinted from Katayama K., et al. Humidity sensitivity of Nb$_2$O$_5$–doped TiO$_2$ ceramics, *Sens Actuators A.*, 24, 55–60, 1990. Copyright 1990, Elsevier.)

powder obtained after hydrolysis was calcined, granulated, pressed, and sintered to obtain porous rectangular wafers 3 mm × 3 mm × 0.4 mm in size. The total porosity was approximately 40%, and the pore size distribution was in the range 5–100 nm, with a small amount of large pores with a radius of 5 μm, and with an average pore radius of 50 nm. RuO_2 electrodes were applied on both faces of the sensing element. No hysteresis was found during RH cycles. The response time was within 10 seconds during water adsorption and 20 seconds during desorption. This sensor showed deterioration in its sensing properties after prolonged exposure to air. To overcome this problem, heat-cleaning treatments for restoring the initial conditions were performed using a far-infrared ceramic heater. A once-a-day cycle of heat cleaning was effective in recovering the sensor's performance in a normal atmosphere. This allowed long, continuous measurement and provided a nearly maintenance-flee ceramic humidity sensor with improved reliability and precision (Yagi and Nakata 1992).

The humidity-sensing properties of mesoporous ZnO-SiO_2 composites synthesized by sol–gel methods and fabricated through screen-printed films with different Si/Zn molar ratios have been investigated by Yuan et al. (2010). By evaluation of the sensors within the range of 11%–95% RH, it was revealed that introducing ZnO improved the humidity sensitivity of the composites, and the specimen with a Si/Zn ratio of 1:1 showed the most promising results among the others. The sensor impedance changed by more than four orders of magnitude over the whole RH range, and low hysteresis also observed. However, the sensor had slightly long response and recovery times of approximately 50 seconds and 100 seconds, respectively. The analysis of the X-ray photoelectron spectroscopy (XPS) data suggested that Si–O–Zn bonds existed in the ZnO-SiO_2 composites, and the sensor was utilizing the protonic conduction mechanism. Later, the same authors (Yuan et al. 2011) established that the sensors based on (Fe_2O_3/SiO_2) composites (Fe/Si with molar ratio of $r = 0.5$) exhibited faster response and recovery times of 20 seconds and 40 seconds, respectively, along with excellent linearity (see Figure 11.32). Based on the complex impedance analysis, the sensing mechanism was assessed to be due to proton transport (proton hopping).

The porous La_2O_3-TiO_2-V_2O_5 glass ceramics also were tested as material for humidity sensors (Shimizu et al. 1987, 1989b). This is a phase-separable glass system, and the interesting feature of this kind of system is that it is possible to control the microstructure of the resulting porous glass by heat treatment to induce phase separation, and subsequent leaching to wash out the soluble phase. Moreover, it is also possible to choose from the various glass-ceramic systems the one that has a suitable intrinsic impedance. Humidity sensitivity up to three orders of magnitude in the changes in impedance, in addition to good linearity of the logarithm of impedance in the whole RH range, have been reported. However, the shortest response time measured was 3 minutes, which is not a satisfactory value (Shimizu et al. 1989b). This is probably due to the fact that the pores for this sensor are smaller than 5 nm.

Yagi and Saiki (1991) have shown that NASICON-based thick-film humidity sensors can also be used in ordinary air-conditioning. NASICON, ($Na_{1+x}Zr_2Si_xP_{3-x}O_{12}$,

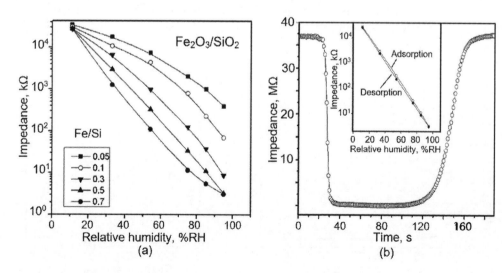

FIGURE 11.32 (a) The dependence of impedance on the relative humidity (RH) for mesoporous Fe_2O_3/SiO_2 samples with different Fe/Si ratios operated under the 1 V ac voltage, applied on a 0.15 mm distance and with frequency of 100 Hz and (b) Response and recovery characteristic of Fe_2O_3/SiO_2 composite ($r = 0.5$). *Insert*: Hysteresis characteristic of Fe_2O_3/SiO_2 composite ($r = 0.5$). (Reprinted with permission from Yuan, Q. et al., Humidity sensing properties of mesoporous iron oxide/silica composite prepared via hydrothermal process, *Sens. Actuators B*, 160, 334–340, 2011. Copyright 2011, Elsevier.)

$0 < x < 3$), which is a high ionic conductor (Sadaoka and Sakai 1985), was selected in order to solve the problems generally posed by ceramic humidity sensors (i.e., high resistance at low humidity and the need for periodic heat cleaning). NASICON was synthesized by the sol–gel method to obtain ultrafine porous powders, which were used for the preparation of thick films by screen printing. Among the NASICON materials, the phosphorus-free material showed comparatively low resistance at low RH and improved linearity. The relation between the logarithm of resistance and RH was nearly linear and hysteresis free in the RH range 20%–90% and at operating temperatures from 0°C to 60°C, as shown in Figure 11.33. When the RH was increasing, the response time was approximately 90 seconds, while it was approximately 130 seconds when the humidity was decreasing.

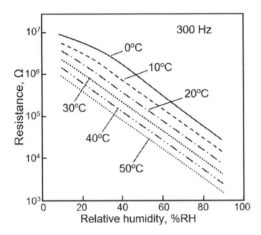

FIGURE 11.33 The relative humidity (RH) dependence of the resistance of the P-free NASICON thick-film sensor at different temperatures. (Reprinted with permission from Yagi, H. and Saiki, T., Humidity sensor using NASICON not containing phosphorus, *Sens. Actuators B*, 5, 135–138, 1991. Copyright 1991, Elsevier.)

There was very little change in the characteristics of the sensors not containing phosphorus during the operation, permitting use without heat-cleaning cycles (Yagi and Saiki (1991).

As follows from the presented results, many sensors have similar performances, and often their parameters are more dependent on the manufacturing technology than on the nature of the metal oxide used. Sensors manufactured on the base of the same material, depending on the technology used, may differ drastically.

For example, in Table 11.4, the results are given for ZnO-based humidity sensors, and Figure 11.34 shows SEM images of SnO_2 nanoparticles, synthesized by various methods, and the resistance response to air humidity of sensors fabricated using these materials. It can be seen how strongly the characteristics of sensors based on the same material may differ. Therefore, it is very difficult to evolve the oxides that are of greatest interest for practical applications, since we do not know how optimal the composition was and what the sensor manufacturing parameters were.

In this regard, when choosing a metal oxide for humidity sensors, the requirement for high stability of the parameters of the material used is of primary importance. For example, it is difficult to expect high stability, especially in an aggressive atmosphere, in sensors based on CuO (Xu et al. 2009). The appearance of H_2S or SO_2 in the atmosphere will be accompanied by chemical interaction with all the ensuing consequences for CuO-based sensors. However, as a rule, numerous studies related to the study of metal oxide–based humidity sensors do not pay attention to these issues. Considering the state of research in this field, in the future we will focus on the problems that are common to all sensors, regardless of the nature of the metal oxide used in the manufacture of sensors.

TABLE 11.4

Comparison of ZnO-Based Humidity Sensors' Performances

Preparative Method	Response Time, s	Recovery Time, s	Max. Response	Ref.
Arc discharge method	–	–	87,900	Fang et al. (2009)
Vapor phase transport	3	20	180	Zhang et al. (2005)
	3	30	5400	Zhang et al. (2005)
	89	175	0.61	Yawale et al. (2007)
	40–70	80–150	230	Zhang et al. (2008)
Precipitation method	–	–	7	Yadav et al. (2008)
	80	–	5.4	Yadav et al. (2009)
dc reactive magnetron sputtering	3	12	42,700	Kannan et al. (2010)
Sol-gel	6	20	~20	Peng et al. (2012)

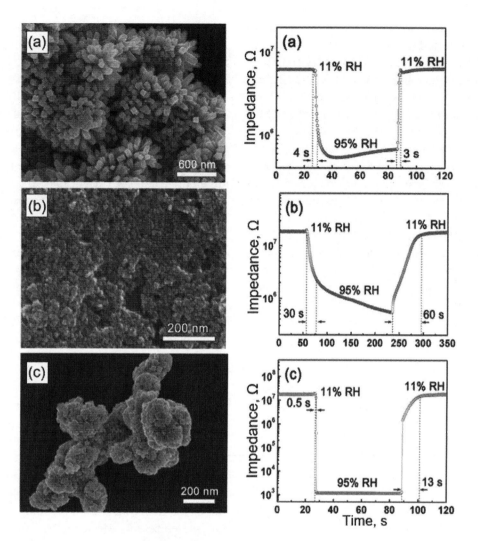

FIGURE 11.34 SEM images and humidity response and recovery curves for (a) three-dimensional (3D), hierarchical SnO_2 nanorods (NRs); (b) SnO_2 nanoparticles (NPs); and (c) 3D hierarchical SnO_2 dodecahedral nanocrystals (DNCs). (Reprinted with permission from Feng, H. et al., Three–dimensional hierarchical SnO_2 dodecahedral nanocrystals with enhanced humidity sensing properties, *Sens. Actuators B*, 243, 704–714, 2017. Copyright 2017, Elsevier.)

11.3.4 Mechanism of Sensor Response and the Role of Structural Factor in the Sensor Performance

One should note that, during the whole period of the humidity sensors' development, attempts were made to understand from the theoretical standpoint the relative weight of the influence of water physisorption and water condensation within the porous structure, leading to electrolytic conduction, on the electrical response of RH sensors. The first theoretical model for the calculation of impedance-humidity characteristics based on the use of the Kelvin equation was proposed by Shimizu et al. (1985b). Knowledge of the pore size distribution, the intrinsic impedance of the sensor element, and the conductivity of condensed water permits the calculation of the impedance-humidity characteristics. In these calculations, the contribution of the thickness of the multilayer of adsorbed water on the impedance was neglected (i.e., the RH sensitivity was attributed only to the capillary condensation). This model was applied to spinel-type oxides (Shimizu et al. 1985a, 1985b). But it was established that this model did not give satisfactory results for these samples (Traversa 1995). This gives possibility to conclude that, in addition to capillary condensation and to the intrinsic resistance of the materials, the surface conduction in a multilayered adsorbed water cannot be ignored in calculating the electrical response of humidity-sensitive ceramic oxides. Furthermore, recent results showed the humidity dependence of the surface resistances of $LiNbO_3$ and $LiTaO_3$ single crystals (Maeda et al. 1992). It was found that the logarithm of the surface electrical resistance decreased linearly with increasing RH. The maximum sensitivity over the RH range 30%–90% was nearly two orders of magnitude for $LiNbO_3$. The humidity dependence of the electrical parameters was interpreted in terms of the water multilayer adsorption mechanism, with conduction due to the Grotthuss chain reaction. For single crystals, water condensation in capillary pores is to be excluded, again confirming the importance of the multilayered water to the humidity sensitivity of ceramics.

Another theoretical model has been proposed by Sun et al. (1989), which seems to follow better the experimental results (Traversa 1995). This model is based on the consideration that electronic and ionic conduction exist simultaneously in porous ceramics at any humidity. The resistance of the sensor element is given by the sum of electronic-type conduction for crystal grains and of ionic-type conduction from proton hopping between water molecules adsorbed on the grain surfaces (Sun and Wu 1988). This model takes into consideration the Grotthuss proton-transfer mechanism and the intrinsic resistivity of the materials. Other attempts to explain humidity-sensing characteristics of various metal oxides can be found in Nitta and Hayakawa (1980), Sadaoka et al. (1987), Garcia-Belmonte et al. (2003), Wang and Virkar (2004), Wagiran and Zaki (2005), and Wang et al. (2009a). For example, based on the study of the impedance of TiO_2 thick porous films, Garcia-Belmonte et al. (2003) have concluded that the conductivity at low frequency in the humidity range from 2% to 60% RH can be described by the fundamental percolation theory of conduction through percolation network, formed by adsorbed H_2O molecules. Wang et al. (2009a) have reported that the conduction mechanism in ZrO_2 films in low RH ranges was mainly dominated by polarization of the grains, while for a higher range of RH, the process occurred by dissociation and polarization of water molecules. The same conclusion was made by Xie et al. (2015) while studying CeO_2-based humidity sensors. Other researchers believe that, in many cases, observed effects should be explained, considering the presence of two types of carriers, electrons and ions (Yeh and Tseng 1989; Wang et al. 2011b). At low RH, only electrons act as the charge carrier, and the increase in conductivity in this region is associated with the release of electrons captured by chemisorbed oxygen, which occurs as a result of the chemisorption of water on the surface of the metal oxide. With increasing RH, the metal oxides can effectively absorb water molecules on their surface to form serial water layers. The ions (H^+, H_3O^+, etc.), which appear as a result of this process, begin to participate in the transport along with electrons, and as RH increases, become dominating in the transport process.

Thus, summarizing previous consideration, one can conclude that—as the metal oxide conductivity at room temperature is strongly dependent on the presence on their surface of monolayer and multilayer of water vapor molecules formed by chemisorption, physisorption, and capillary condensation—metal oxide films designed for humidity sensors should have a specific microstructure, which could be characterized by a specific surface area, porosity, and size and volume of the pores. This means that microstructure control of the porous humidity-sensitive elements is highly determinative of sensor efficiency (Traversa 1995).

For example, Gusmano et al. (1992, 1993a, 1993b) have studied the relationship between the microstructure and the electrical properties of porous $MgAl_2O_4$ and $MgFe_2O_4$ spinel compacts. It was observed that both total open porosity and pore-size distribution had a great influence on the resistance versus humidity sensitivity of spinel compacts. The best results were shown by the samples with a wide distribution of pore sizes in the range 20–500 nm (Gusmano et al. 1993c). The response time was also influenced by the microstructure. Comparison of sensors with metal oxides, having micro- and macro-pore structures, has shown that the presence of micropores caused a slow response time, while the absence of micropores resulted in a fast response time. It was concluded that this effect takes place due to the easy adsorption and desorption of water molecules in large pores (Gusmano et al. 1993b). The same results were reported by Wang and Virkar (2004). They studied a conductometric humidity sensor based on the proton-conducting perovskite oxide $Ba_3Ca_{1.18}Nb_{1.82}O_{9-\delta}$ and found that porous material had a 5–6 times faster time response than that for the dense material. The control of dimension, shape, and distribution of pores in metal oxides was also crucial for the fabrication of the mesostructured thin-film humidity sensors based on $BaTiO_3$ (Yuk and Troczynski 2003), $PbSr_xZr_{0.3}Ti_{0.7}O_3$ (Ansari et al. 2004), and $Cr_{2-x}TixO_3$ (Neri et al. 2004). Varghese et al. (2002) reported the effect of pore size and uniformity on sensing performance for nanoporous Al_2O_3 films. It was reported that the response of the material to humidity was a strong function of pore size. A well-behaved change in impedance more than three orders of magnitude variation over 20% to 90% RH was reported for nanoporous Al_2O_3 films with an average pore size of 13.6 nm. Chang and Tseng (1990) have also found that the humidity conductance and sensitivity characteristics of calcium titanate perovskite sintered at different temperatures is strongly dependent on the microstructure of the sintered ceramics. Viviani et al. (2001) have studied the humidity-sensing behavior of porous barium molybdenum oxide ($BaMO_3$) and reported that the highest sensitivity was found in highly porous $BaTiO_3$:La specimens. In other research, conducted by Hwang and Choi (1997), the electrical characteristics of porous lanthanum (La)-doped $BaTiO_3$ have been studied by means of complex impedance spectroscopy as a function of different sintered densities and humidity conditions. It was revealed that the samples with lower density and higher resistivity exhibited large and nearly linear conductivity in response to humidity changes. Similar conclusion was made by Chou et al. (1999), who established that the sensitivity of the Al_2O_3-SiO_2-ZrO_2-based sensor in the range of high humidity was controlled by the mesopore volume in this material (Figure 11.35a) and in the range of low RH by the specific surface area (Figure 11.35b). According to Yuan et al. (2011), the excellent performance of the Fe_2O_3/SiO_2-based humidity sensor is also attributable to the microstructure properties, such as high Brunauer-Emmett-Teller (BET) surface area with large volume of pores, and a highly ordered porous structure. Remember that the size of crystallites also plays an important role in humidity-sensitive effects (Korotcenkov et al. 2009; Pavelko 2012). As a rule, with sufficient gas permeability of humidity-sensitive material, a decrease in the size of crystallites stimulates an increase in the sensitivity to humidity (see Figure 11.36).

Thus, summarizing conducted analysis, one can conclude that high porosity and a large surface area are desirable to enhance sensitivity, insofar as they do not compromise mechanical stability (Nitta et al. 1980; Shimizu et al. 1985a, 1985b; Drazic and Trontelj 1989; Xu and Miyazaki 1993). In order to obtain a good humidity-sensitive response, the resistivity of the materials must also be high in nearly dry

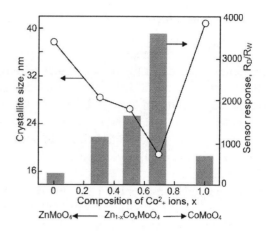

FIGURE 11.36 Graph of average crystalline size (D) and sensitivity to humidity versus composition of Co^{2+} ions in $Zn_{1-x}Co_xMoO_4$ (x = 0, 0.3, 0.5, 0.7, 1) composites. (Data extracted from Jeseentharani, V. et al., *Solid State Sci.*, 67, 46–58, 2017.)

environments, and their pore-size distribution must be controlled (Shimizu et al. 1985b). In particular, according to (Yamazoe and Shimizu 1986), the ceramic sensing elements in RT humidity sensors, in addition to high porosity, should have a wide pore-size distribution over all the size ranges (i.e., micro-, meso and macro-pores, up to radii in which capillary condensation cannot take place at any humidity at their operating temperatures).

From the empirical point of view, it becomes clear that the requirements mentioned above are difficult, but feasible. The experiment showed that in technological techniques capable of controlling the structure of metal oxides, one can use (1) the introduction into the metal oxide matrix of various additives (Shimizu et al. 1989c;

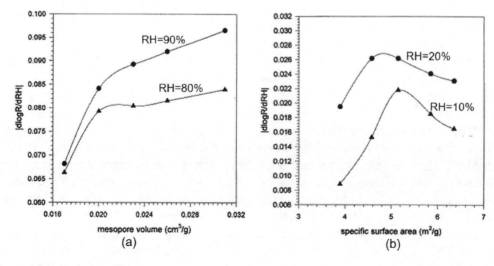

FIGURE 11.35 Correlations between impedance sensitivity and (a) mesopore volume under high relative humidity (RH) and (b) The BET-specific surface area under low RH. (Reprinted with permission from Chou, K.-S. et al., Sensing mechanism of a porous ceramic as humidity sensor, *Sens. Actuators B*, 56, 106–111, 1999. Copyright 1999, Elsevier.)

Yang and Wu 1991; Kim et al. 1992; Liang et al. 2012), (2) the change in the ratio of used components (Arshak and Twomey 2002; Raj et al. 2002; Yuan et al. 2011; Jeseentharani et al. 2017; Sabarilakshmi and Janaki 2017), and (3) the change in the sintering conditions (Golonka et al. 1997; Klym et al. 2016). In particular, the addition of the second phase alters the growth conditions of the main oxide phase and affects the surface activity, and the concentration of adsorption centers on its surface. The main regularities of the effect of the second phase on the structural properties of metal oxides can be found in Traversa (1995), Korotcenkov and Cho (2017), and Farahani et al. (2014). It is important to note that most humidity sensors described earlier have been developed using this approach to the formation of humidity-sensitive materials.

For example, Sun et al. (1988, 1989) have proposed to use graphite powder as the pore former in the humidity-sensitive ceramics during their preparation. Graphite powders can be removed from ceramics during sintering. It was found that when graphite powder was added for the fabrication of porous Co-Fe spinel, the porosity of the specimen increased remarkably, thereby resulting in increased resistance versus RH sensitivity (Sun et al. 1989). These results are presented in Figure 11.37.

Arshak and Twomey (2002) have shown that the stability of the humidity sensors can be significantly improved due to the correct composition (see Figure 11.38). However, these studies have established that improving stability is not always accompanied by an improvement in other parameters. For example, in their case, the most stable sensors had the most nonlinear dependence of resistance on humidity.

As was shown earlier, in resistive humidity sensors based on metal oxides, the main processes responsible for the change in conductivity occur on the surface of the metal oxide and are not associated with capillary condensation of water vapor, as is the case in the humidity sensors of permittivity type (see Chapter 10). Thus, in such sensors, the use of materials with a wider variation in the structure is allowed. For example, in recent years, great interest has been shown in the development of humidity sensors based on hierarchical nanostructures (Xu et al. 2009; Zhen et al. 2016; Feng et al. 2017), nanowires (Wan et al. 2004; Zhang et al. 2005; Hu et al. 2014; Hsu et al. 2017), and nanofibers (Wang et al. 2011b; Hayat et al. 2016; Ganeshkumar et al. 2017), which in principle do not have nanopores, where such condensation is possible. At the same time, such materials are the most gas permeable to water vapor (see Figure 11.39), providing, in addition to high sensitivity (see Figure 11.40), a fast response and recovery. As can be seen from the results shown in Figure 11.40 and Table 11.5, the sensor response to humidity can reach 10^4–10^5. It is understood that this effect is being achieved with a certain combination of the nature of the metal oxide and its structure.

11.3.5 The Role of Alkaline Ions in Sensing Effect Observed in Metal Oxide-Based Humidity Sensors

Studies have shown that alkaline ions introduced into the lattice of metal oxides are among the most effective additives capable of exerting a comprehensive influence on the parameters of humidity sensors. Kim et al. (1992) have established that the addition of alkaline oxides of Na_2O and K_2O (5, 10, 20 wt.%) to $MgCr_2O_4$-TiO_2 help to reduce its pore size distribution, therefore improving its humidity-sensing characteristics at different RH levels (Kim et al. 1992). Joanni and Baptista (1993) have investigated the addition of Li_2O to ZnO

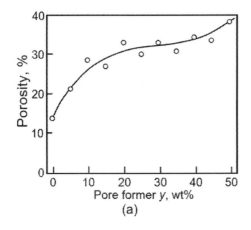
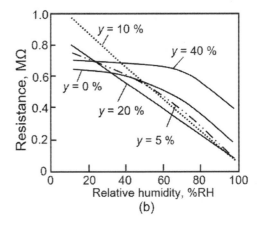

FIGURE 11.37 (a) Relationship between the porosity of the porous ceramic and the amount of added pore-farming agent and (b) Characteristics of resistance (R) and relative humidity (RH) of several porous humidity sensitive ceramics. (Reprinted with permission from Sun, H.T. et al., Porosity control of humidity–sensitive ceramics and theoretical model of humidity sensitive characteristics, *Sens. Actuators*, 19, 61–70, 1989. Copyright 1989, Elsevier.)

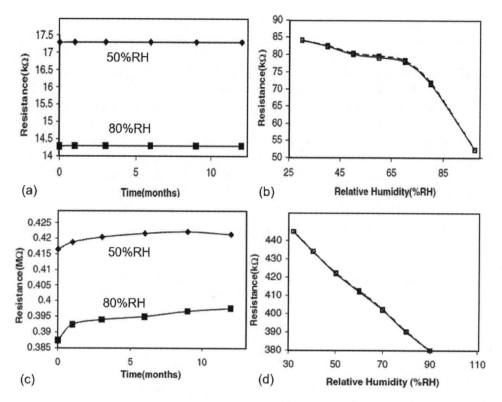

FIGURE 11.38 The effect of fabrication technology on the parameter drift and R = f(RH) dependencies defined for MnO(53%)-ZnO(25%)-Fc$_2$O$_3$(22%) (MZF)-based humidity sensors: (a, b) Air-fired cermet sensor (600–800°C) and (c, d) MZF:IPA (isopropenyl acetate) = 10:1 sensor. (Reprinted with permission from Arshak, K.I. and Twomey, K., Investigation into a novel humidity sensor operating at room temperature, *Microelectronics J.*, 33, 213–220, 2002. Copyright 2002, Elsevier.)

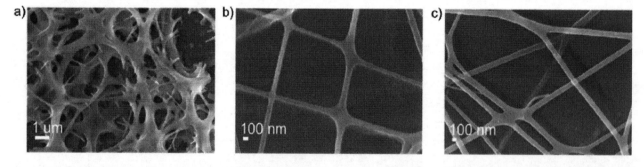

FIGURE 11.39 SEM micrographs of KNbO$_3$ nanofibers annealed at (a) 450°C; (b) 550°C, and (c) 650°C. (Reprinted with permission from Ganeshkumar, R. et al., Annealing temperature and bias voltage dependency of humidity nanosensors based on electrospun KNbO$_3$ nanofibers, *Surf. Interfaces*, 8, 60–64, 2017. Copyright 2017, Elsevier.)

ceramic sensors and found that this additive results in improving sensitivity and longtime stability, as well as in decreasing resistivity and hysteresis. They indicated that Li$_2$O dopants can play the role of sintering agents or liquid glassy phases. The addition of 1 mol% MgO or CrO$_{1.5}$ and FeO$_{1.5}$ to ZrO$_2$-TiO$_2$ increased its conductivity and sensitivity across the humidity range of 20%–90% RH (Yang and Wu 1991). Alkali or other dopant ions can also act as dominant charge carriers instead of protons, and thus lead to higher humidity-sensitive devices (Yamazoe and Shimizu 1986). In particular, Ichinose (1993) observed such effect after addition of 1 mol% K$^+$ to NiWO$_4$, ZnWO$_4$, and MgWO$_4$. Yokomizo et al. (1983) also believed that the hydrated Li$^+$ ions contribute to the charge transport and enhanced the sensitivity to humidity in the ZnCr$_2$O$_4$-LiZnVO$_4$-based sensors. The influence of 0, 1, 2, 5, and 10 mol% of Li, Na and K alkaline earth metal oxides on the humidity-sensitivity response of Nb$_2$O$_5$-doped TiO$_2$ was investigated by Katayama et al. (1990b). Dopant admixtures

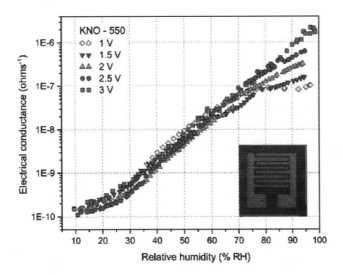

FIGURE 11.40 Electrical conductance versus relative humidity of $KNbO_3$-based nanosensor at different biasing voltages. *Inset*: fabricated humidity nanosensor. (Reprinted with permission from Ganeshkumar, R. et al., Annealing temperature and bias voltage dependency of humidity nanosensors based on electrospun $KNbO_3$ nanofibers, *Surf. Interfaces*, 8, 60–64, 2017. Copyright 2017, Elsevier.)

TABLE 11.5
Parameters of Impedance Humidity Sensors Based on ZnO One-Dimensional Nanostructures

ZnO 1-D Morphology	RH Range	Measurement Results	Max Response	Ref.
Nanowires	25%–95%	Current: $2·10^{-5}$–$2·10^{-3}$ A	10^2	Wan et al. (2004)
	40%–98%	Current: $2·10^{-8}$–$5·10^{-5}$ A	$2·10^3$	Hsu et al. (2017)
	40%–98%	Current: $2·10^{-5}$–$2·10^{-3}$ A	10^2	Hsu et al. (2013)
	15%–66%	Current: $1·10^{-7}$–$3·10^{-7}$ A	3	Hu et al. (2014)
	12%–97%	Resistance: $4·10^8$–$1·10^4$ Ω	$4·10^4$	Zhang et al. (2005)
	1%–60%	Resistance: $8·10^{10}$–$9·10^5$ Ω	$\approx 10^5$	Kiasari et al. (2012)
	25%–90%	Resistance: $6·10^5$–$3·10^5$ Ω	2	Chang et al. (2010)
Nanofibers	11%–95%	Resistance: $4·10^7$–$6·10^3$ Ω	$8·10^3$	Wang et al. (2009b)
	11%–95%	Resistance: $3·10^7$–$3·10^2$ Ω	10^5	Qi et al. (2009b)

1-D, one-dimensional; RH, relative humidity.

---

affected moisture sensitivity as well as the surface microstructure; however, among the dopants, K_2O had the maximum effect on the sensor characteristics, such as sensitivity and response time.

The effects of alkaline earth metal ions from the group II, such as strontium(II), have been emphasized and experimented on the cobalt alumina ($CoAl_2O_4$), zinc alumina ($ZnAl_2O_4$), and barium alumina ($BaAl_2O_4$)-based sensors, which have been studied by Vijaya et al. (2007a, 2007b, 2007c). All the composites were prepared thorough the sol–gel technique and formed by a conventional bulk processing. Similarly, all the specimens were fired under the same conditions of 900°C for 5 hours and the DC electrical properties characterized at RH in the range of 5%–98% RH. It was shown that the addition of Sr, as well as increases of its molar ratio, have led to enhancements of the sensitivity factor. The greatest sensitivity was observed for the composites consisting of 0.8 mol% strontium, while the undoped composites possessed the lowest sensitivity. It was established that these sensors had maximum specific surface area and maximum total pore volume. In all three experiments, the mentioned highly sensitive specimens have shown response and recovery times between 120 seconds and 50 seconds.

Desirable microstructure features of the films for moisture-sensing applications, such as a large surface area and porosity, have been achieved in TiO_2-Cu_2O screen-printed thick films by addition of amounts of sodium oxide. The sensors showed the resistance change of three orders of magnitude between 20% and 95% RH

(Kim and Gong 2005). Song et al. (2009) also strongly improved a performance of humidity sensors via doping with alkaline ions. They fabricated a fast, simple, and reliable thick-film, screen-printed humidity sensor of potassium chloride (KCl)-doped SnO_2 nanofibers based on an interdigitated structure. The sensor showed very low hysteresis and fast and very close response/recovery times of 5 seconds and 6 seconds, respectively, for a range of 11%–95% RH (see Figure 11.41). The sensor impedance was changed by more than five orders of magnitude for this RH range. According to Song et al. (2009), the excellent humidity-sensing characteristics of the sensor described here can be attributed to its structures, such as macropores in the fiber network, the rough porous structure, and large surface area of KCl-doped SnO_2 nanofibers, which easily enable water molecules to adsorb on the surfaces of SnO_2 nanofibers. Moreover, the addition of alkali ions can provide more active sites for water molecules, resulting in the improvement of the sensitivity, as reported in Jain et al. (1999). The same effects were observed for aluminium-doped (Md Sin et al. 2011) and indium-doped ZnO (Liang et al. 2012), K_2O-doped $(Ba_{0.5}Sr_{0.5})TiO_3$-based sensors (Yeh and Tseng 1989), and many other sensors (Bonavita et al. 2001).

Thus, summing up this review, we can conclude that the addition of properly selected concentration of alkalis, as well as many other additives, can strongly affect the microstructure of the humidity-sensitive layers, and thus control the RH sensitivity. Such doping can be accompanied by optimizing effects, such as an increase in open porosity, an increase in the total pore volume, and an increase in the concentration of charge carriers, including ions and electrons. However, it must be borne in mind that doping can give an optimizing effect not only by optimizing the structure of the humidity-sensitive materials. For example, Yeh and Tseng (1989) showed that the porosity and the specific surface area of pure $Ba_{0.5}Sr_{0.5}TiO_3$ were larger than those of K_2O-doped $Ba_{0.5}Sr_{0.5}TiO_3$, but the conductance humidity sensitivity of the former was much smaller than that of the latter. This is because potassium oxide can create more surface-defect lattice sites or oxygen vacancies, resulting in an increased number of efficient adsorption sites for water-vapor adsorption.

11.3.6 HEATED SENSORS WITH SEMICONDUCTING AND IONIC METAL OXIDES

In principle, all the arguments of the previous sections regarding the structure of humidity-sensitive materials are also acceptable for this type of humidity sensor. The only difference is that processes such as capillary condensation, physisorption of water vapors, and the Grotthuss mechanism of conductivity cannot be realized in such sensors. Another difference is that the number of studies aimed at developing exactly such resistive humidity sensors is very limited. Of course, there are a huge number of articles that pay attention to the effect of humidity on the characteristics of heated semiconducting gas sensors, but they are mainly aimed at suppressing this influence, and not at strengthening.

As noted earlier, metal oxides with a semiconductor (Taguchi et al. 1980; Nitta et al. 1983) and ionic conductivity (Johnson and Biefeid 1979; Arai et al. 1983) are suitable for manufacturing heated humidity sensors. However, the sensitivity of electronic-type humidity sensors is much lower than that of ionic-type sensors. At that, it was observed that n-type semiconductors are more suitable than p-type ones as humidity sensors,

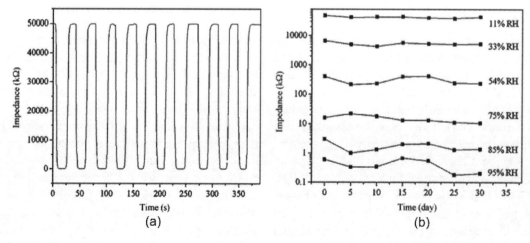

FIGURE 11.41 (a) Response and recovery curve of the 15% KCl-doped SnO_2 nanofibers measured at 100 Hz for 10 cycles. (b) Stability of the sensor after exposure in air for 30 days. (Reprinted with permission from Song, X. et al., Humidity sensor based on KCl–doped SnO_2 nanofibers, *Sens. Actuators B*, 138, 368–373, 2009. Copyright 2009, Elsevier.)

because the conductivity changes upon water-vapor contact are greater for *n*-type than for *p*-type oxides (Seiyama et al. 1983). In addition, the water vapor interaction with *n*-type semiconductors is accompanied by the decrease in resistance that provides better conditions for measurement.

An electronic-type humidity sensor based on ZrO_2-MgO has been developed by Nitta et al. (1983). When this sensor is exposed to an atmosphere containing a given gas at high temperature in the range 400–700°C, reversible chemisorption becomes dominant and the sensor's electrical conduction changes with gas chemisorption. Its electrical resistance is reduced due to adsorption of water vapor in the air, given that ZrO_2-MgO is an *n*-type semiconductor. If this sensor operates between 450°C and 700°C, the influence of interfering reducing gases becomes negligible, because of the enhanced reaction of the reducing gases with oxygen in the air in this temperature range. However, the response time of these sensors was rather slow.

Among the tested heated humidity sensors, the most thoroughly studied are the sensors developed on the basis of $SrSnO_3$. It was established that among perovskite-type oxides (ABO_3), exhibiting a humidity sensitivity in the temperature range between 300°C and 500°C (Arai et al. 1983), $SrSnO_3$ showed the best results. Enhancement of their humidity sensitivity was obtained by partially substituting the A-site element of $SrSnO_3$ with La^{3+} ions. The best results were obtained for the composition $Sr_{0.95}La_{0.05}SnO_3$ (see Figure 11.42) (Shimizu et al. 1985c, 1989a). Further improvement of the sensitivity of $Sr_{0.95}La_{0.05}SnO_3$ was obtained by using Pt electrodes and by adding Pt particles in metal oxide matrix (Shimizu et al. 1989c), in a similar way to that observed for semiconductor gas sensors (Matsushima et al. 1988). It was assumed that dispersed Pt or PtO_2 fine particles apparently act as sensitizers; adsorption of oxygen on the dispersed Pt particles induces a decrease in electron concentration in the positively charged space region, and an increase in the height of the potential barrier at grain boundaries (Shimizu et al. 1989c). In other words, the greater the amount of adsorbed oxygen, the higher is the sensitivity to water vapor. The other important problem (i.e., the interference of reducing gases on the humidity sensitivity), was overcome for humidity sensors based on $Sr_{1-x}La_xSnO_3$ by coating the sensing element with a Pt/Al_2O_3 catalyst (Shimizu et al. 1984).

The effect of the sintering temperature on the humidity sensitivity of $Sr_{0.95}La_{0.05}SnO_3$ was reported (Shimizu et al. 1989a). The higher the sintering temperature, the higher is the conductivity of the materials for a given water vapor pressure. This was explained in terms of a decrease in a contact resistance between adjacent grains due to the sintering process. The decrease in the surface area leads to a reduced number of active adsorption sites for chemisorbed water, thereby resulting in the observed reduction in the humidity sensitivity with increasing sintering temperatures (Figure 11.43) (Shimizu et al. 1989a). The mechanism of humidity sensing is thus related to surface phenomena. Thus, the influence of the microstructure on the humidity sensitivity of heated sensors also takes place, and the control of the microstructure of the sensing element is also one of the important factors for the realization of improved sensitivity.

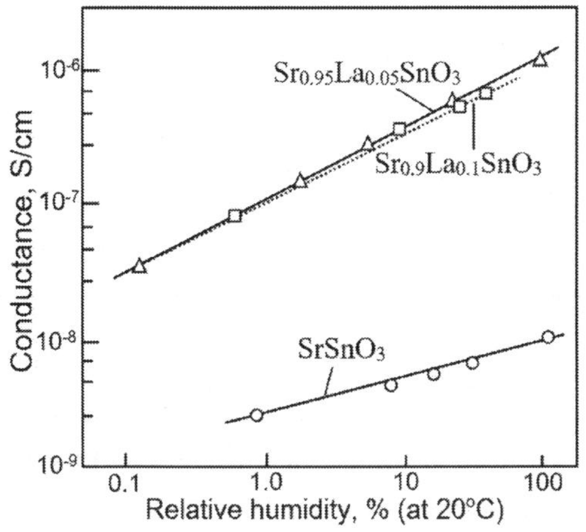

FIGURE 11.42 Effects of partial substitution of La^{3+} on humidity-sensitive characteristics. (o, $SrSnO_3$; Δ, $Sr_{0.95}La_{0.05}SnO_3$; □, $Sr_{0.9}La_{0.1}SnO_3$.) (Adapted with permission from Shimizu, Y. et al., *J. Electrochem. Soc.*, 136, 1206–1210. Copyright 1989, Electrochemical Society.)

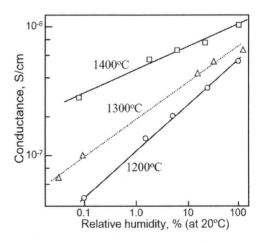

FIGURE 11.43 Effects of the sintering conditions of $Sr_{0.95}La_{0.05}SnO_3$ elements on their humidity-sensitive characteristics. (O, 1200°C; Δ, 1300°C; □, 1400°C.) (Adapted with permission from Shimizu, Y. et al., *J. Electrochem. Soc.*, 136, 1206–1210. Copyright 1989, Electrochemical Society.)

11.3.7 Summary

As it follows from our discussions, ceramic materials possess unique properties, which makes them suitable for humidity sensors when they have a controlled composition and microstructure. Ceramics for application in humidity sensors mainly are porous sintered bodies formed in order to allow water vapor to pass easily through the pores and water condensation in the capillary-like pores between the grain surfaces. Most of the ceramic materials used in humidity sensors have a porosity range from 25% to 50%. Compacts with a given microstructure can be prepared by a traditional ceramic processing. The modification of the microstructure and the chemical composition of ceramic materials using well-known methods permit a performance optimization in the sensors, exploiting their electrical properties and tailoring to specific requirements.

The main advantages of these sensors are easy fabrication, simple operation, and low production cost, which means that well-engineered metal–oxide humidity sensors can be mass produced at reasonable cost. Moreover, these sensors are compact and durable. These sensors are also characterized by high sensitivity. If the humidity content changes within 10%–90% RH, the impedance and resistance of the sensors can vary from 3 to 5 orders of magnitude. This means that even small changes in humidity can be detected. In addition, they can operate at high temperatures and in an aggressive atmosphere. As a result, they can be applied for in situ monitoring in special applications.

Unfortunately, several disadvantages limit more widespread sensor utilization and continue to challenge sensor development. It is important to note that these are the same disadvantages that were noted in metal oxide sensors of capacitive type. The main problem for ceramic humidity sensors refers to the need of their periodic regeneration by a heat cleaning to recover their humidity-sensitive properties. As has been found in ceramic sensors of this type, the base resistance of the element increases gradually during operation (Yamazoe and Shimizu 1986). This drift seems to be connected with the formation of stable, chemisorbed OH groups on the surface of metal oxides, which impede the Grotthuss-type proton conduction. As it was established, the surface hydroxyl ions are removed by heating to temperatures higher than 400°C (Morimoto et al. 1969; Egashira et al. 1978, 1981). This means that the heat cleaning should be at temperatures above 400°C. Moreover, humidity sensors are usually exposed to atmospheres that contain a number of impurities, such as dust, dirt, oil, smoke, alcohol, and solvents. The adhesion or adsorption of these compounds on the ceramic surface causes irreversible changes in the sensor's response. The high surface reactivity of metal oxides makes these materials less resistant than polymers to surface contaminants. Contaminants act in the same way as chemisorbed water and may be removed by heating, too. Therefore, several commercial humidity sensors based on ceramic sensing elements were equipped with a heater for regeneration (Nitta 1981). Yagi and Nakata (1992) have shown that, in many cases, a once-a-day cycle of heat cleaning is enough for recovering the sensor's performance in a normal atmosphere. This allows long continuous measurement and provides a nearly maintenance-free ceramic humidity sensor with improved reliability and precision.

Nonlinear behavior of humidity-resistance characteristics is also a disadvantage of this type of humidity sensors (Traversa 1995). This complicates the processing of results and the construction of a measuring circuit. Another drawback of resistive sensors is their significant temperature dependencies when installed in an environment with large temperature fluctuations. This means that simultaneous temperature compensation is required for accuracy (Wang et al. 1993; Qu and Meyer 1997).

11.4 OTHER MATERIALS ACCEPTABLE FOR APPLICATION IN RESISTIVE HUMIDITY SENSORS

Experimentation has shown that a huge number of different materials can be used in resistive humidity sensors. For example, Chen et al. (2016) reported the application of ultrasonic exfoliated black phosphorus (BP) nanosheets as the sensitive material of a humidity sensor. BP materials have attracted considerable attention owing to their ultrasensitive humidity-sensing characteristics because of the natural absorption of water (H_2O) molecules on the BP surface caused by the specific 2D layer-crystalline structure. Chen et al. (2016) have shown that BP-based sensors had acceptable resistance response in the range of RH from 40% to 70%. As the nanosheet layer was approximately 77 nm, the response time of this sensor was very short. Moreover, Zhu et al. (2016) formed BP quantum dots (BPQDs) and established that, as the humidity changes from 10% to 90%, the resistance of the BPQDs film varied by approximately 4 orders of magnitude. However, the same studies established (see Figure 11.44a) that the BP-based humidity sensors are less repeatable due to the instability of BP with water molecules (Yasaei et al. 2015; Phan et al. 2017). BP reacts naturally with water molecules, limiting its long-term stability in a humid environment. Phan et al. (2017) have shown that this limitation of the BP-based humidity sensor can be overcome

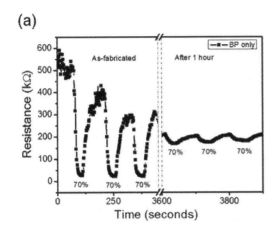
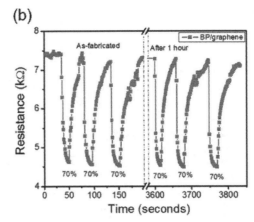

FIGURE 11.44 Transient response and estimated stability of the black phosphorus (BP)-based humidity sensor after 1 hour based on (a) BP only and (b) BP/graphene heterojunction. (Reprinted with permission from Phan, D.-T. et al., Black P/graphene hybrid: A fast response humidity sensor with good reversibility and stability, *Sci. Rep.*, 7, 10561, 2017. Copyright 2017, Springer Science+Business Media.)

by preparing a BP/graphene hybrid. According to Phan et al. (2017), the BP/graphene-based sensors had improved reversibility and stability, and a linear response within the RH range of 15%–70% (see Figure 11.44b). The response/recovery speed of the humidity sensor was extremely fast—within few seconds. The estimated response and recovery time of the sensor was only 9 and 30 seconds at RH of 70% at room temperature. However, the sensor response of such BP/graphene sensors was only 43.4% at RH of 70%. As for other materials used in resistive humidity sensors, such as carbon-based materials (carbon nanotube and graphene oxide) (Phan and Chung 2015) and chalcogenides (Kunakova et al. 2015), their consideration will be continued in Volume 3 (Chapters 4–10).

REFERENCES

Abraham J.K., Philip B., Witchurch A., Varadan V.K., Channa R.C. (2004) A compact wireless gas sensor using a carbon nanotube/PMMA thin film chemiresistor. *Smart Mater. Struct.* 13 (5), 1045–1049.

Agmon N. (1995) The Grotthuss mechanism. *Chem. Phys. Lett.* 244, 456–462.

Agrawal C., Pandey G.P. (2008) Solid polymer electrolytes: Materials designing and all–solid–state battery applications: An overview. *J. Phys. D: Appl. Phys.* 41, 223001.

Ahmadipour M., Ain M.F., Ahmad Z.A. (2016) Effect of thickness on surface morphology, optical and humidity-sensing properties of RF magnetron sputtered CCTO thin films. *Appl. Surf. Sci.* 385, 182–190.

An H., Hirata M., Yosomiya R., Xin Y. (1987) Humidity dependence of the electric characteristics of quaternized 4–vinylpyridine–styrene copolymer/perchlorate complexes. *Die Angew. Makromolekulare Chem.* 150, 33–44.

Anbia M., Fard S.E.M. (2011) Improving humidity sensing properties of nanoporous TiO_2–10 mol % SnO_2 thin film by co–doping with La^+ and K^+. *Sens. Actuators B* 160, 215–221.

Anbia M., Moosavi Fard S.E., Shafiei K., Hassanzadeh M.A., Mayahipour A. (2012) Humidity sensing properties of the sensor based on V–doped nanoporous $Ti_{0.9}Sn_{0.1}O_2$ thin film. *Chin. J. Chem.* 30, 842–846.

Anchisini R., Faglia G., Gallazzi M.C., Sberveglieri G., Zerbi G. (1996) Polyphosphazene membrane as a very sensitive resistive and capacitive humidity sensor. *Sens. Actuators B* 35(36), 99–102.

Anderson J.H., Parks G.A. (1968) The electrical conductivity of silica gel in the presence of adsorbed water. *Z Phys. Chem.* 72, 3362–3368.

Angelopoulos M., Ray A., Mcdiarmid A.G. (1987) Polyaniline: Processability from aqueous solutions and effect of water vapor on conductivity. *Synth. Metals* 21, 21–30.

Ansari Z.A., Ko T.G., Oh J.-H. (2004) Humidity sensing behavior of thick films of strontium–doped lead–zirconium-titanate. *Surf. Coating Technol.* 179, 182–187.

Antoniadis H., Abkowitz M.A., Osaheni J.A., Jenekhe S.A., Stolka M. (1994) Effects of humidity on the dark conductivity and dielectric properties of Poly(benzimidazobenzophenanthroline) thin films. *Chem. Mater.* 6, 63–66.

Arai H., Ezaki S., Shimizu Y., Shippo O., Seiyama T. (1983) Semiconductive humidity sensor of perovskite–type oxides, In: Seiyama T., Fueki K., Shiokawa J., Suzuki S. (eds.) *Analytical Chemistry Symposia Series: Chemical Sensors, Vol. 17.* Kodansha, Tokyo/Elsevier, Amsterdam, the Netherlands, pp. 393–398.

Arshak K.I., Twomey K. (2002) Investigation into a novel humidity sensor operating at room temperature. *Microelectronics J.* 33, 213–220.

Arshak K., Twomey K., Egan D. (2002) A ceramic thick film humidity sensor based on MnZn ferrite. *Sensors* 2, 50–61.

Avani G.N., Nanis L. (1981) Effects of humidity on hydrogen sulfide detection by SnO_2 solid state gas sensors. *Sens. Actuators B* 2, 201–206.

Barkauskas J. (1997) Investigation of conductometric humidity sensors. *Talanta* 44, 1107–1112.

Barsan N., Ionescu R. (1983) The mechanism of the interaction between CO and an SnO_2 surface: The role of water vapour. *Sens. Actuators B* 12, 71–75.

Bauskar D., Kale B.B., Patil P. (2012) Synthesis and humidity sensing properties of $ZnSnO_3$ cubic crystallites. *Sens. Actuators B* 161, 396–400.

Bearzotti A., Bertolo J.M., Innocenzi P., Falcaro P., Traversa E. (2004) Humidity sensors based on mesoporous silica thin films synthesised by block copolymers. *J. Eur. Cer. Soc.* 24, 1969–1972.

Belmaraes M., Blanco M., Goddard W.A. II, Ross R.B., Caldwell G., Chou S.-H., Pham J., Olofson P.M., Thomas C. (2004) Hildebrand and Hansen solubility parameters from molecular dynamics with applications to electronic nose polymer sensors. *J. Comput. Chem.* 25, 1814–1826.

Bonavita A., Caddemi A., Donato N., Accordino P., Galvagno S., Neri G. (2001) Electrical characterization and modeling of this–film humidity sensors. In: *Proceedings of IEEE Conference on Electronics, Circuits and Systems, ICECS 2001*, September 2–5, Malta, Europe, pp. 673–676.

Boyle J.F., Jones K.A. (1977) The effects of CO, water vapor and surface temperature on the conductivity of a SnO_2 gas sensor. *J. Electron. Mater* 6, 717–733.

Caballero A.C., Villegas M., Fernández J.F., Viviani M., Buscaglia M.T., Leoni M. (1999) Effect of humidity on the electrical response of porous $BaTiO_3$ ceramics. *J. Mater. Sci. Lett.* 18, 1297–1299.

Casalbore-Micelia G., Camaioni N., Li Y., Martelli A., Yang M.J., Zanelli A. (2005) Water sorption in polymer electrolytes: kinetics of the conductance variation. *Sens. Actuators B* 105, 351–359.

Cha J.-R., Gong M.-S. (2013) AC complex impedance study on the resistive humidity sensors with ammonium salt–containing polyelectrolyte using a different electrode pattern. *Bull. Korean Chem. Soc.* 34 (9), 2781–2786.

Chakraborty S. (1995) The humidity dependent conductance of $Al_2(SO_4) \cdot 16H_2O$. *Smart Mater. Struct.* 4, 368–369.

Chakraborty S., Nemoto K., Hara K., Lai P.T. (1998) Humidity sensors using a mixture of ammonium paratungstate pentahydrate and aluminium sulphate. *Smart Mater. Struct.* 7, 569–571.

Chaluvaraju B.V., Ganiger S.K., Murugendrappa M.V. (2016) Thermo–electric power and humidity sensing studies of the polypyrrole/tantalum pentoxide composites. *Mater. Sci. Mater. Electron.* 27, 1044–1055.

Chatzandroulis S., Tegou E., Goustouridis D., Polymenakos S., Tsoukalas D. (2004) Capacitive-type chemical sensors using thin silicon-polymer bimorph membranes. *Sens. Actuators B* 103, 392–396.

Chang D.A., Tseng T.Y. (1990) Humidity–sensitivity characteristics of $CaTiO_3$ porous ceramics. *J. Mater. Sci. Lett.* 9, 943–944.

Chang S.P., Chang S.J., Lu C.Y., Li M.J., Hsu C.L., Chiou Y.Z., Hsueh T.J., Chen I.C. (2010) A ZnO nanowire-based humidity sensor. *Superlattices Microstruct.* 47, 772–778.

Chen Z., Lu C. (2005) Humidity sensors: A review of materials and mechanisms. *Sensor Lett.* 3, 274–295.

Chen W.H., Huang J.Q., Zhu C.Y., Huang Q.A. (2016) A black phosphorus humidity sensor with high sensitivity and fast response. In: *Proceedings of IEEE Sensor conference, SENSORS 2016*, October 30–November 3, Orlando, FL. ID: 16597219. doi:10.1109/ICSENS.2016.7808462.

Chou K.-S., Lee T.-K., Liu F.-J. (1999) Sensing mechanism of a porous ceramic as humidity sensor. *Sens. Actuators B* 56, 106–111.

Cho S.M., Kim Y.J., Kim Y.S., Yang Y., Ha S.-C. (2004) The application of carbon nanotube—polymer composite as gas sensing materials. In: *Proceedings of IEEE sensors conference, Sensors 2004*, October 24–27, Vienna, Austria, Vol. 2, pp. 701–704.

Çiğil A.B., Cankurtaran H., Kahraman M.V. (2017) Photo–crosslinked thiolene based hybrid polymeric sensor for humidity detection. *React. Funct. Polymers* 114, 75–85.

Conway B.E., Bockris J.O., Linton H. (1956) Proton conductance and the existence of the H_3O Ion. *J. Chem. Phys.* 24, 834.

Drazic G., Trontelj M. (1989) Preparation and properties of ceramic sensor elements based on $MgCr_2O_4$. *Sens. Actuators* 18, 407–414.

Dunmore F.W. (1938) An electrometer and its application to radio meteorography. *J. Res. Nat. Bur. Std.* 20, 723–744.

Dunmore F.W. (1939) An improved electric hygrometer. *J. Res. Nat. Bur. Std.* 23, 701–714.

Egashira M., Kawasumi S., Kagawa S., Seiyama T. (1978) Temperature programmed desorption study of water absorbed on metal oxides. 1. *Bull. Chem. Soc. Jpn.* 51, 3144–3149.

Egashira M., Nakashima M., Kawasumi S., Seiyama T. (1981) Temperature programmed desorption study of water absorbed on metal oxides. 2. Tin oxide surfaces. *Z. Phys. Chem.* 85, 4125–4130.

Ertug B., Boyraz T., Addemir O. (2010) Humidity sensitivity characteristics of $BaTiO_3$ ceramics with pmma additive at various working temperatures. *J. Ceram. Process. Res.* 11, 443–447.

Fagan J.G., Amarakoon V.R.W. (1913) Reliability and reproducibility of ceramic sensors: Part III, Humidity sensors. *Am. Ceram. Soc. Bull.* 72, 119–130.

Fang F., Futter J., Markwitz A., Kennedy J. (2009) UV and humidity sensing properties of ZnO nanorods prepared by the arc discharge method. *Nanotechnology* 20, 245502–245509.

Farahani H., Wagiran R., Hamidon M.N. (2014) Humidity sensors principle, mechanism, and fabrication technologies: A comprehensive review. *Sensors* 14, 7881–7939.

Fei T., Jiang K., Liu S., Zhang T. (2014a) Humidity sensors based on Li–loaded nanoporous polymers. *Sens. Actuators B* 190, 523–528.

Fei T., Jiang K., Liu S., Zhang T. (2014b) Humidity sensor based on a cross–linked porous polymer with unexpectedly good properties. *RSC Adv.* 4, 21429–21434.

Fei T., Dai J., Jiang K., Zhao H., Zhang T. (2016) Stable cross–linked amphiphilic polymers from a one–pot reaction for application in humidity sensors. *Sens. Actuators B* 227, 649–654.

Feng C.D., Sun S.L., Wang H., Segre C.U., Stetter J.R. (1997) Humidity sensing properties of Nafion and sol–gel derived SiO_2/Nafion composite thin films. *Sens. Actuators B* 40, 217–222.

Feng C.D., Sun S.L., Wang H., Segre C.U., Stetter J.R. (1998) Humidity sensitive properties of Nafion® and sol–gel derived SiO_2/Nafion composite thin films. *Sens. Actuators B* 40, 217–222.

Feng H., Li C., Li T., Diao F., Xin T., Liu B., Wang Y. (2017) Three–dimensional hierarchical SnO_2 dodecahedral nanocrystals with enhanced humidity sensing properties. *Sens. Actuators B* 243, 704–714.

Fleming W.J. (1981) A physical understanding of solid state humidity sensors. *Soc. Automot. Eng. Trans. Sec.* 2, 90, 1656–1667.

Foster A.G. (1932) The sorption of condensible vapours by porous solids. Part I. The applicability of the capillary theory. *Trans. Faraday Soc.* 28, 645–657.

Fratoddi I., Altamura P., Bearzotti A., Furlani A., Russo M.V. (2004) Electrical and morphological characterization of poly(monosubstituted)acetylene based membranes: Application as humidity and organic vapors sensors. *Thin Solid Films* 458, 292–298.

Fratoddi I., Bearzotti A., Venditti I., Cametti C., Russo M.V. (2016) Role of nanostructured polymers on the improvement of electrical response–based relative humidity sensors. *Sens. Actuators B* 225, 96–108.

Furlani A., Iucci G., Russo M.V., Bearzotti A., D'Amico A. (1992a) Iodine–doped polyphenylacetylene thin film as a humidity sensor. *Sens. Actuators* B 8, 123–126.

Furlani A., Iucci G., Russo M.V., Bearzotti A., D'Amico A. (1992b) Thin films of iodine–polyphenylacetylene as starting materials for humidity sensors. *Sens. Actuators B* 7, 447–450.

Ganeshkumar R., Cheah C.W., Zhao R. (2017) Annealing temperature and bias voltage dependency of humidity nanosensors based on electrospun $KNbO_3$ nanofibers. *Surf. Interfaces* 8, 60–64.

Garcia-Belmonte G., Kytin V., Dittrich T., Bisquert J. (2003) Effect of humidity on the AC conductivity of nanoporous TiO_2. *J. Appl. Phys.* 94, 5261.

Goldenberg L.M., Krinichnyi V.I. (1993) Water sensitive sensor based on modified poly(vinyl chloride). *Synth. Met.* 53, 403–407.

Golonka L.J., Licznerski B.W., Nitsch K., Teterycz H. (1997) Thick–film humidity sensors. *Meas. Sci. Technol.* 8, 92–98.

Gong M.S., Lee M.H., Rhee H.W. (2001) Humidity sensor using cross–linked copolymers containing viologen moiety. *Sens. Actuators B* 73, 185–191.

Gong M.S., Park J.S., Lee M.H., Rhee H.W. (2002) Humidity sensor using cross–linked polyelectrolyte prepared from mutually reactive copolymers containing phosphonium salt. *Sens. Actuators B* 86, 160–167.

Gong M.-S., Kim J.-U., Kim J.-G. (2010) Preparation of water–durable humidity sensor by attachment of polyelectrolyte membrane to electrode substrate by photochemical cross-linking reaction. *Sens. Actuators B* 147, 539–547.

Grotthuss C.J.D. (1806) Sur la décomposition de l'eau et des corps qu'elle tient en dissolution à l'aide de l'électricité galvanique. *Ann. Chim. (Paris)* 58, 54–73.

Gusmano G., Montesperelli G., Nunziante P., Traversa E. (1992) The electrical behaviour of $MgAl_2O_4$ pellets as a function of relative humidity. *Mater. Eng.* 3, 417–434.

Gusmano G., Montesperelli G., Nunziante P., Traversa E. (1993a) Microstructure and electrical properties of $MgAl_2O_4$ and $MgFe_2O_4$ spinel porous compacts for use in humidity sensors. *Br. Ceram. Trans.* 92, 104–108.

Gusmano G., Montesperelli G., Nunziante P., Traversa E. (1993b) Humidity–sensitive electrical response of sintered $MgFe_2O_4$. *J. Mater. Sci.* 28, 6195–6198.

Gusmano G., Montesperelli G., Nunziante P., Traversa E. (1993c) Study of the conduction mechanism of $MgAl_2O_4$ at different environmental humidities. *Electrochim. Acta* 38, 2617–2621.

Hallinan Jr. D.T., Balsara N.P. (2013) Polymer electrolytes. *Annu. Rev. Mater. Res.* 43, 503–525.

Harsanyi G. (1995) *Polymer Films in Sensor Applications*. Technomic Publishing Company, Basel, Switzerland.

Hatchett D.W., Josowicz M. (2008) Composites of intrinsically conducting polymers as sensing nanomaterials. *Chem. Rev.* 108, 746–769.

Hayat K., Niaz F., Ali S., Iqbal M.J., Ajmal M., Ali M., Iqbal Y. (2016) Thermoelectric performance and humidity sensing characteristics of La_2CuO_4 nanofibers. *Sens. Actuators B* 231, 102–109.

Henderson M.A. (2002) The interaction of water with solid surfaces: Fundamental aspects revisited. *Surf. Sci. Rep.* 46, 1–308.

Hertl W., Hair M.L. (1968) Hydrogen bonding between adsorbed gases and surface hydroxyl groups on silica. *J. Phys. Chem.* 72, 4676–4682.

Hijikagawa M., Miyoshi S., Sugihara T., Jinda A. (1983) A thin–film resistance humidity sensor. *Sens. Actuators* 4, 307–315.

Hsu C.L., Li H.H., Hsueh T.J. (2013) Water– and humidity–enhanced UV detector by using p–type La–doped ZnO nanowires on flexible polyimide substrate. *ACS Appl. Mater. Interfaces* 5, 11142–11151.

Hsu C.-L., Su I.-L., Hsueh T.-J. (2017) Tunable Schottky contact humidity sensor based on S–doped ZnO nanowires on flexible PET substrate with piezotronic effect. *J. Alloys Comp.* 705, 722–733.

Hu G., Zhao C., Zhang S., Yang M., Wang Z. (2006) Low percolation thresholds of electrical conductivity and rheology in poly(ethylene terephthalate) through the networks of multi-walled carbon nanotubes. *Polymer* 47, 480–488.

Hu G.F., Zhou R.R., Yu R.M., Dong L., Pan C.F., Wang Z.L. (2014) Piezotronic effect enhanced Schottky–contact ZnO micro/nanowire humidity sensors. *Nano Res.* 7, 1083–1091.

Huang P.H. (1985) Electrical and thermodynamic characterization of water vapor/polymeric film system for humidity sensing. In: *Proceedings of 3rd International Conference on Solid–State Sensors and Actuators* (*Transducers' 85*), June 11–14, Philadelphia, PA, pp. 206–208.

Huang H., Dasgupta P.K., Ronchinsky S. (1991) Perfluorosulfonate ionomer–phosphorus pentoxide composite thin films as amperometric sensors for water. *Anal. Chem.* 63, 1570–1573.

Hwang L.S., Ko J.M., Rhee H.W., Kim C.Y. (1993) A polymer humidity sensor. *Synth. Met.* 55–57, 3671–3676.

Hwang T.J., Choi G.M. (1997) Electrical characterization of porous $BaTiO_3$ using impedance spectroscopy in humid condition. *Sens. Actuators B* 40, 187–191.

Ichinose N. (1985) Electronic ceramics for sensors. *Am. Ceram. Soc. Bull.* 64, 1581–1585.

Ichinose N. (1993) Humidity sensitive characteristics of the $MO-WO_3$ (M = Mg, Zn, Ni, Mn) system. *Sens. Actuators B* 13, 100–103.

Ishida T., Kobayashi T., Kuwahara K., Hatanaka H. (1978) Dew sensor composed of resin and carbon. *Nat. Tech. Rep.* 24, 436–444.

Jadkar V., Pawbake A., Waykar R., Jadhavar A., Date A., Late D., et al. (2017) Synthesis of $\gamma-WO_3$ thin films by hot wire–CVD and investigation of its humidity sensing properties. *Phys. Status Solidi A* 214 (5), 1600717.

Jain M.K., Bhatnagar M.C., Sharma G.L. (1999) Effect of Li^+ doping on ZrO_2–TiO_2 humidity sensor, *Sens. Actuators B* 55, 180–185.

Jeon Y.-M., Gong M.-S. (2009) Polymeric humidity sensor using polyelectrolyte derived from poly(amide–sulfone)s. *Macromol. Res.* 17 (4), 227–231.

Jeseentharani V., Dayalan A., Nagaraja K.S. (2017) Co–precipitation synthesis, humidity sensing and photoluminescence properties of nanocrystalline Co^{2+} substituted zinc(II)molybdate ($Zn_{1-x}Co_xMoO_4$; x = 0, 0.3, 0.5, 0.7, 1). *Solid State Sci.* 67, 46–58.

Jingbo L., Wenchao L., Yanxi Z., Zhimin W. (2001) Preparation and characterization of Li+–modified $Ca_xPb_{1-x}TiO_3$ film for humidity sensor. *Sens. Actuators B* 75, 11–17.

Joanni E., Baptista J.L. (1993) ZnO–Li_2O humidity sensors. *Sens. Actuators B* 17, 69–75.

Johnson Jr. R.T., Biefeid R.M. (1979) Ionic conductivity of Li_5AlO_4 and Li_5GeO_4 in moist environment: potential humidity sensors. *Mater. Res. Bull.* 14, 537–542.

Kannan P.K., Saraswathi R., Balaguru Rayappan J.B. (2010) A highly sensitive humidity sensor based on DC reactive magnetron sputtered zinc oxide thin film. *Sens. Actuators A* 164, 8–14.

Katayama K., Akiba T., Yanagida H. (1983) Rutile humidity sensor, In: Seiyama T., Fueki K., Shiokawa J., Suzuki S. (Eds.) *Analytical Chemistry Symposia Series; Chemical Sensors, Vol. 17*. Kodansha, Tokyo/Elsevier, Amsterdam, the Netherlands, pp. 433–438.

Katayama K., Hasegawa K., Takahashi T., Akiba T., Yanagida H. (1990a) Humidity sensitivity of Nb_2O_5–doped TiO_2 ceramics. *Sens. Actuators A* 24, 55–60.

Katayama K., Hasegawa H., Noda T., Akiba T., Yanagida H. (1990b) Effect of alkaline oxide addition on the humidity sensitivity of Nb_2O_5–doped TiO_2. *Sens. Actuators B* 2, 143–149.

Kazemzadeh A., Hessary F.A., Jafari N. (2008) New solid state sensor for detection of humidity, based on Ni, Co, and Mn oxide nano composite doped with Lithium. *Sens. Transducers J.* 94 (7), 161–169.

Kiasari N.M., Soltanian S., Gholamkhass B., Servati P. (2012) Room temperature ultra–sensitive resistive humidity sensor based on single zinc oxide nanowire. *Sens. Actuators A* 182, 101–105.

Kim T.Y., Lee D.H., Shim Y.C., Bu J.U., Kim S.T. (1992) Effects of alkaline oxide additives on the microstructure and humidity sensitivity of $MgCr_2O_4$–TiO_2. *Sens. Actuators B* 9, 221–225.

Kim J.-G. (2002) Electrical properties and fabrication of porous $BaTiO_3$–based ceramics. *J. Mater. Sci. Lett.* 21, 477–479.

Kim D.-U., Gong M.-S. (2005) Thick films of copper–titanate resistive humidity sensor. *Sens. Actuators B* 110, 321–326.

Kim E., Kim S.Y., Jo G., Kim S., Park M.J. (2012) Colorimetric and resistive polymer electrolyte thin films for real–time humidity sensors. *ACS Appl. Mater. Interfaces* 4, 5179–5187.

Kinjo N., Ohara O., Sugawara T., Tsuchitani T. (1983) Changes in electrical resistance of ionic copolymers caused by moisture sorption and desorption. *Polymer J.* 15, 621–623.

Klym H., Ingram A., Shpotyuk O., Hadzaman I., Hotra O., Kostiv Yu. (2016) Nanostructural free–volume effects in humidity–sensitive MgO–Al_2O_3 ceramics for sensor applications. *J. Mater. Eng. Perform., JMEPEG* 25, 866–873.

Korotcenkov G., Brynzari V., Dmitriev S. (1999) Electrical behavior of SnO_2 thin films in humid atmosphere. *Sens. Actuators B* 54, 197–201.

Korotcenkov G. (Ed.) (2010–2011) *Chemical Sensors. Vols. 1–6.* Momentum Press, New York.

Korotcenkov G., Cho B.K. (2009) Thin film SnO_2–based gas sensors: Film thickness influence. *Sens. Actuators B* 142, 321–330.

Korotcenkov G., Cho B.K. (2017) Metal oxide based composites in conductometric gas sensors: achievements and challenges. *Sens. Actuators B* 244, 182–210.

Korotcenkov G., Han S.D., Cho B.K., Brinzari V. (2009) Grain size effects in sensor response of nanostructured SnO_2– and In_2O_3–based conductometric gas sensor. *Crit. Rev. Sol. St. Mater. Sci.* 34 (1–2), 1–17.

Korotcenkov G. (2013) *Handbook of Gas Sensor Materials*. Springer, New York.

Kotnala R.K., Shah J., Singh B., Kishan H., Singh S., Dhawan S.K., Sengupta A. (2008) Humidity response of Li–substituted magnesium ferrite. *Sens. Actuators B* 129, 909–914.

Kulwicki B.M. (1984) Ceramic sensors and transducers. *J. Phys. Chem. Solids* 45, 1015–1031.

Kunakova G., Meija R., Bite I., Prikulis J., Kosmaca J., Varghese J., Holmes J.D., Erts D. (2015) Sensing properties of assembled Bi_2S_3 nanowire array. *Phys. Scr.* 90, 094017.

Kunte G.V., Shivashankar S.A., Umarji A.M. (2009) Humidity sensing characteristics of hydrotungstite thin films. *Bull. Mater. Sci.* 31, 835–839.

Lee D., Yuk J., Lee N., Uchino K. (1994) Humidity–sensitive properties of Nb_2O_5 –doped $Pb(Zr, Ti)O_3$. *Sens Mater.* 5 (4), 231–240.

Lee C.-F., Chiu W.-Y. (2000) Humidity sensitive properties of restructural films prepared from polymer latex with carboxylic acid functional groups. *Polymer J.* 32 (3), 192–197.

Lee C.W., Choi B.K., Gong M.S. (2004) Humidity sensitive properties of alkoxysilane-crosslinked polyelectrolyte using sol-gel process. *Analyst* 129(7), 651–656.

Lee C.W., Rhee H.W., Gong M.S. (1999) Humidity sensitive properties of copolymers containing phosphonium salts. *Syn. Met.* 106, 177–182.

Lee C.W., Rhee H.W., Gong M.S. (2001) Humidity sensor using epoxy resin containing quaternary ammonium salts. *Sens. Actuators B* 73, 124–129.

Lee C.W., Kim Y., Joo S.W., Gong M.S. (2003a) Resistive humidity sensor using polyelectrolytes based on new–type mutually cross–linkable copolymers. *Sens. Actuators B* 88, 21–29.

Lee C.W., Kim O., Gong M.S. (2003b) Humidity–sensitive properties of new polyelectrolytes based on the copolymers containing phosphonium salt and phosphine function. *J. Appl. Polymer Sci.* 89, 1062–1070.

Lee C.-Y., Lee G.-B. (2005) Humidity sensors: A review. *Sensor Lett.* 3, 1–14.

Lee C.-W., Joo S.-W., Gong M.-S. (2005a) Polymeric humidity sensor using polyelectrolytes derived from alkoxysilane cross-linker. *Sens. Actuators B* 105, 150–158.

Lee C.-W., Park H.-S., Kim J.-G., Choi B.-K., Joo S.-W., Gong M.-S. (2005b) Polymeric humidity sensor using organic/inorganic hybrid polyelectrolytes. *Sens. Actuators B* 109, 315–322.

Lee C.-H., Chuang W.-Y., Lin S.-H., Wu W.-J., Lin C.-T. (2013) A printable humidity sensing material based on conductive polymer and nanoparticles composites. *Jpn. J. Appl. Phys.* 52, 05DA08.

Li Y., Yang M.J., Camaioni N., Casalbore–Miceli G. (2001) Humidity sensors based on polymer solid electrolytes: investigation on the capacitive and resistive devices construction. *Sens. Actuators B* 77, 625–631.

Li Y., Yang M. (2002) Humidity sensitive properties of a novel soluble conjugated copolymer: ethynylbenzene–co–propargyl alcohol. *Sens. Actuators B* 85, 73–7.

Li Y., Yang M., Casalbore–Micelib G., Camaioni N. (2002) Humidity sensitive properties and sensing mechanism of π–conjugated polymer p–diethynylbenzene–co–propargyl alcohol. *Synth. Metals* 128, 293–298.

Li Y., Yang M.J., She Y. (2004) Humidity sensors using in situ synthesized sodium polystyrenesulfonate/ZnO nanocomposites. *Talanta* 62, 707–712.

Li Y., Yang M.J., She Y. (2005) Humidity sensitive properties of crosslinked and quaternized poly(4–vinylpyridine-co–butyl methacrylate). *Sens. Actuators B* 107, 252–257.

Li Z.Y., Zhang H.N., Zheng W., Wang W., Huang H.M., Wang C., MacDiarmid A.G., Wei Y. (2008) Highly sensitive and stable humidity nanosensors based on LiCl doped TiO_2 electrospun nanofibers. *J. Amer. Chem. Soc.* 130, 5036–5037.

Li D.M., Zhang J.J., Shen L., Dong W., Feng C.H., Liu C.X., Ruan S.P. (2015) Humidity sensing properties of $SrTiO_3$ nanospheres with high sensitivity and rapid response. *RSC Adv.* 5, 22879–22883.

Li Z., Haidry A.A., Gao B., Wang T., Yao Z. (2017) The effect of Co–doping on the humidity sensing properties of ordered mesoporous TiO_2. *Appl. Surf. Sci.* 412, 638–647.

Liang Q., Xu H., Zhao J., Gao S. (2012) Micro humidity sensors based on $ZnO-In_2O_3$ thin films with high performances. *Sens. Actuators B* 165, 76–81.

Lin W.D., Chang H.M., Wu R.J. (2013) Applied novel sensing material graphene/polypyrrole for humidity sensor. *Sens. Actuators B* 181, 326–331.

Lin W., Liao C., Chang T., Chen S., Wu R. (2015) Humidity sensing properties of novel graphene/TiO_2 composites by sol–gel process. *Sens. Actuators B* 209, 555–561.

Liu J., Agarwal M., Varahramyan K., Berney E.S., Hodo W.D. (2008) Polymer based microsensor for soil moisture measurement. *Sens. Actuators B* 129, 599–604.

Llobet E. (2013) Gas sensors using carbon nanomaterials: A review. *Sens. Actuators B* 179, 32–45.

Londono J.D., Annis B.K., Habenschuss A., Borodin O., Smith S.D., Turner J.Z., Soper A.K. (1997) Cation environment in molten lithium iodide doped poly(ethylene oxide). *Macromolecules* 30 (23), 7151–7157.

Lu J., Cheng L., Zhang Y., Huang J., Li C. (2017) Effect of the seed layer on surface morphology and humidity sensing property of $CoTiO_3$ nanocrystalline film. *Ceram. Intern.* 43, 5823–5827.

Maeda M., Suzuki I., Sakiyama K. (1992) Humidity dependence of surface resistances of $LiNbO_3$ and $LiTaO_3$ single crystals. *Jpn. J. Appl. Phys.* 31, 3229–3231.

Mamishev A.V., Sundara–Rajan K., Zahn M. (2004) Interdigital sensors and transducers. *Proc. IEEE* 92, 808–845.

Matsushima S., Teraoka Y., Miura N., Yamazoe N. (1988) Electronic interaction between metal, additives and tin dioxide–based gas sensors. *Jpn. J. Appl. Phys.* 27, 1798–1802.

Matsuguchi M., Sadaoka Y., Sakai Y., Kuroiwa T., Ito A. (1991) A capacitive-type humidity sensor using cross-linked poly(methylmethacrylate) thin films. *J. Electrochem. Soc.* 138, 1862–1865.

McCafferty E., Zettlemoyer A.C. (1971) Adsorption of water vapor on α–Fe_2O_3. *Discuss. Faraday Soc.* 52, 239–263.

Meanna Perez J.M., Freyre C. (1997) A poly (ethyleneterephthalate)-based humidity sensor. *Sens. Actuators B*. 42, 27–30.

Mogera U., Sagade A.A., George S.J., Kulkarni G.U. (2014) Ultrafast response humidity sensor using supramolecular nanofibre and its application in monitoring breath humidity and flow. *Sci. Rep.* 4, 4103.

Moneyron J.E., de Roy A., Besse J.P. (1991) Realisation of a humidity sensor based on the protonic conductor $Zn_2Al(OH)_6Cl \cdot nH_2O$. *Microelectron. Int.* 8, 26–31.

Morimoto T., Nagao M., Fukuda F. (1969) The relation between the amounts of chemisorbed and physisorbed water on metal oxides. *J. Phys. Chem.* 73, 243–248.

Morrison S.R. (1982) Semiconductor gas sensors. *Sens. Actuators* 2, 329–341.

Munn R.W., Miniewicz A., Kuchta B. (1997) *Electrical and Related Properties of Organic Solids*. Kluwer Academic Publishers, Dordrecht, the Netherlands.

Najjar R., Nematdoust S. (2016) A resistive–type humidity sensor based on polypyrrole and ZnO nanoparticles: hybrid polymers vis–a–vis nanocomposites. *RSC Adv.* 6, 112129–112139.

Nalwa E.H.S. (1997) *Handbook of Organic Conductive Molecules and Polymers, Vol. 4*. John Wiley & Sons, Hoboken, NJ.

Nechtschein M., Santier C., Travers J.P., Chroboczek J., Alix A., Ripert M. (1987) Water effects in polyaniline: NMR and transport properties. *Synth. Metals* 18, 311–316.

Neri D.G., Bona vita A., Rizzo G., Galvagno S. (2004) Low temperature sol–gel synthesis and humidity sensing properties of $Cr_{2-x}Ti_xO_3$. *J. Eur. Cer. Soc.* 24, 1435.

Niranjan R.S., Sathaye S.D., Mulla I.S. (2001) Bilayered tin oxide: Zirconia thin film as a humidity sensor. *Sens. Actuators B* 81, 64–67.

Nitta T., Hayakawa S. (1980) Ceramic humidity sensors. *IEEE Trans. Comport., Hybrids, Manuf. Technol.*, CHMT–3, 237–241.

Nitta T., Terada Z., Hayakawa S. (1980) Humidity–sensitive electrical conduction of $MgCr_2O_4$–TiO_2 porous ceramics. *Z Am. Ceram. Soc.* 63, 295–300.

Nitta T. (1981) Ceramic humidity sensor. *Ind. Eng. Chem. Prod. Res. Dev.* 20, 669–674.

Nitta T., Fukushima F., Matsuo Y. (1983) Water vapor gas sensor using ZrO_2–MgO ceramic body, In: Seiyama T., Fueki K., Shiokawa J., Suzuki S. (Eds.) *Analytical Chemistry Symposia Series; Chemical Sensors, Vol. 17*. Kodansha, Tokyo/Elsevier, Amsterdam, the Netherlands, pp. 387–392.

Nitta T. (1988) Development and application of ceramic humidity sensors, In: Seiyama T. (Ed.), *Chemical Sensor Technology, Vol. 1*. Kodansha, Tokyo/Elsevier, Amsterdam, the Netherlands pp. 57–78.

Noguchi H., Uchida Y., Nomura A., Mori S.–I. (1989) A highly reliable humidity sensor using ionene polymers. *J. Mater. Sci. Lett.* 8, 1278–1280.

Md Sin N.D., Fuad Kamel M., Alip R.I., Mohamad Z., Rusop M. (2011) The electrical characteristics of Aluminium doped Zinc Oxide thin film for humidity sensor applications. *Adv. Mater. Sci. Eng.* 2011, 1–5.

Ogura K., Shiigi H., Nakayama M. (1996) A new humidity sensor using the composite film derived from poly(o-phenylenediamine) and poly(vinyl alcohol). *J. Electrochem. Soc.* 143, 2925–2930.

Ogura K., Saino T., Nakayama M., Shiigi H. (1997) The humidity dependence of the electrical conductivity of a soluble polyaniline–poly(vinyl alcohol) composite film. *J. Mater. Chem.* 7, 2363–2366.

Ogura K., Shiigi H., Nakayama M., Ogawa A. (1999) Thermal properties of poly(anthranilic acid) (PANA) and humidity–sensitive composites derived from heat–treated PANA and poly(vinyl alcohol). *J. Polym. Sci.: Part A: Polym. Chem.* 37, 4458–4465.

Ogura K., Patil R.C., Shiigi H., Tonosaki T., Nakayama M. (2000) Response of protonic acid–doped poly(o–anisidine)/poly(vinyl alcohol) composites to relative humidity and role of dopant anions. *J. Polym. Sci.: Part A: Polym. Chem.* 38, 4343–4352.

Ogura K., Tonosaki T., Shiigi H. (2001) AC impedance spectroscopy of humidity sensor using Poly(o–phenylenediamine)/Poly(vinyl alcohol) composite film. *J. Electrochem. Soc.* 148, H21–H27.

Osada Y., De Rossi D.E. (Eds.) (2000) *Polymer Sensors and Actuators*. Springer, Berlin, Germany.

Otsuka K., Kmokl S., Usul T. (1980) Organic polymer humidity sensor, Denshz–zowyo (September 1980) 68–73 (in Japanese).

Park S.H., Park J.S., Lee C.W., Gong M.S. (2002) Humidity sensor using gel polyelectrolyte prepared from mutually reactive copolymers. *Sens. Actuators B* 86, 68–74.

Park K.-J., Gong M.-S. (2017) A water durable resistive humidity sensor based on rigid sulfonated polybenzimidazole and their properties. *Sens. Actuators B* 246, 53–60.

Parvatikar N., Jain S., Khasim S., Revansiddappa M., Bhoraskar S.V., Ambika Prasad M.V.N. (2006) Electrical and humidity sensing properties of polyaniline/WO_3 composites. *Sens. Actuators B* 114, 599–603.

Pavelko R.G. (2012) The influence of water vapor on the gas–sensing phenomenon of tin dioxide–based gas sensors, In: Korotcenkov G. (Ed.) *Chemical Sensors: Simulation and Modeling. Vol. 2: Conductometric Gas Sensors*. Momentum Press, New York, pp. 297–338.

Pawar M.S., Bankar P.K., More M.A., Late D.J. (2015) Ultra–thin V_2O_5 nanosheet based humidity sensor, photodetector and its enhanced field emission properties. *RSC Adv.* 5, 88796–88804.

Peng X., Chu J., Yang B., Feng P.X. (2012) Mn–doped zinc oxide nanopowders for humidity sensors. *Sens. Actuators B* 174, 258–262.

Phan D.-T., Chung G.-S. (2015) Effects of rapid thermal annealing on humidity sensor based on graphene oxide thin films. *Sens. Actuators B* 220, 1050–1055.

Phan D.-T., Park I., Park A.-R., Park C.-M., Jeon K.-J. (2017) Black P/graphene hybrid: A fast response humidity sensor with good reversibility and stability. *Sci. Rep.* 7, 10561.

Potts J.R., Dreyer D.R., Bielawski C.W., Ruof R.S. (2011) Graphene–based polymer nanocomposites. *Polymer* 52 (1), 5–25.

Qi Q., Fenga Y., Zhanga T., Zheng X., Lu G. (2009a) Influence of crystallographic structure on the humidity sensing properties of KCl–doped TiO_2 nanofibers. *Sens Actuators B* 139, 611–617.

Qi Q., Zhang T., Wang S.J., Zheng X.J. (2009b) Humidity sensing properties of KCl doped ZnO nanofibers with super–rapid response and recovery. *Sens. Actuator B* 137, 649–655.

Qi D., Zhao C., Zhuang Z., Li G., Na H. (2016) Novel humidity sensitive materials derived from naphthalene–based poly (arylene ether ketone) containing sulfobutyl pendant groups. *Electrochim. Acta.* 197, 39–49.

Qu W., Meyer J.-U. (1997) Thick–film humidity sensor based on porous $MnWO_4$ material. *Meas. Sci. Technol.* 8, 593–600.

Qu W., Wlodarski W., Meyer J.U. (2000) Comparative study on micromorphology and humidity sensitive properties of thin–film and thick–film humidity sensors based on semiconducting $MnWO_4$. *Sens. Actuators B* 64, 76–82.

Racheva T.M., Stambolova I.D., Donchev T. (1994) Humidity–sensitive characteristics of SnO_2–Fe_2O_3 thin films prepared by spray pyrolysis. *J. Mater. Sci.* 29, 281–284.

Raj A.M.E.S., Mallika C., Swaminathan K., Sreedgaran O.M., Nagaraja K.S. (2002) Zinc(II) oxide–zinc(II) molybdate composite humidity sensor. *Sens. Actuators B* 81, 229–236.

Ramkumar S., Rajarajan G. (2017) A comparative study of humidity sensing and photocatalytic applications of pure and nickel (Ni)–doped WO_3 thin films. *Appl. Phys. A* 123, 401.

Ratner M.A., Johansson P., Shriver D.F. (2000) Polymer electrolytes: Ionic transport mechanisms and relaxation coupling. *MRS Bull.* 45, 31–35.

Rauen K.L., Smith D.A., Heineman W.R., Johnson J., Seguin R., Stoughton P. (1993) Humidity sensor based on conductivity measurements of a poly(dimethyldiallylammonium chloride) polymer film. *Sens. Actuators B* 17, 61–68.

Rausen K.L., Smith D.A., Heineman W.R., Johnson J., Seguin R., Stoughton P. (1993) Humidity sensor based on conductivity measurements of a poly(dimethyldiallylammonium chloride) polymer film. *Sens. Actuators B* 17, 61–68.

Rittersma Z.M. (2002) Recent achievements in miniaturized humidity sensors – a review of transduction techniques. *Sens. Actuators A* 96, 196–210.

Rubinger C.P.L., Calado H.D.R., Rubinger R.M., Oliveira H., Donnici C.L. (2013) Characterization of a sulfonated polycarbonate resistive humidity sensor. *Sensors* 13, 2023–2032.

Sabarilakshmi M., Janaki K. (2017) Effect of Mg concentration on structural, optical and humidity sensing performance of SnO_2 nanoparticles prepared by one step facile route. *J. Mater. Sci. Mater. Electron.* 28, 8101–8107.

Sadaoka Y., Sakai Y. (1984) Humidity sensor using lithium doped poly(ethylene oxide) thin film. *Denki Kagaku* 52, 132–133.

Sadaoka Y., Sakai Y. (1985) Humidity sensor using porous NASICON. *Denld Kagaku* 53, 395–399.

Sadaoka Y., Sakai Y. (1986) A humidity sensor using alkali saltpoly(ethylene oxide) hybrid films. *J. Mater. Sci.* 21, 235–240.

Sadaoka Y, Matsuguchi M., Sakai Y., Aono H., Nakayama S., Kuroshima H. (1987) Humidity sensors using KH_2PO_4–doped porous (Pb, La)(Zr, $Ti)O_3$. *J. Mater. Sci.* 22, 3685–3692.

Sakai Y., Sadaoka Y., Ikeuchi K. (1985) Humidity sensors composed of graft copolymers. In: *Digest of Technical Papers, 3rd International Conference on Solid–State Sensors and Actuators (Transducers' 85)*, Philadelphia, PA, pp. 213–216.

Sakai Y., Rao V.L., Sadaoka Y., Matsuguchi M. (1987) Humidity sensor composed of a microporous film of polyethylene–graft–poly–(2–acrylamido–2–methylpropane sulfonate). *Polymer Bull.* 18, 501–506.

Sakai Y., Sadaoka Y., Fukumoto H. (1988) Humidity–sensitive and water–resistive polymeric materials. *Sens. Actuators* 13, 243–250.

Sakai Y., Sadaoka Y., Matsuguchi M., Moriga N., Shimada M. (1989) Humidity sensors based on organopolysiloxane having hydrophilic groups. *Sens. Actuators* 16, 359–367.

Sakai Y., Sadaoka Y., Matsuguchi M., Sakai H. (1995) Humidity sensor durable at high humidity using simultaneously crosslinked and quaternized poly(chloromethyl styrene). *Sens. Actuators B* 24, 689–691.

Sakai Y., Sadaoka Y., Matsuguchi M. (1996) Humidity sensors based on polymer thin films. *Sens. Actuators B* 35, 85–90.

Sakai Y. (2000) Ion conducting polymer sensors. In: Osada Y., De Rossi D.E. (Eds.) *Polymer Sensors and Actuators*. Springer, Berlin, Germany, pp. 1–14.

Sakai Y., Matsuguchi M., Hurukawa T. (2000) Humidity sensor using cross–linked poly(chloromethyl styrene). *Sens. Actuators B* 66, 135–138.

Sakai Y., Matsuguchi M., Yonesato N. (2001) Humidity sensor based on alkali salts of poly(2-acrylamido-2-methylpropane sulfonic acid). *Electrochim. Acta.* 46, 1509–1514.

Schopf G., Kobmehl G. (1997) *Polythiophenes–Electrically Conductive Polymers*. Springer, Berlin, Germany.

Seiyama T., Yamazoe N., Arai H. (1983) Ceramic humidity sensors. *Sens. Actuators* 4, 85–96.

Shim Y.-B., Park J.-H. (2000) Humidity sensor using chemically synthesized Poly(1,5–diaminonaphthalene) doped with carbon. *J. Electrochem. Soc.* 147 (1), 381–385.

Shimizu Y., Shippo O., Arai H., Seiyama T. (1984) Influence of the reducing gases on the sensitivity of the semiconductive humidity sensor. *Denki Kagalo A* 5, 849–850.

Shimizu Y., Arai H., Seiyama T. (1985a) Theoretical studies on the impedance–humidity characteristics of ceramic humidity sensors. *Sens. Actuators* 7, 11–22.

Shimizu Y., Ichinose H., Arai H., Seiyama T. (1985b) Ceramic humidity sensors. Microstructure and simulation of humidity sensitive characteristics. *J. Chem. Soc. Jpn.* 1985, 1270–1277.

Shimizu Y., Shimabukuro M., Arai H., Seiyama T. (1985c) Enhancement of humidity–sensitivity for perovskite–type oxides having semiconductivity. *Chem. Lea.* 1985, 917–920.

Shimizu Y., Okada H., Arai H. (1987) Application of La_2O_3–TiO_2 porous glass–ceramic system to a humidity sensor. *J. Ceram. Soc. Jpn., Int. Ed.* 95, 683–687.

Shimizu Y., Shimabukuro M., Arai H., Seiyama T. (1989a) Humidity– sensitive characteristics of La^{3+}–doped and undoped $SrSnO_3$. *J. Electrochem. Soc.* 136, 1206–1210.

Shimizu Y., Okada H., Arai H. (1989b) Humidity–sensitive characteristics of porous La–Ti–V–O glass–ceramics. *J. Am. Ceram. Soc.* 72, 436–440.

Shimizu Y., Shimabukuro M., Arai H. (1989c) The sensing mechanism in a semiconducting humidity sensor with Pt electrodes. *J. Electrochem. Soc.* 136, 3868–3871.

Shukla S.K., Vamakshi M., Bharadavaja A., Shekhar A., Tiwari A. (2012) Fabrication of electro–chemical humidity sensor based on zinc oxide/polyaniline nanocomposites. *Adv. Mat. Lett.* 3 (5), 421–425.

Soloman S. (2010) *Sensor Handbook, 2nd edn.* McGraw Hill, New York.

Son S.Y., Gong M.S. (2002) Polymeric humidity sensor using phosphonium salt–containing polymers. *Sens. Actuators B* 86, 168–173.

Song X., Qi Q., Zhang T., Wang C.A (2009) Humidity sensor based on KCl–doped SnO_2 nanofibers. *Sens. Actuators B* 138, 368–373.

Squillaci M.A., Ferlauto L., Zagranyarski Y., Milita S., Müllen K., Samorì P. (2015) Self–assembly of an amphiphilic π–conjugated dyad into fibers: Ultrafast and ultrasensitive humidity sensor. *Adv. Mater.* 27, 3170–3174.

Su P.G., Chen I.C., Wu R.J. (2001) Use of poly(2–acrylamido–2–methylpropane sulfonate) modified with tetraethyl orthosilicate as sensing material for measurement of humidity. *Anal. Chim. Acta.* 449, 103–109.

Su P.G., Tsai W.Y. (2004) Humidity sensing and electrical properties of a composite material of nano–sized SiO_2 and poly(2–acrylamido–2–methylpropane sulfonate). *Sens. Actuators B* 100, 417–422.

Su P.-G., Wang C.-P. (2008) Flexible humidity sensor based on TiO_2 nanoparticles–polypyrrole–Poly–[3–(methacrylamino)propyl] trimethyl ammonium chloride composite materials. *Sens. Actuators B* 129, 538–543.

Sun H.T., Wu M.T. (1988) Theoretical studies on resistance humidity characteristics of porous ceramic sensors. *Mater. Sci. Prog.* 2, 66–70.

Sun H.T., Yao X., Wu M.T., Li P. (1988) Preparation of porous ceramic humidity sensors by adding graphite powder. *Mater. Sci. Prog.* 2, 61–65.

Sun H.T., Wu M.T., Li P., Yao X. (1989) Porosity control of humidity–sensitive ceramics and theoretical model of humidity sensitive characteristics. *Sens. Actuators* 19, 61–70.

Sun A., Huang L., Li Y. (2009) Study on humidity sensing property based on TiO_2 porous film and Polystyrene Sulfonic Sodium. *Sens. Actuators B* 139, 543–547.

Suna A., Wanga Y., Li Y. (2010) Stability and water-resistance of humidity sensors using crosslinked and quaternized polyelectrolytes films. *Sens. Actuators B* 145, 680–684.

Sundaram R., Nagaraja K.S. (2004a) Solid state electrical conductivity and humidity sensing studies on metal molybdate–molybdenum trioxide composites (M = Ni^{2+}, Cu^{2+} and Pb^{2+}). *Sens. Actuators B* 101, 353–360.

Sundaram R., Nagaraja K.S. (2004b) Electrical and humidity sensing properties of lead(II) tungstate–tungsten(VI) oxide and zinc(II) tungstate–tungsten(VI) oxide composites. *Mater. Res. Bull.* 39, 581–590.

Taguchi H., Takahashi Y., Matsumoto C. (1980) The effect of water adsorption on $(La_{1-x}Sr_x)MnO_3$. *Yogyo–kyokai–shi* 88, 566–570.

Tai W.P., Oh J.H. (2002) Humidity sensing behaviors of nanocrystalline Al–doped ZnO thin films prepared by sol-gel process. *J. Mater. Sci.: Mater. Electron.* 13, 391–394.

Tai W.-P., Kim J.-G., Oh J.-H., Lee C., Park D.-W., Ahn W.-S. (2005) Humidity sensing properties of nanostructured–bilayered potassium tantalate: Titania tilms. *J. Mater. Sci. Mater. Electron.* 16, 517–521.

Tailoka F., Fray D.J., Kumar R.V. (2003) Application of Nafion electrolytes for the detection of humidity in a corrosive atmosphere. *Solid State Ionics* 161, 267–277.

Taka T. (1993) Humidity dependency of electrical conductivity of doped polyaniline. *Synth. Met.* 55–57, 5014–5019.

Takamuku S., Wohlfarth A., Manhart A., Räder P., Jannasch P. (2015) Hyper sulfonated polyelectrolytes: preparation, stability and conductivity. *Polym. Chem.* 6, 1267–1274.

Takaoka Y., Maebashi Y., Kobayashi S., Usui T. (1983) Humidity sensor. *Jpn. Patent* 58, 16467.

Tandon R.P., Tripathy M.R., Arora A.K., Hotchandani S. (2006) Gas and humidity response of Iron oxide—Polypyrrole nanocomposites. *Sens. Actuators B* 114, 768–773.

Thiel P.A., Madey T.E. (1987) The interaction of water with solid surface: Fundamental aspects. *Surf. Sci. Rep.* 7, 211–385.

Tian Y., Xu M., Huang P., Lei E., Hao H., Cui C. (2005) The abnormal sensitivity and its mechanism of $(Ba, Pb)TiO_3$ semiconductor ceramics. *Chinese Sci. Bull.* 50 (9), 936–939.

Tierney M.J. (1996) Practical examples of polymer–based chemical sensors, Chapter 13, In: Taylor R.F., Schultz J.S. (Eds.) *Handbook of Chemical and Biological Sensors.* IOP Publishing, Bristol, UK.

Tonosaki T., Oho T., Isomura K., Ogura K. (2002) Effect of the protonation level of poly(o–phenylenediamine) (PoPD) on the ac impedance of humidity–sensitive PoPD/poly(vinyl alcohol) composite film. *J. Electroanal. Chem.* 520, 89–93.

Traversa E. (1995) Ceramic sensors for humidity detection: the state–of–the–art and future developments. *Sens. Actuators B* 23, 135–156.

Traversa J.P., Nechtschein M. (1987) Water effects in polyaniline: A new conduction process. *Synth. Metals* 21, 135–141.

Traversa E., Sadaoka Y., Carotta M.C., Martinelli G. (2000) Environmental monitoring field tests using screen–printed thick–film sensors based on semiconducting oxides. *Sens. Actuators B* 65, 181–185.

Tsubokawa N., Tsuchida M., Chen J., Nakazawa Y. (2001) A novel contamination sensor in solution: the response of the electric resistance of a composite based on crystalline polymer–grafted carbon black. *Sens. Actuators B* 79, 92–97.

Tsuchitani S., Sugawara T., Kinjo H., Ohara S. (1985) Humidity sensor using ionic copolymer. In:. *Digest of technical papers, 3rd International conference on solid-state sensors and actuators (Transducers' 85)*, June 11–14, Philadelphia, PA, USA, pp. 210–212.

Tsuchitani S., Sugawara T., Kinjo N., Ohara S., Tsunoda T. (1988) A humidity sensor using ionic copolymer and its application to a humidity–temperature sensor module. *Sens. Actuators* 15, 375–386.

Uno S., Harata M., Hiraki H., Sakuma K., Yokomizo Y. (1983) $ZnCr_2O_4$–$LiZnVO_4$ ceramic humidity sensor, In: Seiyama T., Fueki K., Shiokawa J., Suzuki S. (Eds.) *Analytical Chemistry Symposia Series: Chemical Sensors, Vol. 17*. Kodansha, Tokyo/Elsevier, Amsterdam, the Netherlands, pp. 375–380.

Upadhyay S. (2008) Humidity–sensitive characteristic of $Ba_{0.99}La_{0.01}SnO_3$. *Phys. Status Solidi* 205, 1113–1119.

Varghese O.K., Gong D., Paulose M., Ong K.G., Grimes C.A., Dickey E.C. (2002) Extreme changes in the electrical resistance of titania nanotubes with hydrogen exposure. *J. Mater. Res.* 17, 1162–1171.

Vijaya J.J., Kennedy L.J., Sekaran G., Jeyaraj B., Nagaraja K.S. (2007a) Effect of Sr addition on the humidity sensing properties of $CoAl_2O_4$ composites. *Sens. Actuators B* 123, 211–217.

Vijaya J.J., Kennedy L.J., Meenakshisundaram A., Sekaran G., Nagaraja K.S. (2007b) Humidity sensing characteristics of sol–gel derived Sr(II)–added $ZnAl_2O_4$ composites. *Sens. Actuators B* 127, 619–624.

Vijaya J.J., Kennedy L.J., Sekaran G., Nagaraja K.S. (2007c) Synthesis, characterization and humidity sensing properties of Sr(II)–added $BaAl_2O_4$ composites. *Sens. Actuators B* 124, 542–548.

Viviani M., Buscaglia M., Buscaglia V., Leoni M., Nanni P. (2001) Barium perovskites as humidity sensing materials. *J. Eur. Ceram. Soc.* 21, 1981–1984.

Wagiran R., Zaki W.W. (2005) Characterization of screen printed $BaTiO_3$ thick film humidity sensor. *Int. J. Eng. Technol.* 2, 22–26.

Wan Q., Li Q.H., Chen Y.J., Wang T.H., He X.L., Gao X.G., Li J.P. (2004) Positive temperature coefficient resistance and humidity sensing properties of Cd–doped ZnO nanowires. *Appl. Phys. Lett.* 84, 3085–3087.

Wang H., Xu T., Zhang S., Lu T. (1993) A multi–channel temperature and humidity monitor. *Meas. Sci. Technol.* 4, 164–169.

Wang H., Feng C.D., Suin S.L., Segre C.U., Stetter J.R. (1998) Comparison of conductometric humidity-sensitive polymers. *Sens. Actuators B* 40, 211–216.

Wang J., Xu B., Liu G., Zhang J., Zhang T. (2000) Improvement of nanocrystalline $BaTiO_3$ humidity sensing properties. *Sens. Actuators B* 66, 159–160.

Wang J., Lin Q., Zhou R., Xu B. (2002) Humidity sensors based on composite material of nano–$BaTiO_3$ and polymer RMX. *Sens. Actuators B* 81, 248–253.

Wang J., Wan H., Lin Q. (2003a) Properties of a nanocrystalline barium titanate on silicon humidity sensor. *Meas. Sci. Technol.* 14, 172–175.

Wang J., Xu B.K., Ruan S.P., Wang S.P. (2003b) Preparation and electrical properties of humidity sensing films of $BaTiO_3$/polystrene sulfonic sodium. *Mater. Chem. Phys.* 78, 746–750.

Wang W., Virkar A.V. (2004) A conductimetric humidity sensor based on proton conducting perovskite oxides. *Sens. Actuators B* 98, 282–290.

Wang J., Su M.-Y., Qi J.-Q., Chang L.-Q. (2009a) Sensitivity and complex impedance of nanometer zirconia thick film humidity sensors. *Sens. Actuators B* 139, 418–424.

Wang W., Li Z.Y., Liu L., Zhang H.N., Zheng W., Wang Y., Huang H.M., Wang Z.J., Wang C. (2009b) Humidity sensor based on LiCl–doped ZnO electrospun nanofibers. *Sens. Actuator B* 141, 404–409.

Wang Z., Shi L., Wu F., Yuan S., Zhao Y., Zhang M. (2011a) The sol–gel template synthesis of porous TiO_2 for a high performance humidity sensor. *Nature Nanotech.* 22, 275502.

Wang L., He Y., Hu J., Qi Q., Zhang T. (2011b) DC humidity sensing properties of $BaTiO_3$ nanofiber sensors with different electrode materials. *Sens. Actuators B* 153, 460–464.

Wang X., Li J., Li Y., Liu L., Guan W. (2016) Emulsion–templated fully three–dimensional interconnected porous titania ceramics with excellent humidity sensing properties. *Sens. Actuators B* 237, 894–898.

Wilson J.S. (Ed.) (2005) *Sensor Technology Handbook*. Elsevier, Burlington, NJ.

Wraight C.A. (2006) Chance and design—proton transfer in water, channels and bioenergetic proteins. *Biochim. Biophys. Acta* 1757, 886–912.

Wu T.-F., Hong J.-D. (2016) Humidity sensing properties of transferable polyaniline thin films formed at the air–water interface. *RSC Adv.* 6, 96935–96941.

Xie W., Liu B., Xiao S., Li H., Wang Y., Cai D., Wang D. et al. (2015) High performance humidity sensors based on CeO_2 nanoparticles. *Sens. Actuators B* 215, 125–132.

Xin Y., Wang S. (1994) An investigation of sulfonated polysuifone humidity–sensitive materials. *Sens. Actuators A* 40, 147–149.

Xu J., Yu K., Wu J., Shang D., Li L., Xu Y., Zhu Z. (2009) Synthesis, field emission and humidity sensing characteristics of honeycomb–like CuO. *J. Phys. D Appl. Phys.* 42, 075417.

Xu C.-N., Miyazaki K. (1993) Humidity sensor with manganese oxide for room temperature use. *Sens. Actuators B* 13–14, 523–524.

Yadav B.C., Srivastava R., Dwivedi C.D., Pramanik P. (2008) Moisture sensor based on ZnO nanomaterial synthesized through oxalate route. *Sens. Actuators B* 131, 216–222.

Yadav B.C., Srivastava R., Dwivedi C.D., Pramanik P. (2009) Synthesis of nano–sized ZnO using drop wise method and its performance as moisture sensor. *Sens. Actuators A* 53, 137–141.

Yadav B.C., Sharma P., Khanna P.K. (2011) Morphological and humidity sensing characteristics of SnO_2–CuO, SnO_2–Fe_2O_3 and SnO_2–SbO_2 nanocooxides. *Bull. Mater. Sci.* 34, 689–698.

Yagi H., Saiki T. (1991) Humidity sensor using NASICON not containing phosphorus. *Sens. Actuators B* 5, 135–138.

Yagi H., Nakata M. (1992) Humidity sensor using Al_2O_3, TiO_2 and SnO_2 prepared by sol–gel method. *J. Ceram. Soc. Jpn.* 100, 152–156.

Yamamoto T., Murakami K. (1989) Humidity sensor using TiO_2–SnO_2 ceramics, In: Seiyama T. (Ed.) *Chemical Sensor Technology, Vol. 2*. Kodansha, Tokyo/Elsevier, Amsterdam, the Netherland, pp. 133–149.

Yamazoe N., Fuchigami J., Kishikawa M., Seiyama T. (1979) Interactions of tin oxide surface with O_2, H_2O and H_2. *Surf. Sci.* 86, 335–344.

Yamazoe N., Shimizu Y. (1986) Humidity sensors. Principles and applications. *Sens. Actuators* 10, 379–398.

Yang S.-L., Wu J.-M. (1991) ZrO_2–TiO_2 ceramic humidity sensors. *J. Mater. Sci.* 26, 631–636.

Yang M., Li Y., Zhan X., Ling M. (1999) A novel resistive–type humidity sensor based on poly(p–diethynylbenzene). *J. Appl. Polym. Sci.* 74, 2010–2015.

Yang M.J., Li Y., Camaioni N., Casalbore–Miceli G., Martelli A., Ridolfi G. (2002) Polymer electrolytes as humidity sensors: Progress in improving an impedance device. *Sens. Actuators B* 86, 229–234.

Yang B., Aksak B., Lin Q., Sitti M. (2006) Compliant and low-cost humidity nanosensors using nanoporous polymer membranes. *Sens. Actuators B* 114, 254–262.

Yasaei P., Behranginia A., Foroozan T., Asadi M., Kim K., Khalili-Araghi F., Salehi-Khojin A. (2015) Stable and selective humidity sensing using stacked black phosphorus flakes. *ACS Nano* 9, 9898–9905.

Yatsuzuka K., Higashiyama Y., Asano K. (1993) Fundamental characteristics of hydrophilic polymer (polyether block amide) as a humidity sensor. *Jpn. J. Appl. Phys.* 32, L461–L463.

Yawale S.P., Yawale S.S., Lamdhade G.T. (2007) Tin oxide and zinc oxide based doped humidity sensors. *Sens. Actuators B* 135, 388–393.

Yeh Y.C., Tseng T.Y. (1989) Analysis of the d.c. and a.c. properties of K_2O–doped porous $Ba_{0.5}Sr_{0.5}TiO_3$ ceramic humidity sensor. *J. Mater. Sci.* 24, 2739–2745.

Yeh Y.C., Tseng T.Y., Chang D.A. (1989) Electrical properties of porous titania ceramic humidity sensors. *J. Am. Ceram. Soc.* 72, 1472–1475.

Yin H., Ni J., Jiang W., Zhang Z., Yu K. (2011) Synthesis, field emission and humidity sensing characteristics of monoclinic VO_2 nanostructures. *Physica E* 43, 1720–1725.

Ying J., Wan C., He P. (2000) Sol–gel processed TiO_2–K_2O–$LiZnVO_4$ ceramic thin films as innovative humidity sensors. *Sens Actuators B* 63, 165–170.

Yokomizo Y., Uno S., Harata M., Hiraki H. (1983) Microstructure and humidity–sensitive properties of $ZnCr_2O_4$–$LiZnVO_4$ ceramic sensors. *Sens. Actuators* 4, 599–606.

Yoo K.P., Lim L.T., Lee M.J., Lee C.J., Park C.W. (2010) Novel resistive type humidity sensor based on multiwall carbon nanotube–polyimide composite film. *Sens. Actuators B* 145, 120–125.

Yuan Q., Li N., Tu J., Li X., Wang R., Zhang T., Shao C. (2010) Preparation and humidity sensitive property of mesoporous ZnO–SiO_2 composite. *Sens. Actuators B* 149, 413–419.

Yuan Q., Li N., Geng W., Chi Y., Tu J., Li X., Shao C. (2011) Humidity sensing properties of mesoporous iron oxide/silica composite prepared via hydrothermal process. *Sens. Actuators B* 160, 334–340.

Yuk J., Troczynski T. (2003) Sol–gel $BaTiO_3$ thin film for humidity sensors. *Sens. Actuators B* 94, 290–293.

Zafar Q., Sulaiman K. (2016) Utility of PCDTBT polymer for the superior sensing parameters of electrical response based relative humidity sensor. *React. Funct. Polymers* 105, 45–51.

Zhang Y., Yu K., Jiang D., Zhu Z., Geng H., Luo L. (2005) Zinc oxide nanorod and nanowire for humidity sensor. *Appl. Surf. Sci.* 242, 212–217.

Zhang N., Yu K., Zhu Z., Jiang D. (2008) Synthesis and humidity sensing properties of feather like ZnO nanostructures with macroscale in shape. *Sens. Actuators A* 143, 245–250.

Zhang Q., Smith J.R., Saraf L.V., Hua F. (2009) Transparent humidity sensor using cross–linked polyelectrolyte membrane. *IEEE Sensors J.* 9 (7), 854–857.

Zhang D., Jiang C., Sun Y., Zhou Q. (2017) Layer–by–layer self–assembly of tricobalt tetroxide–polymer nanocomposite toward high–performance humidity–sensing. *J. Alloys Compounds* 711, 652–658.

Zhen Y., Sun F.-H., Zhang M., Jia K., Li L., Xue Q. (2016) Ultrafast breathing humidity sensing properties of low–dimensional Fe–doped SnO_2 flower–like spheres. *RSC Adv.* 6, 27008–27015.

Zhu C., Xu F., Zhang L., Li M., Chen J., Xu S., Huang G. et al. (2016) Ultrafast preparation of black phosphorus quantum dots for efficient humidity sensing. *Chem. Eur. J.* 22 (22), 7357–7362.

Zor F.D., Cankurtaran H. (2016) Impedimetric humidity sensor based on nanohybrid composite of conducting poly(diphenylamine sulfonic acid). *J. Sensors* 2016, 5479092.

12 Gravimetric Humidity Sensors

12.1 INTRODUCTION

Gravimetric or *mass-sensitive* humidity sensors include a large group of devices (see Figures 12.1 and 12.2) such as quartz crystal microbalance (QCM) or bulk acoustic wave (BAW) devices (Gerlach and Sager 1994), surface acoustic wave (SAW) sensors (Hoyt et al. 1998), and either microbeam or membrane resonators (Glück et al. 1994; Schroth et al. 1996). As is known, from the measurement point of view, the determination of mass is called *gravimetry* (Lu and Czaderna 1984). At the beginning of their development, mass sensors were synonymous with the QCM. Because of their small size, high sensitivity and stability, piezoelectric crystals have been used as microbalances, namely in the determination of the thicknesses of thin layers and in general gas sorption studies (King 1964; Grate et al. 1993a, 1993b). Only later, other oscillators, such as the SAW devices, vibrating beams, and cantilevers, were added. Virtually all acoustic wave (AW)-based devices use a piezoelectric material

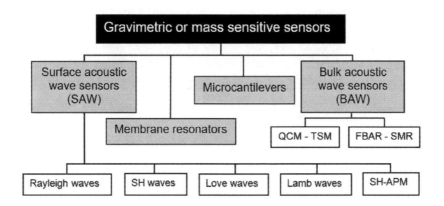

FIGURE 12.1 Different types of acoustic devices that can be used for sensing application: FBAR, film bulk acoustic resonators; QCM, quartz crystal microbalance; SH-APM, shear-horizontal acoustic plate mode; SMR, solidly mounted resonators; TSM, thickness shear mode.

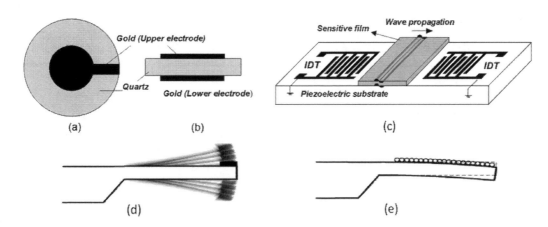

FIGURE 12.2 Schematic diagrams of mass-sensitive gas sensors: (a, b) Quartz crystal microbalance (QCM) device; (c) Surface acoustic wave (SAW) device; (d) Microcantilever in dynamic mode: Absorption of analyte molecules in a sensor layer leads to a shift in resonance-frequency; and (e) Microcantilever in static mode: The cantilever bends owing to adsorption of analyte molecules and the change of the surface stress at the cantilever surface. (d, e - Reprinted with permission from Battiston, F.M. et al., A chemical sensor based on a microfabricated cantilever array with simultaneous resonance-frequency and bending readout, *Sens. Actuators B*, 77, 122–131, 2001. Copyright 2001, Elsevier.)

to generate the AW, which propagates along the surface in SAW devices or throughout the bulk of the structure in BAW devices. Piezoelectricity involves the ability of certain crystals to couple mechanical strain to electrical polarization and will only occur in crystals that lack a center of inversion symmetry (Ballantine et al. 1996). All gravimetric sensors differ constructively, but they have a similar humidity-sensitive mechanism. These *mass-sensitive sensors* rely on disturbances and changes to the mass of the sensor surface during interaction with chemicals (Fanget et al. 2011). In other words, mass-sensitive devices transform the mass change at a specially modified surface into a change in some property of the support material. For example, the mass changes can be monitored by either deflecting a micromechanical structure due to stress changes or mass loading, or by assessing the frequency characteristics of a resonating structure or a traveling AW upon mass loading. In sensors with such architectures, the mechanical amplitude of the vibrating structure is usually kept constant to ensure the stability of the system. Indeed, the loss of the loop is compensated by an amplifier included in the loop. Without any control of the mechanical amplitudes, the vibrations would then become larger and larger because of the gain, and the sensors may be broken.

It is clear that, to increase the sensitivity of such sensors, the active surface of mass-sensitive sensors must be covered with a special adsorption layer that is sensitive to the substance being detected. In particular, when detecting water vapor, the sensing element should be coated with a hygroscopic layer. In the presence of an analyte species, the waves' properties become perturbed in a measurable way that can be correlated to the analyte concentration (Ippolito et al. 2009). Knowing the chemical affinity of the layer with the gas molecules, the frequency shift can then be related to the gas concentration. All this indicates that the sensitivity and selectivity of *mass-sensitive sensors* are determined not only by the design features of the device, but also by the adsorption properties of the sensing materials used.

The optimum domain of applicability of mass sensors can be evaluated by considering some general aspects of behavior of these devices, from the point of view of the basic species–sensor interaction (Janata 2009). Clearly, a mass-related signal will be obtained only if the species–sensor interaction results in a net change of mass of the chemically selective layer attached to the device. Thus, an equilibrium binding will yield a measurable signal. On the other hand, if the interaction is just a displacement of one species with another (i.e., exchange or catalytic reaction), the sensor surface will be only a temporary host to the interacting species, and the net change in mass would be very small. Therefore, mass sensors rely predominantly on equilibrium, rather than on kinetically based selectivity.

Let's now consider these devices in more detail in relation to humidity measurement. As for the description of the general principles of designing and manufacturing gravimetric-based sensors, they can be found in the published books and reviews (Ballantine et al. 1997; Ippolito et al. 2009; Afzal and Dickert 2011; Fanget et al. 2011; Vashist and Korotcenkov 2011; Voinova and Jonson 2011; Bo et al. 2016).

It should be noted that cantilevers and membrane-based mass-sensitive humidity sensors will not be considered in this chapter. It was decided to consider such sensors in a special chapter (Chapter 13). This was done on the basis that (1) these mass-sensitive humidity sensors are fundamentally different from AW sensors in design and operation, and (2) cantilevers and membranes are the basis for the development of humidity sensors, using other properties of humidity-sensitive materials.

12.2 QCM-BASED SENSORS

12.2.1 General Consideration

This category of devices includes the QCM and thin-film resonators (TFRs), the latter encapsulating thin-film bulk acoustic resonator (TFBAR) and solidly mounted resonator (SMR) structures. The conventional QCM setup contains a piezoelectric quartz crystal wafer with deposited metal electrodes (see Figure 12.3), a quartz wafer holder (a metal base), an oscillator, a digital counter for the frequency measurements, and a computer (Voinova and Jonson 2011).

QCM devices or piezoelectric transducers utilize longitudinal standing waves, known as *thickness shear mode*, that propagate in the bulk of material. This is in contrast to surface AWs, which resonate as a standing wave within a thin surface layer of a piezoelectric substrate (see Section 12.3). Usually, QCM sensors consist of a quartz disk, sandwiched between two thin film gold electrodes, deposited onto the quartz substrate. Commercial crystals have a "flat," which marks the orientation of the crystal with respect to the crystallographic coordinates. The electrodes are deposited directly on the quartz crystal, although, in principle, a noncontact activation of the crystal is also possible. One of the electrodes is smaller in order to eliminate the "edge effect"—that is, fringing electric field. Also, the metal is deposited in a "keyhole" pattern, with each

Gravimetric Humidity Sensors

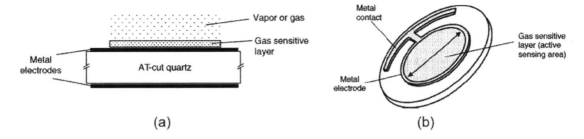

FIGURE 12.3 (a) Cross section and (b) graphical projection of a quartz crystal microbalance (QCM), illustrating a gas-sensitive layer on one electrode.

electrode rotated 180° with respect to the other (Janshoff et al. 2000). This geometry is needed in order to make the field as cylindrical as possible. Both electrodes are covered with a hygroscopic sensing film that can adsorb analyte, including water vapor.

Oscillation of the crystal is due to the AC electric field applied across the crystal from metal electrodes. Typical electric fields are quite low, 10–20 V cm^{-1}. When an electric field is applied to the gold electrodes, it produces an increase or decrease in the thickness of the quartz disk, depending on whether the field is parallel or perpendicular to the internal electrical polarization of the piezoelectric material. The oscillation frequency of the quartz crystal depends on the total mass of the crystal substrate and sensitive coating applied to the crystal surfaces. In particular, the fundamental frequency of the crystal depends on its thickness. The fundamental resonant frequency of a piezoelectric crystal vibrating in thickness longitudinal mode is given by:

$$f_p = \frac{V_p}{2I_p} \quad (12.1)$$

where f_p is the fundamental resonant frequency before the application of the mass, V_p is the acoustic phase velocity of the mode of resonance in the piezoelectric film, and I_p is the thickness of the piezoelectric layer. Thus, a 5 MHz quartz crystal is 330 μm thick, and a 30 MHz crystal has a thickness of 55 μm (Janata 2009). The thickness of the metal electrode also plays an important role. There is no piezoelectric displacement outside the electric field. This means that, at the boundary between the metal and the metal-free region, the energy is reflected, resulting in the formation of a standing wave in the excited part of the crystal, beneath the metal electrodes. This effect is called *energy trapping*. It occurs when the metal is at least 50 nm thick. On the other hand, if the metal is too thick, its mass exceeds the Sauerbrey limit, and the crystal ceases to oscillate.

In principle, there are two modes excited by the AC field, longitudinal and transverse. For crystals in the 100–300 μm thickness range, only the transverse standing wave needs to be considered (Janshoff et al. 2000). The actual lateral displacement of a point on the crystal surface (and therefore the mass sensitivity) is the Gaussian function of the radial distance from the center of the electrode. It also depends on the amplitude of the applied electric field and ranges from few nm/V in water to tens of nm/V in air or in a vacuum. The direction of the displacement comes from the periodic boundary condition of the solution of the wave equation. It is normal to the "flat" of the crystal (Martin and Hager 1989a, 1989b; Hillier and Ward 1992). The above considerations make the description of the physics of the QCM far more complicated than is apparent from the simple Sauerbrey equation (Janata 2009). However, for our consideration, the use of Sauerbrey equation is enough.

Water vapor adsorbed onto the sensitive film adds mass to the system and decreases the resonance frequency in proportion to the mass of the adsorbed gas. If we assume that the adsorbed mass is of infinitely small thickness with negligible phase change across the film thickness, and there is no acoustic energy dissipation, then it is easy to derive the Sauerbrey equation (Sauerbrey 1959; Cheeke and Wang 1999):

$$\Delta f = -\frac{2}{A}\frac{f_0^2}{\sqrt{\mu\rho}}\Delta m \text{ or } \frac{\Delta f}{f_0} = -\frac{2}{A}\frac{f_0 \Delta m}{\sqrt{\mu\rho}} \quad (12.2)$$

where A represents the area of surface, where f_0 is the resonance frequency, μ is the substrate shear stiffness, ρ is the substrate mass density, λ_0 is the acoustic wavelength of the resonator, and Δm is the area mass loading on the surface of the resonator, mass change due to absorption of moisture in our case. Using the Rayleigh hypothesis, which assumes that the added mass layer does not affect the peak kinetic and potential energies, the mass sensitivity, S_m, of a QCM per unit area, A, of added mass $(m_0+\Delta m)$ can be defined as (Wenzel and White 1989):

$$S_m = \lim_{\Delta m \to 0}\left(\frac{\Delta f}{f_0}\cdot\frac{A}{\Delta m}\right) \quad (12.3)$$

where $f_0 \sim v_0/2h_0$, with v_0 and h_0 being the acoustic velocity and thickness of the unloaded resonator, respectively, and $m_0 = \rho A h_0$ is the initial mass of the resonator.

Thus, the Sauerbrey equation indicates that the change in frequency is proportional to the square of the fundamental frequency and inversely proportional to the surface area. This result is the basis of QCM microgravimetry measurements in vacuum and air. In particular, early chemical applications of the quartz crystal microbalance include the detection of volatile compounds in various gases. One of the first papers to suggest chemical detection of analytes by means of a quartz crystal resonator was the short but important publication of King (1964).

Thereby, the QCM sensors are electromechanical oscillators, and the detection consists of measuring the frequency shift with respect to the mass loading, as shown in Figure 12.4. Knowing the chemical affinity of the layer with the gas molecules, the frequency shift can then be related to the gas concentration. However, in practice, the device must be calibrated.

QCMs usually operate at a frequency lower than a few hundred MHz due to the limitation on how a quartz crystal can be thinned down. The relatively low operating frequency of QCMs is the main reason for their mass sensitivity limitation (Fanget et al. 2011). However, due to the relatively low operating frequency, a significant advantage of QCM-based sensors is that inexpensive drive electronics can be employed for routine sensing applications in the gas phase. Therefore, QCM-based gas sensors are widely utilized as a result of their robust nature, availability, and affordable interface electronics.

12.2.2 FBAR Sensors

As shown before, a decrease in the thickness of the piezoelectric layer gives a noticeable increase in sensitivity. But, at the same time as the thickness decreases, the mechanical strength of the self-appliance decreases significantly. To combine high sensitivity and acceptable mechanical strength, QCM devices are made in the form of a membrane on a silicon substrate (see Figure 12.5), using for this purpose various approaches developed for micromachining technology (Zhang et al. 2010). In this case, the resonator can operate up to approximately 1 GHz (O'Toole et al. 1992). In the literature, sensors manufactured using such technology have abbreviations TFR (a thin-film resonator) or TFBAR (thin-film, bulk acoustic resonator). Compared with QCMs, film bulk acoustic resonators (FBARs) are much more compact.

There are several typical configurations of such devices. The first TFBAR or TFR structure is shown in Figure 12.6. The electrodes and thin piezoelectric film are deposited on the top of an optional insulating layer (typically silicon nitride) supported by a substrate. A portion of the substrate is removed, typically by a wet chemical etching process, thereby defining the resonator. The second TFBAR structure is shown in Figure 12.6b.

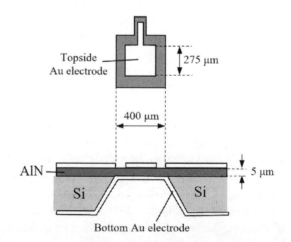

FIGURE 12.5 Schematic diagrams of a thin-film resonator. The upper portion is a top view in which the Au electrode is shown in white and the underlying AlN membrane is shown by the hatch markings. The lower portion is a side view that illustrates the layered structure of the composite material. (Reprinted with permission from O'Toole, R.P. et al., Thin aluminum nitride film resonators: Miniaturized high sensitivity mass sensors, Anal. Chem., 64, 1289–1294, 1992, Copyright 1992, American Chemical Society.)

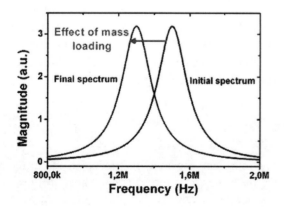

FIGURE 12.4 Principle of operation of mass sensors by the frequency shift measurement. (Reprinted with permission from Fanget, S. et al., Gas sensors based on gravimetric detection—A review, Sens. Actuators B, 160, 804–821, 2001. Copyright 2001, Elsevier.)

Gravimetric Humidity Sensors

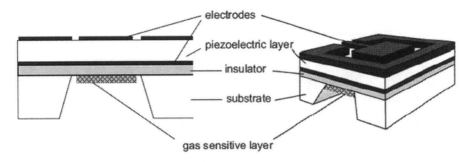

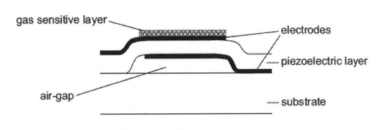

FIGURE 12.6 (a) Cross-section and diagram of a thin-film bulk acoustic resonator (TFBAR) with etched supporting substrate and (b) diagram of TFBAR having the vibrating part suspended over an air-gap. (Reprinted with permission from Ippolito, S. J. et al., Acoustic wave gas and vapor sensors, In: Comini, E. et al. (Eds.), *Solid State Gas Sensing*, pp. 261–304, Springer, New York, 2009. Copyright 2009, Springer Science+Business Media.)

The vibrating membrane is suspended over an air gap. However, for gas-sensing applications, the first configuration is the most widely utilized. In most gas-sensing applications, a sensitive layer is deposited on one side of the structure. For the supporting substrate with the etched cavity, the sensitive layer is typically deposited in the region of this etched cavity, as shown in Figure 12.6a. However, for the air-gap-based structure, the sensitive layer is deposited on the top side. An example of the implementation of the sensor with the configuration shown in Figure 12.6b is shown in Figure 12.7.

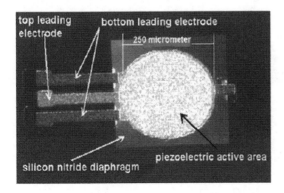

FIGURE 12.7 Photo of a typical ZnO film bulk acoustic resonator (FBAR). (Reprinted from Fanget, S. et al., Gas sensors based on gravimetric detection—A review, *Sens. Actuators B*, 160, 804–821, 2011. Copyright 2011, Elsevier.)

Just as with QCMs, the sensitivity of TFR structure increases proportionally to the square of the resonant frequency, and hence inversely proportionally to the square of the thickness. Therefore, due to the increased operating frequency of a TFR, and since $\Delta f/f^2_0$, the sensitivity of a TFR should be orders of magnitude greater when compared with that of a QCM. However, it should be noted that the detection limit is dependent on the noise and stability of the system (Ippolito 2009). These parameters are worse for TFR-based sensors. In particular, the decrease in the thickness of quartz increases the fragility of the oscillating membrane and hence increases the risk of damage. In addition, it has been found that the sensitivity did not necessarily increase proportionally to the square of the resonant frequency. However, at the very least, sensitivity increases linearly with operating frequency (Gizeli 2002; Rey-Mermet et al. 2006). It has also been suggested that the detection limit is the product of the sensitivity and the Q factor of the structure, which is in TFR structure generally lower than that of a QCM (Rey-Mermet et al. 2006). However, it should be noted that, despite this, many developments have occurred in the field of TFRs since their initial discovery. For example, 2 GHz SMRs based on ZnO were developed by Gabl and co-workers (Gabl et al. 2003; Reichl et al. 2004) for sensing humidity levels in gaseous N_2. Using polyimide as a water vapor–absorbing layer, they observed the mass sensitivity three orders of magnitude larger than for a typical QCM. To improve the mechanical strength in such structures, it is proposed to use piezoelectric materials such as ZnO and AlN, which have higher density and acoustic velocity, and therefore can have thickness of less than 1 μm without the risk of damage. The same approach was used by Liu et al. (2017). They established that polyethylene terephthalate (PET)-coated FBAR exhibited excellent humidity sensitivity of 2202.20 Hz/ppm, which was five orders of magnitude higher than QCM. Comparison of FBAR and QCM is shown in Figure 12.8.

12.2.3 SMR Structures

The structure of SMRs differs considerably from that of a TFBAR, and a typical cross-section is shown in Figure 12.9. The piezoelectric layer is "solidly" mounted onto substrate with acoustic mirrors, that excludes the need for an etched supporting substrate or an air-gap. The reflector stack acts as a Bragg grating, which causes the wave amplitude to diminish within the depth of the reflector, hence confining the mechanical energy to the piezoelectric layer. First described by Newell (1965), the SMRs comprise nominally quarter wavelength–thick layers of materials having alternating high/low acoustic impedance values that are stacked on top of each other. The number of layers required to obtain a satisfactory reflection coefficient strongly depends on the mechanical impedance contrast between layers, and to a lesser extent on the substrate (Lakin 2003; Villa-López et al. 2016). Compared to TFBARs and TFR, SMRs suffer from a lower electromechanical coupling coefficient due to energy being stored in the acoustic reflectors (Lakin 2004). For sensing applications, a sensitive layer is deposited over the top electrode of the structure. However, the manufacture of such sensors can encounter significant difficulties. At present, humidity sensors based on these structures have not been developed, although model samples of SMRs are available. For example, Villa-López et al. (2016) have developed SMRs on the base of a ZnO/SiO$_2$/Mo/SiO$_2$/Mo/ SiO$_2$/Mo/Si structure with a resonant frequency of approximately 1.5 GHz.

To date, neither FBAR nor SMR devices have been demonstrated to be superior to the other; hence the choice between them depends primarily on the users'

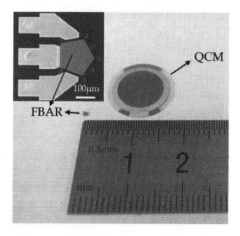

FIGURE 12.8 A comparison of the size between film bulk acoustic resonator (FBAR) and quartz crystal microbalance (QCM). The top-left inset shows the top-view scanning electron microscopy (SEM) image of FBAR. (Reprinted with permission from Liu, W. et al., A highly sensitive humidity sensor based on ultrahigh-frequency microelectromechanical resonator coated with nano-assembled polyelectrolyte thin films, *Micromachines*, 8, 116, 2017. Copyright 2017, MDPI as open access.)

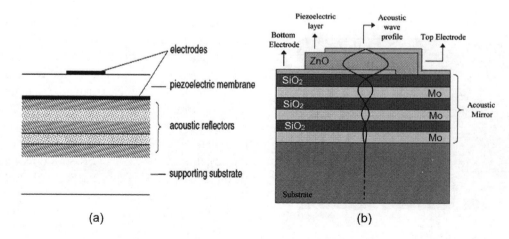

FIGURE 12.9 (a) Cross section of a typical solidly mounted resonator (SMR) and (b) Wave propagation through the SMR and acoustic mirror designed by Villa-López et al. (Reprinted with permission from Villa-López F.H. et al., Design and modelling of solidly mounted resonators for low-cost particle sensing. *Meas. Sci. Technol.*, 27, 025101, 2016. Copyright 2016, IOP.)

ability to design/fabricate membranes and/or Bragg reflectors. However, García-Gancedo et al. (2013) compared identically designed ZnO-based FBAR and SMR devices and showed that, in addition to a more sophisticated manufacturing technology, the SMRs are less responsive than the FBARs. According to García-Gancedo et al. (2013), for the specific device design and resonant frequency ($\approx$ 2 GHz), the FBARs' mass responsivity is approximately 20% greater than that of the SMRs', and although this value is not universal for all possible device designs, it clearly shows that FBAR devices should be favored over SMRs in gravimetric sensing applications where the FBARs' fragility is not an issue. Numerical calculations based on Mason's model offer an insight into the physical mechanisms behind the greater FBARs' responsivity. It was shown that the Bragg reflector has an effect on the acoustic load at one of the facets of the piezoelectric films, which is, in turn, responsible for the SMRs' lower responsivity to mass loadings.

12.3 SURFACE ACOUSTIC WAVE SENSORS

The SAW devices also utilize piezoelectric crystal resonators to generate AWs (Afzal and Dickert 2011; Fanget et al. 2011). A SAW is a kind of AW that propagates along the surface of a material. In other words, a SAW is a mechanical wave in which acoustic energy is confined to the surface of an isotropic single crystal. For the first time, the possibility of propagation of AWs along a solid surface was predicted by Rayleigh in 1885 (Ristic 1983; Grate et al. 1993). These waves, which are sometimes called *Rayleigh waves*, have considerable importance in the areas as diverse as structural testing, telecommunications, and signal processing.

The schematic and working principle of a typical SAW-based humidity sensor is shown in Figure 12.10. Usually, the SAW devices are used as frequency control elements in the feedback path of an oscillator circuit (D'Amico et al. 1982/83; Ippolito et al. 2009). The SAW is generated by interdigital transducer (IDT) electrodes (White and Voltmer 1965), which are periodic metallic bars deposited on a piezoelectric material. The application of voltage between differently polarized transducer fingers of a transmitter IDT generates a periodic electric field that produces a wavelike mechanical deformation of the substrate surface. Just like tiny earthquakes, an acoustic surface wave then exits the transmitter IDT on both sides and travels parallel to the surface in both directions.

SAW devices, used as a sensor platform, mainly have two kinds of structures. The first is a two-port device,

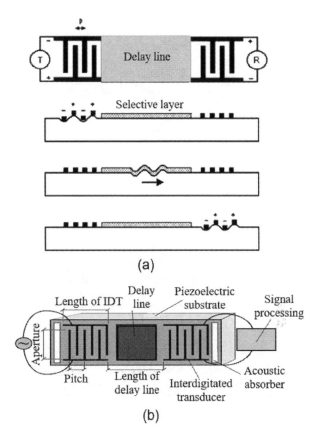

FIGURE 12.10 (a) The schematic and principle of the surface acoustic wave (SAW)-based humidity sensor. A humidity-sensitive layer, usually hygroscopic material, is deposited along the acoustic wave propagation path of the SAW device. The physical water adsorption by the sensing film modulates the phase velocity of the SAW propagating along the device, and as a result, the target relative humidity (RH) can be characterized by the oscillation frequency shift and (b) Diagram illustrating the mechanism of SAW oscillator operation. ([a] – Reprinted with permission from Janata, J., *Principles of Chemical Sensors*, New York, Springer, 2009. Copyright 2009, Springer Science+Business Media. [b] Reprinted with permission from Ippolito, S.J. et al., Acoustic wave gas and vapor sensors, In: Comini E., et al., (Eds.), *Solid State Gas Sensing*, pp. 261–304, Springer, New York, 2009. Copyright 2009, Springer Science+Business Media.)

also called a *delay line*, shown in Figure 12.11a, in which there are two sets of IDTs and a delay line between them. The signal is applied on the first set of IDTs, or the input IDT, which generates the surface wave. This wave is then transmitted over a delay line to the other set of IDTs, or the output IDT, where mechanical wave is converted back into an electric signal by direct piezoelectric effect and analyzed (Ballantine et al. 1997). As is known, the piezoelectric effect is reversible, which means that it works in both directions: The mechanical stress induces electrical polarization charge, and the

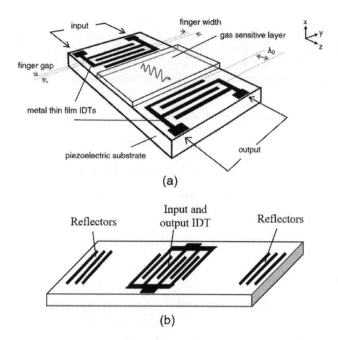

FIGURE 12.11 Two kinds of surface acoustic wave (SAW) devices: (a) schematic of a two-port device, (b) schematic of a one-port device. IDT, interdigital transducer. (Reprinted with permission from Atashbar, M.Z. et al., Basic principles of chemical sensors operation, In: G. Korotcenkov (Ed.) *Chemical Sensors: Fundamentals of Sensing Materials,* Vol. 1, *General Approaches,* New York, pp. 1–62, 2010. Copyright 2010, Momentum Press.)

applied electric field stresses the piezoelectric crystal. As the SAW energy is mainly confined in the surface region, a slight change in the surface condition makes the signal received from the output IDT different, and that lays the foundation of SAW devices' working as high-sensitivity sensors. The sensitive layer normally deposited between the two IDT ports. The second kind of SAW structure is a one-port device, also called a *resonator,* shown in Figure 12.11b. It has only one IDT set. Besides the IDT, there are also grating reflectors that can trap the surface wave, forming a resonating cavity in which the SAW effect is enhanced. Similarly, the change at the surface of the SAW resonator can change the property of SAW, making the SAW resonator a sensitive detector. Moreover, if the signal from the output IDT is amplified by an amplifier and then fed back to the input IDT of the delay line structure, then this will be another type of SAW resonator that has been widely used for sensors application.

The usual substrate materials, used for SAW devices fabrication, are ST-cut (stress- and temperature-compensated) quartz, ZnO, AlN or $LiNbO_3$, which have a high piezoelectric coefficient (Jakubik et al. 2002). Information about these materials can be found in Volume 3 (Chapter 17).

12.3.1 Propagation Modes of the Acoustic Wave

There are many different types of SAWs that are used for sensor applications (Ballantine et al. 1997; Mortet et al. 2008; Ippolito et al. 2009). It is the mode of propagation of the AW that distinguishes different types of AW resonators. Rayleigh waves have longitudinal and vertical shear components. Longitudinal waves have particle displacements parallel to the direction of wave propagation, while shear waves have particle displacements normal to the direction of wave propagation. The particle displacements in shear waves are either normal to the sensing surface (vertical shear wave) or parallel to the sensing surface (horizontal shear wave). Because of the shear component that couples with the medium in contact with the device's surface, SAW devices operating with Rayleigh waves cannot operate in liquids. They are generally used to make high-frequency gas sensors. One should note that the nature of AWs generated in piezoelectric materials is determined by the piezoelectric material orientation, as well as the metal electrodes configuration, employed to generate the electric field that induces AWs by converse piezoelectric effect.

Shear-horizontal surface acoustic wave (SH-SAW) sensors belong to the category of SAW sensors in which the application of a potential to the IDTs produces an AW having particle displacement perpendicular to the direction of wave motion and in the plane of the crystal (shear displacement). As in SAW devices, the acoustic wavelength is determined by the transducer periodicity. Acoustic devices operating with horizontal shear waves are of particular interest, since there are no acoustic losses when operated in liquids compared with vertical shear waves. Thus, they are adequate for sensing applications in liquids (Nomura et al. 2003; Martin et al. 2004).

Shear-horizontal acoustic plate mode (SH-APM) sensors are essentially Rayleigh sensors with the difference that their substrate thickness is of the order of a few acoustic wavelengths (Déjous et al. 1995; Esteban et al. 2000). Under such conditions, IDTs in addition to the Rayleigh waves also generate the shear horizontal waves. These waves are not confined to the surface only but travel through the bulk, being reflected between the top and the bottom surfaces of the substrate that now acts as an acoustic waveguide (see Figure 11.12). The particle displacement for SH-APM devices is parallel to the surface (in plane) and transverse to the direction of wave propagation. The Rayleigh waves typically propagate at much lower velocities than those of the SH-APM modes. The frequency of operation is determined by the material properties, the thickness-to-wavelength ratio of the

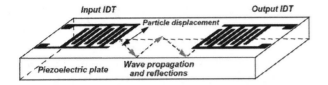

FIGURE 12.12 Schematic view of a SH-APM device structure. (Reprinted with permission from Atashbar, M. Z. et al., Basic principles of chemical sensors operation, In: G. Korotcenkov (Ed.) *Chemical Sensors: Fundamentals of Sensing Materials,* Vol. 1, *General Approaches,* New York, pp. 1–62, 2010. Copyright 2010, Momentum Press.)

substrate, and the IDT finger spacing. These sensors are used mainly for liquid sensing and offer the advantage of using the back surface of the plate as the active sensing area (Kelkar et al. 1991; Dahint et al. 1993). However, the various vapors identification via the electrical surface perturbation is also possible.

Love-wave devices are characterized by AWs that propagate in a layered structure consisting of a piezoelectric substrate and a guiding layer (Bender et al. 2000; Kalantar-zadeh et al. 2001) (Figure 12.13). Love waves are formed by the constructive interference of multiple reflections at the thin coating interface layer that has an acoustic velocity lower than the substrate (Haueis et al. 1994; White et al. 1997). They have a pure shear polarization, with the particle displacements perpendicular to the normal of the surface plane. Love waves are dispersive (i.e., the wave velocity is not solely determined by the material constants but also by the ratio between the thickness of the piezoelectric layer and the wave length defined by the IDT's special period). Under such conditions, the elastic waves generated in the substrate are coupled to this surface guiding layer, which traps the acoustic energy near the surface of the device and thereby increases the sensitivity of the device to mass loadings.

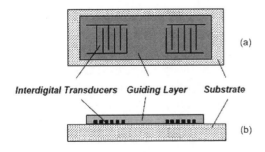

FIGURE 12.13 (a) Top view of Love-wave sensor and (b) Side view of a Love-wave sensor (Reprinted with permission from Atashbar, M. Z. et al., Basic principles of chemical sensors operation, In: G. Korotcenkov (Ed.) *Chemical Sensors: Fundamentals of Sensing Materials,* Vol. 1, *General Approaches,* New York, pp. 1–62, 2010. Copyright 2010, Momentum Press.)

Bulk AW generated by IDTs can be confined between the upper and lower surfaces of a plate that acts as an acoustic waveguide. As a result, both sides of the plate are vibrating. Thus, IDTs can be placed on one side of the plate, and the other side can be used for sensing purpose. The mass sensitivity of a love-mode sensor is determined mainly by the thickness of the guiding layer and its operating frequency. Hence, the guiding layer should be one with low density and low shear velocity. The guiding layer also serves to passivate the IDTs from the contacting liquid (Josse et al. 2001). Love-mode sensors are advantageous for gravimetric sensing applications, such as water and organic vapors absorbed by a thin polymer film (Jakoby et al. 1999).

If the substrate thickness is smaller than the wavelength (membrane), longitudinal waves, called *Lamb waves*, are generated. Lamb-wave sensors are essentially Rayleigh-wave sensors in which the waves propagate in a thin plate (membrane), which has a thickness less than an acoustic wavelength. Particle displacement is transverse to the direction of wave propagation and is parallel and normal to the plane of the surface. Devices based on such waves, also called *flexural plate wave sensors* (FPWs), are of special interest since they are more sensitive than other SAW sensors and can be applied for various applications, including gas and humidity measurement (Wenzel and White 1989; Grate et al. 1991). SAW-based sensors operating in the flexural-plate mode (see Figure 12.14) usually are manufactured using a

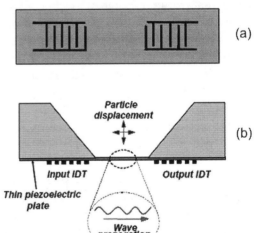

FIGURE 12.14 Configuration of flexural-plate mode surface acoustic wave (SAW) sensor: (a) Top view of a flexural plate wave (FPW) sensor, and (b) Side view of an FPW sensor. IDT, interdigital transducer. (Reprinted with permission from Atashbar, M.Z. et al., Basic principles of chemical sensors operation, In: G. Korotcenkov (Ed.) *Chemical Sensors: Fundamentals of Sensing Materials,* Vol. 1, *General Approaches,* New York, pp. 1–62, 2010. Copyright 2010, Momentum Press.)

membrane technology. The IDTs form a delay line for launching the flexural plate waves that set the whole membrane into motion. The wave velocities and hence the frequency of operation of these devices therefore depends on the plate material and its thickness.

12.3.2 SAW Oscillator as Sensing Element

The SAW oscillator unit is sensitive toward any specific analyte only if delay line is coated with a suitable material that can interact with and distinguish between diverse molecules. As a matter of fact, this sensing material does not have to be even piezoelectric. It is necessary to note that the operation of the SAW and QCM sensors requires temperature and humidity control, because the quartz crystal resonant frequency is affected by variation in the temperature and humidity, and thus affects the frequency shifts during gas molecule collection. The nature of acoustic waves generated in piezoelectric materials is determined by the piezoelectric material orientation, as well as the metal electrodes configuration, employed to generate the electric field that induces acoustic waves by converse piezoelectric effect.

When developing gas sensors, including humidity sensors, it is also necessary to keep in mind that, since piezoelectric substrates are used for AW chemical-sensing applications, an electric field accompanies the propagating wave as it travels through the substrate. This means that perturbations that affect acoustic phase velocity can be caused by many factors, each of which represents a potential sensor response (Ricco et al. 1985; Ricco and Martin 1992). For example, this field can interact with mobile charge carriers in a surface layer/coating and affect both the velocity and amplitude of the wave (Ballantine et al. 1997). This means that the frequency shift in SAW sensors can be not only due to the change in mass, but also due to changes in other physical parameters of the sensitive layer, such as permittivity, electrical conductivity, shear elastic modulus, viscosity, pressure, and density. These interactions change the boundary conditions, producing a measurable shift in the propagating SAW mode's phase velocity. Equation 12.4 illustrates the change in acoustic phase velocity (v) as a result of possible external perturbations, under the assumption that the perturbations are small and linearly combined:

$$\frac{\Delta v}{v_0} \cong \frac{1}{V_p}\left(\begin{array}{l}\frac{\delta v}{\delta T}\Delta T + \frac{\delta v}{\delta m}\Delta m + \frac{\delta v}{\delta \varepsilon}\Delta \varepsilon + \frac{\delta v}{\delta \sigma}\Delta \sigma \\ + \frac{\delta v}{\delta c}\Delta c + \frac{\delta v}{\delta \mu}\Delta \mu + \frac{\delta v}{\delta p}\Delta p + \ldots \end{array}\right) \quad (12.4)$$

where T is the temperature, ε is a permittivity, σ is an electrical conductivity, c is a stiffness, p is a pressure, μ is the shear elastic modulus, and m is a mass. Therefore, the sensor response may be due to a combination of the above parameters. However, during chemical sensing, including humidity measurement, the change of mass, dielectric constant, and conductivity of functionalizing (sensing) layer has the strongest influence. If interaction with an analyte increases the conductivity of the coating film (in the appropriate range), the frequency decreases and the conductivity change enhances frequency shifts due to increased mass ($\Delta f/fr = -\Delta m/M$, where M is the mass of the quartz piezoelectric substrate) or decreased modulus. If it is desired to isolate the conductivity effects from mass and mechanical effects, a reference device can be coated with an initial layer of metal whose sheet conductivity significantly exceeds the window of sensitivity, thereby shorting the electric field (this does not affect sensitivity to mass changes). The chemically sensitive coating material is then applied over the metal film (Ricco et al. 1985). The signal from this device can be compared with that from a sensing device having no metal layer, and the resulting difference in frequency shifts depends only on conductivity changes. The substrate material can also be important in isolating the conductivity effect; $LiNbO_3$ is again the material of choice, with a mass sensitivity less than one-half that of other common substrates, such as quartz and ZnO-on-Si (Ricco et al. 1985).

If the humidity sensitive layer conductivity is unchanged, the effect of adsorbed water on the SAW velocity for a strong piezoelectric substrate can be described by the expression (Dorjin and Simakov 2002)

$$-\frac{\Delta v}{v} = (A + B)\frac{h}{\lambda} \quad (12.5)$$

where A denotes the mechanic and B denotes the electric contribution to the relative velocity change, h represents the adsorbed water layer thickness that is described by an adsorption isotherm (Figure 12.15). It is seen that the influence of both mechanisms on SAW velocity has linear dependence on the thickness of the layer of adsorbed water. The components A and B may be expressed using simple formulae (Dorjin and Simakov 2002):

$$A = \frac{\rho_w}{\rho_s}\sqrt{1 - \frac{v_w^2}{v_s^2}} \quad (12.6)$$

$$B = \frac{\pi \kappa^2 \varepsilon_s}{(\varepsilon_s + \epsilon_0)^2} \cdot \frac{\varepsilon_w^2 + \epsilon_0^2}{\varepsilon_w} \quad (12.7)$$

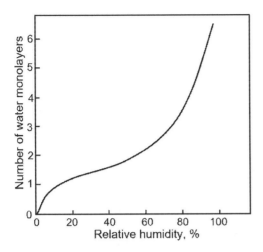

FIGURE 12.15 Water vapor adsorption isotherm for quartz at 23°C. The monolayer thickness calculated from Van der Waals diameter of a water molecule is 0.282 nm. (Reprinted with permission from Awakuni, Y. and Calderwood, J.H., Water vapour adsorption and surface conductivity in solids, *J. Phys. D, Appl. Phys.*, 5, 1038–1045, 1972. Copyright 1972, IOP.)

where ρ_w, ρ_s are densities of water and the piezoelectric substrate, respectively; ε_w is water permittivity; ε_s, ε_0 are substrate and vacuum permittivity; v_w is the sound velocity in water; v_s is SAW velocity; and k^2 is the electromechanical coupling coefficient for the substrate.

Therefore, the design of sensors, depending on the sensitive layer used and its properties, requires judicious selection of the piezoelectric substrate (Ricco and Martin 1992). Changes in the attenuation and velocity of SAWs depend on the square of tile electromechanical coupling coefficient of the substrate, K^2, which varies by orders of magnitude from one material to another. YZ-lithium niobate (LiNbO$_3$), for example, has a K^2 value of 0.048, which is more than 40 times higher than that of ST-quartz ($K^2 = 0.0011$). Thus, LiNbO$_3$ is a more appropriate substrate to choose if this detection mechanism is to be employed. Appropriate temperature controls will be essential, however, since the temperature coefficient of LiNbO$_3$ SAW devices is over 20 times that of comparable ST-quartz devices. If a coating material is used to concentrate the analyte near the sensor surface, it must be a semiconductive material with a sheet conductivity (the product of film thickness and bulk conductivity) that falls within a "window of sensitivity" determined by the substrate. Using values for LiNbO$_3$ (Ricco et al. 1985), maximum sensitivity is obtained in the range of 10^{-7}–10^{-5} Ω^{-1}, and the maximum possible frequency shift ($\Delta f/f_0$) is $K^2/2 = 24{,}000$ ppm. For quartz, the window of sensitivity is an order of magnitude lower (10^{-8}–10^{-6} Ω^{-1}), and the maximum shift obtainable is only 550 ppm; this is, however, more than adequate for the observation of a significant effect (i.e., sensor response) under the proper conditions (Ricco and Martin 1992). In addition to having sheet conductivity within the aforementioned window of sensitivity, the sheet conductivity of the coating material must be altered as a result of a chemical stimulus. As it was shown before, conductive polymers (phthalocyanines, polypyrrole, etc.), metal oxides (ZnO, In$_2$O$_3$, etc.), and metals (Pd) have such properties.

Since many internal and external factors may contribute to the propagation of an AW and, subsequently, to the change in frequency of oscillation, the determination of change in stimulus may be ambiguous and contain errors. An obvious solution is to use a differential technique, in which two identical SAW devices are employed: One device is for sensing the stimulus, and the other is reference. The reference device is shielded from stimulus but subjected to common factors, such as temperature, aging, and so forth. The difference of the frequency changes of both oscillators is sensitive only to variations in the stimulus, thus canceling the effects of spurious factors.

12.4 ADVANTAGES AND SHORTCOMINGS OF ACOUSTIC MASS-SENSITIVE SENSORS

As it follows from our previous discussion, the properties of the considered acoustic sensors depend on many factors, including piezoelectric material, metal electrodes configuration, and sensing material used. However, the clear physical principles underlying the operation of these sensors allow us to estimate detection limits and the relative (S_r) mass sensitivities for different types of gravimetric sensors (Wenzel and White 1978; Bo et al. 2016). This information is presented in Table 12.1. It must be borne in mind that these approximate data do not take into account the latest achievements in the development of these sensors. And this means that the sensitivity can be even higher. The comparison of various types of gravimetric sensors is also presented in Table 12.2.

It is seen that cantilever-based sensors operated in resonant mode are the most sensitive devices. However, among the sensors considered, quartz microbalance-based sensors are the simplest, cheapest, and most reliable. Therefore, it is these sensors, most adapted for wide application, are currently presented on the market of humidity sensors (this information will be presented in Volume 3, Chapter 28). They are inexpensive because they are mass produced for oscillator circuits, and their lifetime is approximately 1 year. QCM sensors have good linearity and can be used for a variety of applications

TABLE 12.1
Comparison of the Relative Sensitivity and Minimum Detectable Mass Density of the Different Acoustic Sensors

Sensors	Sensitivities, cm^2/g	Minimum Detectable Mass Density, ng/cm^2
Flexural plate wave	100–1000	0.5
SH-SAW	65	1
FBAR	1000–10,000	0.1–0.01
Quartz microbalance	10–14	10
	150	1.2
SAW	1000–10,000	0.02–0.04
Microcantilever		

FBAR, film bulk acoustic resonator; SAW, surface acoustic wave; SH-SAW, shear-horizontal surface acoustic wave.

Source: Ward, M.D. and Buttry D.A., *Science*, 249, 1000–1007, 1990; Tipple C.A., Strategies for enhancing the performance of chemical sensors based on microcantilever sensors, PhD thesis, The University of Tennessee, Knoxville, TN, 2003; Mortet, V. et al., *Phys. Stat. Sol.*, 205, 1009–1020, 2008.

TABLE 12.2
Typical Properties of Commonly Utilized Acoustic Wave–Based Sensors

Parameter	QCM	TFR	SAW
Operating frequency	5–30 MHz	500 MHz–20 GHz	40 MHz–1 GHz
range	Yes	Yes	Yes
	No	No	Yes
Sensitivity towards mass	≈ 0.01%	≈ 0.1%	≈ 0.1%
	Up to 10^5	Up to 10^3	Up to 10^4
Sensitivity towards conductivity			
Fractional frequency change to gas			
Quality factor (approx.)			

QCM, quartz crystal microbalance; SAW, surface acoustic wave; TFR, thin film resonator.

Source: Wenzel, S.W. and White R.M., *Appl. Phys. Lett.*, 54, 1976–1978, 1978; Ippolito, S.J. et al., Acoustic wave gas and vapor sensors, In: Comini, E., Faglia, G., Sberveglieri, G. (Eds.), *Solid State Gas Sensing*, Springer, New York, pp. 261–304, 2009.

ranging from solvent detection to food quality analysis, paper and tobacco production, and so on. Many believe that, for the development of dew-point hygrometers based on direct mass measurements of condensation, the SAW sensors are more acceptable.

The major advantages of mass sensors are the simplicity of their construction and operation, low weight, and small power requirements. In addition, their operating principle depends on a highly reliable phenomenon. The measurement of the frequency shift is one of the simplest and most accurate physical measurements. In addition, mass sensors have high sensitivity and can be used with a variety of selective layers for sensing of a very broad range of compounds. Other advantages of such sensors include a relatively short response time, good accuracy, and low drift. All these advantages show that gravimetric sensors are suitable for a wide range of applications. For example, experimental studies have shown that dew-point hygrometers based on direct mass measurements of condensation have the potential to provide more accurate dew-point measurement with high resolution. The results of research carried out in this direction are published by Galipeau et al. (1995), Hoummady et al. (1995), Vetelino et al. (1996), and Jachowicz and Weremczuk (2000). In particular, Hoummady et al. (1995) developed a dew-point detector using an LST-cut quartz SAW sensor. SAW devices increase considerably the accuracy of humidity measurement because of their dual ability to detect dew condensation and to measure the temperature with a great accuracy. Here, the SAW sensor is cooled until condensation occurs on the surface. The appearance of the water layer is accompanied by a sudden change of the delay time or resonance frequency. In comparison with the optical dew point detectors, the accuracy of the SAW devices was improved by a factor of approximately 500 (Hoummady et al. 1995).

It is important to note that SAW devices can be used in a passive mode without the need for batteries (Lee et al. 2007; Kang et al. 2013). An antenna can be added to the input IDT and the signal received by the antenna can then stimulate the SAW, which can be used for sensing application. Thus, SAW devices can be used in the wireless humidity measurement.

Though quite a lot of advantages have been shown for AW sensors, various challenges still exist. The biggest one is the peripheral circuit; the usage of gravimetric sensors requires complicated and expensive driving and detection electronics, including systems for signal processing of sensor outputs. The frequency response and time delay are measured using a network analyzer, which is large and expensive. However, for portable and cheap use of AWs, a better signal processing circuit is needed. Especially for super-high frequency (SHF) SAW, a circuit made using discrete devices is not possible due to the high speed. Another disadvantage of AW sensors is nonlinear characteristics and their relatively high temperature dependence (Bo et al. 2016). The changes in extrinsic variables, such as temperature and pressure, can produce a sensor response, affecting the AW either directly or via changes in the film's intrinsic properties. Therefore, the sensor may not be accurate without adding a temperature compensation part. Another problem is complementary metal–oxide–semiconductor (CMOS) compatibility. SAW is fabricated using piezoelectric materials, and the growth of these materials is not compatible with CMOS technology. This further constrains the integration of a signal-processing circuit with SAW sensors. Therefore, either fabricating traditional piezoelectric materials using CMOS technology is required or finding new CMOS-compatible piezoelectric materials is needed as a crucial issue in the further development and mass production of SAW sensors.

Incorrect interpretation of frequency shifts can also yield flawed results. Proper experimental design can mitigate the temperature effects; however, knowledge of the underlying physics and rigorous verification of the source of the signal variation are the surest protection against misinterpretation of the raw data. Oscillator instability can also pose problems in practical applications of these sensors. AW sensors also suffer from high sensitivity to mechanical strains and vibrations and mechanical stress caused by temperature variations or temperature gradients. Another problem is related to the electrode and mounting geometry, in the sense that the associated mechanical strains may contribute to variations in the measured frequency shifts. For example, more detailed consideration of QCM devices shows that the measured change in resonance frequency is different in the central area of the resonator ($r = 0$) and in the electrode region ($r = R$) due to the mass contribution of the electrodes. The active area of the QCM is approximately half of the disk diameter. Another characteristic feature is the differential mass sensitivity, which demonstrates a maximum in the central part of the resonator surface and drops off in the edge area (Janshoff et al. 2000). The radial distribution of the mass sensitivity depends on the thickness of the disk. For thin crystals, $\Delta f(r) = [f(r) - f_0(0)] \to 0$, so the difference between the resonance frequency in the central and electrode edges regions becomes very small. Another (less unfavorable) aspect is the Q-factor dependence on the crystal thickness (Figure 12.16).

This consideration also testifies that the characteristics of gravimetric sensors, such as response time, sensitivity, and hysteresis, are largely dependent on the type of sensing films used in these devices. Therefore, one should consider that such factors as the history/memory

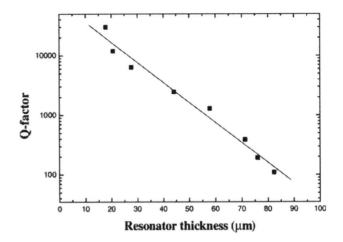

FIGURE 12.16 Q factor of an unloaded quartz crystal microbalance as a function of resonator thickness. (Reprinted with permission from Hung, V.N. et al., High-frequency onechip multichannel quartz crystal microbalance fabricated by deep RIE, *Sens. Actuators A*, 108, 91–96, 2003. Copyright 2003, Elsevier.)

(or nonreversibility) of the sensitive layer, period of exposure, gas concentration, and coexistence of other molecules in the ambient sensing environment can also significantly influence the performance.

12.5 SENSING LAYERS IN GRAVIMETRIC HUMIDITY SENSORS

12.5.1 General Requirements

The detection of air humidity using AW sensors can be based on changes in one or more of the physical characteristics of a thin film or layer in contact with the device surface (Ballantine et al. 1997). Most physical and chemical interactions between analytes and sensor coatings lead to changes in the mass. The surface mass changes can result from sorptive interactions (i.e., adsorption or absorption) or chemical reactions between water vapor and coating. Thus, this sensing mechanism offers the greatest latitude in the selection of sorptive or reactive coating materials (Monreal and Marl 1987). As it follows from general consideration, the working relationship between mass-loading and frequency shift, Δf_m, for AW devices can be written as Equation 12.8:

$$\Delta f_m = -K \cdot S_m \cdot \Delta m_A \qquad (12.8)$$

in which S_m is a device-specific constant that depends upon the factors such as the nature of the piezoelectric substrate, device dimensions, frequency of operation, and the acoustic mode that is utilized; K is a geometric factor for the fraction of the active device area being perturbed; and Δm_A is the change in mass/area on the device surface. It is clear that all other properties being equal, a film having higher surface area results in a larger number of analyte molecules being adsorbed for a given ambient-phase analyte concentration, the consequences of which are enhanced sensitivity and limit of detection. For reactive and (irreversible) adsorptive coatings, a higher surface area translates to a higher capacity and thus greater dynamic range. It is also necessary to consider that, as the plate thickness decreases, the operating frequency and mass sensitivity increase proportionally. The maximal operating frequency usually is limited by the mechanical stability of the substrate.

As it was indicated above, the change of conductivity under the influence of the analyte can also be used for analyte detection using a SAW device (Section 12.3). Some of the intrinsic film properties, such as elastic stiffness (modulus), viscoelasticity, viscosity, and permittivity, can also be utilized for humidity detection. This means that a wide range of materials that change properties under influence of gas can be used for AW-based gas sensor design. Experiments carried out during the last decades confirmed this statement.

However, the ability to use materials with different patterns of interaction with water vapor creates certain difficulties in interpreting the results. For example, an increase in the mass loading alone produces a decrease in the frequency without affecting the attenuation. In contrast, changes in mechanical properties of the coating can produce changes in both the frequency and the attenuation of the AW (Ballantine et al. 1997). Furthermore, these changes can either increase or decrease either or both of the two AW propagation parameters, depending on the details of the relationship the between film thickness, acoustic wavelength, and the complex modulus of the film at the frequency and temperature of operation. Of course, not all detection mechanisms are of practical significance for all types of sensors, and several mechanisms can operate simultaneously (synergistically or antagonistically) to affect a response. In addition, consider that the nature of the sensitive coating—both its inherent physicochemical properties and the physical particulars (thickness, uniformity, etc.) of a specific layer deposited on a specific device—often influences the detection process.

It is worth noting that the performance of a given coating can sometimes be predicted a priori through knowledge of chemical reactions (Ballantine et al. 1997) taking place during interaction with gas surrounding, or mechanical properties of sensing materials used. For example, rigid polymer films with a large value of G move synchronously with the surface of the oscillating resonator, whereas compliant films with lower G do not move synchronously with the surface of the film and tend to lag behind the resonator, creating a viscoelastic response.

As for the requirements for the sensitive layer itself, an attractive feature of AW-based chemical sensors is that they impose relatively few constraints on the materials that can be used as chemically selective coatings. In brief, ideal coatings for QCM applications should be nonvolatile so that the coating stays on the crystal surface and allows rapid and easy diffusion of vapors into and out of the material. The material should also be chemically and physically stable over prolonged use in contact with its working medium (gas and water vapor) and not undergo any hysteresis effects. In addition, the film must be uniform and thin, and it must not electrically short circuit the IDTs. Thickness, uniformity, and other characteristics are affected by the method of deposition, be it painting, dipping, solvent casting, spraying (air-brushing), spin-casting, or subliming. These methods will be detailed shortly in Volume 3 (Chapter 18). Typically, uniformity in film thickness is not crucial but can be important in some circumstances, especially in the case when devices are designed for the sensor market. The selected material must adhere to the device surface in such a manner that it moves synchronously with the AW and must maintain this adhesion in the presence of water vapor and interferants.

The adhesion of thin films to many types of surfaces, including those that are chemically very dissimilar to the coating material, is a much-studied topic outside the sensor field. Often, adhesion-promoting interlayers have been developed for general classes of problems, such as securing a highly nonpolar polymer film to a very polar substrate. Anyone attempting to construct a reliable sensor would do well to examine the relevant literature (Lee 1991). Highly conductive coatings (i.e., most semiconductors and conducting polymers) must, of course, be electrically insulated from IDT electrodes in order to prevent shorting; this is not a concern with the planar electrodes of QCM devices. A final constraint is that the coating be acoustically thin (Ballantine et al. 1997). A somewhat standard rule of thumb is that thicknesses less than 1% of the acoustic wavelength are appropriate. In fact, whether a particular coating thickness is acoustically thin depends critically upon the acoustic properties of that material under the particular set of conditions (temperature, nature and concentration of contacting gaseous species, etc.) being evaluated. In other words, it is the acoustic wavelength in the coating film that is relevant; this can differ appreciably from the acoustic wavelength in the device substrate, particularly in materials such as rubbery polymers, which have vastly different sound velocities than, for example, quartz. In practice, the coating thickness from a few angstroms to several micrometers has been utilized in sensing applications.

When choosing a sensitive layer, Ippolito et al. (2009) also recommend that attention should be paid to a resistance to the influence of external factors (such as fluids that interact with the sensing surface during its operations), environmental stability (humidity, temperature, mechanical shock, and vibration), value of electromechanical coupling with the substrate, ease of processing by the available technologies, and cost of this material.

Certainly, AW devices can work without coating by sensing material. However, experiment has shown that the lack of specificity and low sensitivity are two major drawbacks of sensors with uncoated surfaces.

12.5.2 Humidity-Sensitive Materials Used in Gravimetric Humidity Sensors

12.5.2.1 Polymers

As follows from the name of the sensors under consideration, the change in the mass of the sensitive layer caused by adsorption of water vapor is the most important parameter of this material. Along with the mechanical and adhesive properties that control the longevity of the devices, it is the change in the mass that determines the sensitivity of the humidity sensors. Studies have shown that polymers must fully meet these requirements, including the requirements specified in the previous section, because they are the ideal material for such applications.

Organic polymers are characterized by the capability to reversibly sorb water vapors (Ballantine et al. 1997). Moreover, thin films of many polymeric materials exhibit good adhesive properties and are easily applied to most substrates. In addition, relatively rapid diffusion and a high capacity for organic solutes make amorphous polymers attractive as sensor coatings. The variety of functionalities that can be incorporated into polymers makes it possible to optimize selected sorptive interactions and maximize sensitivity to air humidity (Carey et al. 1986). Analysis of published results testifies that the above-mentioned statements are primarily related to hygroscopic polymers, such as cellulose derivatives (Hu et al. 2011), polyimide (PI) (Schroth et al. 1996), poly-methylmethacrylate (PMMA) (Yoo and Bruckenstein 2013), Nafion (Su et al. 2006), and polyvinyldifluorene (PVDF) (Glück et al. 1994), poly(ethylene glycol methyl ether), phenoxy resin, and poly(vinyl chloride) (Carey et al. 1986), which are widely used in gravimetric humidity sensors. As can be seen from the results presented in Figure 12.17, the change in weight of these polymers with a change in humidity reaches values that allow developing very sensitive sensors. It can be seen that the volume fraction of sorbed water becomes larger as the polymer becomes more hydrophilic. Since less hydrophilic polymers (polyethylene glycol (PEG), ethylcellulose (EC), cellulose acetate butyrate (CAB), and PMMA) do not have highly specific sorption sites, the interaction between a polymer and sorbed water is very small. It is important to note that,

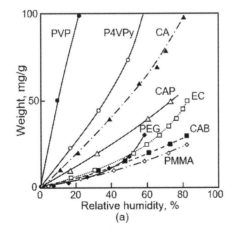
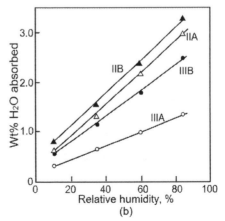

FIGURE 12.17 (a) Sorption isotherms measured at 30°C for PVP, P4VPy, CA, CAP, CAB, EC, PEG, and PMMA. PVP, polyvinylpyrrolidone; P4VPy, poly(4-vinylpyrridine); CA, cellulose acetate; CAP, cellulose acetate phthalate; EC, ethyl cellulose; PEG, polyethylene glycol; PMMA, poly-methylmethacrylate. (b) Water absorption of polyimide derivatives films (IIA, IIIA, IIB, and IIIB) as a function of relative humidity (%RH). ([a] – Reprinted with permission from Sadaoka, Y., Capacitive-type relative humidity sensor with hydrophobic polymer films, In: Comini, E. et al. (Eds.), *Solid State Gas Sensing*, pp. 109–152, Springer, New York, 2009. Copyright 2009, Springer Science+Business Media; [b] – data extracted From Horie, K. and Yamashita, T. *Photosensitive polyimides*, Technomic Publishing Company, Lancaster, PA, 1995.)

along with good adsorption properties, many polymers also have the necessary stability. For example, Nafion is extremely resistant to the chemical compounds attack, so Nafion-based sensors can operate in atmospheres with organic vapors and dust.

It is important to know that the sorption of water depends on many parameters, including methods of forming polymer derivatives (see Figures 12.17b and 12.18). For example, it was established that there is decidedly more water sorbed by cellulose acetate and cellulose acetate hydrogen phthalate films than by cellulose acetate butyrate and ethyl cellulose films in a whole humidity region. The amount of sorbed water in the desiccation process was larger than that in the humidification process. The differences of the sorbed water measured in both processes were more for cellulose acetate and cellulose acetate hydrogen phthalate films than for cellulose acetate butyrate and ethyl cellulose films. For cellulose acetate, the water sorption isotherm observed in the desiccation process is strongly affected by the history of exposure to a humid atmosphere as shown in Figure 12.19.

It should be noted that the sorption ability increases with proceeding of the cross-linking reaction for each polymer. Sorption isotherms measured at 30°C for the films are shown in Figure 12.20. All the curves are almost linear. These polymers were hydrophobic and W_g depended on the polymer species. The difference of sorption ability seems to be derived from the difference of the structure of the films, as well as the difference of the side groups of the polymer.

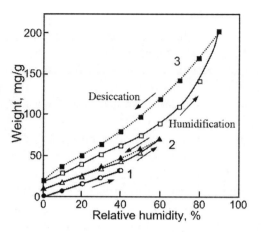

FIGURE 12.19 Water sorption isotherms for cellulose acetate. (1, ○, ●) within 40% RH (2, △, ▲), within 60% RH (3, □, ■), and within 90% RH. *Open symbols*: humidification. *Closed symbols*: desiccation. (Reprinted with permission from Sadaoka, Y., Capacitive-type relative humidity sensor with hydrophobic polymer films, In: Comini, E. et al. (Eds.), *Solid State Gas Sensing*, pp. 109–152, Springer, New York, 2009. Copyright 2009, Springer Science+Business Media.)

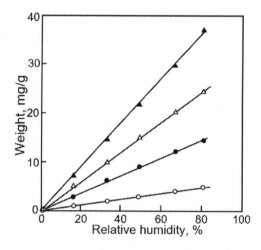

FIGURE 12.20 Sorption isotherms measured at 30°C: (○) poly-VB, (▲) poly-VM, (△) poly-VCr, and (●) poly-VCi. VB, benzoxazine monomer. (Reprinted with permission from Sadaoka, Y., Capacitive-type relative humidity sensor with hydrophobic polymer films, In: Comini, E. et al. (Eds.), *Solid State Gas Sensing*, pp. 109–152, Springer, New York, 2009. Copyright 2009, Springer Science+Business Media.)

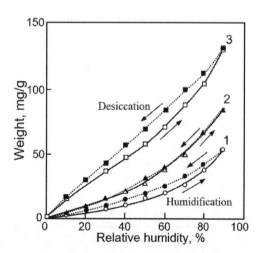

FIGURE 12.18 Water sorption isotherms: (3, □, ■) cellulose acetate hydrogen phthalate. (2, △, ▲) cellulose acetate butyrate, (1, ○, ●) ethyl cellulose. *Open symbols*: humidification. *Closed symbols*: desiccation. (Reprinted with permission from Sadaoka, Y., Capacitive-type relative humidity sensor with hydrophobic polymer films, In: Comini, E. et al. (Eds.), *Solid State Gas Sensing*, pp. 109–152, Springer, New York, 2009. Copyright 2009, Springer Science+Business Media.)

12.5.2.2 Metal Oxides

Metal oxide films can also be applied in QCM and SAW-based humidity sensors, as they also adsorb moisture. At present, SiO_2, ZnO, and TiO_2 have been tested as humidity-sensitive materials for QCM and SAW-based humidity sensors. However, it should be recognized that the use of metal oxides in these sensors is much less common than the use of polymer materials and polymer-based composites. This is due to the fact that the change in the mass of metal oxide films upon adsorption of water is substantially less than for polymers. In addition, metal oxide films cannot provide the same adhesion to the substrate as polymers when using low-temperature processes, which is very important for AW sensors. At the same time, when interacting with water vapor, there is a significant change in electrical conductivity of metal oxides, which can be used to improve the sensitivity of SAW-based humidity sensors, especially in the low-humidity region. Stability of metal oxides in corrosive environments is also a significant advantage of metal oxides.

12.5.2.3 Other Materials

Other hygroscopic materials can also be used in gravimetric humidity sensors. Examples of high-surface area solid adsorbents suitable for humidity sensor coatings are microporous materials such as silica gel, fullerene, graphene, CNTs, porous silicon, and molecular sieves—in particular, zeolites and metal-organic framework materials (MOFs). For most mentioned-above materials, high adsorption capacity arises from the presence of large numbers of micropores and/or mesopores. The total surface area of a single gram of such materials can exceed 1000 m^2 (Cheremisinof and Ellerbusch 1980). Therefore, it is quite understandable why the humidity sensors based on these materials can provide good sensitivity for the detection of vapor-phase species and humidity (King 1964; Sanchez-Pedreno et al. 1986; Bein et al. 1989; Kreno et al. 2012). For example, it was established that MOFs have the ability to adsorb large quantities of water. It is reported that Hong Kong University of Science and Technology (HKUST)-1 can adsorb as much as 40 wt% water (Wang et al. 2002). Of course, this makes them attractive for humidity sensing, which has been demonstrated using a variety of sensing platforms. In particular, Ameloot et al. (2009) demonstrated a water vapor detection with an MOF-coated QCM, using electrochemically synthesized Cu-1,3,5-benzenetricarboxylic acid (Cu-BTC) films grown directly on the device. They showed that changes in relative humidity (RH) can be monitored, and a highly reproducible signal occurs upon cycling between dry and water containing nitrogen flows. The water sorption capacity of the films was found to be 25wt%–30 wt%. However, it is necessary to consider that vapor or organic solvent is also a common interfering gas and must be addressed in the design of MOF-based sensing systems. Experiment has shown that response and recovery times of the above-mentioned sensors depend on the nature, thickness, and structure of used material.

More detailed discussion of some of these materials can be found in Volume 3 of our edition (Chapters 4–13).

12.6 ACOUSTIC HUMIDITY SENSORS' PERFORMANCE

12.6.1 QCM Humidity Sensors

12.6.1.1 Conventional QCM Humidity Sensors

In the development of QCM-based humidity sensors, there are four main directions that are distinguished by the use of humidity-sensitive materials (Table 12.3): (1) polymer-based; (2) inorganic, mainly metal oxides-based; (3) carbon-based; and (4) composite-based sensors, mainly based on polymers. Typically for the manufacture of sensors, AT-cut quartz crystals with a resonant frequency of 10 MHz are used. It can be seen that uncoated QCM-based humidity sensors (covered by Au) have very low sensitivity. The formation of a porous structure in the gold layer increases sensitivity. But this increase is significantly smaller in comparison with the effect observed when using traditional humidity-sensitive materials.

As can be seen from the results presented in Table 12.3, a large number of materials have been tested as humidity-sensitive materials. If we compare the characteristics of sensors using humidity-sensitive materials from different groups, then we see that there is not much difference in parameters for the best representatives. Among the polymeric QCM-based humidity sensors, we can distinguish the results obtained by Neshkova et al. (1996) when using coatings of nitrated polystyrene (NPS) (Figure 12.21). It was established that the humidity sensor in the range 10%–68% RH exhibited a highly sensitive linear frequency vs RH response (up to 30 Hz/%RH) and very good performance reproducibility and stability upon long-term calibration for more than 3 months. The high stability of QCM-based humidity sensors was also observed when using poly (acrylic acid) (PAA) (Figure 12.22) (Wang et al. 2010), cellulose (Randin and Zullig 1987; Hu et al. 2011), PMMA films (Kwon et al. 2006), urea formaldehyde resin (UFR)/SiO$_2$ composite (Lv et al. 2017), graphene (Figure 12.22b) (Yao et al. 2011), and Polyethylenimine-graphene oxide (PEI-GO) (Tai et al. 2016). Neshkova et al. (1996) have established that NPS-based sensors were also characterized by very fast response and recovery times (time response less than 3 s) (Figure 12.21b). In addition, no interference with the frequency response to water vapor in the presence of corroding gases, such as SO$_2$ and NO$_2$, has been observed. Neshkova et al. (1996) have shown that the nonmodified inner interface of this coating provided a

TABLE 12.3
Performances of Quartz Crystal Microbalance–Based Humidity Sensors

Type of Sensing Material	Sensing Material Used	S, Hz/%RH	Res. Time, s	Rec. Time, s	References
Uncoated	Au	0.6			Galatsis et al. (1999)
	Porous Au	2			Galatsis et al. (1999)
Polymers	PVP; PS; PPy; PMMA; Cellulose	1–40	3–140	10–400	Neshkova et al. (1996); Syritski et al. (1999); Kwon et al. (2006); Hu et al. (2011); Yoo and Bruckenstein (2013); Wang et al. (2014); Kosuru et al. 2016)
	PAA; Nafion; PDDA/PSS; PEDOT/PSS	5–23	5–80	5–10	Chen et al. (2005); Su et al. (2006); Wang et al. (2010); Yao and Ma (2014); Muckley et al. (2016)
Inorganic	ZnO; TiO_2; SnO_2/SiO_2;	9–90	1–50	10–16	Erol et al. (2011); Horzum et al. (2011); Asar et al. (2012); Ates et al. (2013); Sakly et al. (2014); Yakuphanoglu (2012); Zhu et al. (2014); Yuan et al. (2016b)
	ND; BP; ZnS	2–46	25–30	3–10	Uzar et al. (2011); Yao et al. (2014a, 2017)
Carbon	MWCNTs; GO; MWCNTs/GO; fullerene	3–25	18–60	3–70	Radeva et al. (1997); Zhang et al. (2005); Yao et al. (2011, 2014b); Li et al. 2013; Su and Kuo (2014); Yuan et al. (2016b); Jin et al. (2017)
Composites	PDDA/GO; ND/GO; Nafion/GO; GO/PEI; GO/ZnO;	27–70	9–50	3–18	Tai et al. (2016); Yuan et al. (2016a, 2016b); Yao and Ma (2014); Yao and Xue (2015); Chen et al. (2016); Yuan et al. (2016b)
	UFR/SiO_2; CNTs/Nafion; PAMAM/AuNPs	16–70	5–50	3–25	Chen et al. (2005); Su et al. (2006); Su and Tzou (2012); Lv et al. (2017)
Other materials	Calix[4]arene; MOF (HKUST-1; [$Cu_3(BTC)_2$])	6–140	300–1700	200–1000	Ameloot et al. (2009); Okura et al. (2010); Su et al. (2013); Kosuru et al. (2016)

BP, black phosphorous; GO, graphene oxide; ND, nanodiamond; PAA, polyacrylic acid; PVA, poly(vinyl alcohol); PAMAM, polyamidoamine; PDDA, poly (dimethyldiallylammonium chloride); PEI, polyethyleneimine; PS, polystyrene; PSS, poly(sodium-*p*-styrenesulfonate); PVP, poly-4-vinylpyridine; Rec., recovery; Res., response; RH, relative humidity; UFR, ureaformaldehyde resin.

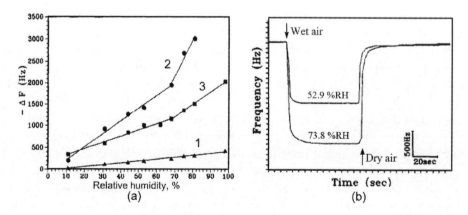

FIGURE 12.21 (a) Frequency change (ΔF) vs relative humidity (%RH) relationships for quartz crystals coated with: (1, ▲) non-modified nitrated polystyrene (NPS), (2, ●) chemically modified nitrated polystyrene (CMNPS), and (3, ■) CMNPS (NPS obtained by in situ nitration of polystyrene [PS] initial coatings); and (b) A typical response and recovery time for CMNPS-coated humidity sensors. (Reprinted from Neshkova, M. et al., Piezoelectric quartz crystal humidity sensor using chemically modified nitrated polystyrene as water sorbing coating, *Anal. Chim. Acta*, 332, 93–103, 1996. Copyright 1996, Elsevier.)

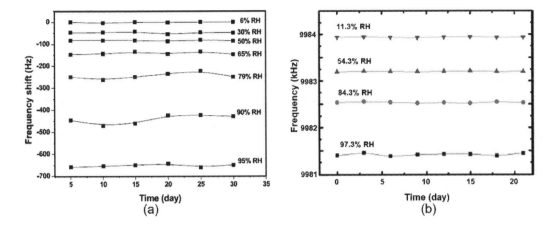

FIGURE 12.22 (a) Stability of the fibrous polyamic acid (PAA) membrane coated quartz crystal microbalance (QCM) sensors with a coating load of 600 Hz after exposure in air for 30 days. (b) Long-term stability of graphene-based QCM humidity sensor. ([a] - From Wang, X. et al., *Nanotechnol.*, 21, 055502, 2010. Copyright 2010, Institute of Physics.) Published by IOP as open access; ([b] - Reprinted from Yao, Y. et al., Graphene oxide thin film coated quartz crystal microbalance for humidity detection, *Appl. Surf. Sci.*, 257, 7778–7782, 2011. Copyright 2011, Elsevier.)

durable, perfectly adhesive and chemically stable with time contact of the coating with the QC electrode surfaces due to the hydrophobicity of NPS. On the other hand, the chemical surface modification yielded an increase in surface hydrophilicity of the coating through stable incorporation of hydrophilic functional groups into its polymer structure, thus bringing about a notable change in its morphology, too. Neshkova et al. (1996) believe that the observed increased sensitivity of the chemically modified nitrated polystyrene (CMNPS)-coated humidity detectors should be attributed to the interrelation between the above two surface coatings' characteristics. In the range from 68% to 98%, the reproducibility becomes worse and considerable hysteresis is observed. However, the abrupt increase in sensitivity, varying from 30 up to 100 Hz/%RH, outweighs the above-mentioned drawbacks, still allowing frequency measurements to be performed with a relative standard deviation (RSD) not surpassing 10% at the worst. It should be noted that an increase in hysteresis and response and recovery time in the high humidity region (i.e., > 70% RH), is common to most polymer-based humidity sensors (Tai et al. 2016).

Among inorganic materials, ZnO-based humidity sensors have the greatest sensitivity (Ates et al. 2013; Sakly et al. 2014). But for the same sensors, there is also the biggest parameters' dispersion. Depending on the synthesis technology and the formation of the humidity-sensitive layer, the sensitivity can vary from 0.5 to 50 Hz/%RH (Erol et al. 2011; Ates et al. 2013). This indicates that the structural factor is one of the most important in the formation of the humidity-sensitive layer. As a rule, sensors based on nanowires (Erol et al. 2011) and nanofibers (Horzum et al. 2011) had the smallest sensitivity, while the largest sensitivity was observed for nanogranulated films (Erol et al. 2010; Sakly et al. 2014). This is understandable, since nanowires and nanofibers with a high specific surface area, calculated per unit weight, have a low specific surface area calculated per unit area. For mass-sensitive devices, this is the determining factor. In addition, in such materials, there are no nanopores, contributing to capillary condensation of water vapor. At the same time, such sensors are the fastest, as in such materials there is practically no diffusion limitation in the penetration of water vapor into the humidity-sensitive layer. For example, Horzum et al. (2011) have reported that their ZnO nanofiber–based humidity sensors had response and recovery time around 0.5 seconds and 1.5 seconds, respectively.

Composites also show good sensing characteristics at the optimum ratio of components (Zhu et al. 2014; Lv et al. 2017). For example, according to Zhu et al. (2014), for SnO_2/SiO_2 composite, the best mixing molar ratio of Sn/Si was found to be 1:1. In this case, sensors exhibited an excellent linear correlation between frequency drift and humidity in the whole range. In addition, sensors developed have shown good stability and short response and recovery time. The optimizing effect when using composites is generally conditioned by the improvement of adsorption properties of humidity-sensitive materials, achieved through optimization of the structural parameters of the material, an increase in the specific surface area and porosity (Yao and Ma 2014; Chen et al. 2016; Tai et al. 2016). It should be borne in mind that not all additives contribute to the improvement of both these parameters. For example, because of the layered structure of graphene, its addition in polymers does not contribute to the growth of gas permeability; therefore, polymer/graphene-based sensors usually have longer response and recovery time.

Concerning calix[4]arene (Okura et al. 2010; Su et al. 2013) and MOFs (Ameloot et al. 2009; Kosuru et al. 2016), characterized by a nanoporous structure, there was not observed a significant optimizing effect in comparison with traditional humidity-sensitive materials. Only when [Cu$_3$(BTC)$_2$] was used, a significant frequency shift was observed when measuring humidity in the range 0%–45% RH (Ameloot et al. 2009). At the same time, large response and recovery times and large hysteresis (Figure 12.23) were typical for such sensors (Okura et al. 2010; Kosuru et al. 2016). This behavior is consistent with the data obtained for other gas sensors based on these materials. The pore size in such materials ensures capillary condensation, even at low humidity levels, ensuring high sensitivity in this range. However, at such sizes of pores, fast establishment of balance with the surrounding atmosphere is impossible. As it is known, the response rate is governed by the rate of guest diffusion within the pores and by MOF particle size. Song et al. (2010) have shown that diffusion times can be shortened by increasing MOF aperture sizes. In this case, the sensitivity will decrease. In addition, the synthesis technology of MOFs is quite complex and expensive. According to experimental studies and theoretical simulations, the best solution to cover the total humidity range (0%–100%) and to obtain acceptable rate of response is to produce all sizes of pores: from nanopores to macropores.

It should be noted that a significant increase in frequency shift at RH higher than 60%–70% is observed for most polymers and many inorganic materials used in the development of humidity sensors (Hu et al. 2011; Chen et al. 2016; Yao et al. 2017). Moreover, for most of these materials, observed exponential dependence is observed, but not linear dependence (Figure 12.24) of the frequency shift from humidity (Wang et al. 2010; Hu et al. 2011; Yao and Xue 2015; Tai et al. 2016; Yao et al. 2017). The dependence close to linear was observed only by Neshkova et al. (1996) for NPS, Su et al. (2006) for CNTs/Nafion, Zhang et al. (2005) for CNTs, and Lv et al. (2017) for UFR/SiO$_2$-based humidity sensors.

The influence of the thickness of the layer on the sensor performances is another important regularity revealed during the research. It was found that an increase in the thickness or mass of the humidity-sensitive layer promotes an increase in the sensitivity of QCM-based sensors (Figures 12.24 and 12.25), regardless of the nature of the material used (Hu et al. 2011; Lv et al. 2017; Yao et al. 2017). It was also established that the increase in the over-the-optimal value increases the hysteresis at higher RH without improving the sensor performances. According to Sakly et al. (2014), 370 nm is the optimal thickness of the ZnO layer, exhibiting high sensitivity, rapid time responses, small hysteresis, best linearity in the range of 9% to 98%, good long-term stability, and reproducibility. It is understood that the optimum thickness of the humidity-sensitive layer is not a universal value and depends on the material used and the formation technology. For example, for nanowires and nanofiber-based films, this thickness can be much larger.

Research has shown that an increase in the sensitivity is also facilitated by an increase in the porosity of the sensing layers (Yoo and Bruckenstein 2013). This allows water vapor to partition into the interior void

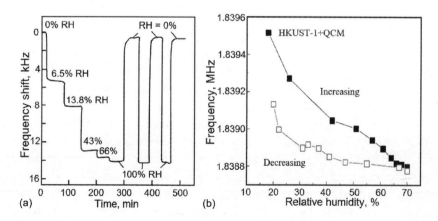

FIGURE 12.23 (a) Signal upon adsorption of water from nitrogen streams at different relative humidity (RH) values, illustrating reversibility and reproducibility of [Cu$_3$(BTC)$_2$]-based humidity sensor. (b) Hysteresis curves of HKUST-1 coated quartz crystal microbalance (QCM) sensors for increasing and decreasing relative humidity. ([a] - Reprinted with permission from Ameloot, R. et al., *Chem. Mater.*, 21, 2580–2582, 2009. Copyright 2009, ACS; [b] - Reprinted from Kosuru, L. et al., *J. Sensors*, 4902790, 2016. Copyright 2016, Hindawi as open access.)

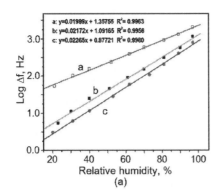
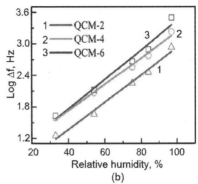

FIGURE 12.24 The dependence of the frequency shifts of (a) polyamic acid (PAA)-based and (b) Black phosphorus (BP)-based quartz crystal microbalance (QCM)-based humidity sensors on relative humidity (RH): 1, 2, and 3 are samples with different thickness or mass. ([a] – Reprinted from Hu, W. et al., Highly stable and sensitive humidity sensors based on quartz crystal microbalance coated with bacterial cellulose membrane, *Sens. Actuators B*, 159, 301–306, 2011. Copyright 2011, Elsevier; [b] – Reprinted from Yao, Y. et al., Novel QCM humidity sensors using stacked black phosphorus nanosheets as sensing film, *Sens. Actuators B*, 244, 259–264, 2017. Copyright 2017, Elsevier.)

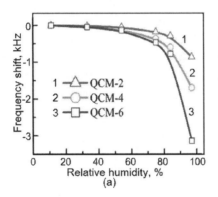
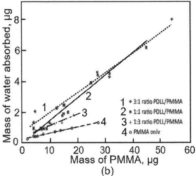

FIGURE 12.25 (a) The frequency response of the black phosphorus (BP)-based quartz crystal microbalance (QCM) humidity sensors as a function of humidity. Deposition mass of BP films: 1, 357 ng; 2, 809 ng; 3, 1102 ng. (b) The water vapor sensitivity dependence on mass of poly(methacrylic acid) (PMAA) in leached mixed polymer films. PDLL, poly(d, l-lactide). ([a] - Reprinted from Yao, Y. et al. Novel QCM humidity sensors using stacked black phosphorus nanosheets as sensing film, *Sens. Actuators B*, 244, 259–264, 2017. Copyright 2017, Elsevier; [b] - Reprinted from Yoo, H.Y. and Bruckenstein, S. A novel quartz crystal microbalance gas sensor based on porous film coatings. A high sensitivity porous poly(methylmethacrylate) water vapor sensor, *Anal. Chim. Acta*, 785, 98–103, 2013. Copyright 2013, Elsevier.)

surfaces rather than partitioning only at the smaller surface of the untreated polymer film (see Figure 12.25b). The increase in the porosity of PMMA films Yoo and Bruckenstein (2013) achieved was by the addition of poly(d,l-lactide) (PDLL), which was then removed from a mixed-polymer film of PMMA/PDLL by using NaOH hydrolysis. In this system, the film sensitivity to water vapor reached a maximum at or near 50% of the leachable component. The slope decreased for a film that had a 75% initial content of PDLL. This effect taken place due to either (1) a collapse the porous structure that is produced by leaching, (2) production of a nonelastic structure, or (3) some combination of both. Another interesting approach to improving sensor parameters was developed by Wang et al. (2014). They suggested to form a humidity-sensitive layer in the form of micropillars (see Figure 12.26). Their modeling and experiment with PMMA micropillars, fabricated using nanoimprint lithography, have shown that ultrahigh QCM sensitivity can be achieved when the height of the micropillar approaches a critical value (15–17 μm).

12.6.1.2 FBAR Humidity Sensors

When considering the basis of the AW sensors' work, it was shown that the sensitivity of such sensors depends on the frequency at which these sensors operate: The higher the frequency was, the higher was the sensitivity.

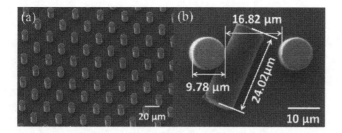

FIGURE 12.26 Scanning electron microscopy (SEM) pictures of pillar-based microstructured surface with a height of 24.02 μm: (a) Side view with a tilt angle of 20° and (b) top view. Pillar is peeled off for the height measurement. (Reprinted with permission from Wang, P. et al., An ultrasensitive quartz crystal microbalance-micropillars based sensor for humidity detection, *J. Appl. Phys.*, 115, 224501, 2014. Copyright 2014, IOP.)

Unfortunately, when using the traditional approach, the increase in the frequency has significant limitations. Typically, the operating frequencies of such sensors do not exceed 10–20 MHz. The experiment showed that micromachining technology can solve this problem. An example of AW sensors manufactured using micromachining technology is FBAR humidity sensors, the parameters of which are given in Table 12.4.

Examples of the configurations and manufacturing technology of FBAR humidity sensors are shown in Figures 12.6 and 12.27a. A 600-nm molybdenum (Mo) and 800-nm high-quality c-axis-oriented aluminium nitride (AlN) were sequentially deposited by the radiofrequency (RF) sputter and patterned to create the bottom electrode and piezoelectric layer. Another 600 nm Mo and 400 nm AlN were sequentially deposited and patterned as the top electrode and passivation layer. A 10 nm/100 nm Cr/Au composite film was then deposited by physical vapor deposition and patterned to form the pads. Finally, phosphosilicate glass (PSG) was etched by diluted hydrofluoric acid to completely release the device. Another version of the FBAR humidity sensor is shown in Figure 12.27b. A sputtered ZnO (1.2 μm) film, deposited by sputtering, acted both as the RH-sensitive layer and the piezoelectric actuation layer for the FBAR sensor. The top and bottom electrodes were made of Au (0.2 μm) and Al (0.2 μm), respectively. The top-view scanning electron microscopy (SEM) image of developed FBAR was shown in the same Figure 12.27b.

From Table 12.4, it can be seen that the use of FBAR structures really increases the sensitivity of humidity sensors by more than 100 times. At that, this sensitivity is achieved even without the use of humidity-sensitive materials (Nagaraju et al. 2014; Zhang et al. 2015). The application of humidity-sensitive materials on the FBAR surface further increases the sensitivity (Liu et al. 2017; Xuan et al. 2017). However, in the case of sensing material that was used by Xuan et al. (2017), this effect was not as significant as in the case of polymers used by Liu et al. (2017). An analysis of the available results shows that, in this case, the same regularities are observed that have been found for traditional QCM-based humidity sensors: Increasing the thickness and specific surface area of the humidity-sensitive materials provides more absorption sites for water molecules and the increase in sensitivity.

It was also found that an increase in the permeability of the upper electrode due to the formation of a microhole array in it also contributes to an increase in the sensitivity of the FBAR sensor (Zhang et al. 2015). The through-hole array provides paths for water vapor to reach the sensitive layer directly. According to Zhang et al. (2015), in this case, sensitivity increased 18 times. As in the case of QCM-based humidity sensors, FBAR humidity sensors show a high temperature dependence (Figure 12.28), and therefore the temperature stabilization during their operation is necessary.

TABLE 12.4
Film Bulk Acoustic Resonator Humidity Sensors' Performance

Sensing Material	Frequency, GHz (platform)	RH Range, %	Sensitivity, kHz/%RH	Res/rec Time, s	References
Uncoated	1.43 (ZnO)	25–85	2.2–8.5	n.a.	Qiu et al. (2010a)
Uncoated	1.5 (ZnO)	15–70	≈ 13	n.a.	Nagaraju et al. (2014)
Uncoated	1.14 (ZnO)	30–85	21	n.a.	Zhang et al. (2015)
GO	1.25 (ZnO)	5–85	5	n.a.	Xuan et al. (2017)
PSS/PDDA	1.44 (AlN)	1–75	42	n.a.	Liu et al. (2017)

GO, oxidized graphene; n.a., not applicable; PDDA, poly(diallyldimethytlammonium choride); PSS, poly(sodium 4-styrenesulfonate); RH, relative humidity.

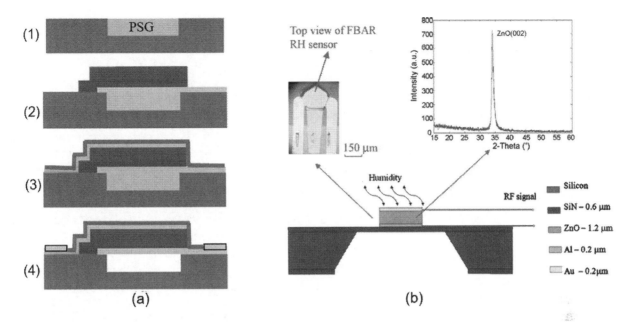

FIGURE 12.27 (a) Schematic illustrating the fabrication process of film bulk acoustic resonator (FBAR): (1) etching of air cavity and deposition of phosphosilicate glass (PSG); (2) deposition of bottom electrode and piezoelectric layer; (3) deposition of top electrode and passivation layer; (4) deposition of Au pads and release of PSG and (b) Schematic cross-sectional structure of the FBAR relative humidity (RH) sensor with a photograph of the top view of a fabricated device on the left and an X-ray diffraction (XRD) trace of the ZnO film illustrating that it had (002) crystal orientation on the right. ([a] - Reprinted with permission from Liu, W. et al., A highly sensitive humidity sensor based on ultrahigh-frequency microelectromechanical resonator coated with nano-assembled polyelectrolyte thin films, *Micromachines*, 8, 116, 2017. Copyright 2017, MDPI as open access; [b] - Reprinted from Qiu, X., et al., Experiment and theoretical analysis of relative humidity sensor based on film bulk acoustic-wave resonator, *Sens. Actuators B*, 147, 381–384, 2010. Copyright 2016, Elsevier.)

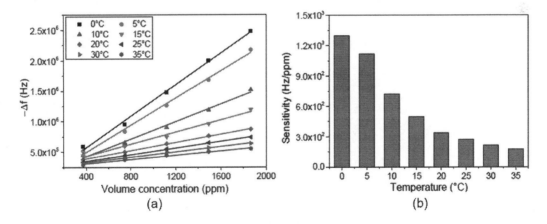

FIGURE 12.28 Temperature dependence of polystyrene sulfonic sodium (PSS)/PDDA-coated film bulk acoustic resonator (FBAR) for humidity sensing: (a) Frequency shift with respect to both the environmental temperature ranging from 0°C to 35°C and the volume concentration of water vapor changing from 373 ppm to 1866 ppm, and (b) sensitivity of poly(vinylidene fluoride) (PET)-coated FBAR with the environmental temperature changing from 0°C to 35°C. (Reprinted with permission from Liu, W. et al., *Micromachines* 8, 116, 2017. Copyright 2017, MDPI as open access.)

12.6.2 SAW-Based Humidity Sensors

12.6.2.1 General Consideration

Humidity-sensitive materials and performances of SAW-based sensors are listed in Table 12.5. It is seen that SAW devices, even without humidity-sensitive films, are sensitive to water vapor (Jasek and Pasternak 2013). According to Dorjin and Simakov (2002), this sensitivity increases with the electromechanical coupling coefficient of piezoelectric substrates. It is also seen that

TABLE 12.5
Performances of Surface Acoustic Wave–Based Humidity Sensors

Type of Sensing Material	Sensing Materials	f_0, MHz	Average Sensitivity f shift kHz/%RH	Average Sensitivity Loss shift dB/%RH	Res. Time, s	Rec. Time, s	Reference
Uncoated	–	120–433	0.02–0.8	0.013	n.a.	n.a.	Kwan and Sit (2012); Balashov et al. (2012, 2015); Hong et al. (2012); Irani and Tunaboylu (2016)
Polymers	PVA	160	460	n.a.	≈ 150	300	Penza and Anisimkin (1999); Penza and Cassano (2000); Chen and Kao (2006); Lieberzeit et al. (2009); Buvailo et al. (2011b); Balashov et al. (2012); Liu and Wang (2014); Lu et al. (2016)
		420–433	0.5–30	0.17–0.7	2–?	5–?	
	BCB; cellulose; PANI; PANi/PVB; PEO; PI; PMMA; PPA; PVB; PVDF; PVP	30–75	0.1–1.3	0.025	10–20	10	Joshi and Brace (1985); Caliendo et al. (1993); Nomura et al. (1994); Buvailo et al. (2011b); Liu et al. (2011); Lin et al. (2012); Wang et al. (2013)
		165–300	0.17–6.1	0.2–0.26	n.a	n.a	
		433	0.65–75	0.7	1.5–?	2.5–?	
Inorganic materials	Ag NPs;	124	0.7–17	0.004–0.04	n.a.	n.a.	Li et al. (2014); Irani and Tunaboylu (2016); Kwan and Sit (2012); Mittal et al. (2015); Tang et al. (2015); Caliendo et al. (1997); Buvailo et al. (2011a); Hong et al. (2012); Guo et al. (2013); Tang et al. (2015); Liu et al. (2015)
	Ag NWs;	433	0.1	n.a			
	SiO_2	123–200	0.1–35	n.a.	10	20	
	Al_2O_3	433	1.2	n.a.	1	3	
	CuO; TiO_2; ZnO; TiO_2/LiCl; NiO;	126–250	0.5–15	1.0	1–10	1–20	
		1540	80	n.a.	23	4	
Carbon	Graphene	160–250	1.5–53	n.a.	1–45	20	Balashov et al. (2012, 2015); Guo et al. (2014); Xuan et al. (2014); Nikolaou et al. (2016)
Composites	BSA-Au; PET-Si;	57–108		0.12–0.16	< 1		Li et al. (2010; 2011b); Rimeika et al. (2017)
	CNTs/Nafion;	433–500	0.1–420	n.a	3–10	3–10	Lei et al. (2011a)
Other	MOF (Cu-BTC)	97	n.a.	n.a.	> 1 h	> 1 h	Robinson et al. (2012)

BCB, benzocyclobutene; CNT, carbon nanotubes; MOF, metal organic framework; NP, nanoparticles; n.a., not applicable; NW, nanowires; PANi, polyaniline; PEO, poly(ethylene oxide); PET, polyelectrolyte; PI, polyimide; PMMA, poly(methyl methacrylate); PPA, polyphenylacetylene; PVA, polyvinyl-alcohol; PVB, polyaniline/poly(vinyl butyral); PVDF, polyvinylidene difluoride; PVP, polyvinyl pyrrolidone; RH, relative humidity.

the deposition of a humidity-sensitive layer strongly increases the response of SAW sensors. Experiment has shown that, for this purpose, one can use the same groups of materials (polymers, inorganic and carbon-based materials, and composites) as in the manufacture of QCM-based sensors. As a piezoelectric platform in the manufacture of humidity sensors, mainly used were $LiNbO_3$ (Penza and Anisimkin 1999; Guo et al. 2013) and quartz (Kwan and Sit 2012). ZnO (Xuan et al. 2014, 2015) and AlN (Hong and Chung 2010; Hong et al. 2012) are also used, but much less often. However, Caliendo (2003) and Hong and Chung (2009) believe that the use of AlN films as piezoelectric substrates has been proven to be beneficial for SAW applications because of

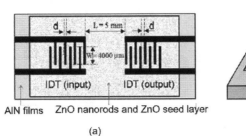
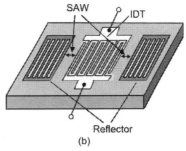

FIGURE 12.29 (a) Schematic diagram of the surface acoustic wave (SAW) humidity sensor ($f_0 = 129$ MHz) developed by Hong et al. (2012). Interdigital transducers (IDTs) of 50 finger pairs had a period of electrodes (d) of 10 μm; the aperture (W) was 100λ = 4000 μm, and the IDT-IDT gap (L) was 5 mm. (b) Schematic figure of the single-port surface acoustic wave resonator (SAWR) structure used by Lei et al. (2011) as the basic component for humidity detection ($f_0 = 433$ MHz): interval of the IDT and reflectors D, 1.8 μm; pairs of IDT fingers, 100; number of reflectors on each side, 300; acoustic aperture W, 730 μm; transmission distance L, 9.1 μm. ([a] - Reprinted from Hong, H.S. et al., High-sensitivity humidity sensors with ZnO nanorods based two-port surface acoustic wave delay line, *Sens. Actuators B*, 171–172, 1283–1287, 2012. Copyright 2012, Elsevier.)

the films' high acoustic velocity, superior temperature, chemical stability, and good piezoelectric properties. In addition, an AlN/Si structure has been found suitable for processes relating to post-heat treatment, which improves the quality of metal oxide sensing layer due to the stability of SAW properties of the thin film at high annealing temperatures (up to 600°C).

Typical configurations of SAW-based humidity sensors are shown in Figure 12.29. The operating frequencies of SAW-based humidity sensors ranged from 30–40 MHz (Nomura et al. 1994; Penza and Anisimkin 1999) to 1.5 GHz (Liu et al. 2015). However, the main research was carried out with sensors designed for frequencies of 160 and 430 MHz. It is also evident that SAW-based humidity sensors, due to the use of higher frequencies, have significantly higher sensitivity in comparison with QCM-based sensors. The experiment also showed that when the RH increased, the resonance frequency decreased, and the resulting peaks became broader and broader. The maximum RH (i.e., the maximum RH that the sensor can detect) corresponds to the condition under which there are no detectable peaks within the interested frequency band at higher RHs. For example, according to data presented in Figure 12.30, the maximum RH that the poly(vinylalcohol) (PVA)-based sensor developed by Lu et al. (2016) can detect is 72.8%.

As was noted earlier, the SAW-based sensors can operate in different modes. At present, Rayleigh-type SAW devices are commonly used. However, the Love wave can also be used for sensing applications (Guo et al. 2013; Liu and Wang 2014; Nikolaou et al. 2016). Unlike the commonly used Rayleigh-type SAW device, the performance of a Love-wave device depends on its

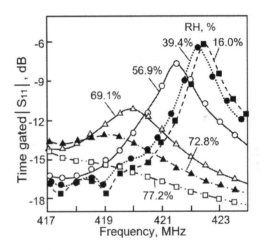

FIGURE 12.30 Measurement results of RH sensor with PVA film loaded on the reflectors of SAW resonator with time gate (Data extracted From Lu, D. et al., *IEEE Sensors J.*, 16, 13–14, 2016.)

guiding layer rather than on IDT structures and substrate characteristics (see Table 12.6). A suitable guiding layer is the key to implement a Love-wave sensor with high mass sensitivity, good temperature stability, and acceptable insertion loss. Due to its low acoustic loss and excellent abrasion resistance, SiO_2 is the most used guiding layer material for Love-wave devices. Unfortunately, a Love-wave sensor incorporating an SiO_2 layer cannot achieve a very high sensitivity, because of the fast transverse AWs in SiO_2. Due to their low shear waves, polymers are often adopted as guiding layers for Love-wave sensors to get a higher sensitivity. The shortcoming of the polymeric layer is the large loss caused by the viscosity of polymers. Based on the perturbation

TABLE 12.6
Examples of Novel and Conventional Surface Acoustic Wave Humidity Sensors, and theirs Characteristics

Authors and Years	Sensors	Guiding Layers	Sensitive Membrane	Production Methods	Resonant Frequency	LOD and Sensitivity	Significance
Takeda and Motozawa (2012)	Quartz ball SAW sensor (Φ 1.0 mm)	–	–	Maskless exposure technology	160 MHz	1.0 µmol/mol; 36–1800 µmol/mol	New and excellent hygrometer candidate
Takayanagi et al. (2014)	Quartz ball SAW sensor (Φ 3.3 mm)	SiO_x	SiO_x film	Subtractive; sol-gel; ion beam deposition	F: 80 MHz; H: 240 MHz	0.035–6.0×10^3 µmol/mol	Detection in a wide range from –95 to 0°C in frost point (<1 µmol/mol)
He et al. (2013)	ZnO/ polyimide (100 µm)	ZnO	ZnO NPs	dc magnetron sputtering (PVD)	132 MHz (12 µm)	5% RH~ 87%RH; 34.7 kHz/10%RH	Great potential for applications in flexible electronics, sensors, and microsystems
Xuan et al. (2014)	GO/ZnO/glass	ZnO	GO	–	225 MHz (12 µm)	0.5% RH~ 5% RH; ~265 kHz/5%RH	Great potential for application in broad humidity range with high-speed detection
Liu and Wang, (2014)	PVA/ST-90°X quartz	PVA layer	PVA layer	Sol-gel	178.15 MHz (28 µm)	$2.845\ 10^{-7}$ kg/m^2; $\Delta f = 5.32$ kHz	Dynamics and humidity response based on polymer-coated Love-wave device
Tang et al. (2015)	SiO_2/ST-cut 42°75' quartz	SiO_2	Porous SiO_2	Sol-gel	197 MHz (16 µm)	520 kHz/63%RH	Beneficial to the practical application due to its good reproducibility and stability
Li et al. (2015)	PNIPAM/128° Y-X $LiNbO_3$	PNIPAM hydrogel	PNIPAM hydrogel	Synthesis	433 MHz	0.13 dB/%RH at 50% RH; 2dB/%RH at 98% RH	Multisensing capabilities and passive wireless features attractive for harsh environmental applications

F, fundamentals; H, harmonic; GO, Graphene oxide; LOD, limit of detection; PNIPAM, poly(N-isopropylacrylamide); RH, relative humidity; ST-cut, stress- and temperature-compensated.

Source: Reprinted from Xu Z. and Yuan Y.J., Implementation of guiding layers of surface acoustic wave devices: A review, Biosens. Bioelectron., 99, 500–512. Copyright 2018, with permission from Elsevier.

approach, Martin et al. (1994) investigated the dynamics and response of polymer-coated SAW devices with acoustically thin and thick layers. They committed that the changes in velocity and attenuation caused by the viscoelastic film were related to the surface mechanical impedances contributed by the film. According to Liu and Wang (2014), for Love waves in a polymer-coated device, the viscosity of the polymeric layer will produce deviations in propagation velocity and mass sensitivity. The real part and imaginary part of the mass sensitivity correspond to the mass velocity sensitivity and mass loss sensitivity, respectively.

Regarding the regularities of the influence of humidity-sensitive materials on the parameters of the sensors detected during the SAW-based sensors testing, they differ little from the regularities established for QCM-based humidity sensors. Let us consider the most important of them.

However, when comparing the parameters of different sensors, it must be borne in mind that, for the SAW sensor coated with thin film, according to mass-loading mechanisms (Sauerbrey 1959; Fan et al. 2012), the frequency shift (Δf) is directly proportional to square of the fundamental frequency of the SAW device (Equation 12.9). This means that even by simply increasing the frequency from 160 to 430 MHz, the sensitivity of SAW sensors can be increased by a factor of 7.

$$\frac{\Delta f}{f_0} = -C_m f_0 \Delta\left(\frac{m}{A}\right) \text{ or } \Delta f \propto f_0^2 \qquad (12.9)$$

where C_m is the coefficients of mass sensitivity of the substrate, respectively, (m/A) is the change in mass per unit area, and f_0 is the fundamental frequency of the SAW device.

12.6.2.2 Polymer SAW-Based Humidity Sensors

Of the polymers, the most studied are PVA. PVA is a hygroscopic polymer of high molecular weight with a glass transition temperature of approximately 70°C, and it has an –OH group bonded to each carbon in the backbone chain (Penza and Cassano 2000). The main conclusion of these studies is that the parameters of the sensors are determined by the technologies of synthesis and the formation of the humidity-sensitive layer. From the data given in the Table 12.5, it can be seen that the sensitivity of PVA-based humidity sensors made in different laboratories can differ by more than two orders of magnitude. The developers note that under optimal conditions of formation of PVA films, the SAW reaction of the sensor shows high sensitivity towards RH, excellent repeatability of the response, quite good water-resistance, and low hysteresis at room temperature. The most sensitive were sensors manufactured by Penza and Cassanor (2000). Sensors with f_0 = 433 MHz when humidity changed from 20% to 75% RH demonstrated a frequency change of approximately 25 MHz (see Figure 12.31a). In addition, the SAW PVA-based sensors showed a negligible cross-sensitivity toward gases such as NH_3, NO_2, CO, and H_2 (Penza and Anisimkin 1999; Buvailo et al. 2011b). This conclusion, of course, does not apply to

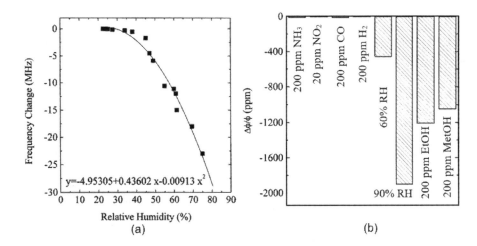

FIGURE 12.31 (a) Surface acoustic wave (SAW) relative humidity (RH) calibration curve of poly(vinylalcohol) (PVA) film, at room temperature: Substrate, ST-cut (stress- and temperature-compensated) quartz; f_0 = 468 MHz; film thickness ≈ 1 μm); and (b) Pattern of the phase response of SAW PVA-based sensor at room temperature (f_0 = 42 MHz). Thickness of PVA film ≈ 1 μm. (Adapted from Penza, M. and Anisimkin, V.I., Surface acoustic wave humidity sensor using polyvinyl-alcohol film, *Sens. Actuators*, 76, 162–166, 1999. Copyright 1999, Elsevier.)

organic vapor molecules (ethanol, methanol) containing OH group (Figure 12.31b). True, high sensitivity to the vapor of the organic solvents is inherent to all polymers.

The 90% response time for sensors developed by Penza and Cassanor (2000) was approximately 3 minutes. Buvailo et al. (2011b) demonstrated that, with a corresponding change in the formation technology and a decrease in the thickness of the PVA layers ($d \approx 510$ nm), the response time can be reduced to 2 seconds. The recovery took a longer period of time, being approximately 5 seconds for 80% recovery of the signal with a long "tail" of an additional 35 seconds until full recovery (Figure 12.32a).

As for other polymers, a limited amount of experimental data does not allow for any fundamental conclusions. In addition, most works do not provide an objective comparison of materials used in identical conditions. With this regard, note the results given by Joshi and Brace (1985), Buvailo et al. (2011b), and Lin et al. (2012). Joshi and Brace (1985) compared polyimide (PI 25451) and cellulose ester polymers and made a conclusion that, in addition to well-established fabrication technology, an adequate humidity sensitivity in all RH range and rapid response (6 s), polyimide had the advantage of high mechanical strength, high temperature capability, and resistance to chemical attack. Unfortunately, a large hysteresis (approximately ± 8% RH) was found in the humidity response of polyimide-coated devices. The use of cellulose ester polymers is found to reduce the hysteresis at the same speed of response. Buvailo et al. (2011b) studied PVA and popyvinylpyrrolidone (PVP)-based humidity sensors in the range of RH from 5% to 95% and concluded that, for the same sensitivity, the PVP-based sensors were faster. These sensors showed the response and recovery time around 1.5 and 2.5 seconds, respectively (Figure 12.32b). This was the fastest response for polymer sensors. However, it should be noted that the humidity-sensitive layer in PVP-based sensors (d[PVP] = 150 nm) was 3.5 times thinner than in PVA-based sensors (d[PVA] = 510 nm). Buvailo et al. (2011b) believe that another reason of fast response of PVP-based sensors is more pronounced hydrophobic nature of PVP polymer compared to PVA that causes weaker water molecule bonding to the polymer surface and, as a result, more "flexible" response behavior. Considering such a strong influence of the thickness of the layer of humidity-sensitive material on the response time, it becomes clear that it is not necessary to expect good performance from sensors, having a humidity-sensitive layer of 10 μm in thickness (Liu et al. 2011).

Lin et al. (2012) have shown that a quick response and high sensitivity can also be achieved by using a nanofiber-based membrane instead of a continuous layer. They compared the performance of SAW humidity sensors developed using electrospun nanofibers of poly(methyl methacrylate) (PMMA), poly(vinyl pyrrolidone) (PVP), poly(ethylene oxide) (PEO), poly(vinylidene fluoride) (PVF), and poly(vinyl butyral) (PVB) and established that, as expected, higher humidity sensitivity was observed for nanofiber sensors based on polymers with high hydrophilicity (i.e., PEO and PVP) (see Figure 12.33a). However, the sensors showed poor sensing linearity and revealed quite low sensitivity at low humidity levels. Moreover, the damaged morphology of the nanofibers after experiencing humid environment suggested their low water durability, and consequently,

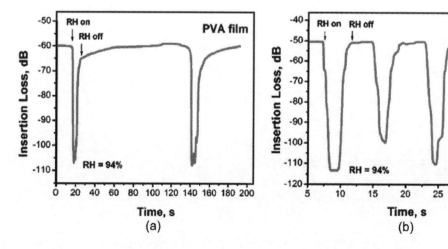

FIGURE 12.32 Surface acoustic wave (SAW) response dynamics of poly(vinylalcohol) (PVA)-based sensor (a) and PVP-based sensor (b), measured for a step from 5% to 94% relative humidity (RH). $f_0 = 250$ MHz, d(PVA) = 510 nm; d(PVP) = 150 nm. (Adapted with permission from Buvailo, A.I., et al., Thin polymer film based rapid surface acoustic wave humidity sensors, *Sens. Actuators B*, 156, 444–449, 2011. Copyright 2011, Elsevier.)

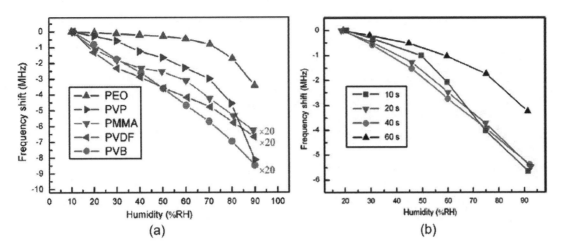

FIGURE 12.33 (a) Frequency response to humidity of surface acoustic wave (SAW) sensors based on electrospun nanofibers of polymethyl methacrylate (PMMA), polyvinyldifluorene (PVDF), PVB, polyethylene oxide (PEO), and PVP; (b) Effect of collection time on the frequency response to humidity of SAW sensors based on electrospun PANi/PVB nanofibers (mass ratio of PANi to PVB: 2/3) (f_0 = 433 MHz; measurement temperature: 30°C). (Reprinted from Lin, Q. et al., Highly sensitive and ultrafast response surface acoustic wave humidity sensor based on electrospun polyaniline/poly(vinyl butyral) nanofibers, *Anal. Chim. Acta*, 748, 73–80, 2012. Copyright 2012, Elsevier.)

unsatisfying stability. In contrast, nanofiber sensors based on relatively hydrophobic PMMA, PVDF, and PVB displayed much lower sensitivity, with frequency shifts of 0.312 MHz, 0.334 MHz, and 0.425 MHz from 90% RH to 20% RH, respectively. But the linearity of the sensing curves was improved. A fairly good linearity, especially in the low-humidity region, was demonstrated by PANi-based humidity sensors (Wang et al. 2013). Lin et al. (2012) supposed that the use of PANi/PVB nanofibers will combine the best qualities of PANi and PVB (Figure 12.33b). PVB has high hydrophilicity, while PANi is an environmentally stable conducting polymer with excellent electrical, magnetic, and optical properties. It has attracted considerable attention over the past 10 years and is generally regarded as a conducting polymer with very high potential in various commercial applications. Studies have confirmed this assumption. PANi/PVB nanofiber–based sensors exhibited very high sensitivity of approximately 75 kHz/%RH from 20% to 90% RH, ultrafast response (1 s and 2 s for humidification and desiccation, respectively) and good sensing linearity. Furthermore, the sensor could detect humidity as low as 0.5% RH. According to Lin et al. (2012), increased sensitivity of PANi/PVB-based sensors in the range of low humidity is conditioned by a change in the conductivity of these films, which can vary by several orders of magnitude (Li et al. 2010a). Thus, the composite nanofibers could exhibit electroacoustic load with humidity variation, while such an effect did not exist in the sensor based on nonconductive nanofibers.

Lei et al. (2011) have shown that even higher sensitivity with excellent linearity of characteristics and good performance can be achieved by using multi-walled carbon nanotubes (MWCNTs)/Nafion nanofibers. Nafion is a sulfonated tetrafluoroethylene based fluoropolymer-copolymer, which is extremely resistant to chemical agent attacks. The sulfonic acid groups ($-SO^{3-}-H^{+}-$) in Nafion have a very high water-of-hydration potential; thus, Nafion absorbs water particles very efficiently and then possesses ionic conductivity. The value of its conductivity is a function of the moisture content and ambient temperature. At the same time, CNT is a material with high surface area for gas adsorption, provided by the central hollow cores and the outside walls. The results of testing the developed sensors are shown in Figure 12.34. The entire response curve showed quite a perfect sensitivity up to 427.6 kHz/%RH with excellent linearity in the wide range from 10% to 80% RH. In addition, the humidity hysteresis of the present sensor was no more than 1% RH. The result indicated that the sensor had quite a short response time (< 3 s for 63% change) in both the adsorption and desorption process. According to Lei et al. (2011), the improvement of sensor performance is the result of the following factors: (1) special electrical properties of Nafion and MWCNTs and conductivity variance due to influence of humidity, and (2) optimization of the structure; the addition of MWCNTs provides a porous 3D structure. With the aid of MWCNTs, the composite nanofibers could have a larger specific surface area and thus a greater number

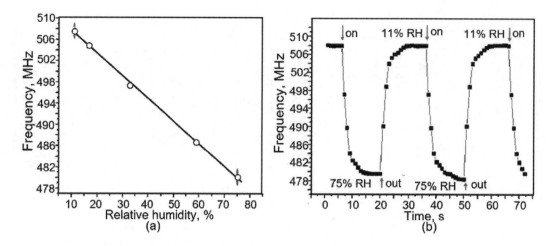

FIGURE 12.34 (a) Sensitivity test result of the MWCNTs/Nafion-based surface acoustic wave (SAW) humidity sensor: Substrate, ST-cut (stress- and temperature-compensated) quartz crystal; pairs of interdigital transducer (IDT) fingers, 100; number of reflectors on each side, 300; acoustic aperture $W = 730$ μm; transmission distance $L = 9.1$ μm ($f_0 = 433$ MHz). The electrospinning method was used to deposit the Nafion/MWCNTs composite nanofibers onto the surface of surface acoustic wave resonators (SAWRs) for humidity sensing; and (b) Dynamic response curve of the present sensors. (Data extracted from Lei, S. et al., *Nanotechnology*, 22, 265504, 2011.)

of active spots for water absorbing. This structure could also make the best use of the Nafion's characteristic of being highly selective and highly permeable to water vapors. However, the MWCNTs must be controlled to be under a relatively small proportion in the composite material, or the sensitivity may decrease, because Nafion in this material was mainly for absorbing water vapor.

12.6.2.3 SAW Humidity Sensors Based on Inorganic Humidity-Sensitive Materials

Inorganic materials used in the development of SAW-based sensors include metal nanoparticles (Li et al. 2014), dielectric oxides, Al_2O_3 and SiO_2 (Kwan and Sit 2012; Mittal et al. 2015; Tang et al. 2015), semiconductor metal oxides such as ZnO, CuO, NiO, TiO_2 (Hong et al. 2012; Tang et al. 2015), and graphene oxide (GO) (Balashov et al. 2012, 2015; Xuan et al. 2014). Just as in the case of PVA, there is a very large sensor's parameter spread, even for the same material (see Table 12.5). For example, for Ag-based films at RH of 80%, the sensitivity varies from 0.08 to 23 kHz/% RH, demonstrating maximum sensitivity for films formed by Ag nanoparticles (Li et al. 2014) and minimal sensitivity for films formed by Ag nanowires (Irani and Tunaboylu 2016). Approximately in the same range, 1.5–53 kHz/% RH, the sensitivity of graphene-based humidity sensors changes. However, for GO-based sensors, it was found that the maximum sensitivity of 53 kHz/% RH ($f_0 = 225$ MHz) was observed when GO was applied by drop-casting on the whole surface, including IDTs area, and a coating thickness of 200–300 nm (Xuan et al. 2014). When the thickness of the coating was reduced to 70–90 nm, the sensitivity decreased to 11 kHz/% RH.

For inorganic materials, as well as for most polymers (Penza and Cassanor 2000), a strong nonlinearity of characteristics (Figure 12.35a) is characteristic when the main changes occur at RH above 30%–40% (Li et al., 2014; Xuan et al, 2014; Mittal et al 2015). This means that these sensors may have insufficient sensitivity for use at low humidity levels (< 10% RH), which is of great interest for certain industrial applications. However, most SAW humidity sensors have high sensitivity, even in the low RH region (Kwan and Sit 2012; Mittal et al. 2015). Moreover, as a rule, in the region of low humidity, a frequency shift has a linear or close-to-linear dependence on humidity (Hong et al., 2012; Irani et al., 2016; Nikolaou et al., 2016). An example of such a relationship is shown in Figure 12.35b. Typically, this behavior is due to the fact that, in this area, the water molecules are primarily physisorbed onto the available active hydrophilic groups and vacancies of the oxide surface through double hydrogen bonding, which are called the *first-layer physisorption* of water. The deviation from linearity begins when the physisorption changes from monolayer to multilayer (read Section 11.3).

In terms of speed of response, the most rapid response and recovery is demonstrated by sensors based on Al_2O_3 (Mittal et al. 2015). Under optimal production conditions, the response and recovery time did not exceed 1 and 3 seconds, correspondingly. Al_2O_3, used in these sensors, was synthesized by sol-gel method. The films were

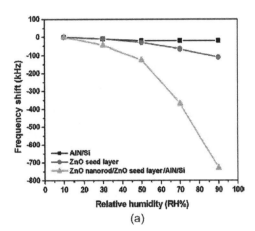 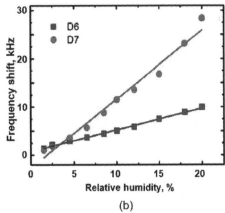

FIGURE 12.35 (a) Relationship between the frequency shift of different surface acoustic wave (SAW) structures on AlN/Si platform and the corresponding relative humidity (10%–90% RH) at room temperature (25°C) ($f_0 = 126$ MHz): 1, ZnO seed layers ($d \approx 0.25$ μm), and 2, -ZnO nanorods/ZnO seed layers ($d \approx 0.8 + 0.25$ μm); and (b) Frequency shift of the SAW humidity sensors with drop-casting graphene oxide (GO) layer at low humidity range from 0.5% to 20% RH: $f_0 = 225$ MHz, drop-casting on the whole surface; D6 – $d \approx 70$–90 nm; D7- $d \approx 100$–130 nm. ([a] – Reprinted with permission from Hong, H. S. et al., High-sensitivity humidity sensors with ZnO nanorods based two-port surface acoustic wave delay line, *Sens. Actuators B*, 171–172, 1283–1287, 2012. Copyright 2012, Elsevier; [b] – Reprinted with permission from Xuan, W. et al., Fast response and high sensitivity ZnO/glass surface acoustic wave humidity sensors using graphene oxide sensing layer. *Sci. Rep.*, 4, 7206, 2014. Copyright 2014, Nature as open access.)

formed by screen-printing technology. They were thin (50 nm) and highly porous. Apparently, this ensured the high speed of these sensors. Tang et al. (2015) showed that other metal oxide sensors with a thickness of 70–80 nm were also high-speed ones (Figure 12.36). At the same time, MOFs-based humidity sensors were the slowest. To establish equilibrium in such sensors, it takes more than an hour (Robinson et al. 2012). The reasons for this behavior were considered earlier. Naturally, such sensors cannot be used for in situ measurements. Heating to certain temperatures allows to quickly remove adsorbed water and regenerate the sensor. But this complicates the operation of the sensor and dramatically increases the power consumption. Most of the other humidity-sensitive materials provided the ability to measure humidity with a response time of 1 to 45 seconds, and no additional heat treatment was required to restore them.

The strong dependence of the sensitivity on the porosity of the films is also inherent to SAW-based humidity sensors. Experiment has shown that a high porosity is beneficial for high sensitivity of the SAW sensors. For example, Tang et al. (2015) have found that, among the sensors developed by them, the sensors with the highest porosity possessed the greatest sensitivity. It is high porosity (53%) that allowed using SiO_2 films with a larger thickness without significant loss in speed (see Figure 12.37a). It should be said that the increase in sensitivity with an increase in the thickness of the humidity-sensitive layer is a generally established regularity for mass-sensitive sensors. Kwan and Sit (2012), by increasing the thickness of SiO_2 films from 200 to 700 nm, increased the sensitivity by a factor of 10 (Figure 12.37b). At 75% RH, the sensitivity for a device with 705 nm of GLAD SiO_2 film was 65 kHz/%RH. It is important to

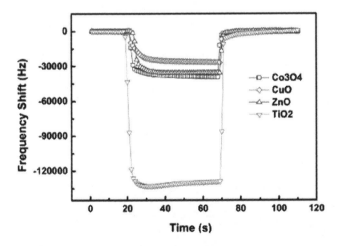

FIGURE 12.36 Dynamic responses of SAW humidity sensors ($f_0 = 200$ MHz) based on semiconducting oxide films when the humidity changed from 30% to 93%. The metal oxides were synthesized by the sol-gel method and had a thickness of 70–80 nm. Semiconducting oxide sols were coated onto the entire quartz substrates using a spin-coating method. (Reprinted with permission from Tang, Y. et al., Highly sensitive surface acoustic wave (SAW) humidity sensors based on sol–gel SiO_2 films: Investigations on the sensing property and mechanism, *Sens. Actuators B*, 215, 283–291, 2015. Copyright 2015, Elsevier.)

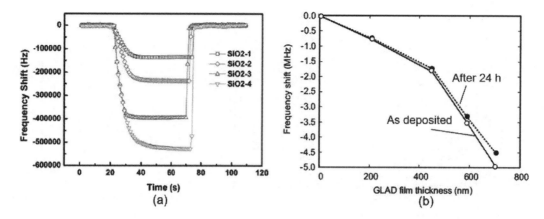

FIGURE 12.37 (a) Dynamic responses of surface acoustic wave (SAW) humidity sensors ($f_0 = 200$ MHz) based on SiO_2 films with different thickness when the humidity changed from 30% to 93%: 1, 70 nm; 2, 130; 3, 200; 4, 260 nm; and (b) Relationship between SiO_2 film thickness and shift in the operating frequency when the humidity changed from 0% to 90% ($f_0 = 120$ MHz). The glancing angle deposition (GLAD) technique was used for deposition of SiO_2 films. ([a] - Reprinted with permission from Tang, Y. et al., Highly sensitive surface acoustic wave (SAW) humidity sensors based on sol–gel SiO_2 films: Investigations on the sensing property and mechanism, *Sens. Actuators B*, 215, 283–291, 2015. Copyright 2015, Elsevier; [b] - Reprinted with permission from Kwan, J.K. and Sit, J.C., High sensitivity Love-wave humidity sensors using glancing angle deposited thin films, *Sens. Actuators B*, 173, 164–168, 2012. Copyright 2012, Elsevier.)

note that this sensitivity was achieved at $f_0 = 120$ MHz. GLAD SiO_2 films, as well as the films used by Tang et al. (2015), had a high porosity (see Figure 12.38). Unfortunately, the authors did not provide data on the dynamics of the response of these sensors, and therefore we do not know whether their speed of response was retained with such an increase in the film thickness (i.e., we do not know how optimal was the thickness of 705 nm in terms of sensitivity/speed-of-response ratio). As it was said earlier, with reference to QCM devices, for each material and technology used, there is an optimal thickness of humidity-sensitive material. For example, with respect to SAW sensors coated with a Cu-BTC-film deposited by a layer-by-layer method, the optimum thickness was 200 nm (Robinson et al. 2012). Robinson et al. (2012) reported that thicker coatings add mass at a distance from the surface that does not couple well with the surface AWs but continue to dampen the energy nonetheless. In addition, 60- and 100-cycle SAWs showed progressively longer equilibration times, with the 100-cycle SAWs failing to level off after 4–6 hours of constant exposure. Guo et al. (2013) have established that the optimal thickness for ZnO films deposited by sputtering was 250 nm. For thicker ZnO films, the frequency shift strongly decreased. Guo et al. (2013) believe that, for Love-wave devices, the sensitivity decreases with increasing thickness due to the decrease in the electromechanical coupling coefficient.

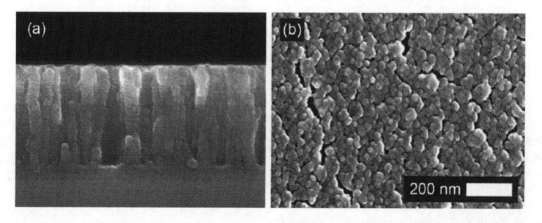

FIGURE 12.38 (a) Cross-sectional and (b) top-down scanning electron microscopy (SEM) images of a 450 nm thick SiO_2 glancing angle deposition (GLAD) film with a 30 nm solid underlayer. Scale bar is applicable to both (a) and (b). (Reprinted with permission from Kwan, J.K. and Sit, J.C., High sensitivity Love-wave humidity sensors using glancing angle deposited thin films, *Sens. Actuators B*, 173, 164–168, 2012. Copyright 2012, Elsevier.)

12.6.2.4 Stability of SAW Humidity Sensors

Many authors (Lieberzeit et al. 2009) indicate a high stability of SAW sensors parameters (see, for example, Figure 12.39). However, some of them, for demonstrating stability, choose conditions with a low level of humidity. For example, such stability test of PVA-based sensors was conducted by Penza and Cassanor (2000) at a RH of 60%, and by Lei et al. (2011b) at 75% RH. Experimental confirmation that the drift of SAW sensors is possible can be found in Kwan and Sit (2012). Kwan and Sit (2012) observed the aging of the SiO_2-based sensor (frequency drift) at high humidity levels (73% RH), while Tang et al (2015) did not observe any changes when using SiO_2 films even at higher levels (93% RH). This means that, in many cases, instability is associated with the features of synthesis and the absence of treatments that stabilize the properties of oxides (Korotcenkov and Cho 2012).

It should be noted that the instability of SAW sensors may be due not only to the instability of humidity-sensitive materials, but also to the instability of the properties of the electrodes used in these devices (Wang et al. 2013). At present, the two-port SAW resonators with aluminum (Al) electrodes are widely used as the frequency feedback element due to their high electrical quality factor (Q) value and low insertion loss over the delay line patterns, resulting in excellent noise immunity and high measurement resolution and accuracy (Chang et al. 1991; Mauder 1995). However, such sensor systems still suffer from a major problem, in that if the sensors are operated in chemically reactive gas-phase environments, the Al electrode structure of the sensor device is easily attacked by the detected gas or gas mixture, which forms corrosive acids or bases with the humidity of the ambient air. The problem is further aggravated if the sensing polymer film on the device surface greatly increased the amount of adsorbed agent and moisture coming in contact with the electrode structure. As a result of that, the sensor performance degrades, and the device electrode structure is easily destroyed. The solution to such problems is the implementation of SAW resonant devices using corrosion-proof electrodes of gold (Au) or platinum (Pt) (Lu et al. 2016). Very impressive results on low-loss resonator filters using heavy metals in their electrode pattern, including Au, have been recently reported by Kadota et al. (2004). Unfortunately, these devices use the shear-horizontal leaky SAW mode, which does not operate as well with the soft polymer films required for high gas and humidity sensitivity as the Rayleigh SAW (RSAW) mode does (Avramov et al. 2002). Recently, an Au-RSAW, two-port SAW resonator, operating at 433 MHz with a typical loaded Q as high as 5,000 and insertion loss in the −8 to −10 dB range in the uncoated state, have been reported for gas sensing (Avramov et al. 2005). However, except for a substantial increase in production cost, much higher velocity perturbation by Au may result in serious distortion of the frequency and phase responses; also, the much larger density of Au compared to Al induces strong excitation of a parasitic surface slimming bulk wave (SSBW) mode. To solve such issues, Wang et al. (2011) presented a new design of an SAW device using a dual-layers electrode structure of Al and a very thin Au film on the top of the Al. According to Wang et al. (2011), first, a thin Au layer not only reduces the cost, but also prevents the attack from the measured gases on the Al electrode,

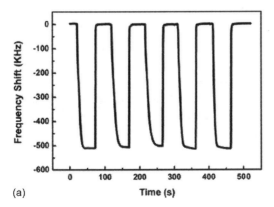
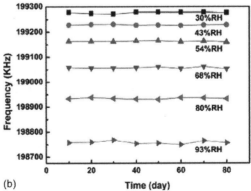

FIGURE 12.39 (a) Frequency response of surface acoustic wave (SAW) sensors based on SiO_2 film when the relative humidity (RH) increases from 30% to 93% for five cycles and (b) the frequency of the sensor in circumstances with various RH values for 80 days. (Reprinted with permission from Tang, Y. et al., Highly sensitive surface acoustic wave (SAW) humidity sensors based on sol–gel SiO_2 films: Investigations on the sensing property and mechanism, *Sens. Actuators B*, 215, 283–291, 2015. Copyright 2015, Elsevier.)

and also, the perturbation from the electrode on the SAW velocity and electromechanical coupling factor is reduced significantly because of the very thick Al film design, leading to performance improvements and technique simplification. Liu and He (2007) characterized the electromechanical coupling factor ($K_2\%$) and reflection coefficient of the Al/Au electrodes by using the theory of acoustic propagation and variational principle of short-circuited grating. It was shown that the Al/Au resonators feature insertion losses, and loaded Q values are comparable with those of SAW resonators with Al metallization, currently used in gas sensor systems. Later, Wang et al. (2013) confirmed that Al/Au electrodes had good corrosion resistance and stable characteristics.

12.6.2.5 New Approaches to Development of SAW-Based Humidity Sensors

The production of SAW-based humidity sensors with increased sensitivity faces some difficulties. First, to increase the sensitivity, it is necessary to increase the operating frequency. However, the dimension of SAW resonators (SAWRs) and thereby the size of the humidity-sensitive area, usually decreased as the operating frequency became higher. This means that the accuracy requirements for the application of humidity-sensitive materials to the SAWR surface are being increased (Caliendo 2006; Sarkara et al. 2006). However, thick-film technology, usually used in the manufacture of humidity sensors, is difficult to adapt to such requirements. The requirements for precision filming instruments and technique greatly increase processing cost of the sensor, which is contrary to the demand for an ideal humidity sensor. Furthermore, until now, the repeatability of this coating method is still not comfortable for mass production. Second, the Al strips on the substrate were very prone to be corroded when exposed in high humidity ambience, which will make a negative influence on the lifespan and stability of the sensor. Last, resonant frequency can be shifted due to dissociation effect in polymer films and movement of conductive ions, especially at high humidity levels.

In order to solve all the problems above, a gold interdigital electrode (IDE) was added in series with original SAWR to form a new sensing topology. By this way, the humidity-sensitive film could be directly deposited on surface of the IDE instead of SAWR. Compared to SAWR, a coating method for IDE, which can have much bigger size, has much better controllability and repeatability, and then the requirements for coating equipment and technique would be much lower. This will greatly reduce the sensors' processing cost and make it more proper for mass production. At the same time, SAWR could be vacuum packaged to be prevented from corrosion and perturbation. This could provide a more stable basic operating frequency of the sensor to improve its stability and to extend its lifespan. For the first time, such an approach was implemented by Shen et al. (1993) in biosensors. Similar devices for humidity detection were developed by Lei et al. (2011b). In the previous design, the SAWR was used as the humidity-sensing component. In the new topological structure, a gold IDE coated with polymer is used as the humidity-sensing element. Resonance frequency of the new topology is mainly determined by parameters of the IDE, such as conductivity and dielectric constant, which are strongly dependent on the humidity. SAWR in this configuration is used as a frequency-selective component to form a positive feedback oscillating circuit with the IDE. A gold IDE, which had been widely used in gas detection, was chosen as the sensitive component due to its low cost and simple fabrication process. A positive feedback oscillation circuit (as shown in Figure 12.40a) was made up to generate an oscillator signal, whose frequency was affected by the ambient moisture content. By measuring the output signal, ambient humidity could be recorded. The performance of the sensor developed by Lei et al. (2011) is shown in Figure 12.40b. It is seen that the new sensitive mechanism is profitable, even for the increase in the sensitivity. For the new sensor configuration, sensitivity at low humidity level (< 30% RH) was approximately 0.4 kHz/%RH, which was only a little larger than in the conventional sensor (Li et al. 2010). However, at a relatively higher humidity level, the sensitivity raised to above 1 kHz/%RH, which was a remarkable improvement compared to the previous one. Lei et al. (2011b) believe that, in the new circuit topology, a resonant frequency was much more sensitive to the conductivity variance of the film but not to the load effect, as in the previous SAWR sensor. For polyelectrolyte material, when absorbing water vapors, its conductivity variance was much more remarkable than its mass variance. At a higher humidity level, the difference was much more obvious. So, the sensitivity improvement was much more remarkable at a higher humidity level.

12.6.3 Dew-Point Hygrometers

For a long time, studies have been conducted to develop dew-point hygrometers using AW sensors as detection elements. These studies were stimulated by the fact that the traditional optical dew-point hygrometers had a number of

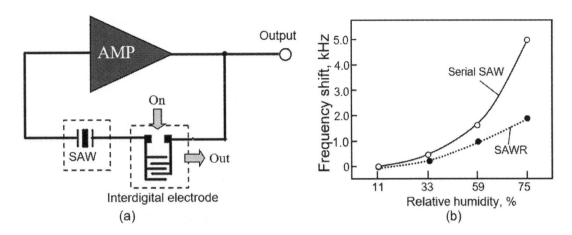

FIGURE 12.40 (a) Schematic of the oscillation circuit typically used for humidity detection and (b) Sensitivity comparison between the surface acoustic wave (SAW) humidity sensor developed by Lei et al. (2011b) and the previous surface acoustic wave resonator (SAWR) sensor ($f_0 = 433$ MHz). Silicon-containing polyelectrolyte was used as humidity-sensitive material. (Reprinted from Lei, S. et al., A novel serial high frequency surface acoustic wave humidity sensor, *Sens. Actuators A*, 167, 231–236, 2011. Copyright 2011, Elsevier.)

limitations associated with surface contamination effects, frost-point transition behavior, and dew-point measurement resolution (see Chapter 5). Considering the high sensitivity of AW devices, it was assumed that their use would allow the development of dew-point hygrometers devoid of the previously mentioned disadvantages. The main difference between the AW-based dew-point hygrometers and AW-humidity sensors is the presence of a Peltier element, integrated with the sensor and a control system, that controls the sensor temperature change according to the program specified (Figure 12.41). The principle of the dew/frost-point measurement is the cooling of the surface of the AW devices until the dew (or frost) point is reached. When it occurs, water (or ice) drops appear on the piezoelectric resonator or the surface of delay line, and the oscillation frequency shift takes place.

Nie et al. (2016) showed that there are also possibilities in which a frequency measurement is not required to measure the dew-point temperature. Figure 12.42 shows the schematic diagram of the dew-point recognition principle. In the Colpitts circuit, there is a quartz crystal resonator (QCR) stuck on the semiconductor refrigeration

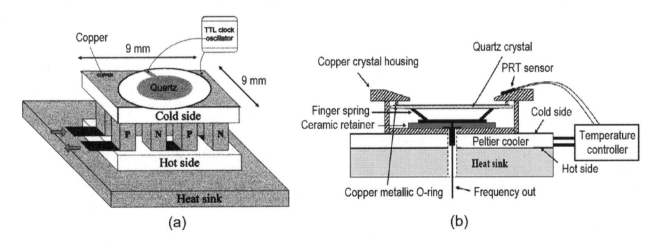

FIGURE 12.41 (a) Schematic diagram of the quartz crystal microbalance (QCM)-based dew-point hygrometer developed by Pascal-Delannoy et al. (2000): active part. Sensor is made up of an AT cut QCM vibrating along the transversal mode (Vig 1992). This quartz does not require any absorbent material, and it is directly stuck on the Peltier element. (b) Holder for a Peltier device system and quartz crystal for measuring dew/frost points at temperatures greater than −60°C. ([a] – Reprinted with permission from Pascal-Delannoy, F. et al., Quartz crystal microbalance (QCM) used as humidity sensor, *Sens. Actuators*, 84, 285–291, 2000. Copyright 2000, Elsevier; [b] – Reprinted with permission from Kwon, S.Y. et al., (2007) Recognition of supercooled dew in a quartz crystal microbalance dew-point sensor by slip phenomena, *Metrologia*, 44, L37–L40, 2007. Copyright 2007, IOP.)

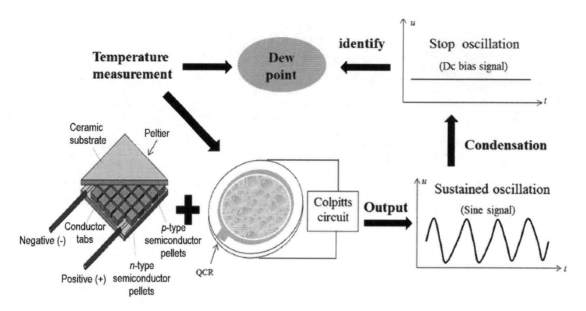

FIGURE 12.42 Schematic diagram of recognition principle of dew-point hygrometer developed by Nie et al. (Reprinted from Nie, J. et al., A new type of fast dew point sensor using quartz crystal without frequency measurement, *Sens. Actuators B*, 236, 749–758, 2016. Copyright 2016, Elsevier.)

device in order to produce condensation on its surface. The principle is described as the following: If there is no condensation, sinusoidal oscillation signals is an output; and the bigger the refrigerating capacity is, more condensation will occur and then the equivalent resistance of QCR will get bigger. As a result, a dc bias signal is the output. It is the signal mutation that can be used for a dew-point recognition. Beyond that, the dew-point temperature could be observed by measuring the surface temperature of QCR with platinum thermometer PT100. Because the Colpitts circuit has a great advantage of high frequency stability, it is commonly used as the drive circuit for the QCM (Gottlieb 1997). Nie et al. (2016) believe that, when using this principle, it is possible to develop a low-cost system, as it neither requires complex steps nor the costly equipment for the fabrication of the sensor. Moreover, it can work in the corrosive gas environment because the sensing layer has a corrosion resistance. The interface electronics circuit simply requires a few components and has digital output that is easy to interface with the microcontroller.

It is important that if the first studies were conducted using SAW devices, then the focus was on using QCM-based devices. SAW-based dew-point hygrometers were developed using $LiNbO_3$ (Vetelino et al. 1995, 1996; Galipeau et al. 1997) and quartz platforms (Hoummady et al. 1995), while for fabrication of QCM-based dew point hygrometers, only quartz platforms were used (Pascal-Delannoy et al. 2000; Kwon et al. 2007, 2008; Lin et al. 2016; Nie et al. 2016). SAW devices operated at frequencies of approximately 100 MHz, and QCM-based sensors at frequencies of 5–6 MHz.

It was established that, for the manufacture of dew point hygrometers, one can use AW sensors without any additional humidity-sensitive layer. However, Vetelino et al. (1996) and Galipeau et al. (1997), based on the example of SAW-based devices, have shown that use of Teflon and polyimide coatings reduced contamination buildup and made dew-point measurements more accurate with improved response time, and these measurements could be carried out at lower water vapor concentrations. As a result of the studies, it was found that the SAW and QCM-based dew-point hygrometers were much less susceptible than the optical technique to stability and resolution degradation caused by surface contamination. The SAW and QCM techniques could also accurately measure surface contamination, which may be used to indicate when cleaning or replacement of the sensor is necessary. In addition, the SAW and QCM-based dew-point instruments had significantly better measurement resolution than the optical dew-point hygrometer. For example, an SAW oscillator dew-point system had a measurement resolution of 0.03°C (Vetelino et al. 1995, 1996; Galipeau et al. 1997), as compared to a resolution of 0.2°C for a commercial optical dew-point hygrometer. Hoummady et al. (1995) believe that the accuracy of humidity measurements by SAW dew-point sensors, in comparison with optical hygrometers, may be improved by a factor of approximately 500. The results obtained by Galipeau et al. (1997) demonstrated that the SAW sensor can detect dew-surface densities more than two orders of magnitude less than those detectable by optical techniques. The response time of SAW and QCM hygrometers varied from 1 to 12 seconds. As a rule, the response time is only limited by the thermal conductivity of the quartz and by the ability of the Peltier element circuit to cool quickly the quartz plate. It was also found that, despite the fact that the frost-point transition, as in the case of the optical dew-point hygrometer, causes a significant instability of dew-point measurements via SAW velocity controlling, the SAW amplitude measurement provides continuous dew-point measurements. It is important to note that the advantages of the considered approach for the development of dew-point hygrometers have been realized, and at present, SAW-based dew-point hygrometers are represented on the market.

With regard to the limitations of this approach to measuring the dew point, it is necessary to distinguish the presence of a strong dependence of the resonant frequency on the temperature. The temperature dependence of the resonant frequency of the quartz resonator is large, and even a small temperature variation causes a significant frequency variation. Thus, it is necessary to somehow compensate this effect of temperature. Kwon et al. (2007) have solved this problem by carrying out additional measurements in dry air and building a temperature-effect compensation curve. Lin et al. (2016) and Nie et al. (2017), to solve the problem, used in a dew-point hygrometer two QCRs. One of these two QCRs only provides a reference frequency, without gas contact, and the other QCR is used for measuring a dew point with gas contact. The frequency difference was calculated between the frequency difference of two QCRs at the initial state and the frequency difference of two QCRs when condensation occurs. The observed value was used as the calibration value of dew-point recognition. When using QCM devices, it also appears that there are difficulties in providing a good and reliable thermal contact with the thermoelectric converter.

12.7 SUMMARY

SAW and QCM devices are small, sensitive, inexpensive, and easy to fabricate, and can work without a local power source. This means that SAW and QCM-based humidity sensors are promising devices for various applications. Regarding the choice of the optimal material for use in QCM and SAW-based humidity sensors, based on an objective comparison of all the parameters of the sensors being developed, there are practically no studies.

TABLE 12.7
Summary of the Main Characteristics of the Coatings

Characteristics	HEM-AMPS	Cellulose acetate	Modified epoxy
Solubility in water	– (b)	++	++ (c)
Sensitivity	++	++	++
Hysteresis	++	–	++
Response time	++	+	++
Selectivity (a)	+	–	–
Ageing	+	+	+

++ = good, + = satisfactory, – = bad; (a) towards CO_2, CH_3OH, C_2H_5OH and (b) working range < 70% RH; (c) does not adhere in hot water; AMPS, 2-acrylamido-2-methylpropane sulfonic acid; HEM, 2-hydroxyethylmethacrylate; RH, relative humidity.
Source: Data extracted from Randin, J.-P. and Zullig, F., *Sens. Actuators*, 11, 319–328, 1978.

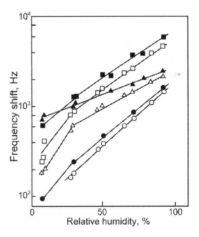

FIGURE 12.43 Response to humidity in nitrogen (*open symbols*) and in the additional presence of 0.83 vol.% ethanol (*solid symbols*) for HEM-AMPS copolymer, $\Delta F_o = 2.9$ kHz (*circles*), cellulose acetate, $\Delta F_o = -22.3$ kHz (*triangles*), and modified epoxy, $\Delta F_o = -60.5$ kHz (*squares*) measured at 25°C: 10% of the nitrogen was saturated in ethanol. HEM, 2-hydroxyethylmethacrylate; AMPS, 2-acrylamido-2-methylpropane sulfonic acid. (Adapted from *Sens. Actuators*, 11, Randin, J.P. and Zullig, F. (1987) Relative humidity measurements using a coated piezoelectrical quartz crystal sensor, 319–328, Copyright 1987, with permission from Elsevier.)

Each developer justifies the advantages of his or her own material. Only a few papers attempt to make a comparative analysis of various humidity-sensitive materials. Randin and Zullig (1987) studied and compared three coatings (see Table 12.7 and Figure 12.43) and concluded that the water solubility of the sulphonic acid copolymer 2-hydroxyethylmethacrylate-2-Acrylamido-2-methylpropane sulfonic acid (HEM-AMPS) restricts its use when RH is less than 70%. Otherwise, this coating gave the best characteristics. The cellulose acetate has a relatively large hysteresis and a poor selectivity towards CO_2, methanol, ethanol, and acetone. The modified epoxy exhibits good characteristics, except for the selectivity. Thus, the HEM-AMPS coating could be used for applications at RH lower than 70%. In pure environments, the cellulose acetate coating could be used to detect slow changes of RH, whereas the modified epoxy would also be convenient, provided that the sensor is not exposed to a liquid water. True, low selectivity is a typical drawback of almost all humidity-sensitive materials.

There are also studies performed by Joshi and Brace (1985), Buvailo et al. (2011b), and Lin et al. (2012) applied to other polymeric materials. However, they cannot be the basis for a conclusion regarding the material most suitable for use in QCM and SAW devices. The parameters of the devices depend too much on the technologies used, and the requirements of the sensors depend on the conditions of their operation. As indicated in many chapters of our handbook, in the development of a humidity sensor for a given application, numerous factors affecting the performance must be considered. Among these are sensitivity, selectivity, reversibility, response time, dynamic range, stability, reliability, and environmental (e.g., temperature) effects. Some of these concerns dictate the preferred substrate and/or acoustic mode; others are important in the selection of the coating, or in establishing optimal operating conditions. These criteria play a large role in the design of a complete sensor system. For example, materials working perfectly at low humidity, can degrade at high speed in conditions of high humidity. The high degree of selectivity afforded by carefully choosing an analyte/coating chemical interaction makes this class of sensors particularly attractive. Unfortunately, the price of selectivity is often an irreversible response and limited lifetime (Ballantine et al. 1997).

The development of new materials and technologies requires increased accuracy of measurements. As shown in Section 12.1, a higher operating frequency leads to a high sensitivity. Therefore, the development of high-frequency and super-high-frequency (SHF) QCM and SAW sensors is a priority for solving this problem. However, for high-frequency devices, both material and structure, as well as fabrication processes, need to be selected especially carefully.

So, as a sensing platform for SAW devices, it is necessary to use materials that have a high sound velocity.

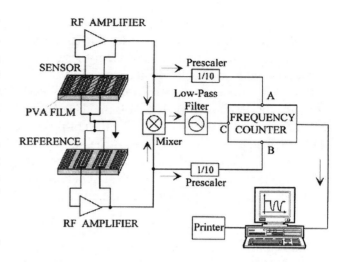

FIGURE 12.44 Scheme of the dual resonator surface acoustic wave (SAW) oscillator for relative humidity (RH) sensing. (Reprinted from Penza, M. and Cassano, G., Relative humidity sensing by PVA-coated dual resonator SAW oscillator, *Sens. Actuators B*, 68, 300–306, 2000. Copyright 2000, Elsevier.)

Such high-frequency piezoelectric materials are ZnO and AlN. ZnO has a sound velocity of more than 4000 m/s, and AlN has a sound velocity of around 5600 m/s, which is almost twice that of $LiNbO_3$. It is also promising to use layered structures. Sapphire has a sound velocity of around 6000 m/s, and diamond has the highest sound velocity, up to more than 10,000 m/s. Optimized thickness of AlN has been deposited onto diamond to acquire a velocity of more than 8000 m/s (Rodríguez-Madrid et al. 2012), and ZnO on diamond has a velocity of 5100 m/s (Shih et al. 2008).

The advantages and shortcomings of AW-based humidity sensors were considered in Section 2.1 and therefore will not be repeated. We note only the strong temperature dependence of the parameters of QCM and SAW-based humidity sensors (see Figure 12.28), which in principle is characteristic for all sensors, based on adsorption effects. Typically, this problem is being solved in a standard way, through temperature stabilization or the use of two QCM or SAW devices and a mixer (Figure 12.44). Figure 12.44 shows that one SAW acts as a reference sensor and the other as a sensing unit. As the two SAW sensors can be placed to experience the same temperature, but only one can sense the pressure change, any interference that can affect the frequency change can be cancelled out after the mixer. This way of temperature compensation is more accurate and is widely used in the sensing field (Nomura et al. 1994; Penza and Cassano 2000; Lei et al. 2011).

REFERENCES

Afzal A., Dickert F.L. (2011) Surface acoustic wave sensors for chemical applications, In: Korotcenkov G. (Ed.) *Chemical Sensors: Comprehensive Sensor Technologies, Vol. 4: Solid State Devices*. Momentum Press, New York, pp. 447–484.

Ameloot R., Stappers L., Fransaer J., Alaerts L., Sels B.F., and De Vos D.E. (2009) Patterned growth of metal-organic framework coatings by electrochemical synthesis. *Chem. Mater.* 21, 2580–2582.

Asar N., Erol A., Okur S., Arikan M.C. (2012) Morphology-dependent humidity adsorption kinetics of ZnO nanostructures. *Sens. Actuators A* 187, 37–42.

Ates T., Tatar C., Yakuphanoglu F. (2013) Preparation of semiconductor ZnO powders by sol–gel method: Humidity sensors. *Sens. Actuators A* 190, 153–160.

Avramov I.D., Rapp M., Kurosawa S., Krawczak P., Radeva E.I. (2002) Gas sensitivity comparison of polymer coated SAW and STW resonators operating at the same acoustic wave length. *IEEE Sensors J.* 2, 150–159.

Avramov I.D., Voigt A., Rapp M. (2005) Rayleigh SAW resonators using gold electrode structure for gas sensor applications in chemically reactive environments. *Electron. Lett.* 41, 450–452.

Awakuni Y., Calderwood J.H. (1972) Water vapour adsorption and surface conductivity in solids. *J. Phys. D, Appl. Phys.* 5, 1038–1045.

Ballantine Jr. D.S, White R. M., Martin S. J., Ricco A.J., Frye G.C., Zellers E.T., and Wohltjen H. (1997) *Acoustic Wave Sensors: Theory, Design, and Physico-Chemical Applications*. Academic Press, San Diego, CA.

Balashov S.M., Balachova O.V., Filho A.P., Bazetto M.C.Q., de Almeida M.G. (2012) Surface acoustic wave humidity sensors based on graphene oxide thin films deposited with the surface acoustic wave atomizer. *ECS Transactions* 49(1), 445–450.

Balashov S.M., Balachova O.V., Braga A.V.U., A Filho A.P., Moshkalev S. (2015) Influence of the deposition parameters of graphene oxide nanofilms on the kinetic characteristics of the SAW humidity sensor. *Sens. Actuators B* 217, 88–91.

Battison F.M., Ramseyer J.-P., Lang H.P., Baller M.K., Gerber C., Gimzewski J.K., Meyer E., and Guntherodt H.-J. (2001) A chemical sensor based on a microfabricated cantilever array with simultaneous resonance-frequency and bending readout. *Sens. Actuators B* 77, 122–131.

Bein T., Brown K., Frye G.C., and Brinker C.J. (1989) Molecular sieve sensors for selective detection at the nanogram level. *J. Am. Chem. Soc.* 111, 7640–7641.

Bender F., Cernosek R.W., and Josse F. (2000) Love wave biosensors using cross-linked polymer waveguide on $LiTaO_3$ substrates. *Electron. Lett.* 36(19), 1–2.

Bo L., Xiao C., Hualin C., Mohammad M.A., Xiangguang T., Luqi T., Yi Y., Tianling R. (2016) Surface acoustic wave devices for sensor applications. *J. Semiconductors* 37 (2), 021001.

Brace J.G. (1985) Measurement of humidity using surface acoustic waves, In: *Proceedings of IEEE Ultrasonics Symposium*, October 16–18, 1, San Francisco, CA, pp. 600–603.

Buvailo A.I., Xing Y., Hines J., Dollahon N., Borguet E. (2011a) TiO_2/LiCl-based nanostructured thin film for humidity sensor applications. *ACS Appl. Mater. Interfaces* 3, 528–533.

Buvailo A.I., Xing Y., Hines J., Borguet E. (2011b) Thin polymer film based rapid surface acoustic wave humidity sensors. *Sens. Actuators B* 156, 444–449.

Caliendo C., Verona E., D'Amico A., Furlani A., Iucci G., Russo M.V. (1993) Surface acoustic wave humidity sensor. *Sens. Actuators B* 15–16, 288–292.

Caliendo C., Verona E., Anisimkin V.I. (1997) Surface acoustic wave humidity sensors: A comparison between different types of sensitive membrane. *Smart Mater. Struct.* 6, 707–715.

Caliendo C., Imperatori P. (2003) High-frequency, high-sensitivity acoustic sensor implemented on AlN/Si substrate, *Appl. Phys. Lett.* 83, 1641–1643.

Caliendo C. (2006) Acoustic method of investigating the material properties and humidity sensing behavior of polymer coated piezoelectric substrates. *J. Appl. Phys.* 100, 054508.

Carey W.P., Beebe K.R., Kowalski B.R., Illman D.L., and Hirschfeld T. (1986) Selection of adsorbates for chemical sensor arrays by pattern recognition. *Anal. Chem.* 58, 149–153.

Chang S.M., Tamiya E., Karube I. (1991) Chemical vapour sensor using a SAW resonator. *Biosens. Bioelectron.* 6, 9–14.

Chen H.W., Wu R.J., Chan K.H., Sun Y.L., Su P.-G. (2005) The application of CNT/Nafion composite material to low humidity sensing measurement. *Sens. Actuators B* 104, 80–84.

Chen Y.T., Kao H.L. (2006) Humidity sensors made on polyvinyl-alcohol film coated SAW devices. *Electron. Lett.* 42 (16), 948–949.

Chen X., Chen X., Li N., Ding X., Zhao X. (2016) A QCM humidity sensors based on GO/Nafion composite films with enhanced sensitivity. *IEEE Sensors J.* 16 (24), 8874–8883.

Cheeke J.D.N., Wang Z. (1999) Acoustic wave gas sensors. *Sens. Actuators* 59, 146–153.

Cheremisinof P.N. and Ellerbusch F. (Eds.) (1980) *Carbon Adsorption Handbook*. Science Publishers, Ann Arbor, MI, pp. 241–279.

Comyn J. (Ed.) (1985) *Permeation of Gases and Vapors in Polymers*. Elsevier, London, UK.

Dahint R., Shana Z.A., Josse F., Riedel S.A., and Grunze M. (1993) Identification of metal ion solutions using acoustic plate mode devices and pattern recognition. *IEEE Trans. Ultrason. Ferroel. Freq. Contr.* 40(2), 114–120.

D'Amico A., Paima A., and Verona E. (1982) Palladium-surface acoustic wave interaction for hydrogen detection. *Appl. Phys. Lett.* 41, 300–301.

Dorjin G.B., Simakov I.G. (2002) Acoustic study of adsorbed liquid layers. *Acoust. Phys.* 48, 436–440.

Erol A., Okur S., Comba B., Mermer O., Arikan M.C. (2010) Humidity sensing properties of ZnO nanoparticles synthesized by sol–gel process. *Sens. Actuators B* 145, 174–180.

Erol A., Okur S., Yagmurcukardes N., Arikan M.C. (2011) Humidity-sensing properties of a ZnO nanowire film as measured with a QCM. *Sens. Actuators B* 152, 115–120.

Fan L., Ge H., Zhang S.-Y., Zhang H., Zhu J. (2012) Optimization of sensitivity induced by surface conductivity and sorbed mass in surface acoustic wave gas sensors. *Sens. Actuators B* 161, 114–123.

Fanget S., Hentz S., Puget P., Arcamone J., Matheron M., Colinet E., Andreucci P., Duraffourg L., Myers E.D., Roukes M.L. (2011) Gas sensors based on gravimetric detection—A review. *Sens. Actuators B* 160, 804–821.

Gabl R., Green E., Schreiter M., Feucht HD., Zeininger H., Primig R., et al. (2003) Novel integrated FBAR sensors: A universal technology platform for bio- and gas-detection. *Proceedings of the IEEE Sensors Conference*. October 22–24 Toronto, Canada, pp. 1184–1188.

Galatsis K., Qu W., Wlodarski W. (1998) Quartz crystal microbalance humidity sensor with porous electrodes, In: *Proceedings of IEEE Conference on Optoelectronic and Microelectronic Materials and Devices*, December 14–16, Perth, WA, Australia, pp. 373–375.

Galipeau D.W., Stroschine J.D., Snow K.A., Vetelino K.A., Hines K.R., Story P.R. (1995) A study of condensation and dew point using a SAW sensor. *Sens. Actuators B* 25, 696–700.

Galipeau D.W., Story P.R., Vetelino K.A., Mileham R.D. (1997) Surface acoustic wave microsensors and applications. *Smart Mater. Struct.* 6, 658–667.

García-Gancedo L., Pedrós J., Iborra E., Clement M., Zhao X.B., Olivares J., et al. (2013) Direct comparison of the gravimetric responsivities of ZnO-based FBARs and SMRs. *Sens. Actuators B* 183, 136–143.

Gizeli E. (2002) Acoustic transducers, In: E. Gizeli, C.R. Lowe (Eds.) *Biomolecular Sensors*. Taylor & Francis Group, London, UK, pp. 176–207.

Glück A., Halder W., Lindner G., Müller H., Weindler P. (1994) PVDF-excited resonance sensors for gas flow and humidity measurements. *Sens. Actuators B* 19, 554–557.

Gottlieb I. (1997) *Practical Oscillator Handbook*. Elsevier, New York.

Grate J.W., Martin S.J., White R.M. (1993a) Acoustic-wave microsensors. Pt. 1. *Anal. Chem.* 65, A940–A948.

Grate J.W., Martin S.J., White R.M. (1993b) Acoustic-wave microsensors. Pt. 2. *Anal. Chem.* 65, A987–A996.

Guo Y.J., Zhang J., Zhao C., Ma J.Y., Pang H.F., Hu P.A., et al. (2013) Characterization and humidity sensing of ZnO/428 YX LiTaO$_3$ Love wave devices with ZnO nanorods. *Mater. Res. Bull.* 48, 5058–5063.

Guo Y.J., Zhang J., Zhao C., Hu P.A., Zu X.T., Fu Y.Q. (2014) Graphene/LiNbO$_3$ surface acoustic wave device based relative humidity sensor. *Optik* 125, 5800–5802.

Haueis R., Vellekoop M.J., Kovacs G., Lubking G.W., and Venema A. (1994) A Love-wave based oscillator for sensing in liquids, In: *Proceedings of the Fifth International Meeting of Chemical Sensors,* Lake Tahoe, CA, Vol. 1, pp. 126–129.

He X.L., Li D.J., Zhou J., Wang W.B., Xuan W.P., Dong S.R., Jin H., Luo J.K. (2013) High sensitivity humidity sensors using flexible surface acoustic wave devices made on nanocrystalline ZnO/polyimide substrates. *J. Mater. Chem. C.* 1, 6210–6215.

Hong H.S., Chung G.S. (2009) Effect of thermal annealing on the SAW properties of AlN films deposited on Si substrate. *J. Korean Phys. Soc.* 54, 1519–1525.

Hong H.S., Chung G.S. (2010) Humidity sensing characteristics of Ga-doped zinc oxide film grown on a polycrystalline AlN thin film based on a surface acoustic wave. *Sens. Actuators B* 150, 681–685.

Hong H.-S., Phan D.-T., Chung G.-S. (2012) High-sensitivity humidity sensors with ZnO nanorods based two-port surface acoustic wave delay line. *Sens. Actuators B* 171–172, 1283–1287.

Horie K., Yamashita T. (1995) *Photosensitive Polyimides*, Technomic Publishing Company, Lancaster, PA.

Horzum N., Tascioglu D., Okur S., Demir M.M. (2011) Humidity sensing properties of ZnO-based fibers by electrospinning. *Talanta* 85, 1105–111.

Hoummady M., Bonjour C., Collin J., Lardet-Vieudrin F., Martin G. (1995) Surface acoustic wave (SAW) dew point sensor: Application to dew point hygrometry. *Sens. Actuators B* 26–27, 315–317.

Hu W., Chen S., Zhou B., Liu L., Ding B., Wang H. (2011) Highly stable and sensitive humidity sensors based on quartz crystal microbalance coated with bacterial cellulose membrane. *Sens. Actuators B* 159, 301–306.

Hung V.N., Takashi Abe T., Minh P.N., Esashi M. (2003) High-frequency one-chip multichannel quartz crystal microbalance fabricated by deep RIE. *Sens. Actuators A* 108, 91–96.

Ippolito S.J., Trinchi A., Powell D.A., Wlodarski W. (2009) Acoustic wave gas and vapor sensors, In: Comini E., Faglia G., Sberveglieri G. (Eds.) *Solid State Gas Sensing*. Springer, New York, pp. 261–304.

Irani F.S., Tunaboylu B. (2016) SAW humidity sensor sensitivity enhancement via electrospraying of silver nanowires. *Sensors* 16, 2024.

Jachowicz R., Weremczuk J. (2000) Sub-cooled water detection in silicon dew point hygrometer. *Sens. Actuators A* 85, 75–83.

Jakoby B., Ismail G.M., Byfield M.P., and Vellekoop M.J. (1999) A novel molecularly imprinted thin film applied to a love wave gas sensor. *Sens. Actuators A* 76, 93–97.

Jakubik W.P., Urbaczyk M.W., Kochowski S., and Bodzenta J. (2002) Bilayer structure for hydrogen detection in a surface acoustic wave sensor system. *Sens. Actuators B* 82, 265–271.

Janata J. (2009) *Principles of Chemical Sensors*. Springer, New York.

Janshoff A., Galla H.-J., and Steinem C. (2000) Piezoelectric mass-sensing devices as biosensors—An alternative to optical biosensors? *Angew. Chem. Int. Ed.* 39, 4004–4032.

Jasek K., Pasternak M. (2013) Influence of humidity on SAW sensor response. *Acta Phys. Polonica* 124 (3), 448–450.

Jin H., Tao X., Feng B., Yu L., Wang D., Dong S., Luo J. (2017) A humidity sensor based on quartz crystal microbalance using graphene oxide as a sensitive layer. *Vacuum* 140, 101–105

Josse F., Bender F., and Cernosek R.W. (2001) Guided shear-horizontal surface acoustic wave sensors for chemical and biochemical detection in liquids. *Anal. Chem.* 73, 5937–5944.

Kadota M., Yoneda T., Fujimoto K., Nakao T. (2004) Resonator filters using shear horizontal-type leaky surface acoustic wave consisting of heavy-metal electrode and quartz substrate. *IEEE Trans. Ultrason. Ferroelectr. Freq. Control* 51(2), 202–210.

Kalantar-Zadeh K., Trinchi A., Wlodarski W., Holland A., and Atashar M.Z. (2001) A novel Love mode device with nanocrystalline ZnO film for gas sensing applications, In: *Proceedings of the 1st IEEE Conference on Nanotechnology (IEEE NANO)*, October 28–30, Maui, Hawaii, pp. 556–561.

Kang A., Zhang C., Ji X., Hana T., Li R., Li X. (2013) SAW-RFID enabled temperature sensor. *Sens. Actuators A* 201, 105–113.

Kelkar U.R., Josse F., Haworth D.T., and Shana Z.A. (1991) Acoustic plate waves for measurements of electrical properties of liquids. *Microchem. J.* 43, 155–164.

King W.H. (1964) Piezoelectric sorption detector. *Anal. Chem.* 36, 1735–1739.

Korotcenkov G. (Ed.) (2011) *Chemical Sensors: Comprehensive Sensor Technologies. Vol. 4. Solid State Devices*. Momentum Press, New York.

Korotcenkov G., Cho B.K. (2012) The role of grain size on the thermal stability of nanostructured metal oxides used in gas sensor applications and approaches for grain-size stabilization, *Prog. Crystal. Growth* 58, 167–208.

Kosuru L., Bouchaala A., Jaber N., Younis M.I. (2016) Humidity detection using metal organic framework coated on QCM. *J. Sensors* 2016, 4902790.

Kreno L.E., Leong K., Farha O.K., Allendorf M., Van Duyne R.P., and Hupp J.T. (2012) Metal-organic framework materials as chemical sensors. *Chem. Rev.* 112, 1105–1125.

Kwan J.K., Sit J.C. (2012) High sensitivity Love-wave humidity sensors using glancing angle deposited thin films. *Sens. Actuators B* 173, 164–168.

Kwon S.Y., Choi B.I., Kim J.C. Nham H.S. (2006) Highly stable quartz crystal microbalance sensor and its application to water vapor measurements. *J. Korean Phys. Soc.* 48 (1), 161–165.

Kwon S.Y., Kim J.C., Choi B.I. (2007) Recognition of supercooled dew in a quartz crystal microbalance dew-point sensor by slip phenomena. *Metrologia* 44, L37–L40.

Kwon S.Y., Kim J.C., Choi B.I. (2008) Accurate dew-point measurement over a wide temperature range using a quartz crystal microbalance dew-point sensor. *Meas. Sci. Technol.* 19, 115206.

Lakin K.M. (2003) A review of thin-film resonator technology. *IEEE Microw. Mag.* 4, 61–67.

Lakin K.M. (2005) Thin film resonator technology. *IEEE Trans. Ultrason. Ferr.* 52, 707–716.

Lee K., Wang W., Kim T., Yang S. (2007) A novel 440 MHz wireless SAW microsensor integrated with pressure–temperature sensors and ID tag. *J. Micromechan. Microeng.* 17(3), 515–523.

Lei S., Deng C., Chen Y., Li Y. (2011a) A novel serial high frequency surface acoustic wave humidity sensor. *Sens. Actuators A* 167, 231–236.

Lei S., Chen D., Chen Y. (2011b) A surface acoustic wave humidity sensor with high sensitivity based on electrospun MWCNT/Nafion nanofiber films. *Nanotechnology* 22, 265504.

Li Y., Ying B.Y., Hong L.J., Yang M.J. (2010a) Water-soluble polyaniline and its composite with poly(vinyl alcohol) for humidity sensing. *Synth. Metal.* 160, 455–461.

Li Y., Li P., Yang M., Lei S., Chen Y., Guo X. (2010b) A surface acoustic wave humidity sensor based on electro-sprayed silicon-containing polyelectrolyte. *Sens. Actuators B* 145, 516–520.

Li X., Chen X., Yao Y., Li N., Chen X., Bi X. (2013) Multi-walled carbon nanotubes/graphene oxide composites for humidity sensing. *EEE Sensors J.* 13(12), 4749–4756.

Li D.J., Zhao C., Fu Y.Q., Luo J.K. (2014) Engineering silver nanostructures for surface acoustic wave humidity sensors sensitivity enhancement. *J. Electrochem. Soc.* 161 (6), B151–B156.

Li B.D., Yassine O., Kosel J. (2015) A surface acoustic wave passive and wireless sensor for magnetic fields, temperature, and humidity. *IEEE Sens. J.* 15, 453–462.

Lieberzeit P.A., Palfinger C., Dickert F.L., Fischerauer G. (2009) SAW RFID-tags for mass-sensitive detection of humidity and vapors. *Sensors* 9, 9805–9815.

Lin Q., Li Y., Yang M. (2012) Highly sensitive and ultrafast response surface acoustic wave humidity sensor based on electrospun polyaniline/poly(vinyl butyral) nanofibers. *Anal. Chim. Acta* 748, 73–80.

Lin N., Meng X., Nie J. (2016) Dew point calibration system using a quartz crystal sensor with a differential frequency method. *Sensors* 16, 1944.

Liu J.S., He S.T. (2007) Fast calculate the parameters of surface acoustic wave coupling-of-modes model. *Chin. J. Acoust.* 26, 269–277.

Liu Y., Wang C.H., Li Y. (2011) BCB film based SAW humidity sensor. *Electron. Lett.* 47 (18), 2011. 2078.

Liu J., Wang L. (2014) Dynamics and response of a humidity sensor based on a Love wave device incorporating a polymeric layer. *Sens. Actuators B* 204, 50–56.

Liu Y., Huang H., Wang L., Liu B., Cai D., Wang D., et al. (2015) Enhanced sensitivity of a GHz surface acoustic wave humidity sensor based on $Ni(SO_4)_{0.3}(OH)_{1.4}$ nanobelts and NiO nanoparticles. *J. Mater. Chem. C* 3, 9902–9909.

Liu W., Qu H., Hu J., Pang W., Zhang H., Duan X. (2017) A highly sensitive humidity sensor based on ultra-high-frequency microelectromechanical resonator coated with nano-assembled polyelectrolyte thin films. *Micromachines* 8, 116.

Lu D., Zheng Y., Penirschke A., Jakoby R. (2016) Humidity sensors based on photolithographically patterned PVA films deposited on SAW resonators. *IEEE Sensors J.* 16, (1), 13–14.

Lv R., Peng J., Chen S., Hu Y., Wang M., Lin J., Zhou X., Zheng X. (2017) A highly linear humidity sensor based on quartz crystal microbalance coated with urea formaldehyde resin/nano silica composite films. *Sens. Actuators B* 250, 721–725.

Martin S.J., Schweizer K.S., Schwartz S.S., and Gunshor R.L. (1984) Vapor sensing by means of a ZnO-on-Si surface acoustic wave resonator, In: *Proceedings of IEEE Ultrasonics Symposium*, November 14–16 Dallas, TX, pp. 207–213.

Martin S.J., Frye G.C., Senturla S.D. (1994) Dynamics and response of polymer-coatedsurface acoustic wave devices: Effect of viscoelastic properties and film resonance. *Anal. Chem.* 66, 2201–2219.

Mauder A. (1995) SAW gas sensor: Comparison between delay line and two port resonator. *Sens. Actuators B* 26, 187–190.

Mittal U., Islam T., Nimal A.T., Sharma M.U. (2015) Novel sol-gel γ-Al_2O_3 thin-film-based rapid SAW humidity sensor. *IEEE Trans. Electron. Dev.* 62(12), 4242–4250.

Mortet V., Williams O.A., and Haenen K. (2008) Diamond: A material for acoustic devices. *Phys. Stat. Sol.* (a) 205(5), 1009–1020.

Muckley E.S., Lynch J., Kumar R., Sumpter B., Ivanov I.N. (2016) PEDOT:PSS/QCM-based multimodal humidity and pressure sensor. *Sens. Actuators B* 236, 91–98.

Nagaraju M., Gu J., Lingley A., Zhang F., Small M., Ruby R., Otis B. (2014) A fully integrated wafer-scale sub-mm^3 FBAR-based wireless mass sensor, In: *Proceedings of IEEE International Frequency Control Symposium (FCS)*, May 19–22, Taipei, Taiwan. doi:10.1109/FCS.2014.6859916

Neshkova M., Petrova R., Petrov V. (1996) Piezoelectric quartz crystal humidity sensor using chemically modified nitrated polystyrene as water sorbing coating. *Anal. Chim. Acta* 332, 93–103.

Newell W.E. (1965) Face-mounted piezoelectric resonators. *Proc. IEEE* 53, 575–581.

Nie J., Liu J., Meng X. (2016) A new type of fast dew point sensor using quartz crystal without frequency measurement. *Sens. Actuators B* 236, 749–758.

Nie J., Liu J., Li N., Meng X. (2017) Dew point measurement using dual quartz crystal resonator sensor. *Sens. Actuators B* 246, 792–799.

Nikolaou I., Hallil H., Conédéra V., Deligeorgis G., Dejous C., Rebiere D. (2016) Inkjet-printed graphene oxide thin layers on Love wave devices for humidity and vapor detection. *IEEE Sensors J.* 16 (21), 7620–7627.

Nomura T, Yasuda T, Furukawa S. (1994) Humidity sensor using surface acoustic waves propagating along layered structures, In: *Proceedings of IEEE MTT-S International Microwave Symposium*, May 23–27, San Diego, CA, pp. 509–512.

Nomura T., Saitoh A., and Miyazaki T. (2003) Liquid sensor probe using reflecting SH- SAW delay line. *Sens. Actuators B* 91(1–3), 298–302.

Okura S., Kus M., Ozelb M., Yılmaz F. (2010) Humidity adsorption kinetics of water soluble calix[4]arene derivatives measured using QCM technique, *Sens. Actuators B* 145, 93–97.

O'Toole R.P., Burns S.G., Bastiaans G.J., Porter M.D. (1992) Thin aluminum nitride film resonators: Miniaturized high sensitivity mass sensors. *Anal. Chem.* 64, 1289–1294.

Pascal-Delannoy F., Sorli B., Boyer A. (2000) Quartz crystal microbalance (QCM) used as humidity sensor. *Sens. Actuators* 84, 285–291.

Penza M., Anisimkin V.I. (1999) Surface acoustic wave humidity sensor using polyvinyl-alcohol film. *Sens. Actuators* 76, 162–166.

Penza M., Cassano G. (2000) Relative humidity sensing by PVA-coated dual resonator SAW oscillator. *Sens. Actuators B* 68, 300–306.

Qiu X., Oiler J., Zhu J., Wang Z., Tang R., Yu C., Yu H. (2010a) Film bulk acoustic-wave resonator based relative humidity sensor using ZnO films. *Electrochem. Solid-State Lett.* 13(5), J65–J67.

Qiu X., Tang R., Zhu J., Oiler J., Yu C., Wang Z., Yu H. (2010b) Experiment and theoretical analysis of relative humidity sensor based on film bulk acoustic-wave resonator. *Sens. Actuators B* 147, 381–384.

Radeva E., Georgiev V., Spassov L., Koprinarov N., Kanev S. (1997) Humidity adsorptive properties of thin fullerene layers studied by means of quartz micro-balance. *Sens. Actuators B* 42, 11–13.

Randin J.-P., Zullig F. (1987) Relative humidity measurements using a coated piezoelectrical quartz crystal sensor. *Sens. Actuators* 11, 319–328.

Reichl W., Runck J., Schreiter M., Greert E., Gabl R. (2004) Novel gas sensors based on thin film bulk acoustic resonators, In: *Proceedings of the IEEE Sensors Conference*, October 24–27, Vienna, Austria, pp. 1504–1505.

Rey-Mermet S., Lanz R., Muralt P. (2006) Bulk acoustic wave resonator operating at 8 GHz for gravimetric sensing of organic films. *Sens. Actuators B* 114, 681–686.

Ricco A.J., Martin S.J., and Zipperian T.E. (1985) Surface acoustic wave gas sensor based on film conductivity changes. *Sens. Actuators* 8, 319–333.

Ricco A.J. and Martin S.J. (1992) Thin metal film characterization and chemical sensors: Monitoring electronic conductivity, mass loading and mechanical properties with surface acoustic wave devices. *Thin Solid Films* 206, 94–101.

Rimeika R., Ciplys D., Poderys V., Rotomskis R., Shur M.S. (2017) Fast-response and low-loss surface acoustic wave humidity sensor based on bovine serum albumin-gold nanoclusters film. *Sens. Actuators B* 239, 352–357.

Ristic V.M. (1983) *Principles of Acoustic Devices*. Wiley, Oxford, UK.

Robinson A.L., Stavila V., Zeitler T.R. White M.I., Thornberg S.M., Greathouse J.A., Allendorf M.D. (2012) Ultrasensitive humidity detection using metal–organic framework-coated microsensors. *Anal. Chem.* 84, 7043–7051.

Rodríguez-Madrid J.G., Iriarte G.F., Araujo D., Villar M.P., Williams O.A., Müller-Sebert W., Calle F. (2012) Optimization of AlN thin layers on diamond substrates for high frequency SAW resonators. *Mater. Lett.* 66(1), 339–342.

Sadaoka Y. (2009) Capacitive-type relative humidity sensor with hydrophobic polymer films, In: Comini E., Faglia G., Sberveglieri G. (Eds.) *Solid State Gas Sensing*. Springer, New York, pp. 109–152.

Sakly N., Said A.H., Ouada H.B. (2014) Humidity-sensing properties of ZnO QDs coated QCM: Optimization, modeling and kinetic investigations. *Mater. Sci. Semicond. Proces.* 27, 130–139.

Sanchez-Pedreno J.A.O., Drew P.K.P., and Alder J.F. (1986) The investigation of coating materials for the detection of nitrobenzene with coated quartz piezoelectric crystals. *Anal. Chim. Acta* 182, 285–291.

Sarkara S., Levita N., Tepper G. (2006) Deposition of polymer coatings onto SAW resonators using AC electrospray. *Sens. Actuators B* 114, 756–761.

Sauerbrey G. (1959) Verwendung von schwingquarzen-zur wagungdunner schichten und zur mikrowagung. *Zeitschrift Physik* 155(2), 206–222.

Schroth A., Sager K., Gerlach G., Hiberli A., Boltshauser T., H. Baltes H. (1996) A resonant poliyimide-based humidity sensor. *Sens. Actuators B* 34, 301–304.

Shen D.Z., Zhu W.H., Nie L.H., Yao S.Z. (1993) Behaviour of a series piezoelectric sensor in electrolyte solution: Part I theory. *Anal. Chim. Acta* 276, 87–97.

Shih W., Wang M., Lin I.N. (2008) Characteristics of ZnO thin film surface acoustic wave devices fabricated using nanocrystalline diamond film on silicon substrates. *Diam. Relat. Mater* 17 (3), 390–395.

Song F., Wang C., Falkowski J.M., Ma L., Lin W. (2010) Isoreticular chiral metal−organic frameworks for asymmetric alkene epoxidation: Tuning catalytic activity by controlling framework catenation and varying open channel sizes. *J. Am. Chem. Soc.* 132, 15390–15398.

Su P.G., Sun Y.L., Lin C.C. (2006) A low humidity sensor made of quartz crystal microbalance coated with multi-walled carbon nanotubes/Nafion composite material films. *Sens. Actuators B* 115, 338–343.

Su P.G., Tzou W.H. (2012) Low-humidity sensing properties of PAMAM dendrimer and PAMAM–Au nanoparticles measured by a quartz-crystal microbalance. *Sens. Actuators A* 179, 44–49.

Su P.G., Lin L.G., Tzou W.H. (2013) Humidity sensing properties of calix[4]arene and functionalized calix[4]arene measured using a quartz-crystal microbalance. *Sens. Actuators B* 181, 795–801.

Su P.G., Kuo X.R. (2014) Low-humidity sensing properties of carboxylic acid functionalized carbon nanomaterials measured by a quartz crystal microbalance. *Sens. Actuators A* 205, 126–132.

Syritski V., Reut J., Orlik A., Idla K. (1999) Environmental QCM sensors coated with polypyrrole. *Synth. Met.* 102, 1326–1327.

Tai H., Zhen Y., Liu C., Ye Z., Xie G., Du X., Jiang Y. (2016) Facile development of high performance QCM humidity sensor based on protonated polyethylenimine-graphene oxide nanocomposite thin film. *Sens. Actuators B* 230, 501–509.

Takeda N., Motozawa M. (2012) Extremely fast 1 µmol mol^{-1} water-vapor measurement by a 1 mm diameter spherical SAW device. *Int. J. Thermophys.* 33, 1642–1649.

Takayanagi K., Akao S., Yanagisawa T., Nakaso N., Tsukahara Y., Hagihara S., Oizumi T., Takeda N., Tsuji T., Yamanaka K. (2014) Detection of trace water vapor using SiO_x-coated ball SAW sensor. *Mater. Trans.* 55, 988–993.

Tang Y., Li Z., Ma J., Wang L., Yang J., Du B., Yu Q., Zu X. (2015) Highly sensitive surface acoustic wave (SAW) humidity sensors based on sol–gel SiO_2 films: Investigations on the sensing property and mechanism. *Sens. Actuators B* 215, 283–291.

Tipple C.A. (2003) Strategies for enhancing the performance of chemical sensors based on microcantilever sensors. PhD thesis, The University of Tennessee, Knoxville, TN.

Uzar N., Okur S., Arıkan M.C. (2011) Investigation of humidity sensing properties of ZnS nanowires synthesized by vapor liquid solid (VLS) technique. *Sens. Actuators A* 167, 188–193.

Vashist S.K., Korotcenkov G. (2011) Microcantilever-based chemical sensors, In: Korotcenkov G. (Ed.) *Chemical Sensors: Comprehensive Sensor Technologies, Vol. 4: Solid State Devices*. Momentum Press, New York, pp. 321–376.

Vetelino K.A., Story P.R., Galipeau D.W. (1995) A comparison of SAW and optical dew point measurement techniques, In: *Proceedings of IEEE Ultrasonics Symposium*, November 7–10, Seattle, WA, pp. 551–554.

Vetelino K.A., Story P.R., Mileham R.D., Galipeau D.W. (1996) Improved dew point measurements based on a SAW sensor. *Sens. Actuators B* 35–36, 91–98.

Vig J.R. (1992) Introduction to quartz frequency standards, United States Army Research and Development Technical Report *SLCET-TR-92-1*.

Villa-López F.H., Rughoobur G., Thomas S., Flewitt A.J., Cole M., Gardner J.W. (2016) Design and modelling of solidly mounted resonators for low-cost particle sensing. *Meas. Sci. Technol.* 27, 025101.

Voinova M., Jonson M. (2011) The quartz crystal microbalance, In: Korotcenkov G. (Ed.) *Chemical Sensors: Comprehensive Sensor Technologies*, Vol. 4: *Solid State Devices*. Momentum Press, New York, pp. 377–446.

Wang Q.M., Shen D.M., Bulow M., Lau M.L., Deng S.G., Fitch F.R., Lemcoff N.O., and Semanscin J. (2002) Metallo-organic molecular sieve for gas separation and purification. *Micropor. Mesopor. Mat.* 55, 217–230.

Wang X., Ding B., Yu J., Wang M., Pan F. (2010) A highly sensitive humidity sensor based on a nanofibrous membrane coated quartz crystal microbalance. *Nanotechnol.* 21, 055502.

Wang W., He S.T., Liu M.H., Pan Y. (2011) Advances in SXFA-coated SAW chemical sensors for organophosphorous compound detection. *Sensors* 11, 1526–1541.

Wang W., Xie X., He S. (2013) Optimal design of a polyaniline-coated surface acoustic wave based humidity sensor, *Sensors* 13, 16816–16828.

Wang P., Su J., Su C.-F., Dai W., Cernigliaro G., Sun H. (2014) An ultrasensitive quartz crystal microbalance-micropillars based sensor for humidity detection. *J. Appl. Phys.* 115, 224501.

Ward M.D. and Buttry D.A. (1990) In situ interfacial mass detection with piezoelectric transducers. *Science* 249, 1000–1007.

Wenzel S.W., White R.M. (1978) Analytic comparison of the sensitivities of bulk-wave, surface-wave, and flexural plate-wave ultrasonic gravimetric sensors. *Appl. Phys. Lett.* 54, 1976–1978.

White R.M., Martin S.J., Ricco A.J., Zellers E.T., Frye G.C., and Wohltjen H. (1997) *Acoustic Wave Sensors: Theory, Design, and Physico-Chemical Application*. Academic Press, San Diego, CA.

Xu Z., Yuan Y.J. (2018) Implementation of guiding layers of surface acoustic wave devices: A review. *Biosens. Bioelectron* 99, 500–512.

Xuan W., He M., Meng N., He X., Wang W., Chen J., et al. (2014) Fast response and high sensitivity ZnO/glass surface acoustic wave humidity sensors using graphene oxide sensing layer. *Sci. Reports* 4, 7206, 9.

Xuan W., He X., Chen J., Wang W., Wang X., Xu Y., et al. (2015) High sensitivity flexible Lamb-wave humidity sensors with a graphene oxide sensing layer. *Nanoscale* 7, 7430–7436.

Xuan W., Cole M., Gardner J.W., Thomas S., Villa-López F.-H., Wang X., Dong S., Luo J. (2017) A film bulk acoustic resonator oscillator based humidity sensor with graphene oxide as the sensitive layer. *J. Micromech. Microeng.* 27, 055017.

Yakuphanoglu F. (2012) Semiconducting and quartz microbalance (QCM) humidity sensor properties of TiO_2 by sol gel calcination method. *Solid State Sci.* 14, 673–676.

Yao Y., Chen X., Guo H., Wu Z. (2011) Graphene oxide thin film coated quartz crystal microbalance for humidity detection. *Appl. Surf. Sci.* 257, 7778–7782.

Yao Y., Ma W. (2014) Self-assembly of polyelectrolytic/graphene oxide multilayer thin films on quartz crystal microbalance for humidity detection. *IEEE Sensors J.* 14(11), 4078–4084.

Yao Y., Chen X., Ma W., Ling W. (2014a) Quartz crystal microbalance humidity sensors based on nanodiamond sensing films. *IEEE Trans. Nanotechnol.* 13(2), 386–393.

Yao Y., Chen X., Li X., Chen X., Li N. (2014) Investigation of the stability of QCM humidity sensor using graphene oxide as sensing films. *Sens. Actuators B* 191, 779–783.

Yao Y., Xue Y. (2015) Impedance analysis of quartz crystal microbalance humidity sensors based on nanodiamond/graphene oxide nanocomposite film. *Sens. Actuators B* 211, 52–58.

Yao Y., Zhang H., Sun J., Ma W., Li L., Li W., Du J. (2017) Novel QCM humidity sensors using stacked black phosphorus nanosheets as sensing film. *Sens. Actuators B* 244, 259–264.

Yoo H.Y., Bruckenstein S. (2013) A novel quartz crystal microbalance gas sensor based on porous film coatings. A high sensitivity porous poly(methylmethacrylate) water vapor sensor. *Anal. Chim. Acta* 785, 98–103.

Yuan Z., Tai H., Ye Z., Liu C., Xie G., Du X., Jiang Y. (2016a) Novel highly sensitive QCM humidity sensor with low hysteresis based on graphene oxide (GO)/poly(ethyleneimine) layered film. *Sens. Actuators B* 234, 145–154.

Yuan Z., Tai H., Bao X., Liu C., Ye Z., Jiang Y. (2016b) Enhanced humidity-sensing properties of novel grapheme oxide/zinc oxide nanoparticles layered thin film QCM sensor. *Mater. Lett.* 174, 28–31.

Zhang Y., Yu K., Xu R., Jiang D., Luo L., Zhu Z. (2005) Quartz crystal microbalance coated with carbon nanotube films used as humidity sensor. *Sens. Actuators A* 120, 142–146.

Zhang M., Du L., Fang Z., Zhao Z. (2015) Micro through-hole array in top electrode of film bulk Acoustic resonator for sensitivity improving as humidity sensor. *Procedia Eng.* 120, 663–666.

Zhu Y., Chen J., Li H., Zhu Y., Xu J. (2014) Synthesis of mesoporous SnO_2–SiO_2 composites and their application as quartz crystal microbalance humidity sensor. *Sens. Actuators B* 193, 320–325.

13 Cantilever- and Membrane-Based Humidity Sensors

13.1 GENERAL CONSIDERATION

As follows from the title of this chapter, the main element of the considered humidity sensors is the membrane (diaphragm) (Morten et al. 1993; Schroth et al. 1996; Buchhold et al. 1998a, 1998b) and cantilever (Gluck et al. 1994; Battiston et al. 2001). Schematic diagrams of such sensors are shown in Figure 13.1. Sensors of this type are being intensively developed in recent years (Lang 2009; Vashist and Korotcenkov 2011). The increased interest in sensors of this type is due to the fact that microcantilever- and membrane-based sensors offer a very promising future for the development of novel physical, chemical, and biological sensors. For example, it was established that advanced analyte detection systems with lower detection limits than in the most advanced techniques, currently employed, can be designed on the base of microcantilevers. For example, microcantilever sensitivities can be as much as 10 times greater than in the quartz crystal microbalance (QCM) and surface acoustic wave (SAW) techniques (see Chapter 12).

13.1.1 Membrane-Based Sensors

Membrane-based humidity sensors represent a further development of the approach used in the development of pressure sensors (Eaton and Smith 1997). In principle, the humidity sensor is a pressure sensor with humidity sensitive film deposited on the surface of the membrane. Many years of operation of pressure sensors having a similar design showed that the sensors of this type are stable and effective, and have a long lifetime. Membranes, as a rule, are made on the basis of silicon or polysilicon. The membrane usually has a size of 1×1 mm^2, and the thickness of the membrane is in the range of 3–15 μm. The use of micromachining technology is the best approach to the fabrication of such sensors (Lang 1996; Madou and Gottehrer 2000; Lavrik et al. 2004). Detailed description of these technologies will be done in Volume 3 (Chapters 15 and 16) of our issue. Typical steps of complementary metal–oxide–semiconductor (CMOS) technology, which can be performed on the wafer during sensor fabrication, include chemical vapor deposition, oxidation, doping, diffusion, metallization, and etching. Photolithography and chemical wet etching are used to pattern and form the silicon platform and measurement structures of the sensor.

Tight dimensional control of the diaphragm features ensures a consistent device sensitivity and overall sensor performance (Kumar and Pant 2014). For example, thin diaphragms are more sensitive than thicker ones, but are difficult to realize. Bigger diaphragms help in achieving higher sensitivity but increase the size of the device. Smaller sensors reduce the cost, but it is often difficult to fabricate the diaphragm with precise thickness. Reducing the diaphragm thickness and increasing the size also leads to degradation in linearity. Square diaphragms are preferred in humidity sensors, as they have a better sensitivity than circular diaphragms with the same thickness h. However, the reverse is true for nonlinearity. Also, it is easier to fabricate square pressure sensors at low cost using anisotropic wet etchants.

13.1.2 Microcantilevers

A microcantilever is the miniaturized counterpart of a diving board that moves up and down at regular intervals. However, there is a large difference in dimension,

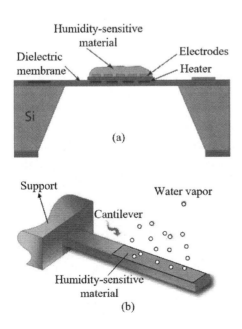

FIGURE 13.1 Schematic diagram of membrane- (a) and cantilever-based (b) humidity sensors.

as microcantilevers have dimensions in micrometers (i.e., a million times smaller than the diving board). Thus, a microcantilever can be modeled as a cantilever beam that is built in (fixed) at one of its ends. Microcantilevers, generally used for sensing applications, are some hundreds of micrometers long and 20–100 μm wide. The thickness can be varied from 1 μm to some tens of micrometers. However, highly sensitive cantilevers with thickness $h = 500$ nm and length $L = 500$ μm are also commercially available (www.concentris.ch; www.micromotive.de), and ultrathin cantilevers with h below 200 nm have been fabricated (Yang et al. 2000; Ramos et al. 2009). Microcantilevers can also have various shapes, as shown in Figure 13.2. The shape and dimensions of cantilevers are design parameters and vary from application to application. Typically, they are made up of silicon, silicon derivatives such as silicon nitride, or polymers. Cantilevers, fabricated with these classical materials, can be operated in a large range of temperatures and environmental conditions. However, developments related to the use of other materials also appear. For example, Boytsova et al. (2015) have fabricated cantilever arrays based on porous anodic alumina. It should be noted that the use of polymer-based microbeam instead of silicon and polysilicon cantilevers gives an increase in the sensitivity of the humidity sensors (Nordstrom et al. 2008; Chatzandroulis et al. 2011). Such an approach in the development of humidity sensors was used by Schmid et al. (2008, 2009) and Patil et al. (2014). As a rule, for the manufacture of such cantilever-based sensors, SU-8 2002 is used.

Research efforts in the field of microcantilever sensors have grown exponentially in the last decade. As has been the case with other types of mass sensors, they were originally developed for an entirely different application: as the tips for scanning probe microscopies, namely the atomic force microscopy (Binning et al. 1986). Gimzewski (1993) discovered that both static and dynamic properties of these tips change with the chemical environment in which they operate.

Currently, microcantilevers are considered as a very promising sensor platform with diverse applications in the fields of biomedical technology, health care, environmental monitoring, and industrial and clinical analysis (Baller et al. 2000; Battiston et al. 2001; Carrascosa et al. 2006; Fantner et al. 2009; Bausells 2015). It is believed that a cantilever-based sensing has several advantages over conventional analytical techniques in terms of high sensitivity, low cost based on mass production, less analyte required, nonhazardous label-free procedures, low energy consumption, rapid response, robustness, and high-throughput analyte detection (Lavrik et al. 2004; Lang and Gerber 2008; Goeders et al. 2008; Lang, 2009). For fabrication of microcantilevers, as well as membranes, the micromachining technologies described in Volume 3 (Chapters 15 and 16) are commonly used.

The principles of cantilever- and membrane-based humidity sensors are identical. But cantilevers are much more often used to develop different sensors, including humidity sensors.

In general, a bimorph cantilever is more sensitive than a similar bimorph membrane because the latter is clamped peripherally instead of having one clamped edge, as in the case of a cantilever. Therefore, in the future, our attention will mainly be focused on the description of the work of cantilever-based sensors.

The description of the technology of manufacturing silicon and polymer cantilevers, their comparative characteristics, and approaches used to optimize the parameters of sensors based on them can be found in Volume 3 of our issue, and also in the reviews of Yi et al. (2002), Mutyala et al. (2009), Boisen et al. (2011), and Bausells (2015). For example, according to Boisen et al. (2011), regardless of the cantilever material, there are some requirements to the final cantilever structure designed for sensor applications.

1. For increased surface stress sensitivity, the cantilever should be as thin and long as possible. This requires processing of suspended fragile structures.
2. For mass sensing, the clamping of the cantilever should minimize clamping losses. Furthermore, the material should have low internal damping, and the cantilever geometry should allow for a high Q factor.
3. For all purposes, the geometries of the cantilevers should be controlled with a high accuracy, since the dimensions have a huge influence on the sensitivity and thus the uniformity of the sensors. For example, precise control of the geometries of reference and measurement cantilevers is crucial to avoid measurement errors.
4. For optical read-out, the surface of the cantilever needs to be reflecting and of high optical quality.

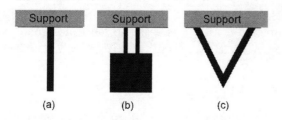

FIGURE 13.2 Different types of microcantilevers (top view): (a) rectangular, (b) double-legged, and (c) triangular.

The surface should not be rough and thereby scatter the light in all directions.

5. The cantilevers should ideally have no initial bending. Initial bending complicates the optical alignment and makes the cantilevers more prone to spurious signals due to changes in temperature, refractive index, etc.

13.2 SORPTION-INDUCED EFFECTS AND THEIR INFLUENCE ON CANTILEVER OPERATION

13.2.1 Sorption Models

Application of the cantilever as a chemical sensor is naturally based on the cantilever's interaction with its environment. The interaction includes thermal effects, viscous damping, and a range of sorption induced effects, such as mass loading, modulation of mechanical stresses, and changes in elasticity of the cantilever. Adsorption-induced modulation of mechanical properties of thin plates has been considered and described elsewhere (Zhou et al. 2005). These sorption-induced effects play a principal role in nano-mechanical gas sensing and recognition. The theory of molecular adsorption on cantilevers has been investigated in several publications (Bottomley et al. 2004; Ryu et al. 2004; Khaled and Vafai 2004; Zhang et al. 2004), whereas the bending, calibration, and curvature of cantilevers have been studied experimentally (Abermann and Martinz 1984; Cherian and Thundat 2002; Tang et al. 2004; Hu et al. 2004).

In order to understand how different modifying coatings provide responses of cantilever sensors, it is useful to consider three distinctive models (Lavrik et al. 2004).

1. The first sorption model is most adequate when interactions between the cantilever and its environment are predominantly surface phenomena (see Figure 13.3a). This model is realized in uncoated cantilever or smooth cantilever surfaces, coated by gold or other metals with very small thickness in comparison to the cantilever. Depending on the nature of adsorption, the analyte species adsorbed on the surface can polarize the surface, creating induced dipole, or molecules can combine chemically with the surface atoms. Exchange coupling, ionic or coordination bonds, electrostatic interactions between molecules, and changes in the hydrophobicity of the surface may occur in this case. All these processes can be accompanied by the change in the surface stress. In general, changes in surface stresses can be largely attributed to changes in the Gibbs free energy. Since spontaneous adsorption processes are driven by an excess of the interfacial free energy, they are typically accompanied by a reduction of the

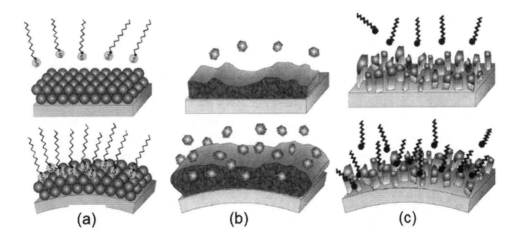

FIGURE 13.3 (a) Schematic depiction of chemisorption of straight-chain thiol molecules on a gold-coated cantilever. Spontaneous adsorption processes are driven by an excess in the interfacial free energy and accompanied by reduction of the interfacial stress; (b) Schematic depiction of analyte-induced cantilever deformation when the surface is modified with a thicker, analyte-permeable coating. Interactions of the analyte molecules with the bulk of the responsive phase lead to coating swelling and can be evaluated using approaches employed in colloidal and polymer science; and (c) Schematic depiction of analyte-induced cantilever deformation in the case of a structured modifying phase. Analyte-induced deflections of cantilevers with structured phases combine mechanisms of bulk, surface, and intersurface interactions. (Reprinted with permission from Lavrik N.V. et al, Cantilever transducers as a platform for chemical and biological sensors, *Rev Sci Instrum.*, 75, 2229–2253, 2004. Copyright 2004, American Institute of Physics.)

interfacial stress. In other words, surfaces usually tend to expand as a result of adsorptive processes. The humidity-induced strain can be approximated with (Buchhold et al. 1998a)

$$\sigma_h(r_h) = -\upsilon(r_h) \frac{E_\parallel^B}{1 + \frac{\varepsilon_\perp}{\varepsilon_\parallel(r_h)}} N(r_h) \quad (13.1)$$

where $\varepsilon_\perp$ and $\varepsilon_\parallel$ are the out-of-plane and in-plane strains, respectively, $E_\parallel^B$ is the bulk modulus, $\upsilon(r_h)$ is the Poisson ratio, and $N(r_h)$ is the number of adsorbed vapor molecules. The latter is typically a nonlinear function of r_h:

$$N(r_h) \approx N_0 \cdot r_h^n \quad (13.2)$$

This type of the surface stress change is defined as compressive, referring to a possibility of return of the surface into the original compressed state. The larger the initial surface free energy of the substrate is, the greater is the possible change in the surface stress resulting from spontaneous adsorption processes. Compressive surface stresses due to adsorption were experimentally observed on the gold side of gold-coated cantilevers exposed to vapor-phase alkanethiols (Berger et al. 1997).

In the bending mode, the bending, caused by the change of the surface energy (stress), can be calculated according to Stoney's law

$$R_B = \frac{Et^2}{6\sigma(1-\xi_P)} \quad (13.3)$$

where R_B is the bending radius, σ is stress caused by interaction with analyte, ξ_P is the Poisson ratio equalled to $\Delta A/\Delta l$, t is thickness, and E is the Young's modulus. The deformation parameters for silicon are $E = 1.7 \times 10^{11} N/m^2$, and $\xi_P = 0.25$ (Janata 2009). The analytical information is obtained from the calibration curve. Taking into account the principle of operation, it is clear that the sensitivity of cantilever-based sensors to additional mass loadings can be increased by reducing the inherent active mass of the resonator, which is the reason for the ongoing research toward nanoelectromechanical systems (NEMS).

One should note that Stoney's equation is based on many assumptions (Freund et al. 1999). As a result, Stoney's formula does not agree well with the experimental data (Freund et al. 1999; Klein 2000) for structures under large deflection of thick films. Therefore, there have been numerous attempts to propose models that better describe the experimental results (Miyatani and Fujihira 1997; Freund et al. 1999; Godin et al. 2001; Sader 2002; Ngo et al. 2006).

2. The second sorption model (see Figure 13.3b) of analyte-induced stresses and cantilever deformation is applicable for a cantilever, modified with a much-thicker-than-monolayer analyte-permeable coating, such as a polymer layer (Betts et al. 2000; Fagan et al. 2000). Taking into account interactions of the analyte molecules with the sensitive material coating the cantilever, a predominant mechanism of cantilever deflection can be described as deformation due to the change in the mass of the coating layer or analyte-induced swelling of the coating. Swelling processes can be quantified using approaches developed in colloidal and polymer science (i.e., by evaluating molecular forces acting in the coating and between the coating and the analyte species). In general, dispersion, electrostatic, steric, osmotic, and solvation forces acting within the coating can be altered by absorbed analytes. In this case, all free surfaces of the film undergo a displacement parallel and normal to the adjacent microbeam surfaces, with the exception of the film surface that is bonded to each cantilever beam. The bond constrains this film surface, preventing it from displacing. This behavior, known as a *full-shear constraint*, produces shear stresses at the film-beam interface that causes the cantilever beams to deflect. It is important to note that the magnitude of apparent surface stress scales up in proportion with the thickness of the responsive phase.

3. The third sorption model (see Figure 13.3c) reflects an analyte-induced cantilever deformation in the case of a structured modifying phase. It is most relevant to nanostructured interfaces and coatings, such as porous metal oxides, nanozeolites, surface-immobilized colloids, and other materials with porous structure that can adsorb the analyte. It is worth noting that grain boundaries, voids, and impurities have been long known as being responsible for high intrinsic stresses in disordered, amorphous, and polycrystalline films (Koch 1994). Analyte-induced deflections of cantilevers with structured phases combine

mechanisms of bulk, surface, and intersurface interactions (Israelachvili 1991). A combination of these mechanisms facilitates efficient conversion of the energy of receptor-analyte interactions into mechanical energy of cantilever bending. Recent studies have demonstrated that up to two orders of magnitude increases in cantilever responses can be obtained when receptor molecules are immobilized on nanostructured instead of smooth gold surfaces (Lavrik et al. 2001a, 2001b; Tipple et al. 2002).

13.2.2 MICROCANTILEVERS AND THEIR MODES OF OPERATION

As indicated before, the adsorption is accompanied by the changes in the mass of the cantilever and the surface stress. All these processes cause changes in its vibrational frequency (dynamic mode) and induce the bending of the microcantilever (static mode) (Lavrik et al. 2004; Goeders et al. 2008). Schematic illustrations of dynamic and static modes of operation are shown in Figure 13.4.

13.2.2.1 Static Mode

In the *static mode* of operation, the analyte produces a deflection of the cantilever that is detected and converted to an electrical signal (Wu et al. 2001). The bending of microcantilevers has been a subject of study in the context of evolution of fundamental material parameters, such as surface stress and the adsorption on exposure to analytes (Ibach 1997; Liu et al. 2002; Vasiljevic et al. 2004; Zhang et al. 2004; McFarland et al. 2005; Goeders et al. 2008). These experiments hinge strongly on the

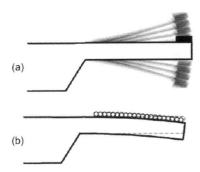

FIGURE 13.4 (a) Absorption of analyte molecules in a sensor layer leads to shift in resonance frequency and (b) Static mode: The cantilever bends owing to adsorption of analyte molecules and the change of the surface stress at the cantilever surface. (Reprinted with permission from Battiston F.M. et al., A chemical sensor based on a microfabricated cantilever array with simultaneous resonance-frequency and bending readout. *Sens Actuators B*, 77, 122–131, 2001. Copyright 2001, Elsevier.)

basis of Stoney's equation, in which the bending of the microcantilevers is associated with the magnitude of the change in the surface stress (Lavrik et al. 2001b). It was established that this deflection is mainly a result of adsorption stress induced by molecular adsorption on just one side of the cantilever. The more analyte is adsorbed, the more the microcantilever will deflect. Depending on the nature of the chemical bonding in the molecule, the deflection can be upward or downward. It is clear that the bending response of a microcantilever is greatly enhanced by passivating a layer (Chen et al. 1995) or tailoring the surface topology (Lavrik et al. 2001b). Lavrik et al. (2001b) compared the deflection sensitivity of microcantilevers with different surface roughness and concluded that it enhances when surface roughness increases. However, Godin et al. (2004) found that a microcantilever with a rough surface deflects less compared to a smooth surface and attributed it to the formation of incomplete mono layers during molecular adsorption. In a similar study, Desikan et al. (2006) did not observe any significant increase in the surface stress, and thus deflection of microcantilevers, due to an increase in the surface roughness. The main advantage of this static mode is its simplicity, since no actuation system is needed.

13.2.2.2 Dynamic Mode

The *dynamic mode* of operation, also called the *resonant mode*, is another method of detecting molecular adsorption using microcantilever- and membrane-based sensors (Oden et al. 1996). In this mode of operation, mass changes during adsorption or desorption can be determined accurately by tracking the eigenfrequency of the microcantilever (Barnes et al. 1994; Schmid et al. 2009; Pustan et al. 2011). Thus, the microcantilever in the sensors is used as a microbalance. The effect of the surface stress on the resonant frequency of microcantilevers was predicted by a model of McFarland et al. (2005). Since the surface stress tends to stretch or compress the beam, the resonant frequency for any given bending mode will either decrease or increase. With this model, one can compute the surface stress that acts on microcantilevers by measuring resonant frequencies. In humidity sensors, upon addition of mass on the cantilever surface, the cantilever's eigenfrequency shifts to a lower value. Resonant frequency shifts of several SU-8-based microcantilevers (microstrings) are shown in Figure 13.5. It is worth noting that the microcantilever-based technique is very sensitive. For example, for the cantilever sensors discussed by Battison et al. (2001), a change of 1 Hz in a resonance frequency corresponded roughly to a mass change of 1 pg. In the dynamic mode, in addition to the resonance

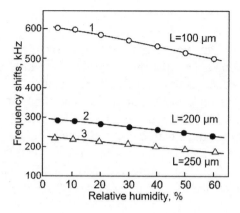

FIGURE 13.5 Resonant frequency shifts of the SU-8 microstrings with length from 100 to 250 μm for a varying relative humidity (RH) measured at 24°C. (Data extracted from Schmid, S. et al., *J. Micromech. Microeng.*, 19, 065018, 2009.)

frequency, the analyte can modify other characteristics of the cantilever oscillation (i.e., resonance amplitude, phase at resonance, or quality factor).

As for the shortcomings of the *dynamic mode*, the main one is the need of an excitation or actuation system in order to excite the cantilever at resonance. For these purposes, one can use the following techniques (Filenko 2008; Martorelli 2008; Vashist and Korotcenkov 2011):

- ***Electrostatic***. This method consists of applying an ac voltage between the two electrodes that constitute a capacitor (one of these electrodes is the resonator), which produces an attractive electrostatic force that bends the resonator harmonically toward the other electrode. This technique is very useful when working in vacuum, since air would damp the motion on the resonator. Otherwise, one has to ensure easy air flow around the resonator. Electrostatic excitation is especially attractive in combination with a capacitive detection method.
- ***Electromagnetic or dielectric excitation***. This method also uses two electrodes that are separated by a dielectric material, thus the whole structure is the cantilever. Here, just as with electrostatic actuation, one applies an alternating voltage between the upper- and bottom-layer electrodes. This voltage will pull the two layers together, thus deforming the dielectric layer and causing lateral stress, which will actuate the cantilever. An important advantage of this method is its avoidance of any external structure, because the actuator is fully integrated onto the cantilever. However, the resonator has to be multilayered, which increases the thickness and complexity of the resonator and thus the minimum detectable mass.
- ***Piezoelectric***. This method uses the dual property of the piezoelectric material. The application of an ac voltage produces a mechanical stress on the piezoelectric material that bends the resonator (Yi et al. 2002). In an *integrated piezoelectric exciter*, there is a sandwich, including piezoelectric material, such as single-crystalline quartz or ZnO (Lee and White 1996), between two metal contacts, which is then placed on a silicon cantilever. By applying an alternating voltage across the piezoelectric material, one can actuate the cantilever. The drawback of this technique is compatibility issues with conventional silicon technology. Modeling of integrated piezoelectric excitation technique is presented by Zhou et al. (2005). *External piezoelectric excitation* utilizes the cantilever holder, which is placed (usually glued) on a commercial quartz crystal. The drawback is that on-chip integration is not provided.
- ***Electrothermal or resistive heating excitation***. This method consists of heating the resonator by means of an ac current through an integrated resistance on the resonator. This method implies expansion of the cantilever material due to the stress exerted by local rise of the temperature around the integrated heating resistor. Usually, the resistor is an integrated diffused resistor or poly-Si that is deposited on the resonator. *Bimorph-effect excitation* is similar to the resistive heating, but the cantilever bends mostly because of the different thermal expansion coefficients of the cantilever material and the heater.
- ***Photothermal***. This method implies thermal expansion of the cantilever material by heating with a focused laser beam or light source. The spot size of a typical optical laser is around 10 μm, which limits the minimum size of the resonator.

The last three methods above use the bimorph effect—that is, the different temperature coefficients or mechanical stress coefficients of the various layer materials forming the cantilever (Fritz et al. 2000b; Lang et al. 1999; Jesenius et al. 2000). This difference in material properties gives rise to a cantilever deflection upon heating or applying mechanical forces. Periodic heating pulses in the cantilever base thus can be used to thermally excite the cantilever in its resonance mode at 10–500 kHz (Baltes et al. 1998;

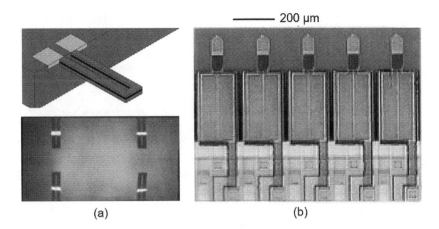

FIGURE 13.6 (a) Microcantilever-NMC60 fabricated by NanoSniff (http://nanosniff.com). The cantilever stack comprises of piezoresistive, boron-doped polysilicon, encapsulated by layers of silicon dioxide. The thickness of the microcantilever stack is 650 nm, and its length is 200 μm. It can be used in chemical and biochemical sensing applications. (b) Piezoelectrically actuated cantilever array. ([a] Reprinted from http://nanosniff.com and [b] Reprinted with permission from Minne, S.C. et al., *Appl. Phys. Lett.*, 72, 2340–2342, 1998. Copyright 1998, AIP Publishing LLC.)

Hierlemann et al. 2000). The cantilevers in which various methods of an excitation are implemented are shown in Figures 13.6 and 13.7.

According to Lange et al. (2002), dynamic mode is preferable in terms of integration of electronics and simplicity of the setup (self-excitation using an amplifying feedback loop). The merit compared to static mode is that the dynamic mode gives absolute information regarding the analyte without a need for careful calibration. The cost in the higher complexity of the dynamic mode compared to the static mode is compensated by much better sensitivity of the sensor.

It is necessary to note that the static and dynamic methods impose completely different constraints on the cantilever design for optimum sensitivity (Lange et al. 2002). The static method requires long and soft cantilevers to achieve large deflections, whereas the dynamic mode requires short and stiff cantilevers to achieve high operation frequencies.

13.3 MICROCANTILEVER DEFLECTION DETECTION METHODS

A scheme for signal readout is critical for real-time measurement, accuracy, and the possibility of integrating microcantilever-based sensors, so it is crucial to implement a readout system capable of monitoring changes with subnanometer accuracy. There are several methods for measuring the deflection of the microcantilever, as discussed below (Lavrik et al. 2004; Goeders et al. 2008; Boisen et al. 2011; Vashist and Korotcenkov 2011).

13.3.1 Optical Methods

Optical readout is one of the most common schemes for detecting the movement of microcantilevers, as derived from standard atomic force microscopy. The optical method (Meyer and Amer 1988), as shown in Figure 13.8, employs a very-low-power laser beam and a position-sensitive detector (PSD). The power of the laser beam is of an order that does not affect the biomolecules coated on the surface of the microcantilever. The laser beam falls on the cantilever and is

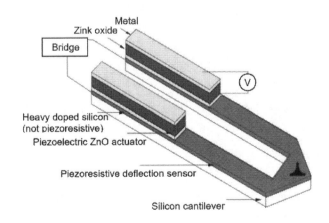

FIGURE 13.7 Schematic of a typical piezoelectric microcantilever (MCL). (Reprinted with permission from Mutyala, M.S.K. et al., Mechanical and electronic approaches to improve the sensitivity of microcantilever sensors, *Acta Mech. Sin.*, 25, 1–12, 2009. Copyright 2009, Springer Science+Business Media.)

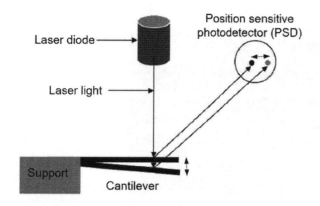

FIGURE 13.8 Schematic of optical detection system for detecting microcantilever deflection. The reflected laser light from the deflected microcantilever falls at different positions on the PSD. Depending on the distance between the two positions of the laser beam on the PSD, the deflection of the microcantilever is determined. (Reprinted from Vashist, S.K., *J. Nanotechnol.*, Online 3, 1–15, 2007. Copyright 2007, AZoiomo as open access.)

reflected, as the gold layer coated on the surface of the cantilever gives it an almost mirrorlike finish. The reflected beam falls on the PSD. When the cantilever is undeflected (i.e., it is not coated with any molecule), the laser beam will fall at a particular spot on the PSD. However, the position of the beam changes as the cantilever deflects.

The optical method is simple and sensitive. The main advantage of this method is the ability to detect deflection in the subnanometer range. However, the method does have disadvantages. This method is critical for small cantilevers based on the diffraction limit of the light used, but the presence of a focused laser beam in a liquid cell environment can result in additional thermal management issues, giving rise to extraneous readings. The alignment system is costly and requires great precision, which can ultimately raise the cost of the sensor device and reduce its portability. In addition, optical techniques are sensitive to changes in the optical density of the sample and can also be subject to artifacts due to changes in the optical properties of the medium surrounding the cantilever, which can move the laser spot on the photodetector surface (Carrascosa et al. 2006). Also, implementation of an optical method for readout of arrays is technologically challenging, as it requires an array of laser sources with the same number of elements as the cantilever array. This technique is employed in optically based commercial array platforms, but sequential switching, on and off, of each laser source is necessary to avoid overlap of the reflected beams on the photodetector.

This problem can be elegantly solved using a scanning laser source, with the laser beam scanned along the array in order to illuminate the free ends of each microcantilever sequentially (Tamayo et al. 2004).

13.3.2 Capacitive Method

The capacitive method is based on the principle that, when the cantilever deflects due to the adsorption of the analyte, the capacitance of a plane capacitor is changed (Blanc et al. 1996). Here, the microcantilever is one of two capacitor plates: a rigid beam with an electrode and a flexible cantilever with another electrode (Goddenhenrich et al. 1990; Brugger et al. 1992). Two electrodes are insulated from each other. This deflection technique is highly sensitive and provides absolute displacement. Capacitance sensing has low temperature coefficients, low-power dissipation, low noise, low cost of fabrication, and compatibility with very large-scale integration (VLSI) technology scaling. For these reasons, capacitive sensing has received the most attention and has been the most used in microelectromechanical systems (MEMS) products. The main problem of this technique in microcantilevers is very small capacitance to measure. Therefore, such sensors are not very sensitive. In addition, the use of this technique is limited in nonconductive environments. In addition, this technique is not suitable to measure large displacements. Therefore, it is limited in its sensing applications.

13.3.3 Piezoelectric Method

The piezoelectric technique requires deposition of a piezoelectric material, such as ZnO, on the cantilever. Due to the piezoelectric effect, transient charges are induced in the piezoelectric layer when the cantilever is deformed. The main disadvantage of this technique is that, in order to obtain large output signals, it requires the thickness of the piezoelectric film to be well above the values that correspond to optimal mechanical characteristics.

13.3.4 Interferometry Method

The interferometry method is an optical detection technique (Erlandsson et al. 1988; Rugar et al. 1989) based on the interference of a reference laser beam with the laser beam reflected by the cantilever. The cleaved end of an optical fiber is brought close to the cantilever surface. One part of the light is reflected at the interface between the fiber and the surrounding medium, while the other part is reflected at the cantilever back into the fiber. The two beams interfere inside the fiber, and the interference signal can be measured with a photodiode. This is a highly sensitive method and gives a direct and absolute measurement of displacement. In this method, light has to be brought close to the cantilever surface to get enough reflected light. Optical fiber is a few micrometers away from the free end of the microcantilever.

13.3.5 OPTICAL DIFFRACTION GRATING METHOD

The reflected laser light from the interdigitated cantilevers forms a diffraction pattern in which the intensity is proportional to the cantilever deflection (Manalis et al. 1996). This can be used for atomic force microscopy, infrared detection, and chemical sensing.

13.3.6 CCD DETECTION METHOD

A charge-coupled device (CCD) camera was used by Kim et al. (2003) to measure the deflection of the cantilever in response to the analyte. The position-sensitive detector, here is the CCD camera that records the laser beam deflected from the cantilever, can measure deflection in the 0.001 nm range. However, the positioning of the fibers is a difficult task. The method works well for small displacement.

13.3.7 HARD-CONTACT/TUNNELING

One of the important parameters of a read-out system is the signal-to-noise ratio. One way to obtain a large signal-to-noise ratio is to use a system with a highly nonlinear response to changes in deflection. Tunneling and hard contact read-out are two such techniques. A tunneling displacement sensor was used as the read-out system in the very first atomic force microscopy (AFM) and in 1991 was fully integrated on a chip to measure static cantilever deflections (Kenny et al 1991). Recently, the detection of resonant frequencies of nanoscale cantilevers has been performed by tunneling (Scheible et al. 2002) and by hard-contact read-out (Dohn et al. 2006). In tunneling read-out, the cantilever is placed in close proximity to a counter electrode, and the tunneling current between the electrode and the cantilever is measured. The signal is, in principle, extremely sensitive, but the fabrication and operation are complex, since very small electrode-cantilever gaps need to be realized. The gap spacing needs to be adjustable, and the measured tunneling current is in the pA regime, which requires high-quality signal amplification. In hard-contact read-out, the cantilever is allowed to touch the electrode and the current ($\approx$ 10 nA) running through the system measured.

13.3.8 PIEZORESISTIVE METHOD

The piezoresistive method (Meyer and Amer 1988; Thaysen et al. 2000; Yang et al. 2003; Bausells 2015) involves embedding a piezoresistive material near the top surface of the cantilever to record the stress change occurring at the surface of the cantilever. Piezoresistance is an intrinsic property of certain materials that causes the change in their electrical resistance in response to an externally applied stress (Ristic 1994). As the microcantilever deflects, it undergoes a stress change that applies strain on the piezoresistor, thereby causing the change in the resistance, which can be measured by electronic means. Thus, the piezoresistor (a mechanical sensitive element) transforms the stress change into an electrical signal. The advantage of this method is the integration of the readout system on the chip. The piezoresistive method is compatible with microfabrication and miniaturization. In addition, the temperature control can be easily implemented.

The piezoresistive approach has been widely studied as a signal transduction method for various applications, but it has several disadvantages. The main disadvantage is the intrinsic noise level, which directly affects the resolution and the sensitivity compared to optically detected cantilevers (Carrascosa et al. 2006), although a reducing of the thickness of piezoresistive cantilevers might increase sensitivity. However, the cross-sectional structure of a piezoresistive cantilever is complex. A piezoresistor has to be embedded in the cantilever, which makes the fabrication of such a cantilever with a composite structure more complicated. Therefore, there are technological limitations on fabricating thin, highly sensitive cantilevers. As a result, the deflection resolution for the piezoresistive readout system is only 1 nm, in comparison to 1 A by optical detection. Moreover, piezoelectric readout requires electrical connections to the cantilever and their isolation from the solution. This method will be considered in more detail in Section 13.6.3.

The piezoresistors are placed where the bending of cantilever (membrane) is maximum and, hence, the signal is large (i.e., near the ridges) (see Figure 13.9).

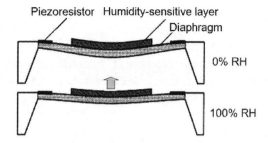

FIGURE 13.9 General layouts of piezoresistive humidity sensors. The membrane-type device uses piezoresistors placed on top of the diaphragm. As the humidity increases, the resulting humidity-dependent volume change of the polyimide layer prompts a deformation of the polyimide bridge substrate bimorph and leads to a bending of the plate, which is then transformed into an output voltage by an integrated piezoresistive bridge. (Reprinted with permission from Buchhold, R. et al., Design studies on piezoresistive humidity sensors, *Sens. Actuators B*, 53, 1–7, 1998. Copyright 1998, Elsevier.)

It has been shown that the sensitivity of piezoresistive humidity sensors with resistors located at the periphery of the diaphragm does not depend significantly on the shape of the diaphragm (Buchhold et al. 1998a, 1998b). Hence, the specification of the diaphragm shape can be based solely on technological considerations.

The deflection is measured as a resistance change in the embedded strain gauges, and is linearly proportional to the shear stress. The relative change in resistance of the piezoresistive material as a function of applied strain is represented by Equation 13.4:

$$\frac{\Delta R}{R} = K_l \cdot \delta_l \cdot K_t \cdot \delta_l \quad (13.4)$$

where δ is the strain in the material and K denotes the gauge factor (GF), which is a material parameter. Subscripts l and t refer to the longitudinal and transverse parts of the gauge factor.

The sensitivity of a piezoresistor varies in proportion to the thickness t and the radius of curvature. The gauge factor is proportional to Young's modulus E, which is an intrinsic characteristic of the material. The gauge factor can also be calculated directly by straining the cantilevers and measuring the resistance change according to Equation 13.5:

$$GF \cdot \delta = \frac{\Delta R}{R} \quad (13.5)$$

where δ is the strain in the material and R is the resistance. For a sensitive device, the gage factor should be of the order of 100.

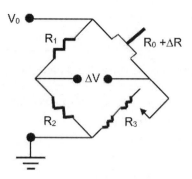

FIGURE 13.10 The Wheatstone bridge circuit used for the piezoresistive microcantilever. (Reprinted from Vashist, S.K., *J. Nanotechnol. Online* 3, 1–15, 2007. Copyright 2007, AZojomo Com Pty Ltd. as open access.)

The piezoresistive cantilever beam can be used as an arm of a Wheatstone bridge circuit, as shown in Figure 13.10. The resistance of the variable-resistance arm ($R_0 + \Delta R$) in Figure 13.10 can be determined using the common voltage divider formula, shown in Equation 13.6:

$$\Delta V = V_0 \left[\frac{R_2}{(R_1 + R_2)} - \frac{R_3}{(R_0 + \Delta R + R_3)} \right] \mapsto R_0$$
$$+ \Delta R = R_3 \left[\frac{V_0(R_1 + R_2)}{R_2 V_0 - \Delta V(R_1 + R_2)} - 1 \right] \quad (13.6)$$

There will be a resistance change whenever the cantilever is subjected to a deflection. Buchhold et al. (1998a, 1998b) have shown that, in the case of using a piezoresistor's bridge, the output voltage can be presented in the form of equation:

$$V_0 = k(\sigma_0 + \sigma_h + \sigma_{th}), \quad (13.7)$$

where k is a constant, and σ_0, σ_h and σ_{th}, respectively, denote the intrinsic-, thermal- and humidity-dependent stress.

As follows from an examination of methods suitable for measuring the deflection of the microcantilever, some methods are robust and well established, but rather bulky, whereas other techniques are a bit more immature but with the promise of becoming miniaturized. Advantages and limitations of presented read-out methods are summarized in Table 13.1.

TABLE 13.1
Summary of Read-Out Methods

Read-Out Method	Advantages	Shortcomings
Optical leverage	Simple read-out method, known from AFM; can be used on any cantilever with a good optical quality	Difficult to apply for large arrays; prone to optical artifacts mass, such as the change in refractive index; not suitable for nanometer-sized cantilevers
Capacitive	Useful for nanometer-sized cantilevers; read-out does not affect the mechanical properties of the cantilever	Stray capacitances make pre-amplifications and CMOS integration necessary; this complicates the fabrication
Piezoelectric	The principle can be used for actuation as well as read-out	Many piezoelectric materials are not cleanroom compatible; many piezoelectrical materials are only suitable for dynamic measurements
Piezoresistive	Facilitates large arrays and system integration; works in all media and in all modes of operation	A piezoresistive layer needs to be integrated into the cantilever, which affects the mechanical performance; for operation in liquid, care has to be taken in order to insulate the resistor
Hard contact	Offers a digital read-out in which a signal is only generated when the cantilever is in resonance; high signal-to-noise ratio	Wear of the counter electrode is a challenge; works only in the air
Tunneling	Potentially very sensitive detection of cantilever bending	Complicated operation and only works in the air
Integrated optical methods (waveguides)	Suitable for large-scale arrays with the same sensitivity as optical leverage	More complicated fabrication and packaging; prone to optical artifacts, such as changes in refractive index

Source: Data extracted from Boisen, A. et al., *Rep. Prog. Phys.*, 74, 036101, 2011.
CMOS, complementary metal–oxide–semiconductor.

13.4 RESONANT OPERATING MODE

13.4.1 MECHANICAL PROPERTIES OF MICROCANTILEVERS

The basic mechanical parameters of a cantilever are the spring constant and the resonance frequency (Cleveland et al. 1993). As specified by the Hooke's law equation (13.8), the spring constant k is the proportionality factor between the applied force, F, and the resulting bending of the cantilever, z:

$$F = -k \cdot z \quad (13.8)$$

The spring constant determines the stiffness of the cantilever. For a rectangular cantilever of length l, the spring constant is given by Equation 13.9:

$$k = \frac{3E \cdot I}{l^3} \quad (13.9)$$

where E is Young's modulus and I is the moment of inertia. A typical spring constant for a stress-sensitive cantilever is in the range of 1 mN/m to 1 N/m.

The resonance frequency f_{res} for a simple rectangular cantilever can be expressed as

$$f_{res} = 0.162 \cdot \frac{\sqrt{E}}{\sqrt{\rho}} \frac{h^3 w}{l^2} \quad (13.10)$$

where ρ is the mass density, and h and w denote the height and width of the cantilever, respectively. The moment of inertia for a rectangular cantilever can be written as

$$I = \frac{wh^3}{l^3} \quad (13.11)$$

A simpler expression for the resonance frequency can be written as a function of the spring constant (Cleveland et al. 1993):

$$f_{res} = 0.32 \cdot \frac{\sqrt{k}}{\sqrt{m}} \quad (13.12)$$

where k is the spring constant and m is effective mass of the cantilever, $m = \rho h l w$. The relation shows that the resonance frequency increases as a function of increasing spring constant and decreasing cantilever mass. As can be seen from the above equation, an increase in mass will be characterized by a decrease in resonance frequency. This result is very important in differentiating mass loading–related events from changes in the material properties of the microcantilever.

The relationship between the change in f_{res} and the change in mass can be seen in Equation 13.13,

$$\Delta m_a = \frac{k}{4n\pi^2}\left(\frac{1}{f_1^2} - \frac{1}{f_0^2}\right) \quad (13.13)$$

where n is a geometric factor ($n = 0.24$ for rectangular microcantilevers) and f_0 and f_1 are the resonance frequencies before and after the mass has been added, respectively.

As can be seen from Equation 13.13, any change in the spring constant of the microcantilever will also lead to a direct effect on the resonance frequency of the microcantilever. There are certain circumstances in which the spring constant of the microcantilever can change during a chemical measurement. For example, if the material properties of the selective film or metallic film applied to the microcantilever alter its spring constant appreciably, a change in resonance frequency will be observed. This can occur when the thickness of the selective film or metallic layer approaches the thickness of the microcantilever, or when these films are innately stiff (Chen et al. 1995). Thus, in general, the change of the resonance frequency is due to a combination of mass loading and changes in the spring constant of the microcantilever.

However, by measuring the bending and resonance frequency simultaneously, any changes in spring constant can be quantified. The mass sensitivity of frequency-based microcantilever measurements, S_m, can be calculated using Equation 13.14:

$$S_m = \frac{1}{f} \cdot \frac{\Delta f}{f \cdot \Delta m} \quad (13.14)$$

where Δm is normalized to the active sensing area of the device (Thundat et al. 1997). Another related measure of the sensitivity is the minimum detectable surface mass density, Δm_s^{min}, given by Equation 13.15:

$$\Delta m_s^{min} = \frac{1}{S_m} \cdot \frac{\Delta f}{f} \quad (13.15)$$

which is the minimum detectable mass over the active sensing area of the sensor (Datskos and Sauers 1999).

13.4.2 Mass Resolution Limitations

Several factors influence the cantilever transducer sensitivity (Ekinci et al. 2003), including variations of the effective mass and the resonance frequency instability. The latter is conditioned by the presence of extrinsic noises of the transducer and readout, and intrinsic fundamental noises of the resonator. In turn, the former is defined by the material and geometry of the resonator, and the latter is governed by fluctuations of quantum nature. Obviously, better mass sensitivity can be attained with lower effective mass of the resonator. However, nanoresonators are more susceptible to fundamental noises as the device geometry is shrunk down to the sizes of detectable particles. Generally speaking, intrinsic fundamental noises due to thermomechanical fluctuations, temperature fluctuations, adsorption–desorption, and momentum exchange are sufficient only for nano-sized cantilevers and Dalton-range sensitivity limits. Therefore, sensing of single atoms would be hardly resolvable without device optimization to diminish the influence of intrinsic noise. Real piezoresistive cantilevers are much more susceptible to readout unit noises, such as Johnson noise and $1/f$ noise. $1/f$ noise can be neglected by operating the cantilever in dynamic mode at frequencies higher than 10 kHz.

13.5 HUMIDITY-SENSITIVE MATERIALS

Hydrophilic polymers are generally used as a hygroscopic material in piezoresistive humidity sensors. As is known, hydrophilic polymers are good adsorbents, and they tend to swell for an increasing moisture content. At the same time, polyimide is a polymer, possessing such properties while maintaining high stability and therefore most often used for these purposes.

13.5.1 Polyimide

Polyimide films are developed for electronic industries to be used as interlayer dielectrics, passivation layers, and stress buffers (Khan et al. 1988; Horie and Yamashita 1995). There are several reasons that make polyimide films applicable in microelectronics industry. The polyimide is chemically inert in its cured form and thermally stable material up to temperatures around 450°C. Polyimides have a good resistance to chemical corrosion (Delapierre et al. 1983). In addition, polyimide is fully compatible with a silicon processing technology (Qui et al. 2001); it is a perfect planarizer used to planarize irregular surfaces (Ralston et al. 1991). It also has a low relative permittivity and high breakdown voltage.

It is the set of these parameters that made polyimide the most widely used humidity-sensitive material in humidity sensors.

Humidity responses of polyimide films have been investigated in many works (Denton et al. 1985, 1990; Ralston et al. 1996; Matsuguchi et al. 1998; Jachowicz and Weremczuk 2000). Experiments show that the dielectric constant of polyimide films change from approximately 3 to 4 as the relative humidity (RH) changes from 0% to 100% RH. Moreover, the dielectric constant change with respect to the humidity change is almost linear, especially within the 20% to 70% RH range (Kang et al. 2000; Dokmeci and Najafi 2002; Laville and Pellet 2002). At that, polyimide films absorb much more water than ceramic oxides, approximately 3% by weight on the average (see Figure 12.17b). Moreover, this absorption is reversible with little hysteresis or its absence.

In polyimides, water molecules are either chemically bound to the polymer matrix or are condensed in microvoids, depending on the humidity level. According to Melcher et al. (1989), water molecules in polyimides are bound either to the carbonyl group or to the oxygen of the ether linkage. By measuring the equilibrium moisture content in a material as a function of the water vapor pressure (the plot of this measurement result is called the *equilibrium moisture content isotherm*), it was shown that moisture begins to condense as the RH level becomes higher (Yang et al. 1985, 1986). Experimental results show that the isotherm curve is concave with respect to the vapor pressure axis at the lower vapor pressures, while it is convex to the axis at higher pressures. The transition from the concave to the convex form is explained by the clustering of water molecules inside the material; the transition point, at which moisture clustering begins, depends on the material and the temperature.

An equivalent expansion coefficient for polyimide, called the *humidity-fraction coefficient*, is reported with typical values of 2.2–$8.5 \cdot 10^{-5}$/%RH (Gerlach et al. 1994; Schroth et al. 1996). The experiments showed that the swelling behavior of polyimide layers is also fairly linear. It was found that the humidity-dependent swelling mechanism of polyimide layers consisted of a very complex mixture of adsorption, absorption, diffusion, and capillary condensation processes. Despite this, the response of humidity-dependent swelling can be calculated approximately and described by Fick's second law (Crank 1976). It is worth noting that humidity extension coefficient of polyimide (CHE) strongly depends on the polyimide type used and its fabrication history, especially on the deposition and curing conditions (see Figure 13.11). It is seen that humidity extension coefficients are approximately 60–80 ppm/%RH.

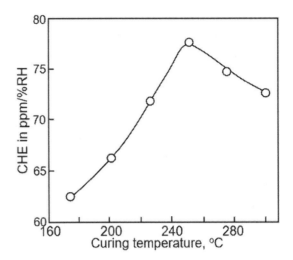

FIGURE 13.11 The humidity extension coefficient of polyimide vs curing temperature. Curing time, 30 min. (Reprinted with permission from Sager, K. et al., Humiditydependent mechanical properties of polyimide films and their use for IC-compatible humidity sensors, *Sens. Actuators A*, 53, 330–334, 1996. Copyright 1996, Elsevier.)

The extension coefficient increases up to approximately 250°C, while above 250°C, a decrease in the humidity extension coefficient is observed.

Water transport inside a material occurs not only by diffusion, but also by reaction. At higher temperature, however, the reaction occurs at a very slow rate so that the transport due to this mechanism can be neglected. In the temperature range at which humidity sensors are normally operated (> −50°C), moisture transport is due to diffusion through the microvoids. The experiment has shown that the diffusion constant of water in polyimides is often very large, leading to the fast response time. According to simulations carried out by Sager et al. (1996), a diffusion coefficient (normalized by the square of thickness) is of approximately $2.2 \cdot 10^{-3}$ s^{-1}. According to experimental estimations, the diffusion coefficient of water in the polyimide films ranges from 3×10^{-8} to 5×10^{-9} cm^2/sec, depending on the type of polyimide (Ree et al. 1991).

Polyimide deposition can be performed at the end of a CMOS fabrication process, and therefore humidity sensors, which are monolithically integrated with the read-out circuit, can be obtained (Qui et al. 2001). Polyimide films can be divided into two groups according to the method that they are processed in microfabrication. Photosensitive polyimide films are processed like photoresists (Bowden 1988; Cech et al. 1991). They are masked, exposed to UV light, and patterned by their own developers. Nonphotosensitive polyimide films are patterned by dry or wet etching like a regular layer in microelectronics, so an additional photolithography step is required in the fabrication. On the other hand, using photosensitive polyimide films considerably reduces the number of processing steps and makes the process easier. Furthermore, since they are directly masked and developed, the resulting pattern has better defined features (Horie and Yamaskito 1995).

Major drawbacks of polyimide films can be stated as long-term stability and chemical durability problems (Ralston et al. 1991). In harsh environments, humidity responses of polyimide films may drift in time. In addition, the presence of water for a long time may cause the failure of the device. However, recent studies have significantly improved the stability of polyimide films in harsh environments.

For cantilever- and membrane-based sensors, the adhesion of the coating and the absence of stresses play a great role. Low-stress materials with high elongation are prerequisites to avoid cracks. The experiment showed that the polyimide has excellent adhesion to metals (and vice versa) and to the cured polyimide films. Difficulties arise in meeting both demands (i.e., good mechanical properties and good adhesion, since polyimides with rigid and linear molecular structure show low stress but have weaker adhesive forces than flexible ones, which, in turn, have poorer mechanical properties) (Numata et al. 1991). Adhesive forces can be increased by using special adhesion prompters for various kinds of polyimides and substrates (e.g., amino-organosilane), or by plasma treatment of the substrate surface (Toray 1992). Some new types of commercially available polyimides already contain an integrated adhesion promoter: Pimel G-X Grade (Asahi 1994), Pyralin PI 2700 (DuPont 1994), and probimide 7000, 7500 (OCG 1994).

Normally, the exposure of polyimide layers to water at elevated temperatures lowers the adhesion of polyimides, because water can hydrolyze chemical bonds between the polyimide and the substrate surface. Recently, the adhesive forces have been significantly improved. New generations of polyimides have passed the tape test, even when wafers coated with the polyimide on a silicon nitride surface were boiled at 121°C and 2 atm for 400 hours (DuPont 1994) and 500 hours (Asahi 1994; OCG 1994). Excellent adhesion strength values of up to 70 MPa have been reported (Asahi 1994). Of course, these data depend on the interface polyimide/substrate.

13.5.2 Other Humidity-Sensitive Polymers

Besides polyimide (Buchhold et al. 1998a, 1998b, 1998c), the volume expansion with respect to the moisture content has been found to be linear to a good approximation for different epoxies (Vanlandingham et al. 1999; Ardebili et al. 2003), cellulose (Fenner 1995),

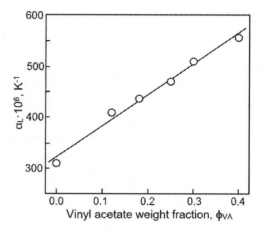

FIGURE 13.12 Coefficient of linear thermal expansion as a function of the vinyl acetate weight fraction in EVA—the composition of poly(ethylene-co-vinylacetate) (PEVA) and polyvinylacetate (PVAc). (Reproduced with permission from Gonzalez-Benito, C. et al., Determination of the linear coefficient of thermal expansion in polymer films at the nanoscale: Influence of the composition of EVA copolymers and the molecular weight of PMMA, *Phys. Chem. Chem. Phys.*, 17, 18495, 2015. Copyright 2015, The Royal Society of Chemistry.)

poly(ethylene-co-vinylacetate), PEVA, (Gonzalez-Benito et al. 2015), poly(methyl methacrylate), PMMA, (Gonzalez-Benito et al. 2015), and also for SU-8 (Feng and Farris 2003). It is important to note that the linear coefficient of thermal expansion, as it was shown by Gonzalez-Benito et al. (2015), can be controlled through a change in the technology of synthesis and composition of polymers (Figure 13.12).

With regard to actual use of the polymers in the development of piezoresistive humidity sensors, then in addition to polyimide films (Gerlach et al. 1994; Sager et al. 1944; Buchhold et al. 1998b; Waber et al. 2014; Huang et al. 2015), the poly(vinyl acetate) (PVA) (Battiston et al. 2001; Gunter et al. 2005), Poly(3,4-ethylenedioxythiophene) (PEDOT)/PSS (Sappat et al. 2011), SU-8 (Patil et al. 2014), cellulose acetate butyrate (CAB) (Loizeau et al. 2012), carboxymethylcellulose (CMC) (Battiston et al. 2001), polyelectrolyte multilayers (PEMs) consisting of poly(allylamine hydrochloride) (Toda et al. 2014), and PANI layers (Patil et al. 2014) were used. However, polyimide-based sensors still prevail.

13.6 HUMIDITY SENSOR IMPLEMENTATION

13.6.1 Functionalization Methods

To apply polymers to the surface of membranes and cantilevers, many methods have been tried (Vashist and Korotcenkov 2011). The coating method used for the surface functionalizing of nanocantilevers should be fast, reproducible, and reliable, and allow one or both cantilever surfaces to be coated separately (Lang 2009). Thin metallic or ceramic films are applied to the desired surfaces of the cantilever using standard film deposition techniques, such as thermal evaporation or sputtering. To prevent delamination, an intermediate adhesion-promoting layer is often employed. For example, during fabrication of cantilevers with a gold covering deposited by evaporation, a thin underlayer of titanium or chromium is used to promote adhesion of the gold film onto the silicon cantilever. The disadvantage of all these methods is that they are suitable only for coating large areas, not individual cantilevers in an array, unless shadow masks are used. Such masks need to be accurately aligned to the cantilever structures, which is a time-consuming process.

Other methods to coat cantilevers use manual placement of particles onto the cantilever, which requires skillful handling of tiny samples. Cantilevers can also be coated by directly pipetting solutions of the probe molecules onto the cantilevers or by employing airbrush spraying and shadow masks to coat the cantilevers separately. Microdropping may also be used for deposition of different sensing layers (Urbiztondo et al. 2009). All these methods, however, have only limited reproducibility and are very time-consuming if a large number of cantilever arrays has to be coated.

A convenient method to coat microcantilevers with probe molecules is to insert the cantilever array into an array of dimension-matched disposable glass capillaries (Lang and Gerber 2008). For example, Baller et al. (2000) and Fritz et al. (2000a) used glass capillaries with an outer diameter of 240 μm. The inner diameter was 150 μm, allowing sufficient room to insert the cantilevers (width, 100 μm) safely. This method has been successfully applied for the deposition onto cantilevers of a variety of materials such as polymer solutions and self-assembled monolayers (Lang and Gerber 2008). Incubation of the microcantilever array in the microcapillaries takes from a few seconds (self-assembly of alkanethiol monolayers) to several tens of minutes (coating with protein solutions).

It was found, however, that for coating with polymer layers, inkjet spotting is the most acceptable and controlled technology (Calvert 2001). In addition, it is possible to coat only the upper or lower surface. The method is also appropriate for coating many cantilever sensor arrays in a rapid and reliable way (Bietsch et al. 2004a, 2004b; Toda et al. 2014). Inkjet spotting allows a cantilever to be coated within seconds and yields very homogeneous, reproducibly deposited layers of well-controlled thickness (Lange et al. 2002; Savran et al. 2003). Figure 13.13 shows a basic print head consisting of a piezoelectric diaphragm above or on the

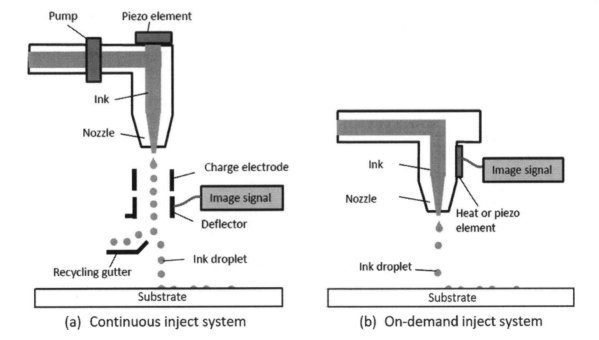

FIGURE 13.13 Schematics of inject printing technology: (a) a continuous inkjet printer and (b) an on-demand inkjet printer. A positioning system allows accurate placement of single droplets onto selected cantilevers. When deposited with a small pitch, the droplets merge into a continuous layer, covering the entire cantilever length. (Reprinted from http://www.dp3project.org/technologies/digital-printing/inkjet.)

side of an ink channel right above the nozzle (Le 1998). Dynamic deflection of the pulsed-voltage activated piezoelectric diaphragm can generate a pressure wave to eject the ink out of the nozzle, squeezing out a liquid jet that breaks into droplets due to the liquid's surface tension. A continuous mode (Le 1998; Wallace et al. 2007) of ink printing has a faster rate of 0.5 µL droplet generation for 80–100 kHz. A deflection plate is electrically controllable to aim the droplets onto a substrate. The excess droplets are recirculated from the gutter. This technique has also been applied to functionalize a polymer-coated microcantilever array for chemical vapor detection experiments (Bietsch et al. 2004b). For these purposes, one can use already available devices on the market, such as an Autodrop MD-P-802 inkjet spotter from Microdrop Technologies that allows depositing a defined number of droplets at the same location. For localized inkjet spotting, one can also use the formation of silicon wafers of hydrophobic layers that prevent spreading over the surface of the surface modification (Figure 13.14). For example, Loizeau et al. (2015) used for these purposes a Teflon-based polymer deposited by molecular vapor deposition.

Because the cantilever beam shows a decrease in resonance frequency due to not only the absorption of an analyte in the sensitive layer, but also the deposition of the sensitive layer itself, the polymer thickness must be monitored during the coating procedure. As can be anticipated, the polymer coating leads to a decrease of the vibrational amplitude and to a reduction of the quality factor (Lange et al. 2002). One example of such influence on the parameters of microcantilevers is presented in Figure 13.15.

It should be noted that a linear model for the decrease of the resonance frequency with increasing polymer thickness is not valid for the thickness of more than 1 µm (modulus contribution). With increasing thickness, the decrease of the cantilever resonance frequency is less than proportional. In fact, for thicknesses greater than 30 µm, the resonance frequency will increase again, since the system has now changed from a polymer-loaded silicon cantilever to a silicon-loaded polymeric cantilever, with the consequence that the resonance frequency again increases linearly with increasing polymer thickness (Lange et al. 2002).

The sensitivity of the cantilever to adsorption also depends on the thickness of the sensing layer. For example, Lange et al. (2002) found that, for cantilevers coated with polymer films, the sensitivity increases linearly with increasing layer thickness in the range of thin polymer layers (2–4 µm) and then shows saturation-like behavior at larger thickness (> 4 µm), as seen in Figure 13.16.

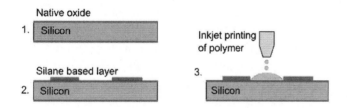

FIGURE 13.14 Fabrication of defined hydrophobic zones on silicon wafers for localized inkjet spotting. (Reprinted from *Sens. Actuators A*, 228, Loizeau, F. et al., Comparing membrane- and cantilever-based surface stress sensors for reproducibility, 9–15. Copyright 2015, with permission from Elsevier.)

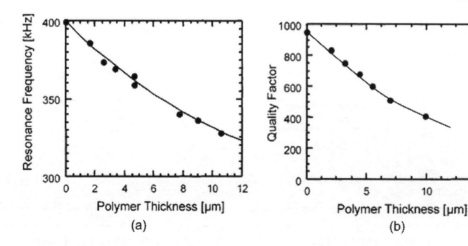

FIGURE 13.15 (a) Resonance frequency shift (a) and quality factor of a cantilever (b) as a function of the polymer thickness (PEUT). The 150 µm-long resonant cantilever was fabricated using complementary metal–oxide–semiconductor (CMOS) technology. (Reprinted with permission from Lange, D. et al., Complementary metal–oxide–semiconductor cantilever arrays on a single chip: Mass-sensitive detection of volatile organic compounds, *Anal. Chem.*, 74, 3084–3095, 2002. Copyright 2002, American Chemical Society.)

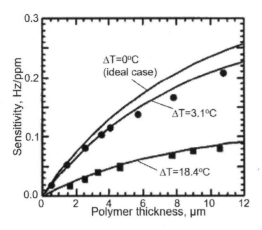

FIGURE 13.16 Measured and modeled (*solid lines*) n-octane sensitivity vs polymer thickness for different cantilever temperatures. ΔT refers to the ambient temperature (gas phase and sensor chip). The 150 μm-long resonant cantilever was fabricated using complementary metal–oxide–semiconductor (CMOS) technology. (Reprinted with permission from Lange, D. et al., Complementary metal–oxide–semiconductor cantilever arrays on a single chip: Mass-sensitive detection of volatile organic compounds, *Anal. Chem.*, 74, 3084–3095, 2002. Copyright 2002, American Chemical Society.)

The reason for this behavior is a mechanical one and is not a consequence of the involved physicochemistry. At small polymer thickness, the mass increase upon the gas absorption and the corresponding frequency shift are linearly correlated with the polymer thickness. The thicker the polymer is, the more is the gas absorbed, and the higher is the absolute frequency change. This does not hold for thicker polymer layers. Here, the fundamental resonance frequency, f_0, of the coated cantilever is considerably reduced as a consequence of the increased added mass of the thick polymer layer. The analyte-induced frequency shift (absolute value) upon gas exposure is also reduced due to the lower starting resonance frequency without analyte dosing. The sensor signal and hence the sensitivity are significantly lower than would be expected from a linear extrapolation of the behavior for the small polymer layer thickness.

13.6.2 Capacitive Humidity Sensors

Cantilever and membrane-based humidity sensors of a capacitive type are devices in which the effect of air humidity on the distance d between the two plates or the overlapping area A between the two plates is realized. Usually, they are the bimorph bending sensors (Britton et al. 2000; Chatzandroulis et al. 2002; Lavirk et al. 2004; Govardhan and Alex 2005; Lang et al 2005; Lee and Lee 2005; Li et al. 2006; Goeders et al. 2008; Nordstrom et al. 2008; Muniraj 2011). The analyte absorption/adsorption on the selective layer material (usually a polyimide) induces mass and the surface stress changes that result in a change in the device resonance frequency and/or bending, which are then detected as a change in capacitance. In capacitive readout, the capacitance is measured between a flexible and a fixed electrode, which are separated by a small gap. The flexible electrode is a conductor on the cantilever or membrane, and the fixed electrode is a conductor on the substrate (Chatzandroulis et al. 2011). As the cantilever or diaphragm deflects, the distance between the two plates changes, and this changes the capacitance of the system. Typical capacitive humidity sensors are shown in Figures 13.17 and 13.18. Figure 13.17 illustrates a diaphragm-based sensor fabricated using bulk-micromachining technology, and Figure 13.18 shows a cantilever-based capacitive humidity sensor. Such a sensor was developed by Chatzandroulis et al. (2002). The Si cantilevers had square-shaped ends for increased capacitance and various lengths (500–2500 μm). The absorption of water molecules causes the upper layer of the cantilever to expand, and hence induces a surface tensile stress. This causes an upward bending of the suspended

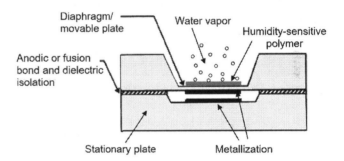

FIGURE 13.17 A cross-section schematic diagram of a bulk-micromachined, diaphragm-based capacitive humidity sensor.

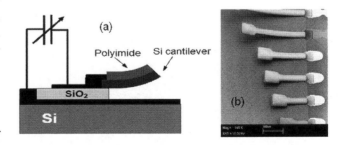

FIGURE 13.18 (a) Schematic diagram and (b) top-surface view of cantilevers covered with polyimide and with a square-shaped end for increased capacitance. (Reprinted with permission from Chatzandroulis S. et al., Fabrication of single crystal Si cantilevers using a dry release process and application in a capacitive-type humidity sensor, *Microelectron. Eng.*, 61–62, 955–961, 2002. Copyright 2002, Elsevier.)

structure away from the glass substrate, which changes the capacitance between the two structures. Thus, in our case, an increase in humidity is accompanied by a decrease in the capacity of the sensor.

The main advantages of capacitive microsensors are low-power, simple structures, low temperature sensitivity, and low-cost integration of the read-out to the portable systems. It outdraws peizoresistive systems because of its low power consumption. The drawback is that the bending amplitude is limited by the space between the microcantilever and the bottom electrode. Excessive signal loss from parasitic capacitance is another serious disadvantage, which hindered the development of miniaturized capacitive sensors until on-chip circuitry could be fabricated.

As shown in Figures 13.17 and 13.18, there are many options for manufacturing capacitive sensors. Of course, each of them has its advantages and disadvantages. For example, Lee and Lee (2003) compared the sensitivity of three humidity sensors operated in a bending mode: single-beam, double-beam, and microbridge-based ones, as shown in Figure 13.19. Three sensors were tested in a closed chamber with humidity level ranging from 45% to 95% RH. Figure 13.19 shows that the single-beam sensor provides a high degree of humidity sensitivity. Although the results suggest that the microbridge type of humidity sensor provides a lower sensitivity, it is a more robust device than the single-beam-based sensor. This means that the configuration of the capacitive sensor should depend on the application and the requirements for the sensor. The operation mode also affects the requirements to cantilever. For resonance frequency measurement, the sensitivity improvement can be realized by reducing the microcantilever dimensions in order to increase stiffness and subsequent resonant frequency. However, for bending mode, higher sensitivity will be achieved with a lower-stiffness microcantilever and larger electrode area (Lee and Lee 2003).

It is understood that, at present, there are no strict requirements for the configuration and technology of manufacturing cantilevers and membranes intended for capacitive humidity sensors. Each developer tries to offer his or her own version and approach to the manufacture of sensors, justifying this by improving the parameters of the devices being developed. For example, Lee and Lee (2003) have developed a cantilever-based humidity sensor with configuration shown in Figure 13.20. In addition to the humidity sensor, the sensor also includes a temperature sensor based on Pt resistor. The integrated microtemperature sensor is required for temperature compensation of the humidity measurement by providing temperature signals to the humidity transducers. The second electrode of the capacitor was fixed on a glass substrate. A gold layer was electron-beam evaporated directly on both the moveable cantilever part of the sensor and on the stationary glass substrate. Lee and Lee (2003) state that, having compensation of the temperature drift, the humidity value can be precisely determined. According to Lee and Lee (2003), sensors had high stability (< ±0.8%), low

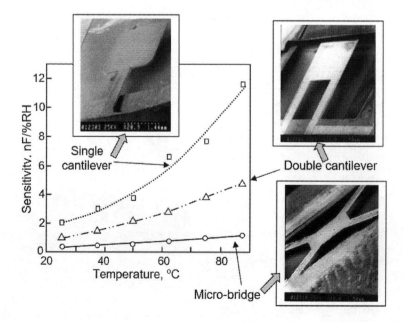

FIGURE 13.19 Temperature effect on humidity sensitivity of the three types of humidity sensors. The total microcantilever beam thickness was 20 μm (nitride: 1 μm/silicon: 19 μm) and all cantilevers exhibit upward bending due to the stress relaxation caused by the shrinking of the cured polyimide. (Data extracted Lee, C.Y. and Lee, G.B., *J. Micromech. Microeng.*, 13, 620–627, 2003.)

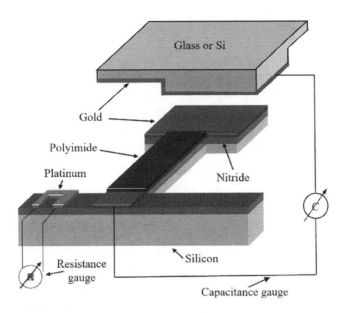

FIGURE 13.20 Schematic representation of the integrated capacitance humidity sensor and temperature sensor. Absorption of water molecules causes the upper layer of the cantilever to expand, inducing a surface tensile stress. As a result, the cantilever bends up, giving rise to a measurable change in the capacitance. (Idea from Chatzandroulis, S. et al., *Microelectron. Eng.*, 61–62, 955–961, 2002; Lee, C.Y. and Lee, G.B. *J. Micromech. Microeng.*, 13, 620–627, 2003; Bombieri, N. et al. (2018) Smart systems design methodologies and tools. In: Bosse S., Lehmhus D., Lang W. (Eds.), *Material-Integrated Intelligent Systems: Technology and Applications*, Wiley, Weinheim, Germany, pp. 55–80.)

hysteresis (1.9% RH) at high humidity, high sensitivity (2.0 nF/%RH), and rapid response time (1.10 s). Govardhan and Alex (2005) used the same approach to developing cantilevers-based humidity sensors. A sensor developed by Govardhan and Alex (2005) exhibited a good linearity between 10% and 40% RH. The capacitance change was in order of Pico Farads to Femto Farads.

Afrang et al. (2015) decided to improve the conventional approach to the development of cantilever-based sensors. They suggested adding a second beam, so-called fixed-fixed beam (see Figure 13.21). Afrang et al. (2015) believe that the added second beam increases the tunability and decreases the maximum applied voltage. They have also shown that the type and the length of the added second beam affected the tunability and linearity, and the applied voltage of the structure.

It should be recognized that a cantilever-based humidity sensing is not one of the more common ones used because of a number of limitations (Goeders et al. 2008). To accurately record cantilever deflection, the dielectric

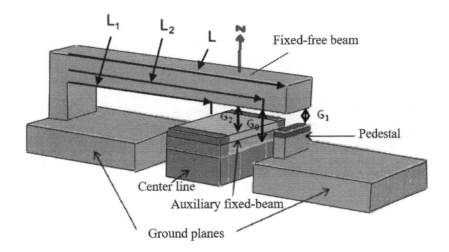

FIGURE 13.21 The schematic diagram of the proposed variable capacitor. (Reprinted with permission from Afrang, S. et al., A new MEMS based variable capacitor with wide tunability, high linearity and low actuation voltage, *Microelectron. J.*, 146, 191–197, 2015. Copyright 2015, Elsevier.)

material between the conductive plates must be constant throughout the experiment. The presence of analyte within the gap often changes its effective dielectric constant. Additionally, if the parallel plates are brought in too close a proximity, they may stick together, which terminates the collection of useful data until they become separated. This phenomenon is frequently encountered when the analyte condenses onto the cantilever surfaces. Although the capacitive cantilevers can be integrated onto a microchip, scaling down the size of the capacitive cantilever will lower its overall sensitivity, because the capacitance of a capacitor is directly proportional to its surface area. This means that microcantilever-based sensors cannot work in the range of low humidity level (< 10% RH).

13.6.3 Piezoresistive Humidity Sensors

Piezoresistive humidity sensors form the largest group of cantilever and membrane-based humidity sensors. As was shown earlier, the main element that transforms the deflection of cantilever and diaphragm into an electrical signal is the piezoresistor. However, it should be noted that other resistive methods for converting a cantilever's deflection into an electrical signal are also offered. For example, Chen et al. (2006, 2008) and Lee et al. (2007) suggested measuring the change in the resistance of the platinum band. An example of such a humidity sensor is shown in Figure 13.22. The polyimide layer is the moisture-sensing layer. A platinum resistor layer is underneath the sensing layer, and an example layout of this platinum layer is shown. The sensing layer (polyimide) sorbs moisture from the surrounding air, which induces stress on the beam. This causes the beam to deflect or deform, resulting in a change in the effective length of the platinum layer. This change can be measured as a change in the resistance of the platinum resistor. Chen et al. (2006, 2008) reported that the sensitivity of these sensors increased with the length of the platinum resistor (length of beam). For a length of 4450 μm, the resistance change of approximately 150 kΩ was reported over 40%–85% RH with polyimide sensing layer, hysteresis of approximately 2%, and the response time of approximately 0.9 seconds (Chen et al. 2008). Cantilever-based resistive humidity sensors usually had nonlinear behavior and lower detection limits below 40% RH.

13.6.3.1 Piezoresistors

In most piezoresistive humidity sensors, the semiconductor resistors, mainly silicon, are used as piezoresistors. Diffusion, ion implantation, and epitaxy are the most common impurity-doping techniques for introducing dopants into a silicon substrate for piezoresistor fabrication (Barlian et al. 2009). These techniques result in different doping profiles. The fabrication of piezoresistors using diffusion involves a predeposition and a drive-in step. During the predeposition step, wafers may be placed in a high-temperature furnace with a gas-phase (iborane [B_2H_6], phosphine [PH_3], or arsine [AsH_3]), or a solid-phase dopant source. The source can be removed and dopants "driven-in" deeper with high temperature annealing (900°C–1300°C).

Ion implantation gained wide use in the 1980s and remains the preferred method today. In ion implantation, dopant ions are accelerated at high energy (keV to MeV) into the substrate. Any layer thick or dense enough to block the implanted ions, such as photoresist, silicon oxide, silicon nitride, or metal, can be used for masking. Typical silicon piezoresistor doses range from 1×10^{14} to 5×10^{16} cm^{-2}, with energy ranges from 30 to 150 keV. One

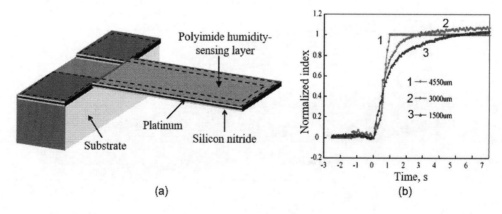

FIGURE 13.22 (a) Cantilever-based resistive humidity sensor with platinum layer and (b) Time-response curve of the humidity sensors with different thickness of humidity-sensitive layer (from 65% to 85% relative humidity [RH]). (Reprinted with permission from Chen L.T. et al., MEMS-based humidity sensor with integrated temperature compensation mechanism, *Sens. Actuators A*, 147, 552–528, 2008. Copyright 2008, Elsevier.)

major disadvantage of ion implantation is a significant damage to the crystal. Lattice order is mostly restored by high-temperature dopant activation and annealing. However, shallow junctions are difficult to obtain with high crystal quality. Parameters that affect the junction depth include the acceleration energy, ion mass, and stopping power of the material (Gibbons 1972).

The most qualitative piezoresistors are formed when using epitaxy. Epitaxy is the growth of atomic layers on single-crystal materials that conform to the crystal structure arrangement on the surface of the crystalline substrate. Chemical vapor deposition (CVD) by decomposing silane (SiH_4) or by reacting silicon chloride ($SiCl_4$) with hydrogen is the most used techniques to deposit epitaxial silicon. Epitaxial silicon films may be doped during the deposition by introducing appropriate dopant source gases. It is important that epitaxial piezoresistors require no annealing and have a uniform dopant profile. In addition, epitaxy has enabled fabrication of ultra-thin piezoresistive layers with increased force sensitivity (Tortonese et al. 1993). Harley and Kenny (1999) and Liang et al. (2001) demonstrated the use of epitaxially grown doped silicon to form piezoresistors in ultra-thin cantilevers (less than 100 nm). This is a practical method for such thin piezoresistive cantilevers, especially given the difficulties of implanting shallow junction depths (less than 50 nm), activating dopant atoms, and restoring lattice quality. A complete review of doping techniques is available elsewhere (Plummer et al. 2000).

However, despite the advantages of epitaxial manufacturing technology for piezoresistors, it must be recognized that ion implantation is the most common method of fabricating piezoresistors (Barlian et al. 2009). Processing complexity and equipment costs and availability are the main drawbacks to epitaxy. Advantages of ion implantation include precise control of dopant concentration and depth. Diffusion has the advantage of batch processing but suffers from poor dopant depth and concentration control. Table 13.2 compares ion implantation, diffusion, and epitaxy techniques. Thus, each resistor fabricated using indicated methods is monolithic and fully integrated into its respective beam. In contrast to conventional glued-on metal foil strain gauges, piezoresistive strain gauges exhibit no mechanical defects from overload or aging (Konrad and Ashauer 1999).

It is worth noting that polysilicon-based piezoresistors can also be used. Moreover, the use of polysilicon allows us to solve the problem of leakage current. At lower temperature (below 125°C), the silicon-based piezoresistors are isolated from the bulk by a p-n junction and the sensor works properly. But the performance of the sensor degrades at higher temperature, due to higher leakage current (Li et al. 2012). In the case of polysilicon-based piezoresistors, the problem of leakage current is absent because piezoresistors are isolated from each other and from the bulk by an oxide film. In spite of this advantage, polysilicon resistors are not preferred because of low sensitivity and problems associated with obtaining controlled and reproducible properties of polysilicon (Xiansong et al. 2004). More detailed information on alternative piezoresistive materials (see Table 13.3), manufacturing techniques for piezoresistive

TABLE 13.2
Comparisons of Doping Methods

Parameter	Process		
	Diffusion	Ion Implantation	Epitaxy
Process conditions	High temperature, batch process	Room temperature, vacuum, batch process	High temperature, low pressure, single wafer
Damage	None	Significant, requires annealing, enhances diffusion	None
Doping concentration control	Acceptable	Excellent	Good
Dopant depth control	Not good	Good	Very good
Typical range of doses or concentration	Concentration is limited to solid solubility	10^{11}–10^{16} cm^{-2}	10^{14}–10^{17} cm^{-2}
Masking	Hard mask	Photoresist or hard mask (silicon oxide, silicon nitride, metal, etc.)	Oxide mask and selective deposition

Source: Plummer, J.D. et al., *Silicon VLSI Technology Fundamentals, Practice, and Modeling*, Prentice Hall, Upper Saddle River, NJ, 2000.

TABLE 13.3
Properties of Piezoresistive Materials

Properties	Si	SiC	Diamond	GaN
Energy gap, eV	1.12	2.3–3.4	5.5	3.4
Breakdown voltage, V/cm	$3–6 \cdot 10^5$	$4 \cdot 10^6$	10^7	$3 \cdot 10^6$
Electron mobility, cm²/V·s	1500	1000	2200	900
Hole mobility, cm²/V·s	100–500	40–100	1600	150
Young's modulus, GPa	130–180	300–500	1000	200–300
Melting point, °C	1410	2830	1400[a]	2400
Thermal conductivity, W/cm·K	1.5	5	20	1.3
Chemical inertness	Poor	Excellent	Good	Good
MEMS compatibility	Excellent	Good	Poor	Fair
Availability/cost	Excellent	Fair	Poor	Fair

Source: Data extracted from Phan, H.-P. et al., *J. Microelectromech. Syst.*, 24, 1663–1677, 2015.

[a] Sublimation temperature.

sensors, and their advantages and disadvantages can be found in numerous reviews already published (Gerlach et al. 1998; Lavrik et al. 2004; Barlian et al. 2009; Mutyala et al. 2009; Park et al. 2010; Kumar and Pant 2014; Bausells 2015; Phan et al. 2015; Liu et al. 2016; Yang and Xu 2017). There, one can find the history of the appearance of piezoresistive sensors, piezoresistance fundamentals, and a description of approaches to optimizing the parameters of these devices. For example, for piezoresistive sensors, there are four parameter types that should be optimized: cantilever and membrane dimensions, piezoresistor dimensions, bias voltage, and fabrication process parameters (implant dose and energy, dopant atom, and annealing time and temperature). According to Mutyala et al. (2009), Park et al. (2010), and Kumar and Pant 2014, the optimization of piezoresistive cantilever- and membrane-based sensors should follow the following steps:

- The thickness of cantilever and membrane is based on the fabrication process constraints. Thin diaphragms and cantilevers are more sensitive than thicker ones but are difficult to realize. For example, fabrication of cantilevers less than a micrometer thick using ion implantation is challenging because of the diffusion during the annealing required to electrically activate the dopant atoms and reduce lattice damage. Submicrometer cantilevers can be fabricated only using epitaxial growth or diffusion.

- Once cantilever and membrane thickness are chosen, the measurement bandwidth and desired stiffness of the cantilever and membrane determine the cantilever length and width. For example, the maximum bandwidth is usually limited by the resvonant frequency (f_0) of the cantilever. Bigger diaphragms and cantilevers help in achieving higher sensitivity but increase the size of the device.

- The length and width of the piezoresistor are chosen to optimize the force resolution. Force resolution is inversely proportional to the piezoresistor width. Therefore, the maximum possible width of the piezoresistor is selected. The choice of piezoresistor length is not straightforward: While a longer piezoresistor is better for $1/f$ noise, a shorter piezoresistor is better for Johnson noise and force sensitivity. The optimal ratio of the cantilever and piezoresistor length ($a = l_p/l_c$) can be found by differentiating the force resolution with respect to a.

- From the technology point of view, larger piezoresistors can be matched with the high-stress regions easily, but at the same time, larger resistors lead to a reduced sensitivity due to the stress averaging effect (Clark and Wise 1979). Smaller resistors can be placed completely within high-stress regions and maximize sensitivity. However, they pose the problem of exactly positioning them at the right location, especially while fabricating the diaphragm through wet bulk micromachining, where there is an uncertainty in position due to undercutting.

- Once the cantilever, membrane, and piezoresistor dimensions are determined, one can calculate the piezoresistor resistance and choose a bridge voltage based on the maximum power dissipation (i.e., $V_{\text{bridge}} = 2\sqrt{W_{\max}R}$). Resolution improves with power dissipation, so one should choose the maximum bias voltage possible. In some design cases, however, the resistance of the piezoresistor is high enough that the bias voltage is limited by the voltage source rather than power dissipation.

- When the cantilever and membrane dimensions, piezoresistor dimensions, and bias voltage are set, one can calculate the force resolution for a variety of process conditions and simulation results to choose the process conditions that achieve the optimal resolution. If the force resolution was not sufficient for the measurement application, the cantilever thickness can be reduced, and the process repeated. In the case of cantilever-based devices for these purposes, one can use Equation 13.16 (Park et al. 2010)

$$F_{min} = \frac{\sqrt{\frac{\alpha V_{bridge}^2}{2 l_p w_p N_z} \cdot ln\left(\frac{f_{max}}{f_{min}}\right) + 8 k_b T R_s \frac{l_p}{w_p}(f_{max} - f_{min})}}{\frac{3(l_c - 0.5 l_p)\pi_{l.max}}{2 w_c t_c^2} \gamma V_{bridge} \beta^*} \quad (13.16)$$

where the numerator is the root-mean-square voltage noise and the denominator is the force sensitivity. Force resolution has several factors: cantilever dimensions (l_c, w_c, and t_c), piezoresistor dimensions (l_p, w_p, t_p), fabrication process parameters (N_z, R_s, α, β, and γ), and operating parameters (V_{bridge}, T, f_{min}, and f_{max}).

- The theoretically achievable resolution for a piezoresistive sensor is dependent on the noise limit set by the fundamental noise. These are two important sources of electrical noise in piezoresistors. Thermal noise exists in all resistors and cannot be eliminated. The origin of 1/f noise still remains an active topic of research, but it is dependent on the implantation dose and the annealing treatment of the resistors. According to Hooge et al. (1981), the empirical 1/f noise model is shown by

$$V_{1/f} = V_b \cdot \sqrt{\frac{\alpha}{N \cdot f}} \quad (13.17)$$

where $V_{1/f}$ is the voltage noise density, V_b is the bias voltage, N is the total number of carriers in the resistor volume, f is the frequency, and α is a fitting parameter attributed to the quality of the lattice. To reduce the noise, N and α can be altered. N can be increased by increasing the doping concentration by giving a higher dose implant. α is dependent on the crystal lattice perfection and the lattice quality, and can be improved by annealing at higher temperature for a longer time. Thus, the noise in piezoresistors can be reduced by high doping and long annealing at higher temperatures.

- Annealing parameters should also be optimized. Thinner device layer and higher annealing time can be used to obtain a very flat doping profile in piezoresistors formed using silicon on insulator (SOI) technology. A longer annealing time also helps in reducing the noise in piezoresistors, as discussed earlier. When the piezoresistors are fabricated on the bulk silicon either by ion implantation or diffusion, there is diffusion profile inside the bulk, leading to nonuniform doping. Consequently, it is difficult to accurately define the uniform doping and junction depth of the resistor. But it is possible to fabricate piezoresistors with very uniform doping using SOI.

13.6.3.2 Sensor Performance

13.6.3.2.1 Membrane-Based Humidity Sensors

The first piezoresistive humidity sensors were developed by Gerlach and co-workers (Gerlach and Sager 1994; Gerlach et al. 1994; Sager et al. 1994, 1995, 1996; Nakladal et al. 1995; Schroth et al. 1996; Buchhold et al. 1998b, 1998c, 1998d) on the base of polyimide thin film deposited on a micromachined silicon diaphragm that is similar to the diaphragm in a piezoresistive pressure sensor. Such membranes have been used for more than 20 years in piezoresistive pressure sensors, with over 200 million such devices deployed in the automotive industry alone (Fenner et al. 2001). The diaphragms were (0.6–1.0) × (1.0–1.8) mm² in area, with an average membrane thickness of 8–12 μm and average polyimide layer thickness of 8–12 μm. The first samples attempted to control the moisture content by changing the eigenfrequency (Schroth et al. 1996). However, they eventually concluded that the static method is more suitable for the development of humidity sensors. Therefore, similar to silicon-based pressure sensors, four piezoresistors were placed along the periphery of the diaphragm (alternatively perpendicular and parallel to the diaphragm edge) and are connected in a Wheatstone bridge configuration (Figure 13.23). In this arrangement, the bridge output voltage U_A is a linear function of the mechanical stress in the resistors perpendicular σ_L and parallel σ_Q to the diaphragm edge (Equation 13.18):

$$U_A = -\frac{\pi_{44}}{2} U_{SS}(\sigma_L - \sigma_Q) = -\frac{\pi_{44}}{2} U_{SS} \sigma_{Res} \quad (13.18)$$

where π_{44} denotes the piezoresistive coefficient for (100)-silicon wafer along the <110> crystal orientations, and U_{SS} the operating voltage.

Technologically induced tensile stress in the polymer layer produces a negative offset voltage U_0, while the thermal expansion or humidity-induced swelling of the polymer layer increases the output voltage U_A. The dependence of U_A on the RH is described by the slope B_{RH}.

$$U_A(\text{RH}) = U_0(\text{RH} = 0\%) + B_{RH} \cdot \text{RH} \quad (13.19)$$

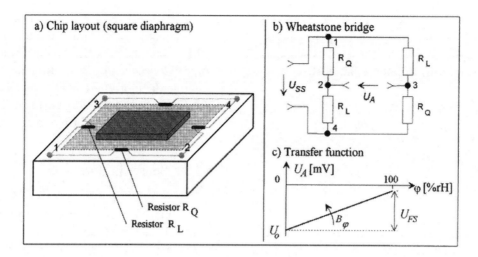

FIGURE 13.23 Layout and transfer function of piezoresistive humidity sensors. (Reprinted with permission from Buchhold, R. et al., Design studies on piezoresistive humidity sensors, *Sens. Actuators B*, 53, 1–7, 1998. Copyright 1998, Elsevier.)

A design developed by Gerlach's group incorporated a thermal oxide and silicon nitride passivation layer between the sensing film and underlying silicon diaphragm to electrically isolate the piezoresistor strain gauges and polysilicon bridge circuit traces (Figure 13.24). The chip semiconductor temperature sensor (i.e., thermistor) was also included. The sensor output is measured as a bridge voltage that is linearly proportional to the excitation voltage and ranges from 0 to 30 mV full scale (FS) for relative humidity values from 0% to 100%, respectively.

The main conclusions of the studies carried out by Gerlach's groups can be formulated as follows (Buchhold et al. 1998b, 1998d):

- Sensitivity of the piezoresistive humidity sensors is nearly independent of the shape of diaphragm (square, rectangular, or circular).
- Sensitivity increases with decreasing diaphragm thickness. The minimum diaphragm thickness is determined by the processing tolerances.

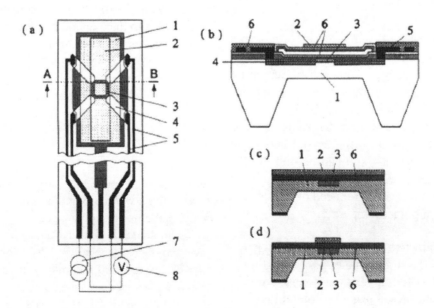

FIGURE 13.24 Top view (a) and cross-sections (b–d) of a humidity sensor based on straining of membrane with polyimide layer: 1, Silicon membrane; 2, polyimide layer; 3, piezoresistors; 4, implanted connections; 5, metallized connections; 6, passivation layer system; 7, current supply; 8, sensor output voltage. (Reprinted with permission from Gerlach G. and Sager K., A piezoresistive humidity sensor, *Sens. Actuators A*, 43, 181–184, Copyright 1994, Elsevier.)

- Sensitivity increases with increasing polymer thickness. However, the dynamic response time of the sensor decreases as the polymer thickness is increased, since diffusion times become longer (Yang et al. 1985).
- Decreased sensitivity is caused by a large, technologically induced stress s_2 in connection with large, thin diaphragm geometries. In addition, the stress dominates the offset voltage of the sensor.
- Polymer coverage x (the ratio between the polymer covered diaphragm width to the total diaphragm width) should be around 0.8 for maximum sensitivity.
- Polymers with a large stress coefficient provide for the highest sensitivity. Therefore, polymers combining a stress coefficient with good long-term stability properties should be chosen.
- The initial drift of sensor parameters is attributed to the relaxation of technologically induced stress in the polymer layer. Artificial aging using empirically developed aging regimes can accelerate the relaxation process significantly.

One should note that Hygrometrix has developed and commercialized the RH sensor of similar design to Gerlach's group. In the sensor market, this sensor has the designation HMX 2000 (read Volume 3). Later on, piezoresistive humidity sensors also were developed and tested by Waber et al. (2014). The starting point of the development was a 1 × 1 mm² piezoresistive pressure sensor chip (Figure 13.25a) from EPCOS (T5400 pressure sensor module) (Waber et al. 2013), in which the membrane was covered with a polyimide layer (Figure 13.25b). The pressure sensor chip was fabricated by the SOI technology. To achieve a high sensitivity, the piezoresistors were placed at the edge of the diaphragm, where the stress is maximal. The resistors were connected as a Wheatstone bridge by highly doped areas and aluminum lines. The measurements showed a sensitivity of 0.25 mV per percent RH (%RH) and a nonlinearity of 3.1%FS (full scale) in the range of 30%–80% RH (Figure 13.26). Time to reach 63% of the saturation output voltage was 21s. The response time could be reduced by decreasing the thickness of the polyimide. However, decreasing the thickness would also lead to an uncritical reduction of the sensitivity.

Piezoresistive humidity sensors with another configuration were developed by Loizeau et al. (2012) using a bulk micromachining process. Graphical representation of these sensor and SEM pictures are shown in Figure 13.27. Two membranes within an array were functionalized by inkjet spotting with CAB diluted in

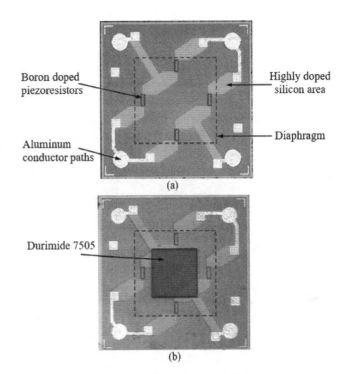

FIGURE 13.25 Images of pressure sensor chip with four piezoresistors interconnected as a Wheatstone bridge (a) before and (b) after deposition of humidity sensitive film. The doping of the piezoresistors is too low to be visible in an optical microscope image. The location of the resistors is therefore indicated by *red boxes*. The footprint is 1 × 1 mm² and the height 150 μm. (Reprinted from Waber, T. et al., *J. Sens. Sens. Syst.*, 3, 167–175, 2014. Copyright 2014, Association of Sensor Technology as open access.)

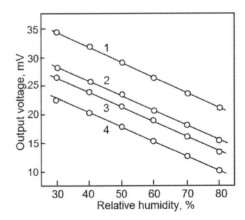

FIGURE 13.26 Relation between relative humidity (RH) and output voltage for four sensors (1, 2, 3, and 4). The sensitivity of the four sensors is between 0.24 mV/%RH and 0.26 mV/%RH. The fact that all four curves are almost parallel is interesting for a one-point calibration. (Reprinted from Waber, T. et al., *J. Sens. Sens. Syst.*, 3, 167–175, 2014. Copyright 2014, Association of Sensor Technology as open access.)

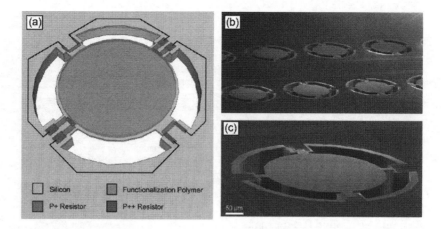

FIGURE 13.27 (a) Graphical representation of a membrane suspended by four constricted beams with integrated piezoresistors connected in a Wheatstone bridge. The membrane is coated with a polymer that reacts to surrounding molecules and (b) Scanning electron microscopy (SEM) pictures of a silicon chip containing two arrays of membrane-type sensors and (c) a close view on one membrane suspended by four constricted beams. Membrane had a diameter of 500 μm and a thickness of 2.5 μm. (Reprinted with permission from Loizeau, F. et al., Membrane-type surface stress sensor with piezoresistive readout, *Procedia Eng.*, 47, 1085–1088, 2012. Copyright 2012, Elsevier as open access.)

hexyl acetate. The results of sensor testing are presented in Figure 13.28. It is seen that the functionalized membranes react quickly to the humidity changes with a rising time constant ($\tau_{62\%}$) of 3.5 seconds and a purging time constant of 1.3 seconds. The difference between the two time constants is due to the mixing of the two gases in the chamber, which is, in reality, not immediate. Therefore, the purging time constant is closer to that of the sensor since it is not limited by the mixing inertia. It is a very good result. In addition, the response of the membrane is linear and shows a sensitivity of 87 mV/%RH. It is the maximal sensitivity observed for the developed piezoresistive humidity sensors. Loizeau et al. (2012) believe that such a sensitivity is the result of the membrane configuration used. By clamping the membrane in the four directions, the whole stress is efficiently transduced on the constricted beams, where the piezoresistors are located. Its sensitivity is therefore increased compared with a standard piezoresistive cantilever with a free end.

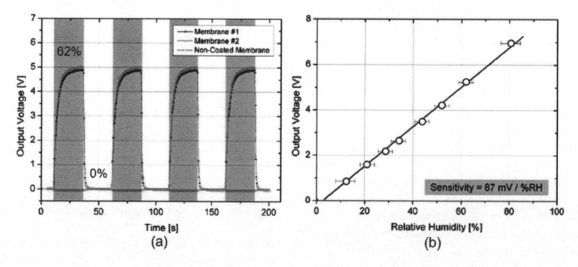

FIGURE 13.28 (a) Dynamic response of three membranes to four humidity pulses of 62%. One of the membranes was used as a reference without the functionalization of the cellulose acetate butyrate (CAB) polymer. The others coated with CAB layers showed a response time of 1.3 s during the purging step. The gain of the amplification stage is 500; and (b) Static responses of membrane #1 to humidity values from 0% to 80%. The response is linear, with a sensitivity of 87 mV/% relative humidity (RH). (Reprinted with permission from Loizeau, F. et al., Membrane-type surface stress sensor with piezoresistive readout, *Procedia Eng.*, 47, 1085–1088, 2012. Copyright 2012, Elsevier as open access.)

13.6.3.2.2 Cantilever-Based Humidity Sensors

Gunter et al. (2005), Sappat et al. (2011), Patil et al. (2014), and Huang et al. (2015) showed that the development of microcantilever-based piezoresistive humidity sensors is also possible. Gunter et al. (2005) used for this purpose cantilevers from Veeco, Inc., fabricated using standard silicon-integrated circuit technology. Each cantilever was approximately 200 μm in length, 40 μm wide, and 1–2 μm thick. These cantilevers had a nominal electrical resistance of 2 kΩ. If the cantilever was bent slightly (strain), even by only a few Angstroms, the resistance of the cantilever changed. This resistance change may be measured using a precision multimeter. UV-crosslinked PVA was used as the active humidity-sensing material. Embedded sensing units were prepared by depositing approximately 50 μL of dissolved PVA onto a clean Si substrate, in the form of a single bead. As a result of testing, it was found that this device responded well to a wide range of humidity values, essentially from 20% up to 100% RH (Figure 13.29), and it was able to measure changes in RH on the order of 1% or less (S ≈ 0.15 Ω/%RH). Unfortunately, in these experiments, the individual sensor elements required as much as 2000–3000 seconds to come to equilibrium (Gunter et al. 2005). Apparently, this is a result of the thickness of the polymer layer in which the piezoresistive cantilever was embedded.

Sappat et al. (2011) have also used commercial microcantilevers, fabricated by PolyMUMP (Multi-User MEMS Processes), which is a MEMS foundry. The piezoresistive microcantilever was designed to be 200 μm long, 50 μm wide, and 3.75 μm thick, residing in a 2 × 2 mm² chip. Each cantilever beam consists of four structural layers of silicon nitride, polysilicon, silicon dioxide, and gold. The polysilicon layer acted as a resistive transducer that would change its resistance when there was bending of microcantilever due to adsorbed molecules on a PEDOT/PSS sensing layer. PEDOT/PSS layers are directly patterned on microcantilever using inkjet printing technology. Experimentation has shown that the water vapor can effectively be adsorbed onto PEDOT/PSS layer on microcantilever surface and induces significant deflection of the cantilever beam. However, the sensitivity of such sensors was only 0.004 Ω/%RH (Sappat et al. 2011). The response and recovery times of resistance were within a few minutes.

A different approach to developing microcantilever-intelligent sensors was proposed by Patil et al. (2014). They suggested using as the main element the polymer nanocomposite microcantilevers, fabricated using the surface micromachining technology. These microcantilevers had length, widths, and thickness of 250, 50, and approximately 3.5 μm, respectively (see Figure 13.30). To obtain a piezoresistive layer, SU-8/carbon black (CB) nanocomposite was prepared by dispersing CB powder approximately 8–9 vol% in SU-8 and SU-8 thinner. This piezoresistive layer was then encapsulated by 1.5 μm SU-8. The microcantilever chip size was 4 mm × 4 mm, and it included four microcantilevers per die. The anchor of the microcantilever was made of approximately 200 μm thick SU-8. Microcantilever surface was modified with PANI nanofibers by drop casting. Since the microcantilevers were made up of SU-8, after PANI drop casting, the other side of the microcantilever was blocked/passivated by depositing a thin layer of Gold, approximately 20 nm to avoid the interference of

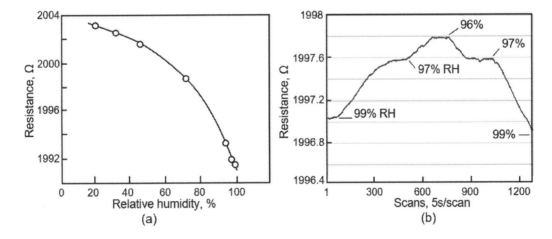

FIGURE 13.29 (a) Plot of resistance vs humidity for the same sensor (4 s/scan). (b) Sensor response to changes in relative humidity (RH) in the range of 96% to 99% (4 s/scan). (Reprinted with permission from Gunter, R.L. et al., Hydration level monitoring using embedded piezoresistive microcantilever sensors, *Medical Eng. Phys.*, 27, 215–220, 2005. Copyright 2005, Elsevier.)

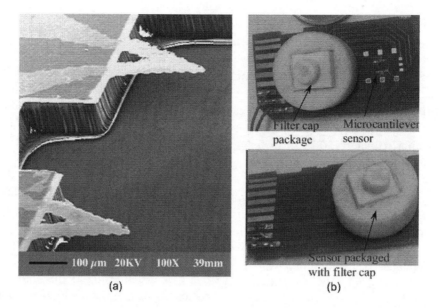

FIGURE 13.30 (a) Scanning electron microscopy (SEM) image of the released polymer nanocomposite microcantilevers; (b) A schematic of the experimental setup for relative humidity (RH) measurement: sensor with and without filter cap. (Reprinted with permission from Patil, S.J. et al., An ultra-sensitive piezoresistive polymer nano-composite microcantilever platform for humidity and soil moisture detection, *Sens. Actuators B*, 203, 165–173, 2014. Copyright 2014, Elsevier.)

structure material during measurement. After nanofiber coating, the individual sensor devices (functionalized and reference nonfunctionalized microcantilevers) were mounted on a printed circuit board (PCB) using conducting silver epoxy (EPO-TEK H20E) and insulating epoxy (EPO-TEK H70E).

As is seen in Figure 13.31, PANI nanofiber-coated microcantilevers showed maximum change of 28 mV at 93% RH. The sensor exhibited high sensitivity (0.4 mV/%RH), linear response, small hysteresis ($\approx$ 1%–2% RH), and fast response and recovery times of 8 and 10 seconds, respectively. These are not bad results. It remains only to wait for the results concerning the stability and durability of these sensors.

Huang et al. (2015) also used surface micromachining technology to manufacture microcantilevers. However, this technology was based on the classical approach, using silicon as the main material of microcantilever and polyimide as a humidity-sensitive material. A typical CMOS process was used to form piezoresistors,

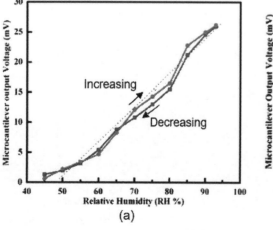
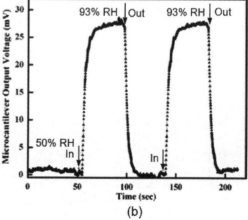

FIGURE 13.31 (a) Hysteresis curve of polyaniline (PANI)-coated microcantilever sensor with change in relative humidity (%RH) and (b) Cyclic response upon repeated exposure to 93% (50% at $\approx$ 0 mV) RH. (Reprinted with permission from Patil, S.J. et al., An ultra-sensitive piezoresistive polymer nano-composite microcantilever platform for humidity and soil moisture detection, *Sens. Actuators B*, 203, 165–173, 2014. Copyright 2014, Elsevier.)

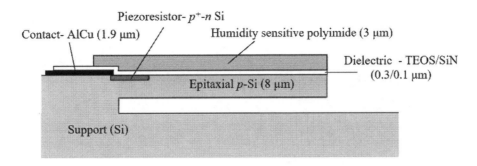

FIGURE 13.32 Schematic of a microcantilever piezoresistive humidity sensor. TEOS, tetraethoxysilane; SiN, silicon nitride.

dielectric and metal wires. In the CMOS process, ion-implanted p^+-type resistors with a sheet resistance of 400 Ω/sq and a junction depth of 1 μm were fabricated as piezoresistors. After the CMOS process, a polyimide layer was spun on the wafer and patterned to form the humidity-sensitive element (Figure 13.32). A hydrophobic porous polytetrafluoroethylene (PTFE) membrane with good humidity penetration was used as a filter to protect the sensor from contaminations of dust and other chemicals. A typical Wheatstone bridge circuit was developed to transform the piezoresistance into an output voltage. As a result of testing the manufactured sensors, it was found that, at 20°C, the sensitivity was 7 mV/%RH and the linearity of the response was 1.9%. The sensor response exhibited some hysteresis (Figure 13.33). However, the experiment showed that the hysteresis property can be improved by increasing operating temperature. Increased temperature promotes the diffusion of the water vapor in polyimide. The recovery time of the sensor was 85 seconds at room temperature (25°C).

13.7 MICRORESONATOR-BASED HUMIDITY SENSORS

Microresonator-based devices are another type of humidity sensors (Gluck et al. 1994; Battiston et al. 2001; Schmid et al. 2008). These sensors operate by monitoring the resonant frequency of cantilever or diaphragm. The resonant beam, which has also been called a *resonating beam force transducer*, acts as a sensitive strain gauge. As the stress state of the diaphragm changes due to water vapor influence, the tension in the embedded structures changes, and so does the resonant frequency. As shown in Section 13.4, there are several mechanisms by which the structures can be driven into resonance while the resonant frequency is sensed. One method is piezoelectric excitation and piezoresistive sensing (Sone et al. 2004): The structure is driven to resonance by ac-applied voltages, and the resonant frequency is measured by piezoresistors (Figure 13.34). In another method, a piezoelectric crystal is used for actuation of the cantilever array, and the cantilever deflection is measured optically via a beam-deflection technique (Battiston et al. 2001). Structures can also be electrostatically excited and sensed by oscillation monitoring with a laser-Doppler vibrometer (Schmid et al. 2008). Humidity sensors with optical excitation by laser and frequency shift control, using the interferometry method, are also developed (Churenkov 2013, 2014). Electrothermal actuation and the output signal measurement using a piezoresistive sensor connected in a Wheatstone bridge circuit can also be used in resonant humidity sensors (Dennis et al. 2015).

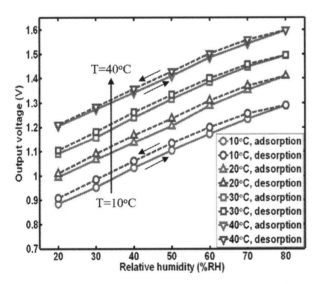

FIGURE 13.33 Relationship between the output voltage of the cantilever-based piezoresistive sensor and the relative humidity (RH) at different temperatures. Cantilevers were fabricated using surface micromachining process and covered by polyimide films. The size of the cantilevers was 400 μm wide by 100 μm long. (Reprinted from Huang, J.Q. et al., *Micromachines*, 6, 1569–1576, 2015. Copyright 2015, MDPI as open access.)

It was established that resonant sensors have been shown to exhibit better sensitivity and lower temperature sensitivity than pure piezoresistive sensors. Furthermore,

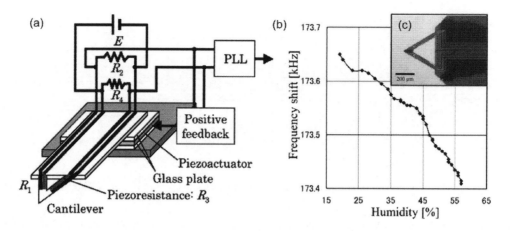

FIGURE 13.34 Harmonic vibration-type mass sensor system and detection circuits using a piezoresistive cantilever: (a) Piezoresistive cantilever and resonance frequency shift detection system; (b) Resonance frequency shift of the cantilever with increasing humidity. As experimental results, a mass sensitivity of 2.2 pg/Hz was obtained from water molecular adsorption on the cantilever; and (c) Optical microscope image of V-shaped piezoresistive cantilever. (Reprinted with permission from Sone, H. et al., *Jpn. J. Appl. Phys.*, 43, 4663–4666, 2004. Copyright 2004, Japan Society of Applied Physics.)

a frequency output is more immune to noise than classical analog piezoresistive and capacitive signals (Eaton and Smith 1997).

Usually, resonator humidity sensors are developed on the base of microcantilevers. Gluck et al. (1994) have developed a cantilever resonator using a piezoelectric polymer material (polyvinyl-difluorene, PVDF). When an electrical signal was applied to the electrode, the cantilever vibrated as a result of expansion or contraction of the beam. However, due to features of configuration, the frequencies of these devices were low, and so were the sensitivities (small frequency shifts over humidity changes). A shift of around 30 Hz was reported for a 0%–90% RH range (Gluck et al. 1994).

A CMOS-MEMS device with embedded microheater for RH sensing via amplitude change, due to mass loading, has been fabricated by Dennis et al. (2015). Figure 13.35a shows the field emission scanning electron microscopy (FESEM) image of the CMOS-MEMS humidity sensor with nanoparticle-sized TiO$_2$ paste successfully deposited on its plate, using a drop-coating method, and Figure 13.35b shows the output voltage of the device for increasing and decreasing humidity levels from 35% to 95% RH. It is seen that the output voltage of the humidity sensor increases from 0.585 mV to 30.580 mV as the humidity is increased from 35% to 95% RH. These changes correspond to the sensitivity of 0.107 mV/%RH in the range of 35%–60% RH and 0.781 mV/%RH in the range

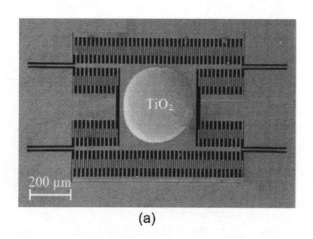
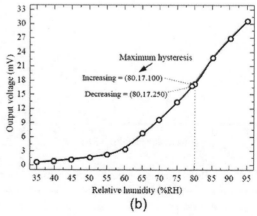

FIGURE 13.35 (a) FESEM image of complementary metal–oxide–semiconductor (CMOS)-MEMS device with TiO$_2$ paste deposited on its plate and (b) Output voltage vs relative humidity (RH) at an operating temperature of 80°C, corresponding to applied voltages of 6 Vpp at a frequency of 4 Hz. (Reprinted with permission from Dennis J.-O. et al., *Sensors*, 15, 16674–16687, 2015. Copyright 2015, MDPI as open access.)

from 60% to 95% RH. Furthermore, the sensitivity of the humidity sensor increased linearly, from 0.102 mV/%RH to 0.501 mV/%RH, with an increase in the temperature from 40°C to 80°C; it was also found to be dependent on the frequency of driving voltage with a maximum sensitivity observed at the frequency of 12 Hz.

However, the highest sensitivity was demonstrated by humidity sensors developed on the basis of all-polymer cantilevers (Schmid et al. 2008, 2009). A string-like, double-clamped microbeam was fabricated using SU-8. A sensitivity of 0.35%–0.78% relative change of the resonant frequency per 1% RH was measured. The frequency resolution was determined by taking twice the standard deviation of the measured signal over a period of 3 minutes, and a value of 6.0 Hz was obtained. This gave an RH resolution of 0.006%, assuming an error-free temperature measurement. The sensitivity is one order of magnitude higher than the sensitivity of a resonant glass cantilever humidity sensor covered with a polymer layer with a sensitivity of 0.024% per % RH (Gluck et al. 1994) and two orders of magnitude higher than the sensitivity of a polymer-coated silicon cantilever with a sensitivity of 0.003% per %RH (Battiston et al. 2001). Such high sensitivity is undoubtedly a significant advantage of the all-polymer design. But the high RH resolution requests an accurate temperature measurement. Other drawbacks are the slower response time (> 8 s) for an increasing humidity and the strong frequency drift after fabrication due to material relaxation. However, Schmid et al. (2008) state that, after storing the samples prior to their usage, the drift due to relaxation can be reduced.

13.8 SUMMARY

Designers of the hygrometric sensors believe that these devices have the following advantages: compared with more conventional sensors, membrane and especially cantilever-based sensors offer improved dynamic response, greatly reduced size, and high precision, and no power supply is required for operation and increased reliability. In addition, sensors have high linearity of sensor signal, long-term stability, and simple readout circuit, and they provide a primary rather than a secondary measurement of dew point. High accuracy, high repeatability, wide dew-point range, wide range of operation temperatures, low hysteresis, and small drift are also attributed to the advantages of these sensors. For example, it is stated that the sensor can measure full RH levels, including 100% RH with water condensation, and they can be stored for a long period without damage from −40°C to 125°C. It was reported that typical sensor response time to a humidity step change is less than 5 seconds (Fenner 1995; Hygrometrix 2000). However, these statements should be treated with caution, since these advantages can be attributed to all electronic humidity sensors, since all these parameters are largely determined by the properties of the hygroscopic material used. In addition, it has become well-established that cantilevers, used in piezoresistive sensors, respond to small changes in viscosity, and ionic strength of the medium in which they are immersed, as well as to the flow dynamics of the cell that houses the cantilever chip (Goeders et al. 2008). Therefore, in many instances, the interpretation of the results of present-day research involving single-cantilever experimentation is based largely on the assumption of fixed conditions between sequential experiments, which reduces the accuracy of measurements. Goeders et al. (2008) believe that microcantilever arrays are the preferred format for sensor design. They enable control experiments to be performed simultaneously with analyses and provide more reliable control of empirical factors, such as thermal drift, changes in viscosity, and solution flow dynamics.

Additional difficulty in designing piezoresistive humidity sensors lies in providing an adequate thermal coupling to ensure a precise temperature control. Piezoresistive sensors are inherently sensitive to changes in the ambient temperature (see Figure 13.36), making it difficult to decouple humidity effects from temperature effects when interpreting the sensor output (Buchhold et al. 1998d). Therefore, as in other sensor types, temperature drift is a significant problem for piezoresistive

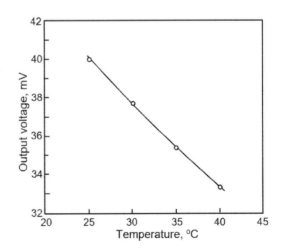

FIGURE 13.36 Bridge output voltage of the sensor at the relative humidity (RH) of 30% RH and at different temperatures. As can be seen, the sensor has a strong cross-sensitivity to temperature. By changing the temperature by 5°C, the output voltage alters by around 2.5mV. (Reprinted with permission from Waber, T. et al., *J. Sens. Sens. Syst.*, 3, 167–175, 2014. Copyright 2014, Association of Sensor Technology as open access.)

humidity sensors. This means that accurate and reproducible operation of piezoresistive humidity sensors requires elaboration of temperature compensation, thus necessitating an additional temperature sensor and, for example, a microcontroller. Buchhold et al. (1998d) believe that this cross-sensitivity may be reduced by employing polymers with large humidity-stress coupling factors and small thermal ones. But this does not solve the problem itself. Therefore, as in most electronic humidity sensors, the temperature drift should be compensated by a circuit or by temperature stabilization (Phan et al. 2003). A description of the approaches that make it possible to compensate the temperature dependence of piezoresistive sensors can be found in Barlian et al. (2009), Han et al. (2012), and Liu et al. (2016).

A significant disadvantage of these sensors is also their complex structure, which requires the use of special expensive equipment. Therefore, manufacturing costs for piezoresistive sensor elements are relatively high compared with other humidity sensors, such as conventional capacitive and resistive ones (Buchhold et al. 1998d). The narrow flow rate range that can be used, periodic cleaning, sensitivity to pressure (Chao et al. 2005), and high uncertainty ($\pm$ 2%–5%) also refer to the shortcomings of piezoresistive humidity sensors. The packaging is another crucial aspect of cantilever- and membrane-based sensors that must be considered at the outset of any application (Kovacs et al. 1996). The numerous advantages of micromachined components can be degraded or might not even be attainable without proper, cost-effective packaging. This problem is intrinsic to micromachining technology simply because the parts are often very small and require comparably minute ports to the outside world for access to gases. These packages are not currently produced in an amount comparable to the number of cases used for packaging conventional integrated circuits, and therefore they are potentially expensive. In addition, microsensors are particularly sensitive to extraneous stresses that can be induced during the process of packaging. Therefore, stress control by mechanical design of the package and by choice of materials is crucial to the practical use of micromachined devices. Cantilever-based devices also require other testing, such an application of pressure or acceleration forces, in addition to electrical testing. This means that the costs of packaging and testing micromachined transducers are also generally many times greater than those of manufacturing the devices themselves. This fact is often overlooked when the "low cost" of batch-fabricated micromachined devices is quoted. The economies of scale for traditional analog and digital integrated circuits are tremendous, because low-cost, commercial-grade packaging is available, preshipment testing is carried out on the entire wafer, and the testing is purely electrical (Kovacs et al. 1996).

REFERENCES

Abermann R., Martinz H.P. (1984) Gas adsorption on thin films of chromium studied by internal stress measurements. *Thin Solid Films* 1, 303–311.

Afrang S., Mobki H., Sadeghi M.H., Rezazadeh G. (2015) A new MEMS based variable capacitor with wide tunability, high linearity and low actuation voltage. *Microelectron. J.* 146, 191–197.

Ardebili H., Wong E.H., Pecht M. (2003) Hygroscopic swelling and sorption characteristics of epoxy molding compounds used in electronic packaging. *IEEE Trans. Compon. Packaging Technol.* 26, 206–214.

Asahi (1994) Tokyo, Japan, Technical data sheet for Asahi Chemical "Pimel TL-500, G-7000 and IX Grade, Asahi Chemical Co., Ltd.," Functional Products Division.

Baller M.K., Lang H.P., Fritz J., Gerber Ch., Gimzewski J.K., Drechsler U., Rothuizen H. et al., (2000) A cantilever array-based artificial nose. *Ultramicroscopy* 82, 1–9.

Baltes H., Lange D., Koll A. (1998) The electronic nose in Lilliput. *IEEE Spectrum* 9, 35–38.

Barlian A.A., Park W.-T., Mallon Jr. J.R., Rastegar A.J., Pruitt B.L. (2009) Review: Semiconductor piezoresistance for microsystems. *Proc. IEEE* 97(3), 513–552.

Barnes J.R., Stephenson R.J., Welland M.E., Gerber Ch., Gimzewski J.K. (1994) Photothermal spectroscopy with femtojoule sensitivity using a micromechanical device. *Nature* 372, 79–81.

Battison F.M., Ramseyer J.-P., Lang H.P., Baller M.K., Gerber C., Gimzewski J.K., Meyer E. et.al. (2001) A chemical sensor based on a microfabricated cantilever array with simultaneous resonance-frequency and bending readout. *Sens. Actuators B* 77, 122–131.

Bausells J. (2015) Piezoresistive cantilevers for nanomechanical sensing. *Microelectron. Eng.* 145, 9–20.

Berger R., Delamarche E., Lang H.P., Gerber Ch., Gimzewski J.K., Meyer E., Guntherodt H.-J. (1997) Surface stress in the self-assembly of alkanethiols on gold. *Science* 276(5321), 2021–2024.

Betts T.A., Tipple C.A., Sepaniak M.J., Datskos P.G. (2000) Selectivity of chemical sensors based on microcantilevers coated with thin polymer films. *Anal. Chim. Acta* 422, 89–99.

Bietsch A., Hegner M., Lang H.P., Gerber C. (2004a) Inkjet deposition of alkanethiolate monolayers and DNA oligonucleotides on gold: Evaluation of spot uniformity by wet etching. *Langmuir* 20, 5119–5122.

Bietsch A., Zhang J., Hegner M., Lang H.P., Gerber C. (2004b) Rapid functionalization of cantilever array sensors by inkjet printing. *Nanotechnology* 15, 873–880.

Binning G., Quate C.F., Gerber C. (1986) Atomic force microscope. *Phys. Rev. Lett.* 56(9), 930–933.

Blanc N., Brugger J., Rooij N.F.D., Durig U. (1996) Scanning force microscopy in the dynamic mode using microfabricated capacitive sensors. *J. Vac. Sci. Technol. B* 14(2), 901–905.

Boisen A., Dohn S., Keller S.S., Schmid S., Tenje M. (2011) Cantilever-like micromechanical sensors. *Rep. Prog. Phys.* 74, 036101.

Bombieri N., Fummi F., Gangemi G., Grosso M., Macii E., Poncino M., Rinaudo S. (2018) Smart systems design methodologies and tools, In: Bosse S., Lehmhus D., Lang W. (Eds.) *Material-Integrated Intelligent Systems: Technology and Applications*, Wiley, Weinheim, Germany, pp. 55–80.

Bottomley L.A., Poggi M.A., Shen S.X. (2004) Impact of nano- and mesoscale particles on the performance of microcantilever-based sensors. *Anal. Chem.* 76(19), 5685–5689.

Bowden M.J (1988) Polymers for electronic and photonic applications, In: Bowden M.J., Turner S.R. (Eds.) *Electronic and Photonic Applications of Polymers*. American Chemical Soc. (Advances in Chemical Series, Vol. 218), American Chemical Society, Washington, DC, pp. 1–73.

Boytsova O., Klimenko A., Lebedev V., Lukashin A., Eliseev A. (2015) Nanomechanical humidity detection through porous alumina cantilevers. *Beilstein J. Nanotechnol.* 6, 1332–1337.

Britton C.L. Jr, Jones R.L., Oden P.I., Hu Z., Warmack R.J., Smith S.F., Bryan W.L. et.al., (2000) Multiple-input microcantilever sensors. *Ultramicroscopy* 82, 17–21.

Brugger J., Buser R.A., and de Rooij N.F. (1992) Micromachined atomic force microprobe with integrated capacitive read-out. *J. Micromech. Microeng.* 2, 218–220.

Buchhold R., Nakladal A., Gerlach G., Sahre K., Eichhorn K.-J., Muller M. (1998a) Reduction of mechanical stress in micromachined components caused by humidity-induced volume expansion of polymer layers. *Microsyst. Technol.* 5, 3–12.

Buchhold R., Nakladal A., Gerlach G., Neumann P. (1998b) Design studies on piezoresistive humidity sensors. *Sens. Actuators B* 53, 1–7.

Buchhold R., Nakladal A., Gerlach G., Sahre K., Eiclhorn K.-J. (1998c) Mechanical stress in micromachined components caused by humidity-induced in-plane expansion of thin polymer films. *Thin Solid Films* 312, 232–239.

Buchhold R., Nakladal A., Buttner U., Gerlach G. (1998d) The metrological behaviour of bimorphic piezoresistive humidity sensors. *Meas. Sci. Technol.* 9, 354–359.

Calvert P. (2001) Inkjet printing for materials and devices. *Chem. Mater.* 13, 3299–3305.

Carrascosa L.G., Moreno M., Alvarez M., Lechuga L.M. (2006) Nanomechanical biosensors: A new sensing tool. *Trends Anal. Chem.* 25, 196–206.

Cech J.M., Burnett A.F., Knapp L. (1991) Pre-imidized photo-imageable polyimide as a dielectric for high density multichip modules. *Polymers Eng. Sci.* 32(21), 1646–1652.

Chao L.P., Lin C.W., Lau Y.D. (2005) A study on the effects of humidity, temperature, and pressure sensor on the piezoresistive film co-structure. In: *Proceedings of the International Conference on MEMS, NANO and Smart Systems* (ICMENS'05), 24–27 July, Banff, Alta, Canada, (doi:10.1109/ICMENS.2005.19).

Chatzandroulis S., Tserepi A., Goustouridis D., Normand P., Tsoukalas D. (2002) Fabrication of single crystal Si cantilevers using a dry release process and application in a capacitive-type humidity sensor. *Microelectron. Eng.* 61–62, 955–961.

Chatzandroulis S., Tsouti V., Raptis I., Goustouridis D. (2011) Capacitance-type chemical sensors, In: Korotcenkov G. (Ed.) *Chemical Sensors: Comprehensive Sensor Technologies, Vol. 4: Solid State Devices*. Momentum Press, New York, pp. 229–260.

Chen G.Y., Thundat T., Wachter E.A., and Warmack R.J. (1995) Adsorption-induced surface stress and its effects on resonance frequency of microcantilevers. *J. Appl. Phys.* 77(8), 3618–3622.

Chen R., Chu C.H., Lee C.Y., Chen H.J., Cheng W.H. (2006) A novel simple humidity sensor constructed by sandwiched cantilever, In: *Proceedings of IEEE Sensors Conference*. October 22–25, Daegu, Korea, pp. 952–955.

Chen L.T., Lee C.Y., Cheng W.H. (2008) MEMS-based humidity sensor with integrated temperature compensation mechanism. *Sens. Actuators A* 147, 552–528.

Cherian S., Thundata T. (2002) Determination of adsorption-induced variation in the spring constant of a microcantilever. *Appl. Phys. Lett.* 80(12), 2219–2221.

Churenkov A.V. (2013) Microresonator interference fiber-optic sensor of relative air humidity. *Techn. Phys. Lett.* 39(8), 723–725.

Churenkov A.V. (2014) Resonant micromechanical fiber optic sensor of relative humidity. *Measurement* 55, 33–38.

Clark S.K., Wise K.D. (1979) Pressure sensitivity in anisotropically etched thin-diaphragm pressure sensors. *IEEE Trans. Electron. Devices* 26(12), 1887–1896.

Cleveland J.P., Manne S., Bocek D., Hansma P.K. (1993) A non-destructive method for determining the spring constant of cantilevers for scanning force microscopy. *Rev. Sci. Instrum.* 64, 403–405.

Crank J. (1976) *The Mathematics of Diffusion*, 2nd edn. Clarendon Press, Oxford, UK.

Datskos P.G., Sauers I. (1999) Detection of 2-mercaptoethanol using gold-coated micromachined cantilevers. *Sens. Actuators B* 61, 75–82.

Delapierre G., Grange H., Chambaz B., Destannes L. (1983) Polymer-based capacitive humidity sensor: Characteristics and experimental results. *Sens. Actuators A* 4, 97–104.

Dennis J.O., Ahmed A.Y., Khir M.-H. (2015) Fabrication and characterization of a CMOS-MEMS humidity sensor. *Sensors* 15, 16674–16687.

Denton D.D., Day D.R., Priore D.F., Senturia S.D. (1985) Moisture diffusion in polyimide films in integrated circuits. *J. Electron. Mater.* 14, 119–136.

Denton D.D., Jaafar M.A.S., Ralston A.R.K. (1990) The long term reliability of a switched capacitor relative humidity sensor system, In: *Proceedings of IEEE 33rd Midwest Symp. on Circuits and Systems*, New York, pp. 854–857.

Desikan R., Lee I., Thundat T. (2006) Effect of nanometer surface morphology on surface stress and adsorption kinetics of alkanethiol self-assembled monolayers. *Ultramicroscopy* 106, 795–799.

Dohn S., Hansen O., Boisen A. (2006) Cantilever based mass sensor with hard contact readout. *Appl. Phys. Lett.* 88, 264104.

Dokmeci M., Najafi K. (2002) A high-sensitivity polyimide capacitive relative humidity sensor for monitoring anodically bonded hermetic micropackages. *IEEE J. Microelectromech. Syst.* 10, 197–204.

DuPont (1994) Technical data sheet for DuPont "Pyralin PI 2700" Series, Dupont, Wilmington, DE.

Eaton W.P., Smith J.H. (1997) Micromachined pressure sensors: Review and recent developments. *Smart Mater. Struct.* 6, 530–539.

Ekinci K.L., Yang Y.T., Roukes M.L. (2003) Ultimate limits to inertial mass sensing based upon nano-electromechanical systems. *J. Appl. Phys.* 95(5), 2682–2689.

Erlandsson R., McClelland G.M., Mate C.M., Chiang S. (1988) Atomic force microscopy using optical interferometry. *J. Vac. Sci. Technol. A* 6(2), 266–270.

Fagan B., Tipple C.A., Xue B., Datskos P., Sepaniak M. (2000) Modification of micro-cantilever sensors with sol-gels to enhance performance and immobilize chemically selective phases. *Talanta* 53, 599–608.

Fantner G.E., Schumann W., Barbero R.J., Deutschinger A., Todorov V., Gray D.S. et al. (2009) Use of self-actuating and self-sensing cantilevers for imaging biological samples in fluid. *Nanotechnology* 20, 434003.

Feng R., Farris R.J. (2003) Influence of processing conditions on the thermal and mechanical properties of SU8 negative photoresist coatings. *J. Micromech. Microeng.* 13, 80–88.

Fenner R.L. (1995) Cellulose crystallite strain gage hygrometry. *Sensors* (Peterborough, NH) 12(5), 51–59.

Fenner R., Zdankiewicz E. (2001) Micromachined water vapor sensors: A review of sensing technologies. *IEEE Sensors J.* 1(4), 309–317.

Filenko D. (2008) Chemical gas sensors based on functionalized self-actuated piezo-resistive cantilevers. PhD thesis, University of Kassel, Kassel, Germany.

Freund L.B., Floro J.A., Chason E. (1999) Extension of the Stoney formula for substrate curvature to configurations with thin substrates or large deformations. *Appl. Phys. Lett.* 74, 1987–1989.

Fritz J., Baller M.K., Lang H.P., Strunz T., Meyer E., Güntherodt H.J., Delamarche E. et al., (2000a) Stress at the solid–liquid interface of self-assembled monolayers on gold investigated with a nanomechanical sensor. *Langmuir* 16, 9694–9696.

Fritz J., Baller M.K., Lang H.P., Rothuizen H., Vettiger P., Meyer E., Guntherodt H.J. et. al., (2000b) Translating biomolecular recognition into nanomechanics. *Science* 288, 316–318.

Gerlach G., Sager K. (1994) A piezoresistive humidity sensor. *Sens. Actuators A* 43, 181–184.

Gerlach G., Sager K., Schroth A. (1994) Simulation of a humidity-sensitive double-layer system, *Sens. Actuators B* 18–19, 303–307.

Gerlach G., Nakladal A., Buchhold R., Baumann K. (1998) Piezoresistive effect: Stable enough for high-accuracy sensor applications? *SPIE* 3514, 377–385.

Gibbons J.F. (1972) Ion implantation in semiconductors: Part II Damage production and annealing. *Proc. IEEE* 60, 1062–1096.

Gimzewski J.K. (1993) Scanning tunneling microscopy. *J. De Phys.* 3(7), 41–48.

Gluck A., Halder W., Lindner G., Muller H., Weindler P. (1994) PVDF-excited resonance sensors for gas flow and humidity measurements. *Sens. Actuators B* 18/19, 554–557.

Goddenhenrich T., Lemke H., Hartmann U., Heiden C. (1990) Force microscope with capacitive displacement detection. *J. Vac. Sci. Technol. A* 8, 383–387.

Godin M., Tabard-Cossa V., Grutter P., Williams P. (2001) Quantitative surface stress measurements using a microcantilever. *Appl. Phys. Lett.* 79, 551–553.

Godin M., Williams P.J., Tabard-Cossa V., Laroche O., Beaulieu L.Y., Lennox R.B., Grutter P. (2004) Surface stress, kinetics, and structure of alkanethiol self-assembled monolayers. *Langmuir* 20, 7090–7096.

Goeders K.M., Colton J.S., Bottomley L.A. (2008) Microcantilevers: Sensing chemical interactions via mechanical motion. *Chem. Rev.* 108, 522–542.

Gonzalez-Benito C., Castillo E., Cruz-Caldito J.F. (2015) Determination of the linear coefficient of thermal expansion in polymer films at the nanoscale: Influence of the composition of EVA copolymers and the molecular weight of PMMA. *Phys. Chem. Chem. Phys.* 17, 18495.

Govardhan K., Alex Z.C. (2005) MEMs based humidity sensor, In: *Proceedings of International Conference on Smart Materials Structures and Systems*, July 28–30, Bangalore, India, pp. SE-20–27.

Gunter R.L., Delinger W.D., Porter T.L., Stewart R., Reed J. (2005) Hydration level monitoring using embedded piezoresistive microcantilever sensors. *Medical Eng. Phys.* 27, 215–220.

Han J., Wang X., Yan T., Li Y., Song M. (2012) A novel method of temperature compensation for piezoresistive microcantilever-based sensors. *Rev. Sci. Instrum.* 83, 035002.

Harley J.A., Kenny T.W. (1999) High-sensitivity piezoresistive cantilevers under 1000 A thick. *App. Phys. Lett.* 75, 289–291.

Hierlemann A., Lange D., Hagleitner C., Kerness N., Koll A., Brand O., and Baltes H. (2000) Application-specific sensor systems based on CMOS chemical microsensors. *Sens. Actuators B* 70, 2–11.

Hooge F.N., Kleinpenning T.G.M., Vandamme L.K.J. (1981) Experimental studies on 1/f noise. *Rep. Prog. Phys.* 44, 479–532.

Horie K., Yamashita T. (1995) *Photosensitive Polyimides*. Technomic Publishing Company, Lancaster, PA.

Hu, Z.Y., Seeley T., Kossek S., Thundat T. (2004) Calibration of optical cantilever deflection readers. *Rev. Sci. Instrum.* 75(2), 400–404.

Huang J.-Q., Li F., Zhao M., Wang K. (2015) A surface micromachined CMOS MEMS humidity sensor. *Micromachines* 6, 1569–1576.

Hygrometrix Inc. (2000) DSHMX2000, HMX2000 Relative Humidity/Moisture Sensor, Product Data Sheet, Hygrometrix, Alpine, CA.

Ibach H. (1997) The role of surface stress in reconstruction, epitaxial growth and stabilization of mesoscopic structures. *Surf. Sci. Rep.* 29, 195–263.

Israelachvili J. (1991) *Intermolecular and Surface Forces*, 2nd edn. Academic Press, San Diego, CA.

Jachowicz R., Weremczuk J. (2000) Sub-cooled water detection in silicon dew point hygrometer. *Sens. Actuators A* 85, 75–83.

Janata J. (2009) *Principles of Chemical Sensors*, 2nd edn. Springer, New York.

Jesenius H., Thaysen J., Rasmussen A.A., Veje L.H., Hansen O., and Boisen A. (2000) A microcantilever-based alcohol vapor sensor-application and response model. *Appl. Phys. Lett.* 76, 2615–2617.

Kang U., Wise K.D. (2000) A high speed capacitive humidity sensor with on-chip thermal reset. *IEEE Trans. Electron Dev.* 47, 702–710.

Kenny T.W., Waltman S.B., Reynolds J.K., Kaiser W.J. (1991) Micromachined silicon tunnel sensor for motion detection. *Appl. Phys. Lett.* 58, 100–102.

Khaled A.R.A. and Vafai K. (2004) Optimization modeling of analyte adhesion over an inclined microcantileverbased biosensor. *J. Micromech. Microeng.* 14(8), 1220–1229.

Khan M.M., Tarter T.S., Fatemi H. (1988) Stress relief in plastic encapsulated integrated circuit devices by die coating with photosensitive polyimide, In: *Digest of the 38th Electronic Components Conference*, Los Angeles, CA, pp. 425–431.

Kim B.H., Mader O., Weimar U., Brock R., Kern D.P. (2003) Detection of antibody peptide interaction using microcantilevers as surface stress sensors. *J. Vac. Sci. Technol. B* 21(4), 1472–1475.

Klein C.A. (2000) How accurate are Stoney's equation and recent modifications? *J. Appl. Phys.* 89, 5487–5489.

Koch R. (1994) The intrinsic stress of polycrystalline and epitaxial thin metal films. *J. Phys.: Condens. Matter* 6, 9519–9550.

Konrad B., Ashauer M. (1999) Demystifying piezoresistive pressure sensors. *Sensors* 16(7), 12–25.

Kovacs G.T.A., Petersen K., Albin M. (1996) Silicon micromachining: Sensors to systems. *Anal. Chemi.* 68(13), 407A–412A.

Kumar S.S., Pant B.D. (2014) Design principles and considerations for the 'ideal' silicon piezoresistive pressure sensor: A focused review. *Microsyst. Technol.* 20, 1213–1247.

Lang W. (1996) Silicon microstructuring technology. *Mater. Sci. Eng. R* 17, 1–55.

Lang H.P. (2009) Cantilever-based gas sensing. In: Comini E., Faglia G., and Sberveglieri G. (eds.), *Solid State Gas Sensing*, Springer Science+Business Media, New York, pp. 305–328.

Lang H.P., Gerber Ch. (2008) Microcantilever sensors. *Top. Curr. Chem.* 285, 1–27

Lang H.P., Baller M.K., Berger R., Gerber C., Gimzewski J.K., Battiston F., Fornaro P. et al., (1999) An artificial nose based on a micromechanical cantilever array. *Anal. Chim. Acta* 393, 59–65.

Lang H.P., Hegner M., Gerber C. (2005) Cantilever array sensors. *Mater. Today* 8, 30–36.

Lang H.P., Baller M.K., Berger R., Gerber Ch., Gimzewski J.K., Battiston F.M. et al. (2002) Adsorption kinetics and mechanical properties of thiolmodified DNA-oligos on gold investigated by microcantilever sensors. *Ultramicroscopy* 91, 29–36.

Lange D., Hagleitner C., Hierlemann A., Brand O., Baltes H. (2002) Complementary metal oxide semiconductor cantilever arrays on a single chip: Mass-sensitive detection of volatile organic compounds. *Anal. Chem.* 74, 3084–3095.

Laville C., Pellet C. (2002) Interdigitated humidity sensors for a portable clinical microsystem. *IEEE Trans. Biomed. Eng.* 49, 1162–1167.

Lavrik N.V., Tipple C.A., Sepaniak M.J., Datskos P.G. (2001a) Gold nano-structures for transduction of biomolecular interactions into micrometer scale movements. *Biomed. Microdevices* 3(1), 35–44.

Lavrik N.V., Tipple C.A., Sepaniak M.J., Datskos P.G. (2001b) Enhanced chemi-mechanical transduction at nanostructured interfaces. *Chem. Phys. Lett.* 336, 371–376.

Lavrik N.V., Sepaniak M.J., Datskosa P.G. (2004) Cantilever transducers as a platform for chemical and biological sensors. *Rev. Sci. Instrum.* 75(7), 2229–2253.

Le H.P. (1998) Progress and trends in ink-jet printing technology. *J. Imaging Sci. Technol.* 42, 49–62.

Lee C.Y., Lee G.B. (2003) Micromachine-based humidity sensors with integrated temperature sensors for signal drift compensation. *J. Micromech. Microeng.* 13, 620–627.

Lee C.Y., Lee G.B. (2005) Humidity sensors: A review. *Sensor Lett.* 3, 1–14

Lee S.S. White R.M. (1996) Self-excited piezoelectric cantilever oscillators. *Sens. Actuators A* 52, 41–45.

Lee C.Y., Ma R.H., Wang Y.H., Chou P.C., Fu L.M. (2007) Microcantilever-based weather station for temperature, humidity and wind velocity measurement, In: *Proceeding of Symposium on Design, Test, Integration and Packaging of MEMS/MOEMS*, 25–27 April, Stresa, Italy.

Li Y.C., Ho M.H., Hung S.J., Chen M.H., Lu M.S.C. (2006) CMOS micromachined capacitive cantilevers for mass sensing. *J. Micromech. Microeng.* 16, 2659–2665.

Li X., Liu Q., Pang S., Xu K., Tang H., Sun C. (2012) High-temperature piezoresistive pressure sensor based on implantation of oxygen into silicon wafer. *Sens. Actuators A* 179, 277–282.

Liang Y.A., Ueng S.-W., Kenny T.W. (2001) Performance characterization of ultra-thin n-type piezoresistive cantilevers, In: *Proceedings of 11th International Conference on Solid-State Sensors and Actuators*, June 10–14, Munich, Germany, pp. 998–1001.

Liu F., Rugheimer P., Mateeva E., Savage D.E., Legally M.G. (2002) Response of a strained semiconductor structure. *Nature* 416, 498.

Liu Y., Wang H., Zhao W., Qin H., Fang X. (2016) Thermal-performance instability in piezoresistive sensors: Inducement and improvement. *Sensors* 16(12), 1984.

Loizeau F., Akiyama T., Gautsch S., Vettiger P., Yoshikawa G., de Rooij N. (2012) Membrane-type surface stress sensor with piezoresistive readout. *Procedia Eng.* 47, 1085–1088.

Loizeau F., Akiyama T., Gautsch S., Vettiger P., Yoshikawa G., de Rooij N.F. (2015) Comparing membrane- and cantilever-based surface stress sensors for reproducibility. *Sens. Actuators A* 228, 9–15.

Madou M., Gottehrer P. (Eds.) (2000) *Fundamentals of Microfabrication*, 2nd edn. CRC Press, Boca Raton, FL.

Manalis S.R., Minne S.C., Atalar A. and Quate C.F. (1996) Interdigital cantilevers for atomic force microscopy. *Appl. Phys. Lett.* 69, 3944–3946.

Martorelli J.V. (2008) Monolithic CMOS-MEMS resonant beams for ultrasensitive mass detechion. Ph. D. thesis, Universitat Autonoma de Barcelona, Spain.

Matsuguchi M., Kuroiwa T., Miyagishi T., Suzuki S., Ogura T., Sakai Y. (1998) Stability and reliability of polyimide humidity sensors using crosslinked polyimide films. *Sens. Actuators B* 52, 53–57.

McFarland A.W., Poggi M.A., Doyle M.J., Bottomley L.A. (2005) Influence of surface stress on the resonance behavior of microcantilevers. *Appl. Phys. Lett.* 87, 053505.

Melcher J., Daben Y., Arlt G. (1989) Dielectric effects of moisture in polyimide, *IEEE Trans. Electron. Insulation* 24(1), 31–38.

Meyer G., Amer N.M. (1988) Novel optical approach to atomic force microscopy. *Appl. Phys. Lett.* 53(12), 1045–1047.

Minne S.C., Yaralioglu G., Manalis S.R., Adams J.D., Zesch J., Atalar A., Quate C.F. (1998) Automated parallel high-speed atomic force microscopy. *Appl. Phys. Lett.* 72, 2340–2342.

Miyatani T., Fujihira M. (1997) Calibration of surface stress measurements with atomic force microscopy. *J. Appl. Phys.* 81, 7099–7115.

Morten B., De Cicco G., Prudenziati M. (1993) A thick-film resonant sensor for humidity measurements, *Sens. Actuators A* 37/38, 337–342.

Muniraj N.J.R. (2011) MEMS based humidity sensor using Si cantilever beam for harsh environmental conditions. *Microsyst. Technol.* 17, 27–29.

Mutyala M.S.K., Bandhanadham D., Pan L., Pendyala V.R., Ji H.-F. (2009) Mechanical and electronic approaches to improve the sensitivity of microcantilever sensors. *Acta Mech. Sin.* 25, 1–12.

Nakladal A., Sager K., Gerlach G. (1995) Influences of humidity and moisture on the long-term stability of piezoresistive pressure sensors. *Measurement* 16, 21–29.

Ngo D., Huang Y., Rosakis A.J., Feng X. (2006) Spatially non-uniform, isotropic Misfit strain in thin films bonded on plate substrates: The relation between non-uniform film stresses and system curvatures. *Thin Solid Films* 515, 2220–2229.

Nordstrom M., Keller S., Lillemose M., Johansson A., Dohn S., Haefliger D., Blagoi G. et al. (2008) SU-8 cantilevers for bio/chemical sensing; fabrication, characterisation and development of novel read-out methods. *Sensors* 8, 1595–1612.

Numata S., Tawata R., Ikeda T., Fujisaki K., Shimanoki H., Miwa T. (1991) Preparation of aromatic acid anhydride complexes as cross linking agents. *Jpn. Patent JP* 03090076.

OCG (1994) *Technical data sheet for Ciba Geigy "Probimide 7000, 7500 and 400", OCG Microelectronics Materials AG*, Basel, Switzerland.

Oden P.I., Chen G.Y., Steele R.A., Warmack R.J., and Thundat T. (1996) Viscous drag measurements utilizing microfabricated cantilevers. *Appl. Phys. Lett.* 68, 3814–3816.

Park S.J., Doll J.C., Rastegar A.J., Pruitt B.L. (2010) Piezoresistive cantilever performance—Part II: Optimization. *J. Microelectromech. Syst.* 19, 149–161.

Patil S.J., Adhikari A., Baghini M.S., Rao V.R. (2014) An ultra-sensitive piezoresistive polymer nano-composite microcantilever platform for humidity and soil moisture detection. *Sens. Actuators B* 203, 165–173.

Phan L.P. Hoa, G. Suchaneck, G. Gerlach (2003) Measurement uncertainty of piezoresistive beam-type humidity sensors. In: *Proceedings of 26th International Spring Seminar on Electronics Technology*, May 8–11, Stara Lesna, Slovak Republic, pp. 408–411.

Phan H.P., Dao D.V., Nakamura K., Dimitrijev S., Nguyen N.T. (2015) The piezoresistive effect of SiC for MEMS sensors at high temperatures: A review. *J. Microelectromech. Syst.* 24(6), 1663–1677.

Plummer J.D., Deal M.D., Griffin P.B. (2000) *Silicon VLSI Technology Fundamentals, Practice, and Modeling*. Prentice Hall, NJ.

Pustan M., Belcin O., Golinval J.C. (2011) Dynamic investigations of paddle MEMs cantilevers used in mass sensing applications, In: *Proceedings of the 3rd International Conference on Advanced Engineering in Mechanical Systems (ADEMS'11)*, 21–23 September., Cluj-Napoca, Romania, Acta Technica Napocensis of the TU of Cluj-Napoca, Vol. 54, pp. 117–122.

Qui Y.Y., Azeredo-Leme C., Alcacer L.R., Franca J.E. (2001) A CMOS humidity sensor with on-chip calibration. *Sens. Actuators A* 92, 80–87.

Ralston A.R.K., Buncick M.C., Denton D.D. (1991) Effects of aging on polyimide: A model for dielectric behavior, In: *Proceedings of the 6th International Conference on Solid-State Sensors and Actuators (TRANSDUCERS'91)*, 24–27 June, San Francisco, CA, pp. 759–763.

Ralston A.R.K., Klein C.F., Thoma P.E., Denton D.D. (1996) A model for the relative environmental stability of a series of polyimide capacitance humidity sensors. *Sens. Actuators B* 34, 343–348.

Ramos D., Arroyo-Hernandez M., Gil-Santos E., Tong H.D., Van Rijn C., Calleja M., Tamayo J. (2009) Arrays of dual nanomechanical resonators for selective biological detection. *Anal. Chem.* 81, 2274–2279.

Ree M., Swanson S., Volksen W. (1991) Residual stress and its relaxation behavior of high-temperature polyimides: Effect of precursor origin, In: *Proceedings of 4th Int. Conf. on Polyimides*, October. 30–November 1, New York, pp. 601–617.

Ristic L. (1994) *Sensor Technology and Devices* (Motorola Series in Solid-State Electronics, Vol. 51). Artech House, Boston, MA, pp. 524–428.

Rugar D., Mamin H.J., and Guethner P. (1989) Improved fiber-optic interferometer for atomic force microscopy. *Appl. Phys. Lett.* 55(25), 2588–2590.

Ryu W.H., Chung Y.C., Choi D.K., Yoon C.S., Kim C.K., Kim Y.H. (2004) Computer simulation of the resonance characteristics and the sensitivity of cantilever-shaped Al/PZT/RuO$_2$ biosensor. *Sens. Actuators B* 97, 98–102.

Sader J.E. (2002) Surface stress induced deflections of cantilever plates with applications to the atomic force microscope: Rectangular plates. *J. Appl. Phys.* 89, 2911–2921.

Sager K., Gerlach G., and Schroth A. (1994) A humidity sensor of a new type. *Sens. Actuators B* 18, 85–88.

Sager K., Gerlach G., Nakladal A., Schroth A. (1995) Ambient humidity and moisture—A decisive failure source in piezoresistive sensors. *Sens. Actuators A* 46–47, 171–175.

Sager K., Schroth A., Nakladal A., Gerlach G. (1996) Humidity-dependent mechanical properties of polyimide films and their use for IC-compatible humidity sensors. *Sens. Actuators A* 53, 330–334.

Sappat A., Wisitsoraat A., Sriprachuabwong C., Jaruwongrungsee K., Lomas T., Tuantranont A. (2011) Humidity sensor based on piezoresistive microcantilever with inkjet printed PEDOT/PSS sensing layers, In: *Proceedings of the International 8th Electrical Engineering/Electronics, Computer, Telecommunications and Information Technology (ECTI)*, 17–19 May, Khon Kaen, Thailand, pp. 34–37.

Savran C.A., Burg T.P., Fritz J., Manalis S.R. (2003) Microfabricated mechanical inherently differential readout. *Appl. Phys. Lett.* 83, 1659–1661.

Scheible D.V., Erbe A., Blick R.H. (2002) Tunable coupled nanomechanical resonators for single-electron transport. *New J. Phys.* 4, 86.

Schmid S., Wägli P., Hierold C. (2008) All-polymer microstring resonant humidity sensor with enhanced sensitivity due to change of intrinsic stress, In: *Proceedings of the EUROSENSORS XXII Conference*, September. 7–10, Dresden, Germany, pp. 697–700.

Schmid S., Kuhne S., Hierold C. (2009) Influence of air humidity on polymeric microresonators. *J. Micromech. Microeng.* 19, 065018.

Schroth A., Sager K., Gerlach G., Hiberli A., Boltshauser T., Baltes H. (1996) A resonant poliyimide-based humidity sensor. *Sens. Actuators B* 34, 301–304.

Sone H., Okano H., Hosaka S. (2004) Picogram mass sensor using piezoresistive cantilever for biosensor. *Jpn. J. Appl. Phys.* 43(7B), 4663–4666.

Tamayo J., Alvarez M., Lechuga L.M. (2004) System and method for detecting the displacement of a plurality of micro- and nanomechanical elements, such as microcantilevers. European Patent PCT/EP2005/002356.

Tang Y.J., Fang J., Yan X., Ji H.F. (2004) Fabrication and characterization of SiO$_2$ microcantilever for microsensor application. *Sens. Actuators B* 97, 109–113.

Thaysen J., Boisen A., Hansen O., Bouwstra S. (2000) Atomic force microscopy probe with piezoresistive readout and highly symmetrical Wheatstone bridge arrangement. *Sens. Actuators A* 83, 47–53.

Thundat T., Oden P.I., Warmack R.J. (1997) Microcantilever sensors. *Microscale Thermophys. Eng.* 1, 185–199.

Tipple C.A., Lavrik N.V., Culha M., Headrick J., Datskos P., Sepaniak M.J. (2002) Nanostructured microcantilevers with functionalized cyclodextrin receptor phases: Self-assembled monolayers and vapor-deposited films. *Anal. Chem.* 74, 3118–3126.

Toda M., Chen Y., Nett S.K., Itakura A.N., Gutmann J., Berger R. (2014) Thin polyelectrolyte multilayers made by inkjet printing and their characterization by nanomechanical cantilever sensors. *J. Phys. Chem. C* 118, 8071–8078.

Toray (1992) Technical data sheet for Toray "Photoneece UR-5100", Toray Industry, Tokyo, Japan.

Tortonese M., Barrett R.C., Quate C.F. (1993) Atomic resolution with an atomic force microscope using piezoresistive detection. *App. Phys. Lett.* 62, 834–836.

Vanlandingham M.R., Eduljee R.F., Gillespie J.W. (1999) Moisture diffusion in epoxy systems *J. Appl. Polym. Sci.* 71, 787–798.

Vashist S.K. (2007) A review of microcantilevers for sensing applications. *J. Nanotechnol. Online* 3, 1–15. DOI: 10.2240/azojono0115.

Vashist S.K., Korotcenkov G. (2011) Microcantilever-based chemical sensors, In: Korotcenkov G. (ed.) *Chemical Sensors: Comprehensive Sensor Technologies, Vol. 4: Solid State Devices*, Momentum Press, New York, 2011, pp. 321–376.

Vasiljevic N., Trimble T., Dimitrov N., Sieradzki K. (2004) Electrocapillarity behavior of Au(111) in SO_4^{2-} and F^-. *Langmuir* 20, 6639–6643.

Urbiztondo M.A., Pellejero I., Villarroya M., Sese J., Pina M.P., Dufour I., Santamaria J. (2009) Zeolite-modified cantilevers for the sensing of nitrotoluene vapors. *Sens. Actuators B* 137, 608–616.

Waber T., Pahl W., Schmidt M., Feiertag G., Stufler S., Dudek R., Leidl A. (2013) Flip-chip packaging of piezoresistive barometric pressure sensors. *Proc. SPIE* 8763, 876321,

Waber T., Sax M., Pahl W., Stufler S., Leidl A., Günther M., Feiertag G. (2014) Fabrication and characterization of a piezoresistive humidity sensor with a stress-free package. *J. Sens. Sens. Syst.* 3, 167–175.

Wallace D., Hayes D., Chen T., Shah V., Radulescu D., Cooley P., Wachtler K. et al., (2007) Ink-jet as a MEMS manufacturing tool. In *Proceedings of the First International Conference on Integration and Commercialization of Micro and Nanosystems*, 10–13 January, Hainan, China, pp. 1161–1168.

Wu G., Ji H., Hansen K., Thundat T., Datar R., Cote R., Hagan M.F., Chakraborty A.K., Majumdar A. (2001) Origin of nanomechanical cantilever motion generated from biomolecular interactions. *Proc. Natl. Acad. Sci. USA* 98, 1560–1564.

Xiansong F, Suying Y, Shuzhi H, Wei Z, Yiqiang Z, Shengcai Z., Weixin Z. (2004) Simulation and test of a novel SOI high temperature pressure sensor. In: *Proceedings of the 7th International Conference on Solid-State and Integrated Circuits Technology*, 18–21 October, Beijing, China, pp. 1824–1827.

Yang D., Koros W., Hopfenberg H. (1985) Sorption and transport studies of water in Kapton polyimide. *J. Appl. Polym. Sci.* 30, 1035–1047.

Yang D., Koros W., Hopfenberg H., Stannett V.T. (1986) The effects of morphology and hygrothermal aging on water sorption and transport in Kapton polyimide. *J. Appl. Polymer Sci.*, 31, 1619–1629.

Yang J., Ono T., Esashi M. (2000) Mechanical behavior of ultrathin microcantilever. *Sens. Actuators A* 82, 102–107.

Yang M., Zhang X., Vafai K., and Ozkan C.S. (2003) High sensitivity piezoresistive cantilever design and optimization for analyte-receptor binding. *J. Micromech. Microeng.* 13, 864–872.

Yang S., Xu Q. (2017) A review on actuation and sensing techniques for MEMS-based microgrippers. *J. Micro-Bio. Robot.* 13, 1–4, doi:10.1007/s12213-017-0098-2.

Yi J.W., Shih W.Y., Shih W.-H. (2002) Effect of length, width, and mode on the mass detection sensitivity of piezoelectric unimorph cantilevers. *J. Appl. Phys.* 91, 1680–1686.

Zhang Y., Ren Q., Zhao Y.P. (2004) Modelling analysis of surface stress on a rectangular cantilever beam. *J. Phys. D: Appl. Phys.* 37(15), 2140–2145.

Zhou W., Khaliq A., Tang Y., Ji H., Selmic R.R. (2005) Simulation and design of piezoelectric microcantilever chemical sensors. *Sens. Actuators A* 125, 69–75.

14 Thermal Conductivity-Based Hygrometers

14.1 PRINCIPLE OF OPERATION

The thermal conductivity-based hygrometer method is another approach to measuring humidity (Daynes 1993; Kimura 1996a, 1996b; Roveti 2001). In this humidity-sensing method, the difference between the thermal conductivities of air and those of air containing water vapor is used for humidity measurement. When the air or gas is dry, it has a greater capacity to "sink" heat, as in a desert climate. A desert can be extremely hot during the day, but at night, the temperature rapidly drops due to the dry atmospheric conditions. By comparison, humid climates do not cool so rapidly at night, because heat is retained by water vapor in the atmosphere.

Thermal conduction is when thermal energy transfers from one object to another. Physical contact between two bodies is required for the heat conduction. Kinetic energy is transferred to a cooler body from a warmer body by thermally agitating its particles. As a result, the cooler body gains heat while the warmer body loses heat. For instance, a heat passage through a rod is governed by a law similar to Ohm's law; the heat flow rate is proportional to the thermal gradient across the material (dT/dx) and its cross-sectional area (A), or

$$H = \frac{dQ}{dT} = -\lambda \cdot A \frac{dT}{dx} \quad (14.1)$$

where λ is called the *thermal conductivity*. The minus sign indicates that heat flows in the direction of temperature decrease. A good thermal conductor has a high value of thermal conductivity, whereas thermal insulators have low values of λ.

Thus, knowing the degree of cooling of the controlled object, one can determine the thermal conductivity of the surrounding space. This is the basis of the work of thermal conductivity–based hygrometers. These devices measure the absolute humidity by quantifying the difference between the thermal conductivity of dry air and that of air containing water vapor. As seen in Figure 14.1, when air or gas is dry, at temperatures up to 473 K it has a greater "thermal conductivity," or ability to transfer heat, compared to humid air (Smetana and Unger 2008), thereby increasing its cooling effect.

The appearance of a conversional thermal-conductivity cell with a platinum (Pt) wire is shown in Figure 14.2. The gas chambers in which these wires are stretched consist of cylindrical holes, 10 mm in diameter, drilled lengthwise through a metal bar 12.5 mm square and 125 mm long (Palmer and Weaver 1924). Inlets and outlets are provided for the circulation of gas. Rolled brass and copper are suitable metals when noncorrosive gases are used; cast metal should not be used because of its probable porosity. To minimize the effect on the calibration of any possible displacement of the active wires, and to reduce the power required to produce a given temperature difference between the wire and the walls of the chamber, the diameter of the

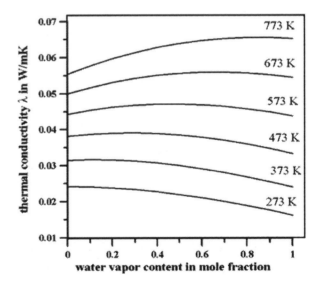

FIGURE 14.1 Thermal conductivity λ_{mix} of air in dependence on water vapor content for different temperatures. (Reprinted with permission from Smetana, W. and Unger, M., Design and characterization of a humidity sensor realized in LTCC-technology, *Microsyst. Technol.*, 14, 979–987, 2008. Copyright 2008, Springer Science+Business Media.)

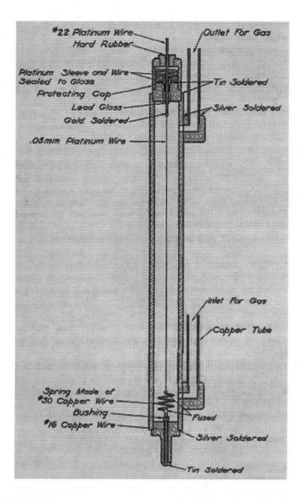

FIGURE 14.2 Section of a conversional thermal-conductivity cell. (Reprinted from Palmer P.E. and Weaver E.R., Thermal-conductivity method for the analysis of gases, *Technological Papers of the Bureau of Standards*, US Bureau of Standards, Washington, DC, 18, 249, 1924.)

cells is made the maximum permissible. Experiments have shown that, when the diameter greatly exceeds 1 cm, convection currents cause unsteady readings (Palmer and Weaver 1924).

14.2 APPROACHES TO THE DESIGN OF THERMAL CONDUCTIVITY-BASED HYGROMETERS

Thermal conductivity-based hygrometers can be manufactured using a variety of approaches. For example, Devoret et al. (1980) suggested a method based on maintaining a constant temperature difference between a "cold" element and a "hot" element. Heat is transferred from the "hot" element to the "cold" element via thermal conduction through the investigated gas. The power needed to heat the "hot" element is therefore a direct measure of the thermal conductivity. In the simplest case, power loss of a single filament by heat conduction via the ambient gas can be expressed as

$$P = k_{TC}\lambda\Delta T \qquad (14.2)$$

where P is the power dissipation of the heater by thermal conduction of the gas, λ is the thermal conductivity of the given gas–gas mixture, k_{TC} is a constant that is characteristic for a given geometry, and ΔT is the temperature difference between the heater and the ambient gas. The precise mechanism is quite complex, because the thermal conductivity of gases varies with temperature and convection, as well as conductivity. Most gases produce a linear output signal, but not all. Heat loss due to convection and heat conduction through the suspensions of the "hot" element need to be minimized by sensor design.

For measuring air humidity, Miura (1985) has used a device, a schematic diagram of which is shown in Figure 14.3. The thermal conductivity humidity sensor (or absolute humidity sensor) consisted of two matched elements (see Figure 14.3a) in a bridge circuit (Figure 14.3b); one is hermetically encapsulated in dry nitrogen, and the other is exposed to the environment. Earlier, platinum filaments were used as elements of thermal conductivity-based hygrometers. However, in modern samples, thermistors with high negative temperature coefficient (NTC) are usually used, which provides a higher sensitivity to the device. Two tiny thermistors ($T1$ and $T2$) are supported by thin wires to minimize the thermal conductivity loss to the housing. Small venting holes are used to expose the thermistor $T1$ to the outside gas, and the thermistor $T2$ is sealed in dry air. Both thermistors are connected into a bridge circuit ($R1$ and $R2$), which is powered by voltage $+E$. Due to the passage of electric current, the thermistors develop self-heating. During operation of the hygrometer, the thermistors heat up to 170°C–200°C over the ambient temperature, using both self-heating and an additional heater. If the sample gas or vapor has a thermal conductivity higher than the reference, the heat is lost from the exposed element and its temperature decreases, whereas if its thermal conductivity is lower than that of the reference, the temperature of the exposed element increases. At that, these temperature changes are proportional to the concentration of the gas present at the sample element. Temperature changes alter the electrical resistance, which is measured using a bridge circuit. To establish a zero reference point initially, the bridge is balanced in dry air. As absolute humidity rises from zero, the output of the sensor gradually increases.

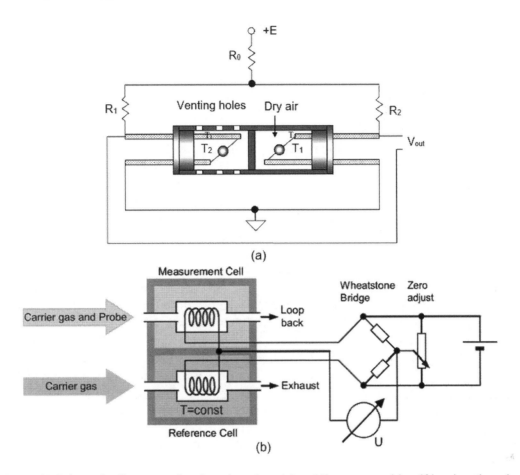

FIGURE 14.3 (a, b) Schematic diagrams of a thermistor-based humidity sensor with self-heating thermistors, showing (a) configuration and (b) principle of operation. (Reprinted with permission from Atashbar, M.Z. et al., Basic principles of chemical sensor operation, In: Korotcenkov, G. (ed.), *Chemical Sensors. Fundamentals of Sensing Materials.*, Vol. 1: *General Approaches*, pp. 1–62, New York, Momentum Press, 2010. Copyright 2010.)

Since the heat dissipated yields different operating temperatures, the difference in the resistance of the thermistors is proportional to the absolute humidity. If the temperature and pressure are known, the absolute humidity easily converts to the relative humidity (RH). Absolute humidity sensors are very durable, operate at temperatures up to 300°C, and are resistant to chemical vapors by virtue of the inert materials used for their construction (i.e., glass, semiconductor material for the thermistors, high-temperature plastics, or aluminum). In general, absolute conductivity–based humidity sensors provide greater resolution at high temperatures (> 90°C–100°C) than do capacitive and resistive sensors, and may be used in applications in which these sensors would not survive. The typical accuracy of an absolute humidity sensor is ±3 g/m³; this converts to approximately ±5% RH at 40°C and ±0.5% RH at 100°C. The improvement in sensitivity with increasing temperature is associated with an increase in the thermal conductivity of the air. In addition, at low temperatures, the moisture content of air is insignificant (see Table 14.1), and therefore the change

TABLE 14.1
Maximum Possible Moisture Content in the Air

Temperature, °C	40	30	20	10	0	−10	−20	−30
Maximum absolute humidity, g/m³	50.67	30.08	17.15	9.36	4.87	2.16	0.88	0.33

in humidity has an insignificant effect on the thermal conductivity of the air.

Bridgeman et al. (2017) suggested a different approach to measuring air humidity. They used two thermistors treated in various ways. A hydrophobically coated thermistor sensor was used as the reference, and the hydrophilically modified thermistor was used as the sensing element. The sensing thermistor was prepared by dip coating a low–thermal mass thermistor with a solution containing a long-chain quaternary ammonium salt and color dye. The salt provides heat upon hydration, and the dye allows for visualization. The measurements were taken at 40 Hz, with an effective temperature resolution of less than 0.05°C.

Figure 14.4 shows the result of the thermistor-based humidity-sensing experiments. The experiments were carried out at room temperature, with insignificant fluctuations in the temperature between the reference and sensing element. Therefore, all temperature changes in the hydrophilically modified thermistor in comparison with the reference element were related to the humidity changes. As seen in all experiments shown in Figure 14.4a, there is a significant increase in temperature with humidity exposure. Since the temperature and flow rates were held constant, for analysis purposes, only the difference between the baseline and peak temperature from humidity exposure was needed. Figure 14.4b shows the relationship between the RH and observed temperature change. It can be seen that there is a proportional correlation between the temperature change and the RH. In the range of > 75% RH, the saturation is observed. However, Bridgeman et al. (2017) believe that the sensor saturation effect could be potentially combated by either changing the chemical coating or heating the thermistor. In addition, no heat release was observed with exposure to acid gases, such as pure carbon dioxide or concentrated hydrochloric acid.

Single-element thermal conductivity–based hygrometers are also being developed. Typically, their work is based on measuring the power needed to maintain a constant temperature of a thermistor in the air flow or to measure the temperature change of a thermistor at a constant current. Such hygrometers are not absolute humidity sensors. To determine the humidity, the sensors must be calibrated.

However, one should note that thermal conductivity humidity sensors respond to any gas that has thermal properties different from those of dry nitrogen (Daynes 1993); this will affect the measurements. The presence of gases that have thermal conductivities relative to air of more than 1 leads to cooling of the exposed thermistor or filament, while gases with thermal conductivities lower than 1 contribute to the heating of the thermistor or filament. At the same time, the greater the difference in thermal conductivity is, the greater is the effect. Gases with thermal conductivities close to 1 cannot be measured by this technique. These include ammonia, carbon monoxide, nitric oxide, oxygen, and nitrogen (Barsony et al. 2009). Thermal conductivities of selected gases are listed in Table 14.2. As seen there, humidity sensors are the most sensitive to the appearance of helium and hydrogen in the atmosphere.

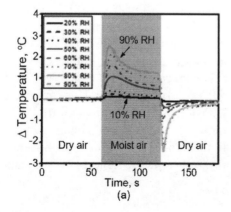 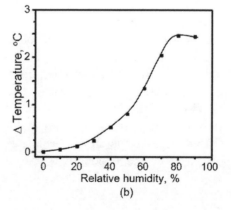

FIGURE 14.4 (a) Test results from alternating between dry and humid air. Absolute temperatures during the tests ranged from approximately 23°C to 26°C. (b) The difference between the baseline and the peak temperature increases as a function of relative humidity (RH). (Reprinted with permission from Bridgeman D. et al., *Sensors*, 17, 1196, 2017. Copyright 2017, MDPI as open access.)

Thermal Conductivity-Based Hygrometers

TABLE 14.2
Thermal Conductivities Relative to Air at 100°C

	Gas	Thermal Conductivity
Thermal conductivity > 1	Helium	5.6
	Hydrogen	6.9
	Methane	1.4
	Neon	1.8
Thermal conductivity ≈ 1	Ammonia	1.12
	Oxygen	1.01
	Nitric oxide	1.003
	Nitrogen	0.97
Thermal conductivity < 1	Argon	0.7
	Butane	0.7
	Carbon dioxide	0.7
	Ethane	0.75
	Freon/Halon	0.4
	Hexane	0.5
	Pentane	0.7
	Propane	0.8
	Water vapor	0.8
	Xenon	0.2

Source: www.engineersedge.com.

14.3 MICROMACHINED HUMIDITY SENSORS

It should be noted that the thermal conductivity-based hygrometer can be manufactured in a variety of designs. For example, Smetana and Unger (2008) have developed a humidity sensor, shown in Figure 14.5a, using a conventional ceramics technology. The setup of the humidity sensor comprises two resistor elements on a substrate, acting as temperature-sensing devices as well as heaters: one of them serves as the humidity-detecting element, while the other as the reference sensor element. The reference sensor element is sealed from the environment ("dry atmosphere" with thermal conductivity λ_1) by attaching a ceramic cap, and the humidity-sensing element is exposed to the environment ("vapor atmosphere" with thermal conductivity λ_2) where it is enclosed by a perforated ceramic cap. Based on the results of calculation shown in Figure 14.1, it becomes mandatory that the operation of the heater elements of the humidity sensor be carried out at a sufficiently high temperature by applying a current pulse in order to achieve a pronounced change of thermal conductivity in dependence on the humidity. Consequently, the heat removal from the resistor, which is exposed to the environment, increases with the increasing amount of water vapor (thermal conductivity λ_2), which results in a decrease of the temperature (T_2) in the sensor element. The heater element in the dry atmosphere (thermal conductivity λ_1), which is acting as reference, is cooled down to a temperature T_1. Since λ_2 is less than λ_1, the relation of the resulting temperatures is T_1 is higher than T_2. The test results are shown in Figure 14.5b. They testify that the sensors are efficient. However, the power consumption during measurements was approximately 3.2 W (Smetana and Unger 2008), which is too much for a humidity sensor, especially when compared to other sensors, such as capacitive or resistive ones. Sensitivity

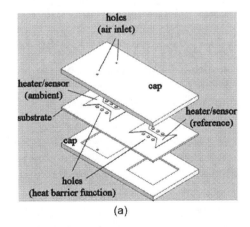 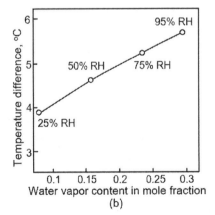

FIGURE 14.5 (a) Exploded view of final version of sensor set-up and (b) characteristic course of a temperature difference determined experimentally at an ambient temperature of 343 K. RH, relative humidity. (Reprinted with permission from Smetana, W. and Unger, M., Design and characterization of a humidity sensor realized in LTCC-technology, *Microsyst. Technol.*, 14, 979–987, 2008. Copyright 2008, Springer Science+Business Media.)

of these sensors is also significantly inferior to most of the developed humidity sensors.

It is understood that a large power consumption significantly narrows the possible field of application of thermal conductivity humidity sensors. Experiment has shown that the application of micromachining solves this problem. The use of membranes isolated from the substrate reduces heat losses and, as a result, significantly reduces the power of the heaters necessary to achieve the required temperature in the core. One of the first thermal conductivity-based sensors fabricated using micromachining technology was the sensor developed by Kimura (1996a, 1996b). Kimura (1996a, 1996b) proposed a thermal humidity sensor with only a single micro-air-bridge heater that was heated up to two temperature levels. Kimura (1996a, 1996b) used specific properties of the air and the water vapor in his developments. The SEM picture of the microbridge heater humidity sensor is shown in Figure 14.6. Thermal conductivities of the air and the water vapor are both increasing functions of temperature, but the change of thermal conductivity of water vapor at high temperatures is stronger (see Figure 14.7). Therefore, Kimura (1996) suggested the following mode of operation: (1) The microheater sensing region is cooled up to a low temperature level, at which the thermal conductivity of the air and the water vapor is almost the same. (2) Then, the same region is heated to a temperature level where

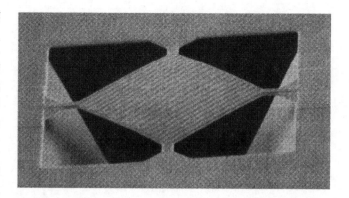

FIGURE 14.7 The SEM picture of the microbridge heater humidity sensor. RH, relative humidity. (Reprinted with permission from Kimura, M., A new method to measure absolute humidity independently of the ambient temperature, *Sens. Actuators B*, 33, 156–160, 1996. Copyright 1996, Elsevier.)

the thermal conductivity difference between the air and water vapor becomes significant. Heating is achieved by applying double pulse currents. The resistance of the microheater changes due to the temperature coefficient of resistance (TCR) property. (3) As a result, the difference of the voltages in two cycles becomes a function of the absolute humidity in the air.

A similar approach was used by Zambrozi and Fruett (2012). They have proposed a humidity sensor in which the sensing element was a single floating resistor of doped polysilicon (Figure 14.8), which basically operated in two phases: thermal actuation and thermal sensing. As thermal actuator, a biasing current is applied to the resistor, leading to its self-heating by Joule effect. As thermal sensor, the biasing current is reduced, and the cooldown process (conduction and convection) is measured through the thermal time constant (τ). Zambrozi and Fruett (2012) assumed that, because the resistor is suspended and cooling is due to heat transfer mechanisms (thermal conduction and thermal convection), the thermal time constant of this process should be dependent on the water vapor quantities in the ambient. The experiment showed that the cooling process on the floating resistor, is really dependent on the amount of water molecules in the environment, and the thermal time constant (τ) is closely related to the RH of the air surrounding the floating resistor (Figure 14.8). The power consumption by a floating resistor during the thermal actuator phase was approximately 5 mW. Earlier, Vass-Várnai et al. (2008), who analyzed the change in thermal conductivity of a porous silicon layer in humid atmosphere, have also shown that the thermal transient methodology could be used to measure humidity. They observed a significant shift in one of the peaks of the

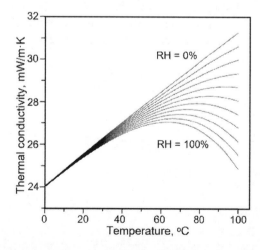

FIGURE 14.6 Results of simulation of the thermal conductivity of moist air as a function of temperature with the relative humidity (RH) as a parameter ranging between dry air (*top curve*, RH = 0%) and saturation conditions (*lower curve*, RH = 100%) in 10% RH steps between curves. (Reprinted with permission from Tsilingiris, P. T., Thermophysical and transport properties of humid air at temperature range between 0°C and 100°C, *Energy Convers. Manage.*, 49, 1098–1110, 2008. Copyright 2008, Elsevier.)

Thermal Conductivity-Based Hygrometers

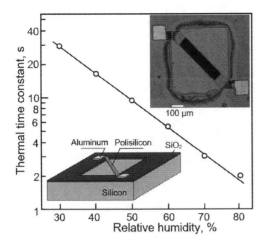

FIGURE 14.8 Thermal time constant (τ) versus relative humidity. Insertions: three-dimensional views of the floating polysilicon resistor and the microscope images of the floating resistor sensor. Sensor was fabricated using a surface micromachining technology. (Reprinted with permission from Zambrozi, Jr. P. and Fruett, F., *J. Integr. Circuits Syst.*, 7, 130–136, 2012. Copyright 2012, Brazilian Microelectronics Society as open access.)

time constant spectrums of the measured device in the driest and wettest state.

A two-element microminiature, thermal conductivity–based humidity sensor manufactured using micromachining technology is shown in Figure 14.9. The sensor was fabricated using a conventional thin-film deposition process and bulk micromachining technology with two masks layout (Lee et al. 2001). In this sensor, resistors are heated to a temperature around 250°C using self-heating effects by applying a constant voltage. One of the resistors is passivated from the humid environment by proper packaging. The other resistor is exposed to the environment, whose resistance changes with respect to the change in RH due to the thermal conductivity difference. In this sensor, the pair of patterned platinum thin-film resistors with meander structure was used as a humidity-sensing element (SE) and temperature-compensating element (CE). Two platinum resistors were located on the top of two different diaphragms. Both resistors were heated electrically to temperatures of around 250°C so that there is a measurable difference in thermal conductivity between dry and humid air. Test results showed that the thermal conductivity–based humidity sensors had a sensitivity of 0.6 mV/% RH, small nonlinearity of 1% of the full-scale (FS) reading (Figure 14.9), and hysteresis of below 2% RH. The response and recovery time to the humidity change were obtained at approximately 25 and 5 seconds, respectively. The fabricated sensor had long-term stability and superior durability in a harsh environment (Lee et al. 2001).

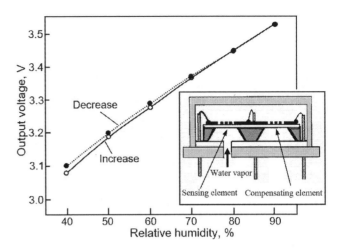

FIGURE 14.9 Output voltage versus relative humidity for the humidity sensor at 20°C. Insertion: The cross-sectional view of the thermal conductivity-based humidity sensor structure. A pair of patterned platinum thin film with meander structure as sensing elements was formed on the thermally isolated thin membranes each other with a pixel size of 240 × 250 μm². (Idea and data from Lee, D. et al., A micromachined robust humidity sensor for harsh environment applications, In: *Proceedings of the 14th IEEE International Conference on Micro Electro Mechanical Systems (MEMS 2001)*, pp. 558–561, 2001.)

In addition, this sensor, due to the high temperature of operation, does not contain any risk of water condensation at high-RH conditions. However, these devices had a lower detection limit of around 40% RH. This is mainly because the difference in thermal conductivities at humidity levels below 40% is small and hard to measure.

The same approach was used by Okcan and Akin (2004, 2007). The schematic diagram of their sensor is shown in Figure 14.10. In the sensor system, the most important fact was to thermally isolate the sensor and the reference in order to achieve high sensitivity. This was achieved by isolating the sensor and the reference from the silicon substrate, which is a good thermal conductor. For thermal isolation, the substrate underneath the sensor and the reference element were etched and structures were suspended. It is important to note that these sensors can be fabricated on the silicon substrate using standard complementary metal–oxide–semiconductor (CMOS) and post-CMOS processes (Okcan and Akin 2004, 2007), which allows the integration of the sensor with the readout circuit monolithically. As the humidity-sensitive structures, *p–n* junction diodes were used. The most important advantage of diodes is that they have high and almost constant temperature sensitivity in all temperature regions. Okcan and Akin (2004, 2007) have also shown that the incorporation of *p–n* junctions as a humidity-sensing element instead of Pt resistors strongly decreased the power dissipated by the sensor.

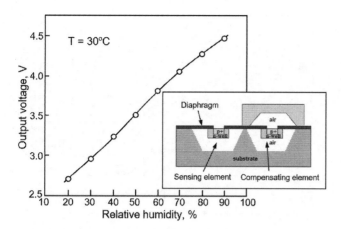

FIGURE 14.10 Output voltage versus relative humidity for the humidity sensor at 30°C. Insertion: A symbolic cross-sectional view of the proposed sensor structure. Measured hysteresis characteristics of the humidity sensor at 20°C and 30°C was less than 1%. (Idea and data from Okcan, B. and Akin, T., A thermal conductivity based humidity sensor in a standard CMOS process, In: *Proceedings of IEEE International Conference on Micro Electro Mechanical Systems*, January 25–29, Maastricht, the Netherlands, pp. 552–555, 2004.)

In addition, they are compatible with the standard CMOS fabrication used. The chip was packaged in such a way that the reference diode did not have any interaction with the humid environment (see Figure 14.10). The sensor provided a sensitivity of 14.3 mV/% RH, 26 mV/% RH, and 46.9 mV/% RH for 20°C, 30°C, and 40°C ambient temperatures, respectively. The sensing diodes were heated to 250°C, so the sensor did not allow water condensation. Total power dissipation of the sensor was only 1.38 mW. Another advantage of using diodes instead of metal resistors is that the lower detection point (20% RH in this case) is lower than that of metal resistor devices (≈40% RH). This is probably due to the higher temperature sensitivities of diodes causing them to be able to detect the small thermal conductivity differences at low humidity levels. Later on, based on the results of the simulations of these sensors, Kouda et al. (2008) proposed a model and corrector, optimizing the work of a thermal-conductivity humidity sensor (THS), operated under a dynamic environment. As a result, even though the sensor characteristics change with temperature and hysteresis, the proposed smart sensor performed quite satisfactorily, irrespective of any change in temperature or hysteresis.

Fang et al. (2011) also developed a two-element thermal conductivity-based humidity sensor based on Micro-Electro-Mechanical Systems (MEMS) technology. However, the micro-air-bridge silicon nitride-platinum (Si_3N_4-Pt) was used as a sensitive element. One micro-air-bridge was used as a reference element for measurement of temperature and another one for measurement of humidity. A Pt thin film about 0.1 μm thick was deposited and shaped by sputter-etching as a zigzag pattern of 15 μm wide stripes and 15 μm spacing for use as a heating and temperature-sensing element. Using Si_3N_4 as covering mask, micro-air-bridge (300 μm × 300 μm in area, 2 μm in thickness) suspended by four narrow 50 μm-wide beams were formed by etching from the back of the silicon wafer. The humidity was measured using an adjustable heater current across a platinum resistor. The RH sensitivity was approximately 0.6 mV/% RH. The response was fast, less than 1 second. Fang et al. (2011) believe that, since the micro-air-bridge in this device has an operating temperature of above 350°C, the surface of its sensing area will be cleaned by the burning out of dust, oil, et cetera, thus guaranteeing long life and stability.

14.4 SUMMARY

It is important to note that thermal conductivity-based sensors are universal devices; they do not exhibit any selectivity and do not require the use of any materials with specific properties. As a result, any thermal conductivity-based gas sensor with appropriate calibration can be used to measure humidity and vice versa. Therefore, when developing humidity sensors, it is necessary to base not only on the results obtained when developing directly humidity sensors, but also on the experience gained in the development of gas sensors (Ohira and Toda 2008; Hautefeuille et al. 2011; De Graaf and Wolffenbuttel 2012; Rastrello et al. 2013; Kommandur et al. 2015; Cai et al. 2016; De Graaf et al. 2016; Struk et al. 2018).

To summarize this consideration of thermal conductivity-based hygrometers, then the advantages of this type of sensor should be attributed to the fact that they (1) are very durable with a long lifetime, (2) work well in corrosive and high-temperature environments up to 300°C, (3) are fast, and (4) have better resolution in the high-temperature range than capacitive and resistive sensors (Wilson 2005).

As for the shortcomings, they should include (1) low selectivity: The sensor responds to any gas with thermal properties different than dry nitrogen, which may affect measurement; (2) a sufficiently large power consumption, especially in the conventional design; (3) low sensitivity due to a small difference in the thermal conductivity of dry and moist air; and (4) a limited temperature range: At low temperatures, the concentration of water vapor in the air is too low to be fixed by this method. The need to control and stabilize the temperature can also be attributed to the disadvantages of this type of sensor.

REFERENCES

Atashbar M.Z., Krishnamurthy S., Korotcenkov G. (2010) Basic principles of chemical sensor operation, In: Korotcenkov G. (ed.) *Chemical Sensors. Fundamentals of Sensing Materials, Vol. 1: General Approaches.* Momentum Press, New York, pp. 1–62.

Barsony I., Ducso C., Furjes P. (2009) Thermometric gas sensing, In: Comini E., Faglia G., Sberveglieri G. (eds.) *Solid State Gas Sensing.* Springer, New York, Chapter 7, pp. 237–260.

Bridgeman D., Tsow F., Xian X., Chang Q., Liu Y., Forzani E. (2017) Thermochemical humidity detection in harsh or non-steady environments. *Sensors* 17, 1196.

Cai Z., van Veldhoven R.H.M., Falepin A., Suy H., Sterckx E., Bitterlich C., Makinwa K.A., Pertijs M.A. (2016) A ratiometric readout circuit for thermal-conductivity-based resistive CO_2 sensors. *IEEE J. Solid-State Circuits* 51 (10), 2463–2474.

Daynes H.S. (1993) *Gas Analysis by Measurement of Thermal Conductivity.* Cambridge Press, New York.

De Graaf G., Prouza A.A., Ghaderi M., Wolffenbuttel R.F. (2016) Micro thermal conductivity detector with flow compensation using a dual MEMS device. *Sens. Actuators A* 249, 186–198.

Devoret M., Sullivan N.S., Esteve D., Deschamps P. (1980) Simple thermal conductivity cell using a miniature thin film printed circuit for analysis of binary gas mixtures. *Rev. Sci. Instrum.* 51(9), 1220–1224.

Fang Z., Zhao Z., Zhang J., Du L., Xu J., Geng D., Shi Y. (2011) A new integrated temperature and humidity. In: *Proceedings of the 6th IEEE International Conference on Nano/Micro Engineered and Molecular Systems,* February 20–23, Kaohsiung, Taiwan, pp. 788–791.

Hautefeuille M., O'Flynn B., Peters F.H., O'Mahony C. (2011) Development of a microelectromechanical system (MEMS)-based multi-sensor platform for environmental monitoring. *Micromachines* 2, 410–430.

Kimura M. (1996a) A new method to measure absolute humidity independently of the ambient temperature. *Sens. Actuators B* 33, 156–160.

Kimura M. (1996b) Absolute-humidity sensing independent of the ambient temperature. *Sens. Actuators A* 55, 7–11.

Kommandur S., Mahdavifar A., Hesketh P.J., Yee S. (2015) A microbridge heater for low power gas sensing based on the 3-Omega technique. *Sens. Actuators A* 233, 231–238.

Kouda S., Dibi Z., Meddour F. (2008) Modeling of thermal-conductivity of smart humidity sensor. In: *Proceedings of 2nd International Conference on Signals, Circuits and Systems (SCS 2008),* November 7–9, Monastir, Tunisia, pp. 1–4.

Lee D., Hong H., Park C., Kim G., Jeon Y., Bu J. (2001) A micromachined robust humidity sensor for harsh environment applications. In: *Proceedings of the 14th IEEE International Conference on Micro Electro Mechanical Systems (MEMS 2001),* January, pp. 558–561.

Miura T. (1985) Thermistor humidity sensor for absolute humidity measurements and their applications, In: Chaddock, J.B. (ed.) *Proceedings of International Symposium on Moisture and Humidity.* ISA, Washington, DC, pp. 555–573.

Ohira S., Toda K. (2008). Micro gas analyzers for environmental and medical applications. *Anal. Chim. Acta* 619, 143–156.

Okcan B., Akin T. (2004) A thermal conductivity based humidity sensor in a standard CMOS process. In: *Proceedings of IEEE International Conference on Micro Electro Mechanical Systems,* January 25–29, Maastricht, the Netherlands, pp. 552–555.

Okcan B., Akin T. (2007) A low-power robust humidity sensor in a standard CMOS process. *IEEE Trans. Electron. Dev.* 54 (11), 3071–3078.

Palmer P.E., Weaver E.R. (1924) Thermal-conductivity method for the analysis of gases. *Technological Papers of the Bureau of Standards,* Vol. 18, No. 249. Bureau of Standards, Washington, DC.

Rastrello F., Placidi P., Scorzoni A., Cozzani E., Messina M., Elmi I., Zampolli S., Cardinali, G.C. (2013) Thermal conductivity detector for gas chromatography: Very wide gain range acquisition system and experimental measurements. *IEEE Trans. Instrum. Measur.* 62 (5), 974–981.

Roveti D.K. (2001) Humidity/moisture: Choosing a humidity sensor: A review of three technologies. *Sensors (Peterborough)* 18 (7), 54–58.

Smetana W., Unger M. (2008) Design and characterization of a humidity sensor realized in LTCC-technology. *Microsyst. Technol.* 14, 979–987.

Struk D., Shirke A., Mahdavifar A., Hesketh P.J., Stetter J.R. (2018) Investigating time-resolved response of micro thermal conductivity sensor under various modes of operation. *Sens. Actuators B* 254, 771–777.

Tsilingiris P.T. (2008) Thermophysical and transport properties of humid air at temperature range between 0°C and 100°C. *Energy Convers. Manage.* 49, 1098–1110.

Vass-Várnai A., Fürjes P., Rencz M. (2008) Possibilities of humidity sensing with thermal transient testing on porous structures. In: *Proceedings of 14th Internashinal Workshop on Thermal Investigations of ICs and Systems, THERMINIC 2008,* September 24–26, Rome, Italy, pp. 200–203.

Wilson J.S. (ed.) (2005) *Sensor Technology Handbook.* Elsevier, Burlington, Amsterdam, the Netherlands.

Zambrozi Jr. P., Fruett F. (2012) Relative-air humidity sensing element based on heat transfer of a single micromachined floating polysilicon resistor. *J. Integr. Circuits Syst.* 7 (2), 130–136.

15 Field Ionization Humidity Sensors

15.1 PRINCIPLES OF OPERATION: CORONA DISCHARGE

All gas sensors, including humidity sensors, can be divided into two types: chemical and physical. All sensors previously discussed in this part of this book belong to the group of chemical-type devices. Field ionization (FI) humidity sensors, discussed in this chapter, are of the physical type. Unlike chemical sensors, physical sensors do not use humidity-sensitive materials, and their work is based on physical principles (i.e., physical properties of the water vapor itself). In particular, FI humidity sensors analyze the ionization characteristics of air containing water vapor. FI is a phenomenon whereby an atom or a molecule is ionized in the presence of a very high electric field, typically of the order 10^{10} V/m (Gomer 1961). This type of sensor can detect gases regardless of their adsorption energies. In particular, Maskell (1970) has investigated the effect of humidity on a corona discharge in air and found that it produced a decrease in the discharge current of approximately 20% for a change in humidity from dry to saturated conditions at atmospheric pressure and room temperature (see Figure 15.1). A corona discharge differs from a uniform field discharge in that one (or both) of its electrodes exhibits sharp curvature, and breakdown occurs first in the high-field region surrounding this electrode. The electrical field in the remainder of the gap is of much smaller magnitude and thus prevents the complete breakdown to a spark. Several electrode configurations will produce this type of discharge, but the most convenient, from both theoretical and experimental points of view, is a coaxial cylindrical system. It is interesting that the larger diameter of wire gave greater sensitivity.

The mechanism of the discharge depends on the polarity of the wire (Maskell 1970). For a positive wire, electrons entering the high-field region produce electron avalanches, which maintain a highly ionized state near the wire. The positive ions produced by these avalanches drift toward the cathode under the influence of a decreasing field and carry most of the current in the region outside the corona envelope. The remainder of the current is carried by electrons that are produced at the cathode or in the gas, mainly by photons from the discharge region. In the case of a negative wire, the cathode is bombarded with high-energy positive ions, which produce the electrons necessary for a self-sustaining discharge. These electrons move toward the anode and constitute the only current in the outer region. In general, they form negative ions by attachment to neutrals. Maskell (1970) established also that the usage of positive wires proved a more stable discharge and a higher and more linear sensitivity to humidity. The sensitivity at high humidity was approximately the same for both polarities, but the monotonic nature of the positive wire response is more suitable for measuring purposes.

Maskell (1970) suggested that this effect can be used to measure humidity in applications in which the fast response is essential, and the need for compensation of the moderate temperature and pressure changes can be tolerated.

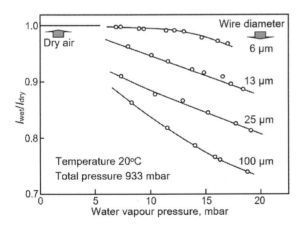

FIGURE 15.1 The effect of humidity on the discharge current for a cylindrical cell with wires (tungsten (W) and platinum (Pt)) of different diameters. The cell was manufactured from copper tube of 20 mm diameter and 100 mm length. (Data extracted from Maskell, B. R., The effect of humidity on a corona discharge in air. Royal Aircraft Establishment. Technical Report 70106, 1970.)

15.2 GAS IONIZATION SENSORS BASED ON 1D STRUCTURES

However, experiments have shown that conventional gas ionization sensors are limited by their huge, bulky architecture, high power consumption, and risky high-voltage operation. For the operation of such sensors, a voltage exceeding several tens of kV is required. Studies conducted in recent decades have shown that indicated disadvantages of gas ionization sensors can be overcome by using one-dimensional (1D) structures (Modi et al. 2003; Hui et al. 2006; Zhang et al. 2010; Wang et al. 2011; Pan et al. 2017). A schematic diagram illustrating the construction of modern gas ionization sensors is shown in Figure 15.2. SEM images of 1D structures used in the FI gas and temperature sensors are shown in Figure 15.3.

It was established that the sharp tips of nanotubes generate very high electric fields at relatively low voltages, lowering breakdown voltages severalfold in comparison to traditional electrodes, and thereby enabling compact, battery-powered, and safe operation of such sensors (Modi et al. 2003; Riley et al. 2003). For example, physical gas sensors based on carbon nanotubes demonstrated low breakdown voltage and showed good sensitivity and selectivity. In many studies, the carbon multi-walled nanotubes (MWNTs) were grown by a chemical vapor deposition (CVD) on the SiO_2 substrate and had 25–30 nm in diameter and lengths from 30 mm (Modi et al. 2003) to 2 µm (Hui et al. 2006). Several curves illustrating operating characteristics of FI humidity sensors are shown in Figure 15.4. It was shown that the nanotube sensor is compact and safe to use, and requires low power to operate. Hui et al. (2006) believe that the simple, low-cost sensors described here could be deployed for a variety of applications, such as environmental monitoring.

Unfortunately, the sensitivity of FI humidity sensors is low in comparison with conductometric or capacitive humidity sensors. For example, below 1% concentration, the breakdown voltage sharply approaches the value for pure air. It was also established that the carbon nanotubes (CNTs)-based ionization gas sensors showed poor stability, because carbon nanotubes could easily be oxidized and degraded in the oxygen-contained atmosphere (Wang et al. 2005; Liao et al. 2008). In addition, despite

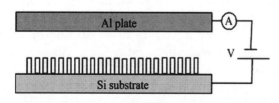

FIGURE 15.2 Schematic diagram of gas-ionized sensor with one-dimensional electrodes. Controlled dc voltage (V) is applied between the anode (vertically aligned nanotube film) and the cathode (Al sheet), which are separated by an insulator. The distance between the two electrodes could be adjusted in a range of 20–200 µm by the thickness of the plastic film or glass. Reprinted with permission from Wang, H. et al., A novel gas ionization sensor using Pd nanoparticle-capped ZnO, *Nanoscale Res. Lett.*, 6, 534, 2011. Copyright 2011, Springer Science+Business Media.)

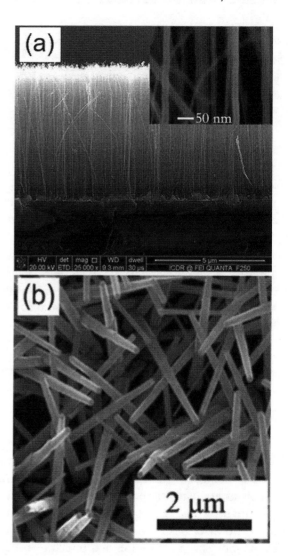

FIGURE 15.3 The cross-sectional SEM image of MWCNTs (a) and ZnO nanorods (b) used in gas-ionization sensors. ([a] Reprinted with permission from Pan, Z. et al., Sensing properties of a novel temperature sensor based on field assisted thermal emission, Sensors, 17, 473, 2017. Copyright 2017, MDPI as open access; [b] Reprinted with permission from Wang, H. et al., A novel gas ionization sensor using Pd nanoparticle-capped ZnO, *Nanoscale Res. Lett.*, 6, 534, 2011. Copyright 2011, Springer Science+Business Media.)

the strong decrease in the gaseous breakdown voltages, up to a range of 100–450 V (Modi et al. 2003; Hui et al. 2006; Wang et al. 2011), such voltages are still hazardous to employ. Moreover, CNTs-based FI sensors operate in a corona discharge mode. Corona discharges are difficult to control, and they generate excessive heat that may destroy sharp and slender CNTs. To resolve this problem, Sadeghian and Kahrizi (2007, 2008) proposed using a sparse array of vertically aligned gold nanorods as substitutes for CNTs. Recently, using this approach, Sadeghian and coworkers Sadeghian and Kahrizi (2007, 2008; Sadeghian and Islam 2011) obtained threshold field-ionization voltages of

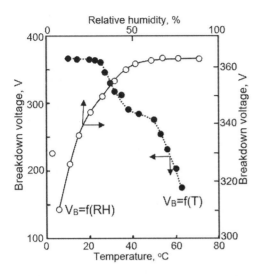

FIGURE 15.4 Effects of environmental factors such as temperature and relative humidity on the operating characteristics of carbon MWNT-based field ionization gas sensors. (Data extracted from Hui, G. et al., *Meas. Sci. Technol.* 17, 2799–2805, 2006.)

0.2–9.0 V (depending on the measured gas) by incorporating whisker-like covering on the cathode (see Figure 15.5). Figure 15.6 is a schematic illustration of such a device. Sadeghian and coworkers (Sadeghian and Kahrizi 2007; Sadeghian and Islam 2008, 2011) believe that, with such an approach, it is possible to make sensors with high sensitivity, high selectivity, long durability, and (ultra)-low-power operation ($P < 10$ μW with ionization currents ≤ 1 μA).

Liao et al. (2008) and Wang et al. (2011) have shown that metal oxide nanowires and nanorods, which are stable at room temperatures, can also be used for stable FI gas sensors instead of CNTs. In particular, Liao et al. (2008) applied ZnO nanowires, and Wang et al. (2011) applied ZnO nanorods. However, it was found that 1D ZnO nanostructures with relatively larger tip radii, compared with CNTs, need higher breakdown voltage (see Figure 15.7). To resolve this problem, Wang et al. (2011) proposed to modify the ZnO nanorod with Pd nanoparticles. The results have shown that the breakdown

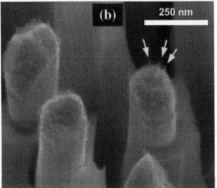

FIGURE 15.5 (a) SEM image of the array of AuNWs grown in porous anodic alumina (PAA) membranes. AuNW tips extruded from the embedding channels can be clearly seen. (b) A close-up view of the AuNW tips. The *arrows* show nanoscale whiskerlike features. (Reprinted with permission from Sadeghian, R.B. and Kahrizi, M., A novel miniature gas ionization sensor based on freestanding gold nanowires, *Sens. Actuators A*, 137, 248–255, 2007. Copyright 2007, Elsevier.)

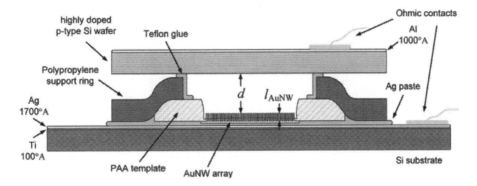

FIGURE 15.6 Cross-sectional schematic illustration of the nanowire-based gas ionization sensor (NWGIS), showing the gold nanowires (AuNW) array at the middle. The polypropylene ring is served as a spacer (not to scale) and is removed at some places to allow access of the gas flow to the measuring cell. (Reprinted with permission from Sadeghian, R.B. and Kahrizi, M., A novel miniature gas ionization sensor based on freestanding gold nanowires, *Sens. Actuators A*, 137, 248–255, 2007. Copyright 2007, Elsevier.)

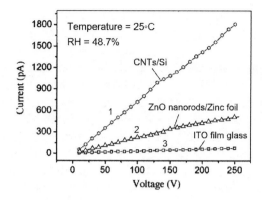

FIGURE 15.7 Effect of work voltages on prebreakdown current in air ($T = 25°C$; 48.7% relative humidity [RH]) for sensors with (1) CNTs and (2) ZnO nanorod electrodes. Indium tin oxide (ITO) films used as other electrode (interelectrode gap is 50 μm). Here, CNTs are 30–40 nm in diameter, and the diameter of ZnO nanorods is 80–120 nm. (Reprinted from Huang, J. et al., A novel highly sensitive gas ionization sensor for ammonia detection, *Sens. Actuators A*, 150, 218–223, 2009. Copyright 2009, Elsevier.)

voltage actually decreased for the Pd/ZnO nanorod-based sensor compared with the bare ZnO nanorod. This means that modification of the surface of 1D ZnO nanostructures for obtaining lower breakdown voltages can be one of the key issues for gas and humidity sensor applications.

15.3 SUMMARY

Nevertheless, the instability of the parameters, need for high-voltage equipment, strong influence of the gas composition, pressure and ambient temperature, and lack of sensitivity significantly limit the field of application of such sensors. In addition, FI humidity sensors are significantly inferior by their parameters to other sensors considered in this book, and therefore it is difficult to expect that they will find a gap in the humidity sensor market.

REFERENCES

Gomer R. (1961) *Field Emission and Field Ionization*. Harvard University Press, Cambridge, MA.

Huang J., Wang J., Gu C., Yu K., Meng F., Liu J. (2009) A novel highly sensitive gas ionization sensor for ammonia detection. *Sens. Actuators A* 150, 218–223.

Hui G., Wu L., Pan M., Chen Y., Li T., Zhang X. (2006) A novel gas-ionization sensor based on aligned multi-walled carbon nanotubes. *Meas. Sci. Technol.* 17, 2799–2805.

Liao L., Lu H.B., Shuai M., Li J.C., Liu Y.L., Liu C., Shen Z.X., Yu T. (2008) A novel gas sensor based on field ionization from ZnO nanowires: Moderate working voltage and high stability. *Nanotechnology* 19, 175501.

Maskell B.R. (1970) The effect of humidity on a corona discharge in air. Royal Aircraft Establishment. Technical Report 70106.

Modi A., Koratkar N., Lass E., Wei B., Ajayan P.M. (2003) Miniaturized gas ionization sensors using carbon nanotubes. *Nature* 424, 171–174.

Pan Z., Zhang Y., Cheng Z., Tong J., Chen Q., Zhang J., Zhang J., Li X., Li Y. (2017) Sensing properties of a novel temperature sensor based on field assisted thermal emission. *Sensors* 17, 473.

Riley D.J., Mann M., MacLaren D.A., Dastoor P.C., Allison W., Teo K.B.K., Amaratunga G.A.J., Milne W. (2003) Helium detection via field ionization from carbon nanotubes. *Nano Lett.* 3, 1455–1458.

Sadeghian R.B., Islam M.S. (2011) Ultralow-voltage field-ionization discharge on whiskered silicon nanowires for gas-sensing applications. *Nature Mater.* 10, 135–140.

Sadeghian R.B., Kahrizi M. (2007) A novel miniature gas ionization sensor based on freestanding gold nanowires. *Sens. Actuators A* 137, 248–255.

Sadeghian R.B., Kahrizi M. (2008) A novel miniature gas ionization sensor based on freestanding gold nanowires. *IEEE Sensors J.* 8 (2), 161–169.

Wang MS, Peng LM, Wang JY, Chen Q (2005) Electron field emission characteristics and field evaporation of a single carbon nanotube. *J. Phys. Chem. B* 109, 110–113.

Wang H., Zou C., Tian C., Zhou L., Wang Z., Fu D. (2011) A novel gas ionization sensor using Pd nanoparticle-capped ZnO. *Nanoscale Res. Lett.* 6, 534.

Zhang Y., Liu J., Zhu C. (2010) Novel gas ionization sensors using carbon nanotubes. *Sensor Lett.* 8 (2), 219–227.

16 Humidity Sensors Based on Thin-Film and Field-Effect Transistors

16.1 THIN-FILM AND FIELD-EFFECT TRANSISTORS

Devices based on the field-effect transistor (FET) and thin-film transistors (TFTs) offer a simple, efficient, and low-cost sensing platform for various applications, in particular for flexible electronics. The experiment showed that these devices can also be used for measuring humidity (Yamazoe and Shimizu 1986; Liao and Yan 2013). For the first time, the idea of using a hygroscopic gate oxide in an FET was proposed by Higikigawa et al. in 1985.

Similar to conventional inorganic FETs, a TFT has a source, drain, and gate electrodes. The channel current between the source and drain electrodes can be modulated by the gate voltage due to the field effect doping. The constructions of TFTs, which can be used for gas and humidity measurements, are shown in Figure 16.1. Figure 16.1a shows the conventional configuration, while Figure 16.1b shows a TFT in which the active semiconductor layer is exposed to the analyte of interest on one side, and the gate electrode is separated from the active layer by an insulator on the other side. It is seen that TFT-based sensors utilize very simple construction. Two electrodes ("source" and "drain") in contact with the semiconductor are used to apply a source–drain voltage and measure the source–drain current that flows through the semiconductor layer, while a third electrode ("gate") is used to modulate the magnitude of the source–drain current. The gate can be used to switch the transistor "on" (high source–drain current) and "off" (negligible source–drain current). In such devices, the channel current could be changed for several orders of magnitude by the gate voltage, indicating that the device is a type of amplifier or a transducer that can convert a voltage/potential signal into a current response (Li et al. 2012). It should be noted that the structure with a highly conductive bottom substrate as a gate is not practical for real-life applications; however, it is very useful for laboratory prototyping printed transistors, especially for investigating semiconducting materials. Usually, such substrates consist of thermally oxidized, highly doped silicon.

In real TFTs—for example, printed TFTs designed for flexible electronics—a gas-permeable gate is usually fabricated from metals, such as gold (Au) or platinum (Pt), but other options are possible. Usually, organic semiconductors (OSCs) are used as the active layer, forming the flow channel (see Section 16.2). However, inorganic semiconductors can also be used (see Section 16.3). The source and drain contacts can be printed using highly conductive polymer, such poly(3,4-ethylenedioxythiophene):poly(styrenesulfonate) (PEDOT:PSS). Of course, other variants are possible.

With zero gate bias, the TFT is a chemiresistor that measures the variation of the film conductivity upon exposure to a chemical species. Depending on the semiconducting material used as the active layer, the mobile charge carriers can either be electrons (n-type material) or holes (p-type material). The source–drain current is controlled by both the gate electrode via an electric field applied across an insulating layer (gate dielectric), and the analyte-semiconductor interaction. It is known that analyte-semiconductor interaction, following an electron transfer between analyte molecules and the sensing layer, is able to change the concentration and mobility of free charge carriers and the work function of the OSC, which causes a response in source–drain current (Janata and Josowicz 1998). In the case of solid-state materials, this interaction is accompanied only by a change in the work function. It has been shown that the interaction occurs through a partial charge transfer between the surface energy states at the semiconductor and the chemisorbed gas. The relationship between the partial pressure of the adsorbing gas and the change of the work function depends on the type of the adsorption isotherm, the conductivity of the solid, and the amount of the transferred charge. It is safe to say that the dynamic range for this type of interaction is limited by the maximum number of the surface adsorption sites and that the response follows some form of a power law given by the adsorption isotherm.

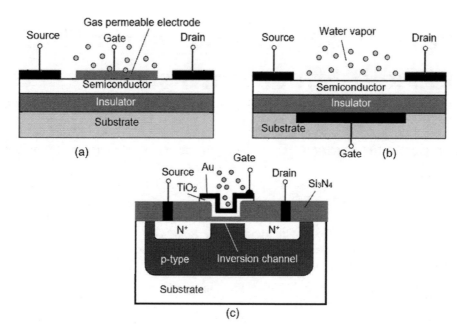

FIGURE 16.1 Schematic cross-sections of field-effect gas-sensing transistors with different configurations: (a, b) Thin-film transistor (TFT) sensors with different position of gates. The analyte of interest is represented by *grey circles*. The applied source–drain voltage V_d and gate voltage V_g are also shown; (c) MOS field-effect transistor (MOSFET) with *n*-channel. The conduction channel is formed due to inversion of the conductivity type.

The variation of the source–drain current, I_D, as a function of the source–drain voltage, V_D, and of the grid voltage, V_G, is a well-established equation (Guillaud et al. 1990).

$$I_D = \frac{Z\mu C_i}{L} V_D \left(V_G - V_T - \frac{V_D}{2} \right) \quad (16.1)$$

where Z is the channel length, L the inter-electrode distance, C_i the capacitance of the insulator per unit surface, and V_T the threshold voltage (Eq. 16.2). In molecular TFTs, V_T has been associated with the presence of traps of structural origin, or to ions or other impurities at the semiconductor–dielectric material interface (Horowitz et al. 1990; Guillaud et al. 1998).

$$V_T = -\frac{ehn_0}{C_i} \quad (16.2)$$

where h is the thickness of the sensing layer and n_0 the density of charge carriers without grid voltage.

Thus, an individual *I–V* curve has a linear region at $V_D \ll (V_G - V_T)$ and a saturation region at $V_D > (V_G - V_T)$. At $V_D = V_G - V_T$, the $I_D = f(V_D)$ curve reaches a maximum, and a pinch off of the channel occurs. For $V_D > V_G - V_T$, the channel is no longer continuous, and the corresponding current is called *saturation current* (Eq. 16.3).

$$I_{DS} = \frac{Z\mu C_i}{2L}(V_G - V_T)^2 \quad (16.3)$$

The curve $\sqrt{I_{DS}} = f(V_G)$ enables deduction of the mobility μ, which is called the *field-effect mobility*. The channel length L (distance between the source and drain electrodes) and the thickness of dielectric layer determine the range of the voltage sweeps. The magnitude of L is also important in minimizing the effect of the contact resistance, which must be kept to a small fraction of the total channel resistance (Torsi and Dodabalapur 2005). It is clear that sensing effect in TFT can be improved by reducing the free charge carrier density and the thickness of the active part of the TFT, and by increasing the mobility of charge carries and capacitance of the insulating material (i.e., decreasing the thickness of insulator) (Bouvet 2006).

As for metal oxide semiconductor (MOS) FETs-based humidity sensors, which are schematically displayed in Figure 16.1c, their operating principle differs somewhat from that described earlier. First, the humidity-sensitive material does not participate in the current transfer and, second, to simplify the structure of the gate, it is desirable that this material has dielectric properties. For such materials, the measure of sensitivity to humidity is a change in capacity, rather than the conductivity of the layer. As a rule, such sensors are manufactured on silicon (Si) substrates and have *n*-channel conductivity formed as a result of inversion of the conductivity type in *p*-Si under the influence of the electric field. According to Lee and Park (1996), the mechanism of sensitivity can be explained by the following way: The adsorption of water vapors in the humidity-sensitive layer results in an increase in their dielectric constant, hence the capacitance of the sensing material that forms the gate. This change in capacitance is accompanied by the threshold voltage changes and the drain current changes in humidity-sensitive FETs (HUSFETs), according to Equation 16.4:

$$V_T = \varnothing'_{MS} + \frac{-Q_{SS} - Q_S - Q_{SD}}{C'_i} + 2\varnothing_F \quad (16.4)$$

where $\varnothing'_{MS}$ is the work function difference between the gate metal and the silicon, Q_{SS} is the fixed positive surface state charge, Q_S is the effective interface charge density, Q_{SD} is the depletion charge density, Φ_F is the Fermi potential of the silicon, and C'_i is the capacitance per unit area in the gate region.

Taking into account that the dielectric in the MOSFET-based humidity sensor shown in Fig. 16.1c has the structure SiO_2-Si_3N_4-TiO_2, where TiO_2 is a humidity-sensitive material (HSM), the capacitance per unit area in the gate region, C'_i, may is expressed by

$$C'_i = \frac{C_o C_n}{\left[(C_o + C_n) + \frac{C_o C_n d_t}{\varepsilon_0 \varepsilon_t}\right]} \quad (16.5)$$

where C_o, C_n and C_t are capacitances per unit area in SiO_2, Si_3N_4, and humidity-sensitive material (HSM), ($C_t = \varepsilon_0 \varepsilon_t / d_t$) respectively, and d_t is the thickness of HSM, ε_0 is permittivity of free space, and ε_t is relative permittivity of HSM. In general, the drain current of metal insulator semiconductor field-effect transistors (MISFETs) is expressed by

$$I_D = \mu \frac{Z}{L} C^i_i \frac{(V_G - 2\varnothing_F - V_D/2)V_D - \left[2(2\varepsilon_S q N_A)^{\frac{1}{2}}\right]/3}{C^i_i \left[(V_D + 2\varnothing_F)^{3/2} - (2\varnothing_F)^{3/2}\right]} \quad (16.6)$$

where Z is the channel depth, L is channel length, and μ is the carrier mobility.

The conductance of the channel in the linear region can be obtained from Equation 16.6, with $V_D \ll (V_O - V_T)$, as the appropriate expression:

$$g_m = \varepsilon_0 \varepsilon_t \frac{1}{d_t + \varepsilon_0 \varepsilon_t \frac{C_o + C_n}{C_o C_n}} \cdot \mu \left(\frac{Z}{L}\right) V_D \quad (16.7)$$

It is seen that, to achieve high sensitivity, it is necessary to use a sufficiently thick layer of humidity-sensitive material ($d_t > \varepsilon_0 \varepsilon_t C_o + C_n / C_o C_n$) with a large change in the dielectric constant under the influence of moisture. In principle, these conditions are similar to those imposed on materials used in capacitive humidity sensors.

16.2 HUMIDITY-SENSING CHARACTERISTICS OF ORGANIC-BASED TRANSISTORS

16.2.1 TFT-Based Sensors

As shown in the previous chapters, in the development of resistive sensors, preference is given to polymers with ionic conductivity. For the production of TFTs, such polymers are unsuitable. Polymers used in TFTs should have semiconductor properties. Research has shown that there are two major overlapping classes of OSCs. Semiconducting properties can be discovered in both single molecules, short-chain (oligomers), so-called organic charge-transfer complexes, and various linear-backbone conductive polymers. Semiconducting small molecules (aromatic hydrocarbons) include the polycyclic aromatic compounds, such as pentacene, anthracene, tetracene, diindenoperylene, perylenediimides, tetracyanoquinodimethane, and rubrene. Polymeric OSCs, conjugated polymers, include poly(3-hexylthiophene) (P3HT), polyfluorene, poly 2,5-thienylene vinylene, and poly(p-phenylene vinylene) (PPV), as well as polyacetylene (PA) and its derivatives.

OSCs, i.e. organic macromolecules with a π-conjugated backbone, are attractive because of their unique electric and electronic properties (Chen et al. 2004; Potje-Kamloth 2010). Almost all organic solids are insulators. But when their constituent molecules have π-conjugate systems, there is a possibility to convert these materials to the conducting state. The charge injection (holes or electrons) into the conjugated backbone of the polymer is called *doping*. It leads to the formation of a self-localized electronic state within the previously forbidden semiconductor bandgap and causes a chain deformation around the charge. The prime source of localization of charge carriers in conducting polymers is a structural disorder, which decreases the conductivity by lowering the charge carrier's mobility. Thus, the added charge on the polymer is not a free electron or hole, but a localized particle connected with a chain deformation, which is known as a *polaron* or *bipolaron state*, depending on the degree of doping. The creation of polaron or bipolaron electronic defects allows the charge transport within a single chain (Bassler and Kohler 2012). Typical current carriers in OSCs are holes and electrons in π-bonds. To match conventions between the physicist and chemist, a polymeric cation-rich material is called *p*-doped, and a polymeric anion-rich material is called *n*-doped. Besides this charge injection, doping in polymers also implies the insertion or the repulsion of counter-ions—that is, dopant anions, to maintain charge neutrality (Potje-Kamloth 2010).

Doping in polymers can be accomplished in a number of ways: chemically, electrochemically, and photochemically, as well as by charge injection at the metal-insulator-semiconductor interface. In the case of electrochemical and/or chemical doping, the neutral polymer chain is oxidized (*p*-doping) or reduced (*n*-doping). Because every monomer is a potential redox site, conjugated polymers can be doped to a high density of charge carriers; the doping level is up to five orders of magnitude greater than that in common inorganic semiconductors. The induced electrical conductivity is permanent, until the charge carriers are purposely removed by undoping—that is, by reversing the (electro)chemical reaction at the redox sites of the polymer chain. Even though a high-density carriers and possible spatial delocalization of charge carriers in OSCs, charge carriers cannot participate in electronic transport similarly to transport in inorganic semiconductors. Electrons in OSCs can move via π-electron cloud overlaps, especially by hopping, tunneling, and related mechanisms (Bassler and Kohler 2012). Polycyclic aromatic hydrocarbons and phthalocyanine salt crystals are the examples of this type of OSCs. At that, it was established that the dopant compounds play an important role in hopping transport. For thermally activated hopping, the charge transport follows the relation $\mu \approx \left(-E_a/kT\right)$. The mobility in most organic thin-film transistors (OTFTs) increases with increasing gate bias, which is attributed to filling of the tail states of the density of states (DOS).

Table 16.1 summarizes band gaps for several of the most used OSCs. It is important that conductivities of OSCs and Fermi levels can be controllably changed at will by chemical and electrochemical doping methods. This means that the characteristics of electronic devices utilizing them can be modified by request. One should note that OSC films, used in TFTs, can be deposited by various fabrication approaches, including thermal evaporation, spin-coating, screen printing, and inkjet printing.

One of the first TFTs designed for humidity measurement was device manufactured by Ohmori et al. (1991). Ohmori et al. (1991) demonstrated that water vapor and chloroform gas also enhanced the source–drain currents of poly(3-butylthiophene)-based thin transistors (TFTs). In each of these interactions, the changes in source–drain currents were observed to be reversible. Since the year 2000, interest in this area has greatly intensified, and numerous review papers are related to this subject of research (Torsi 2000; Torsi and Dodabalapur 2005; Mabeck and Malliaras 2006; Wang et al. 2006; Hasegawa and Takeya 2009; Yamashita 2009; Guo et al. 2010; Klauk 2010; Lin and Yan 2012; Liao and Yan 2013).

Typical operating characteristics of TFT-based humidity sensors are shown in Figure 16.2. These results were obtained by Zhu et al. (2002), who studied pentacene-based TFTs.

Significantly more sensitive pentacene-based sensors are those designed by Park et al. (2013). They have achieved increasing sensitivity by insertion into the transistor of an additional layer of polyelectrolyte (poly[2-(methacryloyloxy)-ethyl trimethylammonium chloride-co-3-(trimethoxysilyl)propyl methacrylate] (poly(METAC-co-TSPM)). Figure 16.3 shows a comparison of the configurations used by Zhu et al. (2002) and Park et al. (2013) The responsive behavior of the new sensor originates from the enhanced migration of mobile ions within the polyelectrolyte upon the absorption of moisture. The authors believe that polyelectrolyte releases free Cl⁻ ions in the electrolyte dielectric layer under humid conditions and

TABLE 16.1
Band Gap of Several Organic Semiconductors Used in Thin-Film Transistor Design

Organic Semiconductor	Abbreviation	Band Gap, eV
Trans-polyacetylene	PA	1.4
Polythiophene	PT	2.0–2.1
Polyaniline	Pani	2.2
Rubrene (5,6,11,12-tetraphenylnaphthacene)		2.2
Poly(p-phenylene vinylene)	PPV	2.5
Poly(p-phenylene)	PPP	2.8–3.0
Polypyrole	PPy	3.2

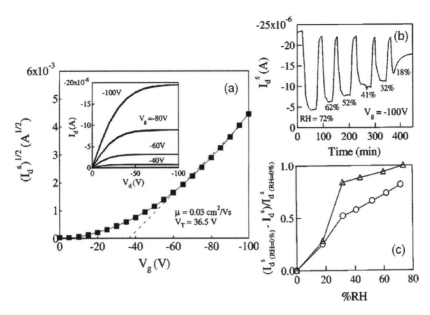

FIGURE 16.2 Operating characteristics of pentacene-based thin-film transistor (TFT). The channel length L is 100 μm and the channel width W is 3 mm. (a) Square root of the saturation current as a function of the gate voltage in a vacuum. The *dashed line* is a fit to obtain the hole mobility. Inset shows the drain current vs drain voltage characteristics measured in a vacuum; (b) Time dependence of the saturation current measured at a gate voltage of −100 V and a drain voltage of −100 V. The OTFT was exposed to wet N_2 gas with different relative humidity (RH) and vacuum, alternatively. (c) The saturation current measured at V_g = −100 V and V_d = −100 V as a function of relative humidity for two OTFTs. The *circles* are for the OTFT with a 50 nm thick pentacene layer; the *triangles* are for the OTFT with a 100 nm thick pentacene layer. (Reprinted with permission from Zhu, Z.T. et al., *Appl. Phys. Lett.*, 81, 4643–4645, 2002. Copyright 2002, American Institute of Physics.)

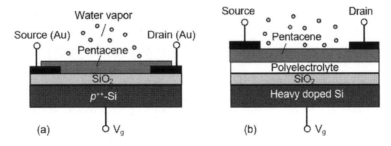

FIGURE 16.3 Configurations of humidity sensors used by Zhu et al. (2002) (a) and Park et al. (2013) (b).

boosts the electrical current in the transistor channel. Figure 16.4 illustrates this mechanism. For efficient operation of the sensor, the polyelectrolyte layer is directly exposed to the humid environment.

Park et al. (2013) showed that application of such an approach to sensor development has led to extreme device sensitivity, such that electrical signal variations exceeding seven orders of magnitude have been achieved in response to a 15% change in the relative humidity (RH) level (see Figure 16.5). In addition, according to Park et al. (2013), the new sensors exhibited a fast responsivity and stable performance toward changes in humidity levels. The transitions between levels occurred within 10 seconds upon exposure to wet N_2 and within 40 seconds upon exposure to dry N_2. It is a good result in comparison with published data. Usually, organic TFT-type humidity sensors have response and recovery times on the order of tens of minutes (Zhu et al. 2002). However, one should recognize that there are articles in which faster response is reported (Subbarao et al. 2016; Wu et al. 2017). In addition, Park et al. (2013) have established that the humidity sensors, mounted on flexible substrates, provided a low voltage (<5 V) operation while preserving the unique ultra-sensitivity and fast responsivity of these devices. The authors believe that the strategy of utilizing the enhanced ion motion in an inserted polyelectrolyte layer of an OTFT structure can potentially improve sensor technologies beyond humidity-responsive systems.

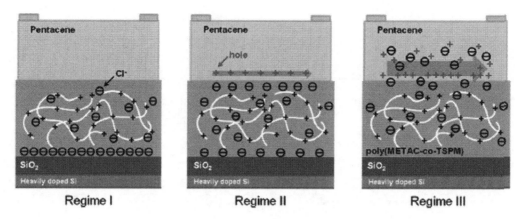

FIGURE 16.4 Schematic diagrams of the proposed working principles under different VG regimes. (Reprinted with permission from Park, Y.D. et al., *ACS Appl. Mater. Interfaces*, 5, 8591–8596, 2013. Copyright 2013, American Chemical Society.)

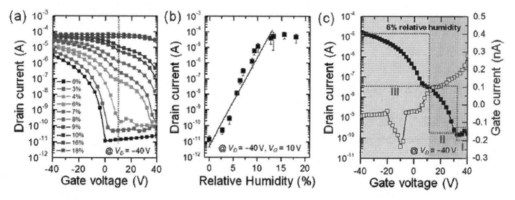

FIGURE 16.5 (a) Transfer characteristics at various humidity levels ($V_D = -40$ V). (b) The plot of I_D at $V_G = 10$ V and $V_D = -40$ V as a function of the humidity level. (c) A representative transfer characteristic plotted with I_G ($V_D = -40$ V). (Reprinted with permission from Park, Y.D. et al., *ACS Appl. Mater. Interfaces*, 5, 8591–8596, 2013. Copyright 2013, American Chemical Society.)

As a further development of the above-described approach to the development of TFT-based sensors, it was proposed to use polyelectrolyte in the formation of the gate of the field-effect transistors (see Figure 16.6). Currently, there is no humidity sensor developed on this principle. However, it has been considered that such a configuration can be a potential platform for effective sensing applications (Dhoot et al. 2006; Kim et al. 2013; Tarabella et al. 2013; Liao and Yan 2013). As is seen in Figure 16.6, being different from the conventional OTFTs, an electrolyte-gated organic thin-film transistor (EG-OTFT) has the OSC channel and the gate electrode that are separated by an electrolyte layer. Even when a small gate voltage is applied (≈ 1 V), a high charge density in the channel of the electrolyte-gated organic field-effect transistor (EG-OFET) can be induced due to the high capacitance (approximately tens of $\mu C/cm^2$) of the electric double layer (EDL) at the electrolyte-OSC interface (Panzer and Frisbie 2008; Said et al. 2008; Larsson et al. 2009; Torsi 2012). Some reports described the underlying operation mechanism of an EG-OTFT as being similar to that of an organic electrochemical transistor (OECT) because of the electrochemical reaction of the active layer in EG-OTFTs (Lee et al. 2009; Yuan et al. 2007). Further work is needed to clarify the working principle of EG-OTFTs. If the modulation of the channel current by the gate voltage in the device was mainly due to electrochemical doping instead of field-effect doping in the active layer, it would be better regarded as a type of OECT.

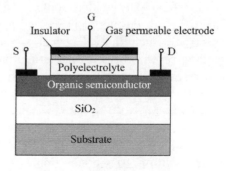

FIGURE 16.6 Configuration of EG-OTFTs.

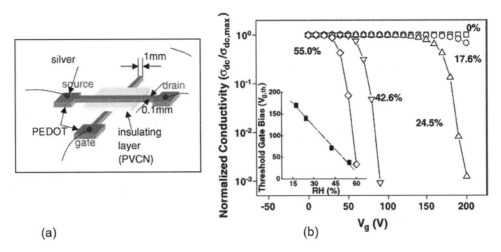

FIGURE 16.7 (a) Tilted view of the poly(3,4-ethylenedioxythiophene) (PEDOT)-based, humidity-dependent field-effect transistor: insulating layer, 400–1000 nm, gate, 500 nm, active layer, 500 nm; and (b) Normalized conductivity (σ_{nor}) vs the gate bias (V_g) at various humidity environments. Inset: threshold gate bias as a function of the relative humidity (RH). (Reprinted with permission from Kang, H.S. et al., Humidity dependent characteristics of thin film poly(3,4-ethylenedioxythiophene) field-effect transistor, *Synth. Met.*, 155, 176–179, 2005. Copyright 2005, Elsevier.)

Of note, the literature describes a configuration significantly different from the traditional configuration of TFTs and FETs. For example, Figure 16.7a shows the humidity sensor configuration proposed by Kang et al. (2005). A thin PEDOT layer (thickness ≈ 500 nm) with the desired pattern was formed as an active humidity-sensitive layer and a gate electrode in such sensors. Figure 16.7b shows the normalized dc conductivity ($\sigma_{nor} = \sigma_{dc}/\sigma_{dc,\,max}$) of the PEDOT active channel as a function of V_g at various humidity environments. Unfortunately, the recovery time of this sensor was too long. From the time-dependent recovery of I_{ds} at $V_g = 0$ V, it was found that the slow increase of I_{ds} with time was mainly caused by redoping of the PEDOT channel by the slow recovery of constituent ions (Kang et al. 2005).

Table 16.2 summarizes OSCs, which were tested as a sensing layer in TFTs designed for water vapor detection. All the above-mentioned devices are based on *p*-type conductivity. N-type TFTs are poorly developed (Yamashita 2009). Water vapor, along with vapors of organic solvents, are the main analytes detected by TFTs. However, as experiments have shown, that such gases as O_2, NO_2, N_2, O_3, H_2 and SO_2 can also be detected (Guillaud et al. 1998; Covington et al. 2001; Kang et al. 2007; Marinelli et al. 2009; Dual et al. 2010; Chen et al. 2012). The optimal thickness of the OSC in TFT-based sensors usually varied in the range from tens of nanometers to several hundreds of nanometers, depending on the material used. For too-thin and too-thick films, the gate modulation was very weak. Moreover, it is impossible to reach the saturation point with great thickness.

Torsi (2000), Mabeck and Malliaras (2006), and Bouvet (2006) have shown that organic TFTs offer a great deal of promise for applications in humidity sensing. According to Yamashita (2009), these sensors have advantages typical for all polymer-based devices and organic electronics. OSCs can be deposited using low-temperature processes on a variety of substrates, including mechanically flexible plastics, and can operate at room temperatures. This means that a low-cost approach, involving the absence of vacuum processing; lithography; and the use of low-cost substrates, can be used for gas sensor fabrication (Chang et al. 2006). For example, for sensor fabrication, one can use novel, low-cost technologies, such as soft lithography and inkjet printing. Thin films of OSCs, deposited using indicated methods, are mechanically robust and flexible. All of the mentioned features testify that OTFTs are promising elements for flexible electronics (Chang et al. 2006; Yamashita 2009). Moreover, easy modification in the conducting polymer structures provides a facile route to sensing materials with different work functions and selectivities to analytes, which ensure the high performance of transistor-configured sensors. Furthermore, it is possible to covalently integrate recognition elements directly to the OSC to provide highly specific interactions with chosen analytes. Miniaturization of these devices is straightforward, so portability, small sample volumes, and arrays with many elements are achievable. In addition, they provide a response (current change) that is easy to measure with simple instrumentation. With these advantages, it is feasible that single-use, disposable sensors could be realized using TFTs.

TABLE 16.2
Summary of Thin-Film Transistor and Field Effect Transistor–Based Humidity Sensors

	Humidity-Sensitive Material	Device	Function	References
Organic semiconductor	Polyaniline	TFT	Active layer	Chao and Wrighton (1987)
		TFT-FET	Hybrid	Barker et al. (1997)
	Polythiophenes	TFT	Active layer	Ohmori et al. (1991)
	Naphthalene tetracarboxylic derivatives	TFT	Active layer	Torsi et al. (2000, 2001)
	Pentacene	TFT	Active layer	Zhu et al. (2002); Li et al (2005); Park et al. (2013)
	PEDOT:PSS	Electrochemical TFT	Active layer	Nilsson et al. (2002)
	PEDOT	FET	Gate and active layer	Kang et al. (2005)
	Oligofluorene derivatives	TFT	Active layer	Li et al (2005)
	PBIBDF-BT	TFT	Active layer	Wu et al. (2017)
	CuPc	TFT	Active layer	Murtaza et al. (2010) Subbarao et al. (2016)
Polymer (isolator)	Crosslinked cellulose acetate	FET	Gate	Higikigawa et al. (1985)
Inorganic semiconductors	Porous TiO_2	FET	Gate	Lee and Park (1996)
	InGaZnO	TFT	Active layer	Park et al. (2008); Kim et al. (2017)
	CNTs	FET	Active layer	Mudimela et al. (2012)
Other materials	Porous Al_2O_3	FET	Gate	Chakraborty et al. (1999a, 1999b)
	CNx	FET	Gate	Lee et al. (2008)
	SiO_2	FET	Active layer	Song et al. (2012)

FET, field-effect transistor; TFT, thin-film transistor.

However, organic TFT-based humidity sensors all have disadvantages peculiar to devices based on semiconducting polymers: bad temporal and thermal stability (Yamashita 2009), and low selectivity to water vapors. Sensors have high sensitivity to vapors of organic solvents (Torsi et al. 2003a, 2003b). For example, the major problem with pentacene, which shows the highest hole mobility of 3 cm^2 V/s in thin-film TFTs, as well as many other organic conductors, is its rapid oxidation in air to form pentacene-quinone (Hasegawa and Takeya 2009). In addition, pentacene has low solubility in solvents. Good solubility is crucial for device fabrication using such solution methods as inkjet printing. Li et al. (2005) studied TFT with different *p*-channel OSCs and various device structures, and established that all of the *p*-channel OSCs investigated showed a degraded transistor performance with increased humidity. Torsi et al. (2002) found that, above 70°C, the grains in alkyl-substituted α-thiophene oligomers films begin to coalesce, and a more regular, terraced morphology with a low surface roughness is observed. This type of morphology reduced the ability of gases to adsorb, decreasing the sensor response. Slow response and recovery also must be attributed to the shortcomings of such sensors (see Figure 16.2). In addition, in many cases, the TFT parameters such as composition, doping, thickness of insulator and sensing layers, grain size, interelectrode distance, etc., are not optimized; therefore, operating characteristics of these devices are far from desired.

One should note that TFTs and chemiresistors based on semiconducting polymers use the same sensing materials and detect the same gases and vapors. This means that the mechanisms of analyte interaction with polymers, proposed as explanation of the sensing characteristics of chemiresistors, can be applied for explanation of the gas-sensing effects observed in OFETs. However, due to complications of this phenomenon and the possibility to use OSCs with different structures in OTFTs, we do not have a unified explanation for all effects observed. Every case requires an individual approach. In addition, it is necessary to consider that a resistor is sensitive to bulk mobility changes, whereas a transistor is sensitive not only to the changes in the channel mobility, but also to the number of trapped or otherwise immobile charges at the insulator-semiconductor interface, which gives rise to a threshold shift (Torsi et al. 2002). This means that the mechanisms of sensitivity for different organic TFTs can have distinctive features. For example, it is known that the response in organic TFTs may depend on both the conductivity and the work function of the polymer. However, experiments have shown that sometimes it is difficult to separate the influence of the various forms of modulation. For example, Polk et al. (2002) found that localized energy states in polyaniline can affect the value of the work function but do not affect the conductivity of the polymer. Torsi et al. (2003a, 2003b) have shown that the modification of the molecular structure of the active material also has a strong influence on the mechanism of sensitivity in TFT humidity sensors. Li et al. (2005) established that in pentacene- and other p-type TFTs, the sensitivity to humidity can be explained in terms of a decrease in the pentacene hole mobility in the presence of water vapor. The presence of polar water molecules can reduce the hole mobility by trapping the charge at grain boundaries and thereby reducing the source–drain current. The change of hole mobility in 1,4,5,8-naphthalene-tetracarboxylic dianhydride (NTCDA) under the influence of water vapor determined by Torsi et al. (2001) is shown in Figure 16.8.

More detailed analysis of the sensing mechanisms of organic TFT-based gas can be found in the reviews prepared by Someya et al. (2010) and Duarte and Dodabalapur (2012). They have shown that environmental molecules ("analytes") can have special effects on OFETs. Noncovalent bonds, such as hydrogen bonds and π-interactions, cause attraction between analytes and OSCs that exceeds simple van der Waals interaction. The interactions can occur on the surfaces of the OSC films, between crystallites that make up a polycrystalline OSC solid, in the free volume of OSCs that are either amorphous or intentionally supplied with flexible side chains, or at interfaces between OSCs and dielectrics or electrodes (Someya et al. 2010). Many analytes, including water, are dipolar, and thus induce local fields that are superimposed on the gate voltage, altering effective charge carrier mobilities. Duarte and Dodabalapur (2012) concluded that organic TFTs are typically injection-limited devices, and the increase in current during interaction with polar analytes most likely occurs due to the dipole interaction at the contact interface with the image charge, which decreases the barrier for hole injection. According to Duarte and Dodabalapur (2012), there are two phenomena that occur in the large-channel devices ($L > 1$ μm) when exposed to a polar analyte. The first

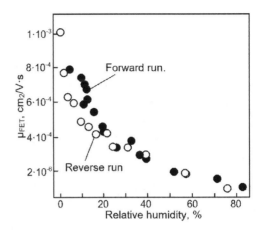

FIGURE 16.8 μ_{FET} dependence in 1,4,5,8-Naphthalene-tetracarboxylic dianhydride (NTCDA) upon H_2O concentration in N_2. (Reprinted with permission from Torsi, L. et al., NTCDA organic thin-film-transistor as humidity sensor: Weaknesses and strengths, *Sens. Actuators B*, 77, 7–11, 2001. Copyright 2001, Elsevier.)

is the percolation of the polar species through the grain boundaries, where they interact with charges to produce trapping effects and increased potential barriers. These effects produce a decrease in drain current based on the polarizability of the OSC. The second effect is the alignment of the analyte dipoles in such a way that they act as added charge, and an increase in drain current occurs. Depending on the carrier density (high–gate bias or low–gate bias operation), either trapping or the added charge phenomena will dominate. An increase in drain current may also occur due to the dielectric interface affecting the dipole interaction at the OSC-analyte interface. Regarding nonpolar analytes, one could say that they also can be a conceivable alternate mobility through induced dipoles and disruption of molecular packing in the OSCs (Someya et al. 2010). Furthermore, depending on the relative positions of OSC and analyte carrier energy levels, analytes can cause the trapping of mobile carriers by localizing them on the analyte molecules, or doping by donating or extracting electrons from the OSC matrix. The effects of doping and trapping on OSC conductivity and interfacial potential barriers can be profound.

It also should be considered that, in nanoscale devices (short-channel devices), the analyte molecules modulate the charge carrier injection at the source and drain electrodes. For a small channel length, the number of grain boundaries within a channel decreases so that the amount of effect of grain boundaries on electrical transport and chemical sensing reduces, and other factors become more important. As a result, when the channel length is comparable with or smaller than the grain size of polycrystalline organic molecules or conjugated polymers, it becomes possible to observe the electrical transport and chemical sensing behavior within the body of grains; the mechanism of this might be different from that in large-scale devices, in which grain boundaries dominate. Wang et al. (2006) believe that electrical transport in very short channel-length organic transistors ($L < 50$ nm) is contact-limited. At that, due to analyte molecules' diffusion near the source/drain contacts through the porous semiconductor layer, the analyte molecules can affect the behavior of the contact. Wang et al. (2006) have shown on the example of P3HT-based TFT that the behavior of nanoscale TFT sensors is remarkably different from that of their microscale counterparts. For organic TFTs of such small dimensions, the drain current typically increases by a factor of more than five in response to the analyte (> 10 for some devices). The smaller the channel length, the stronger this effect is. This means that nanoscale-dimension sensors are more sensitive. This provides a strong argument for employing nanoscale channel-length devices for sensing.

It is necessary to recognize that published data do not make it possible to form a conclusion about OSCs optimal for designing TFT-based humidity sensors. There are only several conclusions related to this topic. For example, Bouvet (2006) concluded that phthalocyanines cannot be used in transistors for humidity and volatile organic compound (VOC) sensing, even though they could be used favorably in such devices. Phthalocyanines better work with oxidizing species, such as ozone and nitride dioxide, which act as true dopants, whereas water and volatile organic compounds act as traps and lead to a decrease in the current. Crone et al. (2001) and Torsi et al. (2002) found that the sensing response increased as the length of the OSC's hydrocarbon end group increased. They believe that this is because of the elongated lamellar morphology and looser molecular packing, which enable greater access of analyte vapor and increased surface area, and change the electronic or spatial barriers between grains. The alkyl chains therefore facilitate adsorption of the analyte molecules by the sensing film. This adsorption mechanism could be a combination of hydrophobic interactions, intercalation to fill defective vacancies, and simple surface binding. All these processes are favored at grain boundaries. The absence of general rules that can be used to select the optimal OSC means that the development of TFT-based humidity sensors is far from complete, and many additional studies are still required to develop devices suitable for the market of humidity sensors. At the moment, one can only say that, as with all types of sensors considered earlier, the composition, structure, and thickness of the humidity-sensitive layer plays an important role in the formation of sensor parameters. For example, Wu et al. (2017) have shown that optimization of the film composition, contributing to the formation of macropores in the humidity-sensitive layer, was accompanied by a significant increase in sensitivity and a decrease in the response time of the sensor (Table 16.3 and Figure 16.9). poly(butyl acrylate) (PBA) was added to poly(3E,7E)-3,7-Bis(2-oxoindolin-3-ylidene)benzo-[1,2-b:4,5-b0]-difuran2,6(3H,7H)-dione- bithiophene (PBIBDF-BT) to prepare the macroporous semiconductor film. The washing of the prepared humidity-sensitive layer with acetone was done to remove any remaining PBA. Through the pores in the semiconductor film, the water molecules are more easily diffused into the charge carrier transport layer. As a result, the contact area between the water molecules and the channel carrier transport layer is increased, resulting in a high sensitivity of the sensor.

TABLE 16.3
Sensing Parameters of OFET-Based Macroporous PBIBDF-BT Films with Different PBIBDF-BT Contents

Parameter	BIBDF content in PBIBDF-BT (BIBDF:PBA)			
	100 wt %	90 wt %	70 wt %	50 wt %
Average pore size, nm	0	82	109	154
Response time, s	10.02	2.40	0.44	0.68
Recovery time, s	67.10	45.57	46.23	45.21
Sensitivity (32%–69% RH)	29	126	269	415

Source: Reprinted with permission from Wu, S. et al., ACS Appl. Mater. Interfaces, 9, 14974–14982, 2017. Copyright 2017, American Chemical Society.

$S = I_{D(gas-off)}/I_{D(gas-on)}$; PBIBDF-BT, (3E,7E)-3,7-bis(2-oxoindolin-3-ylidene)benzo-[1,2-b:4, 5-b0]-difuran-2,6(3H,7H)-dione (BIBDF) with bithiophene; RH, relative humidity.

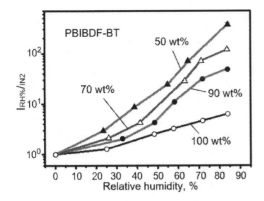

FIGURE 16.9 Relationship between sensitivity and relative humidity (RH) in devices with different PBIBDF content in PBIBDF-BT. Sensors had configuration shown in Figure 16.1b. (Reprinted with permission from Wu, S. et al., *ACS Appl. Mater. Interfaces*, 9, 14974–14982, 2017. Copyright 2017, American Chemical Society.)

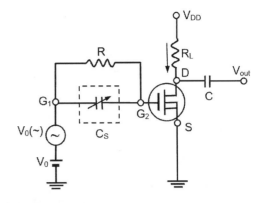

FIGURE 16.11 Equivalent circuit of IGFET humidity sensor. (Reprinted with permission from Yamazoe, N. and Shimizu, Y., Humidity sensors: Principles and applications, *Sens. Actuators*, 10, 379–398, 1986. Copyright 1986, Elsevier.)

16.2.2 MISFET-Based Sensors

As for the organic thin-film, FET-based humidity sensors, the first such devices were reported in the mid-1980s (Ebisawa et al. 1983; Kudo et al. 1984), and by the end of 1980s, it was shown that these devices can be used for gas and water vapor detection (Higikigawa et al. 1985). The FET humidity sensor fabricated by Higikigawa et al. (1985) using standard integral circuit (IC) processing techniques is shown in Figure 16.10. This sensor was integrated with a temperature-sensing diode. The host FET device was an n-channel MISFET with a meandering-gate structure. However, unlike classical FETs, in which an OSC forms a current flow channel, in this FET, crosslinked cellulose acetate 1 μm thick was stacked as a humidity-sensitive membrane between the lower-gate electrode and upper-gate electrode (porous gold electrode, 10–20 nm thick) (see Figure 16.10)—in other words, it formed the duplex gate electrode. This sensor also had a specific measuring circuit (Figure 16.11). The upper and lower gates were electrically connected with a sufficiently large fixed resistance R_B. As a result, it was established that, under adequate conditions, the output voltage, V_{out}, was related to the capacitance of the membrane. In the RH range from 0% to 100%, V_{out} changed from 0.21 to 0.28 V. The authors also reported that the sensor showed good accuracy (hysteresis < 3% RH) with a response time of less than 30 seconds. Long-term stability and a good durability against high RH (90%–95% RH) and dew drops have also been claimed. In devices with one metal electrode in the gate, the classical connection scheme of the field-effect transistor is used for measurement.

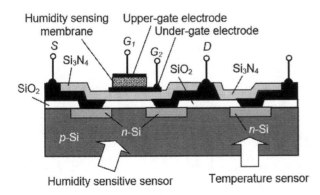

FIGURE 16.10 Schematic cross-section of insulated-gate field effect transistor (IGFET) humidity sensor with duplex-gate electrode designed by Higikigawa et al. in 1985. (Idea from Higikigawa, M. et al., Microchip FET humidity sensor with a long-term stability, In: *Proceedings of the 3rd International Conference on Solid State Sensors and Actuators (Transducers'85)*, Philadelphia, PA, June 11–14, pp. 221–224, 1985.)

16.3 OTHER MATERIALS IN TFT AND FET-BASED HUMIDITY SENSORS

There are almost no studies of the search for other materials suitable for the manufacture of TFT and FET-based humidity sensors. Only a few papers consider TFT and FET-based humidity sensors based on inorganic materials. Currently, the following have been tested as humidity-sensitive materials in such devices: TiO_2

(Lee and Park 1996), amorphous In-Ga-ZnO (a-IGZO) (Takagi et al. 2005; Kamiya et al. 2010), Al_2O_3 (Chakraborty et al. 1999a, 1999b), SiO_2 (Song et al. 2012) and CN_x (Lee et al. 2008).

16.3.1 MISFET-Based Sensors

Lee and Park (1996) presented a HUSFET in 1996. The change of the gate permittivity—TiO_2 with a permeable gold electrode on the top—was used to modulate the threshold voltage V_T. The authors reported a change from 3.0 to 2.4 V, for a change of 30%–90% RH. Lee and Park (1996) used the approach proposed by Higikigawa et al. (1985) (see Figure 16.10), but instead of cellulose, they used the TiO_2 layer.

The construction of MISFET humidity sensors designed by Chakraborty et al. (1999a, 1999b) is shown in Figure 16.12a. It is basically a normally off n-channel metal oxide semiconductor field-effect transistor (MOSFET). SiO_2 and Si_3N_4 layers act as gate insulators. The channel length and width of the FET are 10 and 8300 μm, respectively. The Si_3N_4 layer covers the whole surface area, except contact windows, to keep FET properties constant in ambient condition. Thin films of gold (Au) deposited on Si_3N_4 act as source (S) and drain (D) electrodes. A thin film of Ta (thickness = 2 nm) is deposited on the Si_3N_4 layer, which covers the channel area of the FET, and another thin film of Al (thickness = 200 nm) is deposited on Ta. These films are grown using a dc magnetron sputtering system with the substrate held at room temperature. The porous Al_2O_3 film is obtained by electrochemical anodization of Al film.

The main characteristics of this sensor in shown in Figure 16.12b. It is found that the variation of I_D is consistent with moisture concentration, which indicates the special feature of this structure at low humidity concentration, particularly at the parts per million by volume (ppmv) level. Chakraborty et al. (1999a, 1999b) reported that the sensor had a very small hysteresis loop at higher RH levels (beyond 84% RH level). The response and recovery times were 2 and 6 seconds, respectively, when observations were carried out between 0% and 93% RH levels. The recovery time of the sensor was approximately 10 seconds when observations were carried out between 0% and 100% RH levels.

Lee et al. (2008) have also used this approach for fabrication of HUSFETs, in which the carbon nitride film was applied as sensing material (Figure 16.13). To form hygroscopic CN_x film on the $Si_3N_4/SiO_2/Si$ substrate, a reactive radio frequency (RF) magnetron sputtering method was used. As a result of testing the developed sensors, it was found that the drain current of the n-channel HUSFET increased from 0.85 to 0.99 mA as the RH increased from 20% to 70% (see Figure 16.13b). This corresponded to a sensitivity of 2.8 μA/% RH.

There were also attempts to develop single-walled, carbon nanotube (CNT)-based FETs for humidity measurement (Mudimela et al. 2012). Response of SWCNTs FET to humidity is shown in Figure 16.14. When exposed to humidity, the increase of the current at positive gate voltages and its decrease at negative gate voltages can be explained by the electron-donating nature of water molecules to SWCNTs. However, considering their sensitivity, and especially operation speed, these sensors were significantly inferior to many devices considered earlier. In addition, the technology of their manufacture is complicated, and it is not intended for mass production.

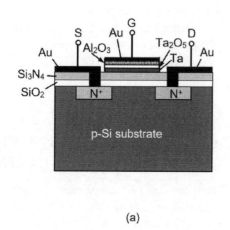
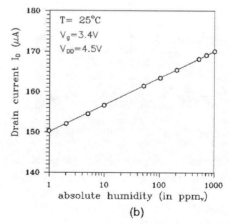

(a) (b)

FIGURE 16.12 (a) Cross-sectional view of the field-effect transistor (FET) sensor and (b) the plot of the drain current with absolute humidity. (Reprinted with permission from Chakraborty, S. et al., *Rev. Sci. Instrum.*, 70, 1565–1567, 1999b. Copyright 1999, American Institute of Physics.)

Humidity Sensors Based on Thin-Film and Field-Effect Transistors

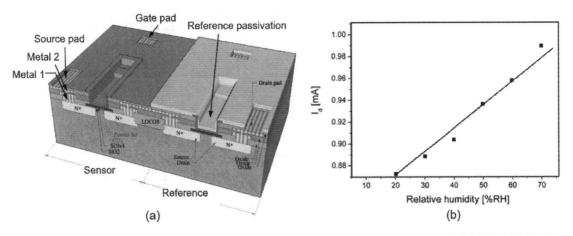

FIGURE 16.13 (a) The structure of n-channel differential HUSFET designed by Lee et al. (2008); (b) The drain current change as a function of relative humidity. (Reprinted with permission from Lee, S.P. et al., *Sensors*, 8, 2662–2672, 2008. Copyright 2008, MDPI as open access.)

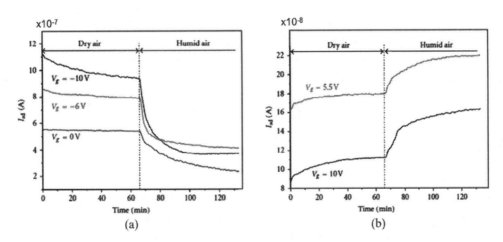

FIGURE 16.14 The source–drain current change of the field-effect transistor (FET) sensor in dry and humid air (relative humidity [RH] = 78%) at (a) negative- and zero-gate voltages and (b) positive-gate voltages. (Reprinted with permission from Mudimela, P.R. et al., *J. Sensors*, 2012, 496–546, 2012. Copyright 2012, Hindawi Publishing Corporation as open access.)

16.3.2 GasFET or FET with Air Gap

Schematically, Gas sensor FETs (GasFETs) are shown in Figure 16.15. The idea of a GasFET was introduced by Janata et al. (Josowicz and Janata 1986; Cassidy et al. 1986) and has been under investigation (Bergstrom et al. 1997; Domansky et al. 1998; Fleischer et al. 2001; Burgmair et al. 2003a, 2003b; Lampe et al. 2005; Stegmeier et al. 2010). The principle of operation of such sensors does not differ from the work of MISFETs, considered earlier: When two chemically different plates are electronically connected, an electric field is created at their interface. This field is proportional to the difference of work functions of the two plate materials (Janata and Josowicz 1998), and both the source–drain current and turn-on voltage are governed by it. Interaction between analyte molecules and the sensing layer is able to change the work function of the sensing material and thus can affect the source–drain current or gate voltage (Janata 1990, 2003). The only difference is that, instead of a dielectric between the gate and the sensitive layer, there is air or another test gas. In a GasFET, the gases come into direct contact with the insulator face of the gate electrode. Therefore, the current in this device in general relies on potential changes upon gas exposure of a sensitive material deposited on the gate electrode. Work function shifts due to the band's bending, or dipoles are directly accessible at the layer-air interface, because air is considered to form an additional gate insulator.

The advantage of GasFETs in comparison with conventional MOSFET- and FET-based sensors is the

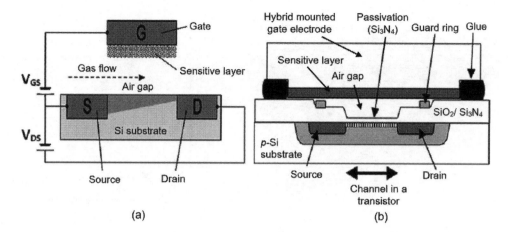

FIGURE 16.15 (a) Schematics of GasFET-type sensor and (b) one of variants of the GasFET realization. The insulation of the transistor channel has to be made of a chemically inert material to avoid additional gas reactions at this surface; otherwise, an additional potential change occurs there. The usage of low-pressure chemical vapor deposition (LPCVD)-deposited Si_3N_4 has been shown to be a good choice. (Reprinted with permission from Burgmair, M. et al., Humidity and low temperature compensation in work function gas sensor FETs, *Sens. Actuators B*, 93, 271–275, 2003. Copyright 2003, Elsevier.)

free choice of the sensing material (Doll and Eisele 1998). The underside of a freestanding gate bridge (see Figure 16.15b) can act as a sensing layer for itself, or it may be covered by sensing material. By using an air-gap FET, all kinds of materials may be used for gas sensing without restrictions to their electrical properties: metals, semiconductors, and thin insulators. However, the materials used in MOSFET and FET-based sensors can be used in GasFETs. In addition, in contrast to classical semiconducting metal–oxide gas and humidity sensors, the gas-sensing properties do not depend on the morphology of the sensing materials. Such a feature gives significant additional freedom in the design.

The disadvantage of the GasFET, in comparison with conventional MOSFET- or FET-based sensor, is that the modulation $\delta I_D/\delta V_G$ is weakened significantly by the air gap (by a factor of approximately 100), necessitating transistors with very steep characteristics. Gergintschew et al. (1996), Fleischer et al. (2001), and Pohle et al. (2003) introduced several options to make more efficient the use of the generated voltage. One significant improvement consists of the formation of a larger area capacitor, built from the suspended gate and a floating gate. The floating gate then transmits the potential coming from the gas-sensitive layer to a small FET device with a short channel (Pohle et al. 2003). This basically minimizes the loss of sensing signal due to a weak coupling via the air gap.

Some examples of the implementation of such sensors follow: Leu et al. (1994) have used a hybrid structure, shown in Figure 16.16. The gate is prepared separately. A view of this hybrid GasFET is shown in Figure 16.17. Manufacturing technology was discussed

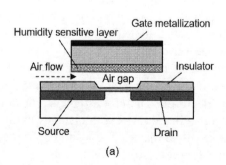
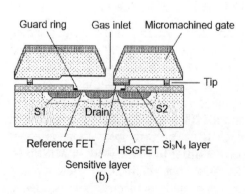

FIGURE 16.16 (a) Schematic plot of a hybrid GasFET. (b) Hybrid GasFET. Schematic construction of the air-gap GasFET fabricated using a micromachined gate. ([a] – Reprinted with permission from Leu, M. et al., Evaluation of gas mixtures with different sensitive layers incorporated in hybrid FET structures, *Sens. Actuators B*, 18–19, 678–681, 1994. Copyright 1994, Elsevier and [b] – Reprinted with permission from Eisele, I. et al., Low power gas detection with FET sensors, *Sens. Actuators B*, 78, 19–25, 2001. Copyright 2001, Elsevier.)

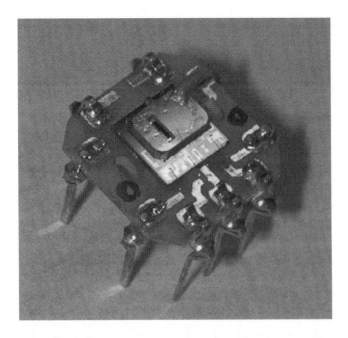

FIGURE 16.17 View of the hybrid GasFET on a TO-8 socket. (Reprinted with permission from Zimmer, M. et al., Gold and platinum as ozone sensing layer in work-function gas sensors, *Sens. Actuators B*, 80, 174–178, 2001. Copyright 2001, Elsevier.)

in Fuchs et al. (1998). Gallium oxide (Ga_2O_3) and platinum (Pt) were tested as sensitive materials were tested Ga_2O_3 and Pt. An interesting aspect of these investigations is the method of obtaining information on the change in the work function of a sensitive material. To measure the contact potential difference (CPD), $\Delta\Phi$, the GasFET operated in the feedback mode. The source–drain current was kept constant by adjusting the gate voltage to compensate the CPD. Therefore, $\Delta\Phi$ results in a shift of the gate voltage U_{gate}, which is directly obtained in millivolts.

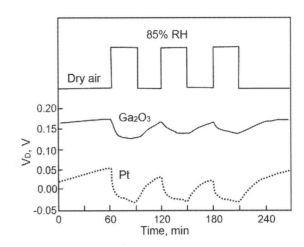

FIGURE 16.18 Influence of water vapor on GasFET at 70°C (shown is 85% relative humidity at room temperature). (Reprinted with permission from Leu, M. et al., Evaluation of gas mixtures with different sensitive layers incorporated in hybrid FET structures, *Sens. Actuators B*, 18–19, 678–681, 1994. Copyright 1994, Elsevier.)

The influence of humidity on the response of sensors modified with Ga_2O_3 and Pt is shown in Figure 16.18. A non-negligible, reversible, and reproducible response is observed for both layers. It is important to note that the sensitivity of these layers to H_2, O_3, and NO_2 was significantly higher (Leu et al. 1994; Zimmer et al. 2001).

Burgmair et al. (2003) also showed that a guard ring (see Figure 16.19) surrounding the area above the FET channel helps to suppress the humidity-induced baseline drifts (Figure 16.19b), by preventing the transport of ions. In this case, the measurement becomes controlled and reproducible. The results, shown in Figure 16.19, are obtained using the standard operating mode of the FET. However, it should be recognized

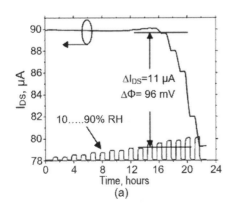 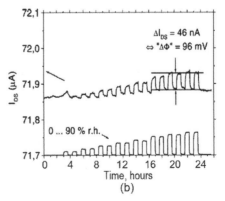

FIGURE 16.19 (a) Influence of humidity pulses (given at 23°C) to an hybrid suspended gate FET (HSGFET) without guard ring ($T_{op} = 30°C$); (b) Optimization of GasFET operation by means of guard ring at ground potential ($T_{op} = 30°C$). (Reprinted from Burgmair, M. et al., Humidity and low temperature compensation in work function gas sensor FETs, *Sens. Actuators B*, 93, 271–275, 2003. Copyright 2003, Elsevier.)

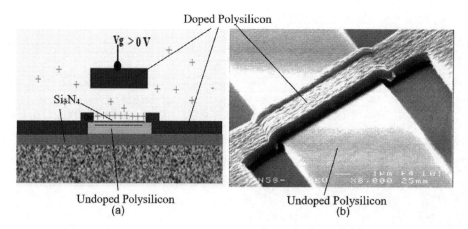

FIGURE 16.20 (a) Structure of SGTFT and (b) polysilicon gate–bridge highlighting the air-gap. (Reprinted with permission from Mahfoz-Kotb, H. et al., Sensing sensibility of surface micromachined suspended gate polysilicon thin film transistors, *Sens. Actuators B*, 118, 243–248, 2006. Copyright 2006, Elsevier.)

that the sensitivity of these sensors is not high, which is apparently due to the large distance between the gate and the channel of the FET.

A similar but simpler version of the humidity sensor based on an FET was proposed by Salaün et al. in 2005. This sensor is a polysilicon suspended-gate TFT (SGTFT), fabricated using a low-temperature surface micromachining process. The structure of this transistor is shown in Figure 16.20. Microtechnology techniques using sacrificial layers were used to fabricate a polysilicon bridge, which acted as the transistor gate. The active layer was polysilicon with passivation by thin film of silicon nitride. Transistors were characterized at various humidity rates, and transfer characteristics showed highly sensitive dependence with humidity (see Figure 16.21).

The humidity-sensing effect was explained in the following way. It is known that silicon nitride contains two different types of surface sites, namely silanol sites (SiOH) and amine sites ($SiNH_2$), on its surface (Niu et al. 1996). These two types of nitride surface sites have a high reactivity to H^+ ions. This means that the properties of the Si_3N_4 insulator and its surface charge directly depend on the surface treatment. In particular, introduction of humidity creates positive charges Qss due to adsorption on the channel surface (Figure 16.20a). These positive charges create a surface potential added to the applied gate voltage that explains the decrease of threshold voltage V_{th}.

It was found that the field effect, due to the voltage applied to the suspended gate, is enough to influence both the distribution of electrical charges present in humidity and the transistor electrical characteristics (i.e., the conductivity of the active layer). The small air gap (0.5 μm) between the gate and the channel explained the amplifying effect of the sensitivity: The threshold voltage shift was more than 17 V when the humidity ratio varies from 20% to 70%. Later, based on these sensors, an suspended gate FET (SGFET)-type sensor array for humidity measurements was developed and manufactured (da Silva Rodrigues et al. 2010)

Another variant of the air-gap FET was proposed and realized by Ahn et al. (2010) and Han et al. (2012). The configuration and technology of fabrication of such a nanogap transistor are shown in Figure 16.22. SEM images are shown in Figure 16.23. Ahn et al. (2010)

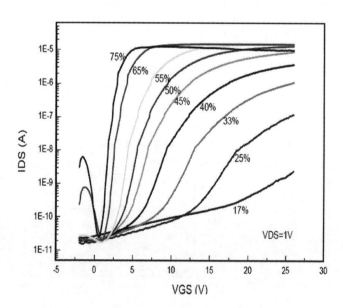

FIGURE 16.21 Transfer characteristic $I_{DS} = f(V_{GS})$, where IDS is the drain–source current and V_{GS} is the gate-source voltage, of an *n*-type suspended polysilicon gate thin-film transistor for different humidity ratios. (Data extracted from Salaün, A.-C. et al., *Proc. SPIE*, 5836, 231–238, 2005.)

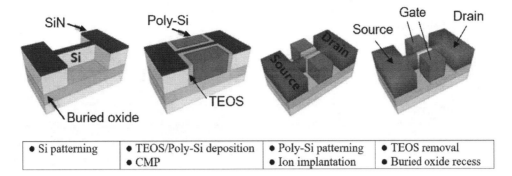

| • Si patterning | • TEOS/Poly-Si deposition
• CMP | • Poly-Si patterning
• Ion implantation | • TEOS removal
• Buried oxide recess |

FIGURE 16.22 Schematics of the process flow. The silicon channel is patterned with the nitride hard mask. The sacrificial SiO_2 with a thickness of 30 nm and the polysilicon with a thickness of 400 nm are subsequently deposited. The polysilicon is then separated by chemical mechanical polishing. Next, the polysilicon gate is patterned, followed by source/drain formation. Finally, the sacrificial SiO_2 and the buried oxide underneath the silicon channel are removed to form suspended silicon. (Reprinted with permission from Han, J.-W. et al., Liquid gate dielectric field effect transistor for a radiation nose, *Sens. Actuators A*, 182, 1–5, 2012. Copyright 2012, Elsevier.)

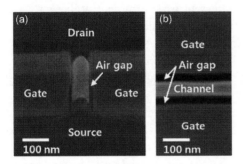

FIGURE 16.23 Color-enhanced scanning electron microscopy images of the fabricated transistor. (a) Bird's eye view and (b) Top view. The two separated gates control the channel, called a *fin*. The air gap has a thickness of ≈ 30 nm. (Reprinted with permission from Han, J.-W. et al., Liquid gate dielectric field effect transistor for a radiation nose, *Sens. Actuators A*, 182, 1–5, 2012. Copyright 2012, Elsevier.)

assumed that water molecules confined in the nanogaps should increase the gate dielectric constant, and as a consequence, it should a shift of the threshold voltage should occur and, accordingly, variation of the drain current. Operating characteristics of these sensors are shown in Figure 16.24.

The same approach to the manufacture of humidity sensors was suggested by Ray (2015). Moreover, the first principles, based on density functional theory (DFT) calculations, were performed to understand the behavior and performance of the device and its usability and detection capability as a humidity sensor. The only difference was that Ray (2015) suggested using an FET with the following parameters: The source and drain electrodes have square cross sections of 0.8 nm × 0.8 nm area, which are separated from the gate electrode via

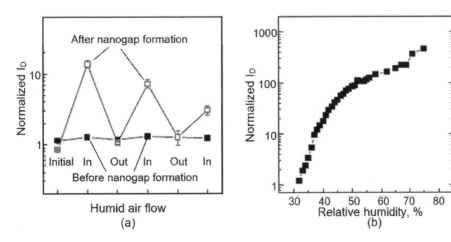

FIGURE 16.24 (a) Humidity response (a) before and (b) after the nanogap formation. The device with the nanogaps is highly sensitive to water molecules. I_D increases after the injection of humid air and returns to its initial baseline after outgassing. (b) Humidity response depending on the relative humidity. (Reprinted with permission from Han, J.-W. et al., Liquid gate dielectric field effect transistor for a radiation nose, *Sens. Actuators A*, 182, 1–5, 2012. Copyright 2012, Elsevier.)

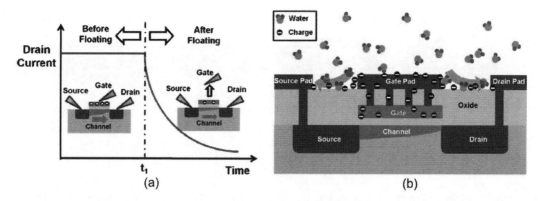

FIGURE 16.25 Schematic illustration of the drain–current drop when the gate bias is electrically floated. (a) Drain–current drops once the gate voltage is floated by lifting up the gate probe due to the dissipation of the gate charges. (b) The gate charges are dissipated to the surrounding electrodes through the oxide surface, helped by the water molecules adhered to the surface. (Reprinted with permission from Song, S.-H. et al., *Appl. Phys. Lett.*, 100, 101603, 2012. Copyright 2012, American Institute Physics.)

dielectric layers of 0.3 nm thickness on both ends of the device. According to Ray (2015), such devices are capable of detecting single water molecules. This is only a model representation of the device, which is not intended for wide application.

A completely different approach to designing the FET-based humidity sensors was proposed by Song et al. (2012). They designed a metal–oxide–semiconductor FET-based humidity sensor that did not use any specific materials to sense the RH (see Figure 16.25a). The principle of operation of their device is based on the surface conductance change of silicon dioxide, deposited by a low-pressure chemical vapor deposition (LPCVD) in humid atmosphere. Many studies have shown that the silicon dioxide surface conductance changes by powers of 10, according to the RH in the surrounding air (Ho et al 1967; Voorthuyzen et al. 1987). According to Song et al. (2012), when the gate in the FET is biased and then floated, the electrical charge in the gate is dissipated through the silicon dioxide's surface to the surrounding ground with a time constant depending on the surface conductance, which, in turn, varies with humidity.

Figure 16.25 also shows the mechanism of humidity measurement in the atmosphere using MOSFET designed by Song et al. (2012). It consists of four steps: Step 1: As shown in the "before floating" part in Figure 16.25a, appropriate voltages to turn on the transistor are applied to the gate, source, and drain by using a probe tip to conduct the drain-to-source current. Step 2: The gate is electrically floated by detaching the probe tip, as demonstrated in the "after floating" part in Figure 16.25a. The gate charges start to dissipate to the drain and source electrodes through the surface of the LPCVD oxide, as shown in Figure 16.25b. Since the drain current is determined by the gate voltage, the drain current starts to decrease as the gate voltage drops. This event is described in the "after floating" part in Figure 16.25a. Step 3: The drop rate of the gate voltage depends on how fast the gate charges are dissipated through the LPCVD oxide's surface from the gate pad to the surrounding electrodes. Step 4: The surface conductance of the LPCVD oxide changes according to the humidity in the atmosphere (Ho et al. 1967; Voorthuyzen et al. 1987). Different humidity in the atmosphere results in different gate voltage drop rates. Thereby, we can sense the humidity change by measuring the drain current drop rate, which is affected by the decreasing rate of the gate voltage.

Figure 16.26 shows the schematic view of the proposed MOSFET humidity sensor and the image of the fabricated MOSFET humidity sensor. The device size is 340 μm by 280 μm. In this special design, we put a ground ring surrounding the gate electrode to gather the dissipated gate charges efficiently. All the electrodes were made of 1 μm thick aluminum, and the probing pad size was 80 μm by 80 μm. The separation gap between the gate electrode and the surrounded ground ring was 15 μm, and the circumference of the ground ring was 340 μm. The gate-to-channel overlap area was 50 μm by 20 μm, and 30 nm thickness of thermal silicon dioxide was used as the gate oxide.

Song et al. (2012) have shown that, with this method, extremely high sensitivity can be achieved. In their experiments, the charge dissipation speed increased a thousand times as the RH increased. Figure 16.27 illustrates this result. The drain current dropped as soon as the probe tip is detached from the gate. The decreasing rate of the current varied according to the RH in the air.

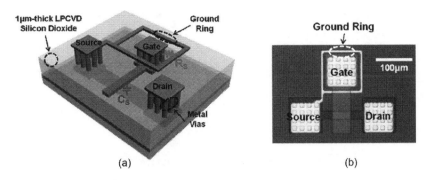

FIGURE 16.26 (a) The device structure, including the ground ring surrounding the gate electrode. (b) Top view of the fabricated MOSFET humidity sensor. (Reprinted with permission from Song, S.-H. et al., *Appl. Phys. Lett.*, 100, 101603, 2012. Copyright 2012, American Institute Physics.)

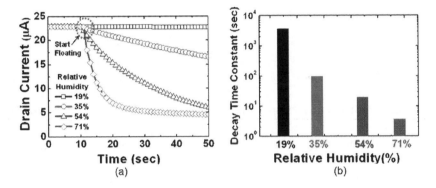

FIGURE 16.27 (a) Measured drain current drop at different relative humidity; (b) Measured decay time constant at different relative humidity. (Reprinted with permission from Song, S.-H. et al., *Appl. Phys. Lett.*, 100, 101603, 2012. Copyright 2012, American Institute Physics.)

16.3.3 TFT-Based Sensors

At present, many TFTs have been developed based on inorganic materials (Nomura et al. 2004, 2006; Mitzi et al. 2004; Kumomi et al. 2008; Seon et al. 2009; Salas-Villasenor et al. 2010; Fortunato et al. 2012: Park et al. 2012). However, as a rule, these devices are designed for other applications and are not used for humidity measurement. In addition, traditional metal oxide materials used in various gas sensors have significant limitations on use in such devices. Studies have shown that, with the use of amorphous a-IGZO (In-Ga-ZnO), there are some prospects for the development of TFT-based humidity sensors. It was found that the Zn-based amorphous oxide semiconductors (AOSs), such as amorphous In-Ga-ZnO (a-IGZO), meet all the requirements of materials aimed for application in TFT-based humidity sensors (Takagi et al. 2005; Kamiya et al. 2010). As an n-type semiconductor, amorphous $InGaZnO_4$ is transparent with an optical bandgap of 3.1–3.3 eV (Nomura 2008). The main advantages of Zn-based AOS for TFT design are high electron mobility (from 5 to 39 cm^2/Vs), even in amorphous form; scalability; uniformity; and availability of low-temperature processing (Nomura et al. 2004; Kamiya and Hosono 2010). Thin films of amorphous $InGaZnO_4$ are usually obtained from a polycrystalline ceramic target by using physical deposition techniques, such as pulsed laser deposition and radio frequency sputtering at room temperature (Hosono and Nomura 2008). Formation of amorphous $InGaZnO_4$ requires an extremely high cooling rate (i.e., amorphous a-IGZO can be fabricated only in the form of a thin film, but not in a bulk form). The amorphous nature of a-IGZO films has been reported to remain stable up to relatively high temperatures ($\approx 500°C–600°C$) (Yang et al. 2012). Also, they have a large process window in the choice of gate insulators. It was also found that the properties of the AOS-based channel layers are easily influenced by various environmental conditions. In particular, it was established that the conductivity of amorphous IGZO is sensitive to oxygen partial pressure and humidity (Kang et al. 2007; Park et al. 2008). Kang et al. (2007) realized

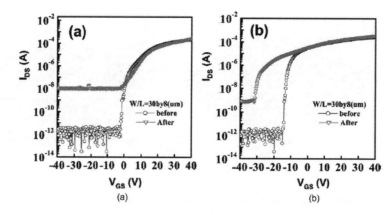

FIGURE 16.28 The comparison of the transfer curves before and after the water exposure for the device with (a) 35 nm thick and (b) 150 nm thick channels, respectively. Thin-film transistors were exposed to 100% relative humidity during 12 h. (Reprinted with permission from Park, J.S. et al., *Appl. Phys. Lett.*, 92, 072104, 2008. Copyright 2008, American Institute Physics.)

IGZO TFT-based gas sensor and showed that IGZO-based TFT had stable characteristics. IGZO TFT sensors were fabricated using the configuration shown in Figure 16.7. Park et al. (2008) found that H_2O adsorption on a-IGZO surfaces significantly influenced the important parameters of the TFTs (see Figure 16.28), including the threshold voltage (V_{th}), subthreshold swing (SS), and off current (I_{off}), due to their charge transfer.

According to Park et al. (2008), the water molecules reversibly diffused in and out of the a-IGZO thin film. While the adsorbed oxygen depleted the electron carriers on the a-IGZO film, the water molecules induced the appearance of extra electron carriers. In addition, the adsorbed water can act as either electron donor or acceptor, like trap sites, depending on the channel thickness. However, during the following research, it was found that the kinetics of response and especially recovery processes in these sensors after interaction with oxygen, NO_2, or water are very slow. For example, Park et al. (2008) established that, for recovery of the initial state after interaction with oxygen and water, annealing in vacuum at $T = 100°C$ during approximately 300 minutes is required. The response to humidity at room temperature was also not too high. Such parameters create difficulties for real application of IGZO-based TFT humidity sensors.

Kim et al. (2017) used the same approach for fabrication of the flexible IGZO-based, TFT-based humidity sensor. As electrodes, they used transparent conductive oxide, such as indium tin oxide (ITO), and as a substrate, parylene (≈ 10 μm). As a result of optimization of the deposition technology of IGZO and ITO films, sensors have been manufactured and tested, the characteristics of which are shown in Figure 16.29. It was established that the TFT-based sensors showed the highest sensitivity for the low gate bias of $-1 \sim 2$ V, resulting in low

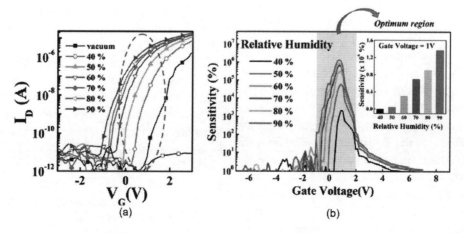

FIGURE 16.29 (a) Variation in transfer characteristics of imperceptible thin-film transistors (TFTs) under various humidity conditions. *B*, Sensitivity values of imperceptible TFTs as a function of gate voltage under different humidity conditions. (Reprinted with permission from Kim, K.S. et al., *Materials*, 10, 530, 2017. Copyright 2017, MDPI as open access.)

power consumption by this sensor. Compared to conventional resistor-type sensors, the oxide TFT sensors exhibited extremely large sensitivities. Considering that these TFTs are ultra-light and imperceptible, Kim et al. (2017) suggested that these sensors are expected to show wide application, because they can be patched or embedded on several objects encountered in everyday life.

16.4 SUMMARY

Analysis of the thin-film and MISFETs for monitoring the atmospheric humidity indicates that, in comparison with chemiresistors, the TFTs and FETs provide more parameters for measurements; thus, they may give more detailed information about the analyte, which can be used for its reliable detection. For example, Torsi et al. (2000) demonstrated that, in addition to measuring the bulk conductivity of the semiconductor upon exposure to chemical species, a threshold voltage and field-effect mobility can also be measured and used for analysis. This means that transistors are, in principle, more selective devices than conventional chemiresistors. Moreover, the detection limit and sensitivity of the sensors based on transistors are better than those of chemiresistors because of the signal amplification of transistor devices (Torsi et al. 2001). In the present case, TFTs and FETs have been compared with conventional chemiresistors based on conducting polymers. Torsi et al. (2001) also demonstrated that transistors, in addition to higher sensitivity, showed a larger dynamic range and superior reversibility in comparison with simple resistors. Bouvet et al. (2001a, 2001b) provided evidence that their TFT-gas sensor performed better than a similar resistor-type sensor in terms of drift, sensitivity, signal-to-noise ratio, response time, lifetime, and operating temperature as well. However, this statement is unfounded, since it is unclear what their comparison was made with. If the comparison was made with devices based on OSCs, then both types of devices used the same materials, and therefore they had the same shortcomings. If we compare polymer chemiresistors with FETs, then in this case, FETs can indeed have much better stability, since they can use cross-linked polymers that do not have such a strong sensitivity to the aggressive environment. Due to their stability, FET-based humidity sensors should not yield to capacitive humidity sensors.

Concerning the technology of manufacturing TFTs and FETs, it should be noted that, as with other semiconductor devices, their preparation is slightly complicated. The characterization of a transistor is also more difficult than a chemiresistor. However, if we compare organic TFT and FET-based sensors with inorganic devices based on silicon, graphene, carbon nanotube, or oxide semiconductors, polymer sensors are undoubtedly advantageous due to the low requirements of processing conditions.

Most inorganic semiconductor devices are prepared on the high-purity crystalline substrates at high temperature, normally relying on the utilization of expensive facilities in clean rooms. In contrast, organic electronic devices can be fabricated by the convenient solution-processable techniques, including spin-coating, and inkjet printing, or screen printing or thermal evaporation at room temperature, enabling the integration of devices on low-cost substrates such as glass, flexible plastics, metal foils, and even papers (Liu et al. 2009). All above-mentioned techniques are cost-effective methods for mass production. Therefore, organic TFTs and MISFETs are the ideal candidates for high-performance and disposable sensors (Cotrone et al. 2012).

Regarding the selectivity of TFT and FET humidity sensors, as well as all previous sensors, these sensors are not selective to water vapor. This is undoubtedly a drawback. For example, experiments carried out by Wu et al. (2017) have shown that PBIBDF-BT-based OTFT humidity sensors are sensitive to the vapors of all organic solvents. The parameters of some of these gases are listed in the Table 16.4. It can be observed that the devices showed a significant response to polar protic molecules compared with both polar aprotic and nonpolar molecules. It is known that polar molecules located on a semiconductor surface and grain boundaries are more likely to interact with semiconductors to induce charge–dipole interactions (Dunlap et al. 1996). As a result, the polar protic solvent (water) exhibited the largest I_{DS} change in the transfer characteristic curves, whereas other chemical vapors showed relatively smaller variations. The protic molecules (water, methanol, ethanol, isopropanol) have relatively larger polarities, which play the role of traps in the electronic transmission. According to Subbarao et al. (2016), the good selectivity for humidity is mainly due to the high polarity and high dielectric constant of the water molecules, and the smaller molecular volume is a secondary factor, which more easily diffuses into the grain boundary. Overall, the device showed a lower response to aprotic or nonpolar molecules, such as diethyl ether, acetone, carbon fluoride (CF), ethyl acetate, and hexane. However, toluene exhibited a higher sensitivity to aprotic molecules. Wu et al. (2017) believe that the reason for this could be related to the π-conjugated structure in the benzene ring, which forms a charge trap that reduces a charge density.

TABLE 16.4
Chemical Nature of Gas Molecules Affected the Sensor Response of PBIBDF-BT–Based OTFT Humidity Sensors

Gas Molecules	Dipole Moment (D)	Dielectric Constant (k)	Polarity	Sensitivity[a]
Water (H^+)	1.86	78.5	10.2	269.8
Methanol (H^+)	1.71	32.6	6.6	6.4
Ethanol (H^+)	1.69	32.6	4.3	15.9
Isopropanol (H^+)	1.70	24.3	4.3	46.6
Diethyl ether	1.30	4.3	2.9	3.85
Toluene	0.40	2.4	2.4	94.7
Acetone	2.88	20.7	5.4	11.1
CF	1.10	4.7	4.4	27.3
Ethyl acetate	1.90	6.03	4.3	15.7
Hexane	0	1.88	0.06	3.64
Anisole	1.29	4.3	2.8	27.6

Source: Reprinted with permission from Wu, S. et al., *ACS Appl. Mater. Interfaces*, 9, 14974–14982, 2017. Copyright 2017, American Chemical Society.

[a] During sensitivity measurement ($S = I_{D,\text{wet}}/I_{D,\text{dry}}$), the concentration of water changed from 32% relative humidity (RH) (9146 ppm) to 69% RH (20 036 ppm), whereas the concentration of other organic solvents ranged from 0 to 50 000 ppm.

CF, specialty solvent, which is a patented blend of hydrofluorocarbons Vertrel® XF (2,3-dihydrodecafluoropentane) and HFC-365mfc(1,1,1,3,3-pentafluorobutane).

REFERENCES

Ahn J.-H., Kim J.-Y., Im M., Han J.W., Choi Y.-K. (2010) A nanogap-embedded nanowire field effect transistor for sensor applications: Immunosensor and humidity sensor. In: *14th International Conference on Miniaturized Systems for Chemistry and Life Sciences*, October 3–7, Groningen, the Netherlands, pp. 1301–1303.

Barker P.S., Monkman A.P., Petty M.C., Pride R. (1997) A polyaniline/silicon hybrid field effect transistor humidity sensor. *Synth. Metals* 85, 1365–1366.

Bassler H., Kohler A. (2012) Charge transport in organic semiconductors. *Top. Curr. Chem.* 312, 1–66.

Bergstrom P.L., Patel N.Y., Schwank J.W., Wise K.D. (1997) A micromachined surface work-function gas sensor for low pressure oxygen detection. *Sens. Actuators B* 42, 195–204.

Bouvet M. (2006) Phthalocyanine-based field-effect transistors as gas sensors. *Anal. Bioanal. Chem.* 384, 366–373.

Bouvet M., Guillaud G., Leroy A., Maillard A., Spirkovitch S. Tournilhac F.G. (2001b) Phthalocyanine-based field-effect transistor as ozone sensor. *Sens. Actuators B* 73, 63–70.

Bouvet M., Leroy A., Simon J., Tournilhac F., Guillaud G., Lessnick P., Maillard A. et al. (2001a) Detection and titration of ozone using metallophthalocyanine based field effect transistors. *Sens. Actuators B* 72, 86–93.

Burgmair M., Zimmer M., Eisele I. (2003) Humidity and temperature compensation in work function gas sensor FETs. *Sens. Actuators B* 93, 271–275.

Cassidy J., Pons S., Janata J. (1986) Hydrogen response of palladium coated gate field effect transistor. *Anal. Chem.* 58, 1757–1761.

Chakraborty S., Hara K., Lai P.T. (1999b) New microhumidity field-effect transistor sensor in ppmv level. *Rev. Sci. Instrum.* 70 (2), 1565–1567.

Chakraborty S., Nemoto K., Hara K., Lai P.T. (1999a) Moisture sensitive field effect transistors using $SiO_2/Si_3N_4/Al_2O_3$ gate structure. *Smart Mater. Struct.* 8, 274–277.

Chang J.B., Liu V., Subramanian V., Sivula K., Luscombe C., Murphy A., Liu J.S., Frechet J.M.J. (2006) Printable polythiophene gas sensor array for low-cost electronic noses. *J. Appl. Phys.* 100, 014506.

Chao S., Wrighton M.S. (1987) Characterization of a solid-state polyaniline-based transistor: Water vapor dependent characteristics of a device employing a poly(vinyl alcohol)/phosphoric acid solid-state electrolyte. *J. Am. Chem. Soc.* 109, 6627–6631.

Chen Y.-C., Chang T.-C., Li H.-W., Chung W.-F., Wu C.-P., Chen S.-C., et al. (2012) High-stability oxygen sensor based on amorphous zinc tin oxide thin film transistor. *Appl. Phys. Lett.* 100, 262908.

Chen H., Josowicz M., Janata J., Potje-Kamloth K. (2004) Chemical effects in organic electronics. *Chem. Mater.* 16, 4728–4735.

Cotrone S., Cafagna D., Cometa S., Giglio E., De Magliulo M., Torsi L., Sabbatini L. (2012) Microcantilevers and organic transistors: Two promising classes of label-free biosensing devices which can be integrated in electronic circuits. *Anal. Bioanal. Chem.* 402, 1799–1811.

Covington J.A., Gardner J.W., Briand D., de Rooij N.F. (2001) A polymer gate FET sensor array for detecting organic vapours. *Sens. Actuators B* 77, 155–162.

Crone B., Dodabalapur A., Gelperin A., Torsi L., Katz H.E., Lovinger A.J., Bao Z. (2001) Electronic sensing of vapors with organic transistors. *Appl. Phys. Lett.* 78, 2229–2231.

Da Silva Rodrigues B., De Sagazan O., Salaün A.-A., Crand S., Le Bihan F., Mohammed-Brahim T., Bonnaud O., Morimoto N.I. (2010) Humidity sensor thanks array of suspended gate field effect transistor. *ECS Trans.* 31 (1), 441–448.

Dhoot A.S., Yuen J.D., Heeney M., McCulloch I., Moses D., Heeger A.J. (2006) Beyond the metal-insulator transition in polymer electrolyte gated polymer field-effect transistors. *Proc. Natl. Acad. Sci. U. S. A.* 103, 11834–11837.

Doll T., Eisele I. (1998) Gas detection with work function sensors. *SPIE* 3539, 96–105.

Domansky K., Baldwin D.L., Grate J.W., Hall T.B., Li J., Josowicz M., Janata J. (1998) Development and calibration of field effect transistor based sensor array for measurement of hydrogen and ammonia gas mixtures in humid air. *Anal. Chem.* 70, 473–481.

Dual V., Surwade S.P., Ammu S., Agnihotra S.R., Jain S., Roberts K.E., et al. (2010) All-organic vapor sensor using inkjet-printed reduced graphene oxide. *Angew Chem. Int. Ed.* 49, 2154–2157.

Duarte D., Dodabalapur A. (2012) Investigation of the physics of sensing in organic field effect transistor based sensors. *J. Appl. Phys.* 111, 044509.

Dunlap D.H., Parris P.E., Kenkre V.M. (1996) Charge-dipole model for the universal field dependence of mobilities in molecularly doped polymers. *Phys. Rev. Lett.* 77, 542–545.

Ebisawa F., Kurokawa T., Nara S. (1983) Electrical properties of polyacetylene/polysiloxane interface. *J. Appl. Phys.* 54, 3255–3259.

Eisele I., Doll T., Burgmair M. (2001) Low power gas detection with FET sensors. *Sens. Actuators B* 78 (2001), 19–25.

Fleischer M., Ostrick B., Pohle R., Simon E., Meixner H., Bilger C., Daeche F. (2001) Low-power gas sensors based on work-function measurement in low-cost hybrid flip-chip technology. *Sens. Actuators B* 80, 169–173.

Fortunato E., Barquinha P., Martins R. (2012) Oxide semiconductor thin-film transistors: A review of recent advances. *Adv. Mater.* 24, 2945–2986.

Fuchs A., Doll T., Eisele I. (1998) Flip-chip mounting of hybrid FET gas sensors with air gap. In: *Proceedings of International conference; 6th, Micro electro, opto, mechanical systems and components*, Micro System Technologies, VDE-Verlag, Potsdam, Germany, pp. 147–152.

Gergintschew Z., Kornetzky P., Schipanski D. (1996) The capacitively controlled field effect transistor (CCFET) as a new low power gas sensor. *Sens. Actuators B* 35–36, 285–289.

Guillaud G., Al Sadoun M., Maitrot M., Simon J., Bouvet M. (1990) Field-effect transistors based on intrinsic molecular semiconductors. *Chem. Phys. Lett.* 167, 503–506.

Guillaud G., Simon J., Germain J.-P. (1998) Metallophthalocyanines: Gas sensors, resistors and field effect transistors. *Coord. Chem. Rev.* 178, 1433–1484.

Guo Y., Yu G., Liu Y. (2010) Functional organic field-effect transistors. *Adv. Mater.* 22, 4427–4447.

Han J.-W., Meyyappan M., Ahn J.-H., Choi Y.-K. (2012) Liquid gate dielectric field effect transistor for a radiation nose. *Sens. Actuators A* 182, 1–5.

Hasegawa T., Takeya J. (2009) Organic field-effect transistors using single crystals. *Sci. Technol. Adv. Mater.* 10 (2), 024314.

Higikigawa M., Suihara T., Tanaka J., Watanabe M., (1985) Microchip FET humidity sensor with a long-term stability. In: *Proceedings of the 3rd International Conference on Solid State Sensors and Actuators (Transducers'85)*, Philadelphia, PA, June 11–14, pp. 221–224.

Ho P.O., Lehovec K., Fedotowsky L. (1967) Charge motion on silicon oxide surfaces. *Surf. Sci.* 6, 440–460.

Horowitz G., Peng X., Fichou D., Garnier F. (1990) The oligothiophene-based field-effect transistor: How it works and how to improve it? *J. Appl. Phys.* 67, 528–532.

Hosono H., Nomura K. (2008) Factors controlling electron transport properties in transient amorphous oxide semiconductors. *J. Non-Crystal. Solids* 354, 2796–2800.

Janata J. (1990) Potentiometric microsensors. *Chem. Rev.* 90, 691–703.

Janata J. (2003) Electrochemical microsensors. *Proc. IEEE* 91, 864–869.

Janata J., Josowicz M. (1998) Chemical modulation of work function as a transduction mechanism for chemical sensors. *Acc. Chem. Res.* 31, 241–248.

Josowicz M., Janata J. (1986) Suspended gate field effect transistor modified with polypyrrole as alcohol sensor. *Anal. Chem.* 58, 514–517.

Kamiya T., Hosono H. (2010) Material characteristics and applications of transparent amorphous oxide semiconductors. *NPG Asia Mater* 2, 15–22.

Kamiya T., Nomura K., Hosono H. (2010) Present status of amorphous In–Ga–Zn–O thin-film transistors. *Sci. Technol. Adv. Mater* 11, 044305.

Kang D., Lim H., Kim C., Song I. (2007) Amorphous gallium indium zinc oxide thin film transistors: Sensitive to oxygen molecules. *Appl. Phys. Lett.* 90, 192101.

Kang H.S., Lee J.K., Lee J.W., Joo J., Ko J.M., Kim M.S., Lee J.Y. (2005) Humidity dependent characteristics of thin film poly(3,4-ethylenedioxythiophene) field-effect transistor. *Synth. Met.* 155, 176–179.

Kim K.S., Ahn C.H., Kang W.J., Cho S.W., Jung S.H., Yoon D.H., Cho H.K. (2017) An all oxide-based imperceptible thin-film transistor with humidity sensing properties. *Materials* 10, 530.

Kim S.H., Hong K., Xie W., Lee K.H., Zhang S., Lodge T.P., Frisbie C.D. (2013) Electrolyte gated transistors for organic and printed electronics. *Adv. Mater.* 25, 1822–1846.

Klauk H. (2010) Organic thin-film transistors. *Chem. Soc. Rev.* 39, 2643–2666.

Kudo K., Yamashina M., Moriizumi T. (1984) Field effect measurement of organic dye films. *Jpn. J. Appl. Phys.* 23, 130–130.

Kumomi H., Nomura K., Kamiya T., Hosono H. (2008) Amorphous oxide channel TFTs. *Thin Solid Films* 516, 1516–1522.

Lampe U., Simon E., Pohle R., Fleischer M., Meixner H., Frerichs H.-P., Lehmann M., Kiss G. (2005) GasFETs for the detection of reducing gases. *Sens. Actuators B* 111–112, 106–110.

Larsson O., Said E., Berggren M., Crispin X. (2009) Insulator polarization mechanisms in polyelectrolyte-gated organic field-effect transistors. *Adv. Funct. Mater.* 19, 3334–3341.

Lee J., Kaake L.G., Cho J.H., Zhu X.Y., Lodge T.P., Frisbie C.D. (2009) Ion gel-gated polymer thin-film transistors: Operating mechanism and characterization of gate dielectric capacitance, switching speed, and stability. *J. Phys. Chem. C* 113, 8972–8981.

Lee S.P., Lee J.G., Chowdhury S. (2008) CMOS Humidity sensor system using carbon nitride film as sensing materials. *Sensors* 8, 2662–2672.

Lee S.P., Park K.J. (1996) Humidity sensitive field effect transistors. *Sens. Actuators B* 35/36, 80–84.

Leu M., Doll T., Flietner B., Lechner J., Eisele I. (1994) Evaluation of gas mixtures with different sensitive layers incorporated in hybrid FET structures. *Sens. Actuators B* 18–19, 678–681.

Li D., Borkent E.-J., Nortrup R., Moon H., Katz H., Bao Z. (2005) Humidity effect on electrical performance of organic thin-film transistors. *Appl. Phys. Lett.* 86, 042105.

Li J., Liu D., Miao Q., Yan F. (2012) The application of a high-k polymer in flexible low-voltage organic thin-film transistors. *J. Mater. Chem.* 22, 15998–16004.

Liao C., Yan F. (2013) Organic semiconductors in organic thin-film transistor-based chemical and biological sensors. *Polymer Rev.* 53, 352–406.

Lin P., Yan F. (2012) Organic thin-film transistors for chemical and biological sensing. *Adv. Mater* 24, 34–51.

Liu S., Wang W.M., Briseno A.L., Mannsfeld S.C.B., Bao Z. (2009) Controlled deposition of crystalline organic semiconductors for field-effect-transistor applications. *Adv. Mater.* 21, 1217–1232.

Mabeck J.T., Malliaras G.G. (2006) Chemical and biological sensors based on organic thin-film transistors. *Anal. Bioanal. Chem.* 384, 343–353.

Mahfoz-Kotb H., Salaun A.C., Bendriaa F., Le Bihan F., Mohammed-Brahim T., Morante J.R. (2006) Sensing sensibility of surface micromachined suspended gate polysilicon thin film transistors. *Sens. Actuators B* 118, 243–248.

Marinelli F., Dell'Aquila A., Torsi L., Tey J., Suranna G.P., Mastrorilli P., et al. (2009) An organic field effect transistor as a selective NOx sensor operated at room temperature. *Sens. Actuators B* 140, 445–450.

Mitzi D.B., Kosbar L.L., Murray C.E., Copel M., Afzali A. (2004) High-mobility ultrathin semiconducting films prepared by spin coating. *Nature* 428, 299–303.

Mudimela P.R., Grigoras K., Anoshkin I.V., Varpula A., Ermolov V., Anisimov A.S., Nasibulin A.G., Novikov S., Kauppinen E. (2012) Single-walled carbon nanotube network field effect transistor as a humidity sensor. *J. Sensors* 2012, 496546.

Murtaza I., Karimov Kh. S., Ahmad Z., Qazi I., Mahroof-Tahir M., Khan T.A., Amin T. (2010) Humidity sensitive organic field effect transistor. *J. Semiconductors* 31 (5), 054001.

Nilsson D., Kugler T., Svensson P.O., Berggren M. (2002) An all-organic sensor–transistor based on a novel electrochemical transducer concept printed electrochemical sensors on paper. *Sens. Actuators B* 86, 193–197.

Niu M.-N., Ding X.-F., Tong Q.-Y. (1996) Effect of two types of surface sites on the characteristics of Si_3N_4-gate pHISFETS. *Sens. Actuators B* 37, 13–17.

Nomura K., Ohta H., Takagi A., Kamiya T., Hirano M., Hosono H. (2004) Room temperature fabrication of transparent flexible thin-film transistors using amorphous oxide semiconductors. *Nature* 432, 488–492.

Nomura K., Takagi A., Kamiya T., Ohta H., Hirano M., Hosono H. (2006) Amorphous oxide semiconductors for high-performance flexible thin-film transistors. *Jpn. J. Appl. Phys.* 45, 4303–4308.

Nomura K. (2008) Bandgap states in transparent amorphous oxide semiconductor, In-Ga-Zn-O, observed by bulk sensitive X-ray photoelectron spectroscopy. *Appl. Phys. Lett.* 92, 202117.

Ohmori Y., Takahashi H., Muro K., Uchida M., Kawai T., Yoshino K. (1991) Gas-sensitive Schottky gated field effect transistors utilizing poly(3-alkylthiophene) films. *Jpn. J. Appl. Phys.* 30, L1247–L1249.

Panzer M.J., Frisbie C.D. (2008) Exploiting ionic coupling in electronic devices: Electrolyte-gated organic field-effect transistors. *Adv. Mater* 20, 3177–3180.

Park J.S., Jeong J.K., Chung H.J., Mo Y.G., Kim H.D. (2008) Electronic transport properties of amorphous indium–gallium–zinc oxide semiconductor upon exposure to water. *Appl. Phys. Lett.* 92, 072104.

Park J.S., Maeng W.-J., Kim H.-S., Park J.-S. (2012) Review of recent developments in amorphous oxide semiconductor thin-film transistor devices. *Thin Solid Films* 520, 1679–1693.

Park Y.D., Kang B., Lim H.S., Cho K., Kang M.S., Ho Cho J.H. (2013) Polyelectrolyte interlayer for ultra-sensitive organic transistor humidity sensors. *ACS Appl. Mater. Interfaces* 5, 8591–8596.

Pohle R., Simon E., Fleischer M., Meixner H., Frerichs H.-P., Lehmann M., Verhoeven H. (2003) Realisation of a new sensor concept: Improved CCFET and SGFET type gas sensors in hybrid flip-chip technology. In: *Proceedings of the 12th international conference on solid-state sensors, actuators and microsystems, TRANSDUCERS 2003*, June 8–12, Boston Massachusetts, USA, pp. 135–138.

Polk B.J., Potje-Kamloth K., Josowicz M., Janata J. (2002) Role of protonic and charge transfer doping in solid-state polyaniline. *J. Phys. Chem.* 106, 11457–11462.

Potje-Kamloth K. (2010) Gas sensing with conducting polymers, In: Cosnier S., Karyakin A. (eds.) *Electropolymerization: Concepts, Materials and Applications*. Wiley-VCH Verlag, New York, pp. 153–171.

Ray S.J. (2015) Humidity sensor using a single molecular transistor. *J. Appl. Phys.* 118, 044307.

Said E., Larsson O., Berggren M., Crispin X. (2008) Effects of the ionic currents in electrolyte-gated organic field-effect transistors. *Adv. Funct. Mater.* 18, 3529–3536.

Salas-Villasenor A.L., Mejia I., Hovarth J., Alshareef H.N., Cha D.K., Ramirez-Bon R., Gnade B.E., Quevedo-Lopez M.A. (2010) Impact of gate dielectric in carrier mobility in low temperature chalcogenide thin film transistors for flexible electronics. *Electrochem. Solid-State Lett.* 13 (9), H313–H316.

Salaün A.-C., Kotb H.M., Mohammed-Brahim T., Le Bihan F., Lhermite H., Bendriaa F. (2005) Suspended-gate thin film transistor as highly sensitive humidity sensor. *Proc. SPIE* 5836, 231–238.

Seon J.-B., Lee S., Kim J.M., Jeong H.-D. (2009) Spin-coated CdS thin films for *n*-channel thin film transistors. *Chem. Mater* 21, 604–611.

Someya T., Dodabalapur A., Huang J., See K.C., Katz H.E. (2010) Chemical and physical sensing by organic field-effect transistors and related devices. *Adv. Mater* 22, 3799–3811.

Song S.-H., Yang H.-H., Han C.-H., Ko S.-D., Lee S.-H., Yoon J.-B. (2012) Metal-oxide-semiconductor field effect transistor humidity sensor using surface conductance. *Appl. Phys. Lett.* 100, 101603.

Stegmeier S., Fleischer M., Hauptmann P. (2010) Thermally activated platinum as VOC sensing material for work function type gas sensors. *Sens. Actuators B* 144, 418–424.

Subbarao N.V.V., Gedda M., Iyer P.K., Goswami D.K. (2016) Organic field-effect transistors as high performance humidity sensors with rapid response, recovery time and remarkable ambient stability. *Org. Electron.* 32, 169–178.

Takagi A., Nomura K., Ohta H., Yanagi H., Kamiya T., Hirano M., Hosono H. (2005) Carrier transport and electronic structure in amorphous oxide semiconductor, *a*-InGaZnO$_4$. *Thin Solid Films* 486, 38–41.

Tarabella G., Mohammadi F.M., Coppedè N., Barbero F., Iannotta S., Santato C., Cicoira F. (2013) New opportunities for organic electronics and bioelectronics: Ions in action. *Chem. Sci.* 4, 1395–1409.

Torsi L. (2000) Novel applications of organic based thin film transistors. *Microelectron. Reliab.* 40, 779–782.

Torsi L. (2012) Phospholipid film in electrolyte-gated organic field-effect transistors. *Org. Electron.* 13, 638–644.

Torsi L., Dodabalapur A. (2005) Organic thin-film transistors as plastic analytical sensors. *Anal. Chem.* 77, 381–387.

Torsi L., Dodabalapur A., Cioffi N., Sabbatini L., Zambonin P.G. (2001) NTCDA organic thin-film-transistor as humidity sensor: Weaknesses and strengths. *Sens. Actuators B* 77, 7–11.

Torsi L., Dodabalapur A., Sabbatini L., Zambonin P.G. (2000) Multi-parameter gas sensors based on organic thin-film-transistors. *Sens. Actuators B* 67, 312–316.

Torsi L., Lovinger A.J., Crone B., Someya T., Dodabalapur A., Katz H.E., Gelperin A. (2002) Correlation between oligothiophene thin film transistor morphology and vapor responses. *J. Phys. Chem. B* 106, 12563–12568.

Torsi L., Tafuri A., Cioffi N., Gallazzi M.C., Sassella A., Sabbatini L., Zambonin P.G. (2003b) Regioregular polythiophene field-effect transistors employed as chemical sensors. *Sens. Actuators B* 93, 257–262.

Torsi L., Tanese M., Cioffi N., Gallazzi M., Sabbatini L., Zambonin P., Raos G., Meille S., Giangregorio M.M. (2003a) Side-chain role in chemically sensing conducting polymer field-effect transistors. *J. Phys. Chem. B* 107, 7589–7594.

Voorthuyzen J.A., Keskin K., Bergveld P. (1987) Investigations of the surface conductivity of silicon dioxide and methods to reduce it. *Surf. Sci.* 187, 201–211.

Wang L., Fine D., Sharma D., Torsi L., Dodabalapur A. (2006) Nanoscale organic and polymeric field-effect transistors as chemical sensors. *Anal. Bioanal. Chem.* 384, 310–321.

Wu S., Guiheng G., Xue Z., Ge F., Zhang G., Lu H., Qiu L. (2017) Organic field-effect transistors with macroporous semiconductor films as high-performance humidity sensors. *ACS Appl. Mater. Interfaces* 9, 14974–14982.

Yamashita Y. (2009) Organic semiconductors for organic field-effect transistors. *Sci. Technol. Adv. Mater.* 10 (2), 024313.

Yamazoe N., Shimizu Y. (1986) humidity sensors. Principles and applications. *Sens. Actuators* 10, 379–398.

Yang D.J., Whitfield G.C., Cho N.G., Cho P.-S., Kim I.-D., Saltsburg H.M., Tuller H.L. (2012) Amorphous InGaZnO$_4$ films: Gas sensor response and stability. *Sens. Actuators B* 171–172, 1166–1171.

Yuan J.D., Dhoot A.S., Namdas E.B., Coates N.E., Heeney M., McCulloch I., Moses D., Heeger A.J. (2007) Electrochemical doping in electrolyte-gated polymer transistors. *J. Am. Chem. Soc.* 129, 14367–14371.

Zhu Z.T., Mason J.T., Dieckmann R., Malliaras G.G. (2002) Humidity sensors based on pentacene thin-film transistors. *Appl. Phys. Lett.* 81, 4643–4645.

Zimmer M., Burgmair M., Scharnagel K., Karthigeyan A., Goll T., Eisele I. (2001) Gold and platinum as ozone sensing layer in work-function gas sensors. *Sens. Actuators B* 80, 174–178.

17 Hetero-Junction-Based Humidity Sensors

17.1 SCHOTTKY BARRIER-BASED HUMIDITY SENSORS

17.1.1 Principles of Operation

The Schottky diode is a semiconductor-based diode that consists of a metal–semiconductor junction instead of a semiconductor–semiconductor p–n junction as in conventional diodes (Rhoderick 1978; Sharma 1984). Schottky diodes use the metal–semiconductor junction as the Schottky barrier. Since the barrier height is lower in metal–semiconductor junctions than in conventional p–n junctions, Schottky diodes have lower forward voltage drop. Based on the thermionic field emission conduction mechanism of the Schottky diode, the I–V characteristic of the diode for forward bias voltage, exceeding 3 kT, is given by (Rhoderick 1978)

$$I = I_{sat} \cdot exp\left(\frac{qV}{nkT}\right) \quad (17.1)$$

where k is the Boltzmann constant, T is the temperature in Kelvin, n is the ideality factor, and I_{SAT} is the saturation current, defined as (Rhoderick 1978)

$$I_{sat} = SA^{**}T^2 \cdot exp\left(-\frac{q\varphi_b}{kT}\right) \quad (17.2)$$

where S is the area of the junction, A^{**} is the effective Richardson constant, and Φ_b is the barrier height. The ideality factor can be extrapolated from Equation 17.1, and the barrier height can be calculated from Equation 17.2.

As it follows from Equation 17.2, these diodes can be used as humidity sensors if the water vapor changes the electrical characteristics of the Schottky diode, i.e., the potential barriers at the Me-semiconductor interface. As it was shown earlier, these changes are possible. (Trinchi et al. 2004; Potje-Kamloth 2008). It is known that the resulting dipole layer can produce an abrupt rise in the surface potential at the metal–oxide and metal-semiconductor interfaces (read Chapters 16 and 18). It follows that the development of a humidity sensor should ensure that water vapor is freely accessible to the interface. At present, methods to solve this problem include approaches such as (1) reducing the thickness of the metal contact until the metal film becomes transparent to water vapor, (2) perforating the metal film by photolithography methods or other methods suitable for this, and (3) formation of a Schottky barrier (SB) on the underside of a humidity-sensitive semiconductor. A cross-section of the last variant is shown in Figure 17.1.

Schottky diodes are advantageous for gas-sensing and humidity-sensing applications due to the simple electrical circuitry required for their operation. The most studied are Schottky diode-hydrogen sensors (Trinchi et al. 2004; Chen and Chou 2004; Lin et al. 2004). Sensors of humidity in the development of these sensors are given much less attention (Traversa and Bearzotti 1995a, 1995b; Salehi et al. 2006).

17.1.2 Sensor Performance

First developments of Schottky diode-based humidity sensors were completed by Tsurumi et al. (1987). The diodes

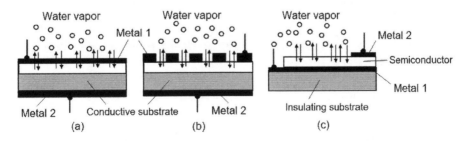

FIGURE 17.1 Cross-sectional view of a typical geometric arrangements (a, b, c) of a Schottky barrier diode used for chemical sensing applications. The Schottky diode comprises two contacts, or junction areas, by definition: (1) between Metal 1 and the semiconductor, forming the Schottky barrier junction, which is the origin of the sensor signal, and (2) between Metal 2 and the semiconductor, forming the ohmic contact, which has to be inactive and hence should not contribute to the sensor response. In the case of (a), the Metal 1 must be permeable to water vapor (i.e., it must be thin). If there is a perforation (case [(b)]), the metal film can be thick.

were fabricated using approach (1), indicated before. The humidity sensitivity of the rectifying *I-V* characteristics of Palladium (Pd)/zinc oxide (ZnO) diodes fabricated by RF (radio frequencies) sputtering has been reported. With increasing humidity, the current under forward bias increased remarkably with physisorption of water at the junction interface (Tsurumi et al. 1987). Later, sensing elements of perovskite-type oxides in contact with a metal needle were investigated by Lukaszewicz (1991). A diode-type *I-V* dependence was clearly observed on application of Al or Fe needles to $LaCrO_3$-sintered bodies (see Figure 17.2). Operating at temperatures between 20°C and 30°C, an increase in the current flow at the same potential was recorded. This was interpreted in terms of a lowering of the potential barrier at the metal–oxide junction due to water adsorption, which facilitated electron transportation through the barrier and increased the current values (Lukaszewicz 1991).

Traversa and Bearzotti (1995a, 1995b), based on the results presented by Tsurumi et al. (1987), carried out several tests of ZnO-dense pellets. It was observed that the introduction of a gold SB made ZnO-dense pellets sensitive to relative humidity (RH) at room temperature (Figure 17.3). The dense ZnO pellets with two ohmic electrodes were insensitive to RH. Au/ZnO samples showed rectifying *I-V* characteristics. The RH sensitivity was bias dependent in the forward region. A maximum sensitivity (I/I_{RH}) of approximately three was measured at 85% RH and 10 V forward bias. A nearly linear forward current versus RH variation was observed in the RH range of 4%–85%. Traversa and Bearzotti (1995b)

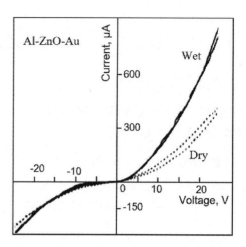

FIGURE 17.3 Dark I-V characteristics of ZnO pellets with one Al and one Au electrode, in dry and wet environments at room temperature. (Reprinted with permission from Traversa, E., Ceramic sensors for humidity detection: The state-of-the-art and future developments, *Sens. Actuators B*, 23, 135–156, 1995. Copyright 1995, Elsevier.)

have found that the behavior of Au/ZnO is closer to that of *p–n* hetero-contacts. The presence of a rectifying *I-V* characteristic, coming from a SB in this case, or from *p–n* junctions, is responsible for the humidity sensitivity. The rectifying character may be related to the presence of surface/interface states. The neutralization of these states by means of physisorbed water may be the cause of the forward-current enhancement with RH. The linear relation of the forward current with RH, which shows that there is a direct proportion between forward current and the number of physisorbed water molecules, is consistent with the proposed humidity-detection mechanism of metal/semiconductor Schottky-barrier interfaces (Traversa and Bearzotti 1995a, 1995b). However, one should note that the current versus RH sensitivity of Au/ZnO, as well as other tested Schottky barriers, was very low for practical applications. However, Traversa and Bearzotti (1995b) believe that it is important to understand the behavior of Schottky diode-based humidity sensors, because it is possible to foresee an increase in the sensitivity of electronic-type humidity sensors by using a Schottky-barrier electrode, in a similar way to that observed for gas sensors.

It is important that, in the case of semiconductor materials such as GaAs and Si, the effect of the humidity influence on the characteristics of Schottky diodes is much more pronounced (see Figure 17.4). Salehi et al. (2006) have studied Pd/porous GaAs and Pd/GaAs Schottky contacts. The pore formation in GaAs was achieved by anodization, which involved immersing the wafer in HF

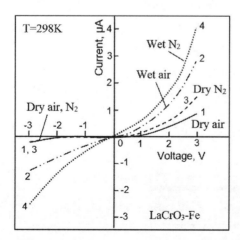

FIGURE 17.2 The current-voltage curves for the $LaCrO_3$-Fe sensor measured at 25°C in dry (1,3) and wet (100% relative humidity [RH]) (2,4) air (1,2) and nitrogen (3,4). (Reprinted with permission from Lukaszewicz, J.P., Diode–type humidity sensor using perovskite-type oxides operable at room temperature, *Sens. Actuators B*, 4, 227–232, 1991. Copyright 1991, Elsevier.)

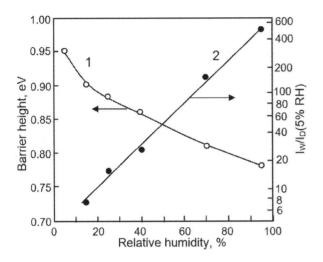

FIGURE 17.4 Humidity influence on the parameters of GaAs-based Schottky diodes: 1-changes in the barrier height of Pd/porous GaAs sample as a function of relative humidity (RH); 2- changes in the sensor response (i.e., $S = I_w/I_D$ of Pd/porous GaAs sample as a function of relative humidity operated at room temperature). The measurements were carried out at 0.2 V of direct inclusion. I_D met the measurements with 5% RH. (Data extracted from Salehi A. et al., (2006) Highly sensitive humidity sensor using Pd/porous GaAs Schottky contact. *IEEE Sensors J.*, 6, 1415–1421.)

acid and then passing an anodic current through it. It was found that the porosity on GaAs wafers promoted the sensing properties of the contact. Pd/porous GaAs contacts, operating at room temperature, demonstrated high sensitivity to air humidity. On the contrary, the Pd/GaAs contacts exhibited negligible sensitivity to RH. When humidity changed from 5% RH to 95% RH, the saturation current, I_{sat}, of *I-V* characteristics of Pd/porous GaAs contacts increased from $8.5 \cdot 10^{-10}$ ampere to $3.0 \cdot 10^{-7}$ ampere. This is a more than two orders of magnitude increase in a saturation current compared to dry conditions. This corresponds to a decrease in the height of the potential barrier by 0.16 eV. In addition, the Pd/porous GaAs sensor showed a fast response time of 2 seconds. The total recovery time of the Pd/porous GaAs sample operated at room temperature was found to be around 10 seconds. According to Salehi et al. (2006), the observed changes in the properties of Pd/porous GaAs contacts are associated with the adsorption and condensation of water vapor in the pores of GaAs. The same situation was observed in the metal/porous silicon/*n*-Si structures (Yarkin 2003). The dissociation of water with Pd is also possible in the Pd/porous GaAs contacts. In light of the above, it becomes clear that the porosity of the humidity-sensitive Schottky contact is important in obtaining quick and high response of such sensors.

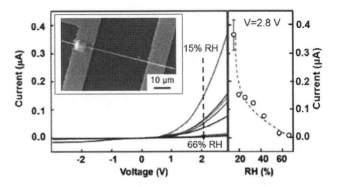

FIGURE 17.5 *I–V* curves of the Pt-ZnO humidity sensor at different air humidity and SEM image (*inset*) of a Schottky-contact sensor. (Reprinted with permission from Yu, R. et al., Piezotronic effect enhanced performance of Schottky-contacted optical, gas, chemical and biological nano sensors, *Nano Energy*, 14, 312–339, 2015. Copyright 2015, Elsevier.)

A clearly pronounced effect associated with the SB was also observed by Hu et al. (2014) and Yu et al. (2015) when using ZnO individual nanowires (NWs) (see Figure 17.5). The Schottky-contacted sensor was fabricated by forming a SB at local ZnO–Pt interface. An important point of these studies is that, unlike the results obtained using granulated ZnO (Traversa and Bearzotti 1995a, 1995b), when studying the SB formed on ZnO at individual nanowires, a decrease in the current (Figure 17.5) was observed, rather than its increase with increasing a humidity level. Apparently, this is exactly the actual behavior of the SB of Me-ZnO in the humid atmosphere, since in this case there is no effect of intergrain contacts, masking the real change in the height of the potential barrier in a humid atmosphere.

Herran et al. (2012), while studying the Ag-ZnO contacts (see Figure 17.6), observed the same effect. As to the nature of this behavior, unfortunately Hu et al. (2014) and Yu et al. (2015) do not offer an acceptable explanation.

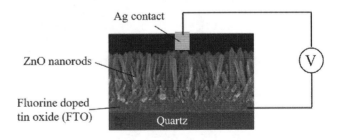

FIGURE 17.6 Scheme of the device and morphology of the ZnO nanorod. The ZnO nanorods were synthesized by electrochemical techniques. The response time was around 5 seconds. (Reprinted from Herran, J. et al., Schottky diodes based on electrodeposited ZnO nanorod arrays for humidity sensing at room temperature, *Sens. Actuators B*, 174, 274–278, 2012. Copyright 2012, Elsevier.)

Herran et al. (2012) and Hu et al. (2014) attempted to relate the decrease in current with the adsorption of water molecules on the surface and the decrease in the carrier concentrations in the ZnO nanowire. But this does not agree with the generally accepted idea of the behavior of water on the surface of semiconductors of *n*-type conductivity.

As for other studies (Hsu et al. 2013, 2017), in which the authors also claim that they use Schottky contact with individual ZnO nanowires, an analysis of these results suggests that the observed effects are not related to the SBs' behavior.

17.2 *p–n* HETERO-CONTACT-TYPE HUMIDITY SENSORS

p–n junctions formed by different metal oxides are another type of hetero-junction-based humidity sensors. The possibility to control humidity with the help of such devices was discovered by Kawakami and Yanagida (1979) in the study of hetero-contact between *p*-type NiO and *n*-type ZnO. Later, the same effect was found in the study of *p–n* hetero-contacts between bulk sintered CuO and ZnO (Toyoshima et al. 1983; Nakamura et al. 1985; Yoo et al. 1992). However, the mechanical contact used in these studies does not provide the necessary reliability; therefore, to overcome this drawback, Ushio et al. (1993) produced open *p–n* hetero-junctions of CuO/ZnO, using the principles of thin-film technology. The process of manufacturing such structures is shown in Figure 17.7. Using the open junction configuration significantly increased the sensitivity of the sensors (Figure 17.8).

The testing of the CuO/ZnO structures showed that increasing humidity results in a substantial increase in the current, flowing across the junction under a forward bias. The result is a highly rectifying device at high levels of RH, with polarity coincident with an ordinary *p–n* junction. These devices operate at room temperature. The observed behavior of hetero-junctions in a humid atmosphere was explained using the mechanism based on electrolysis of adsorbed water, taking place at *p–n* contact sites (Yanagida 1990). The amount of hydrolysis increases at higher humidity, because the higher the humidity is, the higher is the amount of water adsorption. In the proximity of *p–n* junctions, the electron holes are injected from the *p*-type semiconductor into the adsorbed water molecules, giving rise to protons in the adsorbed water phase. The positive charge is liberated at the surface of the *n*-type semiconductor, thereby resulting in electrolysis of the adsorbed water. The process of electrolysis provides a cleaning process during work, with a self-cleaning/self-recovery mechanism. In this respect, the *p–n* hetero-junctions can be regarded as intelligent materials (Yanagida 1988). However, no definitive proof is given in the relevant literature to confirm this mechanism.

Another humidity-sensitive hetero-structure, lanthanum copper oxide (La_2CuO_4)/zinc oxide (ZnO), has been investigated by Traversa et al. (1994). The heterojunctions

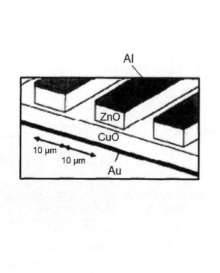
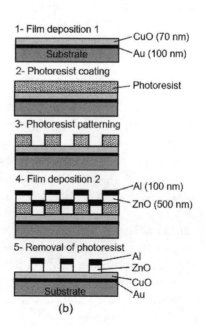

FIGURE 17.7 (a) Structure of a thin-film open junction; (b) A protective layer of SiO_2; (c) Fabrication process of an open junction. (Reprinted from Ushio, Y. et al., Fabrication of thin film CuO/ZnO heterojunction and its humidity–sensing properties, *Sens. Actuators B*, 12, 135–139, 1993. Copyright 1993, Elsevier.)

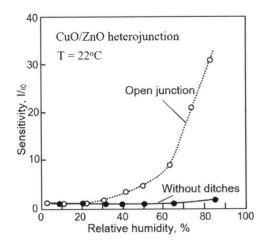

FIGURE 17.8 Comparison of humidity sensitivity between an open junction and a nonpatterned junction. Applied voltage is 2 V. (Reprinted from Ushio, Y. et al., Fabrication of thin film CuO/ZnO heterojunction and its humidity–sensing properties, *Sens. Actuators B*, 12, 135–139, 1993. Copyright 1993, Elsevier.)

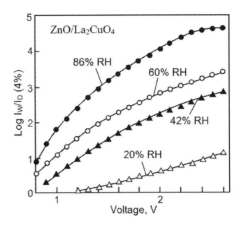

FIGURE 17.10 The bias-dependence of the relative humidity (RH) sensitivity ($I/I_{4\%}$) of ZnO/La$_2$CuO$_4$ heterocontacts at different relative humidities, evaluated from *I-V* curves. (Reprinted with permission from Traversa, E. and Bearzotti, A., A novel humidity detection mechanism for ZnO dense pellets, *Sens. Actuators B*, 23, 181–186, 1995. Copyright 1995, Elsevier.)

were prepared by mechanically pressing sintered discs of the two oxides. Figure 17.9 shows the *I-V* curves of La$_2$CuO$_4$/ZnO heterocontacts measured at different RH values, which are typical for a *p–n* diode behavior. The rectifying character is enhanced with RH. Very high sensitivity, expressed as $I/I_{4\%}$, where $I_{4\%}$ is the current, recorded at 4% RH, was measured for the samples, up to approximately five orders of magnitude at 2.5 V and 86% RH. The RH sensitivity was bias dependent, and very large increases in the current were recorded only with forward bias (Figure 17.10), while with reverse bias, the samples were insensitive to RH. The response time

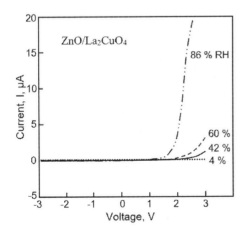

FIGURE 17.9 Dark *I-V* characteristics of ZnO/La$_2$CuO$_4$ heterocontacts at different relative humidity (RH) values at 40°C. (Reprinted with permission from Traversa, E. and Bearzotti, A. A novel humidity detection mechanism for ZnO dense pellets, *Sens. Actuators B*, 23, 181–186, 1995. Copyright 1995, Elsevier.)

at 1 V dc was evaluated at less than 20 seconds during adsorption and less than 10 seconds during desorption. The response time was bias dependent, too: At 0.5 V, it was less than 5 seconds during absorption and approximately 1 second during desorption. This fast response time probably indicates that the RH sensitivity relies on water physisorption.

Traversa (1995) established that the barrier at the La$_2$CuO$_4$/ZnO interface is not a physical diffusion barrier, but a chemical-like barrier due to the presence of a high density of interface states of external origin. These results are largely consistent with those obtained for CuO/ZnO thin films (Ushio et al. 1994). For these samples, it was determined that the presence of interface states is fundamental for their sensing properties. It is likely that, at high RH, the saturation of the original interface states by physisorbed water leads to the release of trapped electrons, resulting in an increase in the forward current. This phenomenon may be assisted by water electrolysis, allowed by a continuous supply of water molecules.

A humidity-sensitive effect was also observed in the *p*-ZnO/*n*-Si hetero-structures by Majumdar and Banerji (2009). The thin-film hetero-junction shown in Figure 17.11 was fabricated on an *n*-type Si substrate by pulsed laser deposition (PLD). As seen in Figure 17.11, the *p*-ZnO/n-Si hetero-structures had a diode-like characteristic similar to a conventional *p–n* junction (Sze 1981, 2001). Majumdar and Banerji (2009) established that the nitrogen doped *p*-ZnO/*n*-Si thin film heterojunction showed almost linear variation of the resistance with RH in the range of 30%–97% at normal

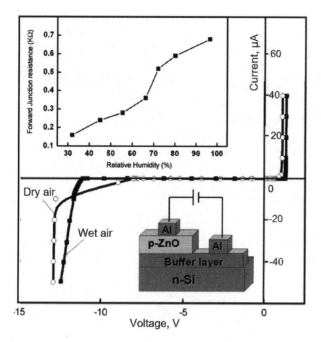

FIGURE 17.11 Current–voltage characteristics of the sensor device in the presence and absence of moisture. *Inset, bottom*: Schematic representation of the device. *Inset, upper left*: Variation of the change in the device resistance as a function of relative humidity. (Reprinted with permission from Majumdar, S. and Banerji, P., Moisture sensitivity of p–ZnO/n–Si heterostructure, *Sens. Actuators B*, 140, 134–138, 2009. Copyright 2009, Elsevier.)

the base of reactions proposed by Bai et al. (2008) and Heiland and Kohl (1988):

$$(H_2O)gas + 2Zn_{Zn} + O_O = 2(Zn_{Zn}^+ - OH^-) + V_O^{2+} + 2e^- \quad (17.3)$$

According to this explanation, an electric field is developed between the positively charged acceptor ions and the negatively charged hydroxyl ions (OH$^-$) on the surface. The acceptors are more and more compensated by donors like (OH$^-$) ions. The (OH$^-$) ions donate an electron to the system to get neutralized according to Equation 17.3, and this electron compensates the hole of the nitrogen-doped ZnO. As the hole concentration is reduced in the presence of moisture so that the current conduction is also reduced for the p-type doped film, hence $\Delta V/\Delta I$ increases from 0.17 to 0.7 kΩ. It is important that the series resistance in p-ZnO (R_p) decreases, but the series resistance in n-Si (R_n) remains the same with adsorption of moisture. Hence, the junction voltage (V) decreases, and this decrease of V lowers the level of current injection so that the injection occurs more slowly with increased bias (Streetman and Banerjee 2001). The reduction of hole concentration causes the increase of depletion width, hence the space charge-limited current (SCLC) is observed at higher voltage. It is also found that in the presence of moisture, there was a slight decrease in the potential barrier.

atmospheric temperature and pressure (Figure 17.11), with a response and recovery time of 12 seconds and 36 seconds, respectively.

The sensing mechanism of p-ZnO/n-Si heterostructures was explained by Majumdar and Banerji (2009) on

Other diode-type humidity sensors used the junction between different carbon-type semiconductors (see insertion in Figure 17.12b) (Lukaszewicz 1992a, 1992b). To obtain thin carbon layers, the spray pyrolysis method was applied. The obtained films were subjected to carbonization at different temperature.

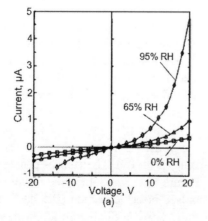
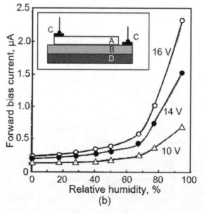

FIGURE 17.12 (a) Current–voltage curves of the low-temperature (LT)-high-temperature (HT) diode measured at 25°C in dry and wet air; (b) Forward bias current vs relative humidity (RH) function for LT-HT diode at 25°C. *Insert*: Schematic view of the diodes tested: *A*, LT carbon film; *B*, HT carbon film; *C*, ohmic contacts; *D*, substrate: *LT*, low-temperature carbon, slightly conductive; *HT*, high-temperature carbon with high electrical conductivity. A protective layer of SiO$_2$. (Reprinted with permission from Lukaszewicz, J.P., An application of carbon–type semiconductors for the construction of a humidity–sensitive diode, *Sens. Actuators B*, 6, 61–65, 1992. Copyright 1992, Elsevier.)

Operating characteristics of these sensors are shown in Figure 17.12. According to Lukaszewicz (1992a), a humidity-sensing effect takes place due to the presence of water molecules at the interface of the films deposited at high temperature (HT) and low temperature (LT). Considering the need for diffusion penetration of water vapor to the HT-LT interface to reduce the response time, it is proposed to make LT carbon films more "transparent" for water molecules.

17.3 SUMMARY

To summarize, one could conclude that Schottky barrier and p–n junction-type humidity sensors, as well as field-effect transistor - and thin-film transistor-based sensors, may have prospects for use. In addition, this type of sensors is most suitable for microminiaturization.

However, if we look at the characteristics of the metal oxide-based Schottky structures and p–n structures, then the prospects for using such structures for humidity control seem doubtful, because in fact there are the same metal oxide–resistive sensors, but with additional rectifying contacts. At the same time, the rectifying contact itself, as it seems, in most cases does not participate in the humidity-sensitive effect. Otherwise, the humidity should have an effect on both direct and reverse I-V characteristics. However, this does not happen. Apparently, in the opposite direction of the switching, the resistance of the contacts is substantially greater than the resistance of the metal oxide forming the given structure; therefore, the change in the metal oxide resistance under the influence of moisture does not affect the overall resistance of the diode.

Only with direct connection, when the contact resistance drops and approaches the series resistance of the diode, is it possible to create conditions under which the interaction of the metal oxide with water vapor can have a significant effect on the I-V characteristics. Therefore, the absence of a noticeable interest in developing such sensors should be considered quite natural. At the present time, based on SBs, only hydrogen sensors are being developed that use hydrogen and palladium (Korotcenkov 2013).

As for SBs and p–n junctions on the base of conventional porous semiconductors, such as Si and GaAs, the development of humidity sensors based on them seems more promising. However, in this case, the inadequate inefficient stability of the characteristics of these sensors, associated with the oxidation of semiconductors when they are in the oxygen atmosphere, comes to the fore. Besides, when using porous semiconductors, an additional uncertainty associated with the technology of porosification appears, which does not provide the required reproducibility of the nanostructure of porous semiconductor.

It should also be noted that, due to the lack of detailed studies, the understanding of the mechanism of the barrier formation at the metal-semiconductor interface and other hetero-structures in different atmospheres is still far from complete. Some generalization with respect to the results of the investigation of the gas-sensitive properties of Schottky contacts and p–n heterostructures can be found only in Sharma (1984), Fonash and Li (1986), Miller et al. (2014), Korotcenkov and Cho (2010, 2017), and Lee (2017). As a result, there are significant problems with the explanation of the nature of phenomena observed in the sensors based on them.

REFERENCES

Bai Z., Xie C., Hub M., Zhang S., Zeng D. (2008) Effect of humidity on the gas sensing property of the tetrapod–shaped ZnO nanopowder sensor. *Mater. Sci. Eng. B* 149, 12–17.

Chen H.I., Chou Y.I. (2004) Evaluation of the perfection of the Pd–InP Schottky interface from the energy viewpoint of hydrogen adsorbates. *Semicond. Sci. Tech.* 19, 39–44.

Fonash S.J., Li Z. (1986) Schottky–barrier diode and metal–oxide–semiconductor capacitor gas sensors. Comparison and performance, In: Schuetzle D. and Hammerle R. (eds.) *Fundamentals and Applications of Chemical Sensors, ACS Symposium Series*, Vol. 309. ACS, Washington DC, pp. 177–202.

Heiland G., Kohl D. (1988) Physical and chemical aspects of oxidic semiconductor gas sensors, In: Seiyama T. (ed.) *Chemical Sensor Technology*, Kodansha, Tokyo, Japan, Chapter 2, pp. 15–38.

Herran J., Fernandez I., Tena–Zaera R., Ochoteco E., Cabanero G., and Grande H. (2012) Schottky diodes based on electrodeposited ZnO nanorod arrays for humidity sensing at room temperature *Sens. Actuators B* 174, 274–278.

Hsu C.L., Li H.H., Hsueh T.J. (2013) Water– and humidity-enhanced UV detector by using p-type La–Doped ZnO nanowires on flexible polyimide substrate. *ACS Appl. Mater. Inter* 5, 11142–11151.

Hsu C.L., Su I.L., Hsueh T.J. (2017) Tunable Schottky contact humidity sensor based on S–doped ZnO nanowires on flexible PET substrate with piezotronic effect. *J. Alloys Compd* 705, 722–733.

Hu G., Zhou R., Yu R., Dong L., Pan C., Wang Z.L. (2014) Piezotronic effect enhanced Schottky–contact ZnO micro/nanowire humidity sensors. *Nano Res.* 7 (7), 1083–1091.

Kawakami K., Yanagida H. (1979) Effects of water vapor on the electrical conductivity of the interface of semiconductor ceramic–ceramic contacts. *Z. Cerant. Soc. Jpn.* 87, 112–115.

Korotcenkov G. (2013) *Handbook of Gas Sensor Materials*, Vol. 1. Springer, New York.

Korotcenkov G., Cho B.K. (2010) Porous semiconductors: Advanced material for gas sensor applications. *Crit. Rev. Sol. St. Mater. Sci.* 35 (1), 1–37.

Korotcenkov G., Cho B.K. (2017) Metal oxide based composites in conductometric gas sensors: Achievements and challenges. *Sens. Actuators B* 244, 182–210.

Lee S.P. (2017) Electrodes for semiconductor gas sensors. *Sensors* 17, 683.

Lin K.W., Chen H.I., Chuang H.M., Chen C.Y., Lu C.T., Cheng C.C., Liu W.C. (2004) Characteristics of Pd/InGaP Schottky diodes hydrogen sensors. *IEEE Sens. J.* 4 (1), 72–79.

Lukaszewicz J.P. (1991) Diode–type humidity sensor using perovskite-type oxides operable at room temperature. *Sens. Actuators B, 4* (1991) 227–232.

Lukaszewicz J.P. (1992a) An application of carbon–type semiconductors for the construction of a humidity-sensitive diode. *Sens. Actuators B* 6, 61–65.

Lukaszewicz J.P. (1992b) Humidity sensitive diode fabricated by means of laser beam, In: *Tech. Digest, 4th Int. Meet. Chem. Sensors,* Tokyo, Japan, September 13–17, pp. 294–297.

Majumdar S., Banerji P. (2009) Moisture sensitivity of p–ZnO/n–Si heterostructure. *Sens. Actuators B* 140, 134–138.

Miller D.R., Akbar S.A., Morris P.A. (2014) Nanoscale metal oxide–based hetero–junctions for gas sensing: A review. *Sens. Actuators B* 204, 250–272.

Nakamura Y., Ikejiri M., Miyayama M., Koumoto K., Yanagida H. (1985) The current–voltage characteristics of CuO/ZnO heterojunctions. *Z. Chem. Soc. Jpn.* 1985, 1154–1159.

Potje–Kamloth K. (2008) Semiconductor junction gas sensors. *Chem. Rev.* 108, 367–399.

Rhoderick E.H. (1978) *Metal–Semiconductor Contacts.* Clarendon Press, Oxford, UK.

Salehi A., Nikfarjam A., Kalantari D.J. (2006) Highly sensitive humidity sensor using Pd/porous GaAs Schottky contact. *IEEE Sensors J.* 6 (6), 1415–1421.

Sharma B. (ed.) (1984) *Metal–Semiconductor Schottky Barrier Junction and Their Application.* Plenum, New York.

Streetman B.G., Banerjee S. (2001) *Solid State Electronic Devices.* Pearson Education Limited, New Delhi, India, Chapter 5, p. 217.

Sze S. (2002) *Semiconductor Devices.* John Wiley & Sons, New York.

Sze S.M. (1981) *Physics of Semiconductor Devices.* John Wiley & Sons, New York.

Toyoshima Y., Miyayama M., Yanagida H., Koumoto K. (1983) Effect of relative humidity on current–voltage characteristics of Li–doped CuO/ZnO junction. *Jpn. Z. Appl. Phys. 22,* 1933–1933.

Traversa E. (1995) Ceramic sensors for humidity detection: The state-of-the-art and future developments. *Sens. Actuators B* 23, 135–156.

Traversa E., Bearzotti A. (1995a) Humidity sensitive electrical properties of dense ZnO with non-Ohmic electrode. *J. Ceram. Soc. Jpn.* 103 (1193), 11–15.

Traversa E., Bearzotti A. (1995b) A novel humidity detection mechanism for ZnO dense pellets. *Sens. Actuators B* 23, 181–186.

Traversa E., Bianco A., Montesperelli G., Gusmano O., Bearzotti A., Miyayama M., Yanagida H. (1994) ZnO/La_2CuO_4 hetero–contacts as humidity sensors, In: Nair I.C.M., Lloydand I.K., Bhalla A.S. (eds.), *Ceramic Transactions, Vol. 43, Ferroic Materials." Design, Preparation, and Characteristics.* American Ceramic Society, Westerville, OH, pp. 375–384.

Trinchi A., Wlodarski W., Li V.X. (2004) Hydrogen sensitive Ga_2O_3 Schottky diode sensor based on SiC. *Sens. Actuators B* 100, 94–98.

Tsurumi S., Mogi K., Noda J. (1987) Humidity sensors of Pd–ZnO diodes, In: *Tech. Digest, 4th Int. Conf. Solid–State Sensors and Actuators (Transducers '87),* Tokyo, Japan, June 7–10, pp. 661–664.

Ushio Y., Miyayama M., Yanagida H. (1993) Fabrication of thin film CuO/ZnO heterojunction and its humidity-sensing properties. *Sens. Actuators B* 12, 135–139.

Ushio Y., Miyayama M., Yanagida H. (1994) Effects of interface states on gas–sensing properties of a CuO/ZnO thin–film heterojunction. *Sens. Actuators B* 17, 221–226.

Yanagida H. (1988) Intelligent materials—A new frontier. *Angew. Chem.* 100, 1443–1446.

Yanagida H. (1990) Intelligent ceramics. *Ferroelectrics* 102, 251–257.

Yarkin D.G. (2003) Impedance of humidity sensitive metal/porous silicon/n–Si structure. *Sens. Actuators A* 107 (1), 1–6.

Yoo D.J., Song K.H., Park S.J. (1992) Humidity sensing characteristics of CuO/ZnO heterojunctions. In: *Tech. Digest, 4th Int. Meet. Chem. Sensors,* Tokyo, Japan, September. 13–17, pp. 108–111.

Yu R., Niu S., Pan C., Wang Z.L. (2015) Piezotronic effect enhanced performance of Schottky–contacted optical, gas, chemical and biological nano sensors. *Nano Energy* 14, 312–339.

18 Kelvin Probe as a Humidity Sensor

The operation principle of Kelvin probe-based humidity sensors is also based on changing the work function of humidity-sensitive material under the influence of water vapour, and therefore, these sensors could be included in the group of field-effect transistor-based humidity sensors described in previous Chapter 16. Moreover, their operation principle is based on the direct measurement of the change in the work function, and not on the measurement of other parameters of the material associated with the work function, as is done in field-effect transistors or Schottky barriers. Therefore, Kelvin probe-based humidity sensors are highlighted as a separate type of humidity sensors.

18.1 WORK FUNCTION

In solid-state physics, the *work function* (Φ) of a material is defined as the minimum energy required by an electron to escape from the Fermi level of the bulk material to the vacuum level (Streetman 1990).

$$\Phi = -q_\theta - E_F \quad (18.1)$$

The Fermi energy (E_F) level corresponds to the electrochemical potential of an electron. At a given temperature, the Fermi level is the highest occupied energy level in the band gap. Thus, the energy difference between E_F and the vacuum (Volta) energy level ($-q_\theta$) is the work function (Janata and Josowicz 2002; Anh et al. 2004). In real situations, work functions Φ of semiconductors contain three contributions: the energy difference between the Fermi level and conduction band in the bulk $(E_C - E_F)_b$, band bending qV_S, and electron affinity χ (see Figure 18.1):

$$\Phi = (E_C - E_F)_b + qV_S + \chi \quad (18.2)$$

In principle, all three contributions may, in principle, change upon gas and water vapor exposure due to their chemisorption or adsorption at the semiconductor surface with forming a surface dipole. It is known that surfaces do attract dipoles. The amount of dipoles adsorbed on a surface depends on the dipole moment of the adsorbed species and the electronic properties of the adsorbing species (Heber 1987). This means that the work function is not a characteristic of a bulk material, but rather a property of the surface of the material (the topmost

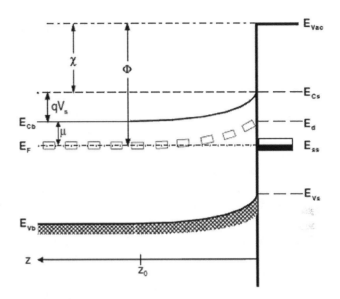

FIGURE 18.1 Band bending of semiconductor in real conditions after chemisorption of charged species (e.g., ionosorption of oxygen on E_{SS} levels). Φ denotes the work function, χ is the electron affinity, and μ is the electrochemical potential. (Reprinted from *Sens. Actuators B*, 118, Sahm, T. et al., (2006) Basics of oxygen and SnO$_2$ interaction; work function change and conductivity measurements, 78–83, Copyright 2006, with permission from Elsevier.)

layers of atoms or molecules). Therefore, by measuring the work function, one can obtain information about the changes in the surface concentration of even such species that are not carrying a net charge—for example, dipoles. This is of special interest when one studies the effect of water vapor, because water interaction with solid can lead to the appearance of such surface species. As is known, water is a polar molecule with strong dipole moment, and adsorption of water is accompanied by the formation of the dipole layer at the solid surface.

18.2 KELVIN PROBE

In 1898, Sir William Thomson, later known as Lord Kelvin, used a gold-leaf electroscope and found that copper and zinc plates placed on insulating shafts produced a charge when they were allowed to come into electrical contact and then move apart. This discovery can be explained with respect to work-function variations. Figure 18.2 shows a diagram of energy and charge, illustrating the principle of the Kelvin probe method.

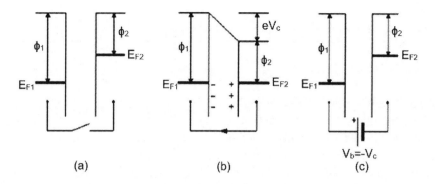

FIGURE 18.2 (a, b, c) Energy and charge diagrams illustrating the Kelvin probe technique principle.

If two different conductors have been brought into electrical contact through an external wire contact, electrons will travel from one conductor with a lower work function to the one with a higher work function, balancing the Fermi energies (Figure 18.2b). If the conductors are made into a parallel plate capacitor, the surfaces will be induced with equal and opposite charges. The potential determined between these two surfaces is known as the *contact potential difference* (CPD), surface potential, or contact potential, which equalizes the work-function variation of the two materials. As a result, CPD measurement becomes relatively simple. The capacitor is applied with an external potential or backing potential until the surface charges vanish completely (Figure 18.2c). If the voltage is chosen such that the electric field is eliminated (the flat vacuum condition), then

$$e\Delta V_{sp} = \Phi_S - \Phi_P \qquad (18.3)$$

where Φ_S and Φ_P are the work function of a sample material and probe material, respectively.

There were many suggestions of how to define this charge-free state. But the most successful was the proposal of William Zisman from Harvard University, who introduced the *nulling concept* and *vibrating electrode method* in 1932. Schematically, a setup of a Kelvin probe is shown in Figure 18.3. One of the two surfaces oscillates periodically with respect to the other. When the tip is vertically vibrated across a sample, the capacitance will differ as the distance changes. This oscillation induces a charge across the surface $Q(t)$. This induced charge can be expressed in terms of the capacitance $C(t)$ and the work-function difference V_{12} between the sensing plate and the reference plate, as (Bergstrom et al. 1997)

$$Q(t) = C(t) \cdot (V_{12} + V_b) = \frac{\varepsilon_0 A}{d(t)} \cdot (V_{12} + V_b) \qquad (18.4)$$

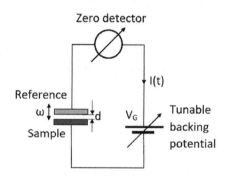

FIGURE 18.3 The setup of a Kelvin probe. A metallic grid oscillates over a sample. The counter voltage, V_G, is adjusted so that the current, i, vanishes. (Reprinted from Oprea, A. et al., Work function changes in gas sensitive materials: Fundamentals and applications, *Sens. Actuators B*, 142, 470–493, 2009. Copyright 2009, Elsevier.)

where V_b is the backing potential which is used to nullify the charge when $V_{12} = -V_b$, A is the area of the plates, and $d(t)$ is the time-varying distance between the plates, which is given by

$$d(t) = d_0 + d_1 \cdot \sin\omega t \qquad (18.5)$$

where d_0 is the plate spacing, d_1 is the displacement, and ω is the frequency of oscillation. The oscillating plate capacitively generates a sinusoidal ac current, which can be described by Equation 18.6. The backing potential, at which ac current is 0 (nulled) or, at a minimum, is observed equal to the CPD. This approach results in the development of systems that automatically monitor the shifts in the contact potential as a result of changes in the sample's work function.

$$i(t) = \frac{\partial Q}{\partial t} = \frac{\partial C(t)}{\partial t}(V_{12} + V_b) + C(t)\frac{\partial V}{\partial t} \qquad (18.6)$$

The work-function measurement using the Kelvin method requires a very sensitive ac current measurement, or a current to voltage impedance transformation prior to the ac measurement (Rossi 1992a). The magnitude of the ac current signal is fed into a control loop, which applies V_b to the sensing film or reference plate to control the current signal at the minimum value, which correlates to $V_{12} = -V_b$. The resolution of the current measurement system and stability of the applied backing potential determine the overall resolution of the system (Baikie and Venderbosch 1991; Rossi 1992b; Bergstrom et al. 1997). For a given work function difference, resolution improves with the magnitude of the current. More detailed description of the Kelvin oscillator method can be found in Riviere (1969), Engelhardt et al. (1970), Surplice and D'Arcy (1970), Belier et al. (1995), Zanoria et al. (1997), Janata and Josowicz (1997, 1998), Doll et al. (1998), Klein et al. (2003), Korotcenkov (2013).

It is important to note that the Kelvin probe method is a relative method because it measures the contact potential difference between the sample and the reference electrode. Although the Kelvin probe technique only measures a work-function difference, it is possible to obtain an absolute work function by first calibrating the probe against the reference material (with known work function), then using the same probe to measure a desired sample.

Effect of water on the work function can be measured by direct methods that use electron emission from the sample induced by photon absorption (photoemission), by high temperature (thermionic emission), due to an electric field (field electron emission), or by using the radiation processing and electron tunneling (Koch et al. 2007; Novikov and Timoshenkov 2010; Musumeci and Pollack 2012). However, these methods, which are even more expensive and experimentally more complex than Kelvin probe method, are more suitable for research than for measuring humidity.

18.3 SENSOR PERFORMANCE

The use of the Kelvin probe in the study of the interaction of metals, metal oxides, and polymers with water vapor has a long history, and many scientific groups are involved in this process (Fuenzalida et al. 1999; Sahoo et al. 2011; Zamarreño et al. 2012; Pohle et al. 2013; etc.). For example, Fuenzalida et al. (1999) studied a relative humidity influence on the work function of $BaTiO_3$ and found that the work function follows the humidity changes closely (see Figure 18.4a), providing a comparatively fast response. The same behavior was observer by Pohle et al. (2013) for doped ZnO-based screen printed layers. Changes of relative humidity (RH) in the range from 33% RH to 56% RH caused the change in the signal response of the Kelvin probe from 25 to 5 mV (Figure 18.4b).

Tanvira et al. (2017) have used the Kelvin probe for study of gas-sensing behavior of the CuO films (see Figure 18.5). They have found that, as the temperature exceeds 50°C, a decrease in the work-function change to the water vapor exposure was detected. This is a natural process for phenomena associated with physisorption.

Zamarreño et al. (2012) measured the work function of conductive polypyrrole (PPy) at room temperature and showed that this work function (CPD) depends not only on the presence of vapors of organic solvents in

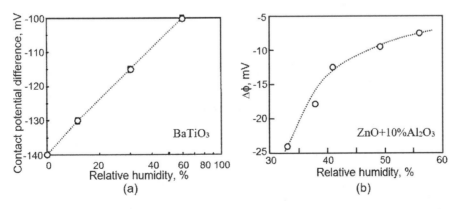

FIGURE 18.4 (a) Contact potential difference (CPD) variation of a hydrothermal $BaTiO_3$ film as a function of the relative humidity. The sensitivity is larger at lower humidity; (b) Response of a ZnO layer with 5% glass and 10% Al to varying humidity levels. Films were deposited using screen printing. $P_{CO2} = 400$ ppm. (Data extracted from Pohle, R. et al., CO_2 sensing by work function readout of ZnO based screen printed films, In: *Proceedings of Transducers 2013*, Barcelona, Spain, Europe, 2048–2050, 2013; Fuenzalida, V.M. et al., *Vacuum* 55, 81–83, 1999.)

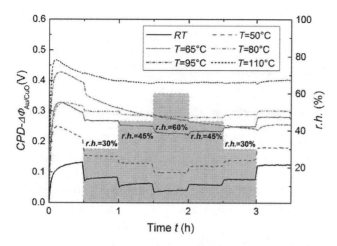

FIGURE 18.5 Kelvin probe measurements with CuO-NPs sensing layers using stepwise changing humidity levels from 0%RH to 60%RH at different temperatures (25°C ≤ T ≤ 110°C). r.h., relative humidity. (Reprinted from Tanvira, N.B. et al., Investigation of low temperature effects on work function based CO_2 gas sensing of nanoparticulate CuO films, *Sens. Actuators B*, 247, 968–974, 2017. Copyright 2017, Elsevier.)

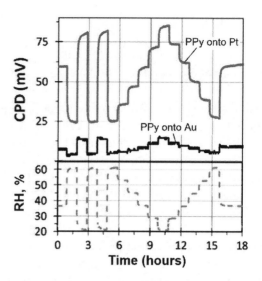

FIGURE 18.7 Kelvin probe measurements of PPy coatings fabricated onto Au (*dark line*) and Pt (*red line*) at different relative humidity (RH) %. Temperature = 23°C. (Data extracted from Zamarreño, C. R. et al., (2012) Work function gas sensors at room temperature by means of conductive polypyrrole thin-films, In: *Proceedings of The 14th International Meeting on Chemical Sensors, IMCS 2012*, 1137–1140, 2012.)

the atmosphere, but even on the type of substrate used to form the film by the in situ polymerization method. In this case, the effect of solvent vapors is much stronger than the effect of humidity. However, in the range of 20%–60% RH, polypyrrole coatings have revealed good sensitivity of the work function to RH variations (see Figures 18.6 and 18.7).

Melios et al. (2016), using the Kelvin probe, have studied the effects of humidity on the electronic properties of mono- and bilayers of graphene prepared by

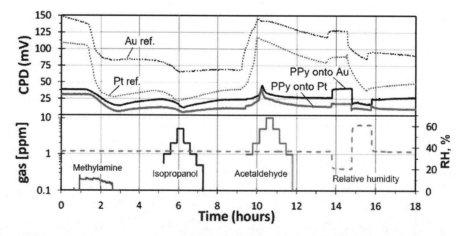

FIGURE 18.6 Kelvin probe measurements of PPy coatings fabricated onto Au and Pt (*solid lines*) and Au and Pt references (*dotted lines*) to different gases (methylamine [0.2 ppm], isopropanol [from 1 to 5 ppm], and acetaldehyde [from 1 to 10 ppm]), and relative humidity (RH). Lines Au_{ref} and Pt_{ref} correspond to the contact potential difference (CPD) measured between a gold reference electrode and the Au (*black dotted line*) and Pt (*red dotted line*) films deposited on alumina substrate, respectively. Temperature = 23°C. (Data extracted from Zamarreño, C.R. et al., Work function gas sensors at room temperature by means of conductive polypyrrole thin-films, In: *Proceedings of The 14th International Meeting on Chemical Sensors, IMCS 2012*, 1137–1140, 2012.)

chemical vapor deposition. It was found that monolayer graphene on SiO_2 was extremely sensitive to water vapor, and water vapor is a strong source of hole doping. The subsequent introduction of controlled humidity resulted in the significant increase of both the hole concentration and the work function, with saturation occurring at approximately 30%–40% RH (see Figure 18.8). It was also concluded that water is not the only dopant present in the ambient air. Other dopants, which are also responsible for the high p-doping of the sample measured in ambient conditions, include O_2, CO, NO_2, and organic compounds.

It is clear that the articles considered above constitute only a small portion of the literature in which an attempt is made to use the Kelvin probe to study the effect of water on surface properties. But they fully reflect the state of research in this field and present the results important for developers of humidity sensors. As for real Kelvin probe-based sensors that could be used to measure humidity, there has been only one attempt to develop such a device. It was made by Bergstrom et al. (1995, 1997) from the K.D. Wise Group (University of Michigan, Ann Arbor, MI).

A schematic diagram of this device is shown in Figure 18.9. The sensor was bulk micromachined from two silicon and one glass wafer, using 13 masking steps to form a three-layer capacitor with an oscillating middle reference plate. The device incorporated a heater to provide temperature control of the sensing film surface with a 7.5°C/mW efficiency in vacuum. The capacitor plates have a 3 μm gap and are 500 μm in diameter. The device can operate from room temperature to over 500°C, with operating frequencies ranging from 10 Hz to 3 kHz. The work-function measurements were enhanced by the measurement of the second harmonic of the Kelvin current, which is not coupled to the electrostatic drive signal and provides the work-function sensitivity of 200 μV/pA.

Chemical testing of the integrated work-function gas sensor has demonstrated repeatable responses to oxygen, water vapor, and pyridine in near-vacuum conditions. However, the sensitivity of the device was low and saturation response time was longer than expected (Bergstrom et al. 1997). Unfortunately, this progress was not developed further.

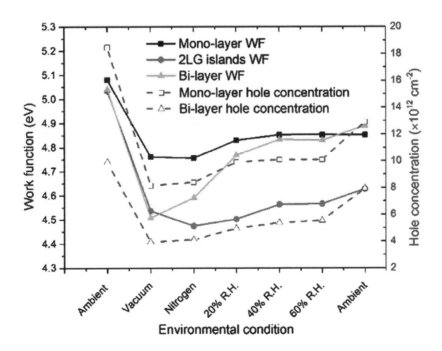

FIGURE 18.8 Summarized results of the environmental measurements of the work function and global hole concentration for monolayer graphene with 2 layer graphene (2LG) islands and bilayer graphene on SiO_2. The environmental steps correspond to initial ambient conditions, vacuum, dry nitrogen, 20%–60% relative humidity (R.H.), and, finally, back to ambient conditions. (Reprinted with permission from Melios, C. et al., Effects of humidity on the electronic properties of graphene prepared by chemical vapour deposition, *Carbon*, 103, 273–280, 2016. Copyright 2016, Elsevier.)

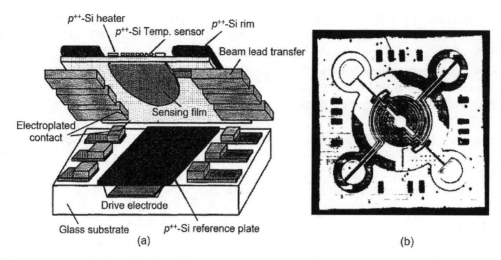

FIGURE 18.9 (a) Three-dimensional perspective view of the work-function gas sensor design. The structure integrates the drive mechanism, reference electrode, and sensing electrode into one microfabricated assembly that is approximately 3.3 mm wide, 2.8 mm high, and 1 mm thick. A 2.5 μm thick silicon reference plate, formed using a boron-diffused etch-stop, is suspended below and electrostatically bonded to the supporting glass substrate. Platinum was used as sensing film. Thus, the Kelvin reference plate is separated by 3 μm from both the drive electrode and the sensing film. The structure also incorporates both the oscillating plate needed in the work-function measurement and the electrostatic drive mechanism to provide oscillation. (b) Top view of an integrated work-function gas sensor. The perforated Kelvin plate is visible through the thermally isolating dielectric window. The sensing film faces the Kelvin plate. (Reprinted with permission from Bergstrom, P.L. et al., A micromachined surface work-function gas sensor for low-pressure oxygen, *Sens. Actuators B*, 42, 195–204, 1997. Copyright 1997, Elsevier.)

18.4 SUMMARY

This review indicates that, in principle, devices based on the Kelvin probe and suitable for controlling air humidity can be developed. This is especially true because the experiment showed that there is a dependence between the work function and humidity that can be used for this purpose. In addition, the process of measuring is not influenced by the electrical or chemical structure of the material. An additional stimulus to develop such sensors could be the sufficient simplicity of the technique and the possibility of its use in many environments.

If we analyze all the possibilities and limitations that arise when using the Kelvin probe for humidity measurement, we can conclude that, for practical sensor applications, Kelvin probes are not suitable. It does not apply to cheap and simple methods of control. Besides that, this method of measurement is complicated in comparison with other methods used in gas sensing and requires specific conditions and equipment. It is too sensitive to the system settings and external influences, such as temperature and vibration. The use of the Kelvin probe also requires qualified specialists for operation and constant calibrations. In addition, this technique has limits to miniaturization. Most probes are rather macroscopic, with a reference electrode the size of a few millimeters or centimeters. The Kelvin probe also cannot be a high-speed device, because its time constant is approximately 1 minute (Fuenzalida et al. 1999). As a result, Kelvin probe-based humidity sensors are not available on the sensor market. The Kelvin probe likely will remain in the laboratories to study the properties of surfaces, since it is an effective instrument for studying gas adsorption and interaction with solids during the study of gas-sensing effect. That is why many organic and inorganic materials have been tested with the Kelvin probe during the past few decades. Some results of these studies were given earlier. In addition, it was found that the Kelvin probe is an effective test tool to screen and characterize possible sensing layers for field-effect devices, such as gas sensor field-effect transistors (GasFETs), in which the materials, demonstrating large work-function changes during interaction with gas, are required (Fleischer 2008). In the meantime, the transferability of the Kelvin probe result to the GasFET has been shown by Burgmair et al. (2002) and Oprea et al. (2005). This means that materials with large changes in the work function measured by Kelvin probes are materials that can be successfully used as a sensing layer in GasFETs.

Another important problem associated with Kelvin probe-based devices is the correct selection of the sample material and probe material. The probe material should be stable, with a minimum change in the work function under the influence of moisture, while the sample material with good stability should adsorb moisture well, but this process must be fast and reversible. This means that if any metal acted as a "reference" electrode, its work function is "assumed" to be constant. Such an "assumption" is necessary, but it cannot be experimentally verified (Janata and Josowicz 1997). For example, as a rule, gold, which has an increased stability, is used as probe material (Fuenzalida et al. 1999; Tanvira et al. 2017). However, even in gold, the work function changes in a humid atmosphere. It is known that the work function of gold decreases under water adsorption (Wells and Fort (1972).

The decrease in the work function of the electron from the metal indicates that, during the adsorption of water on gold, an electron charge is transferred from the H_2O molecule to the metal. Water on the Au surface retains its molecular form. A similar situation is observed when water is adsorbed on Zn, Ru, Rh, Pd, Ag, Ir, and Pt. However, on metals such as Ti, Cr, Fe, Mo, and W, water is adsorbed in a dissociated form. The boundary case is the adsorption of water to Co, Cu, Ni, Cd, and Re. With the change in the nature of the interaction of water with the metal, the nature of the effect of moisture on the work function of the metal also changes. In the presence of other gases or vapors, the situation becomes even more complicated. All this must be taken into account when designing and calibrating the system. This situation is one of the complications in Kelvin probe usage.

It is important to note that, unlike conventional capacitive and resistive sensors, the sample material and probe material should not be porous. This is due to the superficial nature of the processes used in the Kelvin probe. They should have a flat surface and be dense to exclude the possibility of capillary condensation. Otherwise, the values of the work function changes are subject to error due to the history of the sample, as seen in Figure 18.10.

It must also be considered that, as a rule, with an RH of more than 50%, saturation begins, and a further increase in humidity slightly increases the work function of metal oxides (see Figure 18.4B). This is consistent with theoretical models (Wedler 1970) that predict a closed water monolayer at 50% RH—that is, the sample surface is already saturated by OH. High sensitivity to the active gases present in the atmosphere also can be attributed to the shortcomings of the devices under consideration. This is a common drawback of most humidity sensors.

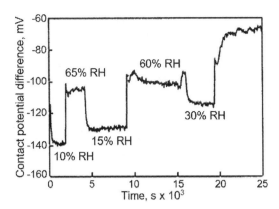

FIGURE 18.10 Time dependence of the contact potential difference (CPD) of a $BaTiO_3$ film, showing fast response. The numbers on the plateaus represent the nominal humidity during the measurement. RH, relative humidity. (Reprinted from Fuenzalida, V.M. et al., Adsorbed water on hydrothermal $BaTiO_3$ films: Work function measurements, Vacuum, 55, 81–83, 1999. Copyright 1999, Elsevier.)

REFERENCES

Anh D.T.V., Olthuis W., Bergveld P. (2004) Work function characterization of electroactive materials using an EMOSFET. *IEEE Sensors J.* 4, 284–287.

Baikie I.D., Venderbosch E. (1991) Analysis of stray capacitance in the Kelvin method, *Rev. Sci. Instr.* 62, 725–735.

Barsan N., Weimar U. (2003) Understanding the fundamental principles of metal oxide based gas sensors; the example of CO sensing with SnO_2 sensors in the presence of humidity. *J. Phys. Condens. Matter* 15, R813–R839.

Belier J.P., Lecoeur J., Koehler C. (1995) Improved Kelvin method for measuring contact potential differences between stepped gold surfaces in ultrahigh vacuum. *Rev. Sci. Instrum.* 66(12), 5544–5547.

Bergstrom P.L., Merchant R., Wise K.D., Schwank J.W. (1995) Dielectric membrane technology for conductivity and work-function gas sensors, In: *Proceedings of the 8th International Conference on Solid-state Sensors and Actuators, and Eurosensors IX*, Stockholm, Sweden, Europe, June 25–29, pp. 993–996.

Bergstrom P.L., Patel N.Y., Schwank J.W., Wise K.D. (1997) A micromachined surface work-function gas sensor for low-pressure oxygen. *Sens. Actuators B* 42, 195–204.

Burgmair M., Wöllenstein J., Böttner H., Karthigeyan A., Anothainart K., Eisele I. (2002) Ti-substituted chromium oxide in work function type sensors: Ammonia detection at room temperature with low humidity cross sensitivity, In: *Proceedings of international workshop on materials and technologies for chemical sensors*, September. 2001, Brescia (Italy). (Paper was not published in a journal; it can be found at http://forschung.unibw.de/berichte/2002/7ut0fvd1htxsoquwk7yl5hkzsvu8aa.pdf.)

Doll T., Scharnagel K., Winter R., Bogner I., Eisele I., Ostrik B., Schoning M. (1998) Work function gas sensors – reference layers and signal analysis, In: *Proceedings of European Conference on Solid state Transducers, Eurosensors XII*, Southampton, UK, 13–16 September, pp. 143–146.

Engelhardt H.A., Feulner P., Pfnür H., Menzel D. (1970) An accurate and versatile vibrating capacitor for surface and adsorption studies. *J. Phys. E* 10, 1133–1136.

Fleischer M. (2008) Advances in application potential of adsorptive-type solid state gas sensors: High-temperature semiconducting oxides and ambient temperature GasFET devices. *Meas. Sci. Technol.* 19, 042001.

Fuenzalida V.M., Pilleux M.E., Eisele I. (1999) Adsorbed water on hydrothermal $BaTiO_3$ films: Work function measurements. *Vacuum* 55, 81–83.

Heber K.V. (1987) Humidity measurement at high temperatures. *Sens. Actuators* 12, 145–157.

Janata J., Josowicz M. (1997) Peer reviewed: A fresh look at some old principles: The Kelvin Probe and the Nernst equation. The combination of two old ideas leads to a new class of chemical sensors. *Anal. Chem.* 69 (9), 293A–296A.

Janata J., Josowicz M. (1998) Chemical modulation of work function as a transduction mechanism for chemical sensors. *Acc. Chem. Res.* 31, 241–248.

Janata J., Josowicz M. (2002) Conducting polymers in electronic chemical sensors. *Nature Mater.* 2, 19–24.

Klein U., Vollmann W., Abatti P.J. (2003) Contact potential differences measurement: Short history and experimental setup for classroom demonstration. *IEEE Trans. Educ.* 46, 338–344.

Koch N., Vollmer A., Elschner A. (2007) Influence of water on the work function of conducting poly(3,4-ethylene-dioxythiophene)/poly(styrenesulfonate). *Appl. Phys. Lett.* 90, 043512.

Korotcenkov G. (2013) Sensing layers in work-function-type gas sensors, In: Korotcenkov G. (ed.) *Handbook of Gas Sensor Materials*. Springer, New York, pp. 377–388.

Melios C., Centeno A., Zurutuza A., Panchal V., Giusca C.E., Spencer S., Silva S.R.P., Kazakova O. (2016) Effects of humidity on the electronic properties of graphene prepared by chemical vapour deposition. *Carbon* 103, 273–280.

Musumeci F., Pollack G.H. (2012) Influence of water on the work function of certain metals. *Chem Phys Lett.* 536, 65–67.

Novikov S.N., Timoshenkov S.P. (2010) A mechanism of changes in the electron work function upon water chemisorption on a Si(100) surface. *Rus. J. Phys. Chem. A.* 84 (7), 1266–1269.

Oprea A., Simon E., Fleischer M., Frerichs H.-P., Wilbertz C., Lehmann M., Weimar U. (2005) Flip-chip suspended gate field effect transistors for ammonia detection. *Sens. Actuators B* 111, 582–586.

Pohle R., Tawil A., Mrotzek C., Fleischer M. (2013) CO_2 sensing by work function readout of ZnO based screen printed films, In: *Proceedings of Transducers 2013*, Barcelona, Spain, Europe, 16–20 June, pp. 2048–2050.

Riviere J.C. (1969) Work function: Measurements and results, In: Green M. (ed.) *Solid State Surface Science*. Dekker, New York, pp. 179–289.

Rossi F. (1992a) Contact potential measurement: The preamplifier. *Rev. Sci. Instr.* 63, 3744–3751.

Rossi F. (1992b) Contact potential measurement: Spacing-dependent errors. *Rev. Sci. Instr.* 63, 4174–4181.

Sahoo P., Oliveira D.S., Cotta M.A., Dhara S., Dash S., Tyagi A.K., Raj B. (2011) Enhanced surface potential variation on nanoprotrusions of GaN microbelt as a probe for humidity sensing. *J. Phys. Chem. C* 115, 5863–5867.

Streetman B.G. (1990) *Solid State Electronic Devices*, 3rd ed. Prentice Hall, Englewood Cliffs, NJ.

Surplice N.A., D'Arcy R.J. (1970) A critique of the Kelvin method of measuring work functions. *J. Phys. E* 3, 477–482.

Tanvira N.B., Yurchenko O., Laubender E., Urban G. (2017) Investigation of low temperature effects on work function based CO_2 gas sensing of nanoparticulate CuO films. *Sens. Actuators B* 247, 968–974.

Wedler G. (1970) *Adsorption. Introduction to Physisorption and Chemisorption*. Wiley, Weinheim, Germany.

Wells R.L., Fort T. (1972) Adsorption of water on clean gold by measurement of work function changes. *Surf. Sci.* 32, 554–560.

Zamarreño C.R., Davydovskaya P., Pohle R., Fleischer M., Matias I.R., Arregui F.J. (2012) Work function gas sensors at room temperature by means of conductive polypyrrole thin-films, In: *Proceedings of The 14th International Meeting on Chemical Sensors, IMCS 2012*, pp. 1137–1140.

Zanoria E.S., Hammall K., Danyluk S., Zharin A.L. (1997) The nonvibrating Kelvin probe and its application for monitoring surface wear. *J. Test. Eval.* 25(2), 233–238.

19 Solid-State Electrochemical Humidity Sensors

19.1 INTRODUCTION

The experiments performed by Nagata et al. (1987) and Iwahara et al. (Iwahara et al. 1988; Iwahara 1990) showed that the ionic electrical conductivity mechanism realized at high temperatures in metal oxides and other materials, so-called *solid electrolytes*, also allows the development and manufacture of humidity sensors. For example, it was established that in metal oxides such as $SrCeO_3$- and $BaCeO_3$-based perovskite-type oxides in the presence of hydrogen and water vapors the electronic conduction decreases and proton conduction appears (Iwahara et al. 1988). Sensors developed on the basis of such materials refer to the electrochemical type (Korotcenkov et al. 2009; Stetter et al. 2011).

Sensors based on solid electrolytes are assigned to a special chapter, since the principle of their operation and the requirements for materials differ from the sensors considered in Chapters 10–18. In particular, the solid electrolytes designed for application in electrochemical humidity sensors, such as potentiometric and amperometric ones, should be dense proton conductors without any pores (Iwahara 1990). This means that the structural factor ceases to be the most important factor controlling the parameters of humidity sensors. In addition, these metal oxides should have a high proton conductivity at a low electronic conductivity at operating temperatures, and good mechanical properties (Traversa 1995). Sensors based on such materials, as a rule, operate at high temperature to achieve a proton conductivity, and therefore good thermal stability is also required. Interest in such sensors is related to the need to control humidity in high-temperature processes, in which standard sensors cannot operate (Greenblatt and Feng 1993).

19.2 PRINCIPLES OF OPERATION

Typically, materials acceptable for application in solid-state electrochemical humidity sensors are subjected to high-temperature sintering to compact the structure. Constructively, electrochemical-based humidity sensors of the potentiometric type are conventional fuel cells that generate the electromotive force at the electrodes of the cell when water vapor appears. The proton transfer occurs from the higher vapor pressure side to the side with lower vapor pressure. The reactions at each electrode are described in Figure 19.1. Experiment has shown that platinum electrodes were the most suitable for this type of application.

As illustrated in Figure 19.1, a gas cell constructed with a proton conductive ceramic as the electrolyte gives rise to an electromotive force (EMF), due to the concentration difference of water vapor, when atmospheres with different humidity are introduced into each electrode compartment. The theoretical EMF, E_0, can be written as

$$E_o = \frac{RT}{2F} ln \frac{P_{H_2O}(I)}{P_{H_2O}(II)} \left(\frac{P_{O_2}(II)}{P_{O_2}(I)} \right)^{1/2} \quad (19.1)$$

where P_{H_2O} and P_{O_2} are the partial pressure of water vapor and oxygen, respectively, and R, F, and T have their usual meanings. When $P_{O_2}(I)$ is close to $P_{O_2}(II)$, as in ambient atmospherics with different humidity, E_0 can be written as

$$E_0 = \frac{RT}{2F} ln \frac{P_{H_2O}(I)}{P_{H_2O}(II)} \quad (19.2)$$

If $P_{H_2O}(II)$ is known and is constant, the humidity in the compartment I (or $P_{H_2O}(I)$) may be estimated from the measured EMF of the cell, wherein this EMF will vary linearly with the logarithm of the water partial pressure. When an electrolyte involves electronic conduction, EMF is lowered to some extent

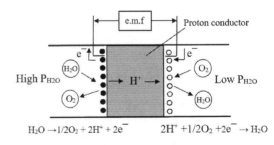

FIGURE 19.1 Concept of the water-vapor concentration cell using a solid proton conductor as the electrolyte. e.m.f., electromotive force. (Idea from Iwahara, H. Use of high temperature proton conductor for gas sensors, In: Yamazoe N. (ed.), *Chemical Sensor Technology,* Vol. 3, Kodansha, Tokyo/Elsevier, Amsterdam, the Netherlands, 117–129, 1990.)

$$E < t'_{H^+} \leq E_0 \quad (19.3)$$

where t'_{H^+} is the transference number of the proton in the solid electrolyte. Even in such a case, we can determine $P_{H_2O}(I)$ from the measured EMF by using a calibration curve.

Of course, the use of other modes of operation of electrochemical humidity sensors, such as amperometric and conductometric ones, is also possible. In particular, Figure 19.2 shows one of the variants of amperometric-based electrochemical sensors. Such a sensor was designed to monitor oxygen and humidity using the ZrO_2-Y_2O_3 membrane. Unlike the potentiometric sensors, amperometric sensors operate under an externally applied voltage. Typically, such sensors have pronounced nonlinear current–voltage relations, which are determined by the electrode kinetics. To eliminate this drawback, it was suggested to introduce a diffusion barrier into the sensor. Therefore, as a rule, amperometric sensors operate in the diffusion-limited mode. Here, each molecule passing the diffusion barrier reacts immediately at the electrode. The corresponding limiting current is a unique function of the geometric parameters of the diffusion barrier. It was established that, in this case, the sensor response (the limiting current I) can be approximated with (Usui et al. 1989a, 1989b)

$$I_{\lim} = \frac{2F \cdot D(H_2O) \cdot S \cdot P}{RTL} pH_2O \quad (19.4)$$

where F is Faraday constant, S is the inner diameter of the capillary, P is the absolute pressure of the analyzed gas, L is the length of the capillary, R is the gas constant, T is the temperature, pH_2O is a partial pressure of water vapors, and $D(H_2O)$ is the diffusion coefficient of the water vapor, which depends on temperature and pressure (Marrero and Mason 1972).

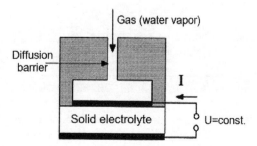

FIGURE 19.2 Experimental setup of a typical amperometric gas sensor based on YSZ electrolyte with a channel-type diffusion barrier.

$$D = D_S \left(\frac{T}{273}\right)^\alpha \frac{1}{P} \quad (19.5)$$

Where D_S is the standard diffusion coefficient and α is a coefficient. It can be seen from Equation 19.4 that the limiting current level depends linearly on pH_2O. This makes it possible to obtain a calibration curve, which can then be used to analyze unknown moisture concentrations in gas mixtures (Medvedev et al. 2017).

To study the sensor characteristics in impedance mode, the configuration, typical for resistive sensors, is being used (see Chapter 11).

19.3 SENSORS PERFORMANCE

It should be noted that a comparatively small amount of solid electrolytes has been tested as a proton conductor in electrochemical humidity sensors (see Table 19.1). However, in principle, all proton conductors used in the development of hydrogen sensors could be used in humidity sensors (Kreuer 1996; Korotcenkov et al. 2009). The classification of proton conductors according to the preparation method, chemical composition, structural dimensionality, mechanism of conduction, et cetera, has been summarized in a comprehensive book on proton conductors (Colomban 1992).

19.3.1 Potentiometric Humidity Sensors

The appropriate humidity-sensing material (i.e., proton-conducting solid electrolyte) is considered to be the key factor for the development of such humidity sensors. The first solid electrolytes used in the development of electrochemical humidity sensors were $SrCeO_3$- and $BaCeO_3$-based metal oxides. $SrCeO_3$- and $BaCeO_3$-based perovskite-type oxides are stable and efficient proton conductors at temperatures above 400°C. These ceramics are only p-type electronic conductors in a hydrogen- or water vapor–free atmosphere. In the presence of these gases, electronic conduction decreases and proton conduction appears (Iwahara et al. 1988). Iwahara et al. (1988) have shown that the protonic conductivity in the Nd_2O_3-doped $BaCeO_3$ was higher than that in the $SrCeO_3$-based proton conductors, which indicates good prospects for the use of these materials in humidity sensors. They are also stable at a high temperature. However, it was suggested that, in the hydrogen-air fuel cell, oxide ions, as well as protons, might be conducting. The conduction mechanism in these cases may be more complex than that in the $SrCeO_3$-based proton conductor. Humidity-sensitive characteristics of these materials have been tested in cells, the concept of which is

TABLE 19.1
Solid Electrolytes Used in Electrochemical Humidity Sensors

Solid Electrolyte	Mode of Operation	Optimal Operation Temperature, °C	Range of Air Humidity Acceptable for Control	References
$SrCe_{0.95}Yb_{0.05}O_{3-\alpha}$	P	300–400	6–100 torr	Iwahara et al. (1983), Nagata et al. (1987), Hassen et al. (2000)
MnO_x	P	RT	> 55% RH	Miyazaki et al. (1994)
MnO_x-based composite	P	RT	0%–90% RH	Miyazaki et al. (1997)
$BaCeO_3$				Iwahara et al. (1988)
β-$Ca(PO_3)_2$	P	580	5–200 torr	Tanase et al. (1989), Tsai et al. (1989), Greenblatt et al. (1990), Greenblatt and Shuk (1996)
NASICON ($HZr_2P_3O_{12}$; ($HZr_2P_3O_{12} \cdot ZrP_2O_7$)	P	450	2–200 torr	Feng and Greenblatt (1992c), Greenblatt and Shuk (1996)
LaF_3	P	30	40%–80% RH	Sun et al. (2011)
ZrO_2-Y_2O_3	A	450–500	0%–75% RH	Usui et al. (1989a. 1989b), Yagi and Hideaki (1995)
$Ag_7I_4AsO_4$	A	RT	10–300 Pa	Lieder et al. (1990)
$La_{0.9}Sr_{0.1}YO_{3-\delta}$	A	650	0.001–0.1 atm	Medvedev et al. (2017)
$Sb_2O_5 \times 2H_2O$	I	RT-150	10%–100%	Ozawa et al. (1983, 1998), Miura et al. (1985)
$HSb(PO_4)_2 \times 2H_2O$	I	150–250	0–4 kPa	Miura et al. (1988)
$SrCe_{0.95}Yb_{0.05}O_3$	I	450	0%–50%	Usui et al. (1989a, 1989b)
NASICON ($HZr_2P_3O_{12} \cdot ZrP_2O_7$)	I	450	10%–100%	Feng and Greenblatt (1993), Greenblatt and Shuk (1996)
NASICON ($HZr_2(PO_4)_3$)	I	25–250	10%–60% RH	Shuk and Greenblatt (1998)
$KMo_3P_6SiO_{25}$	I	180	9–200 torr	Tsai et al. (1989)
α-$Zr(HPO_4)_2 \times H_2O$				
$Na_3HGe_7O_{16}$	I	30–150	0–760 torr	Feng et al. (1992), Feng and Greenblatt (1992a, 1992b)
$Na_3Mo_2P_2O_{11}(OH) \cdot H_2O$	I	25–100	5%–100%	Tsai et al. (1991), Greenblatt and Feng (1993)
$CaZrO_3$:In	I	700	0%–80% RH	Zhou and Ahmad (2008)
$BaZrO_3$:Y	I	400–600	0.05–0.5 atm	Chen et al. (2009a, 2009b)

A, amperometric; I, impedance; P, potentiometric; RT, room temperature.

shown in Figure 19.3. The variation of water-vapor pressure was measured by the electromotive force generated at the electrodes of the cell. Typical constructions of such cells are shown in Figure 19.4. Another variant of galvanic cell–type humidity sensors is shown in Figure 19.5.

It was established that the EMF of galvanic cells, based on $SrCeO_3$, $BaCeO_3$, and more complicated metal oxides, varied linearly with the logarithm of the water partial pressure (Nagata et al. 1987; Iwahara et al. 1988; Hassen et al. 2000). In particular, the EMF of the $SrCe_{0.95}Yb_{0.05}O_3$-based humidity sensor (T_{oper} = 300°C–1000°C) against changes of water-vapor pressure is shown in Figure 19.6a. The EMF of the sensor had good reversibility and responded to the changes of humidity and temperature in approximately 1 minute. However, the EMF was smaller than that expected from the Nernst equation. Moreover, the difference between the theoretical and experimental EMF values became larger with increasing operating temperature. Such behavior was explained by the mixed conductivity of this metal oxide; both protons and positive holes participate in the conductivity, and p-type electronic conduction in the oxide was not negligibly small at high temperatures (Nagata et al. 1987). As shown in Figure 19.6b, the EMF

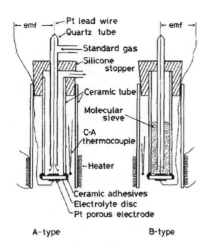

FIGURE 19.3 Schematic illustrations of the galvanic cell–type humidity sensors. The dense sinters obtained were sliced into thin discs (thickness: approximately 0.5 mm, diameter 12 mm) to provide the electrolyte diaphragms for the gas cells. Chromel-Alumel (C-A) thermocouple was used for temperature measurement; emf, electromotive force. (Reprinted with permission from Iwahara, H. et al., Galvanic cell–type humidity sensor using high temperature–type proton conductive solid electrolyte, *J. Appl. Electrochem.*, 13, 365–370, 1983. Copyright 1983, Springer Scinece+Business Media.)

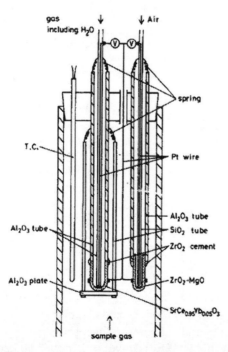

FIGURE 19.4 Construction of humidity sensor composed of the $SrCe_{0.95}Yb_{0.05}O_3$ solid electrolyte. P_{H2O} was estimated by electromotive force (EMF) of the sensor, temperature, and P_{O2}. Temperature was measured using thermocouple (T.C.). P_{O2} was measured by an oxygen sensor constructed by the ZrO_2-MgO solid electrolyte. (Reprinted with permission from Nagata, K. et al., Humidity sensor with $SrCe_{0.95}Yb_{0.05}O_3$ solid electrolyte for high temperature use, *J. Electrochem. Soc.*, 134, 1850–1854, 1987. Copyright 1987, Electrochemical Society.)

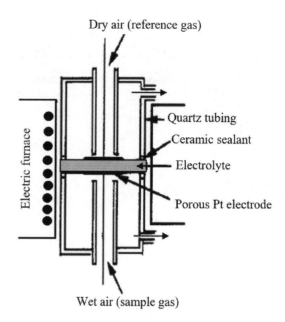

FIGURE 19.5 Schematic illustration of the test galvanic cell–type humidity sensor developed by Feng and Greenblatt. (Reprinted with permission from Feng, S. and Greenblatt, M., Galvanic cell type humidity sensor with NASICON–based material operative at high temperature, *Chem. Mater.*, 4, 1257–1262, 1992. Copyright 1992, American Chemical Society.)

of $SrCe_{0.95}Yb_{0.05}O_3$-based humidity sensors increased with decreasing P_{O2} in sample gas from $1·10^5$ to $5·10^2$ Pa under constant P_{H2O}. The same effect was observed by Hassen et al. (2000). This is consistent with the conclusion of Nagata et al. (1987) that the solid electrolyte $SrCe_{0.95}Yb_{0.05}O_3$ should be also considered as a mixed conductor, whose conductivity is determined by oxygen ions, protons, and positive holes. Experiment showed that the above-mentioned humidity sensors are stable and provide a reliable measurement of the humidity level at high temperatures up to 1000°C (Nagata et al. 1987).

Feng and Greenblatt (1992c) have shown that a much better match with the Nernst equation is performed for Sodium (**Na**) **Su**per **I**onic **CON**ductor (NASICON)-based sensors. It is well-known that NASICON materials ($Na_{1+x}Zr_2Si_xP_{3-x}O_{12}$) are fast sodium ion conductors. The structure of both end members (i.e., $x = 0$ and 3 [$HZr_2P_3O_{12}$ and $HZr_2Si_3O_{12}$, respectively]) is composed of a rigid, three-dimensional network lattice with interconnected cavities. In the phosphate analogue, the sodium ion occupies only one of the cavities. The proton-substituted NASICON, $HZr_2P_3O_{12}$, is a high-temperature proton-conducting material and is thermally stable up to 700°C. Figure 19.7a shows the EMF response of the galvanic cell as a function of the logarithm of the partial pressure of water in the sample compartment at 400°C–550°C. It is seen that in the temperature range

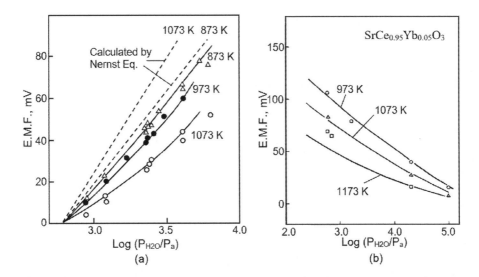

FIGURE 19.6 (a) Humidity dependence of electromotive force (EMF) of $SrCe_{0.95}Yb_{0.05}O_3$-based humidity sensor at 873K, 973K, and 1073K; (b) Relation of EMF of humidity sensor to P_{O2} in O_2, air, N_2, and Ar with P_{H2O} of 2.44×10^3 Pa. $P_{H2O}{}^r$ in air in reference electrode was $6.04 \cdot 10^2$ Po. (Reprinted with permission from Nagata, K. et al., Humidity sensor with $SrCe_{0.95}Yb_{0.05}O_3$ solid electrolyte for high temperature use, *J. Electrochem. Soc.*, 134, 1850–1854, 1987. Copyright 1987, Electrochemical Society.)

425°C–475°C, the EMF behavior of $HZr_2P_3O_{12}$-based cell ideally follows Nernstian behavior over the entire P_{H2O} range examined (2–200 mmHg). The response time of the sensor as a function of temperature on changing P_{H2O} is shown in Figure 19.7b. As the humidity is varied, the EMF responds rapidly and reaches a steady state within a few seconds at all temperatures examined. At 350°C, the response time is 48 seconds; however, the response time is significantly shortened at elevated temperatures; at 450°C, the response time is within 15 seconds (Feng and Greenblatt 1992c).

Potentiometric humidity sensors were also developed on the base of β-$Ca(PO_3)_2$ (Tanase et al. 1989; Tsai et al. 1989; Greenblatt et al. 1990). Complex impedance and dc resistance measurements of the cell under conditions of controlled humidity indicated that both the dc resistance of the cell and the bulk resistance of β-$Ca(PO_3)_2$ obtained by analysis of the complex impedance plots are inversely

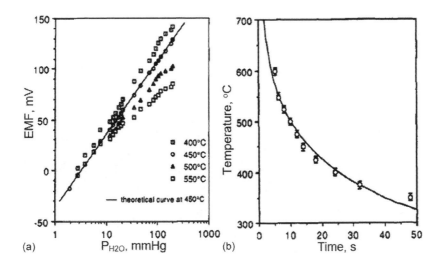

FIGURE 19.7 (a) Humidity dependence of electromotive force (EMF) of sensor at 400°C, 450°C, 500°C, and 550°C; (b) Temperature dependence of the response time for the humidity sensor with the change of partial pressure of water from 6 to 18 mmHg. (Reprinted with permission from Feng, S. and Greenblatt, M., Galvanic cell type humidity sensor with NASICON–based material operative at high temperature, *Chem. Mater.*, 4, 1257–1262, 1992. Copyright 1992, American Chemical Society.)

proportional to the absolute humidity in the cell. These data confirmed the EMF results, which showed that water vapor plays a decisive role in the electrical conductivity of β-Ca(PO$_3$)$_2$ and/or the reversibility of the electrodes on it. Infrared spectrum of β-Ca(PO$_3$)$_2$ showed an absorption band that can be assigned to OH groups, which, according to Abe et al. (1982), may serve as a source of mobile protons and/or produce a channel of proton conduction (Greenblatt et al. 1990). However, experiment has shown that β-Ca(PO$_3$)$_2$-based sensors cannot operate at high temperatures. This material appears to decompose in prolonged application above 650°C. The further investigations to improve the long-term stability, accuracy, and selectivity are also required (Greenblatt and Shuk 1996).

There were also attempts to use Ag$_7$I$_4$As$_5$O$_4$ and certain other silver ion-conducting glasses (Lieder et al. 1990) and LaF$_3$ (Sun et al. 2011) in the cells. However, the LaF$_3$-based cell operated at 30°C had linear Nernst-type response only in the range of 43%–75% relative humidity (RH; Sun et al. 2011), and because of irreversible changes in the electrolyte brought about by the current flow, practical application of Ag$_7$I$_4$As$_5$O$_4$-based cells was limited to low levels of humidity. When using MnOx, the opposite situation was observed by Miyazaki et al. (1994). MnO$_x$-based humidity sensors were sensitive only in the range of RH > 60%. However, later, Miyazaki et al. (1997) showed that through the formation of composites based on MnOx and clay minerals, such as kaolinite, bentonite, activated clay, zeolite, and muscovite, it is possible to effect humidity-sensitive characteristics.

Of these, kaolinite was found to be a good additive to MnO$_x$. As seen in Figure 19.8, the MnO$_x$-kaolinite mixture responded in a wider region, from 10% to 90% RH, and this dependence was close to linear. Miyazaki et al. (1997) assumed that the interlayer water molecules and the capillary condensation of water are responsible for such effect, because the cavity of lamellar structure of

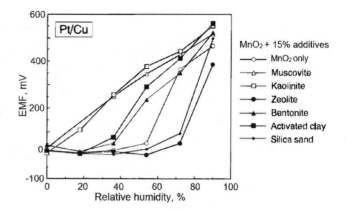

FIGURE 19.8 Typical effects of a series of clay samples, together with silica sand as a standard, as additives to a MnO$_2$ humidity sensor on producing galvanic potentials of Pt/Cu electrodes. EMF, electromotive force. (Reprinted with permission from Miyazaki, K. et al., Development of a novel manganese oxide–clay humidity sensor, *Ind. Eng. Chem. Res. 36*, 88–91, 1997. Copyright 1997, American Chemical Society.)

clay minerals is in the order of several angstroms (Klein and Hurlbut 1993), while the average diameter of micropores of MnO$_x$ ceramics is around 20 Å (Kozawa 1974). The condensation of water molecules from the ambient air into the smaller-sized capillaries will be earlier and smoother than into the larger-sized capillaries, as is dictated by the Kelvin equation for capillary condensation.

19.3.2 Amperometric Humidity Sensors

A solid-electrolyte amperometric humidity sensor, which does not need a reference atmosphere, was designed by Usui et al. (1989a, 1989b) on the basis of yttria-stabilized zirconia. This sensor operated at 450°C and controlled the limiting current, appearing in the diffusion-controlled electrolysis of water vapor. The configuration and appearance of the sensors are shown in Figure 19.9.

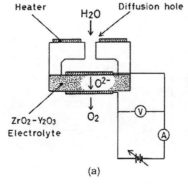 

FIGURE 19.9 (a) Schematic diagram of a gas-polarographic oxygen-humidity sensor using a zirconia electrolyte; (b) Appearance of the sensor. (Reprinted with permission from Usui, T. et al., Gas polarographic multifunctional sensor: Oxygen–humidity sensor, *Sens. Actuators*, 16, 345–358, 1989. Copyright 1989, Elsevier.)

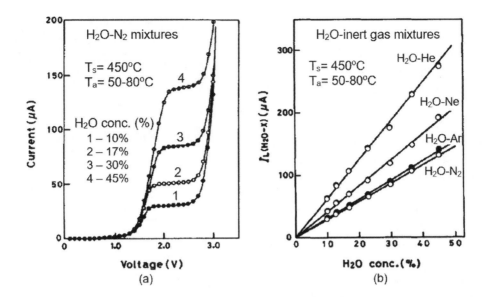

FIGURE 19.10 (a) Voltage-current characteristics of the sensor in H_2O-N_2 gas mixtures. T_s is the temperature of the sensor and T_a is the temperature of the ambient atmosphere; (b) The relationship between the limiting current and the water vapor concentration in H_2O-inert gas mixtures at 450°C. (Reprinted with permission from Usui, T. et al., Gas polarographic multifunctional sensor: Oxygen–humidity sensor, *Sens. Actuators*, 16, 345–358, 1989. Copyright 1989, Elsevier.)

This sensor, designed by Usui et al. (1989a, 1989b), had excellent oxygen- and humidity-sensing characteristics and thus can be applied to accurate and reliable oxygen and humidity control. Figure 19.10 shows the voltage-current characteristics of the sensor in H_2O-N_2 gas mixtures at an ambient temperature T_a of 50°C–80°C when the sensor was heated and kept at 450°C. When the applied voltage was higher than 1.2 V, the output current increased with the applied voltage, and subsequently the limiting current plateaus (saturated current plateaus against the applied voltage) were observed and seen to vary with the water vapor concentration from 10% to 45%. This output current was caused by the fact that oxygen ions, produced by the electrochemical decomposition of water vapor at the cathode of the sensor, passed through the zirconia electrolyte from cathode to anode, as in Equation 19.6

$$H_2O + V_O^{\cdot\cdot} + 2e^- \rightarrow H_2 + O_O^x \quad (19.6)$$

where $V_O^{\cdot\cdot}$ and O_O^x are the oxygen ion vacation with two trapped electrons and the oxygen ion on a normal site in electrolyte, respectively. The experimental threshold voltage of 1.2 V agreed well with the theoretical decomposition voltage of water vapor, estimated as 1.1 V at 450°C (Usui et al. 1989a).

Here, it is important to note that the electrical output of the sensor was quite stable even above 100°C, compared with that of the other types of humidity sensors, and was hardly influenced by the atmospheric temperature (100°C–160°C), as long as the sensor was heated and kept at the temperature of 450°C (see Figure 19.11). Thus, this sensor can be used for monitoring and controlling of humidity at high atmospheric temperatures above 100°C.

Based on this principle, Yagi and Ichikawa (1993) developed sensors with the cathode and anode on the

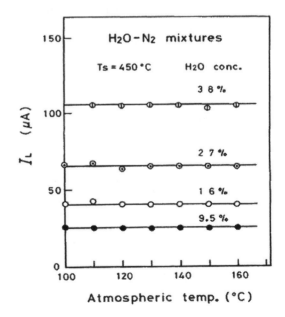

FIGURE 19.11 The relationship between the limiting current and the atmospheric temperatures (100°C–160°C). (Reprinted with permission from Usui T., et al., Gas polarographic sensor usable above 100°C, *Jpn. J. Appl. Phys.*, 28, 2325–2326, 1989. Copyright 1989, Japan Society of Applied Physics.)

same plane; the cathode also acted as the diffusion control hole. The operating temperature range of this sensor was wide, from 20°C to 300°C, and linear detection was available within the humidity range 0–500 mmHg. The response time was approximately 1 minute during adsorption and 2 minutes during desorption. Yagi and Ichikawa (1993, 1995) have also shown that the incorporation of an additional cathode gave possibility to fabricate a simple planar, multifunctional sensor able to measure oxygen and humidity simultaneously (see Figure 19.12). For designing such sensors, they used the ZrO_2 raw material stabilized with Y_2O_3 (8% mol). Sensors can operate at temperatures of 0°C–300°C and detect humidity in the range of 0%–78% vol. The H_2O (10%–80% RH at 0°C–50°C) time of response was approximately 20 seconds, and repeatability ± 1% of full scale (FS).

A different approach to developing amperometric humidity sensors was proposed by Medvedev et al. (2017). Figure 19.13 depicts the operating principles of the as-fabricated electrochemical cell. It can be seen that the cell consists of two ceramic electrolytes, Yttria-Stabilized-Zirconia (YSZ) and Strontium-doped Lanthanum Yttrate (LSY), and a capillary tube. Commercially available YSZ was chosen as a well-known electrolyte providing oxygen-ion transport over a wide temperature range, excellent mechanical properties (high mechanical strength and toughness, low thermal expansion coefficient value), and good thermodynamic stability (Mahato et al. 2015). The proton-conducting electrolyte ($La_{0.9}Sr_{0.1}YO_{3-\delta}$, LSY) was chosen on the basis of the results reported in Okuyama et al. (2014) and Danilov et al. (2016). In particular, this material exhibits a wide electrolytic domain boundary, a low contribution of hole conductivity in oxidizing atmospheres, and a thermal expansion coefficient close to that of YSZ (Danilov et al. 2016). In reducing atmospheres, materials based on $LaYO_3$ exhibit almost protonic transport over a wide temperature range (Okuyama et al. 2014). Porous platinum electrodes were formed on both sides of each

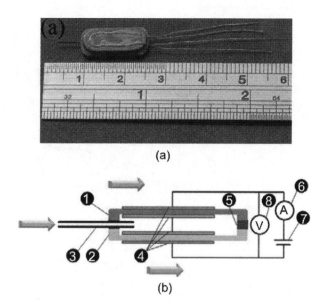

FIGURE 19.13 (a) The general view and (b) The working scheme of the electrochemical cell based on combined oxygen- and proton-conducting electrolytes: 1, YSZ electrolyte; 2, LSY electrolyte; 3, capillary; 4, platinum electrodes; 5, high-temperature glass sealant; 6, amperometer; 7, DC current source; 8, voltmeter. (Reprinted with permission from Medvedev, D. et al., Electrochemical moisture analysis by combining oxygen– and proton–conducting ceramic electrolytes, *Electrochem. Commun.*, 76, 55–58, 2017. Copyright 2017, Elsevier.)

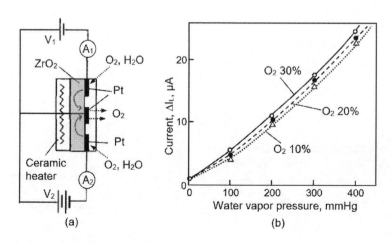

FIGURE 19.12 (a) The structure of the multi-sensor using the planar oxygen sensor; (b) Relation between ΔI_L ($I_{L2}-I_{L1}$) and vapor pressure with changing O_2 concentration in O_2-H_2O-N_2 gas. (Reprinted with permission from Yagi, H. and Ichikawa, K., Humidity sensing characteristics of a limiting current type planar oxygen sensor for high temperatures, *Sens Actuators B*, 13, 92–95, 1993. Copyright 1993, Elsevier.)

individual ceramic material; this was followed by high-temperature annealing (1200°C, 1 h) to achieve a good contact between the electrodes and the electrolyte.

The distinctive feature of the as-fabricated electrochemical cell is the fact that the inner electrodes were shorted together, forming a common electrical circuit (Medvedev et al. 2017). A DC voltage (U) was then applied externally to the outer electrodes of the cell in such a manner that "plus" corresponded to the outer electrode of the oxygen-ion electrolyte and "minus" to that of the proton-conducting electrolyte. The decomposition of water mixed with N_2 is initiated by the direct current supply. As a result, the following electrochemical reactions take place on the inner electrode of YSZ:

$$H_2O_{gas} + 2e'_{Pt\,electrode} \rightarrow O^{2-}_{(YSZ)} + H_{2(gas)} \quad (19.7)$$

and on the inner electrode of LSY:

$$H_2O_{gas} \rightarrow 2H^+_{(ISY)} + 2e'_{Pt\,electrode} + 1/2O_{2(gas)} \quad (19.8)$$

The as-formed molecular hydrogen and oxygen can also be pumped out as follows:

$$H_{2(gas)} \rightarrow 2H^+_{(ISY)} + 2e'_{Pt\,electrode} \quad (19.10)$$

$$1/2O_{2(gas)} + 2e'_{Pt\,electrode} \rightarrow O^{2-}_{(YSZ)} \quad (19.11)$$

In this way, the water flow, electrochemically pumped out from the internal space of the cell, is compensated for equilibrium reasons by the $H_2O + N_2$ flow permeating through the diffusion barrier from the external space to the internal one. According to Faraday's law, this flow corresponds to an electrical current, the amount of which firstly increases and then stabilizes with the gradual increase of the applied voltage. The unchanged value of the current (limiting current) with varying applied voltage corresponds to the steady-state condition, when the concentration of water inside the electrochemical cell becomes negligible. The limiting current (I_{lim}) in the case of low partial pressure of the analyzed component (less than 0.1 atm) can theoretically be found using the Equation 19.4.

Figure 19.14a shows the typical volt-ampere characteristics of the electrochemical cell operating in $N_2 + H_2O$ gas atmospheres. These results demonstrate the efficiency of the sensor developed by Medvedev et al. (2017). As can clearly be seen in Figure 19.14b, the limiting current depends linearly on pH_2O, corresponding qualitatively to Equation 19.4. Since this equation includes known or easily calculated parameters, it is possible to demonstrate the quantitative agreement of the experimental data with the calculated ones. The electrochemical cell can successfully operate as a humidity amperometric sensor at a temperature range of 400°C–700°C. The low temperature boundary is limited by the electrocatalytic activity of the platinum electrodes, while the high temperature boundary is limited by the increasing undesirable oxygen-ion conductivity and its contribution to the total conductivity of the LSY electrolyte. The pH_2O values in the analyzed atmosphere seem to be optimal at 0.001–0.1 atm, because of both the linear form of the concentration dependence of the limiting current in this range and the high precision of the detectable current levels. The response time (t_{90}), taken to reach a reading 90% of the final value, is estimated to be close to 10 minutes (Medvedev et al. 2017). The rather slow dynamic characteristics (in comparison with those of potentiometric-type cells [Liu et al. 2014; Okuyama et al. 2016]) are caused by the presence of the capillary

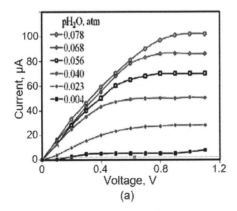
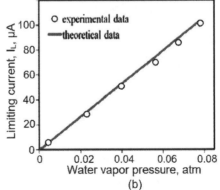

FIGURE 19.14 (a) Volt-ampere characteristics of the electrochemical cell for different water vapor pressure values in nitrogen; (b) Limiting current as a function of water vapor pressure; all the data correspond to the operating temperature of 650°C. (Reprinted with permission from Medvedev, D. et al., Electrochemical moisture analysis by combining oxygen– and proton– conducting ceramic electrolytes, *Electrochem. Commun.*, 76, 55–58, 2017. Copyright 2017, Elsevier.)

acting as a diffusion barrier for gas exchange between the gases of the internal and external parts of the electrochemical cell. Medvedev et al. (2017) believe that when using these sensors, it is possible to determine the water vapor partial pressure in unknown gas mixtures and even to estimate the mutual diffusion coefficients.

19.3.3 Impedance Sensors

Impedance sensors form the third group of humidity sensors, which can be developed on the basis of solid-state electrolytes. The *impedance* is a measure of humidity in such devices. To develop such sensors, Ozawa et al. (1983, 1998) and Miura et al. (1985) proposed to use antimonic acid ($Sb_2O_5 \times 2H_2O$), which is also a proton conductor (Figure 19.15). For this material, sorbed water molecules can remain stable in the bulk up to approximately 150°C; therefore, operating temperature of these sensors could not be above this temperature. The response time of the sensor, which was approximately 20 minutes, was reduced to 3 minutes by the addition of 2 wt% organic binder because of an increase in porosity (Miura et al. 1985). Chandra and Hashmi (1990) have found that proton-conducting solid electrolyte, such as ammonium paratungstate pentahydrate $[(NH_4)_{10}W_{12}O_{41} \cdot 5H_2O]$ (APT·$5H_2O$), can also be used for humidity sensor development. Testing showed that hydrotungstite ($H_2WO_4 \cdot H_2O$) with proton conductivity can also be used as a humidity-sensitive material in the room temperature humidity sensors (Kunte et al. 2008). Sensor operated in the range of 20%–85% RH. The response time was found to be of the order of 45 seconds.

An increase in the operating temperature was obtained by using antimony phosphate ($HSb[PO_4]_2 \times 2H_2O$), which is a proton conductor up to 200°C–300°C due to its thermally stable crystal water (Miura et al. 1988). The response time was approximately 2 minutes without the addition of binders. Some other characteristics of these sensors are shown in Figure 19.16. It was also found that the impedance was negligibly affected by other coexistent-reducing gases, such as H_2, CO, CH_4 (Miura et al. 1988). According to Miura et al. (1988), this may indicate that the present sensor is superior to the humidity sensors using semiconductors or solid electrolytes, such as $SrCe_{0.95}Yb_{0.05}O_3$, in respect to humidity selectivity, since the latter sensors can be seriously interfered by coexisting reducing gases.

The same approach was used by Feng and Greenblatt (1993) and Shuk and Greenblatt (1998) for design of NASICON-based humidity sensors. Shuk and Greenblatt (1998) have designed an impedance-type humidity sensor based on porous $HZr_2P_3O_{12}$ or $HZr_2(PO_4)_3$ using thick-film technology. The authors tested sensors at temperatures up to 250°C and commented on a good and reproducible response. The reported response time was

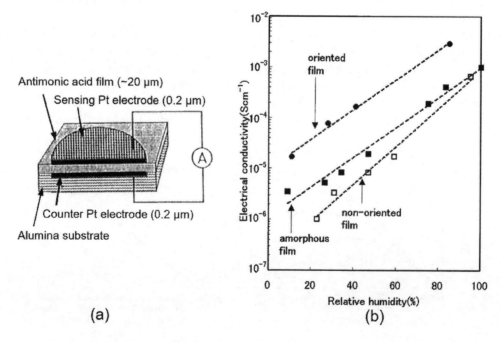

FIGURE 19.15 (a) The structure of the thick-film type, antimonic acid–based, amperometric humidity sensor; (b) Relative humidity (RH) dependence of the electrical conductivity of the (111)-oriented and nonoriented polycrystalline films of cubic $Sb_2O_5 \times nH_2O$, and the amorphous $Sb_2O_5 \times nH_2O$ film. The conductivity was measured at 20°C. (Reprinted with permission from Ozawa K. et al., Preparation and electrical conductivity of three types of antimonic acid films, *J. Mater. Res.*, 13, 830–833, 1998. Copyright 1998, Cambridge University Press.)

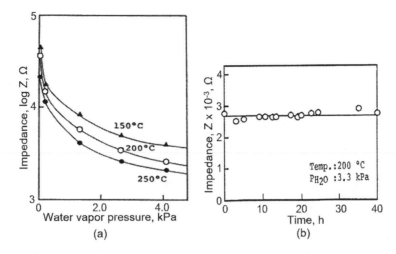

FIGURE 19.16 (a) Dependence of impedance of HSb(PO$_4$)$_2$·2H$_2$O-based sensor on the water vapor pressure at various temperatures; (b) Stability of impedance of the humidity sensor at 200°C. (Reprinted with permission from Miura, N. et al., Humidity sensor using antimony phosphate operative at a medium temperature of 150°C–250°C, *Jpn. J. Appl. Phys.*, 27, L931–L933, 1988. Copyright 1988, Japan Society of Applied Physics.)

10–15 seconds. The sensor signal was dependent on the film thickness of the sensing material, as well as on the microstructure of the film. With increasing film thickness, the conductivity of the film increases and becomes more sensitive to changes of RH (see Figure 19.17). Long-term tests of the film humidity sensors showed relatively good agreement with the commercial humidity sensor, OMEGA-RH-20C, at room temperature over several weeks. In addition, the prototype film sensor was successfully tested in the climate chamber in a wide temperature (25°C–150°C) and RH (2%–100%) range, and results were in good agreement with a commercial humidity sensor by Vaisala (www.vaisala.com). The tests of the film sensor in the climate chamber at higher temperatures (175°C–250°C) also showed good and reproducible response to the change from dry air to steam. One should note that no commercial humidity sensor is available in this temperature range for inline humidity measurements.

Feng and Greenblatt (1993) proposed to use for these purposes HZr$_2$P$_3$O$_{12}$·ZrP$_2$O$_7$. They established that proton motion in the HZr$_2$P$_3$O$_{12}$·ZrP$_2$O$_7$ composite occurs at approximately 450°C, and the dramatic increase of the conductivity in moist conditions takes place at just this temperature. This means that 450°C is the optimal temperature for humidity sensing in this material. The nearly linear behavior of conductivity dependence on the humidity is the main advantage of these sensors (see Figure 19.18). As can be seen in Figure 19.18, deviation from linearity is observed only in the region of low humidity (RH < 20%). On the basis of the experimental evidence, Feng and Greenblatt (1993) assumed that the protonation mechanism takes place via a charge-transfer reaction at the electrode-electrolyte interface, according to the reaction

$$H_2O \text{ (gas)} \rightarrow 2H^+ \text{ (electrolyte)} + 1/2O_2 + 2e^- \quad (19.12)$$

Feng and Greenblatt (1993) have also found that the variation of the real part of the impedance (Z′) with humidity was independent of the frequency in the range 10–20 Hz, which is an additional advantage for application of this composite ceramic material in the humidity-sensing device.

Another example of impedance humidity sensors is devices developed on the basis of Na$_3$HGe$_7$O$_{16}$ (Feng et al. 1992; Feng and Greenblatt 1992a, 1992b).

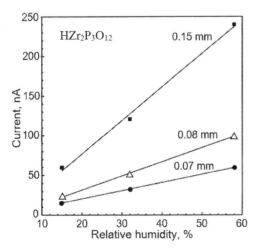

FIGURE 19.17 The sensor response as a function of the thickness of HZr$_2$P$_3$O$_{12}$ film. (Reprinted with permission from Shuk, P. and Greenblatt, M., Solid electrolyte film humidity sensor, *Solid State Ionics*, 113–115, 229–233, 1998. Copyright 1998, Elsevier.)

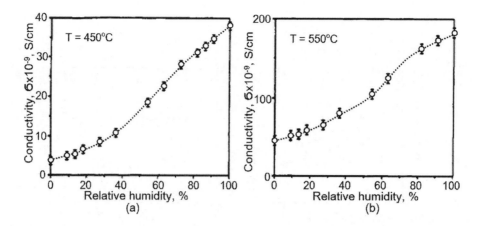

FIGURE 19.18 Humidity dependence of conductivity of the $HZr_2P_3O_{12} \cdot ZrP_2O_7$, determined from ac impedance measurements at different temperatures: (a) 450°C and (b) 550°C. (Reprinted with permission from Feng S. and Greenblatt M., Proton conductivity and humidity–sensing properties at high temperature of the NASICON–based composite material $HZr_2P_3O_{12} \cdot ZrP_2O_7$, *Chem. Mater.*, 5, 1277–1282, 1993. Copyright 1993, American Chemical Society.)

It was found that, for the hydrated samples, the extent of ionization of water molecules (protonation) by the cations and the water and/or ammonia content (formation of conducting path) dramatically affect the magnitude of proton conductivity. In addition, compared with aluminosilicate zeolites, the germinates studied here show higher ionic conductivities and lower activation energies. Sodium ion conductivity in dehydrated $Na_3HGe_7O_{16}$ was observed with conductivities from approximately 10^{-8} $(\Omega \cdot cm)^{-1}$ at 125°C to approximately 10^{-3} $(\Omega \cdot cm)^{-1}$ at 500°C with $E_a = 0.64$ eV. Humidity-sensing measurement on the Na-germanates indicated that the conductivity was sensitive to humidity in the temperature range 50°C–120°C. At 50°C, the variation of conductivity is approximately 2.5 orders of magnitude from $5.3 \cdot 10^{-7}$ $(\Omega \cdot cm)^{-1}$ at P_{H_2O} of approximately 50 mmHg to $2.1 \cdot 10^{-5}$ $(\Omega \cdot cm)^{-1}$ at P_{H_2O} of approximately 450 mmHg. It was suggested that the conductivity at 120°C was dominated by the intergranular contribution. In addition, Feng et al. (1992) concluded that, in the temperature range 50°C–120°C, the mobile species contributing to the conductivity with the change of humidity are protons, or H_3O^+ ions, rather than sodium ions associated with water molecules.

Tsai et al. (1991) believe that $Na_3Mo_2P_2O_{11}(OH)$ is also promising material for humidity sensors. Figure 19.19 plots logσ versus log(RH) at 100°C for this material. The resulting slope is 1.5 above log(RH) of approximately 1.5. According to Tsai et al. (1991), this implies that three surface water molecules (solvation number = 3) are needed to form one proton above RH = 32%. Moreover, since the experimental solvation number is less than the theoretical primary solvation number (≈ 4.5) typically required for Na^+ (Rabo 1979), the mobile ion species are likely H^+ ions rather than Na^+ (H_2O) ions (Bockris and Saluja 1972). Tsai et al. (1991) have also found by thermogravimetric analysis (TGA) that 2 moles of water are rapidly incorporated as water of crystallization per formula $Na_3Mo_2P_2O_{11}(OH)$ at RH of approximately 12%. The conductivity increases slowly from RH = 12% to RH = 32%, typical of type-II Brunauer behavior. Therefore, Tsai et al. (1991) suggested that the conductivity mechanism of $Na_3Mo_2P_2O_{11}(OH) \cdot 2H_2O$ at 12% < RH < 32% is attributed to mixed particle hydrate and framework hydrate mechanism. At RH > 32%, the variation of logσ versus log(RH) clearly shows a particle hydrate behavior. However, $Na_3Mo_2P_2O_{11}(OH) \cdot 2H_2O$ is not sufficiently stable material and has a limited temperature

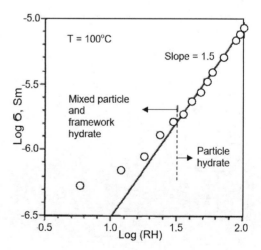

FIGURE 19.19 Profile of logσ versus log(relative humidity [RH]) for pelletized pressed sample of $Na_3Mo_2P_2O_{11}(OH)$ at 100°C. (Reprinted with permission from Tsai, M. et al., Ionic conductivity in layered $Na_3Mo_2P_2O_{11}(OH) \cdot 2H_2O$, *Solid State Ionics*, 47, 305–311, 1991. Copyright 1991, Elsevier.)

range for applications. $Na_3Mo_2P_2O_{11}(OH)\cdot 2H_2O$ is a proton conductor at low temperatures < 200°C and a Na^+ ion conductor in the range 260°C–420°C. At 420°C, the compound decomposes.

As for the most high-temperature-impedance humidity sensors, they were developed on the basis of $CaZrO_3$ (Zhou and Ahmad 2008). Zhou and Ahmad (2008) established that, by doping with indium oxide, the oxygen ion conductivity of the In-doped $CaZrO_3$ increased dramatically, and the resulting sensor became sensitive to humidity and hydrogen (in the presence of oxygen) at high temperatures (e.g., 700°C). The impedance plots of the $CaZrO_3$ at different humidity and of the 10 mol% In-doped $CaZrO_3$ at 700°C in 20% humidity and at different oxygen concentrations are shown in Figure 19.20. This indicates the In-doped $CaZrO_3$ is also a good protonic conductor in the presence of oxygen. The results also show that the evolution of protonic conduction occurs at the expense of oxygen ion conduction. Zhou and Ahmad (2008) believe that, although the sensor is not selective for humidity and oxygen due to the ion conducting mechanism, the results indicate that this sensor can be used to detect humidity in an environment where oxygen and other gas components are maintained constant.

Chen et al. (2009a, 2009b) have shown that $BaZrO_3$-based impedance sensors can also operate at high temperatures. $BaZrO_3$, like $CaZrO_3$, is a ceramic material with Perovskite structure that has been reported to have high chemical stability at high operating temperatures (i.e., it does not [permanently] change its basic chemical composition and structure when exposed to other gaseous species) and high proton conductivity when doped appropriately. At temperatures (T) > 500°C, H_2O dissociative adsorption is accompanied by "filling in" the oxygen vacancies and releasing protons into the structure. Sensors were fabricated using Y-doped $BaZrO_3$ thin films deposited by pulsed laser ablation technique (Chen et al. (2009a) and by sputtering (Chen et al. 2009b). It was established that T = 500°C is an optimal temperature for these sensors' operation. The reduction in sensitivity with increasing film thickness indicated that absorption of H_2O into the lattice occurs primarily in the surface layers. This is consistent with the results of the structural characterization of the $BaZrO_3$:Y films, which showed that the films had a dense structure. Response time at T = 500°C was in the range of 100–250 seconds. Desorption times were larger than absorption times, indicating a slower removal process of humidity. As for the stability of the parameters of these sensors, the same experiments showed that the sensors are subject to aging (see Figure 19.21). This means that further research in this direction is necessary.

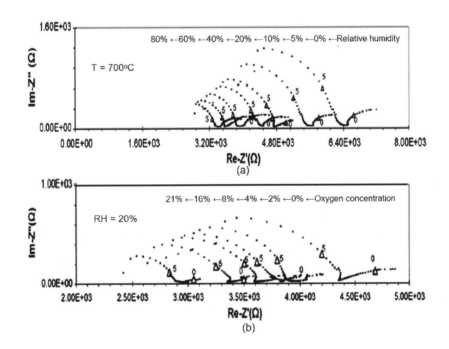

FIGURE 19.20 (a) Complex impedance plots of the sintered In-doped $CaZrO_3$ at 700°C in nitrogen with different humidity; (b) The influence of oxygen concentration on the complex impedance plots of the 10 mol% In-doped $CaZrO_3$ at 700°C with 20% humidity. RH, relative humidity. (Reprinted with permission from Zhou, M. and Ahmad, A., Sol–gel processing of In–doped $CaZrO_3$ solid electrolyte and the impedimetric sensing characteristics of humidity and hydrogen, *Sens. Actuators B*, 129, 285–291, 2008. Copyright 2008, Elsevier.)

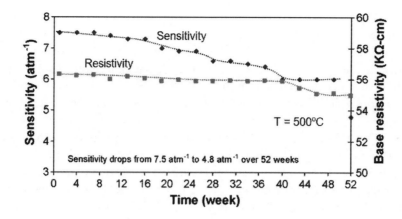

FIGURE 19.21 The stability test of the humidity sensors with Y–doped BaZrO$_3$ 200 nm pulsed laser deposited (PLD) film during 52 weeks, the sensitivity range drops from 7.5 atm^{-1} to 4.8 atm^{-1} with the corresponding base resistivity at 50°C when partial pressure of water at 0.058 atm. The base resistivity (i.e., at 0 atm water vapor) of the sample drops but remains comparably stable. (Reprinted with permission from Chen, X. X., et al., Pulsed laser deposited Y–doped BaZrO$_3$ thin films for high temperature humidity sensors, *Sens. Actuators B*, 142, 166–174, 2009. Copyright 2009, Elsevier.)

19.4 SUMMARY

As follows from this review, the solid-state electrochemical humidity sensors are reliable and stable devices that allow monitoring humidity in extreme conditions that are not suitable for the operation of most known humidity sensors. As shown, the sensors are capable of operating even at temperatures of approximately 1000°C. Moreover, their indications are predictable, which is an essential advantage of such devices. The main disadvantage of potentiometric-type humidity sensors is that they need a reference atmosphere, and this requires a complex structure for the measuring device.

In addition, for reliable operation, it is necessary to monitor the concentration of oxygen and hydrogen in the gas mixture. As can be seen from the results presented in this chapter, the most reliable and stable operation of humidity sensors is observed in an oxygen-free atmosphere.

The large power consumption and significant limitations on microminiaturization also significantly limit the wide use of such sensors. Therefore, despite the advantages, the field of use of such sensors is limited by specific high-temperature applications, such as in situ source emissions control in power plants during combustion processes.

REFERENCES

Abe Y., Shimakawa H., Hench L.L. (1982) Protonic conduction in alkaline earth metaphosphate glasses containing water. *J. Non–Cryst. Solids* 51, 357–365.

Bockris J. O'M., Saluja P.P.S. (1972) Ionic solvation numbers from compressibilities and ionic vibration potentials measurements. *J. Phys. Chem.* 76, 2140–2151.

Chandra S., Hashmi S.A. (1990) Humidity sensor using a proton conductor ammonium paratungstate pentahydrate (APT·5H$_2$O). *Solid State Ionics* 40/41, 460–462.

Chen X.X., Rieth L., Miller M.S., Solzbacher F. (2009a) Pulsed laser deposited Y–doped BaZrO$_3$ thin films for high temperature humidity sensors. *Sens. Actuators B* 142, 166–174.

Chen X.X., Rieth L., Miller M.S., Solzbacher F. (2009b) High temperature humidity sensors based on sputtered Y–doped BaZrO$_3$ thin films. *Sens. Actuators B* 137, 578–585.

Colomban P. (ed.) (1992) *Proton Conductors*. Cambridge University Press, New York.

Danilov N., Vdovin G., Reznitskikh O., Medvedev D., Demin A., Tsiakaras P. (2016) Physicochemical characterization and transport features of proton–conducting Sr–doped LaYO$_3$ electrolyte ceramics, *J. Eur. Ceram. Soc.* 36, 2795–2800.

Feng S., Greenblatt M. (1992a) Preparation, characterization, and ionic conductivity of novel crystalline, microporous Germanates, M$_3$HGe$_7$O$_{16}$·xH$_2$O, M = NH^{4+}, Li$^+$, K$^+$, Rb$^+$, Cs$^+$; x = 4–6. 2. *Chem. Mater.* 4, 462–468.

Feng S., Greenblatt M. (1992b) Preparation, characterization, and ionic conductivity of novel crystalline, microporous silicogermanates, M$_3$HGe$_{7-x}$Si$_x$O$_{16-12x}$·H$_2$O, M = K$^+$, Rb$^+$, Cs$^+$; 0 < m < 3; x = 0–4. 3. *Chem. Mater.* 4, 468–472.

Feng S., Greenblatt M. (1992c) Galvanic cell type humidity sensor with NASICON–based material operative at high temperature. *Chem. Mater.* 4, 1257–1262.

Feng S., Greenblatt M. (1993) Proton conductivity and humidity–sensing properties at high temperature of the NASICON–based composite material HZr$_2$P$_3$O$_{12}$·ZrP$_2$O$_7$. *Chem. Mater.* 5, 1277–1282.

Feng S., Tsai M., Greenblatt M. (1992) Preparation, ionic conductivity, and humidity–sensing property of novel, crystalline microporous Germanates, Na$_3$HGe$_7$O$_{16}$·H$_2$O, x = 0–6. 1. *Chem. Mater.* 4, 388–393.

Greenblatt M., Feng S. (1993) Proton conducting solid electrolytes for high temperature humidity sensing. *Mat. Res. Symp. Proc.* 293, 283–294.

Greenblatt M., Shuk P. (1996) Solid–state humidity sensors. *Solid State Ionics* 86–88, 995–1000.

Greenblatt M., Tsai P.P., Kodama T., Tanase S. (1990) Humidity sensors with sintered β–Ca(PO)$_4$ for high temperature use. *Solid State Ionics* 40/41, 444–447.

Hassen M.A., Clarke A.G., Swetnam M.A., Kumar R.V., Fray D.J. (2000) High temperature humidity monitoring using doped strontium cerate sensors. *Sens. Actuators B* 69, 138–143.

Iwahara H., Uchida H., Kondo J. (1983) Galvanic cell–type humidity sensor using high temperature–type proton conductive solid electrolyte. *J. Appl. Electrochem.* 13, 365–370.

Iwahara H., Uchida H., Ono K., Ogaki K. (1988) Proton conduction in sintered oxides based on $BaCeO_3$. *J. Electrochem. Soc.* 135, 529–533.

Iwahara H. (1990) Use of high temperature proton conductor for gas sensors, In: Yamazoe N. (Ed.) *Chemical Sensor Technology,* Vol. 3, Kodansha, Tokyo/Elsevier, Amsterdam, the Netherlands pp. 117–129.

Klein C., Hurlbut C.K. Jr. (1993) *Manual of Mineralogy (after James D. Dana)*, 21st edn. John Wiley & Sons, New York.

Korotcenkov G., Han S.D., Stetter J.R. (2009) Review of electrochemical hydrogen sensors. *Chem. Rev.* 109 (3), 1402–1433.

Kozawa A. (1974) Electrochemistry of manganese dioxide and production and properties of electrolytic manganese dioxide (EMD), In: Kordesch K.V. (Ed.) *Batteries. Manganese Dioxide.* Marcel Dekker, New York, Vol. 1, Ch. 3.

Kreuer K.D. (1996) Proton conductivity: Materials and applications. *Chem. Mater.* 8, 610–641.

Lieder M., Rourke F., Vincent C.A. (1990) A solid state amperometric humidity sensor. *J. Appl. Electrochem.* 20, 964–968.

Kunte G.V., Shivashankar S.A., Umarji A.M. (2008) Humidity sensing characteristics of hydrotungstite thin films. *Bull. Mater. Sci.* 31 (6), 835–839.

Liu Y., Parisi J., Sun X., Lei Y. (2014) Solid–state gas sensors for high temperature applications – a review. *J. Mater. Chem. A* 2, 9919–9943.

Mahato N., Banerjee A., Gupta A., Omar S., Balani K. (2015) Progress in material selection for solid oxide fuel cell technology: A review. *Prog. Mater. Sci.* 72, 141–337.

Marrero T.R., Mason E.A. (1972) Gaseous diffusion coefficients. *J. Phys. Chem. Ref. Data* 1, 3–118.

Medvedev D., Kalyakin A., Volkov A., Demin A., Tsiakaras P. (2017) Electrochemical moisture analysis by combining oxygen- and proton-conducting ceramic electrolytes. *Electrochem. Commun.* 76, 55–58.

Miura N., Yashima I., Yamazoe N. (1985) Humidity–sensing characteristics of antimonic acid proton conductor element at medium temperatures. *Z Chem. Soc. Jpn.* 9, 1644–1649 (in Japanese).

Miura N., Mizuno H., Yamazoe N. (1988) Humidity sensor using antimony phosphate operative at a medium temperature of 150°C–250°C. *Jpn. J. Appl. Phys.* 27 (5), L931–L933.

Miyazaki K., Xu C.N., Hieda M. (1994) A new potential–type humidity sensor using EMD–based Manganese oxides as a solid electrolyte. *Electrochem. Soc.* 141 (4), L35–L37.

Miyazaki K., Hieda M., Kato T. (1997) Development of a novel manganese oxide–clay humidity sensor. *Ind. Eng. Chem. Res.* 36, 88–91.

Nagata K., Nishino M., Goto K.S. (1987) Humidity sensor with $SrCe_{0.95}Yb_{0.05}O_3$ solid electrolyte for high temperature use. *J. Electrochem. Soc.* 134, 1850–1854.

Okuyama Y., Kozai T., Ikeda S., Matsuka M., Sakai T., Matsumoto H. (2014) Incorporation and conduction of proton in Sr–doped $LaMO_3$ (M = Al, Sc, In, Yb, Y). *Electrochim. Acta* 125, 443–449.

Okuyama Y., Nagamine S., Nakajima A., Sakai G., Matsunaga N., Takahashi F., Kimata K., Oshima T., Tsuneyoshi K. (2016) Proton–conducting oxide with redox protonation and its application to a hydrogen sensor with a self–standard electrode. *RSC Adv.* 6, 34019–34026.

Ozawa Y., Miura N., Yamazoe N., Seiyama T. (1983) Proton conduction in antimonic acid at medium temperatures in the presence of water vapor. *Chem. Lett.* 12 (10), 1569–1572.

Ozawa K., Sakka Y., Amano M. (1998) Preparation and electrical conductivity of three types of antimonic acid films. *J. Mater. Res.* 13 (4) 830–833.

Rabo J.A. (ed.) (1979) *Zeolite Chemistry and Catalysis*, ACS Monograph 171. American Chemical Society, Washington DC, p. 187.

Shuk P., Greenblatt M. (1998) Solid electrolyte film humidity sensor. *Solid State Ionics* 113–115, 229–233.

Stetter J.R., Korotcenkov G., Zeng X., Liu Y., Tang Y. (2011) Electrochemical gas sensors: Fundamentals, fabrication and parameters. In: Korotcenkov G. (ed.), *Chemical Sensors: Comprehensive Sensor Technologies. Vol. 5. Electrochemical and Optical Sensors.* Momentum Press, New York, pp. 1–89.

Sun G., Wang H., Jiang Z. (2011) Humidity response properties of a potentiometric sensor using LaF_3 thin film as the solid electrolyte. *Rev. Sci. Instrum.* 82, 083901.

Tanase S., Greenblatt M., Tsai P.P. (1989) Apparatus for evaluating humidity-sensing characteristics of solid electrolytes. *Rev. Sci. Instr.* 60, 3809–3811.

Traversa E. (1995) Ceramic sensors for humidity detection: The state–of–the–art and future developments. *Sens. Actuators B* 23, 135–156.

Tsai F.T., Tanase S., Greenblatt M. (1989) High temperature humidity sensing materials. *Mat. Res. Symp. Proc.* 135, 603–608.

Tsai M., Feng S., Greenblatt M., Haushalter R.C. (1991) Ionic conductivity in layered $Na_3Mo_2P_2O_{11}(OH)\cdot 2H_2O$. *Solid State Ionics* 47, 305–311.

Usui T., Kurumiya Y., Nuri K., Nakazawa M. (1989a) Gas polarographic multifunctional sensor: Oxygen–humidity sensor. *Sens. Actuators* 16, 345–358.

Usui T., Kurukiya Y., Ishibashi K., Nakazawa M. (1989b) Gas polarographic sensor usable above 100°C. *Jpn. J. Appl. Phys.* 28 (11), 2325–2326.

Yagi H., Ichikawa K. (1993) Humidity sensing characteristics of a limiting current type planar oxygen sensor for high temperatures. *Sens. Actuators B 13*, 92–95.

Yagi H., Ichikawa K. (1995) High temperature humidity sensor using a limiting–current–type plane multi–oxygen sensor for direct firing system. *Sens. Actuators B* 25, 701–704.

Zhou M., Ahmad A. (2008) Sol–gel processing of In–doped $CaZrO_3$ solid electrolyte and the impedimetric sensing characteristics of humidity and hydrogen. *Sens. Actuators B* 129, 285–291.

Section IV

New Trends and Outlook

20 Microwave-Based Humidity Sensors

20.1 INTRODUCTION

Microwave sensors are based on the interaction between microwaves and matter. Microwave sensors utilize electromagnetic fields and devices internally operating at frequencies starting from approximately 300 MHz up to the terahertz range. This interaction may be in the form of reflection, refraction, scattering, emission, absorption, or change of speed and phase. In Volume 1 (Chapter 5) it was shown that passive and active microwave radiometry can be used for remote monitoring of water vapor in the atmosphere. The microwave energy recorded by a passive sensor can be emitted by the atmosphere, reflected from the surface, emitted from the surface, or transmitted from the subsurface. Active microwave sensors provide their own source of microwave radiation to illuminate the target. In this chapter, it will be shown that the same principle can be used for humidity sensor development. It is important to note that constructively microwave humidity sensors are fundamentally different from those considered earlier. A description of the various approaches to the development of microwave sensors for various purposes can be found in Kraszewski (1980a, 1980b, 1996, 2001), Vainikainen et al. (1986), Nyfors (2000), and Kempka et al. (2006).

For the development of microwave humidity sensors, a microwave sensing based on the transmitter-receiver systems is used. The sensor transmits a microwave (radio) signal through the tested gas and detects the transmitted or reflected signal using the measuring circuit shown in Figure 20.1. As was shown in Volume 1, the water vapor has powerful absorption bands in the microwave region (Table 20.1), which allow estimation of the concentration of water vapor by absorption. In addition, water vapor, especially when condensed, has a strong effect on the dielectric constant of the medium with concomitant effects (Kraszewski 2001). This property is mostly proposed for use in the development of gas sensors, including humidity sensors. The relationship between a microwave parameter of the material (e.g., attenuation), A, and the dielectric permittivity of the mixture may be expressed in a general form as $A = f(\varepsilon)$ (Kraszewski 2001). Similarly to capacitance measurements, this transduction technique estimates the change in the permittivity of the sensitive material at microwave frequencies as a function of the adsorbed quantity of water vapor molecules on the surface of the sensitive layer at room temperature. This approach utilizes some well-known physical theories—for example, the theory of dielectric mixtures and the theory of bound water: adsorption and capillary condensation (see Section 10.3).

The dispersion and dissipation of electromagnetic energy interacting with dielectric material depend upon the dimensions, shape, and relative permittivity (dielectric properties) of the material (Kraszewski 2001). When the moisture content of the material changes, the change is reflected in the wave parameters. Because the relative permittivity of water differs significantly from that of

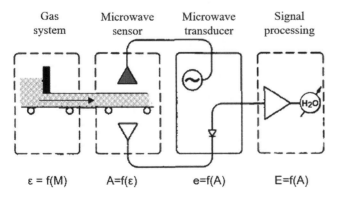

FIGURE 20.1 Block diagram of the microwave moisture content meter.

TABLE 20.1
Frequency of Active Microwave Sensor for Water Vapor Monitoring

Monitoring Target	Frequency (GHz)
Soil moisture	Around 1.4
Soil moisture	Around 2.7
Water vapor, rain	Around 15
Rain, water vapor	Around 18
Water vapor, liquid water	Around 21
Water vapor, liquid water	22.24
Water vapor, liquid water	Around 24
Water vapor, cloud, liquid water	Around 30
Rain, cloud, water vapor	Around 37
Cloud, snow	Around 90
Water vapor	183.31
Water vapor	325.10
Water vapor	380.20

most hygroscopic dielectric materials, its effect can be separated from the effect of the dry dielectric material. In general, this may be expressed in a functional form as

$$\alpha = \Phi_1(m_w, m_d) \text{ and } \beta = \Phi_2(m_w, m_d) \quad (20.1)$$

where α and β are any two descriptive electromagnetic wave parameters. Regardless of the complexity of the analytical expressions described by Equation 20.1, it is generally possible to solve the two equations by separation of variables and to express the mass of water and the mass of dry material in terms of two measured parameters in the form

$$m_w = \Psi_1(\alpha, \beta) \text{ and } m_d = \Psi_2(\alpha, \beta) \quad (20.2)$$

Substituting the analytical expressions corresponding to Equation 20.2 into Equation 20.3,

$$M = \frac{m_w}{m_w + m_d} \times 100, \quad (20.3)$$

where M is the moisture content of the material, m_w is the mass of eater, and m_d is the mass of dry material, then the general expression for the moisture content of material can be written as

$$M = \frac{\Psi_1(\alpha, \beta)}{\Psi_1(\alpha, \beta) + \Psi_2(\alpha, \beta)} \times 100 \quad (20.4)$$

This equation contains only the wave parameters determined experimentally and is totally independent of the material density. Taking into account the direct relationship between the quantity of water, adsorbed by the material and the humidity of the environment, it is possible to go to the determination of air humidity or test gases through calibration.

20.2 MICROWAVE SENSORS

There are several possible ways to arrange a microwave sensor measurement. They all have different characteristics, which make them suitable for different applications. For example, many forms of radiating elements in waveguide, coaxial-line, or stripline configurations can be used as microwave sensors. They may be divided into resonant and aperiodic groups, into open and closed structure, and into reflection and transmission types (Kraszewski 1980a, 1980b; King and Smith 1981; Gardiol 1985; Nyfors and Vainikainen 1989). Classification and some examples of microwave sensors are shown in Figure 20.2. Waveguide and coaxial line transmission measurements represent closed structures, while the free-space transmission measurements and open-ended coaxial-line systems represent open-structure techniques, respectively. Resonant structures can include either closed resonant cavities or open resonant structures operated as two-port devices for transmission

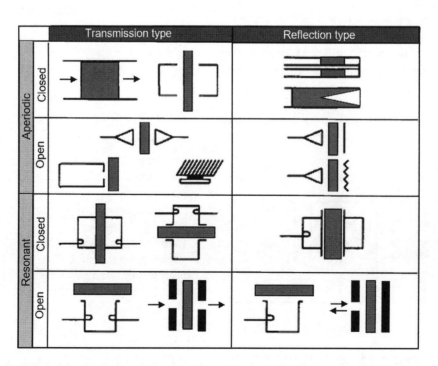

FIGURE 20.2 Classification of microwave sensors used to measure moisture content. (Idea from Kraszewski, A., *IEEE Trans. Microw. Theor. Tech.*, 39, 828–835, 2001.)

measurements or as one-port devices for reflection measurements. A brief overview of the main groups will be briefly given here.

20.2.1 Transmission Sensors

The basic geometrical configuration consists of two antennas: One is transmitting and one is receiving. Those systems are close to their optical equivalents. On the path from the transmitting antenna to the receiving antenna, the microwaves are transmitted through the target object, which affects both the phase and the amplitude of the microwaves. A common application is to measure a gas stream flowing in the pipe. To measure a gas stream in a pipe, the microwaves must be transmitted through dielectric windows on opposite sides of the pipe, with a transmitting antenna on one side and a receiving antenna on the other side (Figure 20.3). On the way between the antennas, the microwave signal penetrates the gas flowing in the pipe. Free-space transmission devices can also be used for moisture monitoring in granulated materials (Narayanan and Vu 2000). Of course, the target object can also be located in the waveguide between the transmitter and receiver (closed system). The permittivity of the target object affects both the phase and the amplitude of the signal. The advantage with this configuration is its simplicity, and the main problem is the sensitivity to reflections in various parts of the system, like the dielectric windows and interfaces inside the material dependent on the flow regime. The reflections in the system cause ripples on the frequency response, and the amplitude is much more affected than the phase. If the sensor is based on measuring only one microwave parameter, a higher accuracy is therefore normally achieved by measuring the phase than the attenuation. It is important to note that the sample geometry is an important factor that needs to be properly controlled for meaningful results (Mladek and Beran 1980). Other requirements for measuring system can be found in Narayanan and Vu (2000).

If the gas flowing through the pipe has a small loss, the pipe will also affect the microwaves. Waveguide modes will be excited that strongly affect the transmission properties. Especially if the measurement is done on a fixed frequency, the errors will be large when the changing permittivity of the flow moves the cut-off frequencies relative to the measurement frequency. By performing a frequency sweep and averaging, or by lining the pipe with a microwave absorber, the error can be decreased. When using a frequency that is much higher than the lowest cut-off frequency, the conditions are similar to the free-space conditions, but there is still the problem with reflections from the walls. For a lossy target object, the influence of the pipe is smaller. In many cases, the best solution is to use the frequency-modulated continuous wave (FMCW) technique, which is often used in radars. The FMCW technique discriminates signals in time, thus being able to exclude the reflections that arrive slightly later than the main signal. The FMCW technique measures the signal delay, which is closely related to the phase measurement in sensors, but lacking the 2p ambiguity in the phase. Both the phase and the signal delay depend on the speed of propagation, which depends on the permittivity.

In some cases, a transmission sensor needs a reference channel for making phase measurements. Normally the reference channel is external to the sensor. If it is directed through the sample instead, but in a different way than the main channel, special features can be achieved.

The approach described above is the simplest approach to the development of humidity sensors, but

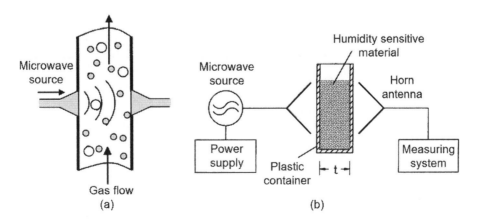

FIGURE 20.3 The basic geometrical configuration of a free-space transmission sensor designed for controlling the gas flow (a) and solid granular material (b).

also the most dimensional. True, in the last decade, significant progress has been made in the development of flat, microstrip, and patch antennas (Pozar and Schaubert 1995), which affected the reduction in the size of such devices.

There are also sensors based on the principle of the reflected wave measurements. They are easier to use, because they allow one-sided sensing and have more robust construction (Kraszewski 2001). It is also clear that the sensitivity of free-space transmission sensors, developed for gas flow control, is lower than in the devices using the humidity-sensitive materials. As was shown in previous chapters, significant changes in the dielectric constant occur only when condensed water appears, which can form in humidity-sensitive materials due to capillary condensation at relatively low humidity.

20.2.2 Resonator Sensors

The resonator-based sensor is another approach to the development of humidity sensors. Microwave resonators can have different configuration—for example, a cavity or metallic chamber, resonating when the operating wavelength exactly matches its dimensions. The measurement frequency range of resonator-based sensor usually is from 50 MHz to more than 100 GHz. Inserting a dielectric object into the cavity changes its electrical dimensions, and the change can be correlated with the object permittivity and then with its moisture content and the humidity of surrounding atmosphere. The output signal versus frequency has the shape of a resonant curve, magnitude of which decreases for increasing moisture content in the material. Thus, the variation in air humidity and water content in a humidity-sensitive material changes the operating frequency of the system. It is important to note that the sensitivity of moisture measurement using resonant structure sensors is significantly higher than when using sensors, discussed above (Nelson 1998; Kraszewski 2001). Table 20.2 shows a general comparison of the microwave measurement systems based on collective information available in the literature and Nelson's (1998) experiences in this field.

20.2.3 Impedance Meters

Impedance-measuring sensors are close to the transmitter-receiver systems described earlier. They differ in that instead of antennas, "applicators" are used to transmit the generated microwave field into and out of the tested sample or object (Barochi et al. 2011; De Fonseca et al. 2015). One or two applicators can be used, as, for example, in Figure 20.4. For example, Barochi et al. (2011) proposed to use the microstrip line. The microstrip line (Kanaya et al. 2005; Marynowski et al. 2010) is one of the most studied propagative structures for high-frequency electronic circuits. It consists of a conductive line of distribution and a ground plan situated on both sides of a substrate (Figure 20.5a). The electromagnetic field orientation is represented on Figure 20.5b. In classic electronic circuits, the substrate is generally an insulating material like glass or epoxy. The characteristics of a microstrip line depend on its structural parameters (line width [w], length, and thickness [t]) and the dielectric constant of the substrate. The characteristic

TABLE 20.2
A General Comparison of the Microwave Dielectric Measurement Systems

Parameter	Slotted Line Reflection System	Guided Wave Transmission System	Free Space Transmission System	Filled Cavity Resonance System	Partial Filled Cavity Resonance System	Probe Reflection System
Frequency	Broad band	Banded	Banded	Single	Single	Broad band
Sample size	Moderate	Moderate	Large	Large	Very small	Small
Temperature control	Difficult	Difficult	Very easy	Very easy	Very easy	Easy
Accuracy for:						
Low-loss material	Very low	Moderate	Moderate	Very high	High	Low
High-loss material	Low	Moderate	Moderate	Does not work	Low	High
Sample preparation	Easy	Difficult	Easy	Very difficult	Very difficult	Easy

Source: Data extracted from Nelson, S.O., Dielectric properties measuring techniques and applications, ASAE Paper No. 983067, ASAE, St. Joseph, MI, 1998.

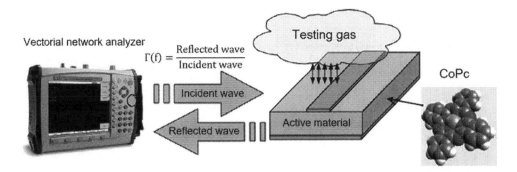

FIGURE 20.4 Scheme of the detection principle by means of a microwave transduction. Γ(f) is the response of vectorial network analyser, CoPc is a cobalt phthalocyanine molecule. The geometry of the propagative structure is a type of microstrip, namely a grounded coplanar waveguide (GCPW), where a sensing material replaces the substrate or is deposited in thin layer on a glass substrate. So, the permittivity variation modifies the characteristic impedance and the propagation constant of the microstrip line that define the sensor. At each frequency, the wave reflected by the sensor is attenuated and out of phase compared to the incident wave, due to specific gas-sensing material interactions. Using a vectorial network analyzer, the sweep in a wide frequency range (30 MHz to 20 GHz) gives spectrum of this gas–material interaction. (Reprinted with permission from Barochi, G. et al., Development of microwave gas sensors, *Sens. Actuators B*, 157, 374–379, 2011. Copyright 2011, Elsevier.)

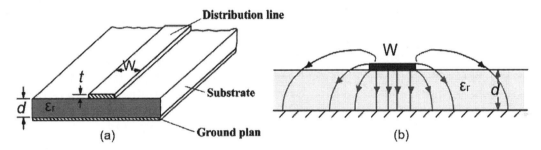

FIGURE 20.5 (a) Representation of a microstrip line and (b) Circulation of the electromagnetic waves between the propagation line and the ground plan. (Reprinted with permission from De Fonseca, B. et al., Microwave signature for gas sensing: 2005 to present, *Urban Climate*, 14, 502–515, 2015. Copyright 2015, Elsevier.)

impedance of a microstrip line and its other characteristics can be calculated through equations, which can be found in the literature (Komarov et al. 2005; Marynowski et al. 2010).

The microwave test instrumentation is often a vector network analyzer, allowing measurement of insertion loss, reflectivity, and time delay over a chosen frequency band. However, this method is laborious, and its use requires a fairly long time for data sampling (Barochi et al. 2011). Instead of such complex and expensive instrumentation, simpler structures using a bridge or a reflectometer can also be designed for particular applications. When such methods are used to determine the moisture content, the frequency used should be above 5 GHz to avoid the influence of ionic conductivity and bound water relaxation (Kraszewski 1996). Using high frequencies can also reduce the size of the device. As is known, the size of microwave components is usually proportional to the wavelength and therefore inversely proportional to the frequency.

20.3 EXAMPLES OF HUMIDITY SENSOR REALIZATION: HUMIDITY SENSOR PERFORMANCE

Microwave sensors developed for humidity measurements are listed in Table 20.3. As is seen, almost all sensors are of resonant and impedance types.

20.3.1 TRANSMISSION SENSORS

Carullo et al. (1998, 1999) tried to develop transmission-based humidity sensors using a holed waveguide filled with air. As indicated before, the air permittivity can be obtained by measuring its effect on the amplitude of a microwave signal that propagates through a holed waveguide. However, the permittivity changes at room temperature, and humidity ranges (20%–90% relative humidity [RH]) are not large enough to be easily detectable in this way (see Figure 20.6). For example, calculations show that the sensitivity of ε_r to the humidity at 20°C does

TABLE 20.3
Microwave Sensors Developed for Humidity Measurements

Configuration	Frequency	Sensing Material	Max. Sensitivity	References
Transmission-Type Sensors				
Holed waveguide	9 GHz	Air	—	Carullo et al. (1998, 1999)
Resonant-Type Sensors				
A circular waveguide	9.5 GHz	Air	—	Vainikainen et al. (1986)
Cylindrical cavity resonator. Differential technique	7.8 GHz	CF_6	0.13 µV/ppm	Rouleau et al. (2000, 2001)
Quasi-spherical resonator	13.5 GHz	Air	—	Cuccaro et al. (2012); Underwood et al. (2012); Gavioso et al. (2014)
Impedance Resonant Sensors				
Microstrip line planar technique	1 GHz	PI	5 kHz/%RH	Bernou et al. (2000)
Planar-inductive coupling with a remote antenna	≈ 18 MHz	PI	16 kHz/%RH	Harpster et al. (2002)
Coplanar waveguide-to-slotline ring resonator	2.9 GHz	PI	108 kHz/%RH	Kim et al. (2006)
Coplanar waveguide line with stepped impedance resonators	1 GHz	PI	0.6 MHz/%RH	Amin and Karmakar (2012); Amin et al. (2013)
		PVA	2.4 MHz/%RH	
Coplanar waveguide line with ELC resonator	6–7 GHz	PI	1.7 MHz/%RH	Amin et al. (2013, 2014)
		PVA	6.8 MHz/%RH	
Coplanar waveguide line with ELC resonator	≈ 200 MHz	PEL paper	0.6 MHz/%RH	Feng et al. (2015)
Loop antenna loaded with interdigital capacitor	2.85 GHz	PVA	5.4 MHz/%RH	Lu et al. (2014)
Interdigital capacitor-controlled oscillator with a phase-locked loop	900 MHz	PI	18.8 fF/%RH	Wu et al. (2015)
Chipless sensor tag with multiple coupled loops resonator	3.3 GHz	Si nanowires	≈ 1 MHz/%RH	Vena et al. (2016)
Coplanar waveguide line	4.6 GHz	Cellulose nanofibers	2.8 MHz/%RH	Eyebe et al. (2017)

ELC, electric field-coupled inductor capacitor; PEL paper, nonorganic coated inkjet paper (PEL Nano P60, Printed Electronics Ltd.); PI, polyimide; PVA, polyvinyl alcohol.

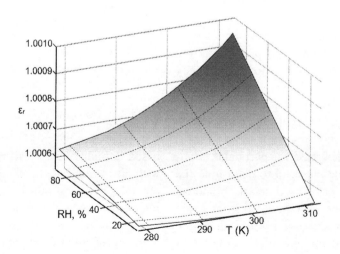

FIGURE 20.6 Theoretical behavior of air relative permittivity. (Data extracted from Carullo, A. et al., Microwave system for relative humidity measurement, In: *Proceedings of 16th IEEE Instrumentation and Technology Conference*, May 24–26, New York, pp. 124–129, 1999.)

not exceed $2 \cdot 10^{-6}$ % RH^{-1}. For this reason, Carullo et al. (1998, 1999) proposed to detect the phase of a microwave signal by means of an interferometer system. The principle scheme of the proposed sensor is shown in Figure 20.7a. However, such solution requires another signal that acts as a reference for the signal that propagates through the waveguide. As reference channel, Carullo et al. (1998, 1999) planned to use either coaxial cable (single waveguide hygrometer [SWH]) (Figure 20.7b) or an additional holed waveguide (twin waveguide hygrometer [TWH]) (Figure 20.7c). The SWH is based on a simple assembly but requires severe specifications for the temperature sensor and the signal generator. On the contrary, the TWH assembly is more complex but allows one to relax the specification of the temperature uncertainty and the frequency stability. The waveguides in experimental instrument had a length of approximately 40 cm, and the working frequency was of 9 GHz. However, these elaborations have not received further development. We can assume that this is due to the low sensitivity and large dimensions of the device, which do not allow us to hope for the possibility of their practical application.

20.3.2 Resonant and Impedance Sensors

The first resonant humidity sensors were developed by Hasegawa and Stokesberry (1975), and these sensors did not assume the use of humidity-sensitive materials. Hasegawa and Stokesberry (1975) have designed a hygrometer aimed to measure the humidity of atmospheric air over the vapor pressure range 3–7400 Pa (0.03–74.00 mbar). The instrument was an adaptation of a microwave refractometer using two cavities operating at 12 GHz. One cavity was exposed to the moist test air, and the other was exposed to the same air sample with all the water vapor removed. Both cavities were maintained at the same fixed temperature in a thermostated oven and at the same total pressure. The difference in frequency between the cavities was automatically nulled by a tuning probe in the sampling cavity. The instrument was calibrated by two independent methods. One involved the measurement of the resonance frequency of the sampling cavity as a function of probe penetration and using it in a theoretically derived equation for vapor pressure. The second method involved the measurement of the probe penetration as an empirical function of known vapor pressure of a test gas. These two methods yielded the results that agreed on the average to better than 0.5% for vapor pressures up to 3050 Pa (30.5 mbar).

Rouleau et al. (2000) have also used a microwave differential system with reference resonator. The microwave differential technique consists of measuring the reflection coefficient difference $\Delta\Gamma$, between the reference and the measuring resonators (Rouleau et al. 2001). The reference resonator was filled with dehydrated gas. This system with cylindrical cavity was used to measure trace moisture levels in SF_6. In their calculations, they proceeded from the assumption that, for a cylindrical cavity resonator of diameter D and length l filled with a gas of permittivity ε, the resonant frequency, f_r, is given by Rizzi (1988)

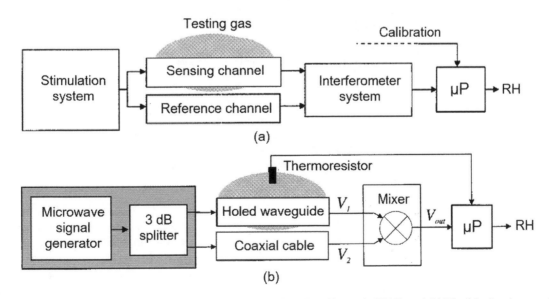

FIGURE 20.7 (a) The principle scheme of the sensor proposed by Carullo et al. (1999) and (b) The block scheme of the single waveguide hygrometer. (Idea from Carullo, A. et al., Microwave system for relative humidity measurement, In: *Proceedings of 16th IEEE Instrumentation and Technology Conference*, May 24–26, New York, pp. 124–129, 1999.)

$$f_r = \frac{c}{\sqrt{\mu_r \varepsilon_r}} \sqrt{\left(\frac{q'_{nm}}{\pi D}\right)^2 + \left(\frac{p}{2l}\right)^2} \quad (20.5)$$

where c is the velocity of light in free space, μ_r the relative permeability, ε_r the relative permittivity, p the number of half wavelengths from one end of the cavity to the other, and q'_{nm} (for transverse electric (TE) mode) is the mth root of the first derivative of the nth order Bessel function J, of the first kind. For any TE mode, the relative permittivity can be found by

$$\varepsilon_r = \left(\frac{f_0}{f_r}\right)^2, \quad (20.6)$$

where f_0 is the resonant frequency of the resonator under vacuum. Experimental instrument had resonators with dimensions $D = 4.925$ cm and $l = 6.300$ cm, and resonant frequency equaled approximately 7.8 GHz. Rouleau et al. (2000) believe that this instrument is able to measure moisture levels lower than 10 ppm, and it can be used on line with an appropriate filter for moisture removal connected to the reference resonator.

Vainikainen et al. (1986) developed a resonance humidity sensor (Figure 20.8b), in which the resonant frequency and the Q-factor of the cavity depended on the medium characteristic (i.e., the humidity) of the air. As the humidity changes, the resonant frequency of the cavity varied (Figure 20.8a). The resolution of the measurement equipment was approximately two parts in 10^6. The measurement time of the equipment was approximately 50 milliseconds, which made fast dynamic measurements possible. The resonant cavity was made of Invar, and it was designed to resonate at 9.5 GHz. It is seen that the device has large dimensions. In addition, it is noteworthy that the device becomes sensitive only with an increase in temperature (see Figure 20.8a), when the concentration of water vapor in the air can reach large values.

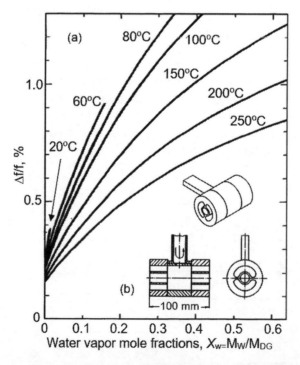

FIGURE 20.8 (a) Negative resonance frequency shift caused by moist air. M_W is the mass of water vapor, and M_{DG} is the mass of dry air. (b) Electromagnetic TE_{011}-mode resonator cavity sensor for humidity measurements developed by Vainikainen et al. (1986). The humidity sensor consists of a circular waveguide that is terminated at each end with a coaxial cylindrical partition. This partition separates a portion of the waveguide into coaxial waveguide plus a smaller circular waveguide, which also serves as the inner conductor for the coaxial waveguide. This partition acts like a perfect reflector to the microwave energy in the TE_{01} mode. The waveguide cavity thus supports the TE_{011} resonant mode. Owing to the construction, the terminations permit the free flow of the medium through the resonator, although they act as electrical short-circuits. The electric field strength of the TE_{011}-mode is zero on the inner surface of the cavity, which makes the sensor fairly insensitive to dirt. Thus, continuous measurement of the dielectric properties of the medium can be made. (Data extracted from Vainikainen, P.-V. et al., *Electron. Lett.*, 22, 985–987, 1986.)

Underwood et al. (2012) also did not use humidity-sensitive materials when developing the microwave resonance dew-point hygrometer (see Figure 20.9a). They have used a quasi-spherical resonator (Figure 20.9b) and detected the onset of condensation by measuring the frequency ratio of selected microwave modes. Based on the simulation, the authors supposed that this technique will be able to detect as little as 0.1 mm^3 of water in the 69 cm^3 resonator, corresponding to a condensed layer of minimum average thickness around 12 nm. The authors believe that, in comparison to optical dew-point hygrometers, the microwave device presented may offer some interesting advantages. Firstly, the same instrumentation may be used in two ways: to measure the water fraction in a gaseous mixture well away from saturation conditions, or alternatively, for a direct measurement of the dew point. Measurement of the water fraction requires the independent knowledge of the permittivity and equation of state of the mixture components. However, the saturation curve or condensation temperature can be found even when this information is unavailable. Secondly, when used as a condensation hygrometer, the method described in this work allows a quantitative estimate of the volume and thickness of the condensate layer.

The same approach and the same microwave resonator were used by Cuccaro et al. (2012). A small-volume (65 cm^3), gold-plated, quasi-spherical microwave resonator has been used to measure the water vapor mole fraction x_w of H$_2$O/N$_2$ and H$_2$O/air mixtures. According to Cuccaro et al. (2012), this experimental technique exploits the high precision, achievable in the determination of the cavity microwave resonance frequencies, and is particularly sensitive to the presence of small concentrations of water vapor as a result of the high polarizability of this substance. For their simulations, they used an assumption that the relative permittivity of the mixture ε_{mix} may be conveniently approximated (Buckingham and Raab 1958) as

$$\varepsilon_{mix} = \frac{2\rho_{mix}\alpha_{mix} + 1}{1 - \rho_{mix}\alpha_{mix}} \quad (20.7)$$

where ρ_{mix} is the density of the mixture, α_{mix} is the molar polarizability of a mixture, which can be calculated using the simple mixing rule $\alpha_{mix} = \sum_i x_i \alpha_i$, α_i is the molar polarizability of pure component i, and x_i is the corresponding mole fraction. The validity of this simplifying assumption, for a mixture whose components differ greatly in volatility, has been discussed by Harvey and Lemmon (2005). From this equation, it is clear that the determination of the mixture composition from a measurement of the dielectric constant requires an estimate of the density and the dielectric virial coefficients of the pure components. For common substances, these quantities are accurately known from experiment and, for the particular case of He, the same quantities are amenable to extremely accurate calculation from the theory (Lach et al. 2004; Mehl 2009).

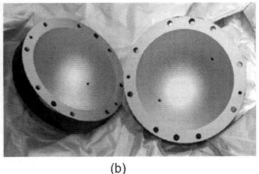

(a) (b)

FIGURE 20.9 (a) The assembled resonator fitted for gas flow, temperature measurement, and control; and (b) The two quasi-hemispheres comprising the 69 cm^3 gold-plated, triaxial, ellipsoid microwave resonator. (Reprinted with permission from Underwood R.J. et al., A microwave resonance dew-point hygrometer, *Meas. Sci. Technol.*, 23, 085905, 2012. Copyright 2012, Institute of Physics.)

As a result of modeling and testing of the developed installation, Cuccaro et al. (2012) concluded that, from the point of view of sensitivity, their microwave technique is limited in its useful operating range by the rapid decrease of the relative permittivity of the humid mixture for water molar fractions below 1 part in 10^4; thus, it cannot compete with spectroscopic techniques (Funke et al. 2003). However, it should be stressed that, in principle, and in contrast to chilled-mirror hygrometers and Al_2O_3 capacitive sensors (Funke et al. 2003), the microwave hygrometer does not need to be calibrated and may be used as a primary humidity standard if it is coupled with a suitable generator (Vega Maza et al. 2012). Cuccaro et al. (2012) believe that the applications of developed experimental methods are mainly related to the accurate determination of high mixing ratios in moist gases at high pressures and temperatures. The same conclusion was made earlier, in relation to a device developed by Vainikainen et al. (1986). In fact, thanks to its simple and rugged design, the microwave hygrometer is suitable for accurate measurements at temperatures up to 500 K and pressures up to 4 MPa, far above the current upper working range of standard humidity generators. Such determinations are of interest for a variety of specific applications, including moisture measurement and control in the operation of certain types of fuel cells (Vega Maza et al. 2012). In later studies, Gavioso et al. (2014) confirmed the possibility of using this technique for analyzing the humidity of gases at elevated pressures. May et al. (2002, 2003) also used a radio frequency re-entrant resonator to determine dielectric constants for gas mixtures. But all measurements were performed at pressures of 5–10 MPa, which far exceed the standard atmospheric pressure.

As we see, the approach to the development of microwave humidity sensors does not provide high sensitivity and small size of devices. Therefore, in recent decades, developments based on other approaches have emerged. For example, Bernou et al. (2000) have developed humidity sensors based on the resonant structure, operated at approximately 1 GHz. However, unlike previous authors for such sensor fabrication, they used the planar technology, and more particularly, the microstrip line technique. Such an approach is much more technologically advanced, and it allows to significantly reduce the dimensions of the device. The humidity sensor was a narrow band-pass filter that consisted of a resonant structure and two ports for supply and measurements. The resonant structure was screen-printed lines on a substrate, and this microwave element selected a resonant frequency. In this microstrip technique, a resonant structure can be a quarter-wavelength stub, a half-wavelength coupled line, or a ring resonator. Bernou et al. (2000) used coupled lines because it is quite simple to study and optimize them. The designed filter consisted of a half-wavelength line coupled on a quarter wavelength with a supply line, so that the resonator had to be bent in a U form. In order to have a band-pass behavior, a gap was done in the middle of the supply line (Figure 20.10a). The sensor was electrically connected to an oscillator, so that the change in electromagnetic (e.m.) parameters of the structure was converted into a frequency variation, which is easier to measure. The sensor was manufactured on an alumina substrate coated with polyimide, a humidity-sensitive material. If the parameters of a polyimide layer change, the propagation is disturbed, and the resonant frequency is going to change. The results of these sensor testing

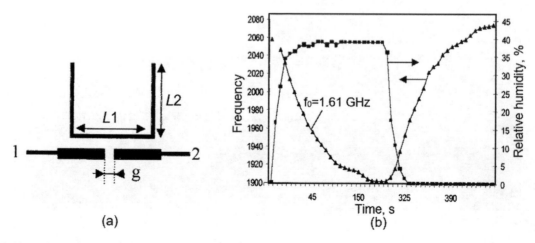

FIGURE 20.10 (a) Final filter design. $L_1 + 2L_2 = \lambda/2$ and (b) Response to a 40% humidity step at 25°C. Polyimide thickness: 6 µm. A H5000 Coreci humidity sensor gives the reference for %RH. (Reprinted with permission from Bernou, C. et al., Microwave sensors: A new sensing principle. Application to humidity detection, *Sens. Actuators B*, 68, 88–93, 2000. Copyright 2000, Elsevier.)

are presented in Figure 20.10b. As seen, the frequency decreases as the RH increases. Taking the gas line delay into account, the rising time (10%–90%) of the response to the RH step was estimated at 68 seconds. The sensor sensitivity was approximately 5 kHz per % RH at room temperature. Big response and recovery time all appear to be associated with the large thickness of the polyimide film used in the sensor.

Amin and coworkers (Amin and Karmakar 2012; Amin et al. 2013) used a similar approach to developing a cheap, printable humidity sensor. The humidity sensor was also based on the planar impedance resonator structure. However, instead of uniform impedance resonators (UIRs), (Amin and Karmakar 2012; Amin et al. 2013) used stepped impedance resonators (SIRs). SIRs are transmission line resonators utilizing quasi-TEM modes. SIRs have advantages over UIR in their wide degree of freedom of design and compact size (Sagawa et al. 1997). The basic structure of a three-element, half-wave SIR is shown in Figure 20.11a. This structure is symmetric at the mid-center 'O' and comprised of two cascaded quarter-wave, tri-step SIRs. Amin et al. (2013, 2014) also tested an electric field-coupled inductor capacitor (ELC) resonator (Figure 20.11b) having a resonant frequency at 6.96 GHz. As a plane wave illuminates the resonator, the middle capacitor-like structure couples to the E field and is connected to two parallel loops, which provide the inductance. Thus, the structure resonates at a frequency determined by its equivalent L and C components (Equation 20.8). Here, the capacitance generated between two split gaps of the ELC resonator has a major influence in the structure resonance frequency. Humidity sensitivity was incorporated via dielectric change of commercially available Kapton HN polyimide (Amin and Karmakar 2012) or polyvinyl alcohol (PVA) (Amin et al. 2013, 2014) films, applied to the surface of the coplanar waveguide (CPW) (Figure 20.11c).

$$f_0 = \frac{1}{2\pi}\sqrt{\frac{2}{LC}} \quad (20.8)$$

The results of testing the developed sensors are given in Table 20.4 and in Figure 20.12. It can be seen that the sensors with ELC resonator have a higher sensitivity in comparison with sensors that used SIR. The use of PVA instead of polyimide also gives rise to sensitivity. It is important to note that this is the highest sensitivity observed in microwave humidity sensors. Approximately the same sensitivity was demonstrated by the microwave humidity sensors developed by Lu et al. (2014) (see Table 20.4). The sensor consisted of a narrow-band-loop antenna loaded with PVA film ($d = 3$ μm)-coated interdigital capacitor (IDC). The sensor operated at 2.85 GHz. The RH sensor generates an amplitude peak at resonance in the radar cross section (RCS), which changes with different RHs. The sensor structure is suitable for printed circuit fabrication technique to lower the cost and can

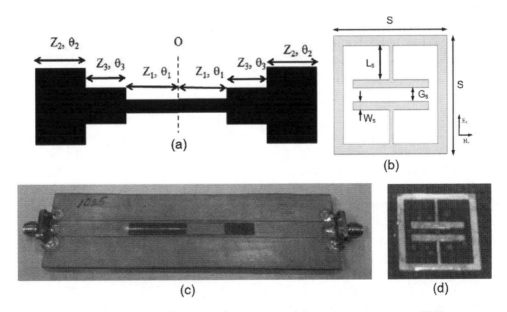

FIGURE 20.11 (a) Layout of proposed half-wave, tri-step stepped impedance resonator (SIR) structure; (b) Schematic diagram and photo of fabricated electric field-coupled inductor capacitor (ELC) resonator. The dimensions are $S = 6$ mm; $L_S = 1.75$ mm; $G_s = 0.7$ mm; $W_s = 0.4$ mm. Substrate Taconic TLX 0; height, $h = 0.5$ mm; $\varepsilon_r = 2.45$; tan $\delta = 0.0019$; (c) Photo of fabricated SIR structure for humidity sensing without Kapton coating. The length of the SIR in sensor prototype, calculated for the operating frequency of 1025 MHz, was 50 mm. (d) Photo of ELC resonator. (Reprinted with permission from Amin, E.M. et al., *Prog. Electromagn. Res. B*, 54, 149–166, 2013. Copyright 2013, JPIER as open access.)

TABLE 20.4
Summary of Humidity Sensitivity Parameter Values for Kapton and Polyvinyl Alcohol

Humidity-Sensitive Polymer	SIR resonator ($f_0 = 1.025$ GHz) RH - 50–90%		ELC resonator ($f_0 = 6.96$ GHz) RH - 35%–85%	
	S_{fr} (MHz/%RH)	$S_{\delta P/\delta f}$ (%/%RH)	S_{fr} (MHz/%RH)	$S_{\delta P/\delta f}$ (%/%RH)
Polyimid (Kapton)	0.63	1.02	1.68	0.7
PVA	2.38	2.1	6.75	1.4

Source: Reprinted from Amin, E.M. et al., *Prog. Electromagn. Res. B*, 54, 149–166, 2013. Published by JPIER as open access.
ELC, electric field-coupled inductor capacitor; PVA, polyvinyl alcohol; RH, relative humidity; SIR, stepped impedance resonator.

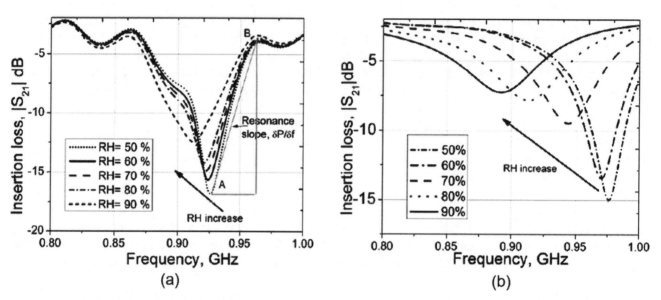

FIGURE 20.12 (a) Magnitude of measured insertion loss (S_{21}) vs frequency for different humidity conditions with Kapton. (b) Magnitude of measured insertion loss (S_{21}) vs frequency for different humidity conditions with polyvinyl-alcohol (PVA). (Reprinted with permission from Amin E.M. et al., *Prog. Electromagn. Res. B*, 54, 149–166, 2013. Copyright 2013, JPIER as open access.)

be realized on flexible substrates to allow mounting on a curved surface. The metallization of the sensor structure was realized on a 17 μm-thick copper layer.

There have also been attempts to use other materials to develop a radio-frequency identification (RFID) humidity sensor tag. Thus, Vena et al. (2016) have used silicon nanowires (SiNWs) as a humidity-sensitive material, while Eyebe et al. (2017) used for this purpose cellulose nanofibers (CNFs). However, these materials are not optimal for these purposes. Sensors based on these materials became sensitive to humidity only at RHs above 70% in the case of using SiNWs, and at RHs above 55% when using CNF (see Figure 20.13). This is understandable, since these materials do not possess the necessary porosity of the structure. These materials have lack of nanopores (silicon nanowires) or they are present in insufficient quantities (CNFs), in which the capillary condensation of water vapor at low RH is possible.

High sensitivity was also demonstrated by sensors developed by Kim et al. (2006) and Feng et al. (2015). Feng et al. (2015), like Amin et al. (2013, 2014), used a printed structure with an ELC resonator. RH sensors developed by Kim et al. (2006) were based on two different microwave resonators in its structure, a coplanar waveguide (CPW)-to-slotline ring resonator and a microstrip patch antenna. In both structures, a polyimide film was used as humidity-sensing material. To integrate a humidity sensor without making an RF system complex, the first harmonic resonant frequency of the resonators, which were designed to be 3.375 and 5.0 GHz for the CPW-to-slotline ring resonator and the microstrip patch antenna, respectively. A sensing structure was fabricated through a "polymer-metal multilayer processing technique" (Kim 1997). A schematic diagram illustrating fabrication technology of CPW-to-slotline ring resonator-based sensor is shown

Microwave-Based Humidity Sensors

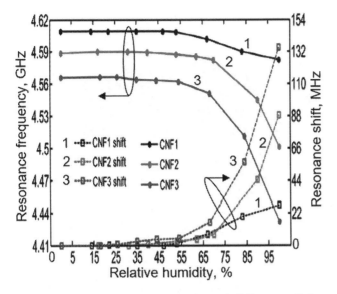

FIGURE 20.13 Resonant frequency of different cellulose nanofiber (CNF) films as a function of relative humidity (RH). (Reprinted with permission from Eyebe, G.A. et al., Environmentally-friendly cellulose nanofibre sheets for humidity sensing in microwave frequencies, *Sens. Actuators B*, 245, 484–492, 2017. Copyright 2017, Elsevier.)

FIGURE 20.15 Sensitivity of the coplanar waveguide (CPW)-to-slotline ring resonator to air humidity in terms of resonant frequency. (Reprinted with permission from Kim, Y.-H. et al., A novel relative humidity sensor based on microwave resonators and a customized polymeric film, *Sens. Actuators B*, 117, 315–322, 2006. Copyright 2006, Elsevier.)

in Figure 20.14. Testing showed that the sensitivity of the RH sensor using the CPW-to-slotline ring resonator was 181 kHz/%RH and 4.95 mdB/%RH in terms of its resonant frequency and insertion loss, respectively (see Figure 20.15). Hysteresis was 0.0013%, and average percent deviation from an average resonant frequency was 0.002% at 25°C. Sensitivity of the antenna RH sensor was 108 kHz/%RH and 5.50 mdB/%RH in terms of resonant frequency and return loss, respectively.

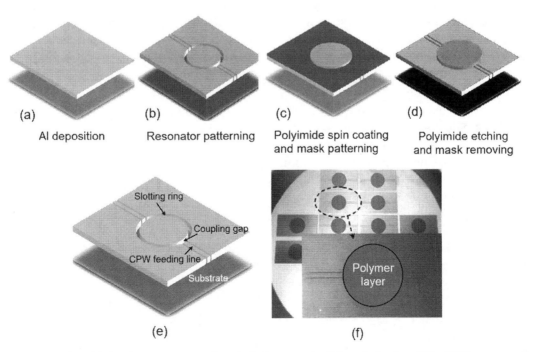

FIGURE 20.14 Simplified fabrication sequence and optical photograph of the relative humidity (RH) sensor using the coplanar waveguide (CPW)-to-slotline ring resonator: (a, b, c, d, e) fabrication sequence, (f) optical photograph. (Adapted from (Reprinted with permission from Kim, Y.-H. et al., A novel relative humidity sensor based on microwave resonators and a customized polymeric film, *Sens. Actuators B*, 117, 315–322, 2006. Copyright 2006, Elsevier.)

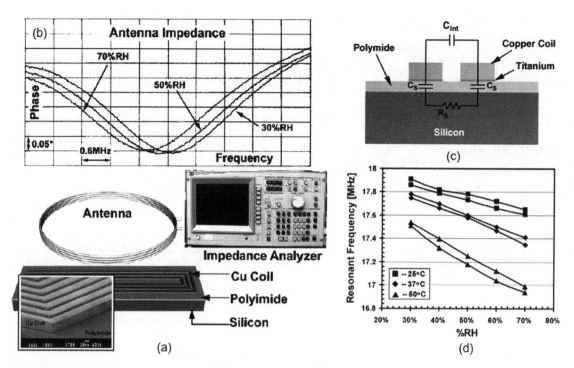

FIGURE 20.16 (a) Passive wireless integrated humidity sensor and scanning electron microscopy (SEM) image of integrated copper coil on polyimide; (b) Air humidity influence on antenna impedance; (c) Capacitive elements of the humidity sensor; (d) Calibration results for two t_p = 560 nm sensors at 25, 37, and 50°C. (Reprinted with permission from Harpster, T.J. et al., A passive wireless integrated humidity sensor, *Sens. Actuators A*, 95, 100–107, 2002. Copyright 2002, Elsevier.)

Another approach to developing humidity sensors was proposed by Harpster et al. (2002). They developed a sensor capable of wireless operation through inductive coupling with remote antenna (see Figure 20.16a). This system consisted of a planar electroplated copper coil (20 μm thick, 30 μm pitch, and 23 μm turns) and a *p*-type 10–20Ω cm silicon substrate separated by 460–560 nm polyimide film. The humidity sensor was modeled as an inductor-capacitor (LC) tank circuit in which the copper coil formed both the inductor and the humidity-sensitive capacitor (Figure 20.16c). The natural resonant frequency of this system changes when the permittivity of the polyimide under the coil changes in response to humidity changes (see Figure 20.16b, 20.16d). To remotely monitor the resonant frequency shift due to humidity changes, a 1.0–1.5 cm diameter loop antenna was used to stimulate the tank circuit as it is shown in Figure 20.27a. As a result of testing such a sensor, it was found that the resonance frequency changed with humidity in the measured sensitive range from 4 to 16 kHz/%RH. Hysteresis in the range of 30%–70% RH measured at T = 25°C did not exceed 4.5% RH (see Figure 20.16d).

It is important that the sensors developed by Harpster et al. (2002), Kim et al. (2006), Amin and Karmakar (2012), Amin et al. (2013, 2014), Lu et al. (2014), and Feng et al. (2015) and reviewed earlier can be used in the RFID sensing systems, interest in which has grown dramatically in recent years (Finkenzeller 2003; Want 2004, 2006; Ahuja and Potti 2010; Meng and Li 2016). It is expected that such systems will be an important element in the realization of ubiquitous environment monitoring and noninvasive control. The RFID operates at a variety of frequencies that are summarized in Table 20.5. Both low-frequency (LF) and high-frequency (HF) RFIDs operate in the near field, and energy transfer is through inductive

TABLE 20.5
Frequencies and Reading Range of Radio-frequency Identification Techniques

RFID Techniques	Operating Frequencies	Free Space Reading Range
LF	125–134.3 kHz	< 10.0 cm
HF	13.56 MHz	< 1.0 m
UHF	860–960 MHz	1.0–12.0 m
SHF	2.45–5.8 GHz	Up to 100.0 m (Active)

Source: Data extracted from Meng, Z. and Li, Z., *Meas. Sci. Rev.*, 16, 305–315, 2016.

HF, high frequency; LF, low frequency; RFID, radio-frequency identification; SHF, super high frequency.

coupling. However, for ultra high frequency (UHF) and higher, like super high frequency (SHF) within the microwave frequency range, the communication and energy transfer is in the far field through backscattering. RFID mechanisms can also be applied to collecting sensed data. Figure 20.17 shows the configuration of the RFID sensing systems. For the monitoring of environmental conditions, as shown in Figure 20.17, the RFID tags with sensors are required so that the tags can send sensor-derived data to readers, although the addition of a sensor component increases the tag size and cost. However, Kim et al. (2006) and Chang et al. (2007) have shown that the physical and functional integration of the antenna and the sensor component is possible, and this integration provides compactness and cost-efficiency for a sensor tag. To produce a modified polyimide-based appliance with good sensitivity, Kim et al. (2006) and Chang et al. (2007) used a microstrip patch antenna as a microwave resonator. Because the microstrip structure has a narrow bandwidth, the patch antenna can serve as a high-resolution sensor. Furthermore, to enhance sensitivity, the patch antenna was meandered. A meandering patch adds a large reactance to the input impedance, and the bandwidth of the antenna becomes narrower, thereby improving the sensitivity of the sensing function.

Jia et al. (2008) developed a simpler and cheaper version of RFID humidity sensor, capable of passive wireless sensing through the far-field backscatter coupling. As is known, RFID tags can be classified into two categories: *active* and *passive* ones. Active tags require a power source, a powered infrastructure or use energy, stored in an integrated battery. Passive RFID tags do not contain a discrete power source but derive their energy from an incident RF signal and reflect the RF carrier back to the reader. Undoubtedly, passive RFID tags are more preferable for a wide range of applications. Jia et al. (2008) resolved the technical task of developing a cheap RFID humidity sensor as follows: A commercial UHF RFID tag was employed as a sensing platform to receive the power and to reflect the sensed data back to the RFID reader, and a humidity-sensitive polyimide film was incorporated onto the top surface of the RFID tag for humidity sensing. The sensor characterization system was a commercial passive EPC-Gen2 RFID system operating in the license-free EU band at 865–868 MHz. In a humid environment, the adsorption of water in the polyimide films causes degradation to the tag's antenna due to dielectric losses and changes the input impedance. These result in a change of the power required to activate the sensor tag. The experimental system allows controlling the antenna output power and the threshold level, necessary to remotely power up the passive wireless humidity sensor tag. The prototype sensor demonstrated that power, required to activate the sensor tag, was a linear function of RH, and the maximum sensing distance between the reader antenna and the RFID humidity sensor tag was approximately 1.5 m. Jia et al. (2008) believe that, due to its unique features of being low cost, battery-less, wireless, maintenance-free, and disposable, the RFID humidity sensor can be integrated into wallpaper for humidity monitoring in a built environment.

Virtanen et al. (2011) also developed a humidity sensor for passive UHF RFID systems. However, to simplify and reduce the cost of manufacturing such sensors, they suggested using flexible substrate and inkjet-printed technology. The sensor tag was a double-sided structure formed by inkjet printing silver nanoparticle ink on 125 μm-thick Kapton 500 HN polyimide film by DuPont. A silver nanoparticle ink was used to form the conductors of the tag. The sensor tag consisted of three main components: (1) an integrated circuit (IC), which provided the basic identification functionality of the tag; (2) inkjet-printed sensor elements; and (3) radiation elements. The IC was a Higgs 3 RFID IC from Alien Technology (www.alientechnology.com). The substrate, Kapton 500 HN, is a flexible, low-loss, and extremely durable polyimide film dielectric. Kapton HN film was selected because of its permittivity dependence on the environmental humidity. This electrical property is the key factor in the sensor's functionality. Furthermore, Kapton HN is able to withstand the high-temperature levels that the printed tag will undergo during sintering as a part of the inkjet printing process. The varying relative permittivity of the Kapton film was transformed into a capacitance, dependent on the environmental humidity using the parallel plate capacitors in the sensor elements. The sensor elements used in the sensor tag occupied both the top and bottom sides of the structure (see Figure 20.18). The radiation element was a short dipole antenna. A dipole-type antenna was chosen due to its omnidirectional radiation pattern. Note that the sensor elements were also a part of the tag antenna. The sensor tag developed and fabricated by Virtanen et al. (2011) is fully passive, and as such, it

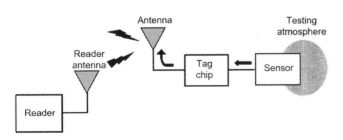

FIGURE 20.17 Configuration of radio-frequency identification (RFID) sensing systems.

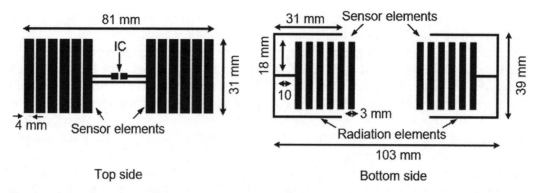

FIGURE 20.18 Printing layout of the sensor tag. The size is indicated in millimeters. (Data extracted from Virtanen J. et al., *IEEE Trans. Instrum. Measur.*, 60, 2768–2777, 2011.)

does not need any power supply of its own. Therefore, the sensor tag does not need any maintenance procedures, and it can be permanently enclosed inside walls, ceilings, and floors for long-term monitoring spanning several years. In addition, the sensor tag is flexible and small in size, allowing it to fit inside various structures.

To increase the accuracy of measurements, Virtanen et al. (2011) suggest using two tags: One acts as a sensor, and the other as a stable reference (see Figure 20.19). This type of humidity measurement setup has been successfully used by Sidén et al. (2007). The reference tag is shielded in a casing of dielectric impervious to humidity. As a result, the power-on-tag of the reference tag is unaffected by the humidity, whereas the power-on-tag of the sensor tag changes according to the ambient humidity level. At that, the reference and sensor tags are used as a whole (i.e., the reference tag is always at a fixed orientation and distance to the actual sensor tag). The RFID testing showed that the humidity sensitivity of the sensor was approximately 198.8 kHz/%RH. The operating frequency range was in the range of 860–930 MHz. The measured standard deviation in the humidity sensitivity allowed humidity measurements within ± 4.0% RH accuracy. The read range of the sensor (i.e., the operating range) was approximately 8 m at various humidity levels (Virtanen et al. 2011). True, these sensors cannot be called *high-speed*, since measurements are performed based on the analysis of frequency characteristics, which requires a certain time.

They proposed to fabricate the passive humidity sensor tag directly on plywood substrate by using brush-painting and photonic sintering of cost-effective silver ink. Earlier, this technology was used in studies by Virkki et al. (2014) and Sipilä et al. (2015). According to Sipilä et al. (2016), brush-painting is a versatile but simple and fast additive manufacturing method. The method not only reduces the process-steps of RFID tag manufacturing, but also minimizes the need of conductive ink material, as the material is dispensed directly to the brush and from the brush directly to the antenna area in the substrate. By brush-painting RFID tags directly on plywood, one can manufacture very thin tags through eliminating the need for additional substrate material. The thickest part of

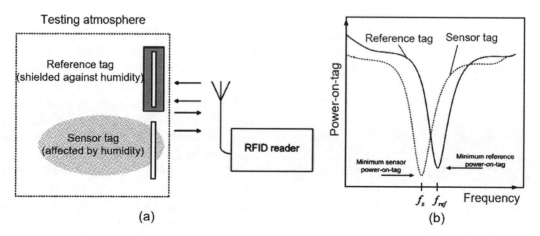

FIGURE 20.19 (a) Measurement setup while measuring ambient humidity; (b) illustration of the power-on-tag curves measured during a humidity level measurement. (Data extracted from Virtanen, J. et al., *IEEE Trans. Instrum. Measur.*, 60, 2768–2777, 2011.)

the tag in this case is the IC, which determines the scale of the thickness of the tag. Sipilä et al. (2016), in their experiments, have used the IC with thickness 120 μm, but the use of even thinner ICs—for example, 75 μm—is possible, meaning that embedding the tags inside versatile products is convenient. In addition, when the tags are embedded as a part of the wooden product, they will be almost impossible to remove from the product without breaking it, as the wooden item itself acts as the substrate of the tag. The prototype of the RFID sensor tag, operated in the range of 800–1000 MHz, is shown in Figure 20.20. The tag antennas were brush-painted through a stencil (50 μm-thick polyimide film) on plywood substrate, by using only one layer of ink. The wood substrate used in this study was 4 mm-thick birch plywood.

Certainly, these sensors were significantly inferior to the sensors considered earlier in their parameters. However, Sipilä et al. (2016) believe that the fabricated RFID-based humidity sensor components have a great potential to be utilized in the humidity sensing applications and also in automatic identification and supply chain control of various wooden products, especially in the packaging and construction industry. These fabrication methods enable fast and cost-effective manufacturing of sensor tags.

It should be noted that, in addition to research aimed at simplifying and reducing the cost of manufacturing the RFID humidity sensor tags, there are developments aimed at improving the parameters of the passive RFID tags. The passive RFID tag collects the radiation energy from the RFID reader as its power supply. Hence, the power dissipation of the passive RFID tag, which determines the maximum operating distance of the tag, is crucial for the design of a passive RFID tag. Recently, with the rapid development of the Internet of Things and sensor technology, research on adding sensing functionality to the RFID tag has become a hot topic. Results of this research can be found in Abad et al. (2009), Oprea et al. (2009), Wei et al. (2011), Zhao et al. (2011), Beriain et al. (2012), Deng et al. (2014), and Wu et al. (2015). The smart RFID sensing tag not only extends the application field of RFID, but also contributes to the reduction in the fabrication cost of RFID systems. For example, Wu et al. (2015) developed a humidity sensor element integrated with the wireless transceiver blocks (see Figure 20.21). The humidity sensor element with the

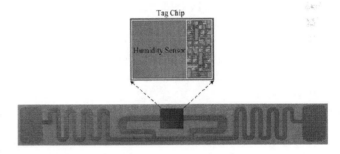

FIGURE 20.21 Photo of the proposed wireless humidity sensor. (Reprinted with permission from Wu, X. et al., Design of a humidity sensor tag for passive wireless applications, *Sensors*, 15, 25564–25576, 2015. Copyright 2015, MDPI as open access.)

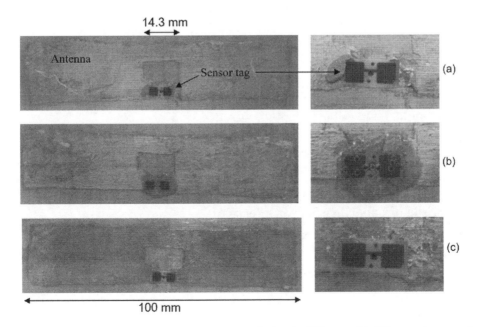

FIGURE 20.20 Noncoated tag and a close-up of the integrated circuit (IC) area (a), IC-coated tag and a close-up of the IC-area (b), and all-coated tag and a close-up of the IC area (c). (Reprinted with permission from Sipilä E. et al., Experimental study on brush-painted passive RFID-based humidity sensors embedded into plywood structures, *Int. J. Antennas Propag.*, 2016, 1203673, 2016. Copyright 2016, Hindawi as open access.)

polyimide humidity-sensitive film was fabricated by the standard complementary metal–oxide–semiconductor (CMOS) technology without any post-processing, which resulted in integration with other tag blocks and low fabrication cost. An ultra-low-power sensor interface was introduced to ensure that the wireless sensor can work in passive mode. A two-stage rectifier used a dynamic bias-voltage generator to boost the effective gate-source voltage of the switches in differential-drive architecture, resulting in a flat power conversion efficiency curve. Wu et al. (2015) reported that the total power dissipation of the sensor tag was 2.5 µW, resulting in a maximum operating distance of 23 m under 4 W RFID reader radiation power conditions. The parameters of integrated humidity sensors developed by Wu et al. (2015), in comparison with parameters of other integrated humidity sensors, are presented in Tables 20.6 and 20.7. Wu et al. (2015) believe that developed smart RFID sensing tag is promising for low-cost applications. For such applications Sipilä et al. (2016) suggested using an even simpler and cheaper technology for manufacturing RFID-based humidity sensors.

Certainly, there are many more possible versions of RFID tags humidity than what was previously considered. Any RFID tag sensors developed for gas detection can be transformed into a humidity sensor by simply changing the gas-sensitive material on the humidity-sensitive material. Therefore, it will be useful for the developers of RFID tag humidity sensors to get to know the results of studies published in Dragoman et al. (2007), Jouhannaud et al. (2007), Penirschke et al. (2007), Rossignol et al. (2010), Hattenhorst et al. (2015), Ali et al. (2016), Bailly et al. (2016a, 2016b), Rydosz et al. (2016), Bahoumina et al. (2017), Staszeka et al. (2017), and Zarifi et al. (2017). For example, Rossignol et al. (2010) suggested a coaxial design for the development of a microwave gas sensor (see Figure 20.22). The sensor represents an open line with an extreme surface in contact with gas. The principle of this sensor operation is the use of sensitive material as a substrate of a transmission coaxial line. Interaction with the gas and the surface induces a dielectric variation of the substrate. Rossignol et al. (2010) have shown that the interaction between sensitive material and each tested gas, in particular water vapor, presents a specific frequency

TABLE 20.6
Comparison of Integrated Humidity Sensors

Sensor Structure	Process, µm	Sensitivity	Fabrication Post-Processing	On-Chip Readout Circuit	References
Interdigitated	3.0	5 fF/%RH	Yes	No	Gu et al. (2004)
Parallel plate	0.5	303 fF/%RH	Yes	No	Kim et al. (2009)
Interdigitated	0.35	0.11 MHz/%RH	Yes	No	Dai and Lu (2010)
Woven mesh	0.15	1.78 mV/%RH	No	Yes	Nizhnik et al. (2012)
Interdigitated	0.16	7.43 fF/%RH	No	Yes	Tan et al. (2013)
Interdigitated	0.6	30 fF/%RH	No	Yes	Cirmirakis et al. (2013)
Interdigitated	0.18	18.75 fF/%RH	No	Yes	Wu et al. (2015)

Source: Data extracted from Wu, X. et al., *Sensors*, 15, 25564–25576, 2015.
RH, relative humidity.

TABLE 20.7
Performance Comparison of Capacitive Sensor Interfaces

Process (µm)	Supply (V)	Area (mm²)	ENOB (bits)	FOM (pJ/conv)	Power (µW)	References
0.32	3.0	0.52	9.8	4.5	84.0	Tan et al. (2012)
0.35	3.0	0.09	9.3	3.4	54.0	Sheu et al. (2012)
0.16	1.2	0.15	12.5	8300	10.3	Tan et al. (2013)
0.09	1.0	N/A	10.4	1.4	3.0	Nguyen and Hafliger (2013)
0.18	0.5	0.01	6.8	1.6	1.1	Wu et al. (2015)

Source: Data extracted from Wu, X. et al., *Sensors*, 15, 25564–25576, 2015.
ENOB, effective number of bits; FOM, figure-of-merit; energy per *conversion* step.

Microwave-Based Humidity Sensors

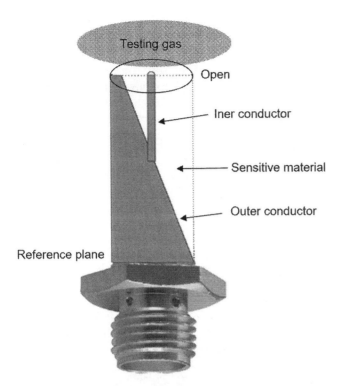

FIGURE 20.22 Exploded scheme of the coaxial gas sensor. (Reprinted with permission from Rossignol J. et al., Broadband microwave gas sensor: A coaxial design, *Microw. Opt. Technol. Lett.*, 52, 1739–1741. Copyright 2010, Wiley-VCH Verlag GmbH & Co, KGaA.)

spectrum of the reflection coefficient at the input of the line. According to Rossignol et al. (2010), zeolite can be used as sensitive material for humidity detection.

20.4 SUMMARY

An analysis of the published results shows that, just as in other humidity sensors, the parameters of the microwave humidity sensors are largely determined by the humidity-sensitive materials used. If you analyze the principles of operation of microwave humidity sensors, you can conclude that these devices have the same requirements for humidity-sensitive materials as capacitive sensors (see Chapter 10). Also, as for capacitive sensors, a reduction in the pore size was accompanied by an increase in sensitivity. For example, De Fonseca et al. (2015) observed that the use of synthesized SnO_2 with a grain size of 5 nm, instead of commercial SnO_2 with particle size equal to 96 nm, gave an increase in the signal (real and imaginary parts) in the impedance microwave sensor of more than 10 times. As is known, the pore size is directly proportional to the grain size. This means that polymers and metal oxides are the most suitable humidity-sensitive materials for these applications (Bernou et al. 2000; De Fonseca et al. 2015).

It is also clear that rather expensive and complicated microwave sensors are not intended for wide application. However, for industrial applications, where a humidity measurement over a wide dynamic range is required, such humidity sensors may be useful.

The benefits and drawbacks of different sensors depend to a large degree on the specific application, but some general remarks can be given. According to Nyfors (2000), the advantages of microwave humidity sensors include the following:

- Microwave sensors sense a very good contrast between water and other materials, which makes them well suited for water content measurements.
- Microwave resonator sensors are inherently stable because the resonant frequency is related to the physical dimensions.
- Microwave sensors are insensitive to environmental conditions, such as dust (contrary to infrared sensors), and high temperatures (contrary to semiconductor sensors).
- In low-frequency capacitive sensors, the dc conductivity often makes measurement difficult; the dc conductivity depends strongly on the temperature and ion content. At microwave frequencies, the influence of the dc conductivity often disappears.

However, according to Nyfors (2000), there may be some disadvantages, which are as follows:

- The higher the frequency is, the more expensive the electronic components are, with the exception of the frequency bands used by radars and other high-volume applications. The price level has generally been falling with time.
- Microwave sensors must be calibrated separately for different humidity-sensitive materials.
- The sensors are often adapted to a specific application, resulting in a low universal applicability.
- The sensors are sensitive to more than one variable. As is known, the air relative permittivity ε_r is a function of temperature T, relative humidity h, and pressure P: $\varepsilon_r = f_\varepsilon(T, h, P)$ (Nyfors and Vainikainen (1989):

$$\varepsilon_r = 1 + A \cdot \frac{P}{T} + B \cdot \frac{\rho_w}{T} + C \cdot \rho_w, \quad (20.9)$$

where P is the atmospheric pressure, T is the absolute temperature, ρ_w is the density of the water vapor in the air, and A, B, and C are three constant parameters whose values are $1.552 \cdot 10^{-6}$ Km²/N, 3.456 Km³/kg, and $-76.57 \cdot 10^{-6}$ m³/kg, respectively (Carullo et al. 1999). According to Figure 20.17d, one can judge the influence of temperature. Therefore, in some cases, additional sensors and systems for compensation are necessary. Other gases and solvent vapors may also affect the readings of the sensors.

- There are also difficulties with microminiaturization and integration. Devices, even with high frequencies, remain cumbersome in comparison with the electronic sensors, discussed in the previous chapters. However, for industrial use, this restriction may not be so significant.

REFERENCES

Abad E., Palacio F., Nuin M., Zárate A., Juarros A., Gómez J.M., Marco S. (2009) RFID smart tag for traceability and cold chain monitoring of foods: Demonstration in an intercontinental fresh fish logistic chain. *J. Food Eng.* 93, 394–399.

Ahuja S., Potti P. (2010) An introduction to RFID technology. *Commun. Network* 2, 183–186.

Ali M.A., Cheng M.M.-C., Chen J.C.-M., Wu C.-T.M. (2016) Microwave gas sensor based on graphene-loaded substrate integrated waveguide cavity resonator. In: *IEEE International Microwave Symposium (IMS)*, May 22–27, San Francisco, CA.

Amin E.M., Karmakar N.C. (2012) Development of a low cost printable humidity sensor for chipless RFID technology. In: *Proceedings of IEEE International Conference on RFID-Technologies and Applications (RFID-TA)*, November 5–7, Nice, France, pp. 165–170.

Amin E.M., Karmakar N., Winther-Jensen B. (2013) Polyvinyl-alcohol (PVA)-based humidity sensor in microwave frequency. *Prog. Electromagn. Res. B* 54, 149–166.

Amin E.M., Bhuiyan M.S., Karmakar N.C., Winther-Jensen B. (2014) Development of a low cost printable humidity sensor for chipless RFID technology. *IEEE Sensors J.* 14 (1), 140–149.

Bahoumina P., Hallil H., Lachaud J.L., Abdelghani A., Frigui K., Bila S., et al. (2017) Microwave flexible gas sensor based on polymer multi wall carbon nanotubes sensitive layer. *Sens. Actuators B* 249, 708–714.

Bailly G., Rossignol J., de Fonseca B., Pribetich P., Stuerga D. (2016a) Microwave gas sensing with hematite: Shape effect on ammonia detection using pseudocubic, rhombohedral, and spindlelike particles. *ACS Sens.* 1, 656–662.

Bailly G., Harrabi A., Rossignol J., Stuerga D., Pribetich P. (2016b) Microwave gas sensing with a microstrip interdigital capacitor: Detection of NH_3 with TiO_2 nanoparticles. *Sens. Actuators B* 236, 554–564.

Barochi G., Rossignol J., Bouvet M. (2011) Development of microwave gas sensors. *Sens. Actuators B* 157, 374–379.

Beriain A., Rebollo I., Fernandez I., Sevillano J.F., Berenguer R. (2012) A passive UHF RFID pressure sensor tag with a 7.27 bit and 5.47 pj capacitive sensor interface. In: *Proceedings of the 2012 IEEE International Microwave Symposium Digest (MTT)*, Montreal, PQ, Canada, June 17–22, pp. 1–3.

Bernou C., Rebiere D., Pistre J. (2000) Microwave sensors: A new sensing principle. Application to humidity detection. *Sens. Actuators B* 68, 88–93.

Buckingham A.D., Raab R.E. (1958) The dielectric constant of a compressed gas mixture. *Trans. Faraday Soc.* 54, 623–628.

Carullo A., Ferrero A., Parvis M. (1998) A microwave interferometer system for humidity measurement, In: *Proceedings of the IEEE Conference on Precision Electromagnetic Measurements*, July 6–10, Washington, DC, pp. 528–529.

Carullo A., Ferrero A., Parvis M. (1999) Microwave system for relative humidity measurement, In: *Proceedings of 16th IEEE Instrumentation and Technology Conference*, May 24–26, New York, pp. 124–129.

Chang K., Kim Y.H., Kim Y.J., Yoon Y.L. (2007) Functional antenna integrated with relative humidity sensor using synthesised polyimide for passive RFID sensing. *Electron. Lett.* 43 (5), 7–8.

Cirmirakis D., Demosthenous A., Saeidi N., Donaldson N. (2013) Humidity-to-frequency sensor in CMOS technology with wireless readout. *IEEE Sens. J.* 13, 900–908.

Cuccaro R., Gavioso R.M., Benedetto G., Madonna Ripa D., Fernicola V., Guianvarc'h C. (2012) Microwave determination of water mole fraction in humid gas mixtures. *Int. J. Thermophys.* 33, 1352–1362.

Dai C.L., Lu D.H. (2010) Fabrication of a micro humidity sensor with polypyrrole using the CMOS process. In: *Proceedings of the 5th IEEE International Conference on Nano/Micro Engineered and Molecular Systems*, Nice, France, January 20–23, pp. 110–113.

De Fonseca B., Rossignol J., Stuerga D., Pribetich P. (2015) Microwave signature for gas sensing: 2005 to present. *Urban Climate* 14, 502–515.

Deng F.M., He Y.G., Zhang C.L., Feng W. (2014) A CMOS humidity sensor for passive RFID sensing applications. *Sensors* 14, 8728–8739.

Dragoman M., Grenier K., Dubuc D., Bary L., Plana R., Foum E., Flahaut E. (2007) Millimeter wave carbon nanotube gas sensor. *J. Appl. Phys.* 101, 106103.

Eyebe G.A., Bideau B., Boubekeur N., Loranger E., Domingue F. (2017) Environmentally-friendly cellulose nanofibre sheets for humidity sensing in microwave frequencies. *Sens. Actuators B* 245, 484–492.

Feng Y., Xie L., Chen Q., Zheng L.R. (2015) Low-cost printed chipless RFID humidity sensor tag for intelligent packaging. *IEEE Sens. J.* 15, 3201–3208.

Finkenzeller K. (2003) *RFID Handbook*, 2nd ed. John Wiley & Sons, Madrid, Spain.

Funke H.H., Grissom B.L., McGrew C.E., Raynor M.W. (2003) Techniques for the measurement of trace moisture in high-purity electronic specialty gases. *Rev. Sci. Instrum.* 74, 3909–3933.

Gardiol F.E. (1985) Open-ended waveguides: Principles and applications, In: *Advances in Electronics and Electron Physics*, Vol. 63. Academic Press, New York, pp. 139–187.

Gavioso R.M., Ripa D.M., Benyon R., Gallegos J.G., Perez-Sanz F., Corbellini S., Avila S., Benito A.M. (2014) Measuring humidity in methane and natural gas with a microwave technique. *Int. J. Thermophys.* 35, 748–766.

Gu L., Huang Q.A., Qin M. (2004) A novel capacitive-type humidity sensor using CMOS fabrication technology. *Sens. Actuators B* 99, 491–498.

Harpster T.J., Stark B., Najafi K. (2002) A passive wireless integrated humidity sensor. *Sens. Actuators A* 95, 100–107.

Harvey A.H., Lemmon E.W. (2005) Method for estimating the dielectric constant of natural gas mixtures. *Int. J. Thermophys.* 26, 31–46.

Hasegawa S., Stokesberry D.P. (1975) Automatic digital microwave hygrometer. *Rev. Sci. Instrum.* 46, 867–873.

Hattenhorst B., Theissen H., Schulz C., Rolfes I., Baer C., Musch T. (2015) Microwave sensor concept for the detection of gas inclusions inside microfluidic channels. In: *IEEE International Microwave Workshop Series on RF and Wireless Technologies for Biomedical and Healthcare Applications* (*IMWS-BIO*), September 21–23, Taipei, Taiwan, pp. 108–109.

Jia Y., Heiß M., Fu Q., Gay N.A. (2008) A prototype RFID humidity sensor for built environment monitoring. In: *Proceedings of IEEE International Workshop on Education Technology and Training and International Workshop on Geoscience and Remote Sensing* (*ETT and GRS 2008*), December 21–22, Shanghai, China, pp. 496–499.

Jouhannaud J., Rossignol J., Stuerga D. (2007) Metal oxide-based gas sensor and microwave broad-band measurements: An innovative approach to gas sensing. *J. Phys Conf. Ser.* 76, 012043.

Kanaya H., Nakamura T., Kawakami K., Yoshida K. (2005) Design of coplanar waveguide matching circuit for RF-CMOS front-end. *Electron. Commun. Jpn.* 88 (7), 19–26.

Kempka T., Kaiser T., Solbach K. (2006) Microwaves in fire detection. *Fire Safety J.* 41, 327–333.

Kim Y.J. (1997) Application of polymer metal multi-layer processing techniques to microelectromechanical system, PhD thesis, Georgia Institute of Technology, Atlanta, GA.

Kim Y.-H., Jang K., Yoon Y.J., Kim Y.-J. (2006) A novel relative humidity sensor based on microwave resonators and a customized polymeric film. *Sens. Actuators B* 117, 315–322.

Kim J.H., Hong S.M., Lee J.S., Moon B.M., Kim K. (2009) High sensitivity capacitive humidity sensor with a novel polyimide design fabricated by MEMS technology. In: *Proceeding of the 4th IEEE International Conference on Nano/Micro Engineered and Molecular Systems* (*NEMS*), January 5–8, Shenzhen, China, pp. 703–706.

King R.W.P., Smith D.S. (1981) *Antennas in Matter*. MIT Press, Cambridge, MA.

Komarov V., Wang S., Tang J. (2005) *Permittivity and Measurement*. John Wiley & Sons, New York.

Kraszewski A. (1980a) Microwave aquametry—a review. *J. Microwave Power* 15 (4), 209–220.

Kraszewski A. (1980b) Microwave aquametry—a bibliography. *J. Microwave Power* 15 (4), 298–310.

Kraszewski A. (Ed.) (1996) *Microwave Aquametry: Electromagnetic Wave Interaction with Water-Containing Materials*. IEEE Press, Washington, DC.

Kraszewski A. (2001) Microwave aquametry–needs and perspectives. *IEEE Trans. Microw. Theory Tech.* 39 (5), 828–835.

Lach G., Jeziorski B., Szalewicz K. (2004) Radiative corrections to the polarizability of helium. *Phys. Rev. Lett.* 92, 233001.

Lu D., Zheng Y., Schussler M., Penirschke A., Jakoby R. (2014) Highly sensitive chipless wireless relative humidity sensor based on polyvinyl-alcohol film. In: *Proceedings of IEEE International Symposium of Antennas and Propagation Society* (*APSURSI*), July 6–11, Memphis, TN, pp. 1612–1613.

Marynowski W., Kowalczyk P., Mazur J. (2010) On the characteristic impedance definition in microstrip and coplanar lines. *Prog. Electromagnet. Res.* 110, 219–235.

May E.F., Miller R.C., Goodwin A.R.H. (2002) Dielectric constants and molar polarizabilities for vapor mixtures of methane + propane and methane + propane + hexane obtained with a radio frequency reentrant cavity. *J. Chem. Eng. Data* 47, 102–105.

May E.F., Edwards T.J., Mann A.G., Edwards C. (2003) Dew point, liquid volume, and dielectric constant measurements in a vapor mixture of methane + propane using microwave apparatus. *Int. J. Thermophys.* 24, 1509–1525.

Mehl J.B. (2009) Ab initio properties of gaseous helium. *C. R. Phys.* 10, 859–865.

Meng Z., Li Z. (2016) RFID tag as a sensor–A review on the innovative designs and applications. *Meas. Sci. Rev.* 16 (6), 305–315.

Mladek J., Beran Z. (1980) Sample geometry, temperature and density factors in the microwave measurement of moisture. *J. Microw. Power* 15, 243–250.

Narayanan R.M., Vu K.T. (2000) Free-space microwave measurement of low moisture content in powdered foods. *J. Food Process. Preserv.* 24, 39–56.

Nelson S.O. (1998), Dielectric properties measuring techniques and applications. ASAE Paper No. 983067. ASAE, St. Joseph, MI.

Nguyen T.T., Hafliger P. (2013) An energy efficient inverter based readout circuit for capacitive sensor. In: *Proceedings of the Biomedical Circuits and Systems Conference* (*BioCAS*), November 1–2, Rotterdam, the Netherlands, pp. 326–329.

Nizhnik O., Higuchi K., Maenaka K. (2012) Self-calibrated humidity sensor in CMOS without post-processing. *Sensors* 12, 226–232.

Nyfors E., Vainikainen P. (1989) *Industrial Microwave Sensors*. Artech House, Norwood, MA.

Nyfors E. (2000) Industrial microwave sensors–A review. *Subsurf. Sens. Technol. Appl.* 1 (1), 23–43.

Oprea A., Courbat J., Bârsan N., Briand D., de Rooij N.F., Weimar U. (2009) Temperature, humidity and gas sensors integrated on plastic foil for low power applications. *Sens. Actuators B Chem.* 140, 227–232.

Penirschke A., Schuller M., Jakoby R. (2007) New microwave flow sensor based on a left-handed transmission line resonator. In: *Proceedings of IEEE International Microwave Symposium*, June 3–8, Honolulu, HI, pp. 393–396.

Pozar D.M., Schaubert D.H. (Eds.) (1995) *Microstrip Antennas*. IEEE Press, Piscataway, NJ.

Rizzi P.A. (1988) *Microwave Engineering*. Prentice Hall, Upper Saddle River, NJ.

Rossignol J., Stuerga D., Jouhannaud J. (2010) Broadband microwave gas sensor: A coaxial design. *Microwave Opt. Technol. Lett.* 52 (8), 1739–1741.

Rouleau J.F., Goyette J., Bose T.K. (2000) Performance of a microwave sensor for the precise measurement of water vapor in gases. *IEEE Trans. Dielectr. Electr. Insul.* 7 (6), 825–831.

Rouleau J.F., Goyette J., Bose T.K., Frechette M.F. (2001) Optimization study of a microwave differential technique for humidity measurement in gases. *IEEE Trans. Instrum. Meas.* 50 (3), 839–845.

Rydosz A., Maciak E., Wincza K., Gruszczynski S. (2016) Microwave-based sensors with phthalocyanine films for acetone, ethanol and methanol detection. *Sens. Actuators B* 237, 876–886.

Sagawa M., Makimoto M., Yamashita S. (1997) Geometrical structures and fundamental characteristics of microwave stepped-impedance resonators. *IEEE Trans. Microw. Theory Tech.* 45, 1078–1085.

Sheu M.L., Hsu W.H., Tsao L.J. (2012) A capacitance-ratio-modulated current front-end circuit with pulsewidth modulation output for a capacitive sensor interface. *IEEE Trans. Instrum. Meas.* 61, 447–455.

Sidén J., Xuezhi Z., Unander T., Koptyug A., Nilsson H.E. (2007) Remote moisture sensing utilizing ordinary RFID tags. In: *Proceedings of IEEE Sensors Conference*, October 28–31, Atlanta, GA, pp. 308–311.

Sipila E., Virkki J., Sydanheimo L., Ukkonen L. (2015) Experimental study on brush-painted metallic nanoparticle UHF RFID tags on wood substrates. *IEEE Antennas Wireless Propag. Lett.* 14, 301–304.

Sipilä E., Virkki J., Sydänheimo L., Ukkonen L. (2016) Experimental study on brush-painted passive RFID-based humidity sensors embedded into plywood structures. *Int. J. Antennas Propag.* 2016, 1203673.

Staszeka K., Rydosz A., Maciak E., Wincza K., Gruszczynski S. (2017) Six-port microwave system for volatile organic compounds detection. *Sens. Actuators B* 245, 882–894.

Tan Z., Shalmany S.H., Meijer G.C., Pertijs M.A. (2012) An energy-efficient 15-bit capacitive-sensor interface based on period modulation. *IEEE J. Solid State Circuits* 47, 1703–1711.

Tan Z., Daamen R., Humbert A., Ponomarev Y.V., Chae Y., Pertijs M.A. (2013) A 1.2-V 8.3-nJ CMOS humidity sensor for RFID applications. *IEEE J. Solid State Circuits* 48, 2469–2477.

Vainikainen P.-V., Agarwal R.P., Nyfors E.G., Toropainen A.P. (1986) Electromagnetic humidity sensor for industrial applications. *Electron. Lett.* 22 (19), 985–987.

Vega Maza D., Miller W.W., Ripple D.C., Scace G.E. (2012) A humidity generator for temperatures up to 200°C and pressures up to 1.6 MPa. *Int. J. Thermophys.* 33 (8–9), 1477–1487.

Vena A., Perret E., Kaddour D., Baron T. (2016) Toward a reliable chipless RFID humidity sensor tag based on silicon nanowires. *IEEE Trans. Microw. Theory Tech.* 64 (9), 2977–2985.

Virkki J., Bjorninen T., Sydanheimo L., Ukkonen L. (2014) Brush-painted silver nanoparticle UHF RFID tags on fabric substrates. In: *Proceedings of the Progress in Electromagnetics Research Symposium (PIERS'14)*, August 25–28, Guangzhou, China, pp. 2106–2110.

Virtanen J., Ukkonen L., Björninen T., Elsherbeni A.Z., Sydänheimo L. (2011) Inkjet-printed humidity sensor for passive UHF RFID systems. *IEEE Trans. Instrum. Meas.* 60 (8), 2768–2777.

Underwood R.J., Cuccaro R., Bell S., Gavioso R.M., Ripa D.M., Stevens M., de Podesta M. (2012) A microwave resonance dew-point hygrometer. *Meas. Sci. Technol.* 23, 085905.

Want R. (2004) Enabling ubiquitous sensing with RFID. *Computer* 37 (4), 84–86.

Want R. (2006) An introduction to RFID technology. *IEEE Pervasive Comput.* 5 (1), 25–33.

Wei P., Che W., Bi Z., Wei C., Na Y., Li Q., Min H. (2011) High-efficiency differential RF front-end for a Gen2 RFID tag. *IEEE Trans. Circuits Syst. II* 4, 189–194.

Wu X., Deng F., Hao Y., Fu Z., Zhang L. (2015) Design of a humidity sensor tag for passive wireless applications. *Sensors* 15, 25564–25576.

Zarifi M.H., Shariaty P., Hashisho Z., Daneshmand M. (2017) A non-contact microwave sensor for monitoring the interaction of zeolite 13X with CO_2 and CH_4 in gaseous streams. *Sens. Actuators B* 238, 1240–1247.

Zhao C.L., Qin M., Huang Q.A. (2011) A fully packaged CMOS interdigital capacitive humidity sensor with polysilicon heaters. *IEEE Sens. J.* 11, 2986–2992.

21 Integrated Humidity Sensors

As shown in the previous chapters, the humidity sensor readings are temperature dependent, which means that in order to obtain reliable information, it is necessary to monitor and stabilize the sensor temperature. In addition, in many cases, for the efficient operation of sensors, especially of the metal oxide type, their reactivation by heating them time to time is required. This treatment helps to clean the surface and evaporate condensed vapors. Temporary heating could also be a solution for the drift caused by the formation of chemisorbed OH⁻ groups on the surface of humidity-sensitive materials.

Another problem, especially urgent for capacitive humidity sensors, is the of the joints, which in many cases is commensurate with the capacitance of the measuring element. This forces either to significantly increase the dimensions of the sensor, or to minimize the distance between the sensor and the measuring devices. In this regard, in the development of integral humidity sensors, there are two directions:

- Development of humidity sensors integrated with a temperature sensor and a heater.
- Development of humidity sensors integrated with the readout circuitry.

21.1 HUMIDITY SENSORS INTEGRATED WITH HEATER

To solve the problem of temperature control, several variants of capacitive sensors were suggested (Rittersma 2002). First, humidity sensors with an integrated refresh heater have been presented by Kang and Wise (1999) and Qu and colleague (1998, 2000). The heater was placed on the backside of the sensor to optimize the chip size. An additional heater was also incorporated into the sensor developed by Laville and Pellet (2002a), and by Kim et al. (2009). Laville and Pellet (2002) have shown that the use of a heater is useful since it removes condensed water from the surface. Even if the sensitivity is decreased, the time response is greatly improved. The same results were received by Kim et al. (2009). The integrated humidity sensor developed by Kim et al. (2009) is shown in Figure 21.1. They established that, although heating decreases sensitivity, it improved the performance of such parameters as the nonlinearity, hysteresis, response time, and temperature dependence of Al_2O_3-based sensors.

Other versions of capacitive humidity sensors with an integrated heater are shown in Figure 21.2. These sensors were developed and fabricated by O'Halloran

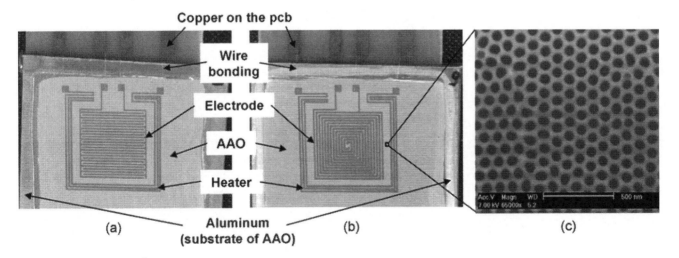

FIGURE 21.1 Humidity sensor devices with (a) interdigitated type and (b) rectangular spiral-shaped type, and (c) scanning electron micrograph of anodic aluminum oxide (AAO). (Reprinted with permission from Kim, Y. et al., Capacitive humidity sensor design based on anodic aluminum oxide, *Sens. Actuators B*, 141, 441–446, 2009. Copyright 2009, Elsevier.)

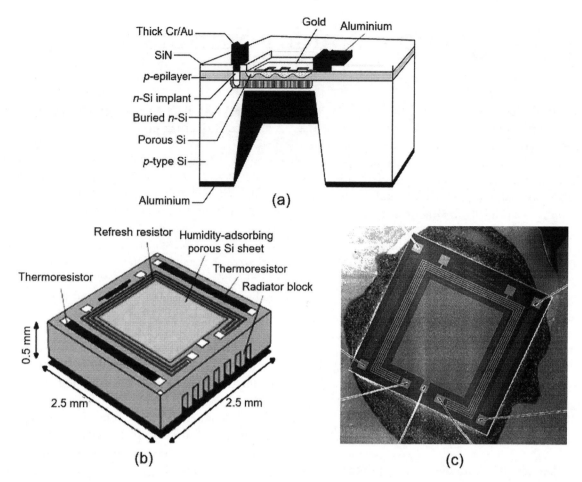

FIGURE 21.2 Two capacitive humidity sensors with integrated heaters: (a) design in French group (Connolly et al. 2002); (b and c) design by Rittersma et al. (2000): (b) schematic diagram and (c) scanning electron microscopy (SEM) image of the porous silicon capacitive humidity sensor. ([a] Reprinted with permission from Connolly, E.J. et al., Comparison of porous silicon. Porous polysilicon and porous silicon carbide as materials for humidity sensing applications, *Sens. Actuators A*, 99, 25–30, 2002. Copyright 2002, Elsevier; [b and c] Reprinted with permission from Rittersma, Z.M. et al., Novel surface-micromachined capacitive porous silicon humidity sensor, *Sens. Actuators B*, 68, 210–217, 2000. Copyright 2000, Elsevier.)

et al. (1998) and Rittersma and Benecke (2000). The first device, shown in Figure 21.2a, consists of a membrane with a buried (silicon) heater, with meshed electrodes on the top: The temperature of the sensor is measured with an integrated diode (O'Halloran 1999). The second sensor (Figure 21.2b and c) is a bulk device, in which both the heater and the temperature sensor consist of metal resistors placed on the front side (Rittersma et al. 2000). This device also has a meshed top electrode to allow for rapid vapor uptake (Rittersma 1999). These sensors were fabricated using methods of micromachining technology.

The development carried out by Regtien (1981) also deserves attention. Regtien (1981) proposed a capacitor in combination with a thermoelectric cooler. Here, an integrated diode was used as a temperature sensor. Upon cooling with a thermoelectric Peltier element, condensation nuclei gradually develop on the capacitor, and the capacitance increases nonlinearly. When the dew-point temperature is reached, the capacitance suddenly increases. And thus, the dew point can be measured. Of course, this sensor requires calibration. The accuracy of this device is limited by the nonlinearity and the fact that the surface needs regular cleaning. The same approach was used by Foucaran et al. (2000). An improvement of Regtien's design with respect to thermal efficiency was presented by Jachowicz and coworkers (Jachowicz 1992; Jachowicz and Weremczuk 2000; Weremczuk et al. 2001). In particular, Figure 21.3 shows the top view of the humidity sensor with integrated heater, elaborated by Jachowicz and Weremczuk (2000). They positioned the capacitor on a silicon membrane of 50 μm thickness, thus reducing the thermal capacitance and improving the recovery time of the sensor. The sensor has been formed by depositing a humidity-sensitive material (polyimide) on interdigitated electrodes. Two heaters and a thermometer have been integrated to the

Integrated Humidity Sensors

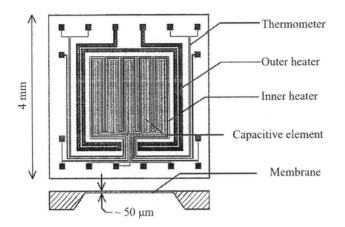

FIGURE 21.3 The top view of the capacitive dew-point detector with integrated heater and temperature sensors. (Reprinted with permission from Jachowicz, R.S., and Weremczuk, J., Sub-cooled water detection in silicon dew point hygrometer, *Sens. Actuators A*, 85, 75–83, 2000. Copyright 2000, Elsevier.)

sensor area using metal resistors. This device is a typical example of the achievements of microsystem technology, because it constitutes a successful attempt to beat the difficulties in hygrometry by integrating different components on a single chip.

21.2 MONOLITHIC INTEGRATION OF THE HUMIDITY SENSORS WITH THE READOUT CIRCUITRY

It should be noted that low-power dissipation and compatibility of capacitive sensor technology with a standard complementary metal–oxide–semiconductor (CMOS) fabrication process allow the monolithic integration of the humidity sensors with the readout circuitry (Hagleitner et al. 2001; Qui et al. 2001; Dai 2007; Eder et al. 2014). The principles of CMOS technology are described in Volume 3 (Chapter 15) of our issue). An example of implementation of humidity sensors in the monolithic integration variant is shown in Figure 21.4. The integrated circuit (IC) contains the sensing element and the calibration circuitry on a single chip. The humidity-sensing property on-chip was obtained by a simple post-processing step after the standard CMOS fabrication whereby a commercial polyimide with thickness 3.8 μm was deposited on the packaged chip using spin-coating process. The fabricated sensor had a linear transfer function and a negligible hysteresis. The resolution of the sensor was 0.06%. No measurable drift was observed after aging the sensor in 60% relative humidity (RH), 26°C for 7 days. The response time of the sensor was approximately 20 seconds (Qui et al. 2001).

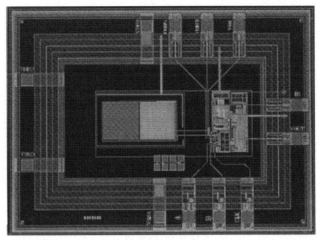

FIGURE 21.4 The capacitive humidity sensor fabricated in a standard complementary metal–oxide–semiconductor (CMOS) process. Polyimide film was used as humidity-sensitive material. (Reprinted with permission from Qui, Y.Y. et al., A CMOS humidity sensor with on-chip calibration, *Sens. Actuators A*, 92, 80–87, 2001. Copyright 2001, Elsevier.)

A simpler version of the humidity sensor's monolithic integration with the readout circuitry was proposed by Dai (2007). Figure 21.5 shows this device. A capacitive humidity sensor integrated with a microheater and a ring oscillator circuit was fabricated using the commercial 0.35 μm CMOS process and the post-process. The post-process used reactive-ion etching (RIE) to etch the sacrificial layers of humidity sensor, and then coated a polyimide film on the interdigital electrodes of the humidity sensor. The ring oscillator circuit was employed to convert the capacitance of the humidity sensor into the oscillation frequency output. The microheater provided the operation of the humidity sensor at temperatures above the ambient temperature, which contributed to the reduction of the sensor signal drift. Experiments showed that the microheater with the voltage of 8V generated the working temperature of approximately 80°C, and the humidity sensitivity of the humidity sensor was 25.5 kHz/%RH at 80°C.

Integrated sensor systems, in addition to the humidity sensor, heater, and thermometer, may contain other sensors, such as pressure sensors (Won et al. 2006), calorimetric sensor, mass-sensitive resonant (Hagleitner et al. 2001), sensors, monitoring corrosion, gas thermal conductivity, gas flow rates (Hautefeuille et al. 2011), and gas sensors (Oprea et al. 2012). An example of such an integrated sensor system is shown in Figure 21.6. As a rule, such integrated sensors are manufactured based on surface and bulk micromachining techniques and CMOS technology. However, other technologies, including standard operations of thin-film technology and flexible substrates, can also be used. For example, Oprea et al. (2012),

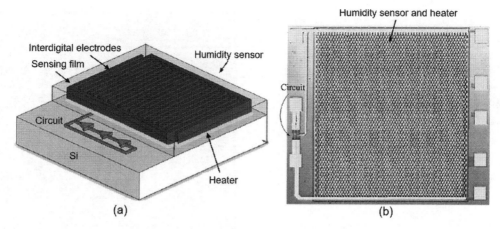

FIGURE 21.5 (a) Schematic structure of the humidity sensor with heater and circuit; (b) Photograph of the integrated chip after the complementary metal–oxide–semiconductor (CMOS) process. (Reprinted with permission from Dai, C.-L., A capacitive humidity sensor integrated with micro heater and ring oscillator circuit fabricated by CMOS–MEMS technique, *Sens. Actuators B*, 122, 375–380, 2007. Copyright 2007, Elsevier.)

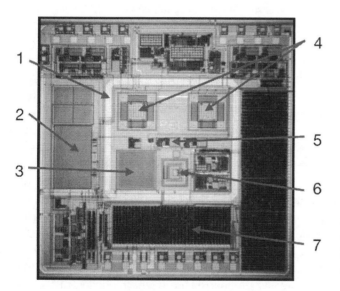

FIGURE 21.6 Micrograph of the gas microsensor system chip (size, 7 mm × 7 mm). The different components include: 1, flip-chip frame; 2, reference capacitor; 3, sensing capacitor; 4, calorimetric sensor and reference; 5, temperature sensor; 6, mass-sensitive resonant cantilever; and 7, digital interface. (Reprinted from Hagleitner, C. et al., Smart single-chip gas sensor microsystem, *Nature*, 414, 293–296, Copyright 2001, with permission from Nature.)

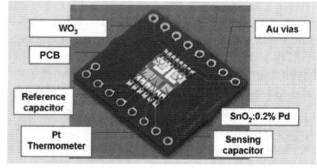

FIGURE 21.7 Optical micrograph of the multi-sensor platform fabricated on polyimide foil and packaged on a printed circuit board (PCB) for its characterization. (Reprinted with permission from Oprea, A. et al., Environmental monitoring with a multisensor platform on polyimide foil, *Sens. Actuators B*, 171–172, 190–197, 2012. Copyright 2012, Elsevier.)

using this approach, produced a multisensor platform that combined three different sensor types on the same polymeric substrate: two metal oxides and two capacitive gases, and humidity-sensing structures, as well as a resistive thermometer (Figure 21.7) (see Chapter 22). Huang et al. (2012) used a different approach to developing a fully integrated humidity sensor system. They proposed to make the pseudo-three-dimensional (3D) sensor system-on-chip (SSoC) by stacking sensing materials directly on the top of a CMOS-fabricated chip (see Figure 21.8), utilizing the microstamping technology. Designed analog and digital circuits were implemented following a standard 0.35 μm CMOS process flow.

Manzan et al. (2005) and Hu et al. (2014) also elaborated the integrated resistive-based humidity sensors. Manzan et al. (2005) integrated an ionic polymer–based resistive sensor with a dedicated signal conditioning circuit implemented in 0.35 μm CMOS. In addition to providing excitation and measurement of the film resistance, the implemented signal conditioning circuit detected the temperature of the sensing film by way of an on-chip temperature sensor in order to perform the necessary compensation for the influence of the temperature

Integrated Humidity Sensors

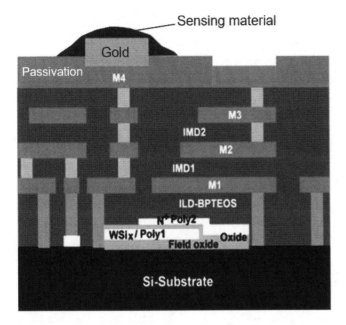

FIGURE 21.8 The cross-section schematic of the developed humidity sensor system-on-chip (SSoC). The circuit elements, such as metals, dielectrics, and transistors, are under the microstamped sensing material. (Reprinted with permission from Huang C.-W. et al., A fully integrated humidity sensor system-on-chip fabricated by micro-stamping technology, *Sensors*, 12, 11592–11600, 2012. Copyright 2012, MDPI as open access.)

upon the measured relative humidity. The authors claim that the sensor developed on the base of poly(ethylene oxide-co-epichlorohydrin) is stable and repetitive, and can operate in the range 25%–85% RH, 10°C–50°C with very low hysteresis. The humidity microsensor integrated with a readout circuit on-a-chip fabricated by Hu et al. (2014), using the commercial 0.18 μm CMOS process, operated in approximately in the same range, from 30% to 90% RH. The titanium dioxide deposited on the interdigitated electrodes was used as a humidity-sensitive material. The resistive humidity sensor had a sensitivity of 4.5 mV/RH% at room temperature.

As studies have shown, a monolithic, integrated sensor system with microcantilevers, operating in resonant (Hagleitner et al. 2001) and deflection modes (Zimmermann et al. 2008), can also be manufactured using micromachining and CMOS technologies. Figure 21.9 shows the micrograph of the cantilever array fabricated by Zimmermann et al. (2008). The sensing cantilevers (second and fourth cantilever in the array from the left in Figure 21.9) have an additional gold layer on the top, while the reference cantilevers (first and third in the same figure) have only silicon nitride as protective layer on the surface. Sensing and reference cantilevers are paired (a gold-coated sensing cantilever and an uncoated reference cantilever) to achieve a dual-cantilever symmetric Wheatstone bridge configuration: Two resistors are located on cantilever 1, two on cantilever 2. When using the appropriate polymer coating, these integrated sensor systems can detect both vapors of volatile organic compounds and water vapor. The electronic circuitry, shown in Figure 21.10, was designed for high-gain amplification of the sensor output signal, while maintaining low-noise characteristics and providing the possibility of sensor offset compensation. The input multiplexer connects one sensor of the sensor array at a time to the analog front-end. A micrograph of the single-chip sensor system is shown in Figure 21.11. The chip size is 5.2 mm × 2.7 mm, including the silicon lugs on each side of the cantilever array. The system consumes approximately 50 mW power at a supply voltage of 5 V. However, summing up their studies, Zimmermann et al. (2008) note that the analyte sensitivity of this system is by a factor of approximately five lower in comparison to a resonant-cantilever system, and, particularly for analytes like humidity or ethanol,

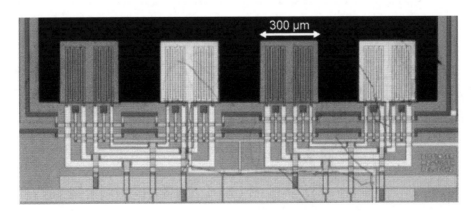

FIGURE 21.9 The sensor array with four microcantilevers. (Reprinted with permission from Zimmermann, M. et al., A CMOS-based integrated-system architecture for a static cantilever array, *Sens. Actuators B*, 131, 254–264, 2008. Copyright 2008, Elsevier.)

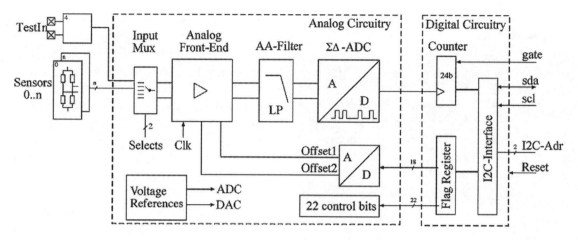

FIGURE 21.10 The block diagram of the sensor system: The sensor array, multiplexer (Input Mux), amplifiers, and filters are integrated on a single chip along with a digital block. The analog signal chain is fully differential. ADC - the analog-to-digital converter; DAC - the digital-to-analog converter; LP - low-pass filter. (Reprinted with permission from Zimmermann, M. et al., A CMOS-based integrated-system architecture for a static cantilever array, *Sens. Actuators B*, 131, 254–264, 2008. Copyright 2008, Elsevier.)

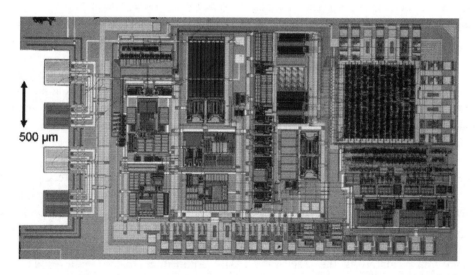

FIGURE 21.11 Micrograph of the single-chip sensor system. (Reprinted with permission from Zimmermann, M. et al., A CMOS-based integrated-system architecture for a static cantilever array, *Sens. Actuators B*, 131, 254–264, 2008. Copyright 2008, Elsevier.)

capacitive microsensors perform pronouncedly better. So, it is a valid question whether it is worth spending efforts on further developing the CMOS-integrated static cantilevers for gas (humidity) sensor applications.

Nagaraju et al. (2014) reported that hybrid integrated humidity sensors can also provide improved sensor parameters. Nagaraju et al. (2014) managed to integrate a thin-film bulk acoustic wave resonator (FBAR)-based humidity sensor with CMOS circuitry, and due to this, the drift in frequency of the FBAR sensor due to temperature, age, and stress was reduced by a factor of 10. The device contained two FBAR sensors, one for measuring humidity, and the other was used as a reference sensor. The integrated FBAR-CMOS (FMOS) circuit package was hermetically sealed and immune to any sensor inputs. To expose the FBAR membrane of the sensing oscillator, 80 μm (diameter) holes into the backside of the wafer stack were etched.

21.3 SUMMARY

Summing up this review, we can state that integrated sensors can really solve problems with temperature control, large parasitic capacitances, and the convenience of processing and transmitting information. However, if integration with temperature sensors can be considered justified, since it does not require the use of any new technologies and materials, integration with the heater and especially monolithic integration with the readout circuitry for sensors intended for widespread use seems redundant. Such sensors become too expensive.

If replacement of the humidity sensor, which must be done periodically, is necessary, the whole system will have to be changed, since the replacement of the sensor in an integrated sensor system is not possible. Besides that, for applying a humidity-sensitive material, technologies that are not used in the manufacture of conventional IC are needed. There are also problems connected with the packaging of such systems and the protection of the electronic part of integrated systems from external influences. Therefore, the development of integrated sensors, especially in monolithic performance, in most cases represents a demonstration of technology capabilities, rather than a prospect for developing real sensor systems for the market.

REFERENCES

Connolly E.J., O'Halloran G.M., Pham H.T.M., Sarro P.M., French P.J. (2002) Comparison of porous silicon. Porous polysilicon and porous silicon carbide as materials for humidity sensing applications. *Sens. Actuators A* 99, 25–30.

Dai C.-L. (2007) A capacitive humidity sensor integrated with micro heater and ring oscillator circuit fabricated by CMOS–MEMS technique. *Sens. Actuators B* 122, 375–380.

Eder C., Valente V., Donaldson N., Demosthenous A. (2014) A CMOS smart temperature and humidity sensor with combined readout. *Sensors* 14, 17192–17211.

Foucaran A., Sorli B., Garcia M., Pascal-Delannoy F., Giani A., Boyer A. (2000) Porous silicon layer couples with a thermoelectric cooler: A humidity sensor. *Sens. Actuators A* 79, 189–193.

Hagleitner C., Hierlemann A., Lange D., Kummer A., Kerness N., Brand O., Baltes H. (2001) Smart single-chip gas sensor microsystem. *Nature* 414, 293–296.

Hautefeuille M., O'Flynn B., Peters F.H., O'Mahony C. (2011) Development of a microelectromechanical system (MEMS)-based multisensor platform for environmental monitoring. *Micromachines* 2, 410–430.

Hu Y.-C., Dai C.-L., Hsu C.-C. (2014) Titanium dioxide nanoparticle humidity microsensors integrated with circuitry on-a-chip. *Sensors* 14, 4177–4188.

Huang C.-W., Huang Y.-J., Lu S.-S., Lin C.-T. (2012) A fully integrated humidity sensor system-on-chip fabricated by micro-stamping technology. *Sensors* 12, 11592–11600.

Jachowicz R.S. (1992) Dew point hygrometer with heat injection-principle of construction and operation. *Sens. Actuators B* 7, 455–459.

Jachowicz R.S., Weremczuk J. (2000) Sub-cooled water detection in silicon dew point hygrometer. *Sens. Actuators A* 85, 75–83.

Kang U., Wise K.D. (1999) A robust high-speed capacitive humidity sensor integrated on a polysilicon heater, In: *Proceedings of Transducers'99*, June 7–10, Sendai, Japan, pp. 1674–1677.

Kim Y., Jung B., Lee H., Kim H., Lee K., Park H. (2009) Capacitive humidity sensor design based on anodic aluminum oxide. *Sens. Actuators B* 141, 441–446.

Laville C., Pellet C. (2002) Interdigitated humidity sensors for a portable clinical microsystem. *IEEE Trans. Biomed. Eng.* 49, 1162–1167.

Manzan D. Jr., Nogueiraf V., De Paolif M.A., dos Reis Filho C.A. (2005) Polymer-electrolyte-film-based humidity sensor with integrated signal conditioner. In: *Proceedings of the IEEE International Conference on Electronics, Circuits, and Systems, ICECS 2005*, December 11–14, Gammarth, Tunisia, Article number 4633609.

Nagaraju M., Gu J., Lingley A., Zhang F., Small M., Ruby R., Otis B. (2014) A fully integrated wafer-scale sub-mm^3 FBAR-based wireless mass sensor. In: *Proceedings of IEEE International Conference on Frequency Control Symposium (FCS)*, May 19–22, Taipei, Taiwan. doi:10.1109/FCS.2014.6859916.

O'Halloran G.M. (1999) *Capacitive Humidity Sensor Based on Porous Silicon*. Delft University Press, Delft, the Netherlands.

O'Halloran G.M., Groeneweg J., Sarro P.M., French P.J. (1998) Porous silicon membrane for humidity sensing applications. In: *Proceedings of Eurosensors XII*, September 13–18, Southampton, UK, pp. 901–904.

Oprea A., Courbat J., Briand D., Bârsan N., Weimar U., de Rooij N.F. (2012) Environmental monitoring with a multisensor platform on polyimide foil. *Sens. Actuators B* 171–172, 190–197.

Qu W., Meyer J.U. (1998) A novel thick-film ceramic humidity sensor. *Sens. Actuators B* 40, 175–182.

Qu W., Wlodarski W. (2000) A thin-film sensing element for ozone, humidity and temperature. *Sens. Actuators B* 64, 42–48.

Qui Y.Y., Azeredo-Leme C., Alcacer L.R., Franca J.E. (2001) A CMOS humidity sensor with on-chip calibration. *Sens. Actuators A* 92, 80–87.

Regtien P.P.L. (1981) Solid-state humidity sensors. *Sens. Actuators A* 2, 85–95.

Rittersma Z.M. (1999) *Microsensor Applications of Porous Silicon. From Humidity-Sensitive Sheet to Sacrificial Layer*. Shaker Publishing, Maastricht, the Netherlands.

Rittersma Z.M. (2002) Recent achievements in miniaturized humidity sensors – A review of transduction techniques. *Sens. Actuators A* 96, 196–210.

Rittersma Z.M., Benecke W. (2000) A humidity sensor featuring a porous silicon capacitor with an integrated refresh resistor. *Sens. Mater.* 12 (1), 35–55.

Rittersma Z.M., Splinter A., Bodecker A., Benecke W. (2000) A novel surface-micromachined capacitive porous silicon humidity sensor. *Sens. Actuators B* 68 (1–3), 210–217.

Weremczuk J., Gniazdowski Z., Lysko J.M., Jachowicz R.S. (2001) Optimization of semiconductor dew point hygrometer mirrors surface temperature homogeneity. *Sens. Actuators A* 92 (1), 10–15.

Won J., Choa S.-H., Yulong Z. (2006) An integrated sensor for pressure, temperature and relative humidity based on MEMS technology. *J. Mech. Sci. Technol.* 20 (4), 506–512.

Zimmermann M., Volden T., Kirstein K.-U., Hafizovic S., Lichtenberg J., Brand O., Hierlemann A. (2008) A CMOS-based integrated-system architecture for a static cantilever array. *Sens. Actuators B* 131, 254–264.

22 Humidity Sensors on Flexible Substrate

22.1 FLEXIBLE ELECTRONICS

Lately, a new trend has appeared with the integration of sensors directly on flexible plastic foils. Their flexibility and simplified processing, targeting the production on large area using roll-to-roll and printed electronics processing (see Figure 22.1), bring new opportunities and will also reduce several technical limitations that characterize the production processes of conventional microelectronics (Logothetidis 2008). In conventional Si microelectronics, patterning is most often done using photolithography, in which the active material is deposited initially over the entire substrate area, and selected areas of it are removed by physical or chemical processes. Despite its high resolution, the photolithographic process is very complex and expensive, uses extremely expensive equipment, requires many steps, and is time consuming, subtractive, and only suitable for patterning of small areas. Moreover, the harsh conditions required for dissolving resists, etching the underlying layers, and removing the photoresist destroys the activity of most organic electronic materials. De Rooij and co-workers (Courbat et al. 2010a, 2010b) believe that the printing technology applied in flexible electronics is experiencing significant growth, and the sensors field can benefit from these developments with the availability of new types of materials and fabrication processes. In particular, the above-mentioned approach to gas and humidity sensor design guarantees reduced price, new functionalities, and a possibility to integrate sensors where it was impossible to imagine them a decade ago. This approach is compatible with a new generation of electronic devices made out of polymer materials (known as *organic electronic devices*), which are the future in bringing fabrication costs down and thus opening the field to a variety of applications, such as monitoring humidity and temperature of goods vulnerable to environmental conditions (e.g., foods, medical supplies, aircraft, and automotive replacement parts) (Chatzandroulis et al. 2011). In addition, flexible substrates are ideal for deployment on curved surfaces, for integration on cloth for application in wearable sensors and smart textiles, or for development of wireless radio frequency identification (RFID) tags for logistics applications (Vergara et al. 2007). In this case, no direct optical contact is required to identify an item. It leads to higher efficiency in goods handling (Angeles 2005). In addition, plastic substrates possess many attractive properties including biocompatibility, light weight, shock resistance, softness, and transparency (Xu 2000). More detailed information on the achievements and prospects in the development of sensors on plastic foil can be read in the reviews prepared by Briand et al. (2011) and Mattana and Briand (2016).

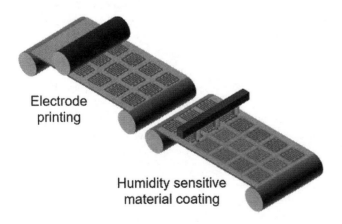

FIGURE 22.1 Schematic drawing of a roll-to-roll production line for chemical gas and humidity sensors on plastic foil. The transducers and coating layers are coated using additive printing techniques, such as the gravure printing of interdigitated electrodes and the local inkjet printing of different sensing layers. (Reprinted with permission from Briand, D. et al., Making environmental sensors on plastic foil, *Mater. Today*, 14, 416–423, 2011. Copyright 2011, Elsevier.)

22.2 FLEXIBLE PLATFORMS

One of the most fundamental difficulties in flexible electronic systems is the substrate, because a material for such substrates would present a unique balance of properties (Crawford 2005). No doubt, substrates should be flexible. Specifically, this means that a substrate must be able to bend but not crack or lose its other properties. Ideally, the substrate could repeatedly bend without a significant long-term degradation. Along with bending, it must be robust and cannot stretch. The substrate must withstand reasonable processing temperatures. This means that the melting temperature of the substrate must be sufficiently high. Additionally, the coefficient of thermal expansion (CTE) must be sufficiently low. If the film expanded or shrank (or both) too much under the heating, the layers deposited on the top (probably a mixture of inorganics that typically have low CTEs) are inclined to cracking or de-adhering, as in the case of external stress. The substrate should be thermally stable as well. Thermal stability is an important factor, primarily for fabrication reasons, because the ability to achieve very low processing temperatures is still the subject of investigation. Stability in aggressive environments is also required. In addition, a flexible substrate must be smooth and sufficiently adhesive, transparent, and, above all, economically viable. There are a number of materials that meet most of these requirements and could possibly function as a flexible electronic substrate. However, the perfect material is not yet found; therefore, at present, when choosing a substrate, depending on its application, compromises are necessary. When making this choice, we need to consider multiple material properties of the material. Plastic substrates that can be used in gas sensor design are listed in Table 22.1.

As shown in Table 22.1, most plastics that can be used as substrates for flexible gas and humidity sensors deform or melt at temperatures of only 100°C–300°C, placing severe limitations on the quality of crystalline semiconductors that can be grown directly on plastic. Therefore, in many cases, developers prefer to use a polymer as sensing material. However, it is known that polymer-based gas and humidity sensors have limitations in many applications. For resolving the above problem with crystalline material deposition, three categories of approaches for overcoming temperature restriction have been proposed (McAlpine et al. 2007). The first approaches are crystallization methods, in which an inferior inorganic semiconductor is vapor, deposited at low temperatures onto plastic and subsequently crystallized. An example is the conversion of amorphous material into polycrystalline one via laser crystallization. The second methods are the wet-transfer, or "bottom-up" methods. Single-crystalline material is prepared at high temperatures; then synthesized material is transferred onto plastic at ambient temperatures in the form of solution or paste, and then it is exposed to annealing at temperature, admissible for

TABLE 22.1
Polymers Acceptable as and Their Parameters

Material	Parameters
Polytetrafluoroethylene (PTFE)	PTFE is most well-known by the DuPont brand name Teflon. It is highly resistant to chemicals and wear. According to DuPont, its melting point is 327°C, but its mechanical properties degrade above 260°C. Therefore, PTFE allows for processing steps in the 250°C range. PTFE has excellent dielectric properties.
Polyimide (PI)	Thermosetting polyimides are known for thermal stability, good chemical resistance, excellent mechanical properties, and characteristic orange/yellow color. Polyimide is available commercially from DuPont as a product called Kapton. Polyimides compounded with graphite or glass fiber reinforcements have excellent flexible properties. Thermoset polyimides exhibit very low creep and high tensile strength. Molded polyimide parts and laminates have very good heat resistance. Polyimides maintain their properties during continuous use to temperatures from cryogenic to 232°C and, for short excursions, as high as 482°C. As a result, polyimides are suitable for processing around 350°C. Polyimides are also inherently resistant to flame combustion and do not usually need to be mixed with flame retardants. Typical polyimide parts are not affected by commonly used solvents and oils—including hydrocarbons, esters, ethers, alcohols, and freons. They also resist weak acids but are not recommended for use in environments that contain alkalis or inorganic acids. The above-mentioned properties make polyimide an excellent choice as a substrate from a stability and processing standpoint.
Polyethylenes (PET) and (PEN)	Two polymers in the polyethylene family showing promise are polyethylene terephthalate (PET) and polyethylene naphthalate (PEN). PET and PEN films are commercially available from DuPont under the names Melinex and Teonex, respectively. As compared to polyimide films, they offer transparency. Depending on its processing and thermal history, PET may exist both as an amorphous (transparent) and as a semicrystalline polymer. The semicrystalline material might appear transparent or opaque and white, depending on its crystal structure and particle size. However, the glass transition temperature of these polymers is only in the range of 100°C–300°C. This limits the processability of these films and requires the design of new techniques in processing. PET can be semirigid to rigid, depending on its thickness, and it is very lightweight. It makes a good gas and fair moisture barrier, as well as a good barrier to alcohol and solvents.

substrate used for sensor fabrication (Oprea et al. 2012). The local inkjet printing is also a promising method for the preparation of different sensing layers (Weng et al. 2010). The third category of techniques are dry-transfer methods, involving the relocation of semiconductor materials or fully fabricated devices from inorganic substrates to plastic, using poly(dimethylsiloxane) (PDMS) stamps or soluble glues. All of the above-mentioned methods can be used for the manufacture of humidity sensors. Examples of flexible humidity sensors designed on the base of plastic substrates are listed in Table 22.2 and shown in Figure 22.2.

A brief analysis showed that no polymer meets all requirements. Right now, the most suitable for the widest number of applications appears to be polyethylene terephthalate (PET), offering curing advantages, or polyethylene naphthalate (PEN), offering thermal and dimensional stability. However, the experiments and results presented in Table 22.2 show that polyimide (PI) offers more options in terms of processing, as long as transparency isn't necessary. For example, Briand et al. (2006) have shown that PI-based flexible micro-hotplates on PI sheets and on silicon for gas-sensing and thermal actuating applications can be realized. Moreover, platinum and aluminium microheating elements on PI exhibited

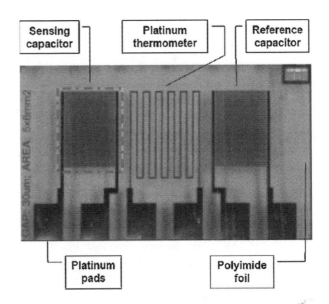

FIGURE 22.2 The sensor platform substrate with Pt thermometer, electrodes, and connection pads. The interdigital electrode structures realize two plane capacitive transducers, a sensing one (left) and a reference one (right). The area reserved for the polymer-sensing layer is surrounded by a *dotted line*. (Reprinted with permission from Oprea, A. et al., Temperature, humidity and gas sensors integrated on plastic foil for low power applications, *Sens. Actuators B*, 140, 227–232, 2009. Copyright 2009, Elsevier.)

TABLE 22.2
Examples of Plastic Substrates Using in Humidity Sensors

Substrate	Sensor Type (T_{oper})	Sensing Material (Deposition Technology)	References
PEN or PI	Capacitance	Cellulose acetate butyrate; polyimide (spray or screen printing)	Oprea et al. (2008)
PI	Capacitance	Cellulose acetate butyrate (drop coating)	Courbat et al. (2009a); Oprea et al. (2012)
PI	Capacitance	Bisbenzocyclobytene (casting)	Pecora et al. (2008)
P	Capacitance	Polyimide (spin coating)	Lee et al. (2008)
PI	Capacitance	Polydimethylsiloxane (PDMS) (drop coating)	Oprea et al. (2009)
PI	Capacitance	Bis(benzo cyclobutene) (BCB)	Zampetti et al. (2009)
PI	Capacitance	Polyimide foil	Rivadeneyra et al. (2014)
PI or PET	Capacitance	Poly (2-hydroxyethyl methacrylate) (pHEMA) (gravure printing)	Reddy et al. (2011)
PET	Capacitance	Cellulose acetate butyrate (inkjet-printing)	Molina-Lopez et al. (2012)
PET	Resistive	PMMA/PMAPTAC (copolymerization)	Su and Wang (2007)
PTFE	Resistive	Poly-tetrafluoroethylene	Miyoshi et al. (2009)
GE	Resistive	MMA/MDBAC (silk-screening printing)	Ahn et al. (2012)
PI	Resistive	MEPTDD/CEMA (copolymerization; screen printing)	Lim et al. (2013)
PET	Resistive	Single-walled CNTs (dip coating)	Parikh et al. (2006)
PDMS	Resistive	R-GO/PU (spin-coating)	Trung et al. (2017)

CEMA, 2-(cinnamyloxy)ethyl methacrylate; GE, glass epoxy; MAPTAC, [3-(methacrylamino)propyl] trimethyl ammonium chloride; MDBAC, [2-(methacryloyloxy)ethyl] dimethyl benzyl ammonium chloride; MEPTDD, 2-(N-Methacryloyloxy)ethyl-N′-propyl-N, N,N′,N′-tetramethylethylene diammonium dibromide; MMA, methyl methacrylate; P, parylene; PDMS, polydimethylsiloxane; PTFE, poly-tetrafluoroethylene; PU, polyurethane; R-GO, reduced graphene oxide.

TABLE 22.3
Properties of Polymer Materials Used in Flexible Substrates

Parameter	Flexible Polymer Substrate			
	PEN	PI	PPS	PEI
T_G	121°C	354°C	92°C	217°C
T_M	269°C	–	285°C	–
CTE	18 ppm/°C	16 ppm/°C	30 ppm/°C	31 ppm/°C
Solvent stability	++	++	++	++
Transparency	transparent	–	translucent	translucent
CHE	n.a.	8 ppm/%RH	1.5 ppm/%RH	n.a.
Water absorption	0.3%	1.8%	0.05%	0.2%

Source: Data extracted from Kinkeldei, T. et al., Influence of flexible substrate materials on the performance of polymer composite gas sensors, In: *Proceedings of The 14th International Meeting on Chemical Sensors, IMCS 2012*, May 20–23, Nuremberg, Germany, pp. 537–540, 2012.

CHE, coefficient of hydroscopic expansion; CTE, coefficient of thermal expansion; n.a., not available; PEI, polyethylenimine; PEN, polyethylene naphthalate; PI, polyimide; PPS, polyphenylene sulfide; T_G, glass transition temperature (temperature at which the amorphous phase of the polymer is converted between rubbery and glassy states); T_M, melting point.

promising characteristics for their integration in the low-power gas sensors and thermal actuators. In particular, a high operating temperature (up to 400°C–500°C) was obtained at a relatively low-power (100–150 mW), and the thermal stability of the structure allowed the annealing of a metal–oxide film to realize metal–oxide gas sensors. Compared to micro-hotplates on silicon with their membranes made of dielectric layers, the fabrication of micro-hotplates with PI-based membranes brings the advantages of simplified processing and an improved robustness and flexibility. Meanwhile, Briand et al. (2011) believe that PI will only be used for applications with specific requirements regarding temperature and the robustness of the substrate: Most devices will be produced on PET and PEN substrates.

Kinkeldei et al. (2012) also compared different polymeric substrates such as PEN, PI, polyphenylene sulfide (PPS), and PEI (see Table 22.3) and concluded that the selection of polymer for sensor platform depends on the working principle of the sensor designed. In case of metal oxides–based sensors, the substrate has to be heated during operation of sensors. This requires temperatures above the melting point of PET/PEN and PPS. This means that these materials are unacceptable for application in metal–oxide chemiresistors. In case of capacitive sensors, the substrate material should be inert against environmental changes so that the signal is only dependent on the gas-sensitive layer. PI has a high water adsorption that affects the sensing performance (Oprea et al. 2007), while PEN/PET, compared to PI, have a lower water absorption coefficient and are less affected from humidity changes (Oprea et al. 2008). Therefore, bare PI and PEN substrates have been used by Oprea et al. (2007, 2008) for capacitive humidity sensors. It is important to note that, in this case, the substrate itself can act as a humidity-sensitive material. Kinkeldei et al. (2012) have found that PEI substrates also had a low water absorption coefficient. The experiments with metal oxide–resistive sensors have shown that the influence of water on the gas sensor performance showed least influences with PEI as substrate materials. This means that, in humidity sensors manufactured on such substrates, the sensor performance will be determined by the humidity-sensitive material, and not by the substrate. According to Kinkeldei et al. (2012), besides the stable measurement results with PEI substrates, the PEI combines several benefits of the other substrates. It is translucent and can be heated up to 200°C without degradation (soldering is possible). Further evaporation of metal films on the substrate showed excellent adhesion already on non-plasma-treated substrates. For comparison, the adhesion was poor on PEN substrates and scratched metal lines were destroyed. Kinkeldei et al. (2012) believe that these properties make PEI a promising substrate material for the resistive-type gas sensor applications.

22.3 HUMIDITY SENSORS ON FLEXIBLE SUBSTRATES

As mentioned earlier, the development of sensors based on flexible substrates is a popular trend in the development of various sensors. So, at present, it is already possible to find a large number of articles describing gas-sensing devices fabricated on plastic substrates (Ki 2006; Huanget et al. 2006; Lee et al. 2008; Oprea et al. 2007, 2008, 2009; Arena et al. 2009; Gu et al. 2009; Miyoshi et al. 2009; Su et al. 2009c; Zampetti et al. 2009; Ahn et al. 2010; Jeong et al. 2010; Wang et al. 2010a, 2010b; Lee et al. 2011; Reddy et al. 2011). It is important to note that a major part of these published works is devoted to flexible humidity sensors. Let's consider some of them.

22.3.1 Capacitive Humidity Sensors on Flexible Substrates

Studies have shown that the development of humidity sensors based on flexible substrate can use the same approaches as for the development of sensors based on conventional substrates. Restrictions can be associated only with the need to ensure high adhesion of humidity-sensitive material and electrodes to the substrate by using sufficiently low-temperature deposition or synthesis of these materials. However, it should be recognized that the configuration with interdigital electrodes (Figure 22.3a) is most suitable for sensors on plastic substrates fabricated using printing technology (Oprea et al. 2007, 2008, 2009; Chatzandroulis et al. 2011). However, one should note that sensors with two parallel metal plates, separated by a thin humidity-sensitive film layer (Figure 20.3b), have also been elaborated (Zampetti et al. 2009). The sensor shown in Figure 22.3b is a capacitor implemented on a flexible PI substrate of approximately 8 μm thick, where a thin layer of bis(benzo cyclobutene) (BCB) is approximately 0.8 μm thick and is used as a dielectric-sensitive material. The bottom electrode (electrode B) is a metal layer shaped as a big island, while the upper electrode is a sequence of square-pattern structure, in order to allow the interaction between BCB and the analyte.

Typically, flexible capacitive sensors are manufactured on the basis of humidity-sensitive polymers. As well as in conventional sensors, for these purposes, we used poly cellulose acetate-butyrate (CAB) (Oprea et al. 2008), polydimethylsiloxane (PDMS) (Oprea et al. 2009), poly (2-hydroxyethyl methacrylate) (pHEMA) (Reddy et al. 2011), bis(benzo cyclobutene) (BCB) (Zampetti et al. 2009), and PI layers (Oprea et al. 2008; Rivadeneyra et al. 2014), which were realized by spray deposition, screen printing from suitable polymer solutions/gels, drop coating, and other methods of thick-film technology.

Oprea et al. (2008) have reported that the devices, manufactured on 30–100 μm PI and PEN substrates, with approximately 1 cm² of active area, 30–40 μm of interdigital traces and gaps, and 4–20 pF of nominal capacitance, had sensitivities of approximately 100–1000 ppm/%RH. PI and PEN substrates were also tested as humidity-sensitive materials. Oprea et al. (2008) showed that the sensitivity of the sensors produced on PI substrates was higher (Figure 22.4). Response and recovery time were within 1–7 minutes and depended on the used humidity-sensitive material. When using a specially applied humidity-sensitive layer, the response time constant was between 1 and 3 minutes, and for foils, it was between 1 and 7 minutes. This is understandable, since the porosity of the deposited layer is much higher than the porosity of the PI foil. This means that humidity sensors fabricated using PI foil as humidity-sensitive material can be applied where the response and recovery times are not critical. Faster sensors require the use of specially deposited humidity-sensitive layers.

Rivadeneyra et al. (2014) also manufactured a humidity sensor using polyimide foil simultaneously as a substrate and as a humidity-sensitive material (see Figure 22.5). The proposed sensor showed a reproducible humidity response from 100 kHz to 10 MHz; its sensitivity was

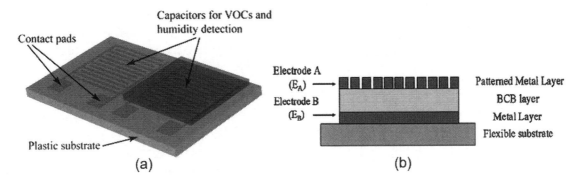

FIGURE 22.3 (a) Schematic view of two capacitors with interdigital electrodes (one uncoated on the left, one coated with a functional material on the right) patterned on a plastic substrate and (b) Schematic view of humidity sensor with two parallel metal plates. (Reprinted with permission from Zampetti, E. et al., Design and optimization of an ultra-thin flexible capacitive humidity sensor, *Sens. Actuators B*, 143, 302–397, 2009. Copyright 2009, Elsevier.)

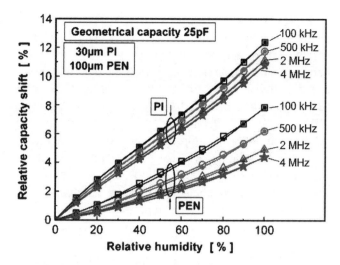

FIGURE 22.4 Comparison of the humidity responses from a 30 μm polyimide (PI) (Kapton®) and 100 μm polyethylene naphthalate (PEN) humidity foil sensors. The *open symbols* stem from exposure events performed during increasing humidity sequences, while the *full symbols* represent the decreasing humidity ones. Measurement performed with HP 4285A, at 25°C and 200 sccm synthetic air. (Reprinted with permission from Oprea, A. et al., Capacitive humidity sensors on flexible RFID labels, *Sens. Actuators B*, 132, 404–410, 2008. Copyright 2008, with permission from Elsevier.)

FIGURE 22.5 The image of the humidity sensor developed by Rivadeneyra et al. (2014) on the base of polyimide foil with thickness 75 μm. Silver electrodes were fabricated by inject printing. The designed interdigitated electrode (IDE) area was 11.65 mm² (L = 1.85 mm; W = 6.3 mm) composed of 63 fingers (32 fingers for one electrode and 31 for the other one), with 50 μm width and interspacing. (Reprinted with permission from Rivadeneyra, A. et al., Design and characterization of a low thermal drift capacitive humidity sensor by inkjet-printing, *Sens. Actuators B*, 195, 123–131, 2014. Copyright 2014, Elsevier.)

(4.5 ± 0.2) fF/%RH at 100 kHz and (4.2 ± 0.2) fF/%RH at 1 MHz. These sensors had stable characteristics. The sensors have been measured once a week for more than 6 months, and data showed a small maximum variation of ± 9.3 fF of the average value. After the fifth month, the humidity sensor was tested for 10 days at fixed RH (30%) and controlled temperature (30°C) every 6 hours. The aging drift was less than 3% RH. However, the response and recovery times were too long (τ~30 min). Of course, such sensors can hardly compete in terms of response time with commercial complementary metal–oxide–semiconductor (CMOS)-based humidity sensors (SHT15, Sensirion AG, Switzerland) with times smaller than 10 seconds. However, Rivadeneyra et al. (2014) believe that this sensor, as it is presented here, could be useful for environmental humidity monitoring, in which the changes are usually gradual and slow. The same conclusion was made earlier by Oprea et al. (2007, 2008). According to Rivadeneyra et al. (2014), the response time can be reduced by using a thinner substrate. Great times for response and recovery (τ> 25 min) were also observed for their sensors by Molina-Lopez et al. (2012). Similarly to Oprea et al. (2007, 2008) and Rivadeneyra et al. (2014), they are sure that this is due to the large thickness of the substrate used. A thick substrate requires a long time to attain stabilization inside the substrate.

Reddy et al. (2011) reported that the maximum percentage change in capacitance of a poly (2-hydroxyethyl methacrylate) (pHEMA)-based sensor was 172% at 80% RH when compared to the base capacitance at 30% RH. Sensors developed by Reddy et al. (2011) had better sensitivity compared to sensors developed by Oprea et al. (2007, 2008), but they had a pronounced nonlinearity and the main growth of sensitivity was observed at RH above 60%.

As noted earlier, the main advantage of sensors made on polymer substrates is their flexibility. The capacitance sensors, developed by Zampetti et al. (2009), fully demonstrate this very property. The fabricated sensor had dimensions of 3 mm × 6 mm and an active area of 9 mm². The devices have shown sensitivity to relative humidity of 0.38%/RH% and a linearity of 0.996 in the range of 10%–90% RH. The response time, estimated from 10% to 90% of the response curve, was 216 seconds. Zampetti et al. (2009) have found that due to the ultra-thin thickness (8 μm) of the polymeric substrate, these sensors can be attached directly to the object surfaces (e.g., to a bottle, plaster, paper, RFID tag) and many others. It was established that the electric characteristic variations were negligible, with a maximum shift of 0.5% for capacitance and 1.5% for conductance, for bending the radius down to 1.4 mm.

22.3.2 Resistive Humidity Sensors on Flexible Substrates

Resistive humidity sensors, as well as capacitive sensors, can be manufactured on flexible substrates such as PI and PET (Su and Wang 2007; Miyoshi et al. 2009; Cha and Gong 2013; Lim et al. 2013; Su et al. 2013; Tarapata et al. 2016; Li et al. 2017). PI may remain too expensive compared to PET foil, which was preferred when less-demanding specifications were required (low temperature during fabrication or operation, no contact with aggressive chemical). The first humidity-sensing devices on PET were developed by Parikh et al. (2006) and Su and Wang (2007). Parikh et al. (2006) have used electronically conductive, single-walled carbon nanotubes (CNTs) films (1–2 μm thick) deposited by dip-coating using aqueous surfactant-supported dispersions. The measurements were carried out using a four-contact method. Sensors developed were very flexible (e.g., they could be bent to diameters as small as 10 mm without significantly compromising sensor function).

Su and Wang (2007), unlike Parikh et al. (2006), used a traditional approach based on the use of interdigital electrodes and polyelectrolyte humidity-sensitive material. Su and Wang (2007) have reported that they successfully fabricated a flexible humidity sensor (see Figure 22.6a), using a methyl methacrylate (MMA) and [3-(methacrylamino) propyl] trimethyl ammonium chloride (MAPTAC) copolymer material, prepared by in situ copolymerization. The sensor resistance in the range of 10%–10% RH varied by more than three orders of magnitude. The response time (humidification from 7% to 78% RH) was approximately 45 seconds, and the recovery time (desiccation from 73% to 5% RH) was approximately 150 seconds. The effect of long-term stability is shown in Figure 22.6b. The impedance of the flexible humidity sensor had no obvious deviation at four testing points—10%, 30%, 60%, and 90% RH—for at least 120 days.

Ahn et al. (2012) have used another polyelectrolyte. A thin, polymeric, resistive humidity-sensitive film of MMA/MDBAC copolymer was formed on the gold electrode (GE) substrate by the silk-screening printing method. The effective area of the humidity-sensitive membrane in the sensor was 25–35 mm^2. Interdigital GEs were made by first printing nanosilver paste (thickness 6–7 μm), followed by consecutive electroless plating of Cu (thickness 5 μm), Ni (thickness 2 μm), and Au (thickness 80 nm) on the silver pattern. Impedance of these printed sensors was found to respond proportionally to changes in RH over a range of 20%–95% RH (Figure 22.7a). Typical response times were approximately 75 seconds for the adsorption process and 95 seconds for the desorption processes. Impedance of the humidity sensor using GE substrate showed no obvious deviation at three testing points—30, 60, and 80% RH—for at least 225 days (Figure 22.7b).

Approximately the same parameters were demonstrated by the sensors developed by Lim et al. (2013) on the base of MEPTDD/CEMA copolymer films, having strong adhesion to the polyimide substrate (Figure 22.8). A resistive-type microhumidity sensor with interdigitated microelectrodes had size 2.0 mm × 3.0 mm. Lim et al. (2013) reported that the microhumidity sensors exhibited high sensitivity, little hysteresis, rapid

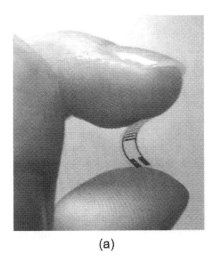

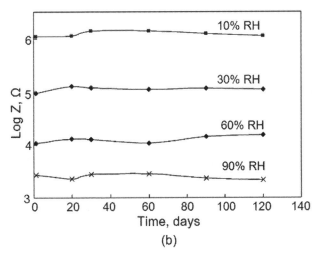

(a) (b)

FIGURE 22.6 (a) The photo of a flexible resistive humidity sensor fabricated on the polyethylene terephthalate (PET) substrate; and (b) The long-term stability of a flexible MMA/MARTAC-based humidity sensor measured at 1 V, 1 kHz, 25°C and different humidity. (Reprinted with permission from Su, P. G. and Wang, C. S., Novel flexible resistive-type humidity sensor, *Sens. Actuators B*, 123, 1071–1076, 2007. Copyright 2007, Elsevier.)

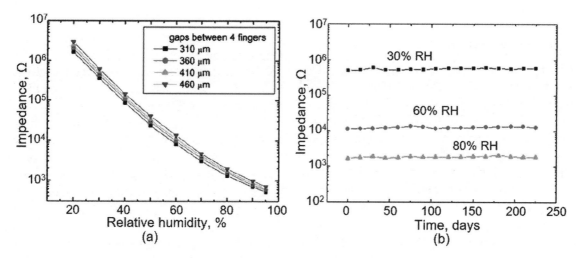

FIGURE 22.7 (a) Impedance changes of humidity sensors obtained from methyl methacrylate/[2-(methacryloyloxy)ethyl] dimethyl benzyl ammonium chloride (MDBAC/MMA) = 70/30 polyelectrolyte ink formed on gold electrodes with different gaps between fingers: 25°C, 1 kHz, and 1 V; and (b) Stability of the sensor obtained from MDBAC/MMA = 70/30 on the electrode with four fingers and a gap of 310 μm at 25°C. (Reprinted with permission from Ahn, H.Y. et al., Preparation of flexible resistive humidity sensors with different electrode gaps by screen printing and their humidity-sensing properties, *Macromolecular Res.*, 20, 174–180, 2012. Copyright 2012, Springer Science+Business Media.)

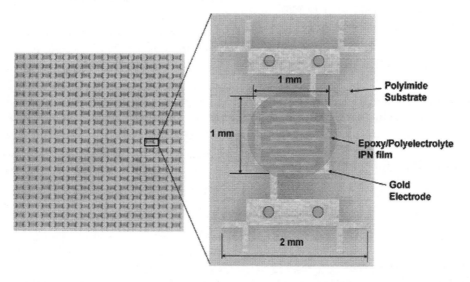

FIGURE 22.8 Photographs of micro-electrode with flexible polyimide substrate developed by Lim et al. (2013). (Reprinted with permission from Lim, D.-I. et al., Preparation of flexible resistive micro-humidity sensors and their humidity-sensing properties, *Sens. Actuators B*, 183, 574–582, 2013. Copyright 2013, Elsevier.)

response, water durability, and stability when exposed to high temperature and humidity.

Miyoshi et al. (2009) have shown that poly-tetrafluoroethylene (PTFE) can also be applied in humidity sensors. They reported a flexible design configuration for a resistive humidity sensor based on a sandwiched porous hydrophilic PTFE membrane structure deposited by a soft-MEMS technique applicable to physiological humidity ranges. The polymer membrane with thickness 80 μm was placed between two gold deposited layers (see Figure 22.9). The device was calibrated at 100 Hz against moist air over the range of 30%–85% RH (Figure 22.10), which includes normal humidity levels in the atmosphere and physiological air, such as breath and evaporating sweat. The response sensitivity of the humidity device was extremely high, even for the recovery to dry air; for example, the response time was less than 1 second for a conductivity shift between humid air of 80% RH and dry air of −60°C dew point.

A completely different approach to the development of humidity sensors was used by Li et al. (2017). They proposed to use in flexible humidity sensors the porous ionic membrane (PIM). The basic humidity-sensing response

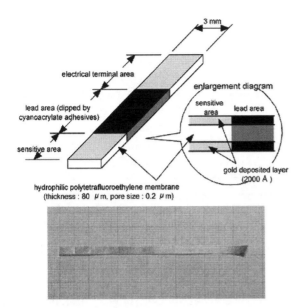

FIGURE 22.9 The structure and photograph of a wearable humidity sensor shaped by a knife into a 3-mm widestrip. (Reprinted with permission from Miyoshi, Y. et al., Flexible humidity sensor in a sandwich configuration with a hydrophilic porous membrane, *Sens. Actuators B*, 142, 28–32, 2009. Copyright 2009, Elsevier.)

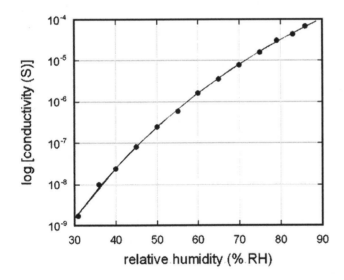

FIGURE 22.10 Calibration curve (semilogarithmic plot) of the wearable polytetrafluoroethylene (PTFE)-based humidity sensor in moisture gases varying humidity. X-axis (electrical conductivity of the humidity sensor) is expressed as a logarithmic scale. (Reprinted with permission from Miyoshi, Y. et al., Flexible humidity sensor in a sandwich configuration with a hydrophilic porous membrane, *Sens. Actuators B*, 142, 28–32, 2009. Copyright 2009, Elsevier.)

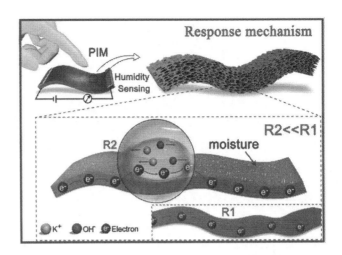

FIGURE 22.11 The humidity-sensing mechanism of the porous ionic membrane (PIM)-based sensor. (Reprinted with permission from Li, T. et al., Porous ionic membrane based flexible humidity sensor and its multifunctional applications, *Adv. Sci.*, 4, 1600404, 2017. Copyright 2017, Wiley-VCH Verlag GmbH & Co. KGaA.)

mechanism of the PIM-based sensor is explained in Figure 22.11. A PIM with well-defined three-dimensional (3D) porous structures was prepared by naturally evaporating a polyvinyl alcohol (PVA)/KOH polymer gel electrolyte, and its humidity-sensing properties showed that a PVA/KOH mass ratio of 6:3 and the thickness of approximately 70 μm were optimal for humidity sensor fabrication. The conductance of the PIM changed more than 70 times as the RH increased from 10.89% to 81.75% at room temperature. The response was fast (< 1 s) and reversible. According to Li et al. (2017), the potassium hydroxide has a higher ability to dissolve in water than in PVA, so, during the adsorption process, some potassium and hydroxide ions are transferred from the PVA to the adsorbed moisture and form a liquid electrolyte; the number of transferred electrons in the adsorbed moisture increases dramatically as a result of the directional migration of potassium and hydroxide ions in the electric field, so that the resistance value of the PIM decreases sharply in humid environments. In addition, Li et al. (2017) state that the PIM-based sensor was insensitive to the temperature (0°C–95°C) and pressure (0–6.8 kPa) change.

An interesting approach to the development of humidity sensors was also proposed by Trung et al. (2017). They proposed a transparent, stretchable humidity sensor with a simple fabrication process, having intrinsically stretchable components that provide high stretchability, sensitivity, and stability along with fast response and relaxation time. As a humidity-sensitive material, Trung et al. (2017) suggested to use a reduced graphene oxide (R-GO)/polyurethane (PU) composite with a thickness of 200 nm, and as a substrate a polydimethylsiloxane (PDMS) (Figure 22.12). The mechanism of operation of such sensors was described in Section 11.2.5. According to the authors, these sensors with fast response have immense potential in a variety of

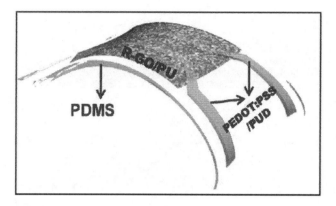

FIGURE 22.12 Schematic of the transparent, stretchable humidity sensor. (Reprinted with permission from Trung, T.Q. et al., Transparent, stretchable, and rapid-response humidity sensor for body-attachable wearable electronics, *Nano Res.*, 10, 2021–2033, 2017. Copyright 2017, Springer Science+Business Media.)

applications, including e-skin, personal health care, monitoring the moisture of the soil during irrigation in agriculture, and gauging the effectiveness of cosmetic products. However, it is not clear how to protect the sensor from extraneous mechanical influences, which will also lead to a change in resistance. This is especially important, since the change in sensor resistance under the influence of moisture does not exceed 10%.

An unusual solution for sensors on flexible substrates was proposed by He et al. (2013). On the base of flexible substrates, the surface acoustic wave (SAW)-based humidity sensors were developed (Figure 22.13a). SAW sensors were fabricated on polyimide substrates by deposition of (0002) oriented columnar ZnO nanocrystals with grain sizes of 50–60 nm. ZnO piezoelectric films were deposited using a direct-current (dc) magnetron sputtering system. The ZnO thickness of the SAW sensors studied was approximately 3.5 μm. SAW devices were made by standard UV-light photolithography and lift-off process. An aluminium (Al) layer of approximately 80 nm thickness was used for the interdigitated transducers (IDTs), which were deposited by thermal evaporation. Figure 22.13b shows the frequency shift, Δf, of the SAW devices as a function of humidity with wavelength as a variable. It is clear that the sensitivity of the SAW sensors increases with the decrease in wavelength, and also increases with humidity for all the devices with different wavelengths, although the sensitivity varies from sample to sample. He et al. (2013) reported that the SAW sensors developed without any surface treatment had a sensitivity of 34.7 kHz/10% RH. This is a good result, even in comparison with devices designed on conventional substrates. Therefore, the authors are confident that such devices have a great potential for applications in flexible electronics, sensors, and microsystems. As for the shortcomings, in addition to the disadvantages inherent in all SAW-based sensors (see Chapter 12), these devices, as well as all humidity sensors using polyimide foil ($d = 100$ μm) as a humidity-sensitive material, had long response and recovery time, more than 10 minutes.

22.3.3 Multiparameter Sensing Platform

In the previous chapter, it was noted that the development of integrated sensors based on micromachining technologies is an important step in the development of sensor technologies. Studies have shown that integrated sensors can be manufactured and based on flexible substrates. Besides, high technologies are not required for their production. For

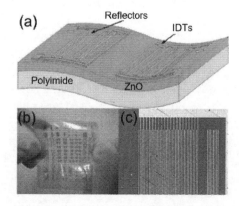
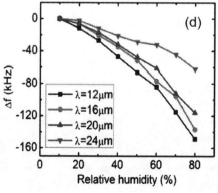

FIGURE 22.13 A schematic drawing of the flexible surface acoustic wave (SAW) device: (a) A photo of the fabricated flexible SAW devices on a polyimide substrate; (b) Microscopy image of the interdigitated transducers (IDTs); (c) The wavelength of the devices is 12 μm; and (d) Resonant frequency shift as a function of relative humidity (RH) for flexible SAW sensors with different wavelengths. The time taken for each measurement is 50 min. (Reprinted with permission from He, X.L. et al., High sensitivity humidity sensors using flexible surface acoustic wave devices made on nanocrystalline ZnO/polyimide substrates, *J. Mater. Chem. C*, 1, 6210–6215, 2013. Copyright 2013, The Royal Society Chemistry.)

example, a multiparameter sensing platform (for volatile organic compounds [VOCs], temperature, humidity, and reducing and oxidizing gases) on the plastic foil, based on standard clean room processes, was already designed (Courbat et al. 2009b; Oprea et al. 2009, 2012; Briand et al. 2011). In particular, Oprea et al. (2012) demonstrated that, by joint operation of the Pt thermoresistive thermometer and polymer-based capacitive and metal oxide conductometric sensors (Figure 22.14), several interfering gases can be detected and, to some extent, separated at hardware level. Thus, the capacitive structures monitored the humidity and ethanol (as representative VOC), while the metal-oxide ones monitored oxidizing and reducing gases (NO_2, CO) as well as the reducing ethanol vapors. Oprea et al. (2012) reported that all the integrated devices were stable and gave reproducible signals for more than 2 months of operation, even when the metal oxide (MOX)-based sensors ran continuously at 300°C. A typical result of the multisensor platform is presented in Figure 22.15. As well as monolithic sensors, to eliminate a long time of response of the microhumidity sensor, flexible sensors can be integrated with a microheater (Lee et al. 2011).

It is important to note that a hybrid approach to design gas-sensing devices on plastic substrates can be also used. In particular, an SAW chip, which requires substrates with strongly different properties, can be transferred onto a plastic substrate (Cobianu et al. 2007). Another interesting and new approach is the coating of passive (no power source on board) conventional RFID tags with chemically sensitive films to form a chemical sensor (Subramanian et al. 2005; Potyrailo and Morris 2007). The detection of several vapors of industrial, health, law enforcement, and security interest (ethanol, methanol, acetonitrile, and water vapors) was demonstrated with a single 13.56 MHz RFID tag coated with a solid polymer electrolyte-sensing film. More details related to such a sensor can be found in Chapter 10.

22.3.4 FEATURES OF FABRICATION TECHNOLOGY

It should be noted that the creation of robust conjunctions between polymers and plastic substrates, frequently PET and PI substrates, is a main concern of current developments in the area of sensors fabricated on flexible substrates (Lin et al. 2012; Su et al. 2013). This challenge has been somewhat mitigated by introducing physical and chemical reactive methods to form a bonding matrix. For example, Su et al. (2013) proposed such an approach as an anchoring of humidity-sensing polyelectrolytes to the electrode surface on plastic substrates. A humidity-sensitive layer of polyelectrolyte, produced by copolymerization of MMA and [3-(methacrylamino)propyl] trimethylammonium chloride, was anchored to an interdigital GE (pretreated with 3-mercaptopropionic acid) on a PET substrate by a peptide chemical protocol. N-(3-dimethylaminopropyl)-N'-ethylcarbodiimide hydrochloride (EDC) was employed as a peptide coupling reagent. The sensors have shown good humidity dependence over wide humidity ranges (20%–90% RH) with good long-term stability. The water stability of the anchored poly-MMA-[3-(methacrylamino)propyl] trimethyl ammonium chloride (MAPTAC) to the 3-mercaptopropionic acid (MPA)/Au contact area was influenced by the amount of added EDC. For improvement of adhesion, Li et al. (2009)

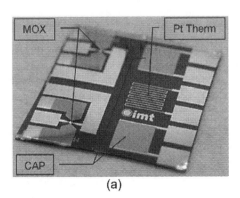
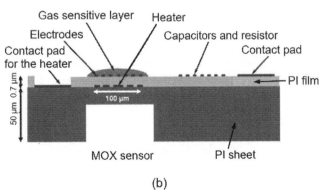

FIGURE 22.14 (a) The multisensor platform micrograph. MOX: nanogranular SnO_2 and WO_3 metal oxide thick films; CAP: interdigital capacitors, one of them coated with poly(ether urethane) (PEUT); Pt Them: Pt-resistance thermometer and (b) Cross-sectional schematic of the multisensor platform on polyimide foil: Capacitive humidity and volatile organic compound (VOC) sensors and resistive temperature sensors. The power consumption of the MOX sensors can be reduced by using a back-etched membrane. ([a] Reprinted with permission from Oprea, A. et al., Environmental monitoring with a multisensor platform on polyimide foil, *Sens. Actuators B*, 171–172, 190–197, 2012, Copyright 2012, Elsevier; [b] Reprinted with permission from Courbat, J. et al., Multi sensor platform on plastic foil for environmental monitoring, *Procedia Chem.*, 1, 597–600, 2009. Copyright 2009, Elsevier as open access.)

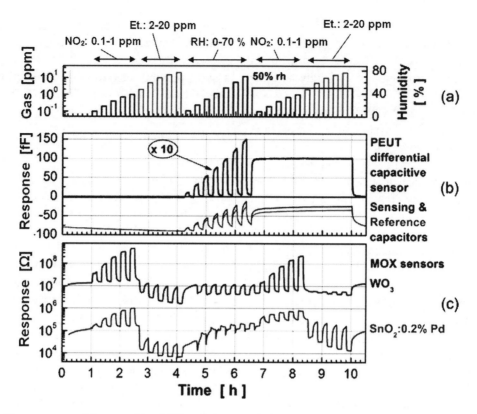

FIGURE 22.15 Gas response of the capacitive (RT) and chemoresistive (300°C) gas sensors to nitrogen dioxide (0.1–1.0 ppm), ethanol (2–20 ppm), and relative humidity (RH) (0%–70%). (a) Gas protocol; (b) Differential PEUT capacitive sensor response (10 times magnified), PEUT sensing capacitor signal, and reference capacitor signal. The PEUT sensing capacitor and reference capacitor signals (in the 2 pF range) have been arbitrarily shifted in order to be displayed on the same panel with the differential capacitive sensor response; and (c) Gas response of the SnO_2:0.2%Pd and WO_3 MOX sensors, respectively. (Reprinted with permission from Oprea, A. et al., Environmental monitoring with a multisensor platform on polyimide foil, *Sens. Actuators B*, 171–172, 190–197, 2012, Copyright 2012, Elsevier.)

have proposed to apply surface modification of the electrode surface with poly (diallyldimethylammonium chloride) (PDDA) before depositing the sensitive layer.

Molina-Lopez et al. (2012) also paid attention to the improvement of the technology of manufacturing humidity sensors based on plastic substrates. They established that the selected silver ink exhibited better adhesion performances on PET than on PI, regardless of pretreatment with oxygen plasma, which, on the contrary, was useful to increase the smoothness of the printed lines' edges. Molina-Lopez et al. (2012) have also found the PET substrate was usually too thick (125 μm) to allow a quick stabilization of the sensor response. In addition, it was shown that passivation of the printed silver electrodes with nickel could be a justification for the improvement in the stability of electroplated sensors. High sensor stability has been obtained only for nickel-electrodeposited devices. Using nickel electrodeposition allowed for increasing the thickness of electrodes up to 15 μm on PET and up to 5 μm on PI. This limitation in the thickness was due to the increasing stress induced in thicker layers, which eventually forced the metallic lines to delaminate from the substrate. Molina-Lopez et al. (2012) believe that further reduction of the device size would yield an important increase in the sensor sensitivity.

Regarding the integration of temperature sensors on plastic-flexible foil, the conventional platinum resistance temperature detector (RTD) has been realized on a flexible polyimide substrate for operation up to 400°C, and resistors made of TaSiN have demonstrated high temperature coefficient of resistance (TCR) values (Moser and Gijs 2007; Chung et al. 2009). Courbat (2010) established that the sputtering of Ti/Pt with the plasma activation of the surface before deposition provided the best adhesion properties to polyimide substrates. The use of Cr led generally to satisfactory adhesion results by evaporation. However, despite its good adhesive properties for Pt, this metal suffers from oxidation when exposed to air and diffuses easily in platinum. They are significant disadvantages for heating elements or chemical sensors for which a high stability over time is required. To counteract these drawbacks, Ta should ideally be used. However, it showed poor adhesion when evaporated.

Some approaches, based on thermosensitive polymers based on graphite or metallic powders in a PDMS matrix, that suffered from nonlinearity and were limited to 100°C have also been evaluated (Chuang and Wereley 2009; Shih et al. 2010). Another approach that has been reported is the screen printing of a polymeric, thermosensitive material on Kapton for textronic applications (e.g., measurement of the temperature of the human body) (Bielska et al. 2009).

22.4 PAPER-BASED HUMIDITY SENSORS

In recent years, there have been studies aimed at studying the possibilities of using paper substrates for the development of various sensors (Liana et al. 2012). It is believed that paper-based sensors are a new alternative technology for fabricating simple, low-cost, portable, and disposable analytical devices for many application areas, including clinical diagnosis, food-quality control, and environmental monitoring. The unique properties of paper, such as its versatility, high abundance, and low cost, and its unique physical properties allow paper to be considered as a promising sensing platform (Martinez et al. 2010). Paper is a highly sophisticated material, as it can be made thin, lightweight, and flexible depending on its pulp processing. In addition, paper is combustible and biodegradable. It can be easily stored, transported, and disposed. Moreover, cellulose fibers can be functionalized, thus changing properties such as hydrophilicity, if desired, as well as its permeability and reactivity (Bracher et al. 2010). The paper surface can be easily manipulated through printing, coating, and impregnation and can be fabricated in large quantities. Depending on the main goal to be achieved in paper-based sensors, the fabrication methods and analysis techniques can be tuned to fulfill the needs of the end-user.

There are a variety of paper materials available to the user, although the choice is based mainly on the fabrication steps required in developing a device and also on the specific application area (Li et al. 2010; Martinez et al. 2010). However, most sensors are based on the filter paper (the Whatman® cellulose range) and glossy paper. Glossy paper is a flexible substrate made of cellulose fiber blended with an inorganic filler. In particular, Arena et al. (2010) used glossy paper for developing a flexible, paper-based sensor for the detection of ethanol vapors in air using indium tin oxide nanoparticulate powder as a sensing material and multiwalled carbon nanotubes as electrodes.

Of course, paper substrates are inferior to plastic substrates for such parameters as mechanical strength, resistance to aggressive media, and manufacturability. The inherent surface roughness and porosity make a device fabrication on paper more complicated. There are methods of processing paper that can significantly improve its surface properties (Jalkanen et al. 2015). For example, many studies have focused on devising methods for improving the surface hydrophobicity of paper, such as laser ablation to modify the surface morphology and/or change the surface energy (Chitnis and Ziaie 2012), and plasma-induced polymerization to create hydrophobic polymer chains on the paper's surface to make it water-repellent (Song et al. 2013). Other interesting examples include the possibility to improve surface hydrophobicity of paper by using a coating of organic or inorganic nanoparticles (Stanssens et al. 2011; Ogihara et al. 2012; Bollstrom et al. 2014). Such treatment makes it possible to reduce the effect of humidity on paper-based electronics. However, even though this process is effective, it requires a specific type of paper, takes longer, and incurs additional costs. Therefore, paper substrates cannot replace plastic ones. However, in specific applications, paper substrates can undoubtedly find applications. For example, the high porosity of the paper makes it possible to incorporate materials that have properties that are important for sensor applications, but which are difficult to fix on plastic substrates. In addition, continuous pore channels, allowing for efficient diffusion of gaseous molecules throughout the film matrix, provide maximal exposure of sensing material to the gaseous analytes and thus make it possible to enhance the sensor signal and accuracy (Xu et al. (2011)

For example, Jalkanen et al. (2015), using this approach, managed to implement paper-based humidity sensors with a humidity-sensitive layer formed by porous silicon (PSi) nanoparticles. Earlier, Jalkanen et al. (2012) showed that the porous silicon nanoparticles are an interesting material for humidity sensor applications, because they host a large internal surface area, and the surface chemistry can be modified to accommodate a large number of environmental and biochemical sensing schemes. The sensing elements, consisting of printed interdigitated silver electrodes and a spray-coated PSi layer, were fabricated on a coated paper substrate by a two-step process. The image of a PSi-based humidity sensor is shown in Figure 22.16. A multilayer coated paper was used as a substrate. pSi particles were dispersed in a toluene suspension, making them solution processable. A broad particle size distribution ($\approx$ 100–2000 nm) was used to improve packing and adhesion of the silicon film on the substrate. Different interdigitated silver electrode patterns with varying gaps (100–500 µm) were produced on the substrate with two techniques—namely, flexography and inkjet printing. Jalkanen et al. (2015) believe that the fabrication steps used for the sensor's fabrication are easily up-scalable, which suggests that the presented method is feasible for

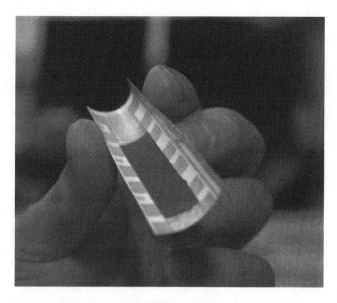

FIGURE 22.16 A photograph showing a spray-coated *p*-Si layer on flexographically printed silver electrodes. (Reprinted with permission from Jalkanen, T. et al., Fabrication of porous silicon based humidity sensing elements on paper, *J. Sensors*, 2015, 927396, 2015. Copyright 2015, Hindawi as open access.)

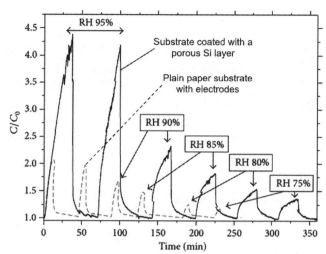

FIGURE 22.17 Relative capacitive response of paper-based sensor to changes in relative humidity (RH) concentration shows that, with proper thickness of the sensing layer, the signal from the sensing layer is clearly stronger than the signal from the substrate. (Reprinted with permission from Jalkanen, T. et al., Fabrication of porous silicon based humidity sensing elements on paper, *J. Sensors*, 2015, 927396, 2015. Copyright 2015, Hindawi as open access.)

mass production of humidity-sensing devices. The results of testing are presented in Figure 22.17. It is seen that in the range of 70%–95% RH, the response increased in accordance with concentration of water vapor, and a concentration change that is smaller than 5 RH% can be detected in this RH range. However, it is also evident that the sensors have a limited sensitivity range (RH > 70%) and long response and recovery time. It was also established that sensors had significant hysteresis characteristics.

Niarchos et al. (2017) also used the paper support for their resistive humidity sensors. For these purposes, they used two paper substrates: a plain printing paper of 80 g/m² basis weight and a glossy, photographic quality paper of a 200 gm⁻² basis weight. Interdigitated electrodes (IDEs), Au/Cr (100 nm/10 nm), were directly patterned on them, using laser ablation. The layer of ZnO nanoparticles deposited on the paper substrates using successive spin-coating was used as a humidity-sensitive element. Figure 22.18. shows the performances of the developed sensors. It is seen that the sensors can be used for humidity monitoring in the range of 20%–70% RH. However, the sensors have extremely great resistance. In addition, the response and recovery time are long enough. At an RH of 70%, the response time reaches 30 minutes for glossy-based and 10 minutes for plain paper–based sensors. This can in part be attributed to differences in the structure and porosity between two papers. The most interesting thing in these studies is that the paper-based humidity sensors without any ZnO coating had a higher response than after applying this layer. These results reveal that the ZnO layer, at least under the experimental conditions used in this work, acts as a passivation layer to humidity, prohibiting water molecules from interacting with the cellulose fibers.

It should be noted that the development of humidity sensors without the use of special humidity-sensitive material is not something specific in the development of paper-based sensors. This approach, in particular, was used by Feng et al. (2015) in the development of RFID humidity sensors (see Chapter 20). The resonators were fabricated by inkjet printing using silver nanoparticle ink. Three types of paper substrates were compared in this work, including nonorganic coated inkjet paper (PEL Nano P60, Printed Electronics Ltd.), commercial inkjet photo paper (Ultra premium, Kodak), commercial UV-coated packaging paper (Korsnäs AB). Testing has shown that the paper-based LC resonators exhibited excellent sensitivity upon exposure to moisture, with reasonable response time ($\tau \sim 10$–15 min). At that, the sample on the packaging paper exhibited the highest sensitivity out of the three samples when RH was below 70% RH, whereas it showed the lowest sensitivity when RH exceeds 70% RH. Feng et al. (2015) believe that using a thinner paper substrate or a higher working temperature can help desorb the water vapor more efficiently if required by applications.

Mraovic et al. (2014) also tested the possibility of using different cellulose substrates, such as recycled paper and

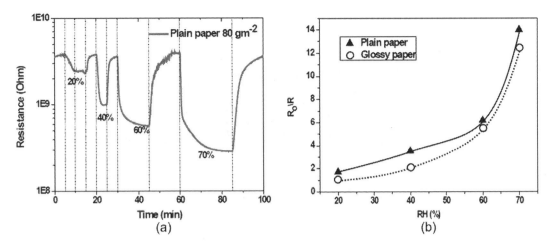

FIGURE 22.18 (a) Humidity response of plain paper-based sensor at controlled relative humidity (RH) levels ranging from 20% to 70%; and (b) Sensing response for plain and glossy paper-based sensors as function of the RH. (Reprinted with permission from Niarchos, G. et al., Humidity sensing properties of paper substrates and their passivation with ZnO nanoparticles for sensor applications, *J. Sensors*, 17, 516, 2017. Copyright 2017, Hindawi as open access.)

cardboard, in the development of humidity sensors. The sensors did not have additional layers of humidity-sensitive materials and were fabricated via the screen-printing method with silver-based conductive ink. Studies have shown that, despite the absence of a humidity-sensitive material, the sensors allowed fixing the humidity changes in the range of 30%–80% RH with a fairly good speed, which was several minutes, but with very poor repeatability. After analyzing the results, the authors also concluded that the sensor response is much more a function of the conductive ink than of the substrate.

The analysis shows that, although paper-based sensors are promising, they suffer from certain limitations, such as accuracy, repeatability, and sensitivity. However, developers hope that in the future, with advances in fabrication and analytical techniques, there will be newer and more innovative developments in the paper-based sensors. This innovative approach was used in developing sensors by Fraiwan et al. (2016). They created a novel, paper-based cantilever sensor array that enabled detection of VOCs, and hence water vapor, with the naked eye (see Figure 22.19). This simple VOC detection method was achieved using

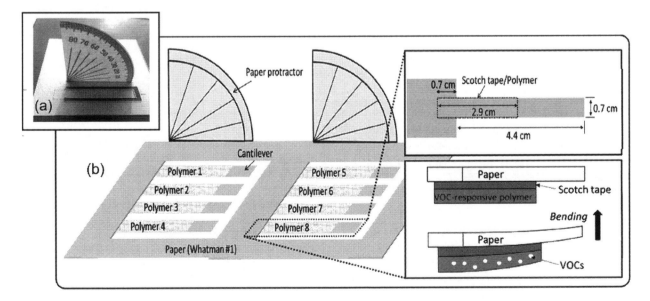

FIGURE 22.19 (a) Demonstrated paper-based volatile organic compound (VOC) sensor unit. A protractor and cantilever were patterned on a paper; and (b) Proposed paper-based VOC sensor array with eight different swellable polymer matrices. (Reprinted with permission from Fraiwan, A. et al., A paper-based cantilever array sensor: Monitoring volatile organic compounds with naked eye, *Talanta*, 158, 57–62, 2016. Copyright 2016, Elsevier.)

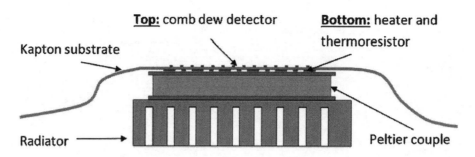

FIGURE 22.20 Cross-section of the hygrometer with the inkjet-printed dew-point detector. (Reprinted with permission from Tarapata, G. et al., Novel dew point hygrometer fabricated with inkjet printing technology, *Sens. Actuators A*, 247, 641–646, 2016. Copyright 2016, Elsevier.)

low-cost paper materials that monitored VOC-induced polymer swelling through the mechanical deflection of the cantilevers, thereby generating a selective angle pattern for a specific VOC. It is clear that such sensors are not very accurate. But, this is an extremely simple method that allows moisture monitoring without external equipment or interface electronics for readout.

22.5 SUMMARY

Experimentation has shown that, in terms of their parameters, flexible sensors as a rule do not differ much from sensors made on conventional substrates (Lim et al. 2013). Adequate sensing properties (enough sensitivity, good selectivity, stability, linearity, and reasonable response and recovery times) have been achieved. This means that such sensors can be used in the same areas where sensors on the solid-state substrates are used. For example, Tarapata et al. (2016) have demonstrated that even dew-point hygrometers (Figure 22.20) can be developed on the basis of flexible resistive humidity sensors. For these hydrometers, only inkjet printing technology can be used. All components of this hygrometer were printed on 12.5 μm flexible polyimide (Kapton) foil. The Kapton film provides adequate resistance against temperature (over 200°C), and its thinness allows minimization of the surface temperature measurement error with the bottom thermoresistor.

REFERENCES

Ahn H., Park J.-H., Kim S.-B., Jee S.H., Yoon Y.S., Kim D.J. (2010) Vertically aligned ZnO nanorod sensor on flexible substrate for ethanol gas monitoring. *Electrochem. Solid-State Lett.* 13 (11), J125–J128.

Ahn H.Y., Kim J.-G., Gong M.-S. (2012) Preparation of flexible resistive humidity sensors with different electrode gaps by screen printing and their humidity-sensing properties. *Macromolecular Res.* 20 (2), 174–180.

Angeles R. (2005) RFID technologies: Supply-chain applications and implementation issues. *Inform. Syst. Management* 22 (1), 51–65.

Arena A., Donato B., Saitta G. (2009) Capacitive humidity sensors based on MWCNTs/polyelectrolyte interfaces deposited on flexible substrates. *Microelectr. J.* 40, 887–890.

Arena A., Donato N., Saitta G., Bonavita A., Rizzo G., Neri G. (2010) Flexible ethanol sensors on glossy paper substrates operating at room temperature. *Sens. Actuators B* 145, 488–494.

Bielska S., Sibinski M., Lukasik A. (2009) Polymer temperature sensor for textronic applications. *Mater. Sci. Eng. B* 165, 50–52.

Bollstrom R., Pettersson F., Dolietis P., Preston J., Osterbacka R., Toivakka M. (2014) Impact of humidity on functionality of on-paper printed electronics. *Nanotechnology* 25, 094003.

Bracher P.J., Gupta M., Whitesides G.M. (2010) Patterning precipitates of reactions in paper. *J. Mater. Chem.* 20, 5117–5122.

Briand D., Colin S., Gangadharaiah A., Vela E., Dubois P., Thiery L., de Rooij N.F. (2006) Micro-hotplates on polyimide for sensors and actuators. *Sens. Actuators A* 132, 317–324.

Briand D., Oprea A., Courbat J., Bârsan N. (2011) Making environmental sensors on plastic foil. *Mater. Today* 14 (9), 416–423.

Cha J.-R., Gong M.-S. (2013) Preparation of epoxy/polyelectrolyte IPNs for flexible polyimide-based humidity sensors and their properties. *Sens. Actuators B* 178, 656–662.

Chatzandroulis S., Tsouti V., Raptis I., Goustouridis D. (2011) Capacitance-type chemical sensors, In: Korotcenkov G. (ed.) *Chemical Sensors: Comprehensive Sensor Technologies, Vol. 4: Solid State Devices*. Momentum Press, New York, pp. 229–260.

Chitnis G., Ziaie B. (2012) Waterproof active paper via laser surface micropatterning of magnetic nanoparticles. *ACS Appl. Mater. Interfaces* 4, 4435–4439.

Chuang H.-S., Wereley S. (2009) Design, fabrication and characterization of a conducting PDMS for microheaters and temperature sensors. *J. Micromech. Microeng.* 19, 45010.

Chung C.K., Chang Y.L., Wu J.C., Jhu J.J., Chen T.S. (2009) Characterization and patterning of novel high-TCR Ta–Si–N thin films for sensor application. *Sens. Actuators A* 156, 323–327.

Cobianu C., Georgescu I., Buiculescu V. (2007) Chip level packaging for wireless surface acoustic wave sensor. *US patent* 0114889, A1.

Courbat J. (2010) Gas sensors on plastic foil with reduced power consumption for wireless applications. PhD Thesis, École Polytechnique Federale de Lausanne, Switzerland.

Courbat J., Briand D., de Rooij N. (2010b) Ink-jet printed colorimetric gas sensors on plastic foil. In: *Proceedings of SPIE Photonic Devices + Applications Conference 2010*, July 31–August 5, San Diego, CA, 77790A. doi:10.1117/12.861142.

Courbat J., Briand D., Oprea A., Bârsan N., Weimar U., de Rooij N.F. (2009a) Multi sensor platform on plastic foil for environmental monitoring. *Procedia Chem.* 1, 597–600.

Courbat J., Briand D., Wöllenstein J., de Rooij N.F. (2009b) Colorimetric gas sensors based on optical waveguides made on plastic foil. *Procedia Chem.* 1, 576–579.

Courbat J., Canonica M., Teyssieux D., Briand D., de Rooij N.F. (2010a) Design and fabrication of micro-hotplates made on a polyimide foil: electrothermal simulation and characterization to achieve power consumption in the low mW range. *J. Micromech. Microeng.* 21 (1), 015014.

Crawford G.P. (ed.) (2005) *Flexible Flat Panel Displays*. John Wiley & Sons, Hoboken, NJ.

Feng Y., Xie L., Chen Q., Zheng L.-R. (2015) Low-cost printed chipless RFID humidity sensor tag for intelligent packaging. *IEEE Sensors J.* 15 (6), 3201–3208.

Fraiwan A., Lee H., Choi S. (2016) A paper-based cantilever array sensor: Monitoring volatile organic compounds with naked eye. *Talanta* 158, 57–62.

Gu P.-G., Lee C.-T., Chou C.-Y., Cheng K.-H., Chuang Y.-S. (2009) Fabrication of flexible NO_2 sensors by layer-by-layer self-assembly of multi-walled carbon nanotubes and their gas sensing properties. *Sens. Actuators B* 139, 488–493.

He X.L., Li D.J., Zhou J., Wang W.B., Xuan W.P., Dong S.R., Jin H., Luo J.K. (2013) High sensitivity humidity sensors using flexible surface acoustic wave devices made on nanocrystalline ZnO/polyimide substrates. *J. Mater. Chem. C* 1, 6210–6215.

Huanget A., Wong V.T.S., Ho C.-M. (2006) Silicone polymer chemical vapor sensors fabricated by direct polymer patterning on substrate technique (DPPOST). *Sens. Actuators B* 116, 2–10.

Jalkanen T., Määttänen A., Mäkilä E., Tuura J., Kaasalainen M., Lehto V.-P., et al. (2015) Fabrication of porous silicon based humidity sensing elements on paper. *J. Sensors* 2015, 927396.

Jalkanen T., Makila E., Maattanen A., Tuura J., Kaasalainen M., Lehto V.-P., et al. (2012) Porous silicon micro- and nanoparticles for printed humidity sensors. *Appl. Phys. Lett.* 101 (26), 263110.

Jeong H.Y., Lee D.-S., Choi H.K., Lee D.H., Kim J.-E., Lee J.L., et al. (2010) Flexible room-temperature NO_2 gas sensors based on carbon nanotubes/reduced graphene hybrid films. *Appl. Phys. Lett.* 96, 213105.

Ki Y.S. (2006) Microheater-integrated single gas sensor array chip fabricated on flexible polyimide substrate. *Sens. Actuators B* 114, 410–417.

Kinkeldei T., Zysset C., Münzenrieder N., Tröster G. (2012) Influence of flexible substrate materials on the performance of polymer composite gas sensors. In: *Proceedings of The 14th International Meeting on Chemical Sensors, IMCS 2012*, May 20–23, Nuremberg, Germany, pp. 537–540.

Lee C.-Y., Fan W.-Y., Chang C.-P. (2011) A novel method for in-situ monitoring of local voltage, temperature and humidity distributions in fuel cells using flexible multifunctional micro sensors. *Sensors* 11, 1418–1432.

Lee C.-Y., Wu G.-W., Hsieh W.-J. (2008) Fabrication of micro sensors on a flexible substrate. *Sens. Actuators A* 147, 173–176.

Li P., Li Y., Ying B., Yang M. (2009) Electrospun nanofibers of polymer composite as a promising humidity sensitive material. *Sens. Actuators B* 141, 390–395.

Li T., Li L., Sun H., Xu Y., Wang X., Luo H., Liu Z., Zhang T. (2017) Porous ionic membrane based flexible humidity sensor and its multifunctional applications. *Adv. Sci.* 4, 1600404.

Li X., Tian J.F., Garnier G., Shen W. (2010) Fabrication of paper-based microfluidic sensors by printing. *Colloid Surf. B* 76, 564–570.

Liana D.D., Raguse B., Gooding J.J., Chow E. (2012) Recent advances in paper-based sensors. *Sensors* 12, 11505–11526.

Lim D.-I., Cha J.-R., Gong M.-S. (2013) Preparation of flexible resistive micro-humidity sensors and their humidity-sensing properties. *Sens. Actuators B* 183, 574–582.

Lin Q., Li Y., Yang M. (2012) Polyaniline nanofiber humidity sensor prepared by electrospinning. *Sens. Actuators B* 161, 967–972.

Logothetidis S. (2008) Flexible organic electronic devices: Materials, process and applications. *Mater. Sci. Eng. B* 152, 96–104.

Martinez A.W., Phillips S.T., Whitesides G.M., Carrilho E. (2010) Diagnostics for the developing world: Microfluidic paper-based analytical devices. *Anal. Chem.* 82, 3–10.

Mattana G., Briand D. (2016) Recent advances in printed sensors on foil. *Mater. Today* 19 (2), 88–99.

McAlpine M.C., Ahmad H., Wang D., Heath J.R. (2007) Highly ordered nanowire arrays on plastic substrates for ultrasensitive flexible chemical sensors. *Nature Mater.* 6, 379–384.

Miyoshi Y., Miyajima K., Saito H., Kudo H., Takeuchi T., Karube I., Mitsubayashi K. (2009) Flexible humidity sensor in a sandwich configuration with a hydrophilic porous membrane. *Sens. Actuators B* 142, 28–32.

Molina-Lopez F., Briand D., de Rooij N.F. (2012) All additive inkjet printed humidity sensors on plastic substrate. *Sens. Actuators B* 166–167, 212–222.

Moser Y., Gijs M.A.M. (2007) Miniaturized flexible temperature sensor. *J. Microelectromech. S* 16 (6), 1349–1354.

Mraovic M., Muck T., Pivar M., Trontelj J., Pleteršek A. (2014) Humidity sensors printed on recycled paper and cardboard. *Sensors* 14, 13628–13643.

Niarchos G., Dubourg G., Afroudakis G., Georgopoulos M., Tsouti V., Makarona E., Crnojevic-Bengin, V., Tsamis, C. (2017) Humidity sensing properties of paper substrates and their passivation with ZnO nanoparticles for sensor applications. *Sensors* 17, 516.

Ogihara H., Xie J., Okagaki J., Saji T. (2012) Simple method for preparing superhydrophobic paper: Spray-deposited hydrophobic silica nanoparticle coatings exhibit high water-repellency and transparency. *Langmuir* 28, 4605–4608.

Oprea A., Barsan N., Weimar U., Bauersfeld M.L., Ebling D., Wollenstein J. (2008) Capacitive humidity sensors on flexible RFID labels. *Sens. Actuators B* 132, 404–410.

Oprea A., Barsan N., Weimar U., Courbat J., Briand D., de Rooij N.F. (2007) Integrated temperature, humidity and gas sensors on flexible substrates for low-power applications. In: *Proceedings of the IEEE Sensors Conference*, October 28–31, Atlanta, GA, pp. 158–161.

Oprea A., Courbat J., Barsan N., Briand D., de Rooij N.F., Weimar U. (2009) Temperature, humidity and gas sensors integrated on plastic foil for low power applications. *Sens. Actuators B* 140, 227–232.

Oprea A., Courbat J., Briand D., Bârsan N., Weimar U., de Rooij N.F. (2012) Environmental monitoring with a multisensor platform on polyimide foil. *Sens. Actuators B* 171–172, 190–197.

Parikh K., Cattanach K., Rao R., Suh D.-S., Wu A., Manohar S.K. (2006) Flexible vapour sensors using single walled carbon nanotubes. *Sens. Actuators B* 113, 55–63.

Pecora A., Zampetti E., Pantalei S., Valletta A., Minotti A., Maiolo L., et al. (2008) Interdigitated sensorial system on flexible substrate. In: *Proceedings of the Seventh IEEE Conference on Sensors, IEEE SENSORS*, October 26–29, Lecce, Italy, pp. 21–24.

Potyrailo R.A., Morris W.G. (2007) Multianalyte chemical identification and quantitation using a single radio frequency identification sensor. *Anal. Chem.* 79, 45–51.

Reddy S.G., Narakathu B.B., Atashbar M.Z., Rebros M., Rebrosova E., Joyce M.K. (2011) Fully printed flexible humidity sensor. *Procedia Eng.* 25, 120–123.

Rivadeneyra A., Fernández-Salmerón J., Agudo M., López-Villanueva J.A., Capitan-Vallvey L.F., Palma A.J. (2014) Design and characterization of a low thermal drift capacitive humiditysensor by inkjet-printing. *Sens. Actuators B* 195, 123–131.

Shih W.-P., Tsao L.-C., Lee C.-W., Cheng M.-Y., Chang C., Yang Y.-J., Fan K.-C. (2010) Flexible temperature sensor array based on a graphite-polydimethylsiloxane composite. *Sensors* 10, 3597–3610.

Song Z., Tang J., Li J., Xiao H. (2013) Plasma-induced polymerization for enhancing paper hydrophobicity. *Carnohydr. Polym.* 92, 928–933.

Stanssens D., Van den Abbeele H., Vonck L., Schoukens G., Deconinck M., Samyn P. (2011) Creating water-repellent and super-hydrophobic cellulose substrates by deposition of organic nanoparticle. *Mater. Lett.* 65, 1781–1784.

Su P.-G., Hsu H.-C., Liu C.-Y. (2013) Layer-by-layer anchoring of copolymer of methyl methacrylate and [3-(methacrylamino)propyl] trimethyl ammonium chloride to gold surface on flexible substrate for sensing humidity. *Sens. Actuators B* 178, 289–295.

Su P.-G., Tseng J.-Y., Huang Y.-C., Pan H.-H., Li P.-C. (2009c) Novel fully transparent and flexible humidity sensor. *Sens. Actuators B* 137, 496–500.

Su P.G., Wang C.S. (2007) Novel flexible resistive-type humidity sensor. *Sens. Actuators B* 123, 1071–1076.

Subramanian V., Chang P., Lee J.B., Molesa S.E., Volkman S.K. (2005) Printed organic transistors for ultra-low-cost RFID applications. *IEEE Trans. Comp. Packag. Technol.* 28 (4), 742–747.

Tarapata G., Marzecki M., Selma R., Paczesny D., Jachowicz R. (2016) Novel dew point hygrometer fabricated with inkjet printing technology. *Sens. Actuators A* 247, 641–646.

Trung T.Q., Duy L.T., Ramasundaram S., Lee N.E. (2017) Transparent, stretchable, and rapid-response humidity sensor for body-attachable wearable electronics. *Nano Res.* 10 (6), 2021–2033.

Vergara A., Llobet E., Raмнrez J.L., Ivanov P., Fonseca L., Zampolli S., Scorzoni A. et al. (2007) An RFID reader with onboard sensing capability for monitoring fruit quality. *Sens. Actuators B* 127, 143–149.

Wang L., Luo J., Yin J., Zhang H., Wu J., Shi X., Crew E. et al. (2010b) Flexible chemiresistor sensors: Thin film assemblies of nanoparticles on a polyethylene terephthalate substrate. *J. Mater. Chem.* 20, 907–915.

Wang Y., Yang Z., Hou Z., Xu D., Wei L., Kong E.S.-W., Zhang Y. (2010a) Flexible gas sensors with assembled carbon nanotube thin films for DMMP vapor detection. *Sens. Actuators B* 150, 708–714.

Weng B., Shepherd R.L., Crowley K., Killard A.J., Wallace G.G. (2010) Printing conducting polymers. *Analyst.* 135, 2779–2789.

Xu J.M. (2000) Plastic electronics and future trends in microelectronics. *Synth. Met.* 115, 1–3.

Xu M., Bunes B.R., Zang L. (2011) Paper-based vapor detection of hydrogen peroxide: Colorimetric sensing with tunable interface. *ACS Appl. Mater. Interfaces* 3, 642–647.

Zampetti E., Pantalei S., Pecora A., Valletta A., Maiolo L., Minotti A., Macagnano A., et al. (2009) Design and optimization of an ultra-thin flexible capacitive humidity sensor. *Sens. Actuators B* 143, 302–397.

23 Nontraditional Approaches to Humidity Measurement

In the previous chapters, we have considered the most popular methods of humidity measurement. However, it should be noted that periodically, there are works in which other approaches to such measurements are proposed. Let us briefly consider some of them.

23.1 HUMIDITY DETECTION USING TRIBOELECTRIC EFFECT

It is known that when friction of various materials occurs, a charge transfer, the so-called *triboelectric charging*, takes place. The description of this phenomenon dates back to ancient Greece, but even today, it is an open debate whether electrons, ions, or the exchange of surface materials causes the net charge transfer from one contacting body to the other (Lowell 1986; Matsusaka et al. 2010). A number of experimental and theoretical studies suggest that water molecules adhered to the surface of the materials play an important role in charge transfer (Nomura et al. 2003; Diaz and Felix-Navarro 2004; Xie et al. 2016). Even in 1902, Knoblauch (1902) hypothesized that the H^+ and OH^- ions dissolved in the water adsorbed to the surface of polar solids would be reasonable charge carriers. Subsequent experimental observations confirmed the idea of the importance of surface water. For instance, it was found that humidity strongly alters the charges that are generated via triboelectricity (Diaz and Felix-Navarro 2004; Xie et al. 2016). Given the direct relationship between the amount of surface water and air humidity, it can be assumed that triboelectric charging can be used to measure the moisture. Measurements performed in different groups confirmed this possibility (Glor and Moritz 2016; Schella et al. 2017). The relationship between magnitude of charge and humidity for different materials is shown in Figure 23.1. It is seen that, up to moderate humidities, all fillings are highly charged. Then, at a material-specific humidity, a crossover to a low-charged regime is found. At very high humidities, only the hydrophobic materials, such as polystyrene (PS) and polytetrafluoroethylene (PTFE), are able to accumulate charges. Thus, it can be seen that in a certain range of humidity there

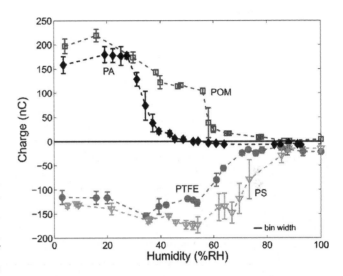

FIGURE 23.1 Magnitude of charge as a function of humidity for each 2000 polyimide (PA), polyoxymethylene (POM), polystyrene (PS), and polytetrafluoroethylene (PTFE) sphere shaken in a steel container. (Reprinted with permission from Schella, A. et al., *Soft Matter.*, 13, 394–401, 2017. Copyright 2017, The Royal Society of Chemistry.)

is a clearly expressed dependence of the magnitude of the charge on the humidity of the air, which, with some calibration, can give information on the humidity of the surrounding atmosphere.

Measurements carried out by Glor and Moritz (2016) have shown that charge transfer from different sample plates after corona charging depends also on the air humidity. Moreover, they established that, in cases of low relative humidity (RH), the charge transfer is for most samples higher with tribo charging, whereas at high RH, where no charge transfer can be measured after tribo charging, a substantial charge transfer can still be measured after corona charging.

However, it must be borne in mind that the measurement results are highly dependent on conditions and materials that are difficult to reproduce. Unfortunately, friction and corona charging are difficult both to define quantitatively to produce a reproducible state of friction and corona charging in different labs. The measurement results also have a widespread (see Figure 23.1) and strong dependence on temperature (Glor and Moritz 2016), which also does not contribute to achieving the required reproducibility during measurements.

23.2 HUMIDITY INFLUENCE ON THE BREAKDOWN VOLTAGE

The effect of humidity on the breakdown voltage of air can also be used for determining air humidity. It was established that, for a uniform gap, the effect of humidity on the breakdown voltage is negligible, while for nonuniform gaps, such as rod-plane like gaps, the influence of humidity is found to be of significance. In particular, for inhomogeneous fields, like in many practical insulation configurations, the switching-impulse breakdown voltage of a given air-gap usually decreases with the decrease of atmospheric humidity (Allen and Phillips 1959; IEC 1989; Wu et al. 2005). Effect of humidity on breakdown voltage of rod gaps is shown in Figure 23.2. However, this method requires a high-voltage installation (Wu et al. 2005; Soni and Choubey 2015), and this method does not have a high measurement accuracy. In addition, the phenomenon behind is complicated and influenced by the gap structures (i.e., the electric field distribution and the type of discharges). A large number of contributions on these subjects exist in literature and have been well covered in the reviews (TF 1991; Allen et al. 1992).

23.3 HUMIDITY MEASUREMENT USING HEAT PIPE

It is known that many hygrometers are not durable at high temperatures and high humidities, or under dew condensation. To resolve this problem, Yamauchi et al. (2016) have developed a phase transition thermally balanced (PTTB) sensor. The PTTB sensor is similar to a psychrometer, in the sense that the latent heat of water is used for measuring humidity. However, whereas a psychrometer uses the heat of evaporation of water vapor on a wet bulb, the PTTB sensor uses the heat of condensation on a heat pipe. Moreover, the surface of a heat pipe, sensing part of the PTTB sensor, is covered with a metal; therefore, it is robust, even at high temperatures and high humidities.

Figure 23.3 shows a schematic diagram of the proposed measurement principle based on a simple model, and Figure 23.4 shows a diagram of the chamber used in the PTTB sensor. The PTTB sensor is divided into three regions: the inside of the chamber (the measurement region), the adiabatic region, and the outside of the chamber (the heat-release region). A heat pipe

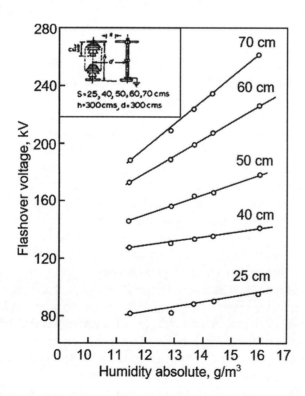

FIGURE 23.2 Effect of humidity on breakdown voltage of rod gaps at 20°C. (Data extracted from Hussian, E. et al., *J. Indian Inst. Sci.*, 56, 481–494, 1974.)

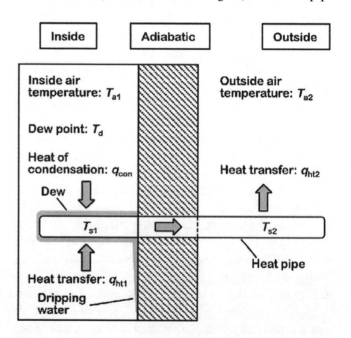

FIGURE 23.3 Schematic diagram of the measurement principle of a phase transition thermally balanced (PTTB) sensor based on a simple model. (Reprinted with permission from Yamauchi, S. et al., Novel humidity sensor using heat pipe: Phase transition thermally balanced sensor designed for measurement of high humidity at high temperature, *Sens. Actuators A*, 250, 1–6, 2016. Copyright 2016, Elsevier.)

Nontraditional Approaches to Humidity Measurement

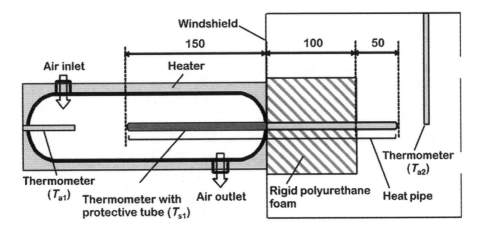

FIGURE 23.4 Schematic diagram of chamber used in a phase transition thermally balanced (PTTB) sensor. The diameter and length of the heat pipe were 6 and 300 mm, respectively. A thermal insulation material made of rigid polyurethane foam was sandwiched between the two ends of the heat pipe to create the adiabatic region in the system. (Reprinted with permission from Yamauchi, S. et al., Novel humidity sensor using heat pipe: Phase transition thermally balanced sensor designed for measurement of high humidity at high temperature, *Sens. Actuators A*, 250, 1–6, 2016. Copyright 2016, Elsevier.)

penetrates into the inside of a chamber from the outside through the adiabatic region. The temperature of the air and that of the surface of the heat pipe are given by T_a and T_s, respectively. The subscripts 1 and 2 used in the symbols in the figure represent the inside and outside of the chamber, respectively. The inside of the chamber is filled with humid air with a dew point of T_d and an air temperature of T_{a1}. If the surface temperature of the heat pipe inside the chamber, T_{s1}, is equal to or lower than T_d, condensation occurs on the surface; the heat pipe receives both condensation heat from water and other heat from the inside. If the surface temperature of the heat pipe outside the chamber, T_{s2}, is lower than T_{s1}, the heat received inside the chamber is conducted through the heat pipe to the outside of the chamber and is released to the outside air (provided that the temperature of the outside air, T_{a2}, is lower than T_{s2}, and the dew point of the outside air is lower than T_{s2}, and no condensation occurs on the surface of the heat pipe outside the chamber). Thus, the difference in the temperatures of the pipe measured inside and outside of the chamber is a measure of air humidity level.

However, experiments carried out by Yamauchi et al. (2016) have shown that the sensitivity of their hygrometer is very low, and the range of humidity that can be measured is very narrow (see Figure 23.5). According to Yamauchi et al. (2016), reliable measurements will be possible if the following problems can be solved: (1) Sufficient heat should be radiated in the heat-release region such that the surface temperature of the heat pipe in the measurement region is always lower than the dew point of the air inside the chamber. Cooling using Peltier devices or refrigerants, instead of cooling used air, will

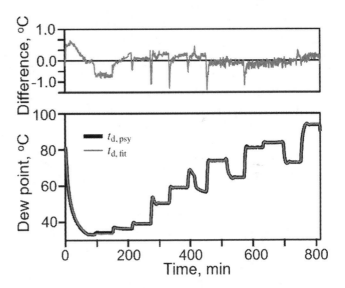

FIGURE 23.5 Comparison between a psychrometer and a phase transition thermally balanced (PTTB) sensor. The *thick black line* is dew point, measured using the psychrometer ($t_{d,\text{psy}}$) and the *gray line* is the fitted value of the PTTB sensor ($t_{d,\text{fit}}$). The value of $t_{d,\text{fit}} - t_{d,\text{psy}}$ is also shown in the *upper graph*. (Reprinted with permission from Yamauchi, S. et al., Novel humidity sensor using heat pipe: Phase transition thermally balanced sensor designed for measurement of high humidity at high temperature, *Sens. Actuators A*, 250, 1–6, 2016. Copyright 2016, Elsevier.)

be a solution for this. (2) The measurement is affected by the ambient temperature, airflow velocity, and radiation heat; thus, measures to suppress these effects may sometimes be required, such as adoption of a shield to reduce the influence of fluctuations of the ambient temperature, air flow, and radiation heat. (3) The sudden re-evaporation of water that has condensed inside the chamber can cause measurement errors. A water droplet dripping from the heat pipe directly to the bottom of the chamber is the main cause of the sudden re-evaporation; therefore, it can be prevented by placing a droplet collector underneath the heat pipe inside the chamber and connecting one end of the droplet collector to the adiabatic wall at a slight angle so that the collected water moves to the wall. (4) To expand the measurement range to a lower humidity, it is necessary to increase the amount of heat radiated from the heat pipe outside the chamber.

23.4 SELF-POWERED ACTIVE HUMIDITY SENSOR

With the rapid development of functional nanodevices, investigating portable, small-size, and sustainable power sources for driving these functional nanodevices is becoming more and more important (Zang et al. 2014). In the past several years, a new self-powered system that integrates an energy generator and functional nanodevice has been proposed, aimed at harvesting energy from the environment to power the functional nanodevice, such as self-powered pH sensors, automobile speedometers, gas sensors, and magnetic sensors (Xu et al. 2010; Hu et al. 2011; Cui et al. 2012; Zhang et al. 2012).

Zang et al. (2014) have shown that ZnO nanowire (NW) arrays can be used to fabricate a piezo-nanogenerator (NG) that can operate as a self-powered, active humidity sensor. It was found that the piezoelectric output of the device can act not only as a power source, but also as a very sensitive response signal to the RH of the ambience. The piezoelectric nanogenerators that can convert mechanical vibration energy into electrical power have exhibited unique advantages compared to the solar cells, and thermoelectric devices because mechanical energy is ubiquitous in our environment. In addition, mechanical energy harvesting is less affected by working conditions than optical and thermal energies.

Figure 23.6a shows the final device structure of the self-powered active humidity sensor, which is composed of three major components: Al-doped ZnO NW arrays on Ti foil, Al layer, and Kapton boards. The vertically aligned Al-doped ZnO NW arrays were synthesized by a seed-assisted, wet-chemical method. Such a device structure is similar to a typical NG without packaging. Ti foil acts as both the substrate for Al-doped ZnO NW arrays and the conductive electrode. A piece of Al foil (thickness ≈ 0.05 mm) positioned on the top of Al-doped ZnO NW arrays acts as the counter electrode. Two terminal copper leads are glued on the two electrodes with silver paste for electrical measurements, respectively. To ensure the stability of the device, the finished device was fixed between two sheets of Kapton boards as the frame.

Figure 23.7a shows the piezoelectric output voltage of Al-doped ZnO NW arrays upon exposure to different RHs at room temperature under constant applied deformation. The compressive force in these measurements remains the same (34 N, 2.7 Hz). It is seen that there is a relationship between the piezoelectric output voltage and the RH. The piezoelectric voltage dramatically decreases as the RH increases while the sensor response increases. Figure 23.7c also shows that the response of Al-doped ZnO NW arrays is much higher than that of undoped ZnO NW arrays. Such high performance can be attributed to the large amount of adsorption sites from the Al doping.

The detailed working mechanism of the self-powered active humidity sensor based on Al-doped ZnO NW arrays, proposed by Zang et al. (2014), is shown in Figure 23.8. Since the change in free-carrier density can affect the piezoelectric output of the NWs, the piezoelectric output of Al-doped, ZnO nanoarray

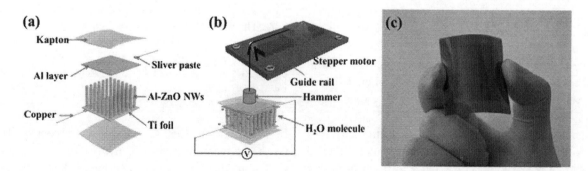

FIGURE 23.6 Fabrication process of the self-powered active humidity sensor based on Al-doped ZnO NW arrays. (a) Schematic diagram showing the structural design of self-powered active ethanol sensor. (b) Schematic image showing the device actively detecting humidity at room temperature. (c) The optical image of the flexible device. (Reprinted with permission from Zang, W. et al., *RSC Adv.*, 4, 56211, 2014. Copyright 2014, The Royal Society of Chemistry.)

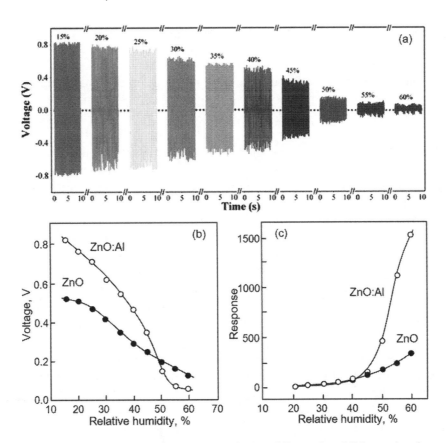

FIGURE 23.7 (a) The piezoelectric output voltage of the device at different humidities under the same applied strain at room temperature; (b) The relationship between the piezoelectric output voltage and relative humidity (RH) of Al-doped ZnO and pure ZnO nanowire (NW) arrays; (c) The response of Al–ZnO NW arrays and pure ZnO NW arrays at different RHs. (Reprinted with permission from Zang, W. et al., *RSC Adv.*, 4, 56211, 2014. Copyright 2014, The Royal Society of Chemistry.)

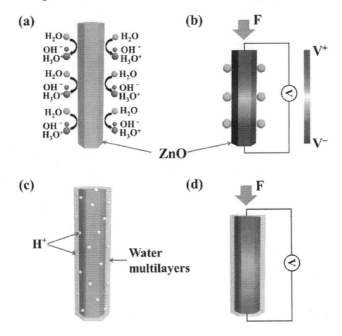

FIGURE 23.8 The working mechanism of the self-powered active humidity sensor based on Al-doped ZnO nanowire (NW) arrays: (a) schematic illustration showing the device based on Al-doped ZnO NW arrays at low relative humidity (RH) without applied force; (b) the piezoelectric output of the device under mechanical deformation at low RH; (c) Al-doped ZnO NW without compression at high RH; (d) the piezoelectric output of the device under mechanical deformation at high RH. (Reprinted with permission from Zang, W. et al., *RSC Adv.*, 4, 56211, 2014. Copyright 2014, The Royal Society of Chemistry.)

NG can act not only as a power source, but also as a response signal to humidity at room temperature. Small inositol of Zn^{2+} and Al^{3+} produces high local charge density and strong electrostatic field and represent good sites for chemisorption of water molecules (Tai and Oh 2002). When the device is at low RH without any applied force (Figure 23.8a), water molecules quickly occupy the available sites under exposure to the atmosphere. Initially, water vapor is chemisorbed on the surface of the Al-doped ZnO NW arrays, and then hydroxyl groups can form on the surface. After the first layer of chemisorbed water forms, subsequent layers of water molecules are physically adsorbed. The physisorbed water dissociates into H_3O^+ and OH^- ions because of the high electrostatic field in the chemisorbed layer. A charge transport by a Grotthuss chain reaction occurs when H_3O^+ releases a proton to a neighboring water molecule, which accepts it while releasing another proton. H_3O^+ appears in the physisorbed water and serves as a charge carrier in H_2O-Al-ZnO NWs (Fu et al. 2007). Under the compressive deformation, both the conductive H_3O^+ in the water layer and the free electrons inside of the Al-doped ZnO NWs can have directional movement and screen the piezoelectric polarization charges in the NWs; thus, the piezoelectric output voltage of the humidity sensor is lowered, as shown in Figure 23.8b. At high RH (Figure 23.8c), continuous water adsorption on the material surface will give rise to physisorbed water multilayers, which are less affected by the underlying chemosorbed layer. Consequently, the protons will gain freedom to move randomly inside the physisorbed water multilayers according to the Grotthuss mechanism (read Section 11.3). When the device is under a compressive strain at high RH (Figure 23.8d), the high density of protons in the water layer play a crucial role in decreasing the piezoelectric output of the device. Thus the piezoelectric output voltage of the device is significantly lowered.

Gu et al. (2016) have noted that, as the free charge carrier density of the ZnO nanowires was sensitive to the surface-absorption of chemicals such as hydrogen, ethanol, CO, and other toxic gases and vapors, those devices do not have selective response to water vapors. Therefore, although the ZnO nanowires can play both the piezoelectric and chemical sensing properties in the devices, the limited piezoelectric constant and poor gas sensing selectivity may limit the practical application of such active sensors. For resolving this problem, Gu et al. (2016) proposed to use $NaNbO_3$ nanofibers (NFs). As reported, the perovskite $NaNbO_3$ NFs have exhibited ultra-high sensitivity and selectivity to water molecules at room temperature. The surface absorption of water molecules leads to a great enhancement of an NF's conductivity (Zhang et al. 2015). Simultaneously, the $NaNbO_3$ nanowires have also been reported to have an outstanding piezoelectric energy-harvesting performance by Jung et al. (2011). This means that the $NaNbO_3$ nanowires have great potential for building high-performance, active humidity sensors.

The schematic fabrication process of the flexible $NaNbO_3$ active humidity sensor developed by Gu et al. (2016) is shown in Figure 23.9. The $NaNbO_3$ NFs were synthesized on the clean silicon substrates through a far-field electrospinning method. The average diameter of the NFs was approximately 78 nm. In order to fabricate

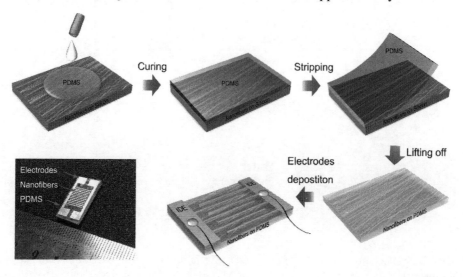

FIGURE 23.9 $NaNbO_3$ nanofiber (NF)-based active sensors fabrication procedure. PDMS, polydimethylsiloxane. (Reprinted with permission from Gu L. et al., Piezoelectric active humidity sensors based on lead-free NaNbO3 piezoelectric nanofibers, *J. Sensors*, 16, 833, 2016. Copyright 2016, MDPI as open access.)

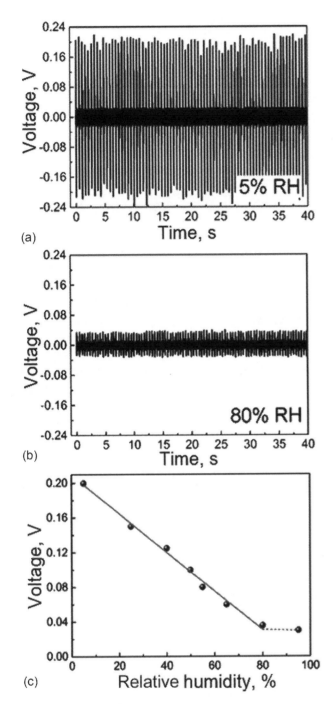

FIGURE 23.10 (a, b) The output voltage generated by the active sensor at different humidity conditions; (c) the relationship between the peak value of output voltage and the humidity. RH, relative humidity. (Reprinted with permission from Gu L. et al., Piezoelectric active humidity sensors based on lead-free NaNbO3 piezoelectric nanofibers, *J. Sensors*, 16, 833, 2016. Copyright 2016, MDPI as open access.)

the flexible nanogenerators, the prepared $NaNbO_3$ NFs were transferred from the silicon substrate onto the flexible polydimethylsiloxane (PDMS) substrate.

Figure 23.10 shows the piezoelectric active humidity response of the $NaNbO_3$ sensors with the voltage signal conditioned by the charge amplifier. The sensing performance was tested by changing the humidity of the environment when the $NaNbO_3$ sensors were continuously vibrating under the knocks of the rotating stick. As shown in Figure 23.10a and b, the sensor could generate distinguishable impulsive voltage with the humidity changing from 5% to 95% RH, where the peak-to-peak value of the voltage decreased from 0.4 to 0.06 V. Figure 23.10c shows the relationship between the voltage and the humidity of the environment. The voltage amplitude exhibited a negative linear correlation, with the humidity varying from 5% to 80% RH, and saturated when the humidity rose higher than 80% RH. Finally, the sensitivity of the active sensor could be obtained as approximately 2 mV/%RH according to the linear result fitting within the humidity range of 5%–80% RH. As indicated by Gu et al. (2016), the operation of these sensors is controlled by the same mechanism discussed above.

It is seen that $NaNbO_3$-based sensors had lower sensor response. According to Gu et al. (2016), the relatively lower sensitivity of this $NaNbO_3$ sensor compared to the reported ZnO sensors should be attributed to the decreased output voltage amplitude after the signal conditioning. However, it is worth noting that the structure of the $NaNbO_3$ NF-based sensors was much more stable than the sensors based on the ZnO nanorod arrays, which was assembled through the stacking of the top electrode and the nanorod arrays. Moreover, the sensor also exhibited fast response speed to the variation in humidity. The response time of the sensors for the humidity changing from 65% to 95% RH is approximately 12 seconds. In addition, it was established that the output voltage did not exhibit any change during the introducing and removing process of ethanol and H_2, which confirmed the excellent selectivity of the $NaNbO_3$-based humidity sensor. Gu et al. (2016) also evaluated the influence of temperature on the sensing result and established that no obvious change in the voltage amplitude could be found. That suggested that the sensor possessed great temperature stability during the testing process. Thus, Gu et al. (2016) believe that the active humidity sensor based on $NaNbO_3$ NFs provides an effective solution for self-powered sensor systems with high sensitivity, simple structure and fabrication processes, and low production costs.

Another variant of self-powered humidity sensors was developed by Su et al. (2017). They have shown that, by coupling a triboelectric nanogenerator (TENG) (Du et al. 2014) with a resistive humidity sensor, the degree of ambient moisture can be actively detected at room temperature. The prepared TENG consisted of polyethylene terephthalate (PET) film, a Cu film, a PTFE film, and an Al foil, where PET film was selected for the substrates at the top and bottom of the TENG due to

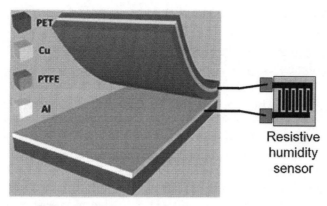

FIGURE 23.11 Schematic diagram of triboelectric nanogenerator (TENG). (Reprinted with permission from Su, Y. et al., Novel high-performance self-powered humidity detection enabled by triboelectric effect, Sens. *Actuators B*, 251, 144–152, 2017. Copyright 2017, Elsevier.)

its excellent flexibility and durability (see Figure 23.11). PTFE film served as the contact layer. The Cu layer was deposited on the back of PTFE film acting as the back electrode. The layer of Al played dual roles of electrode and triboelectric materials. Triggered by a linear motor (JZK-10, Sinoceramics, Inc) with a tunable impact speed, force, and frequency, the TENG was connected to the resistance sensor to form the self-powered triboelectric humidity sensor (THS). A reduced graphene oxide (RGO)-polyvinylpyrrolidone (PVP) composite material was used as sensing material in resistive humidity sensors. Su et al. (2017) reported that the prepared, self-powered THS had a high humidity response of approximately 7 and fast response/recovery times of 2.8 and 3.5 seconds.

23.5 HUMIDITY SENSOR USING A SMT

Ray (2015) proposed a completely different approach to the development of humidity sensors: a single molecular transistor (SMT)-based novel detection methodology was proposed for humidity sensing. A SMT (Ray 2014) is a modified version of a single electron transistor (SET), where the "island" is a single molecule and the device operation can be explained using the theory of Coulomb blockade (Fulton and Dolan 1987). The novel device structure proposed by Ray (2015) is illustrated in Figure 23.12, which is an SMT with the source (S) and the drain (D) electrodes placed on both sides of the gate electrode. The source and the drain electrodes have square cross-sections of 0.8 nm × 0.8 nm

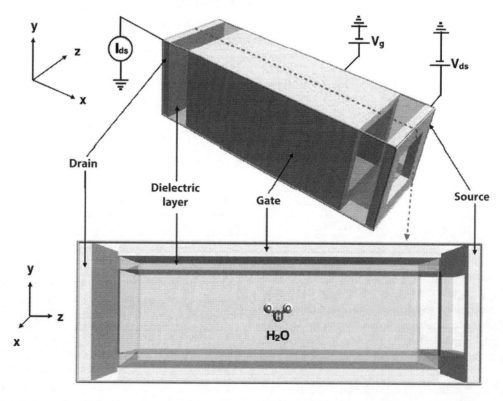

FIGURE 23.12 Schematic of the single molecular transistor (SMT) device structure proposed by Ray S.J. (2015). (*Top*) Side view of the device and (*bottom*) cross-section of the device taken along the *dotted red lines* in the xy-plane with the H_2O molecule inside. (Reprinted with permission from Ray S.J., *J. Appl. Phys.*, 118, 044307, 2015. Copyright 2015, American Institute of Physics.)

area, which are separated from the gate electrode via dielectric layers of 0.3 nm thickness on both ends of the device. The active component of the device is the water molecule, which is introduced into the device through the aperture of the source electrode; it becomes operational on its arrival. For the present case, the water molecule works as the island within the SMT, and the presence of it can be realized from the charge stability diagram. The island is tunnel coupled to the source (S) and the drain (D) electrodes on both sides, through which electrical conduction occurs and the island potential can be controlled locally via using a gate electrode (Clapera et al. 2015). As the other end of the device is closed, the molecule gets trapped within the device once it is inside it. The gate electrode covers the entire channel region (3 nm in length) from all different sides, backed by a dielectric layer of 0.5 nm thickness, which allows perfect electrostatic control over the island region of the device.

The performance and operation of a novel SMT-based humidity sensor were demonstrated using density function theory-based Ab-initio calculations. Ray (2015) have shown that this device has a design that can allow real-time detection through the charge stability diagram. It is found that, due to the large charging energy of the water molecule, the device can work over a large temperature range of operation with extremely high detection sensitivity. Ray (2015) believes that the simplistic design of the device can be useful for practical realization and can offer fast detection possibility. Due to these combined advantages, this can be a completive candidate as a nanoelectronic sensor over solid state and two-dimensional (2D) single crystal-based sensors to achieve better performance and sensitivity.

However, it should be noted that the proposed sensors are not yet implemented and exist only in the form of a theoretical model that is unlikely to be realized in the near future and will be able to appear in the market of humidity sensors.

REFERENCES

Allen K.P., Phillips K. (1959) Effect of humidity on the spark breakdown voltage. *Nature* 183, 174–175.

Allen N.L., Fonseca J.R., Geldenhuys H.J., Zheng J.C. (1992) Influence of air humidity on the dielectric strength of external insulation. Cigré Brochure 72, Guidelines for the evaluation of the dielectric strength of external insulation, Working Group 07 of Study Committee 3, pp. 59–72.

Clapera P., Jehl X., Corna A., Ray S.J., Sanquer M., Valentian A., Barraud S. (2015) Design and cryogenic operation of a hybrid quantum-CMOS circuit. *Phys. Rev. Appl.* 4 (4), 044009.

Cui N.Y., Wu W.W., Zhao Y., Bai S., Meng L.X., Qin Y., Wang Z.L. (2012) Magnetic force driven nanogenerators as a noncontact energy harvester and sensor. *Nano Lett.* 12, 3701–3705.

Diaz A., Felix-Navarro R. (2004) A semi-quantitative tribo-electric series for polymeric materials: The influence of chemical structure and properties. *J. Electrostat.* 62, 277–290.

Du W.M., Han X., Lin L., Chen M.X., Li X.Y., Pan C.F., Wang Z.L. (2014) A three dimensional multi-layered sliding triboelectric nanogenerator. *Adv. Energy Mater.* 4 (2014), 7963–7975.

Fu X.Q., Wang C., Yu H.C., Wang Y.G., Wang T.H. (2007) Fast humidity sensors based on CeO_2 nanowires. *Nanotechnology* 18, 145503.

Fulton T.A., Dolan G.J. (1987) Observation of single-electron charging effects in small tunnel junctions. *Phys. Rev. Lett.* 59, 109.

Glor M., Moritz K. (2016) Effectiveness of increase of relative humidity as a measure to reduce the ignition probability of explosive atmospheres by static electricity. *Chem. Eng. Trans.* 48, 331–336.

Gu L., Zhou D., Cao J.C. (2016) Piezoelectric active humidity sensors based on lead-free $NaNbO_3$ piezoelectric nanofibers. *Sensors* 16, 833.

Hu Y.F., Xu C., Zhang Y., Lin L., Snyder R.L., Wang Z.L. (2011) A Nanogenerator for energy harvesting from a rotating tire and its application as a self-powered pressure/speed sensor. *Adv. Mater.* 23, 4068–4071.

Hussian E., Nandagopal M.R., Prabhakar B.R. (1974) Effect of humidity on breakdown voltages of gaps and insulators. *J. Indian Inst. Sci.* 56 (11), 48–494.

IEC (1989) IEC 60060-1. High-voltage test techniques, Part 1: General definitions and test requirements. 2nd ed., 1989–2011.

Jung J.H., Lee M., Hong J.I., Ding Y., Chen C.Y., Chou L.J., Wang Z.L. (2011) Lead-free $NaNbO_3$ nanowires for a high output piezoelectric nanogenerator. *ACS Nano* 5, 10041–10046.

Knoblauch O. (1902) Versuche uber die beruhrungselektrizitat. *Z. Phys. Chem.* 39, 225–244.

Lowell J. (1986) Constraints on contact charging of insulators. II. Energy constraints. *J. Phys. D: Appl. Phys.* 19 (1), 105–113.

Matsusaka S., Maruyama H., Matsuyama T., Ghadiri M. (2010) Triboelectric charging of powders: A review. *Chem. Eng. Sci.* 65, 5781–5807.

Nomura T., Satoh T., Masuda H. (2003) The environment humidity effect on the tribo-charge of powder. *Powder Technol.* 135–136, 43–49.

Ray S.J. (2014) Single molecule transistor based nanopore for the detection of nicotine. *J. Appl. Phys.* 116, 244307.

Ray S.J. (2015) Humidity sensor using a single molecular transistor. *J. Appl. Phys.* 118, 044307.

Schella A., Herminghaus S., Schroter M. (2017) Influence of humidity on tribo-electric charging and segregation in shaken granular media. *Soft Matter.* 13, 394–401.

Soni N., Choubey H. (2015) Effect of humidity on the impulse breakdown characteristic of air. *Int. J. Recent Res. Electr. Electron. Eng.* 2 (3), 63–65.

Su Y., Xie G., Wang S., Tai H., Zhang Q., Du H. et al. (2017) Novel high-performance self-powered humidity detection enabled by triboelectric effect. *Sens. Actuators B* 251, 144–152.

Tai W.P., Oh J.H. (2002) Humidity sensing behaviors of nanocrystalline Al-doped ZnO thin films prepared by sol–gel process. *J. Mater. Sci. Mater. Electron.* 13, 391–394.

TF (1991) TF 33.07.03. Humidity influence on non-uniform field breakdown in air. *Electra*, No 134, pp. 63–89.

Wu D., Asplund G., Jacobson B., Li M., Sahlen F. (2005) Humidity influence on switching-impulse breakdown voltage of air gaps for indoor high-voltage installations. In: *Proceedings of the 14th International symposium on High Voltage Engineering Tsinghua University*, August 25–29, Beijing, China. doi:10.1109/CEIDP.2005.1560616.

Xie L., Bao N., Jiang Y., Zhou J. (2016) Effect of humidity on contact electrification due to collision between spherical particles. *AIP Adv.* 6, 035117.

Xu S., Qin Y., Xu C., Wei Y.G., Yang R.S., Wang Z.L. (2010) Self-powered nanowire devices. *Nat. Nanotechnol.* 5, 366–373.

Yamauchi S., Akamatsu K., Niwa T., Kitano H., Abe H. (2016) Novel humidity sensor using heat pipe: Phase transition thermally balanced sensor designed for measurement of high humidity at high temperature. *Sens. Actuators A* 250, 1–6.

Zang W., Wang W., Zhu D., Xing L., Xue X. (2014) Humidity-dependent piezoelectric output of Al–ZnO nanowire nanogenerator and its applications as a self-powered active humidity sensor. *RSC Adv.* 2014, 4, 56211.

Zhang R., Lin L., Jing Q.S., Wu W.Z., Zhang Y., Jiao Z.X., Yan L., Han R.P.S., Wang Z.L. (2012) Nanogenerator as an active sensor for vortex capture and ambient wind-velocity detection. *Energy Environ. Sci.* 5, 8528–8533.

Zhang Y., Pan X., Wang Z., Hu Y., Zhou X., Hu Z., Gu H. (2015) Fast and highly sensitive humidity sensors based on $NaNbO_3$ nanofibers. *RSC Adv.* 5, 20453–20458.

24 Summary and Outlook

24.1 SUMMARY

To sum up our consideration of electronic and electrical humidity sensors, recently developed humidity sensors in both university research and industry provide promising performance with a high-impact contribution to accuracy, reliability, and economic efficiency (Rittersma 2002; Chen and Lu 2005; Lee and Lee 2005; Farahani et al. 2014). However, in real, practical environments, challenges remain for enhancing sensor efficiency and response characteristics. In addition, despite the large number of humidity sensors already on the market (read Chapter 28, Volume 3), there is a need to expand the range of manufactured sensors. The development of new technologies and tightening of environmental requirements and requirements for living conditions, work, and storage of various manufactured products require more sensors with improved or application-specific parameters. However, despite the apparent simplicity of the humidity sensors, the selection and design of a suitable sensor for a new application is a difficult task for the design engineer. It is necessary to choose the best method of moisture measurement and a suitable sensing material, sensing platform, and sampling system to optimize the configuration of the sensor and to develop a technology for their manufacture. All these steps are important, because these decisions determine the specificity, sensitivity, response time, and stability of the final device. As shown in this volume, there are many opportunities to develop a humidity sensor and many technological approaches that could be used in their manufacture. Each of these approaches has its advantages and disadvantages, which were considered in the chapters devoted to these instruments and summarized and systematized in Volume 3. Therefore, we will not analyze them.

It is clear that, ideally, it would be desirable to have a highly sensitive, fast-responding, selective, and stable sensor capable of operating in any environment and under any conditions (Korotcenkov and Cho 2011). Of course, the task of developing such a sensor can be solved, but the question arises as to what efforts will be required to solve such a problem and what costs will be. Therefore, in designing a humidity sensor for any new application, the developers have to answer the following questions:

1. Does the application's operating environment require special materials or fabrication procedures? The environment determines appropriate materials and the transducing principle that can be utilized in a sensor. Thus, the first step in the sensor design is to determine the required working conditions, such as expected temperature and humidity changes, possible chemical interferences, and electromagnetic interactions.
2. Does the application require high sensitivity or a broader range of detection?
3. How selective should the sensor be? Can the task be solved using a sensor sensitive to other gases and vapors?
4. Are there any restrictions on the power consumption?
5. How reliable, accurate, and stable will the sensor be? Can it be manufactured with high reproducibility?

The measurement range, response time, and resolution of measurements, as well as the size of the sensor, its weight, safety of operation, and reliability should also be considered. In addition, commercialization of designed sensors actually influences the process of their elaboration. At this stage, questions appear, such as (1) How much will it cost? (2) What is its shelf life? (3) What restrictions must be placed on storage (e.g., refrigeration, desiccation)? (4) Who owns it? (5) Is the technology protected? (6) What is its target market?

It seems that there are many "ready" solutions for given problems; therefore, the careful study of results of previous research may reveal shortages that remain to be solved in the future. For example, considering the complex mechanism of interaction of water vapor with polymers and inorganic materials, and the presence of a large number of factors influencing the sensor performance, it becomes clear the development of a new sensor requires good understanding of the functioning of all types of humidity sensors, their design, features of exploitation, and modern trends in their development. This means that multidisciplinary knowledge and careful approaches are needed to achieve good results (see Figure 24.1). Knowledge of materials science is necessary for elaboration of effective technologies required for fabricating appropriate coatings and membranes, synthesis or deposition of sensitive materials, and their surface functionalization and modification. Without this knowledge, it is also impossible to solve the tasks of increasing the reliability and stability of humidity-sensitive materials (Rittersma 2002). Chemical sciences, such as interfacial and interphase chemistry, electrochemistry, molecular chemistry, analytical chemistry, and so on, are necessary for precise understanding of the processes that are the basis on which sensors function. The physical sciences provide spectroscopic detection methods (optical, mass, etc.), as well as knowledge necessary for understanding the mechanisms of interaction at water vapor–solid or condensed water–solid interfaces that determine the sensor's signal.

There is also no doubt that the role of engineers in this process is as important as the role of scientists. Certainly, during the initial stages of research, interest is generally focused on exploring and proving the principles by which a new sensing technology can be applied to measure a water vapor. However, at the following stages of elaboration, the impact of engineers is significantly increased, especially when they are close to completing a design (Figure 24.2).

Sometimes, the design of the device that will be suitable for commercial marketing requires resolving

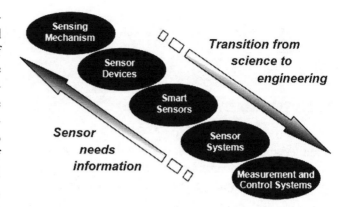

FIGURE 24.2 From idea to devices—from science to engineering. (Reprinted from Korotcenkov, G., and Cho, B.K., Introduction in chemical sensor technologies. in: Korotcenkov G. (Ed.), *Chemical Sensors: Comprehensive Sensor Technologies*, Vol. 4 Solid State Devices, Momentum Press, New York, pp. 1–38, 2011. Copyright 2011, with permission from Momentum Press.)

a number of complex engineering problems that do not necessarily rely on elaboration of basic principles and testing of sensor prototypes (Rittersma 2002). One should note that modern progress in the field of chemical and humidity sensors has, in many aspects, been determined by the successes of engineers working in such areas as microfabrication and signal processing. It is the efforts of these engineers that are currently solving the problem of developing "smart sensors" and "intelligent sensors." We note that the development of such sensors is one of the most promising directions in chemical and humidity sensors design. These sensors are needed to acquire data in order to correct errors in automated process and to adapt the process to changing circumstances (Singh 2005).

24.2 SMART SENSORS

Smart and intelligent sensors are integrated sensors that utilize the transduction properties of one class of sensing materials and the electronic properties of silicon (Van der Horn and Huijsing 1998; Lin et al. 2015). These sensors are defined as sensors with small memory and standardized physical connection to enable the communication with processor and data network (sensor + interfacing circuit = smart sensor). Moreover, smart or intelligent sensors are capable of logic functions, two-way communication, and decision-making. This means that smart or intelligent sensors are sensors with additional functions. The intelligent sensor concept is based on adding the possibility of processing the sensor data and the flexibility to reconfigure embedded functions, as well as to aggregate external sensors' data.

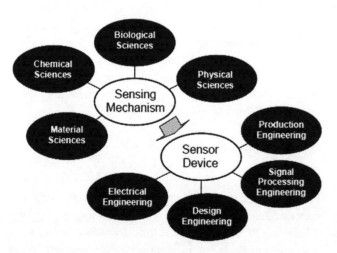

FIGURE 24.1 Scheme illustrating the correlation of different sciences in chemical sensor design. (Reprinted from Korotcenkov, G., and Cho, B.K., Introduction in chemical sensor technologies, in Korotcenkov, G. (Ed.), *Chemical Sensors: Comprehensive Sensor Technologies*, Vol. 4 Solid State Devices, Momentum Press, New York, pp. 1–38, 2011. Copyright 2011, with permission from Momentum Press.)

Summary and Outlook

24.2.1 Architecture of Smart Sensors

The general architecture of smart sensors is shown in Figure 24.3. A classical block diagram of a smart sensor interface is shown in Figure 24.3a. The sensor output signal is first processed in the analog domain by a signal conditioning circuit that generally relies on operational amplifiers (Sifuentes et al. 2017). The main functions of this block are level shifting and amplification, so as to match the sensor output span to the input span of the ensuing analog-to-digital converter (ADC) and, hence, to make good use of the ADC dynamic range. Other common tasks of the signal conditioning circuit are: sensor output-to-voltage conversion, filtering, linearization, and/or demodulation. The resulting analog signal is then digitized via the ADC. Finally, a digital system acquires, stores, processes, controls, communicates (to other devices or systems), and/or displays the digital value with information about the measurand. Nowadays, the most popular digital systems are microcontrollers (μC) and field-programmable gate arrays (FPGAs).

An alternative approach to reading some sensors (e.g., resistive, capacitive, and voltage-output sensors) (Reverter and Casas 2008; Pelegrí-Sebastiá et al. 2012; Chetpattananondh et al. 2014) is shown in Figure 24.3b. This circuit topology is known as a *direct interface circuit* (DIC), since the sensor is directly connected to the digital system without using either the signal conditioning circuit or the ADC (Bengtsson 2012), resulting in a direct sensor-to-μC (Reverter 2012) or to-FPGA interface circuit (Oballe-Peinado et al. 2015). In this topology, the digital system excites the sensor to obtain a time-modulated signal that is directly measured in the digital domain through a digital timer embedded into the digital system. In comparison with the sensor electronic interface shown in Figure 24.3a, a DIC is simpler and needs fewer components.

Thus, a smart sensor is the combination of a sensor, a small microcontroller, the necessary memory (flash, random access memory [RAM], and read-only memory [ROM]), and a communication interface (typically a receiving and transmitting device, i.e., a transceiver) (see Figure 24.4). Since a microcontroller is available, the management and control of external sensors or external devices are now also possible. In addition, the intelligent sensor structure in an application enables the customization of the embedded algorithms to specific applications; moreover, the customization can happen by reprogramming the flash memory. Since it is an evolution of smart sensors, no penalties in term of cost and performance are expected. The communication unit is often composed of a single device, the transceiver, that allows data transmission and reception on the selected media (wired or wireless). According to the International Organization for Standardization (ISO)-Open System Interconnection (OSI) model (Hayyes 1988), the transceiver implements at least the physical interface. Normally, it also provides a connection to higher layers of the communication stack, such as the data link layer. For instance, the transceiver generally manages the media access control, allowing the microcontroller not to care about the state of occupancy of the communication medium.

It should be highlighted that the power supply unit may be the most critical block of a smart sensor, in that it should provide different and very stable power supply

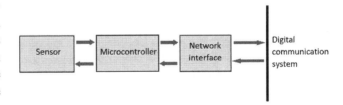

FIGURE 24.4 A smart sensor containing a transducer and processing unit; it may contain a communication interface.

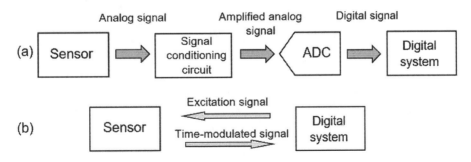

FIGURE 24.3 (a) Classical block diagram of a sensor electronic interface; (b) Direct interface circuit. ADC, analog-to-digital converter. (Reprinted with permission from Sifuentes, E. et al., Measuring dynamic signals with direct sensor-tomicrocontroller interfaces applied to a magnetoresistive sensor, *J. Sensors*, 17, 1150, 2017. Copyright 2017, MDPI as open access.)

voltages and currents to all the units (Flammini and De Vito 2012; Guo and Healy 2014). For instance, the sensing elements and the first conditioning electronic circuits need a very low-noise power supply and ground reference, which may not be easy to implement in the same circuit in which a heater must be supplied. The source of power may be the main supply, batteries (Guo and Healy 2014), or, in some cases, the same cable used for communication, through which power can be recovered (Wall et al. 2005). In this case, data is carried on a conductor that is normally used for electric power transmission by impressing a modulated carrier signal on the wiring system.

As seen in Figure 24.5, a smart chemical sensor can be wired or wireless according to different schemes (Flammini and De Vito 2012). In particular, Figure 24.5a shows a smart chemical sensor with a wired communication interface, also called a *wired chemical sensor*; power supply is derived from mains.

Figure 24.5b represents a wired chemical sensor with power recovered by the wired interface, directly from the data lines or from specific connector pins dedicated to power transmission (e.g., USB). Figure 24.5c is a block diagram of a wired chemical sensor with data coded in the power interface, according, for instance, to power line communication technologies. In Figure 24.5d, the smart chemical sensor has a wireless communication interface, but wires are still necessary for power supply. Such a situation usually is realized in industrial, building, and home applications, where power is present almost anywhere. In this case, advantages with respect to the scheme depicted in Figure 24.5c are in a more reliable and fast communication system and a better integration with information and communication technologies, such as personal computers, internet, and web technologies. A true mobile, smart, chemical sensor is shown in Figure 24.5e, in which a wireless communication interface has been adopted and the sensor

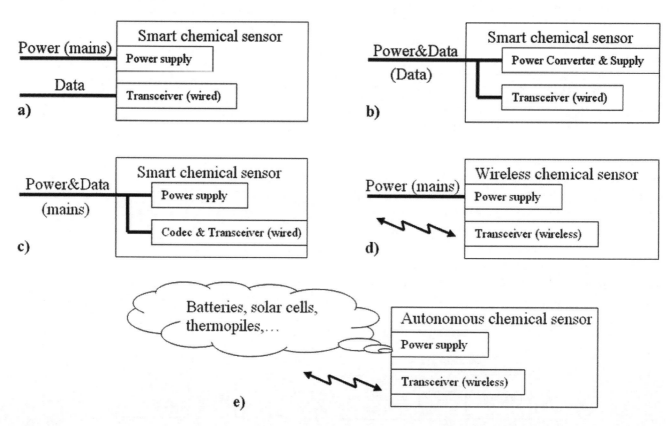

FIGURE 24.5 Possible schemes for a smart chemical sensor: (a) smart chemical sensor with wired communication interface (wired chemical sensor) and power supply from mains; (b) wired chemical sensor with power recovered by this interface (e.g., USB); (c) wired chemical sensor with data coded in the power interface; (d) smart chemical sensor with wireless interface and wired connection for power supply; (e) autonomous chemical sensor with no wires, making the sensor movable. (Reprinted from Flammini, A., and De Vito, S., Wireless chemical sensors, in: Korotcenkov, G. (Ed.), *Chemical Sensors. Comprehensive Sensor Technologies*, Vol. 6, Chemical Sensors Application, pp. 87–126, 2012. Copyright 2012, with permission from Momentum Press.)

is powered without the need for wires, simply using energy stored in batteries or in supercapacitors, powered by solar cells or other forms of energy harvesting (see Chapter 23, Section 23.4). It should be highlighted that passive wireless sensors (Deivasigamani et al. 2013) that do not require additional power supplies are also possible (see Chapter 20).

It is important to note that wireless technologies have been under rapid development during recent years (Wang et al. 2006). An obvious advantage of wireless transmission is a significant reduction and simplification in wiring and harness. Additional savings in overall cost can be obtained by more efficient control of the equipment through effective monitoring of the environment. Wireless sensors allow otherwise impossible sensor applications, such as monitoring dangerous, hazardous, unwired, or remote areas and locations. This technology provides nearly unlimited installation flexibility for sensors and increased network robustness. Furthermore, wireless technology reduces maintenance complexity and costs. Wireless sensor networks allow faster deployment and installation of various types of sensors, because many of these networks provide self-organizing, self-configuring, self-diagnosing, and self-healing capabilities to the sensor nodes. Some of them also allow flexible extension of the network. Another advantage of wireless sensors is their mobility. These sensors can be placed in transporting vehicles to monitor the "on-the-go" environment. They also can be placed on rotating equipment, such as a shaft to measure critical parameters. At present, this technology is widely used in various fields (Wang et al. 2006; Tavares et al. 2008; Deivasigamani et al. 2013), including environmental monitoring and humidity measurement (Shinghal et al. 2011; Borgese et al. 2017). One should note that all wireless sensors, including humidity sensors, are developed on the same principles (Flammini and De Vito 2012).

24.2.2 Advantages and Limitations

An analysis of the measurement capabilities of smart and integrated sensors shows that these sensors provide significant advantages in terms of the overall size and the ability to use small signals from the transduction element. Additional potential advantages of the smart and intelligent sensors concept include the following:

- Packaging or assembling of sensing and actuating, or signal-processing devices, in the proximity of the analyte.
- Lower maintenance.
- Reduced downtime.
- Better reliability.
- Improved resolution and sensitivity, and superior functionality.
- Fault-tolerant systems.
- Broadcast information about its own status.
- Adaptability for self-calibration and compensation.
- Lower cost due to batch fabrication.
- Lower weight.
- Each smart sensor comprising the distributed architecture has communication capability, and its information is available for more than one client. This way, the sensor information can be used for multiple aims.
- Wireless data transfer.
- Fewer interconnections between multiple sensors and control systems.
- The network of smart sensors can adapt to a variable number of RH sensors, improving reliability. If a sensor fails, the network can estimate its response on the basis of previous behavior and the response of the closest sensors.
- Capability to obtain the average, variance, and standard deviation for the set of measurements.
- Less complex system architecture.

These advantages of smart sensors are application-specific. There is certainly justification for many applications in distributing the signal processing throughout a large sensor system so that each sensor has its own calibration, fault diagnostics, signal processing, and communication, thereby creating a hierarchical system. Innovations in sensor technology have generally allowed a greater number of sensors to be networked, more accurate sensors to be developed, and on-chip calibration to be included. Output linearization and compensation of the temperature effect are also possible in smart sensors (Nenov and Ivanov 2007). Thus, in general, new technology has contributed to better performance by increasing the efficiency and accuracy of information distribution and reducing overall costs. These performance enhancements are achieved at the expense of increased complexity and higher cost of individual sensor systems. However, less hardware and reduction of repetitive testing make smart sensors and intelligent sensors cost-effective. The appearance of such sensors in the market of humidity sensors (read Chapter 28, Volume 3) testifies that the development of the technology of manufacturing smart and intelligent humidity sensors is successful; therefore, we expect further progress in this direction.

REFERENCES

Bengtsson L. (2012) Direct analog-to-microcontroller interfacing. *Sens. Actuators A* 179, 105–113.

Borgese M., Dicandia F.A., Costa F., Genovesi S., Manara G. (2017) An inkjet printed chipless RFID sensor for wireless humidity monitoring. *IEEE Sensor J.* 17 (15), 4699–4707.

Chen Z., Lu C. (2005) Humidity sensors: A review of materials and mechanisms. *Sensor Lett.* 3, 274–295.

Chetpattananondh K., Tapoanoi T., Phukpattaranont P., Jindapetch N. (2014) A self-calibration water level measurement using an interdigital capacitive sensor. *Sens. Actuators A* 209, 175–182.

Deivasigamani A., Daliri A., Wang C.H., John S. (2013) A review of passive wireless sensors for structural health monitoring. *Modern Appl. Sci.* 7 (2), 57–76.

Farahani H., Wagiran R., Hamidon M.N. (2014) Humidity sensors principle, mechanism, and fabrication technologies: A comprehensive review. *Sensors* 14, 7881–7939.

Flammini A., De Vito S. (2012) Wireless chemical sensors, In: Korotcenkov G. (Ed.) *Chemical Sensors. Comprehensive Sensor Technologies.* Vol. 6: Chemical Sensors Applications. Momentum Press, New York, pp. 87–126.

Guo W., Healy W.M. (2014) Power supply issues in battery reliant wireless sensor networks: A review. *Int. J. Intell. Contr. Syst.* 19 (1), 15–23.

Hayyes J.P. (1988) *Computer Architecture and Organization*, 2nd ed. McGraw-Hill, New York.

Korotcenkov G., Cho B.K. (2011) Introduction in chemical sensor technologies, In: Korotcenkov G. (Ed.) *Chemical Sensors: Comprehensive Sensor Technologies.* Vol. 4 Solid State Devices. Momentum Press, New York, pp. 1–38.

Lee C.-Y., Lee G.-B. (2005) Humidity sensors: A review. *Sensor Lett.* 3, 1–14.

Lin Y.-L., Kyung C.-M., Yasuura H., Liu Y. (Eds.) (2015) *Smart Sensors and Systems*. Springer, New York.

Nenov T., Ivanov S. (2007) Linearization of characteristics of relative humidity sensor and compensation of temperature impact. *Sens. Mater.* 19 (2), 95–106.

Oballe-Peinado O., Vidal-Verdú F., Sánchez-Durán J.A., Castellanos-Ramos J., Hidalgo-López J.A. (2015) Smart capture modules for direct sensor-to-FPGA interfaces. *Sensors* 2015, 15, 31762–31780.

Pelegrí-Sebastiá J., García-Breijo E., Ibáñez J., Sogorb T., Laguarda-Miro N., Garrigues J. (2012) Low-cost capacitive humidity sensor for application within flexible RFID labels based on microcontroller systems. *IEEE Trans. Instrum. Meas.* 61, 545–553.

Reverter F. (2012) The art of directly interfacing sensors to microcontrollers. *J. Low Power Electron. Appl.* 2, 265–281.

Reverter F., Casas O. (2008) Direct interface circuit for capacitive humidity sensors. *Sens. Actuators A* 143, 315–322.

Rittersma Z.M. (2002) Recent achievements in miniaturized humidity sensors – a review of transduction techniques. *Sens. Actuators A* 96, 196–210.

Shinghal K., Noor A., Srivastava N., Singh R. (2011) Intelligent humidity sensor for wireless sensor network agricultural application. *Int. J. Wirel. Mob. Netw.* 3 (1), 118–128.

Sifuentes E., Gonzalez-Landaeta R., Cota-Ruiz J., Reverter F. (2017) Measuring dynamic signals with direct sensor-to-microcontroller interfaces applied to a magnetoresistive sensor. *Sensors* 17, 1150.

Singh V.R. (2005) Smart sensors: Physics, technology and applications. *Indian J. Pure Appl. Phys.* 43, 7–16.

Tavares J., Velez F.J., Ferro J.M. (2008) Application of wireless sensor networks to automobiles. *Meas. Sci. Rev.* 8 (3), 65–70.

Van der Horn G., Huijsing J.L. (Eds.) (1998) *Integrated Smart Sensors: Design and Calibration.* Springer, Dordrecht, the Netherlands.

Wall R.W., and King B.A. (2005) Power line carrier communications for center pivot irrigation control. In: *Proceedings of International Symposium on Power Line Communications and Its Applications*, April 6–8, Moscow, ID, 109–115.

Wang N., Zhang N., Wang M. (2006) Wireless sensors in agriculture and food industry—Recent development and future perspective. *Comp. Electron. Agric.* 50, 1–14.

Index

Note: Page numbers in italic and bold refer to figures and tables respectively.

absolute humidity sensors 6–7, 245
acid-base interactions 8
acoustic wave (AW) sensors: dew-point hygrometers 194–7; disadvantages 173; love-wave devices 169; minimum detectable mass density **172**; propagation modes 168–70; properties **172**; types *161*
2-acrylamido-2-methylpropane sulfonic acid (AMPS) 118
active microwave sensors 317
active RFID tags 331
ADC (analog-to-digital converter) 377
AFM (atomic force microscopy) 213
airborne mass spectrometer (AIMS)-H_2O 73, *74*
air humidity: gravimetric measurement 18–21; mass spectrometric measurement 73–4
Alberti L.B. 17
alkali ions 135–6
alkaline ions in metal oxides 145–8
aluminum oxide sensor 98
AMETEK, electrolytic cell by 62, *62*
Amontons G. 26
amperometric humidity sensors 304–8, *305*, *306*, *307*
analog-to-digital converter (ADC) 377
aspiration psychrometer 34, *35*, *36*
Assmann R. 31
atomic force microscopy (AFM) 213
AW sensors *see* acoustic wave (AW) sensors

BAW (bulk acoustic wave) devices 161–2
benzo cyclobutene (BCB) 351
bimorph cantilever 206
black phosphorus (BP) sensors 150–1, *151*
Bragg grating sensors 166
Bravais A. 33
bulk acoustic wave (BAW) devices 161–2
bulk polymer-resistive sensors 120

calcium carbide method 71
cantilever-based sensors 171
capacitance-type humidity sensors 79; Al_2O_3-based sensor *98*; $BaTiO_3$ *94*; materials used 85–6; metal oxides in **93**; parallel plate structure *81*, 81–3; permittivity type 79–81; planar capacitive sensors 83–5; time-dependent *101*
capacitance *vs.* RH *89*, *90*

capacitive dew-point detector *341*
capacitive method 212
capacitive sensor interfaces, performance comparison **334**
capillary condensation 8; effect 95
carbon nanotubes (CNTs) 353
CCD (charge-coupled device) 213
CEC (coulometric electrolytic cell) 58
cellulose derivatives–based sensors 91
cellulose nanofibers (CNFs) 328; resonant frequency *329*
ceramic humidity sensors 140, 150
charge-coupled device (CCD) 213
chemical method 71
chemical sensor design *376*
chemisorbed layer 95
chemisorption process 8
chilled-mirror hygrometers 42–6
classic electronic circuits 320
cloud/fog chamber hygrometer 73
CMOS *see* complementary metal–oxide–semiconductor (CMOS)
CNFs *see* cellulose nanofibers (CNFs)
CNTs (carbon nanotubes) 353
coefficient of thermal expansion (CTE) 347–8
cold sensors 6
compensation method 34
complementary metal–oxide–semiconductor (CMOS) 6, 83, 173, 205, 249, 334, 341
conductive polymers 115–17
conductometric gas sensors 113, *113*, 143
constant-pressure hygrometer 71
constant-volume hygrometer 71–2
contact potential difference (CPD) 292, *293*
continuous methods 6
conventional and planar coulometric sensors 61, *61*
conventional hygrometers, humidity measurement in **11**
coordination chemistry interactions 8
coplanar waveguide (CPW) 327; -to-slotline ring resonator-based sensor 328, *329*
corona discharge: environmental factors *255*; gas ionization sensors 253–6; operation principles 253
coulometric electrolytic cell (CEC) 58
coulometric hygrometers 57; equipment *58*; humidity measurement 64; P_2O_5 57–62; sensing elements *58*
coulometric titration 66

counter ions 115
CPD (contact potential difference) 292, *293*
CPW *see* coplanar waveguide (CPW)
crystallization method 348
CTE (coefficient of thermal expansion) 347–8

Daniell J.F. 41, *42*
da Vinci L. 17
delay line 167
density of states (DOS) 260
depression capability 47
de Saussure S.H.B. 23
detection principles 5, **6**
dew cup 43
dew-point hygrometers 41–2; advantages and disadvantage **45**; chilled-mirror *44*; Colpitts circuit 195–6; Lambrecht's 42, *42*; Peltier element *195*; and performances 47–8; planar *53*; QCM-based devices 196–7; recognition principle *196*; SAW devices 196–7; sensor *43*; surface conductive 46–7
DIC (direct interface circuit) 377
diffusion barrier 58
diffusion hygrometer 72–3, *73*
diode-type humidity sensors 288
direct interface circuit (DIC) 377
discrete methods 6
divinylbenzene (DVB) 91
doping 259
DOS (density of states) 260
drift 68
dry- and wet-bulb thermometers 31–2; principle *32*
Dunmore cell 51–2, *51*
durability 10
DVB (divinylbenzene) 91

EG-OTFT (electrolyte-gated organic thin-film transistor) 262, *262*
electrical sensors 5
electric field-coupled inductor capacitor (ELC) resonator 327
electrochemical sensors 5
electrolysis process 57, 286
electrolyte-gated organic thin-film transistor (EG-OTFT) 262, *262*
electrolytic hygrometers 57–63; measurement cell *59*; operation *57*
electromagnetic/dielectric excitation method 210

electromotive force (EMF) 299
electronic expansion hygrometer 28
electronic sensor-based hygrometers 25
electronic-type humidity sensor 135, 149
electrostatic method 210
electrothermal/resistive heating excitation method 210
EMF (electromotive force) 299
energy trapping effect 163
equilibrium moisture content isotherm 87, 217

Faraday's law 57, 60, 307
FBARs *see* film bulk acoustic resonators (FBARs)
Fermi energy (E_F) level 291
FESEM (field emission scanning electron microscopy) 234, *234*
FET *see* field-effect transistor (FET)
field-effect mobility 258
field-effect transistor (FET) 257; GasFET 269–75; humidity sensors 257–77; MISFET 268–9
field emission scanning electron microscopy (FESEM) 234, *234*
field ionization (FI) humidity sensors 253
film bulk acoustic resonators (FBARs) 164–6; fabrication process *183*; humidity sensors 181–3; PSS/PDDA-coated, temperature dependence *183*
film humidity sensors 309
first-layer physisorption 190
Fischer K. 65
flexible substrate, humidity sensors on 350–9, *352*; capacitive 351–2; electronics 347; interdigital electrodes *351*; local inkjet printing 349; micro-electrode *354*; multiparameter sensing platform 356–7; multisensor platform micrograph *357*; paper-based humidity sensors 359–62; paper-based VOC *361*; PIM-based sensor *355*; plastic foil *347*; plastic substrates 349, **349**; platforms 347–50; polymer materials properties **350**; polymers acceptable **348**; RH sensors *353*, 353–6; SAW device *356*
flexural plate wave (FPWs) sensors 169, *169*
FPH (frost-point hygrometer) 45
free-proton mechanism 134
free-space transmission devices 319, *319*
frequency-modulated continuous wave (FMCW) technique 319
frost-point hygrometer (FPH) 45
full-shear constraint 208

galvanic cell–type humidity sensor *302*
GasFETs (gas sensor field-effect transistors) 296

gas-ionized sensor 254
gas-polarographic oxygen-humidity sensor *304*
gas sensor field-effect transistors (GasFETs) 296
gauge factor (GF) 214
"glass"-sensitive element 58
gravimetric humidity sensors: characteristics 173; devices 161; intrinsic film properties 174; mass-loading and frequency shift 173; metal oxides 176–7; MOFs 177; polymers 175–6; sensitive coating 174
gravimetric hygrometer 19; NIST *20*; water-collection tubes in *20*
gravimetric method 18–19, *18*
Grotthuss chain reaction 134
Grotthuss mechanism 133–5

hair hygrometer/hygrographs 23, 25, *27*; humidity control features using 25–6; old retro *23*; operation *23*; variants 24, *24*; working 25
hard-contact read-out technique 213
heated sensors 148–9
heater/circuit, structure humidity sensor *342*
heating, ventilation, and air conditioning (HVAC) systems 3
hierarchically porous polymers (HPPs) 54
hot sensors 6
HPPs (hierarchically porous polymers) 54
HSM *see* humidity-sensitive material (HSM)
humidity-fraction coefficient 217
humidity-induced strain 208
humidity measurement: breakdown voltage 366, *366*; chamber used in PTTB sensor *367*; condensation methods 41–2; coulometric hygrometers 64; fabrication process *368*; heated salt-solution method for 51–4; on KF titration 65–9; magnitude of charge *365*; NF-based active sensor *370*; piezoelectric output voltage *369*; pneumatic bridge method 72, *72*; range of *365*; self-powered active sensor 368–72; single molecular transistor 372–3; using heat pipe 366–8
humidity-sensitive FETs (HUSFETs) 259
humidity-sensitive material (HSM) 259; properties 8
humidity sensor realization: resonant/impedance sensors 323–35; transmission sensors 321–3
humidity sensors 3, 205; absolute 6–7; accuracy of measurements 36–8; applications 4; capacitive humidity sensors 221–4; characteristics **7**; classification

5, 5–7, **6**; development and testing 5; diode-type 288; fabrication 5; functionalization methods 218–21; integrated with heater 339, 339–41, *340*; lithium chloride *51*; materials for application in 7–10, **9**; by mechanical-optoelectronic principle 28; micromachined 6; microresonator-based devices 233–5; monitoring 5; P_2O_5 61–2; piezoresistive humidity sensors 224–33; *p–n* hetero-contact-type 286–9; polystyrene surface conductivity 63; requirements for 10–11; Schottky barrier-based 283–6
HUSFETs (humidity-sensitive FETs) 259
Hutton J. 31
HVAC (heating, ventilation, and air conditioning) systems 3
hybrid integrated humidity sensors 344
hydrogels 114
hydrophilic acrylic polymer 127
hydrophilic polymers 216
hygrographs and hygrothermographs 27, 27–8
hygrometer 17; chilled-mirror 42–6; cloud/fog chamber 73; condensation *41*; constant-pressure 71; constant-volume 71–2; dew-point *see* dew-point hygrometers; diffusion 72–3; electrolytic testing **59**; electronic expansion 28; electronic sensor-based 25; factors 61; with fiberoptic dew-point detector 44, *44*; gravimetric *see* gravimetric hygrometer; gravimetric-based 17–18, *18*; hair tension dial 25; invented/perfected 26; by Leonardo da Vinci *17*; mechanical *see* mechanical hygrometer; old retro 28; Sanctorius 24; string 24; using metal/wood or metal/paper laminate 26; volumetric 26
hygroscopy 24
hysteresis 10; problem 89

IDEs *see* interdigitated electrodes (IDEs)
Idiota de Staticis Experimentis (book) 17
IDTs (interdigitated transducers) 356
IGFET (insulated-gate field effect transistor) humidity sensor *267*
impedance meters 320–1, *321*
impedance sensors 308–12
inorganic salts 117–19
insulated-gate field effect transistor (IGFET) humidity sensor *267*
integrated humidity sensors: comparison **334**; with heater 339, 339–41, *340*; with readout circuitry 341–4, *343*; schematic diagram *340*

Index

interdigital electrode 81
interdigitated electrodes (IDEs) 113; structure 83
interdigitated transducers (IDTs) 356
interferometry method 212–13
International Organization for Standardization (ISO) 67, 377

Karl Fischer (KF) titration method 65; advantages and limitations 67–9; gas analyzer by Metrohm *68*; measurement cell for *65*; by Mettler Toledo *68*; principles 65–7; reagent 66; volumetric and coulometric 66
Kelvin probe as humidity sensor 291–3, *292*; operation principle 291; sensor performance 293–5; work function 291, *291*
Kelvin radius 95
KF *see* Karl Fischer (KF) titration method
killer criteria 8–9
Knudsen diffusion 10

Lambrecht's dew-point hygrometer 42, *42*
lamb-wave sensors 169
laser interferometry method 21
LED (light-emitting diode) 28, 43
LeRoy C. 43
LiCl/HPPMs sensors: durability *54*; impedance modulus *54*
light-emitting diode (LED) 28, 43
linearity 10
lithium chloride humidity sensors *51*; limitations 52–3; planar 53–4
logarithmic response behavior *vs.* RH *102*

mass-sensitive sensors: active surface 162; advantages 172; devices 161; gas 5; schematic diagrams *161*; species–sensor interaction 162
Maxwell-Wagner effect 97
MCL sensors *see* microcantilever (MCL) sensors
mechanical hygrometer; *see also* hair hygrometer/hygrographs: overview 23–5; types 26–7
MEMS (microelectromechanical systems) 83, 212
metal-insulator-semiconductor (MIS) 85
metal insulator semiconductor field-effect transistors (MISFETs) 259
metal-organic framework materials (MOFs) 177, 180
metal oxide–based capacitive humidity sensors 93–4; Al_2O_3 96–100; other oxides 100–3; water interaction mechanism 94–6
metal oxide resistive sensors: advantages 150; with alkali ions 135–6; alkaline ions 145–8; $BaTiO_3$ thin film *vs.* RH *137*; comparison with 143; conductivity in 133–6; disadvantages 150; Grotthuss mechanism 133–5; humidity-sensing characteristics 142–3; $KNbO_3$ nanosensors *146*, *147*; RT humidity sensors 138–42; semiconducting 135; structural properties 144–5; thick-film technology 137; thin-film technology 137; type **136**
metal oxide semiconductor (MOS) 258
microcantilever (MCL) sensors 205–7, *206*; detection methods 211–15, **215**; modes of operation 209–11; NEMS 208; NMC60 *211*; sorption models 207–9
microelectromechanical systems (MEMS) 83, 212
microfabrication 376
micromachined humidity sensors 6, 247–50
microresonator humidity sensor 233–5
microstrip line, representation *321*
microstrip technique 326
microwave dielectric measurement systems **320**
microwave differential technique 323
microwave moisture content meter *317*
microwave sensors 202–3, *318*; for humidity measurements **322**; impedance meters 320–1, *321*; overview **317**, 317–18; resonator sensors 320, **320**; transmission sensors *319*, 319–20
microwave test instrumentation 321
mirror-based dew-point sensors 41
MIS (metal-insulator-semiconductor) 85
MISFETs (metal insulator semiconductor field-effect transistors) 259
mobility 379
modern P_2O_5-based humidity sensors 61–2
MOFs (metal-organic framework materials) 177, 180
moisture absorption 86
moisture-sensing applications 147
MOS (metal oxide semiconductor) 258
multi-sensor platform, optical micrograph *342*
multi-walled carbon nanotubes (MWCNTs) 189–90

nafion 122; based sensors 176, 189
nanocomposite-based humidity sensors: CNTs 127; conductive polymer 128; groups 126; NaPSS 130; PEDOT:PSS 130; percolation threshold 126, *126*; PPy/Ta_2O_5 composites 128, *128*; SiO_2/nafion composite 129, *129*; ZnO/PPy composite 131
nanofiber sensors 188–9
National Institute of Standards and Technology (NIST) 19; gravimetric hygrometer 20
National Oceanic and Atmospheric Administration (NOAA) 45; FPH instrument *46*
n-doped, TFTs 260
Nernst equation 302

OECT (organic electrochemical transistor) 262
Ohm's law 243
Open System Interconnection (OSI) 377
optical diffraction grating method 213
optical gas sensors 5
optical readout 211–12, *212*
organic charge-transfer complexes 259
organic electrochemical transistor (OECT) 262
organic semiconductors (OSCs) 257
organic thin-film transistors (OTFTs) 260
OSI (Open System Interconnection) 377

PAA (polyamic acid) 90
PANI (polyaniline) 115–16, *116*, 189
paper-based humidity sensors 359
passive RFID tags 331
PCB (printed circuit board) 232
PDMS (polydimethylsiloxane) 351, 355, 359
p-doped, TFTs 260
PDVB (poly divinyl benzene) 118–19
PEN (polyethylene naphthalate) 348
percolation threshold 126–8
PE sensitive materials: bulk polymer-resistive sensors 120; categories 119; copolymers 120–1; cross-linking techniques 122–3; hysteresis 124–5, *125*; PoAN doping 121; sodium styrenesulphonate 120; TA film 125
PET (polyethylene terephthalate) 348
phase-separable glass system 140
phase transition thermally balanced (PTTB) sensor 366, *366*
phosphonium salts 119
phosphorous pentoxide (P_2O_5)-based coulometric hygrometers: advantages and limitations 58–61; operation principles 57; sensor configuration 57–8
photolithography 347
photosensitive polyimides 90, *90*
photothermal method 210
physisorbed layer 95
piezoelectricity 162
piezoelectric method 210, 212
piezoresistive method 213–14
planar coulometric humidity sensor 62, *62*
planar lithium chloride humidity sensors 53–4
planar oxygen sensor, multi-sensor *306*
pneumatic bridge method 72, *72*
p–n hetero-contact-type humidity sensors 286–9

polaron/bipolaron state 259
polyamic acid (PAA) 90
polyaniline (PANI) 115–16, *116*, 189
polydimethylsiloxane (PDMS) 351, 355, 359
poly divinyl benzene (PDVB) 118–19
poly(d,l-lactide) (PDLL) 181
polyelectrolytes (PEs) 9
polyethylene naphthalate (PEN) 348
polyethylene terephthalate (PET) 348
poly(hydroxyethymethacrylate) (PHEMA) 125
polyimide (PI) *348*, 349; CNTs *127*; films, diffusion coefficient 88–9; humidity-sensitive materials 216–18
polymer-based capacitance-type humidity sensors 86; implementation of 87–93; water interaction mechanisms 86–7
polymer-based resistive sensors: advantages 131; conductive polymers 114–17; containing inorganic salts 117–19; disadvantages 131–3; nanocomposite 126–31; PE 119–25; type of 114; VTF 115; water-durable 124
polymer SAW sensors 187–90
poly(*p*-diethynylbenzene) (PDEB) 116, *116*
polysalts 119
polytetrafluoroethylene (PTFE) 233, **348**
poly(vinylalcohol) (PVA) 114, 187–8
poly(vinyl butyral) (PVB) 189
polyvinyl chloride acetate (PVCA) 91
pope cell 62–3
popyvinylpyrrolidone (PVP)-based humidity sensors 188
porous chromium electrode 92
porous humidity-sensitive material 143–4
porous platinum electrodes 307
porous polymers 53
position-sensitive detector (PSD) 211
potentiometric humidity sensors 300–4
printed circuit board (PCB) 232
prototype sensor 331
PSD (position-sensitive detector) 211
psychrometer 31, *31*; advantages and limitations **38**, 38–9; coefficient 33, 39; for home use *34*; overview 31; principles of operation 31–3; realization 33–6; sling *34*
PTFE (polytetrafluoroethylene) 233, **348**
PTTB (phase transition thermally balanced) sensor 366, *366*
pulsed laser ablation technique 311
pulsed laser deposition (PLD) 287
PVCA (polyvinyl chloride acetate) 91

QCM *see* quartz crystal microbalance (QCM)
QCR (quartz crystal resonator) 195–7
quartz crystal microbalance (QCM): advantage 164; coatings for 174; edge effect 162; energy trapping effect 163; frequency shifts *181*; graphical projection *163*; humidity sensors 177–81, **178**; metal oxide films 176–7; oscillation frequency 163; PAA coating *179*; piezoelectric crystal, frequency 163; Sauerbrey equation 163–4; schematic diagrams *161*; thickness shear mode 162; *vs.* FBAR *166*
quartz crystal resonator (QCR) 195–7

radio frequency identification (RFID) *331*, 347, 357, 360
Rayleigh waves 167
recombination effect 61
rectangular spiral-shaped type, humidity sensor *339*
reference channel 323
reference tag 332
Regnault H.V. 41
Regnault's hygrometer 41, *42*
relative humidity (RH) 25, 32, 51, 79, 217, 245, 293, 352–3, 360, 369; BaTiO$_3$ thin film *vs.* *137*; capacitance *vs.* 89, *90*; logarithmic response behavior *vs.* *102*; Magnus parameters in **33**; measurements 63; polymer sensor 45; psychrometric chart **33**; sensors 6; U.S. Weather Bureau psychrometric table **37**
reproducibility 10
resistive sensors: advantages 114; humidity-sensitive materials for 113; metal oxide 133–50; polymer-based 114–33; response time for 125
resonant/impedance sensors 323–35
resonant operating mode 209; cantilever, mechanical properties 215–16; mass resolution limitations 216
resonant structures 318
resonating beam force transducer 233
resonator sensors 320, **320**
reversible reactions 8
RFID *see* radio frequency identification (RFID)
RH *see* relative humidity (RH)
room temperature (RT) humidity sensors: NASICON 140–1; phase-separable glass system 140; sol-gel method 139–40; TiO$_2$ 138

sampling system 47, *47*
Sanctorius hygrometer 24
SAPs (superabsorbent polymers) 122
saturated salt sensors 53
saturation current 258
Sauerbrey equation 163–4
SAW resonators (SAWRs) 194

SAW sensors *see* surface acoustic wave (SAW) sensors
scanning electron microscopy (SEM) image *232*, *340*; passive wireless integrated humidity sensor *330*
Schottky barrier-based humidity sensors: principles of operation 283, *283*; sensor performance 283–6, *284*, *285*
Schottky diode 283
SCLC (space charge-limited current) 288
self-calibration 44
semiconductor polymers 115–17
SEM image *see* scanning electron microscopy (SEM) image
sensor system, block diagram *344*
sensor system-on-chip (SSoC) *343*
sensor tag, printing layout *332*
sensor wafer *35*
serpentine electrodes (SREs) 84, *85*
shear-horizontal acoustic plate mode (SH-APM) sensors 168–9, *169*
shear-horizontal surface acoustic wave (SH-SAW) 168
SHF (super-high frequency) 172–3
silicon nanowires (SiNWs) 328
silver nanoparticle ink 331
single molecular transistor (SMT) *372*, 372–3
single waveguide hygrometer (SWH) 323
SiNWs (silicon nanowires) 328
SIRs (stepped impedance resonators) 327
sling psychrometer 33, *34*, 36
smart sensors 376–9; advantages and limitations 379; architecture 377–9; chemical sensor *378*; communication interface *377*; sensor electronic interface *377*
SMR (solidly mounted resonator) 166–7
SMT (single molecular transistor) *372*, 372–3
sodium styrenesulphonate (NaSS) 120
sol–gel method 139–40
solid electrolytes 299
solidly mounted resonator (SMR) 166–7
solid-state devices 3
solid-state electrochemical humidity sensors: overview 299; principles of operation *299*, 299–300, *300*; sensors performance 300–12, **301**
solid-state physics 291
space charge-limited current (SCLC) 288
SrCeO$_3$- and BaCeO$_3$-based metal oxides 300
SREs (serpentine electrodes) 84, *85*
SSoC (sensor system-on-chip) *343*
stepped impedance resonators (SIRs) 327
Stoney's law 208
string hygrometer 24
Strontiumdoped Lanthanum Yttrate (LSY) 306
superabsorbent polymers (SAPs) 122

Index

super-high frequency (SHF) 172–3
surface acoustic wave (SAW) sensors:
AW propagation modes 168–70; characteristics **186**; configurations *185*; Cu-BTC-film 192; delay line 167, *168*; dynamic responses *191*, *192*; frequency response *193*; humidity sensors 183–7; inorganic materials 190–2; love-wave sensor 185; metal oxide films 176–7; oscillator 170–1; passive mode 172; performances **184**; polymer-based sensors 187–90; principle *167*; resonator 168; SAWRs *185*, 194; schematic diagrams *161*, *167*; stability 193–4; working principle *167*
surface conductive dew-point hygrometer 46–7, *47*
surface stress change 208
swelling effect 86
SWH (single waveguide hygrometer) 323

TENG (triboelectric nanogenerator) 372
TFBAR (thin-film bulk acoustic resonator) *164*, *165*
TFR (thin-film resonator) 164–5
TFTs *see* thin-film transistors (TFTs)
thermal conduction 243, *244*

thermal conductivity-based hygrometer method: approaches 244–7; operation principle 243–4
thermal stability 348
thermistor-based humidity sensor *245*
thermistors 39
thermocouples 39
thermometric sensors 5
thick-film technology 99
thin-film bulk acoustic resonator (TFBAR) *164*, *165*
thin-film open junction, structure *286*
thin-film resonator (TFR) 164–5
thin-film transistors (TFTs) 257, *258*; band gap **260**; humidity sensors 257–77; pentacene-based *261*
TiO$_2$-based humidity sensors 138–9, **139**
transduction technique 317
transmission sensors *319*, 319–20
triboelectric charging 365
triboelectric effect 365–6
triboelectric nanogenerator (TENG) *372*
tunneling 213
twin waveguide hygrometer (TWH) 323

uniform impedance resonators (UIRs) 327
unsaturated hydrocarbons 60
upper troposphere and lower stratosphere (UTLS) 73

van der Waals forces 8, 86–7
vehicle mechanism 135
very large-scale integration (VLSI) technology 212
VOC *see* volatile organic compound (VOC)
Vogal–Tamman–Fulcher (VTF) relationship 115
volatile organic compound (VOC) 361; sensing 266
volt-ampere characteristics *307*
VTF (Vogal–Tamman–Fulcher) relationship 115

water sorption isotherms 176, *176*
weather house device 24
wet- and dry-bulb psychrometer *35*
wet-bulb hygrometers *31*
wheatstone bridge circuit 62–3, *214*; *see also* pneumatic bridge method
whirling psychrometer 33
wired chemical sensor 378
work function (Φ) 291
World Meteorological Organization (WMO) 32, 38, 48

Yttria-Stabilized-Zirconia (YSZ) 306

ZnO-based humidity sensors 179; one-dimensional nanostructures **147**; performance comparison **141**